INTEGRATED PEST MANAGEMENT AND PEST CONTROL – CURRENT AND FUTURE TACTICS

Edited by **Marcelo L. Larramendy and Sonia Soloneski**

INTECHOPEN.COM

Integrated Pest Management and Pest Control – Current and Future Tactics
Edited by Marcelo L. Larramendy and Sonia Soloneski

Published by InTech

Janeza Trdine 9, 51000 Rijeka, Croatia

Edition 2014

Publishing Process Manager Maja Kisic

Technical Editor Teodora Smiljanic

Cover Designer InTech Design Team

Additional hard copies can be obtained from orders@intechopen.com

Integrated Pest Management and Pest Control – Current and Future Tactics, Edited by Marcelo L. Larramendy and Sonia Soloneski
p. cm.
ISBN 978-953-51-0050-8

INTECH

open science | open minds

Contents

Preface

Pests have always represented one of the key limiting factors in crop production. For the first time in human history, we are technically able to produce as much food as needed for the ever increasing world population, thus theoretically eliminating the risk of food crisis. Since the beginning of the humanity, pesticides have revolutionized pest control to protect its food source. The first known pesticide was elemental sulfur dusting used in ancient Sumer about 4,500 years ago in ancient Mesopotamia. Toxic chemicals such as arsenic, mercury and lead were being applied to crops to kill pests by the 15th century whereas nicotine sulfate was extracted from tobacco leaves for use as an insecticide during the 17th century. Nowadays, it is commonly accepted that the increase in food production worldwide has led to employ large amounts of pesticides. Although the benefits of conventional food production practices have been immense, they utilize levels of pesticides and fertilizers that can result in a detrimental impact factor on the environment. Pesticides are high volume, widely used environmental chemicals and there is continuous debate concerning their probable role in both acute and chronic human health effects. Latest report of the United States Environmental Protection Agency -USEPA- disclaimed that the world pesticide market estimation usage only for the 2006-2007, was nearly over 5.2 billion pounds. Their application is still the most effective and accepted method for plant and animal protection from a large number of pests, being the environment consequently and inevitably exposed to these compounds. The goal in pesticide investigation and development is identifying the specificity of action of a pesticide toward the organisms it is supposed to kill only the target organisms should be affected by the application of the product. However, because pesticides are designed and selected for their biological activity, toxicity on non-target organisms, including humans, frequently remains a significant potential risk. The benefits in using pesticides must be weighed against their deleterious effects on human health, biological interactions with non-target organisms, pesticide resistance and/or accumulation of these chemicals in the environment.

Integrated Pest Management -IPM- traces its first real beginnings to the late 1960s, where a number of factors were associated to initiate a search for better methods of pest control rather than the reliance on pesticide use. Among these factors are included not only the very well known problems related to pesticide use, abuse, and misuse, but also the rapid development of new technologies enabling more sophisticated approaches, primarily due to the overwhelming advances in communication and computational tools, with the allied new sciences of operations research, systems analysis, and modeling. As defined by the USEPA, IPM is an effective and environmentally sensitive approach to pest management that relies on a combination of common-sense practices. IPM programs use current and comprehensive information on the life cycles of pests and their interaction with the environment. This information, in combination with available pest control methods, is used to manage pest damage by the most economical means, and with the least possible hazard to people, property, and the environment.

IPM is not a single pest control method but, rather, a series of pest management evaluations, decisions and controls. The methodology includes four major steps, namely: a) *set action thresholds*, a point at which pest populations or environmental conditions indicate that pest control action must be taken; b) *monitoring and pest identification*, so that appropriate control decisions can be made in conjunction with action thresholds as well as to remove the possibility that pesticides will be used when they are not really needed or that the wrong kind of pesticide will be used; c) *prevention*, to manage the crop, lawn, or indoor space to prevent pests from becoming a threat; d) *control*, evaluating the proper method for balancing both effectiveness and risk.

Increasing number of scientific reports within the complex IPM came out in the last years. A simple search in a databank as Scopus, displays more than 5,600 reports published in scientific journals from which approximately 4,400 have been reported during the last decade. As developments in this field have been quite rapid, we believe the writing of a new book scoping the subject is fully justified. To tackle among others, related geopolitical, economical and population issues in our modern, cloud computing-economy connected societies, we aim to present a more holistic approach of the matter, in order to appreciate the full scope of the question.

Many researchers have contributed to the publication of this book. The editors hope that this book will continue to meet the expectations and needs of all interested in pest management to minimize the use, abuse and misuse of pesticides.

Marcelo L. Larramendy and **Sonia Soloneski**
Faculty of Natural Sciences and Museum
National University of La Plata
Argentina

Part 1

Integrated Pest Management – Theory and Concepts

1

Principles and Practices of Integrated Pest Management on Cotton in the Lower Rio Grande Valley of Texas

Shoil M. Greenberg, John J. Adamczyk and John S. Armstrong
Kika de la Garza Subtropical Agricultural Research Center, Agricultural Research Service, United States Department of Agriculture, Weslaco USA

1. Introduction

Sustainable agriculture is ecologically sound, economically viable, socially just, and humane. These four goals for sustainability can be applied to all aspects of any agricultural system, from production and marketing, to processing and consumption. Integrated Pest Management (IPM) may be considered a key component of a sustainable agriculture system. This publication reviews recent advances in the development of IPM programs for cotton in the Lower Rio Grande Valley of Texas. We describe annual losses caused by arthropod pests in general and by specific key insect pests, briefly showed sampling of insect populations and cotton growth stages, which importance of the proper timing of scouting procedures and treatments; and economic threshold harmfulness (ETH) for optimizing control and minimizing risk from insects. We describe effectiveness of cotton insecticides; transgenically modified cotton; microbial insecticides; native, most widely-distributed and augmentative releases of beneficial insects; and cultural control techniques for cotton insects. We also show cotton diseases and weed controls. IPM is a process that considers all control options in the proportion shown in the model of a pyramid, and it can be used to demonstrate how growers might productively construct their pest management programs.

2. What is IPM

Integrated Pest Management (IPM) has been defined as a sustainable approach to managing pests by combining biological, cultural, physical, and chemical tools in a way that minimizes economic, health, and environmental risks (ND IPM Homepage, Texas Pest Management Association); IPM has also been defined as a knowledge-based, decision-making process that anticipates limits and eliminates or prevents pest problems, ideally before they have become established. IPM typically combines several strategies to achieve long-term solutions. IPM programs include education, proper waste management, structural repair, maintenance, biological and mechanical control techniques, and pesticide application when necessary (www.PestControlCanada.com). IPM is a pest management strategy that focuses on long-term prevention or suppression of pest problems through a combination of techniques such as 1) monitoring for pest presence and establishing treatment threshold levels, 2) using non-chemical practices to make the habitat less conducive to pest development; improving sanitation; and 3) employing mechanical and physical controls. Pesticides that pose the least possible hazard and are effective in a manner that minimizes risk to people, property, and the environment are used only after careful monitoring indicates they are needed, according to established guidelines and treatment thresholds (California Department of Pesticide Regulation, cdprweb@cdpr.ca.gov). IPM employs approaches, methods, and disciplines to minimize environmental impact, minimize risks, and optimize benefits. An expansion of the IPM concept is the process of Integrated Crop Management (ICM), which includes other agricultural decision-making tasks such as fertilizer and soil water management. An ICM program would include an IPM component to deal with pest management decisions plus address remaining issues applicable to the total crop production process (Ohio Pest Management & Survey Program, http://ohioline.osu.edu/icm-fact/fc-01.html). Thus, IPM is a system of pest management decisions based on ecological, economic, and sociological values.

2.1 Pest management practices and set of IPM principles

It may be classified according to the approach or the method used to deal with a pest problem. In terms of approach, pest management practices may be designed to prevent, suppress, or eradicate problems. Pest management practices are grouped under four categories: biological, chemical, cultural and mechanical, and legal. IPM approaches and methods are used to minimize environmental contamination, minimize risk from harmful organisms, and optimize benefits. It is a systems approach to pest management that utilizes decision making procedures based on either quantitative or qualitative observations of the pest problem and the related host or habitat (Ohio Pest Management & Survey Program, http://ohioline.osu.edu/icm-fact/fc-01.html).

The U.S. Environmental Protection Agency (EPA) has developed a useful set of IPM principles. *Acceptable pest levels* occur when pest population (s) are present but occur at densities too low to cause economic damage. Controls are applied only if pest densities increase to *action thresholds* for that particular crop. *Preventive cultural practices* involve selecting the best varieties for local growing conditions, together with plant quarantine, cultural techniques, and plant sanitation. *Monitoring plant growth* and densities of key and secondary pest species (commonly referred to as *scouting*) is a cornerstone of IPM. *Mechanical controls* include a variety traps, vacuuming, and tillage to disrupt survival and reproduction by various pest species. *Biological controls* involve the use of predators, parasitoids and pathogens to maintain pest populations at densities lower than would occur in their absence (and hopefully at subeconomic levels). *Chemical controls* which involve use of synthetic pesticides only as required and often only at specific times in a pest life cycle (Bennett et al., 2005)

Therefore, setting up an IPM program and designing a monitoring plan for a given crop should be based on the phenology of the plant and population densities of key and secondary pests.

2.1.1 Cotton production and insect diversity

Cotton production in the U. S. occurs on 30,000 farms and covers an average of 14.4 million acres (5.8 m ha) with a mean yield of 683.3 lb of lint per acre (766 kg/ha) (for 2004-2006)

(Williams, 2007). Cotton generates $6.2 billion in cash for farmers, and the total business revenue for the U.S. cotton industry is estimated at $40.2 billion per year. Texas ranks first in cotton production in the U.S., averaging 6.0 million acres (2.4 m ha) and generates $1.6 billion in cash for farmers, thus providing a total economic impact of $5.2 billion (Statistical Highlights of United States Agriculture, 2007; Agricultural Statistics, 2008). In the Lower Rio Grande Valley (LRGV) of Texas, an average of 220,000 acres (88,710 ha) of cotton were planted each year during 2004-2006 and generated an estimated $63.8 million in crop production (Lower Rio Grande Valley Cotton Blue Book, 2006)

Cotton production in the LRGV is challenged with a diversity of pests, and links the North American cotton states with those of Mexico and other South American cotton-producing areas. The most notable pest of Texas cotton production is the boll weevil (BW), *Anthonomus grandis grandis* Boheman, which entered the U.S. near Brownsville, Cameron Co, TX, during the 1890's. Other noted pests of cotton that emerged during the progression of cotton production in the LRGV were numerous lepidopterans (bollworm, *Heliothis zea* (Boddie); tobacco budworm, *Heliothis virescens* (Fabricius); beet armyworm, *Spodoptera exigua* (Hübner); cabbage looper, *Trichoplusia ni* (Hübner); black cutworm, *Agrotis insilon* (Hufnagel); fall armyworm, *Spodoptera frugiperda* (J. E. Smith); pink bollworm, *Pectinophora gossypiella* (Saunders); yellowstriped armyworm, *Spodoptera ornithogalli* (Guenée); and the leaf perforator, *Bucculatrix thurberiella* Busck); the plant sucking cotton aphid, *Aphis gossypii* Glover; stinkbugs; cotton fleahoppers, *Pseudatomoscelis seriatus* (Reuter); whiteflies, *Bemisia tabaci* (Gennadius) biotype B and *Trialeurodes abutilonea* (Haldeman); spider mite, *Tetranychus* spp.; thrips, *Thrips* spp.; cotton leafminer, *Stigmella gossyppi* (Forbes & Leonard); the verde plant bug, *Creontiades signatus* (Distant); Texas leaf cutting ant, *Atta texana*; and lubber grasshopper, *Brachystola magna* (Girard) (Cotton insects and mites: Characterization and management, 1996; French et al., 2006; Armstrong et al., 2007; Castro et al., 2007; Lei et al., 2009; Greenberg et al., 2009a and 2009b)

2.1.2 Cotton losses due to pests

A diversity of harmful organisms challenges the profitable production of agricultural crops and if left unmanaged, can result in significant losses. Estimates of crop losses vary widely by location and by year, but those are about one-third of potential global agricultural production in the form of food and fiber. Total annual losses in the world are estimated at about U.S. $300 billion (FAO, 2005). Average yield loss range from 30 to 40% and are generally much higher in many tropical and subtropical countries.

Cotton is the most important fiber crop in the world and is grown in almost all tropical and subtropical countries. Cotton production is especially threatened by insect attacks (Homoptera, Lepidoptera, Thysanoptera, Coleoptera) and by weed competition during the early stages of development. Pathogens may be harmful in some areas and years. Only recently have viruses reached pest status in South Asia and some states of the U.S. The estimates of the potential worldwide losses of animal pests and weeds averaged 37 and 36%, respectively. Pathogens and viruses added about 9% to total potential loss. The proportional contribution of crop protection in cotton production areas varied from 0.37 in West Africa to 0.65 in Australia where the intensity in cotton production is very high. Despite the actual measures, about 29% of attainable production is lost to pests (Oerke, 2006).

In the U.S. arthropod pests reduced overall cotton yield by $ 406.2 million (the mean for 2004-2006), in Texas - $ 99.3 million, and in the LRGV - $ 5.6 million (Williams 2005-2007) (Table 1).

Insect	Rank by % loss	Bales lost	Rank by % loss	Bales lost	Rank by % loss	Bales lost
	USA		Texas		LRGV of Texas	
Bollworm/Budworm	1	229,186	2	78,826	1	39,063
Lygus	2	171,478	6	10,314	0	0
Thrips	3	145,040	3	65,062	6	1,563
Fleahopper	4	119,745	1	108,057	2	26,042
Aphids	5	80,418	4	61,162	3	5,208
Stinkbugs	6	68,823	5	13,186	0	0
Spider mites	7	60,720	10	2,917	9	163
Bemisia tabaci	8	14,817	8	3,926	4	3,906
Fall armyworm	9	12,071	7	5,404	7	456
Boll weevil	10	3,190	9	3,190	5	3,190
Beet armyworm	11	1,104	12	229	8	228
Cutworms	12	1,100	0	0	0	0
Saltmarsh Caterpillars	13	237	0	0	0	0
Pink bollworm	14	232	13	28	0	0
Grasshopper	15	131	0	0	0	0
Loopers	16	144	0	0	0	0
Green Mirid	17	0	11	685	0	0
Total lost: bales*		908,436		352,985		79,818

*One bale of lint = 200kg
Source: Williams, 2007.

Table 1. Cotton losses in the United States due to insects.

2.1.3 Sampling insect populations

IPM is a process of pest monitoring and sampling to determine the status of a pest, and, when control actions are needed, all control options are considered. Field observation (scouting) is a vital component of cotton insect control. Fields should be checked at least once and preferably twice a week to estimate the species present, the type of damage, and the level of damage which has occurred up to that point in time. Scouting should also include monitoring plant growth, fruiting, weeds, diseases, beneficial insect activity, and the effects of prior pest suppression practices. The number of samples required depends on the field (plot) size and variability. Several different sampling methods are used in IPM programs. Visual observations of plants (generally ranges from 25-100 plants;

preferred method is to examine 5 consecutive plants in 10-20 representative locations within a field); sweep net (5 sweeps per sample, and at least 20 samples per treatment); beat bucket (3-5 plants per bucket, and at least 20 samples per treatment); drop cloth (the standard length - three feet long [=0.9144 m], used if row spacing is 30 inches [=0.762 m] or wider; a minimum of 4-6 drop cloth samples should be taken per field); colored sticky traps; and pheromone traps. Some of the sampling methods are shown in Fig.1. Methods of identification and sampling procedures for cotton insect pests and beneficial are available in some sources (Steyskal et al., 1986; Cotton scouting manual, 1988; Bohmfalk et al., 2002; Spark & Norman, 2003; Greenberg et al., 2005). Scouting is not a suppression tool, but is essential in formulating management decisions. The cost of controlling insects is one of the larger items of the crop production budget, ranging from $70 to over $100 per acre (from $173 to over $247 per ha) (Pest management strategic plan for cotton in the midsouth, 2003).

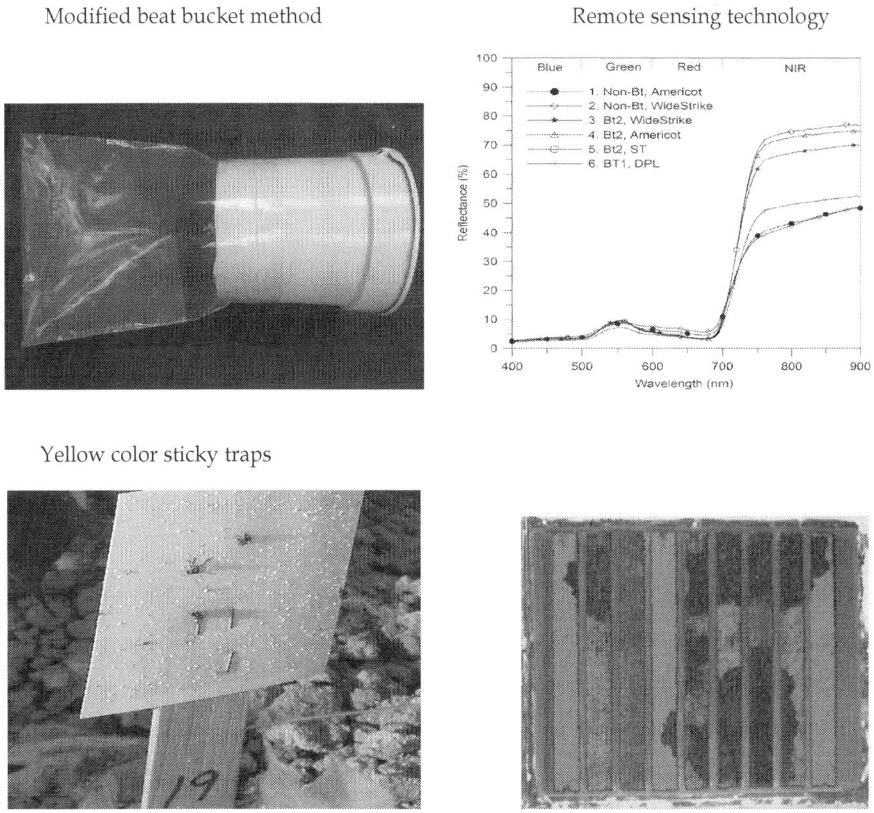

Fig. 1. Examples of sampling methods used in cotton IPM programs.

Knowledge of growth stages is important to the proper timing of scouting procedures and treatments (Table 2).

Developmental period	Calendar days		Accumulated heat units, DD60's	
	Avg.	Range	Avg.	Range
Planting to emergence	7	5-10	43	15-71
Emergence of:				
First true leaf	8	7-9	74	53-96
Six true leaf	25	23-27	239	161-320
Pinhead square	29	27-30	269	192-351
1/3 grown square	43	35-48	400	264-536
Square initiation to bloom	23	20-25	496	382-609
Bloom to: peak bloom	18	14-21	693	525-861
Full grown boll	23	20-25	751	588-912
Open boll	47	40-55	1059	1014-1105

Source: Lower Rio Grande Valley of Texas, Cotton Blue Book, 2006-2008).

Table 2. Cotton development by calendar days and heat units. Accumulated heat units, DD60's measures are in Fahrenheit (F°). Conversion degrees Fahrenheit to Centigrade (C°): C°= F° - (32*5/9).

2.1.4 Economical threshold of harmfulness

Control is needed when a pest population reaches an economic threshold (Table 3) or treatment level at which further increases would result in excessive yield or quality losses. This level is one of the most important indices in IPM for optimizing control and minimizing risk from insects.

Suppression activities are initiated when insect pest populations reach treatment thresholds which are designed to prevent pest population levels from reaching the Economic Injury Level (EIL) when economic losses begin to occur (value of the crop loss exceeds the cost of control).

Insects	Season	Economical Threshold of Harmfulness (ETH)
Boll weevil	Early Mid and Late	40 overwintered boll weevils per acre, 15-20% damage squares from squaring to peak bloom
Thrips	From 50% emergence to 3-4 true leaves	The average number of thrips counted per plant is equal to the number of true leaves at the time of inspection
Fleahoppers (FH)	All	1st-3rd weeks of squaring - 15-25 nymphs and adults per 100 terminals. After 1st bloom – treatment is rarely justified.
Aphids	All	≥50 aphids per leaf
Whiteflies	All	When ≥40% of the 5th node leaves are infested with 3 or more adults
Plant Bugs (*Creontiades* spp.)	During the first 4 to 5 weeks of fruiting	15-25 bugs per 100 sweeps
Spider Mites	All	When 50% of the plants show noticeable reddened leaf damage
Bollworm	Before bloom After boll formation	≥ 30 % of the green squares examined are worm damaged and small larvae are present 10 worms ≤ ¼-inch in length per 100 plants and 10% damage fruit for Non-Bt cottons; or 10 worms >1/4-inch in length per 100 plants with 5% damaged fruit
Beet Armyworm	All	When leaf feeding and small larvae counts exceeded 16-24 larvae per 100 plants and at least 10% of plants examined are infested; when feeding on squares, blooms, or bolls the threshold needs to be 8-12 larvae larger than ¼ inch per 100 plants
Fall Armyworm	Before first bloom	30% of the green squares are damaged
	Bolls are presented	15-25% small larvae are present per 100 plant terminals and 10-15% of squares or bolls are worm damaged

Inch =2.54 cm
Source: Norman & Sparks, 2003; Castro et al., 2007.

Table 3. Economic thresholds for some major cotton insects on cotton in the Lower Rio Grande Valley of Texas.

2.2 Insect control by synthetic chemicals

Synthetic chemicals continue to be the main tool for insect control. The total cost of pesticides applied for pest control is valued at $10 billion annually (Sharma & Ortiz 2000). Conventionally grown cotton uses more insecticides than any other single crop and epitomizes the worst effects of chemically dependent agriculture. Each year, cotton producers around the world use nearly $2.6 billion worth of pesticides, more than 10% of the world's pesticides and nearly 25% of the world's insecticides (http://www.panna.org/

files/conventionalCotton.dv.html). On agricultural crops in the U.S., about 74.1 million kg of insecticides is used. Over half of this amount is applied to cotton fields, corresponding roughly to 7.3 kg/ha of AI per hectare (Gianessi & Reigner 2006). In Texas, the direct insect management treatment cost is $115.6/ha; and, in the LRGV of Texas, the direct cost is $168.9 per hectare (Williams 2005-2007). Insecticides recommended for use on cotton are described in Table 4. Statewide, 46% of insecticides are applied aerially, 46% with ground equipment, and 8% by irrigation. Farmers perform 51% of pesticide application themselves (Lower Rio Grande Valley Cotton Blue Book, 2006-2008). Hollow cone spray nozzles are recommended for insecticide applications because they provide better foliar coverage than flat-fan or flood-jet nozzles. A straight spray boom with two nozzles per row is required for adequate coverage.

2.3 Changes in Texas cotton IPM during recent years

During recent years, there have been significant changes in Texas cotton IPM, and this system continues to evolve rapidly. These changes are occurring because of three major factors: boll weevil (BW) eradication; new and more target-specific insecticides used; and the development and use of transgenic Bt-cotton. The BW is currently the most important key pest of cotton in the LRGV of Texas where it has caused extensive damage since its appearance in 1892. Control of BW is through multiple applications of synthetic insecticides. In 1995, during the initial BW eradication program, farmers in the LRGV lost 13.5 million kg of cotton lint worth $150 million. This loss of 15% of the harvest was due to extensive ULV malathion spraying, mostly by plane, that led to massive secondary pest outbreaks of the beet armyworm (BAW) and areawide natural enemy disruption (http://www.panna.org/files/ conventionalCotton.dv.html; Summy et al., 1996). The BW eradication program in the LRGV was initiated for the second time during 2005. The second attempt at BW eradication did not trigger major secondary pest outbreaks because was initiated in the fall and reduced the heavy malathion use before the following the spring planting of cotton; improved pesticide application techniques (mostly ground rigs, helicopters versus airplane, treatments only edge strip of the fields); preventive activity; availability of target-specific pesticides for lepidopterans. Progress in the U.S. BW eradication effort where BW was successfully eradicated has resulted in a sharp decrease in the number of insecticide applications. The reduction in foliar sprays has also had an indirect effect in reducing outbreaks of secondary pests, such as cotton aphids and beet armyworm.

Cotton IPM in the LRGV of Texas has also improved due to: target specific insecticides such as Tracer and Steward for lepidopterans, (Leonard, 2006); cotton seed treatments with the systemic insecticides Gaucho Grande and Cruiser, which protect cotton from sucking insect damage for 30 days after planting (Greenberg et al., 2009, Zhank et al., 2011); reducing the application rate of insecticides without reducing efficacy of the program, for example, the malathion rate was reduced from 16-oz/ac to 12-oz/ac when oil was added as an adjuvant (Texas Boll Weevil Eradication Foundation, 2011); combination of applications for maintaining and preserving beneficial insects, lessening the environmental impacts, such as early-season spraying of cotton for overwintering BW and fleahoppers; pre-harvest application of the insecticides Karate or Guthion at half-rate with the cotton defoliant Def [synergistic effects] (Greenberg et al., 2004; 2007); termination of insecticide treatments based upon crop maturity; and improved pesticide application techniques (correct nozzle placement, nozzle type, and nozzle pressure) (Leonard et al., 2006; Lopez et al., 2008).

Class	Common name	Brand name	Recommended target pests
OP	Acephate (0.5-1.0)*	Orthene® 90S (generics)	Thrips, cutworms, *Greontiadis* plant bugs, fleahoppers, cutworm, fall armyworm
OP	Dicrotophos (0.25-0.5)	Bidrin	Thrips, plant bugs, fleahoppers, stinkbugs, aphids, boll weevil
OP	Dimethoate (0.11-0.22)	Dimethoate (generics)	Thrips, fleahopper, and *Greontiadis* plant bugs
OP	Malathion (0.61-0.92)	Fufanon ULV9.9	Boll weevil
OP	Methamidophos (0.7-2.2)	Monitor	Thrips, plant bugs, fleahoppers, whiteflies
C	Oxamyl (0.25)	Vydate® 2L	Boll weevil, plant bugs, fleahoppers
C	Methomyl (0.45)	Lannate®2.4LV	Aphids, beet armyworm, fall armyworm, fleahoppers
C	Thiodicarb (0.6-0.9)	Larvin ®3.2	Boll worm, beet armyworm, fall armyworm, tobacco budworm, loopers
CN	Imidacloprid (0.05)	Provado®1.6F	Plant bugs, fleahoppers, aphids, whiteflies
CN	Acetamiprid (0.025-0.05)	Intruder®70WP	Aphids, whiteflies, fleahoppers
CN	Thiamethoxam (0.03-0.06)	Centric® 40WG	Plant bugs, aphids, whiteflies, fleahoppers
IGR	Methoxyfenozid (0.06-0.16)	Intrepid®2F	Beet armyworm, fall armyworm, loopers
OC	Dicofol (0.75-1.5)	Kelthane® MF	Spider mites
P	Bifenthrin (0.37)	Capture or Discipline	Bollworms, fall armyworm, aphids, plant bugs
P	Cyfluthrin (0.01-0.06)	Baythroid® 2E	Cutworm, stinkbug, bollworms, boll weevil, whiteflies
P	Cyhalothrin (0.01-0.04)	Karate-Z	Cutworm, stinkbug, bollworms, boll weevil
P	Deltamelthrin (0.04-0.2)	Decis	Cutworm, stinkbug, bollworms, whiteflies, thrips
	Spiromesifen (0.094-0.25)	Oberon® 2SC	Whiteflies, spider mites
	Plant Growth Regulation	Ethephon (Prep) Mepiquart Clorade	Modified plant growth
	Defoliants	Def, Dropp, Ginstar	For early harvest

*In parentheses - rate AI lb/ac; 1 pound (lb) =0.4536 kg; 1 ac= 0.4047 ha; OP -organophosphate; C - carbamate; CN -chloro-nicotinyl; IGR -insect growth regulator; OC -organochlorine; P -pyrethroid
Source: The Pesticide Manual, 2003.

Table 4. Insecticides recommended for use on cotton in U.S.

2.3.1 Changes in the sucking bug complex – Stinkbugs, plant bugs and the cotton fleahopper

The sucking bug pests of cotton (suborder Heteroptera) have been elevated in pest status within the cotton growing regions of the United States over the past decade. Some of the most notable heteropterans are: tarnished plant bug, *Lygus lineolaris* (Palisot de Beauvois); western tarnished plant bug, *Lygus hesperus* Knight; the stinkbug complex (Pentatomidae); and the cotton fleahopper, *Pseudatomoscelis seriatus* (Reuter). This transition from being considered secondary pests and now elevated to key pest status has also coincidentally followed the functional eradication of the boll weevil from the southeastern and southern cotton belt regions. (Grefenstette and El-Lissy, 2008).

Other reasons often mentioned for increases in bugs infesting cotton with the progression of BW eradication is the adoption of varieties containing the Bt endotoxins that were being released in conjunction with eradication efforts. Over time, the number of BW was reduced, coinciding with a reduction in number of ULV malathion applications within a season, which may have been suppressing the bugs. Because lepidopteran pests were the key target at the time, Bt cotton varieties significantly reduced these pests, and, at the same time, safer, more target-specific insecticides were in development and being applied under full label. These three factors - the progress of BW eradication and the reduction of ULV malathion, the adoption of cotton varieties with BT, and the use of target-specific insecticides for control of lepidopteran pests are most often cited as the reason for changes in shift from lepidopteran management to sucking bug attacking cotton (Layton, 2000; Greene & Capps, 2003).

Some of the cotton growing regions of Texas are in the process of actively eradicating the BW from the LRGV in south Texas and the Winter Garden area (WGA) south and west of San Antonio, near Uvalde. However, the intensity of problems with the sucking bug complex and economic losses they cause varies by production region. For example, the tarnished plant bug, *L. lineolaris* (Palisot de Beavois) has increased in pest status in the southern and mid-south cotton regions following BW eradication (Layton, 2000), and has developed resistance to a wide variety of insecticides (Snodgrass, 1996; 2008). Not all bug complexes have increased or are related to BW eradication. California and Arizona had perennial problems with *L. hesperus* and *L. elisus* Van Duzee (Heteroptera: Miridae) in alfalfa and cotton before and after BW was eradicated from the cotton producing regions of these 2 states (Leigh et al., 1985; Zink & Rosenheim, 2005). Cotton damage from tarnished plant bugs results from feeding on cotton squares (flower buds), with the most significant impact when fruit abscises or drops to the ground (Tugwell et al., 1976). Further to the west in Arizona and California, the western tarnished plant bug causes similar feeding injury to cotton (Leigh T. et al., 1996).

For the last few years, the verde plant bug, *Creontiades signatus* Distant, has been reported infesting cotton grown in the LRGV and the Lower-Coastal Bend regions of south Texas, causing injury to developing lint and seed inside cotton bolls (Armstrong et al., 2009 a, 2010). The verde plant bug has increased in pest status since the initiation of the second attempt to eradicate the BW in the LRGV (2005) and from 1999 to the present in the Upper and Lower Coastal Bend production areas (Texas Boll Weevil Eradication Foundation, 2011). Feeding injury from the verde plant bug is similar to that caused by lygus bugs, but it has thus far been considered a late season pest, injuring and causing abscission in bolls <315 heat units (DD) from anthesis. Molecular and taxonomic work identified *C. signatus* as being native to the Gulf Coast of the U.S. and Mexico (Coleman et al., 2008). Reasons for increases in the densities of this new plant bug pest of south Texas can only be speculated. Some factors that may account for these increases the significant recent increase in the acres of soybean, *Glycine max* (L.) Merr., planted in the LRGV. *C. signatus* can reproduce on soybean and within the seed-head of grain sorghum, *Sorghum bicolor* (L.) Moench. Moreover, several weedy species also serve as reproductive hosts. Cotton may not be the most highly preferred host of the verde plant bug, but the bug survives on the cotton plant and has a preference for oviposition on the petioles of cotton leaves similar to other *Lygus* species (Armsrong & Coleman, 2009, Armstrong et. al, 2009 b, c).

The stinkbugs attacking cotton can be varied and complex. The most frequently encountered species are the southern green stinkbug, *Nezara viridula* (L.), the green stinkbug, *Acrosternum hilare* (Say), and the brown stinkbug, *Euschistus servus* (Say) (Hemiptera: Pentatomidae). These three species are considered the primary targets for a significant number of insecticide applications applied to cotton (Williams, 2008), most notably in the mid-south and southern cotton regions and have also been associated with elevated pest status following BW eradication (Green et al., 1999; Turnipseed et al., 2004; Willrich et al., 2004). However, in Texas, the diversity of species seems to be broader from central Texas to the Lower-Gulf Coast region south of Corpus Christi, and includes the rice stinkbug (RSB), *Oebalus pugnax* (F.); in the LRGV, Winter Garden area, and in far west Texas, there is the Conchuela stinkbug, *Chlorochroa ligata* (Say) (Muegge, 2002). Stinkbugs of all species and localities are noted for being more injurious to small to medium size cotton bolls, and, on a comparative basis, can cause more injury by lacerating thicker boll tissue, resulting in greater injury to the tissues, seed, and lint (Greene et al., 1999; Musser et al., 2009).

The most consistent early season true-bug pest of cotton in the state of Texas is the cotton fleahopper, which prefers feeding on small, primordial squares developing in the upper terminal of plants (Stewart & Sterling, 1989). When injured, the small squares abscise from the plant. However, the cotton plant is noted for compensation, and if management practices are instigated or populations decrease before the EIL is reached, losses due to fleahopper feeding injury may be negligible (Sterling, 1984). The length of the growing season is often associated with compensatory gain because of the delayed fruit set. The historical relationship between the severities of cotton fleahopper infestations with the progress of BW eradication, in the state of Texas is difficult to make, as severe fleahopper outbreaks have been noted before, during, and after an area has been functionally eradicated. The High Plains of Texas was declared functionally eradicated in 2003, but cotton fleahopper populations are as much a threat now as they were before eradication. In south Texas, cotton fleahoppers are still considered a significant pest, and BW eradication has not yet been fully realized.

With the more recent changes in the pest status of heteropteran pests of cotton, there is a greater realization of the pests' feeding injury and association with incidence of boll rot. Cotton fleahopper feeding injury to cotton squares and bolls is important because the wounds allow bacterial and fungal pathogens to enter and invade the interior of the forming fruit. Environmental conditions in the cotton field, mostly in the form of temperature, humidity, and moisture, can prevent or promote the growth of the boll rotting pathogens.

Economic thresholds established for most sucking pests are generally based on direct feeding injury and do not include boll rot as a yield-limiting factor. Square and boll rot may promote the delayed abscission of cotton fruit due to the production of ethylene by the rotting and degradation of fruiting tissue (Duffey & Powell, 1979). Cotton bolls do not normally sustain extensive damage from cotton fleahopper, due to the fact that their mouthparts (stylets) are not long enough to penetrate the wall of the boll. Boll rot pathogens have, however, been associated with direct transmission of common plant pathogen and cottonseed-rotting bacteria, *Pantoea ananatis* (Bell et al., 2006; Bell et al., 2010). The stinkbugs and plant bugs possess stylets that are long and broad enough to cause physical damage from insertion and laceration of the tissue, injection of digestive enzymes, and the ingestion of the enzymatic soup. This subsequently causes loss of boll, lint, and seed tissue, and provides an entry for pathogens that collectively may cause boll rot (Medrano et al., 2009). Even if the cotton fruit, including bolls, does not abscise, the quality and quantity of lint will be reduced.

2.3.2 Improving management options for the integrated approach to control bug pests

The plant bugs as a group have, in the past, been targeted for the discovery of host plant resistance traits that could be integrated into traditional cotton breeding programs. Host plant resistance of the cotton fleahopper and plant bugs have been studied extensively during the last four decades. The three main sources of host plant resistance identified were relatively high gossypol levels (Lukefahr, 1975), smooth (rather than hirsute) genotypes (Lukefahr, 1970), and production of nectar. No active cotton breeding programs have continued with any forms of resistance since Lidell et al. (1986) screened for glabourous, pilose, and nactariless traits. Many of these same traits were screened in cotton for the lygus bugs (Gannaway & Rummel, 1994; Tingey & Pellemer, 1977; Jenkins & Wilson, 1996). No information is available for host plant resistance for stinkbugs in cotton. Treatment thresholds for insecticide applications for these bugs have been provided in several extension-based publications that list the bug pests and insecticides used for their control. Little research-based economic injury levels (EIL) have been provided for the green plant bug, which has, thus far, been considered a late season pest. Late-season injury levels for the green plant bug, based on boll damage parameters such as boll size (diameter) and age from tagged white-blooms, has been reported by Armstrong et al. (2009c, 2010). Early season infestations occurring during the pre-bloom period have not been observed in south Texas. Economic thresholds could improve if the dynamics of confounding factors, such as the relationship of boll rot and injury levels based on bug pest densities are studied. The overwintering biology and ecology of plant bugs and stinkbugs and the means to monitor movement into the agricultural crops would be of significant use for management of stinkbugs.

2.4 Control Lepidopteran by using transgenically modified cotton

Transgenically modified cotton that expresses an insecticidal protein derived from *B. thuringiensis* Berlinger is revolutionizing global agriculture (Head et al., 2005). In 1996, it was introduced as transgenic cotton, Bollgard® (Monsanto Co., St. Louis, MO) encoding the Cry 1Ac insect toxin protein (Layton, 1997); in 2002, Bollgard II® (Monsanto Co., St. Louis, MO), which produced the Cry1Ac and Cry2Ab endotoxins (Sherrick et al., 2003); Dow

AgroSciences, LLC (Indianapolis, IN) introduced their pyramided-gene technology into the market in 2004 as Widestrike™, which produced two Bt endotoxins, Cry1Ac and Cry1Fa (Adamczyk and Gore, 2004). VipCot is new transgenic cotton. The active Bt toxin is Vip 3A, which is an exotoxin produced during vegetative stages of Bt growth (Mascarenhas et al., 2003). In the first year of commercial availability in the United States, Bollgard cotton was planted on 850,000 hectares or 15% of the total cotton area, and, by 2007, expanded to about 2.9 million hectares, or 65.8% of U.S. cotton area. However, adoption of Bt cotton has varied greatly across growing regions in the U.S., and other countries, depending on the availability of suitable varieties and, more importantly, the particular combination of pest control problems. Bollgard cotton varieties have been rapidly accepted by farmers in areas where tobacco budworm-bollworm complex (BBWC) is the primary pest problem, particularly when resistance to chemical pesticides is high. There are many factors which can affect changes in expressing the amount of stacked endotoxins. Individual lepidopteran species vary in their susceptibility to Bt proteins (Luttrell & Mink, 1999), and efficacy can be affected by protein expression levels in different plant structures (Adamczyk et al., 2008) and among different varieties (Adamczyk and Gore, 2004). Differences in susceptibility can also occur based on the geographic location of populations (Luttrell et al., 1999). The LRGV of Texas is dominated by beet armyworm, bollworm, and fall armyworm, and suitable Bt varieties have not been readily available for more rapid increase in the adoption of Bt technology.

Microbial insecticides are environmentally friendly and highly selective. Transgenic plants reduce the need for conventional insecticides, providing benefits for human health and the environment. For example, in U.S. cotton, the average number of insecticide applications used against tobacco budworm [*Heliothis virescens* (Fabricius)]-bollworm [*Helicoverpa zea* (Boddie)] complex decreased from 5.6 in 1990-1995 to 0.63 in 2005-2009 (from Proceedings of Beltwide Cotton Conferences).

Year	Bt cotton, ha	% Bt cotton of total planted	Hectares Bt sprayed	Average number applications
		USA		
2005	2,994,086	51.8	1,234,855	0.54
2006	3,439,604	57.2	1,603,722	0.59
2007	2,877,114	65.8	895,232	0.50
		Texas		
2005	546,898	22.6	75,061	0.78
2006	669,891	27.2	37,823	0.44
2007	929,654	47.5	22,657	0.44
		LRGV of Texas		
2005	3,474	4.7	0	0
2006	2,285	5.8	0	0
2007	8,097	20.0	0	0

Source: (Williams, 2006-2008).

Table 5. Bt cotton area.

Carpenter & Ginanessi (2001) estimated that the average annual reduction in use of pesticides on cotton in the U.S. has been approximately 1,000 tons of AI . Traxler et al. (2003) estimated that the benefits gained from the introduction of Bt cotton fluctuates from year to year but averaged $215 million. The adoption of transgenic Bt-cotton is described in Table 5.

Bt types, traits, and varieties mostly used in the LRGV of Texas for the last five years (2005-2010) are shown in Table 6.

Bt type	Bt trait	Variety	Bt endotoxins	Owner of Bt trait	Owner of variety
None	Non-Bt	DPL 5415RR	None	None	Delta & Pineland
Single	Bollgard	NuCotn 33B	Cry1Ac	Monsanto	Delta & Pineland (Monsanto)
Dual	Bollgard II	DPL424 BGII/RR	Cry1Ac + Cry2Ab	Monsanto	Delta & Pineland
Dual	WideStrike	Phy485 WRF	Cry1Ac + Cry2F	Dow Agroscience	Dow Agroscience

Source: Greenberg & Adamczyk, 2010.

Table 6. Bt cottons used in the LRGV of Texas.

During the 2005-2007 seasons, the average percentage of leaf damage on non-Bt trait varieties was 1.5-fold greater than on Bollgard varieties. Leaf damage was 3.6-fold less on Bollgard II and WideStrike-trait varieties than on non-Bt cotton, and 2.4-fold less than on Bollgard-trait varieties (F = 18.8, df = 3, 36, P = 0.001, 2005; F = 15.6, df = 3, 36, P = 0.001, 2006; and F = 10.2, df = 3, 36, P = 0.009, 2007) (Fig. 2). The same trend was observed for the

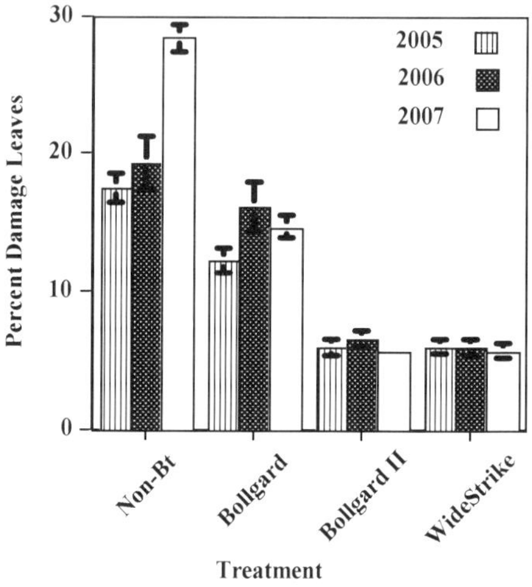

Fig. 2. Percent damage.

proportion of consumed leaves. On non-Bt cotton varieties, the index was 1.6-fold greater than on Bollgard varieties and 2.4-fold greater than on Bollgard II and WideStrike varieties. The proportion of consumed leaves on Bollgard was 1.5-fold greater than on Bollgard II or WideStrike cotton (F = 23.3, df = 3, 36, P = 0.001, 2005; F = 25.8, df = 3, 36, P = 0.002, 2006; F = 23.1, df = 3, 36, P = 0.001, 2007) (Fig. 3). The differences of leaf damage between varieties containing dual Bt endotoxins (Bollgard II and WideStrike) during the cotton-growing seasons were not significant (t = 0.440; P = 0.668) except at the end of the season (110 days of age). The damage to WideStrike cotton (Phy 485 WRF) was 1.4-fold greater than to the Bollgard II variety (ST 4357 BG2RF) (t = 4.332; P = 0.001).

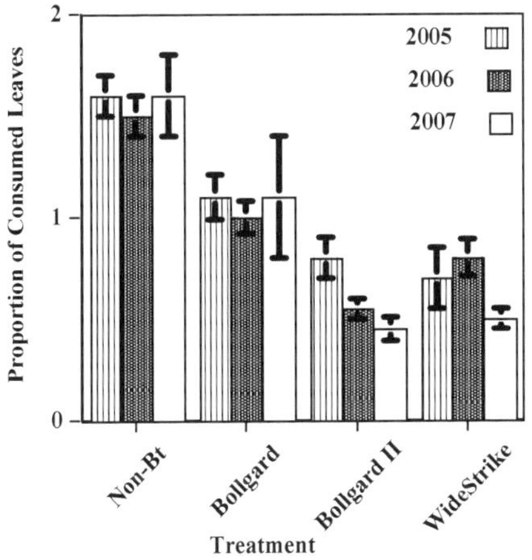

Fig. 3. Proportion of consumed leaves on different Bt trait of cotton.

The seasonal average of damage to fruit on the plant (88.5% attributed to bollworm and, to a lesser extent, beet armyworm) on non-Bt cotton (15.2%) was about 4.6-fold greater than on WideStrike (3.3%), 3.8-fold greater than Bollgard II (4.0%), and 1.7-fold greater than Bollgard (9.0%) (F = 8.9, df = 3, 31, P = 0.001). Damage by noctuids on abscised cotton fruit was 39.0% for non-Bt, 28.5% for Bollgard, 12.6% for Bollgard II, and 8.5% for WideStrike cottons (F = 17.8; df = 3, 16; P = 0.001). In non-Bt cotton, live larvae were 6.2-fold greater than on WideStrike, 4.5-fold greater than on Bollgard II, and only 1.7-fold greater than on Bollgard (F = 11.7; df = 3, 16; P = 0.001). Live larvae in fallen fruit were 92.6% bollworm and 7.4% beet armyworm (Greenberg & Adamczyk, 2010).

Bt cotton has proven itself to be a useful tool in BW eradication zones in minimizing risk of outbreaks of lepidopteran, secondary pest problems; and augmenting activity of beneficial insects.

2.5 Biorational and botanical insecticides

Some registered and produced biorational and botanical insecticides are shown in Tables 7 and 8.

Country	Product name	Based on	Target Insects
U.S.	DiPel DF or ES, Condor, Javelin WG	*Bacillus thuringiensis*	Noctuids
U.S.	Mycotrol	*Beauveria bassiana*	Sucking insects
U.S.	Naturalis	*Beauveria bassiana*	Sucking insects
U.S.	BioBlast	*Metarhizium anisopliae*	Thrips, mites, Coleoptera
U.S.-Europe	PFR-97TM	*Paecilomyces fumosoroseus*	Whiteflies, thrips
U.S.	Spinosad (SpinTor)	*Saccharopolyspora spinosa*	Noctuids, thrips

Source: The Biopesticides Manual, 2001.

Table 7. Registered and produced biorational pesticides.

Common name	Produced	Azadirachtin	Target insects
Neemix™	W.R. Grace & Co. -Conn., Columbia, MD	0.25%	Noctuids, aphids
Neemix®4.5	Certis USA, L.L.C.	4.5	Noctuids, aphids, whiteflies, thrips
Ecozin EC	Amvac, USA, CA	3.0	Noctuids, whiteflies
Agroneem	AgroLogistic Systems, Inc., CA	0.15	Noctuids

Source: Isman, 1999.

Table 8. Registered and produced botanical insecticides.

The effectiveness of some biopesticides based on *B .bassiana* and *M. anisoplia* against sucking insects is not significantly different from synthetic insecticides (Table 9), but *B. thuringiensis* showed satisfactory results against lepidopteran pests (Table 10).

Pesticides	Rate	Mortality, %			
		Young *Bemisia tabaci*	Old *Bemisia tabaci*	*Aphis gossypii*	*Thrips* spp.
B.bassiana	2gr/L	98.8 ± 0.6a	97.6 ± 1.4a	96.4 ± 2.1a	90.4 ± 1.8a
M. anisoplia	5gr/L	90.4 ± 4.8a	91.4 ± 3.1a	91.6 ± 3.6a	98.6 ± 0.8a
Neemix	41.3gr/L	41.6 ± 10.4b	26.0 ± 6.7d	72.1 ± 9.7b	51.4 ± 4.2c
Azadirect	32.3gr/L	68.0 ± 10.2b	64.6 ± 2.5c	90.4 ± 6.5a	46.7 ± 1.8c
QRD	1.3gr/gal	82.1 ± 5.5a	80.3 ± 5.5a	92.4 ±2.7a	69.1 ± 7.7b
Insecticides:					
Fulfil	0.4gr/L	-	-	100a	-
Oberon	0.2gr/L	98.9 ± 0.8a	95.9 ± 3.3a	-	-
Control (H_2O)		6.2 ± 2.0c	1.8 ± 0.8e	4.6 ± 2.0c	1.4 ± 0.9d

Source: Greenberg, unpublished data.

Table 9. Effects of different biorational and botanical pesticides on sucking insects (Greenberg, unpublished data).

Insect Larvae	Pesticides	Mortality, %
Fall armyworm	Spinosad (SpinTor), 12-150 g a.i. per ha	72.3 ± 1.6
Complex (Fall and beet armyworms, bollworm)	Spinosad, 1st spray; DiPel, 2nd spray, 100-300 g a. i. per ha	76.2 ± 3.8
Beet armyworm	DiPel	65.3 ± 3.6
Bollworm	Spinosad	71.3 ± 5.8
Bollworm	DiPel	61.3 ±2.1

Source: Greenberg, unpublished data.

Table 10. Effectiveness of biorational pesticides against lepidopteran.

Three commercial neem-based insecticides, Agroneem, Ecozin, and Neemix, were evaluated for oviposition deterrence of beet armyworm. In controls, the proportion of eggs laid on cotton leaves by beet armyworm was from 2.5 to 9.3-fold higher than neem-based treatments. Neem-based insecticides also deterred feeding by beet armyworm larvae. In controls, the mean percentage of cotton leaves eaten by first instars per day were 3-fold; third instars, 5-fold; and fifth instars,9.3-fold higher than in neem-based treatments, respectively ($P<0.001$). Agroneem, Ecozin, and Neemix caused 78, 77, and 72% beet armyworm egg mortality after direct contact with neem-based insecticides, respectively, while in non-treated controls, only 7.4 % mortality. Survival of beet armyworm larvae fed for 7 days on cotton leaves treated with neem-based insecticides was reduced to 33, 60, and 61% for Ecozin, Agroneem, and Neemix, respectively, compared with 93% in the non-treated controls ($P=0.015$) (Greenberg et al., 2005). Neem-based insecticides could control other lepidopteran, also (Isman, 1999, Ma et al., 2000, Saxena & Rembold, 1984).

2.6 Beneficial insects

Beneficial insects in conventional cotton under BW eradication or intensive pressure of synthetic insecticides can control about 10-15% of harmful insects. Native, most widely-distributed beneficial insects in the LRGV of Texas are described in Table 11.

Beneficial Insects	Target insects
Minute pirate bug, *Orius tristicolor* (White)	Aphids, thrips, whiteflies, mites, and moth eggs and small larvae
Bigeyed bug, *Geocoris uliginosus* (Say)	Mites, whiteflies, thrips, plant bug *Creontiades*, fleahoppers, and moth eggs
Lady beetles, *Hippodamia convergens* (Guerin-Meneville)	Aphids, moth eggs and small larvae
Green lacewings, *Chrysopa rufilabris* (Burmeister)	Immature feed on aphids, spider mites, whiteflies,
Syrphid fly larva	Aphids
Spider, *Hibana futilis* (Banks)	Fleahoppers, *Pseudomatoscelis seritatus* (Reuter), plant bug, *Creontiades signatus* (Distant)
Encarsia pergandiella Howard	Parasites on whiteflies nymphs
Trichogramma spp.	Egg parasite
Bracon spp.	Larva parasite mostly of lepidopteran

Source: Based on Extension Entomologists of LRGV of Texas and authors observations.

Table 11. Native, most widely-distributed beneficial insects in the LRGV of Texas.

We estimated that native parasitoids can control whiteflies in organic cotton (95-100%); sustainable agriculture cotton (80-90 %); Bt cotton (50-60%); conventional cotton (25-30%); and under BW eradication (0-5%).

One of potentially effective strategy for early-season suppression BW involves periodic augmentation an ecto-parasitoid of BW larvae such as *Catolaccus grantis* (Burks) (Summy et al., 1994)

Parasitism of boll weevils by *Catolaccus grandis* in release sites

Site	Date	Percent parasitism
Monte Alto	04.28.93	80.0
	05.05.93	52.8
	0.5.12.93	76.4
	05.19.93	78.3
	05.26.93	74.9
	06.02.93	85.2
Weslaco	05.24.94	83.3
	0.6.02.94	69.2
	06.09.94	62.5
	06.16.94	50.0

Boll weevil used : 3rd instar larva and pupa
Parasites released : Monte Alto - 1,000; Weslaco – 500 females / ac / wk

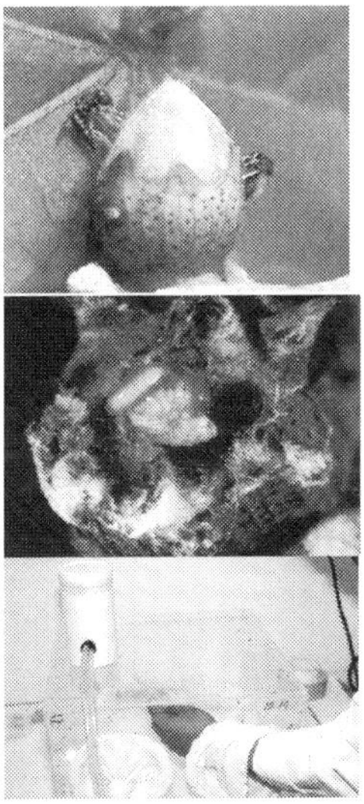

Source: Summy et al., 1994.

Fig. 4. Parasitism of boll weevil larvae by *C. grandis*.

The alternative to chemical control can be propagation and augmentative releases *Trichogramma* spp., an egg parasite of numerous lepidopteran species. *Trichogramma pretiosum* Riley and *T. minutum* Riley are widely use species in the USA. Some lepidopteran species distributed in LRGV, like as beet armyworm and fall armyworm, deposited hair-covering egg masses and protected a portion of eggs from parasitization. But these eggs punctured by *Trichogramma* and rapidly desiccated. The percentage of desiccated eggs tended to increase the total host mortality induced by *Trichogramma* compared with those on bollworm eggs (Greenberg et al., 1998) (Table 12, Fig. 5)

Treatment	Percentage					
	Parasitized eggs		Desiccated eggs		Total mortality	
	BAW	BWTreat	BAW	BW	BAW	BW
T. pretiosum	44.8±4.3	90.3±1.7	24.9±2.1	5.5±1.2	69.7±5.6	95.9±0.5
T. minutum	51.6±3.7	88.9±1.6	29.3±2.2	6.5±1.6	80.9±3.3	95.3±1.6
Control	0	0	5.8±1.2	4.4±1.7	5.8±1.2	4.4±1.7

Source: Greenberg et al., 1998.

Table 12. Effectiveness of *Trichogramma* spp. against noctuids on cotton.

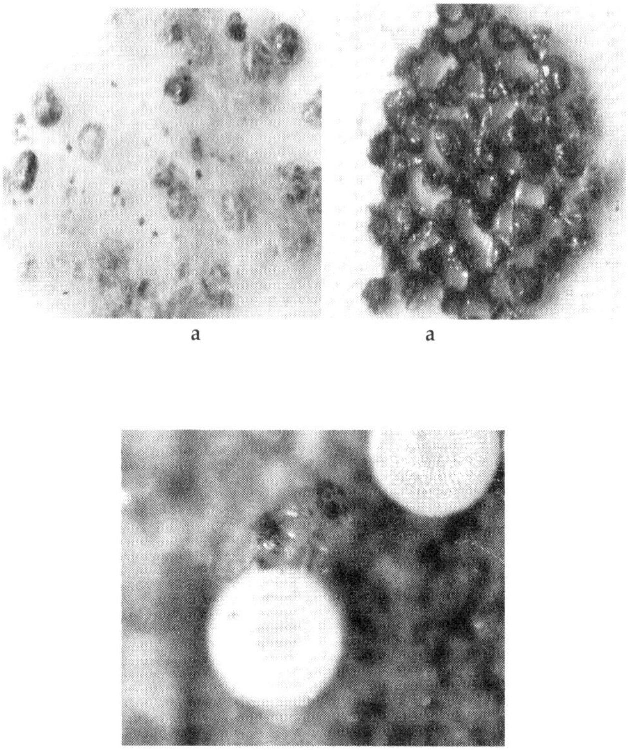

Fig. 5. *Trichogramma* parasitized beet armyworm (a) and bollworm (b) eggs.

3. Cultural control in IPM system

Among the important alternatives to insecticides in cotton are cultural control techniques. Different tillage systems are one of the most important cultural control tools. Conservation tillage has found some acceptance among growers because it reduces soil erosion, conserves soil moisture, and substantially lowers cost of field operations compared to conventionally tilled systems. In the LRGV, 30% of cotton acreage is under conservation tillage. Water availability for irrigation has become a major concern for south Texas. In this case, conservation tillage can be a valuable tool for improving soil moisture. Our results demonstrated that different tillage practices had indirect potentially positive or negative effects on pest and beneficial populations in cotton. The effects are influenced by both abiotic and biotic factors which can be created or manipulated by conventional (cv) and conservation (cs) tillage systems. Tillage operations modify soil habitats where some insect pests and beneficial insects reside during at least part of their life cycles. These modifications can alter survival and development of both soil and foliage-inhabiting insects.

Conventional tillage in dryland cotton increased water stress, causing plants to shed squares and bolls, and allocated more resources into vegetative growth. The conservation tillage cotton responded by fruiting at a higher rate. Increased plant height and number of leaves in conventional tillage provided significantly more light interception and shading of the soil surface between rows. Temperatures in conservation tillage rows were higher than in conventional tillage fields by about 15°C and resulted in increased mortality of insects in fallen fruit (Greenberg et al., 2004, 2010).

Boll Weevil: In dryland cotton, the average number of boll weevils per plant during the 2001 cotton growing season was 2.3-fold (P=0.011) and, in 2002, - 3.5-fold (P=0.019) higher in conventional versus conservation tillage fields (Greenberg et al., 2003).

Aphids: On seedling cotton, numbers of aphids were higher in conventional tillage plots. In late spring and early summer, aphids primarily migrated to conservation tillage cotton where there was higher soil moisture and RH, and plants were more succulent and attractive to aphids than in conventional tillage.

Bollworm and Tobacco Budworm, Beet Armyworm. Fruit fallen on the ground were infested with larvae at 15.7 % higher in conventional than in conservation plots. Numbers of live larvae in infested fruit were 4.7-fold higher in conventional versus conservation tillage plots (69.3% vs. 14.7%). The number of larvae per plant was 5.9-fold higher in conventional than conservation tillage.

Cutworm. Higher infestation densities and plant damage have been observed in conservation tillage fields on seedling cotton (18.3% damaged plants in conservation tillage and 2.7 % in conventional tillage). Conservation tillage promotes the development of weeds that serve as oviposition sites for adults and alternative plant hosts for larval development (Greenberg et al., 2010).

4. Cotton diseases

A plant disease occurs when there is an interaction between a plant host, a pathogen, and the environment. When a virulent pathogen is dispersed onto a susceptible host and the environmental conditions are suitable, then a plant disease develops and symptoms become evident.

Seedling Disease Complex. Seedling disease is caused by a complex of soil fungi which may occur separately or in combinations. These fungi are *Pythium* sp., *Fusarium* sp., *Rhizoctonia solani*, and *Thielaviopsis basicola*. Symptoms include decay of the seed before germination, decay of the seedling before emergence, girdling of the emerged seedling at or near the soil surface, and rotting of root tips. Crop rotation, quality of the seed, timely planting, and the use of fungicides like Captain, Maxim, Nu-Flow ND, Nu-flow M, Vitavax, and Baytan can reduce losses to seedling diseases and are registered for commercial seed and soil treatments (Allen et al., 2010).

Root Rot. This disease, caused by the fungus, *Phymatotrichum omnivorum*, generally becomes evident during the early summer. It causes rapid wilting, followed by death of the plants within a few days. Leaves shrivel, turn brown and die, but they remain attached to the plant. The disease kills plants in circular areas ranging from a few square yards to an acre or more in size. Dead plants will remain standing in the field but can be easily pulled from the soil. Control procedures include: 1) altering the growing environment in the root zone by applying soil amendments to increase organic matter and reducing soil PH by using the chelated element sulfur and in organic trace elements zinc and iron; 2) using winter cover *Brassicae* plants as a cultural control for disease suppression; 3) fumigating infested planting holes will usually only delay the onset of disease in non-infested plants; and 4) applying sulfur in trenches 4 to 6 inches wide and 4 to 6 feet deep around the outside of the drip line of infested plants to prevent the spread of root rot. Incidence and control of cotton root rot is observed with color-infrared imagery by using remote sensing equipment (Matocha et al., 2008, 2009).

Boll Rot. This disease is prevalent in high moisture and heavy plant densities. If excessive stalk growth has occurred, one may encounter boll rot problems. Reducing some of the leaf tissue with the selective use of defoliants may be a practical answer. Good weed and insect management will decrease incidence of boll rot (Allen et al., 2010).

Nematodes. The nematode *Rotylenchus reniformis* Linford & Oliveria is a major problem confronting cotton production in the LRGV of Texas. Root-knot nematode, *Meloidogyne incognita* (Kofoid & White), is prevalent in sandy or sandy clay loam soils. Larvae feed on the root plants causing swellings (galls) on them. Control practices for nematodes include crop rotation and chemical control with nematicides or soil fumigants (Robinson et al., 2008).

5. Weed control

The main winter and spring weeds in cotton are common purslane (*Portulaca oleracea* L.), pigweed (*Amaranthus palmeri* Wats.), wild sunflower (*Helianthus annuus* L.), and Johnsongrass[*Sorghum halepense* (L.) Persoon]. Control is by use of a conventional tillage system, winter cover crops, and selective herbicides. Black oat (*Avena strigosa* Schreb.) and hairy vetch (*Vicia villosa* Roth) suppressed winter weeds to the same extent or more than did winter tillage in no-cover plots. In the spring, soil incorporated black oats cover was slightly more beneficial to cotton than incorporated hairy vetch, but neither cover controlled spring weeds. Two years of winter cover cropping did not obviate the need for cultivation, and hand-weeding for sustainable spring weed management in cotton in the LRGV of Texas. (Moran & Greenberg, 2008).

6. Conclusion – IPM models

The model of a pyramid can be used to demonstrate how growers might construct their pest management programs. There are different models of pyramids, but they are basically similar. In Fig. 6 (Model #1), the foundation of a sound pest and disease management program in an annual cropping system that begins with cultural practices which alter the environment to promote crop health. These include crop rotations that limit the availability of host material used by plant pathogens, judicious use of tillage to disrupt pest and pathogen life cycles, destruction of weeds, and preparation of seed beds. Management of soil fertility and moisture can also limit plant diseases by minimizing plant stress. Environmental control can regulate in terms of temperature, light, moisture, and soil composition. However, the design of such systems cannot wholly eliminate pest problems. The second layer of defense against pests consists of the quality of crop germplasm. Newer technologies that directly incorporate genes into crop genomes, commonly referred to as genetic modification or genetic engineering, are integrating new traits into crop germplasm. The most-widely distributed are the different insecticidal proteins derived from *Bacillus thuringiensis*. Upon these two layers, growers can further reduce pest pressure by considering both biological and chemical inputs (McSpadden Cardener & Fravel, 2002).

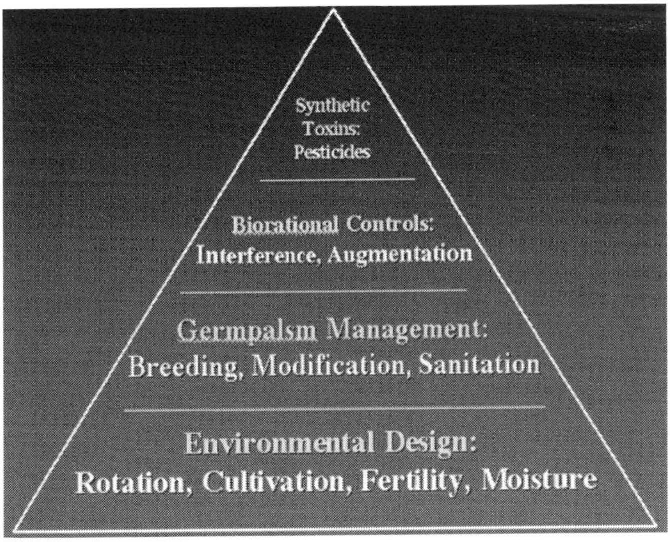

Source: Gardener & Fravel, 2002.

Fig. 6. Model # 1.

High yields of agricultural crops can only be obtained if there is sufficient control of pests. In the mid 20th century, development of chemical pesticides seemed to provide an effective answer, but pests became resistant and, by killing natural beneficial species, resurgence of pest populations occurred. The LRGV played a key role in the acceptance of IPM concept by entomologists. The devastating outbreaks of tobacco budworm (*Heliothis virescens*) in the LRGV of Texas during the late 1960's and early 1970's (and the similar outbreaks of *Heliothis armigera* in Australia during the same period) demonstrated conclusively that unilateral reliance on pesticides for insect control was not sustainable and could lead to economic calamities. This led to the concept of integrated pest management utilizing a range of control tactics in a harmonious way (Fig.7, Model #2 adapted from Naranjo, 2001). The diagram shows the different aspects of IPM - avoidance of pest, then surveillance and finally, if necessary, control using a bio- or chemical pesticide.

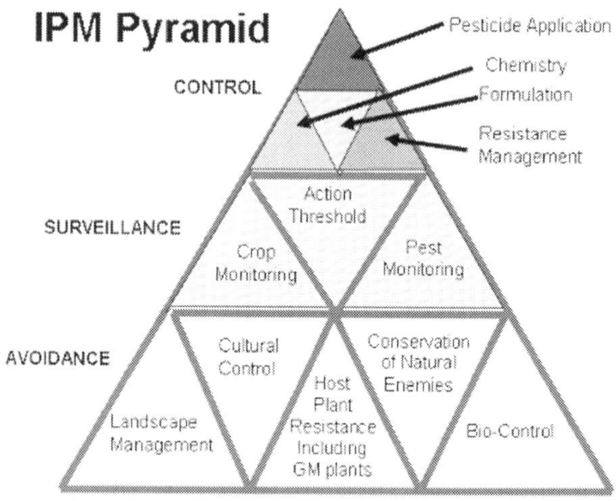

Source: Naranjo, 2001.

Fig. 7. Model #2.

In Texas, IPM implies integration of approaches and methods into a pest management system, which takes into consideration that environmental impacts and economic risks have been minimized.

IPM models (Figs. 8, 9) based on conceptions of Extension Entomologists Texas A&M University System and authors of this article. No single pest control method is relied on in IPM systems. Chemical control is used only when needed (in relation to economic thresholds), and it is important to optimize their application. Nozzles need to be selected to optimize the droplet sizes so that the pesticides can be distributed where the pests are located with minimal spray drift. Monitoring (sampling) of the pest is constantly needed. Mere presence of a pest is not a reason to justify action for control.

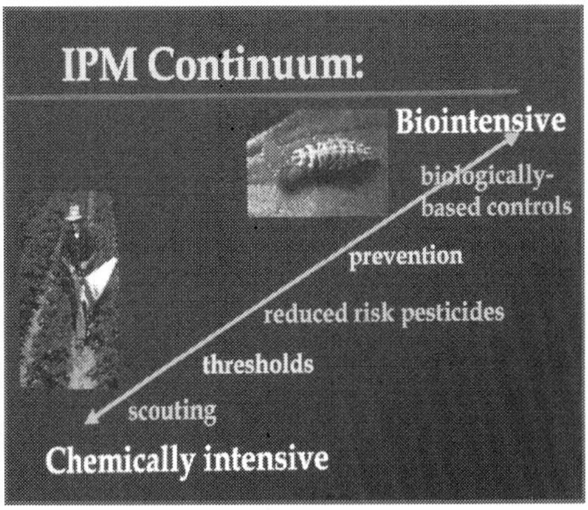

Fig. 8.

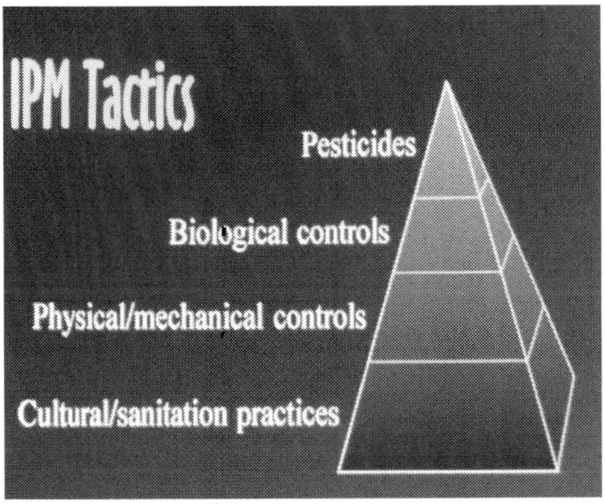

Fig. 9.

In the future, IPM is expected to continue to be dominant in agriculture. This will include increased use of reduced-risk pesticides and genetically-engineered crops. Recent surveys of both conventional and organic growers indicate an interest in using biocontrol products (Van Arsdall & Frantz, 2001). The future success of the biological control industry will depend on innovative business management, product marketing, extension education, and research (Mathre et al., 1999). These will contribute substantially to making the 21st century the age of biotechnology by the development of innovative IPM strategies.

7. Acknowledgment

We thank Prof. K. R. Summy (University of Texas Pan-American), Dr. L. A. Wood (Mission Plant Protection Lab, APHIS, USDA), Assistant Prof. L. Rivera (Texas Agrilife Extension Service, Texas A&M University System) for reviewing this manuscript; and J. Alejandro for technical assistance.

8. References

Adamczyk, J. J., Jr. & Gore, J. (2004). Laboratory and field performance of cotton containing Cry1Ac, Cry1F, and both Cry1Ac and Cry1F (WideStrike) against beet armyworm and fall armyworm larvae (Lepidoptera: Noctuidae). *Florida Entomologist* 87: 427-432.

Adamczyk J. J., Jr., Greenberg, S., Armstrong, S., Mullins, W. J., Braxton, L. B., Lassiter, R. B. & Siebert, M. W. (2008). Evaluation Bollgard II and WideStrike technologies against beet and fall armyworms. *Proceedings, Beltwide Cotton Conferences. National Cotton Council, Memphis, TN*, Nashville, Tennessee, January 2008. CD-ROM.

Agricultural Statistics. (2008). United States Department of Agriculture. United States Government Printing Office. ISBN 978-0-16-079497-1. Washington.

Allen, S. J., Scheikowski, L. J., Gambley, C., Sharman, M. & Maas, S. (2010). Integrated disease management for cotton. *Cotton pest management guide*. Pp. 115-121.

Armstrong, J. S., Coleman, R. J. & Duggan, B. L. (2007). Characterizing the damage and oviposition of a *Creontiades* plant bug to South Texas cotton. *Proceedings, Beltwide Cotton Conferences. National Cotton Council, Memphis, TN*, New Orleans, Louisiana, January, 2007. CD-ROM.

Armstrong, J. S., Adamczyk, J. J. & Coleman, R. J. (2009a). Determining the relationship between boll age and green plant bug feeding injury to South Texas cotton. *Proceedings, Beltwide Cotton Conferences. National Cotton Council, Memphis, TN*, San Antonio, Texas, January, 2009. CD-ROM.

Armstrong, J. S., Medrano, E. G. & Esquivel, J. F. (2009b). Isolating and identifying the microbes associated with green mirid feeding injury to cotton bolls. *Proceedings, Beltwide Cotton Conferences. National Cotton Council, Memphis, TN*, San Antonio, Texas, January, 2009. CD-ROM.

Armstrong, J. S., Coleman, R. J. & Setamou M. (2009c). Oviposition patterns of *Creontiades signatus* Distant (Hemiptera: Miridae) on okra-leaf and normal leaf cotton. Annals Entomological Society of America.102: 196-200.

Armstrong, J. S., Coleman, R. J. & Duggan, B. L. (2010). Actual and simulated injury of *Creontiades signatus* Distant (Hemiptera: Miridae) feeding on cotton bolls. *Journal of Entomological Science* 45: 170-177.

Bell, A. A., Lopez, J. D. & Medrano, E. G. (2006). Frequency and identification of cottonseed-rotting bacteria from cotton fleahoppers. . *Proceedings, Beltwide Cotton Conferences. National Cotton Council, Memphis, TN*, San Antonio, Texas, February, 2006. CD-ROM.

Bell, A. A., Medrano, E. G. & Lopez, J. D. (2010). Transmission and importance of *Pantoea ananatis* during feeding on cotton buds (*Gossypium hirsutum* L.) by cotton fleahoppers (*Pseudatomoscelis seriatus* Reuter). *Proceedings World Cotton Research Conference-4*, Lubbock, Texas, September 2007. CD-ROM.

Bennett, G. W., Owens, J. M. & Corrigan, R. M. (2005). *Truman's Scientific Guide to Pest Management Operations*. 6th Edition: 10-12. Purdue University, Questex.

Bohmfalk, G. T., Frisbie, R. E., Sterling, W. L., Metzer, R. B. & Knutson, A. E. (2002). *Identification, biology, and sampling of cotton insects*. Texas A&M University System, College Station.

Brewer, M., Anderson, D. and Armstrong J. S. (2011) Comparing external and internal symptoms of feeding injury caused by sucking bugs to harvest potential, with emphasis on *Creontiades signatus. Proceedings, Beltwide Cotton Conferences. National Cotton Council, Memphis, TN,* Atlanta, Georgia, January, 2011. CD-ROM

Carpenter, J. E. & Gianessi, H. 2001. *Agricultural biotechnology: Updated benefits, estimates,* National Center for Food and Agricultural Police: Washington, D. C.

Castro, B. A., Cattaneo, M. and Sansone, C. G. (2007). *Managing Cotton Insects in the Lower Rio Grande Valley.* Texas Agricultural Extension Service. Texas A&M University System. Weslaco, Texas.

Coleman, R.J., Hereward, H.P., De Barro, P.J., Frohlich, D.R., Adamczyk Jr, J.J. & Goolsby, J. (2008). Molecular comparison of *Creontiades* plant bugs from South Texas and Australia. *Southwestern Entomologist* 33:111-117.

Copping, L. G. (Fd.) (2001). *The biopesticide manual.* Second Edition. British Crop Protection Council. UK

King, E. G., Phillips, J. R. & Coleman, R. J. (Eds.). (1996). *Cotton insects and mites: characterization and management,* The Cotton Foundation Publisher, ISBN 0-939809-03-6, Memphis, TN.

Duffey, J. E. & Powell, R. D. (1979). Microbial induced ethylene synthesis as a possible factor of square abscission and stunting in cotton infested by cotton fleahopper. *Annals of the Entomological Society of America* 72: 599-601.

FAO (2005). FAOSTAT - Agriculture.

French, J. V., Greenberg, S. M. & Haslem, P. (2006). Leaf cutter ants on the go. *Newsletter of Citrus Center Texas A&M University-Kingsville.* Vol. 24, # 2. Weslaco, Texas.

Gannaway, J. R. (1994). Breeding for insect resistance. In: *Insect pests of cotton,* Matthewsand, G. A. & Tunstall, J. P. (eds.), pp. 431-453, CAB International, Wallingford, UK.

Gianessi, L. and Reigner, N. (2006). Pesticides use in U. S. crop production: 2002. *Insecticides & other pesticides.* CropLife Foundation. Washington, DC.

Greenberg, S. M., Summy, K. R. Roulston, J. R. & Nordlund, D. A. (1998). Parasitism of beet armyworm by *T. pretiosum* and *T. minutum* under laboratory and field conditions. *Southwestern Entomologist.* 23 (2): 183-188.

Greenberg, S. M., Smart, J. R., Bradford, J. M., Sappington, T. W., Norman, J. W., Jr. & Coleman, R. J. (2003). Effects of different tillage systems in dryland cotton on population dynamics of boll weevil (Colioptera: Curculionidae). *Subtropical Plant Science.* 55: 32-39.

Greenberg, S. M., Showler, A. T., Sappington, T. W. & Bradford, J. M. (2004). Effects of burial and soil condition on post-harvest mortality of boll weevils (Colioptera: Curculionidae) in fallen cotton fruit. *Journal of Economic Entomology.* 97 (2): 409-413.

Greenberg, S. M., Showler, A. T. & Liu, T. X. (2005). Effects of neems-based insecticides on beet armyworm (Lepidoptera: Noctuidae). *Insect Science.* 12: 17-23.

Greenberg, S. M., Sappington, T. W., Elzen, G. W., Norman, J. W. & Sparks, A. N. (2004). Effects of insecticides and defoliants applied alone and in combination for control of overwintering boll weevil (*Anthonomus grandis,* Coleoptera: Curculionidae) - laboratory and field studies. *Pest Management Science.* 60: 849-859.

Greenberg, S. M., Yang, C. & Everitt, J. H. (2005). Evaluation effectiveness of some agricultural operations on cotton by using remote sensing technology. *Proceedings 20th Workshop of remote sensing technology,* Weslaco, Texas, October 2005, CD.

Greenberg, S. M., Sparks, A. N., Norman, J. W., Coleman, R., Bradford, J. M., Yang, C., Sappington, T. W. & Showler, A. (2007). Chemical cotton stalk destruction for maintenance of host-free periods for the control of overwintering boll weevil in tropical and subtropical climates. *Pest Management Science.* 63: 372-380.

Greenberg, S. M., Liu, T.-X. & Adamczyk, J. J. (2009a). Thrips (Thysanoptera: Thripidae) on cotton in the lower Rio Grande Valley of Texas: Species composition, Seasonal abundance, Damage, and Control. *Southwestern Entomologist.* 34: 417-430.

Greenberg, S. M., Jones, W. A. & Liu, T.-X. (2009b). Tritrophic interactions among host plants, whiteflies, and parasitoids. *Southwestern Entomologist* 34: 431-445.

Greenberg, S. M. & Adamczyk, J. J. (2010). Effectiveness of transgenic Bt cottons against noctuids in the Lower Rio Grande Valley of Texas. *Southwestern Entomologist* 35: 539-549.

Greenberg, S. M., Bradford, J. M., Adamczyk, J. J., Smart, J.R. & Liu, T.-X. (2010). Insects population trends in different tillage systems of cotton in south Texas. *Subtropical Plant Science.* 62: 1-17.

Greene, J. K., Turnipseed, S. G., Sullivan, M. J. & Herzog, G. A. (1999). Boll damage by southern green stinkbug (Hemiptera: Pentatomidae) and tarnished plant bug (Hemiptera: Miridae) caged on transgenic *Bacillus thuringiensis* cotton. *Journal of Economic Entomology.* 92(4): 941-944.

Grefenstette, W. J & El-Lissy, O. (2008). Boll weevil eradication – *2008. www.aphis.usda.*

Head, G., Moar, M., Eubanks, M., Freeman, B., Ruberson, J., Hagerty, A. & Turnipseed, S. (2005). A multiyear, large-scale comparison of arthropod populations on commercially managed Bt and non-Bt cotton fields. *Environmental Entomology.* 34: 1257-1266.

http://www.panna.org/files/conventionalCotton.dv.html

IPM definitions: ND IPM Homepage, TX Pest Management Association. *www.PestControlCanada.com.* California Department of Pesticide Regulation, *cdprweb@cdpr.ca.gov* Ohio Pest Management & Survey Program, *http://ohioline.osu.edu/icm-fact/fc-01. html* (+ Pest management practices).

Isman, M. B. (1999). Neem and related natural products. In: *Biopesticides: Use and Delivery,* Hall, F. R. & Menn, J. J., Eds, pp. 139-154. Humana Press, Totowa, NJ.

Jenkins, J. N. & Wilson, F. D. (1996). Host Plant Resistance. In: Cotton Insects and Mites: Characterization and management, King, E.C., Phillips, J.R. & Coleman, R.J., Eds., pp. 563-597. The Cotton Foundation Publisher, ISBN 0-939809-03-6, Memphis, TN.

Layton, M. B. (1997). Insect scouting and management in Bt-transgenic cotton. Mississippi Cooperative Extension Service Publication 2108.

Layton, M. B. (2000). Biology and damage of the tarnished plant bug, *Lygus lineolaris,* in cotton. *Southwestern Entomologist.* 23: 7-20.

Lei, Z., Liu, T.-X. & Greenberg, S. M. (2009). Feeding, oviposition, and survival of *Liriomyza trifolii* (Diptera: Agromyzidae) on Bt and non-Bt cottons., 99: 253-261. *Bulletin of Entomological Research,* Cambridge University Press.

Leigh, T.F., Roach, S.H. & Watson, T.F. (1996). Biology and Ecology of Important Insect and Mite Pests of Cotton. In: Cotton Insects and Mites: Characterization and management, King, E.C., Phillips, J.R. & Coleman, R.J., Eds., pp. 17-50. The Cotton Foundation Publisher, ISBN 0-939809-03-6, Memphis, TN.

Leonard, R. 2006. Status of chemical control strategies. *Proceedings, Beltwide Cotton Conferences. National Cotton Council, Memphis, TN,* San Antonio, Texas, February, 2006. CD-ROM.

Leonard, B. R., Gore, J., Temple, J. & Price, P. P. (2006). Insecticide efficacy against cotton insect pests using air induction and hollow cone nozzles, 957961. *Proceedings, Beltwide Cotton Conferences. National Cotton Council, Memphis, TN*, San Antonio, Texas, February, 2006. CD-ROM.

Luttrell R. G., Wan L. & Knighten, K. (1999). Variation in susceptibility of noctuid (Lepidoptera) larvae attacking cotton and soybean to purified endotoxin protein and commercial formulations of *Bacillus thuringiensis*. *Journal of Economic Entomology*. 92: 21-32.

Lopez, J. D., Fritz, B. K., Latheef, M. A., Lan, Y., Martin, D. E. & Hoffmann, W. C. (2008). Evaluation of toxicity of selected insecticides against thrips on cotton in laboratory bioassays. Journal of Cotton Science. 12: 188-194.

Lower Rio Grande Valley Cotton Blue Book. (2006). Texas Cooperative Extension. The Texas A&M University System.

Lower Rio Grande Valley Cotton Blue Book. (2007). Texas Cooperative Extension. The Texas A&M University System.

Lower Rio Grande Valley Cotton Blue Book. 2008. Texas Cooperative Extension. The Texas A&M University System.

Lukefahr, M. J., Cowan, C. B. Jr., Bariola, L. A. & Houghtaling, J. E. (1968). Cotton strains resistant to the cotton fleahopper. *Journal of Economic Entomology*. 61: 661-664.

Lukefahr, M. J., Cowan, C. B., Jr. & Houghtaling, J. E. (1970). Field evaluations of improved cotton strains resistant to the cotton fleahopper. *Journal of Economic Entomology*. 63: 1101-1103.

Luttrell, R. G. & Mink, J. S. (1999). Damage to cotton fruiting structures by the fall armyworm, *Spodoptera frugiperda* (Lepidoptera: Noctuidae). *Journal of Cotton Science*. 3: 35-44.

Leigh, T. F., Grimes, D. W., Dickens, W. L. & Jackson, C. E. (1974). Planting pattern, plant populations, and insect interactions in cotton. *Environmental Entomology*. 3: 492-496.

Ma, D.-L., Gordh, G. & Zalucki, M. P. (2000). Biological effects of azadirachtin on *Helicoverpa armigera* (Hubner) (Lepidoptera: Noctuidae) fed on cotton and artificial diet. *Australian Journal of Entomology*. 39: 301-304.

Mascarenhas, V. J., Shotkoski, F. & Boykin. (2003). Field performance of Vip Cotton against various lepidopteran cotton pests in the U.S., pp. 1316-1322. *Proceedings, Beltwide Cotton Conferences. National Cotton Council, Memphis, TN*, Nashville, Tennessee, January, 2003.

Mathre, D. E., Cook, R. J. & Callan, N. W. (1999). From discovery to use: Traversing the world of commercializing biocontrol agents for plant disease control. *Plant Diseases* 83: 972-983.

Matocha,J. E., Greenberg, S. M., Bradford, J. M. & Wilborn, J. R. (2008). Biofumigation and soil amendment effects on cotton root rot suppression. *Proceedings, Beltwide Cotton Conferences. National Cotton Council, Memphis, TN*, Nashville, Tennessee, January 2008. CD-ROM

Matocha,J. E., Greenberg, S. M., Bradford, J. M., Yang, C., Wilborn, J. R. & Nichols, R. (2009). Controlled release fungicides, soil amendments, and biofumigation effects on cotton root rot suppression. *Proceedings, Beltwide Cotton Conferences. National Cotton Council, Memphis, TN*, San Antonio, Texas, January, 2009. CD-ROM.

McSpadden Gardener, B. B. & Fravel, D. R. (2002). Biological control of plant pathogens: Research, commercialization, and applications in the USA. *Online. Plant Health Progress doi:10.1094/PHP-2002-0510-01-RV.*

Medrano, E.G., Esquivel, J.F., Bell, A.A., Greene, J., Roberts, P., Bachelor, J., Marois, J.J., Wright, D.L., Nichols, R.L. & Lopez, J. (2009). Potential for *Nezara virdula* (Hemiptera: Pentatomidae) to transmit bacterial and fungal pathogens into cotton bolls. *Current Microbiology*. 59:405-412.

Moran, P. J., & Greenberg, S. M. (2008). Winter cover crop and vinegar for early-season weed control in sustainable cotton. *Journal of Sustainable Agriculture*. 32 (3): 483-506.

Muegge, M. 2002. *Far West Texas cotton Insects Update*. 1, (3). Texas AgriLife Extension Service.

Musser, F. R., Knighten, K. S. & Reed. J. T. (2009) Comparison of damage from tarnished plant bug (Hemiptera: Miridae) and southern green stinkbug (Hemiptera: Pentatomidae) adults and nymphs. Midsouthern Entomology. 2: 1-9.

Naranjo, S. E., Ellsworth, P. C., Chu, C. C. & Henneberry, T. J. (2001). Conservation of predatory arthropods in cotton: Role of action thresholds for *Bemisia tabaci*. *Journal of Economic Entomology*. 95: 682- 691.

Norman, J. W. & A. N. Sparks. (2003). Managing Cotton Insects in the Lower Rio Grande Valley. *Texas Agricultural Extension Service*. TX A&M University System.

Oerke, E. C. (2006). Crop losses to pests. *Journal of Agricultural Science*. 144: 31-43.

Pest management strategic plan for cotton in the midsouth. (2003). Arkansas, Louisiana, Mississippi Cooperative Extension.

Robinson, A. F., Westphal, A., Overstreet, C., Padgett, G. B., Greenberg, S. M., Wheeler, T. A. & Stetina, S. R. (2008). Detection of suppressiveness against *Rotylenchulus reniformis* in soil from cotton (*Gossypium hirsutum*) fields in Texas and Louisiana. *Journal of Nematology*. 40 (1): 35-38.

Saxena, K, N. & Rembold, H. (1984). Orientation and ovipositional responses of *Heliothis armigera* to certain neem constituents, pp. 199-210. *Proceedings of the 2nd International Neem Conference*, Rauischholzhausen, Germany.

Sharma H. C. & Ortiz R. (2000). Transgenics, pest management, and the environment. *Current Science*. 79: 421–437.

Sherrick, S. S., Pitts, D., Voth, R. & Mullins, W. 2002. Bollgard II performance in the Southeast, pp. 1034-1049. *Proceedings, Beltwide Cotton Conferences. National Cotton Council, Memphis, TN*, Atlanta, Georgia, January, 2002.

Snodgrass, G. L., Gore, J., Abel, C. A. & Jackson, R. E. (2009). Predicting field control for tarnished plant bug (Hemiptera:Miridae) populations with pyrethroid insecticides by glass-vial bioassays. *Southwestern Entomologist*. 33: 181-189.

Sparks, A. N. & Norman, J. W. (2003). Pest and beneficial arthropods of cotton in the Lower Rio Grande Valley. *Texas Agricultural Extension Service*. The Texas A&M University System.

Statistical Highlights of United States Agriculture. (2007). National Agricultural Statistics Service. U. S. Government Printing Office. Washington.

Sterling, W. L. (1984). Action and inaction levels in pest management. *Texas Agricultural Experimental Station Bulletin B-1480.*

Stewart, S.D. & Sterling, W.L. (1989). Susceptibility of different ages of cotton fruit to insects, pathogens and physical stress. *Journal of Economic Entomology*. 82(2): 593-598.

Steyskal, G. C., Murphy, W. L. & Hoover, E. M. (1986). Insects and mites, 103 pp. *Techniques for collection and preservation*. USDA, ARS, Publication No. 1443

Summy, K. R., Morales-Ramos, J. A., King, E. G., Wolfenbarger, D. A., Coleman, R. J. & S. M. Greenberg. (1994). Integration of boll weevil parasite augmentation into the short-season cotton production system of the Low Rio Grande Valley, pp. 953-964. *Proceedings, Beltwide Cotton Conferences. National Cotton Council, Memphis, TN*, San Diego, January, 1994.

Summy, K. R., Raulston, J. R., Spurgeon, D. & Vargas, J. (1996). An analysis of the beet armyworm outbreak on cotton in the Lower Rio Grande Valley of Texas during the 1995 production season, pp. 837-842. *Proceedings, Beltwide Cotton Conferences. National Cotton Council, Memphis, TN*, Nashville, Tennessee. January, 1996.

Texas Boll Weevil Eradication Foundation. (2011). *txbollweevi.org*.

Tingey, W. M., & Pellemer, E. A. (1977). Lygus bugs: Crop resistance and physiological nature of feeding. *Bulletin Entomological Society of America*. 23:277-287.

Tomlin, C. D. S. (Ed.) (2003). *The pesticide manual*. Thirteenth Edition. British Crop Protection Council. UK.

Traxler, G., Godoy-Avila, S., Falck-Zepeda, J. & Espinoza-Arellano, J. (2003). Transgenic cotton in Mexico: Economic and Environmental impacts, In: *Economic and Environmental Impacts of First Generation Biotechnologies*, Kalaitzandonakes, N., Ed., pp. 2-32. Department of Agricultural Economics, Auburn University, Mexico.

Tugwell, P., Young, S. C., Dumas, B. A. & Phillips, J. R. (1976). Plant bugs in cotton: importance of infestation time, types of cotton injury, and significance of wild hosts near cotton. *Report Series 227*. University of Arkansas Agricultural Experiment Station.

Van Arsdall, R. T., & Frantz, C. (2001). Potential role of farmer cooperatives in reducing pest risk: *Final report. Online*. National Council of Farmer Cooperatives. US EPA. Pesticide Environmental Stewardship Program.

Williams, M. R. (2005). Cotton insect losses. *Proceedings, Beltwide Cotton Conferences. National Cotton Council, Memphis, TN*, New Orleans, Louisiana, January, 2005. CD-ROM.

Williams, M. R. (2006). Cotton insect losses. *Proceedings, Beltwide Cotton Conferences. National Cotton Council, Memphis, TN*, San Antonio, Texas, February, 2006. CD-ROM.

Williams, M. R. (2007). Cotton insect losses. *Proceedings, Beltwide Cotton Conferences. National Cotton Council, Memphis, TN*, New Orleans, Louisiana, January, 2007. CD-ROM.

Williams, M. R. (2008). Cotton insect losses. *Proceedings, Beltwide Cotton Conferences. National Cotton Council, Memphis, TN*, Nashville, Tennessee, January 2008. CD-ROM.

Willrich, M. M., Leonard, B. R., Gable, R. H. & LaMotte, L. R. (2004). Boll injury and yield losses in cotton associated with brown stinkbug (Heteroptera: Pentatomidae) during flowering. Journal of Economic Entomology. 97: 1928-1934.

Zhang, L., Greenberg, S. M., Zhang, Y. & Liu, T.-X. (2011). Effectiveness of systemic insecticides, thiamethoxfam and imidacloprid treated cotton seeds against *Bemisia tabaci* (Homoptera: Aleyrodidae). *Pest Management Science*. 67: 226-232.

Zink, A.W. and Rosenheim, J. (2005). State-dependent feeding behavior by western tarnished plant bugs influences flower bud abscission in cotton. *Entomologia Experimentalis et Applicata* 117: 235-242.

2

Toward the Development of Novel Long-Term Pest Control Strategies Based on Insect Ecological and Evolutionary Dynamics

René Cerritos[1]*, Ana Wegier[2,3] and Valeria Alavez[2]
[1]*Laboratorio de inmunología, Unidad de Medicina experimental, Facultad de Medicina, Universidad Nacional Autónoma de México*
[2]*Instituto de Ecología, Universidad Nacional Autónoma de México,*
[3]*Current institution: CENID-COMEF, Instituto Nacional Investigaciones Forestales, Agrícolas y Pecuarias*
México

1. Introduction

Most of the organisms that have negative impacts on agroecosystems and human health, namely bacteria, arthropods, fungi and weeds, share distinctive traits: short generation times, numerous offspring, and therefore large population sizes. These characteristics allow these organisms to change so fast that control of their population growth is difficult to achieve. However, species with these traits, viewed in an ecological context, are subjected to different selective pressures that impede unlimited growth. For instance, as insects are currently the most devastating group within agroecosystems, producing grave economic losses, farmers have resorted to the use of insecticides—whether natural, synthetic, or expressed in genetically modified organisms—as the main control method used to deal with this problem. While the use of insecticides can aid in the short-term control of insect pests these control methods present six fundamental problems that are environmentally irreversible: 1) the pest evolution of insecticide resistance; 2) eradication of non-target species; 3) elimination of ecological interactions; 4) modifications of the biogeochemical cycles; 5) environmental pollution; and 6) impact on human health. Currently, a great deal of knowledge about ecological and evolutionary processes and dynamics is becoming available; this can help to explain the issues mentioned above. In this chapter, we will analyze these processes to subsequently propose alternatives for a long-term integral pest management system.

1.1 Understanding insect pest populations

Insects represent almost 60% of the total species diversity existent in the planet (Strong *et al.*, 1984; Purvis & Hector, 2000; Gibbs, 2001). It is estimated that 26% of all extant living organisms on Earth (361,000 species) are phytophagous insects, while 31% (430,000 species) are saprophagous or predators (Stong *et al.*, 1984). Insect populations are mostly characterized by presenting early reproduction, small body size, undergoing just one reproductive event in their lifetime (i.e., semelparity), having small progeny, and allocating substantial resources to reproduction (Borror *et al.*, 1992). In general, they lack paternal care (Daly *et al.*, 1978) and individuals produce a great amount of descendants to maintain the stability of the population size (Huffaker *et al.*, 1984). Populations of insects are generally discrete, which means that generations do not overlap (Begon *et al.*, 1996).

The size of insect populations is regulated when intrinsic or environmental forces modify their capacity for survival, reproduction, or migration (Berryman, 1973). Predation, competition, and resource availability (e.g., habitat, food) are factors that regulate populations (Price, 1984). If these factors pose a limitation when a certain population density and growth rate is attained, then they are acknowledged as density-dependent processes (Varley *et al.*, 1973; Price, 1984). On the other hand, external factors such as the weather or the soil type can control the population size independently of population density (i.e., density-independent; Price, 1984). Thus, Huffaker *et al.* (1984) mention that insect populations are basically regulated by the ecological relationships they sustain, being predation and parasitism the most relevant interactions. Taking this into consideration, when natural enemies of an insect species are eliminated, its populations may undergo an accelerated increase, such that it becomes a pest. Any organism that causes economic losses by affecting crops and/or domestic animals or human health is considered a pest (Speight *et al.*, 1999). As a general rule, a species needs to have numerous individuals to be regarded as such. Moreover, pests generally emerge as a consequence of human activities (Uvarov, 1964), given the fact that in the wild these species tend to occur in low densities, although their numbers drastically increase when favorable environmental conditions arise (Dominguez, 1992). Other authors define a pest as a species that causes an economic loss to humans by damaging their food, house, or dress. In nature, there are no such things as pests; therefore, this concept does not have a strict biological meaning—it is derived from human values related to health, economics, and aesthetics (Leyva & Ibarra, 1992).

Pests arise for three main reasons (Uvarov, 1964; Leyva & Ibarra, 1992; Speight *et al.*, 1999):

Increased resources: Human activities provide insect species with the best resources: These are unlimited. A good example comes from the huge extensions of mono-crops, which supply a great amount of food.

Elimination of natural enemies by control methods: Many parasites, parasitoids, and predators control the growth of phytophagous insect populations and limit their distribution.

Introduction of exotic species: When a species is introduced to a new region, it can multiply rapidly due to the lack of natural enemies.

1.2 Recent trends in insect-control methods

Throughout history, human societies have battled pests, sometimes losing against them and thus confronting dramatic losses (Losey & Vaughan, 2006). Yet, we do not have a complete record of these events, as knowledge regarding plagues was somewhat unspecific before better tools (e.g., microscopes) allowed us to characterize them. In addition, scientific interest was less attentive to normal agricultural complications. From the late nineteenth to the early twentieth century, crop protection specialists relied on knowledge of pest biology and cultural practices to produce multitactical control strategies (Gaines, 1957). This approach changed in the early 1940s, when the use of organosynthetic insecticides supplanted virtually all other tactics and became the dominant approach to insect pest control. This period (from the late 1940s through the mid-1960s) was called "the dark age of pest control" (Newsom, 1980), because specialists began to focus on testing chemicals to the detriment of studying pest biology and non-insecticidal control methods. By the late 1950s, however, warnings about the risks of the preponderance of insecticides in pest control began to arise. Reports coming from the workers of cropping areas in North and South America (Dout & Smith, 1971) and Europe (MacPhee & MacLellan, 1971) described early signs of the consequences of insecticides, but did not have much impact, given that pesticides seemed very successful at a relatively low cost, providing long-season crop protection against pests and complementing the benefits of fertilizers and other agricultural production practices. During the decade of the 1960s, the application of insecticides reached its highest exploitation, and the negative consequences became evident in the agricultural yield. Given this problem, in the 1970s, different alternatives were adopted that focused on combining chemical methods with other strategies, for instance, considering biological and traditional agricultural knowledge.

Integrated pest management (IPM) is a concept that arises from the difficulties presented by the unsystematic use of insecticides. IPM has a long history and a broad scope, including the use of chemical insecticides in combination with improved cultural and biologically based techniques with a focus on achieving the most permanent, satisfactory, and economical insect control possible (Kogan, 1998). Yet, although conceived as a strategy friendlier to the environment, even the most successful contemporary IPM programs have been implemented with little consideration for ecosystemic processes. While species and population ecology have been the foundations of those programs because populations are the biological units in which species exist (Geier, 1966; Kogan, 1998), little attention has been paid to understanding the characteristics, processes, and dynamics at all ecosystemic levels (Gliessman, 1990). This information is essential for a scientific analysis of agroecosystems (Risser, 1986). Different definitions of IPM proposed in the last decades tried to incorporate these concepts; some were discussed and adopted in international committees (Kogan, 1998) with varying degrees of success. For instance, the management of rice pests in Southeast Asia was proposed to be based mainly on the restoration of natural controls through the removal of broad-spectrum insecticides (Kenmore, 1996). Furthermore, Kogan (1988) proposed that the four basic hierarchical ecological scales— individual, populations, communities and ecosystems—should serve as the template for IPM integration. This framework will be further discussed within this contribution, as the notion is compatible with our proposal.

Lastly, at the present time, excitement about genetic engineering dominates the literature and global management strategies; nevertheless, nothing will have been learned from past experiences if genetic engineering prevails over all of the other technologies that are also blossoming (Kogan, 1998). Like in the "dark ages," we do not have enough knowledge about the risks that genetic engineered crops could pose for wild plant populations (e.g. through gene flow; Ellstrand, 2003; Andersson & de Vicente, 2010; Dyer *et al.*, 2009), non-target organisms (Dale *et al.*, 2002; Hilbeck & Schmith, 2006), and human health (Schubert, 2002; Finamore *et al.*, 2008; Spiroux de Vendomois *et al.*, 2009).

Despite the control strategy used, the pest evolutionary arms race continues, and the ongoing development of insect resistance to insecticides has become a serious problem. Moreover, other factors linked to human populations have complicated the problem. For instance, policy strategies concerning these practices should be guided firstly by strict scientific knowledge.

1.3 Economic, ecological, and evolutionary costs of insecticide use

Historically, with the advent of agriculture, the human social structure changed and the establishment and growth of human societies began as a result, this practice lies at the very core of human cultures. Agriculture is an activity that clearly benefits from environmental services, since these provide primary sources essential to farming such as soil, biogeochemical cycles, and ecological interactions (e.g., pollinators, predators, nitrogen fixing bacteria). By altering the environmental services that sustain agriculture, we would be jeopardizing not only valuable biological diversity and ecological processes, but also a series of economic, social, and cultural components in a way that the calculation of the costs would involve many different levels. Nevertheless, an important question to ask in a broader sense, not just in the context of agriculture is as follows: How much is an environmental service worth? Some works have revolved around this issue and estimated that the costs of losing even one ecological service rises up to billions of dollars (Losey & Vaughan, 2006, 2008; Pimentel 1992, 2005).

At first glance, the use of pesticides may appear to be an advantageous and low-priced pest control option; nevertheless, major environmental complications follow this practice, finally resulting in economic, ecological, and evolutionary costs. As summarized in Figure 1, the alteration of ecological dynamics has short-term (ecological) and long-term (evolutionary) consequences that may or may not be reversible according to the magnitude of the damage. All of the biome processes at different levels—namely the individual, population, community, and the biome itself—are affected by the drastic conditions insecticides impose when applied irresponsibly. This practice affects the evolutionary processes in agroecosystems and jeopardizes the environmental sustainability necessary for future generations to survive, sometimes altering biological processes that are irreplaceable. Within lower ecological levels, insecticides compromise the life span, growth, reproductive potential, and behavior of individuals, thus reducing their fitness, and with time, modifying the evolution of life history traits. At the population level, increases in the mortality rate and alteration of the age structure could lead to the reduction of genetic and phenotypic diversity, decreasing the fitness of surviving individuals and increasing the potential for species extinction. At higher ecological levels, the affections inflicted by insecticides must not only be considered according to their ecological and evolutionary consequences, but also evaluated in relation to the ecological services that are being affected, since they are mostly provided by a complex of species through network interactions. Indiscriminate insecticide use damages the services that ecosystems inherently provide through their proper functioning, and insects, being one of the most diverse and successful animal groups, are involved in performing important ecological tasks such as pollination, pest control, suppression of weeds and exotic herbivorous species, decomposition, and soil improvement (Losey & Vaughan 2006, 2008).

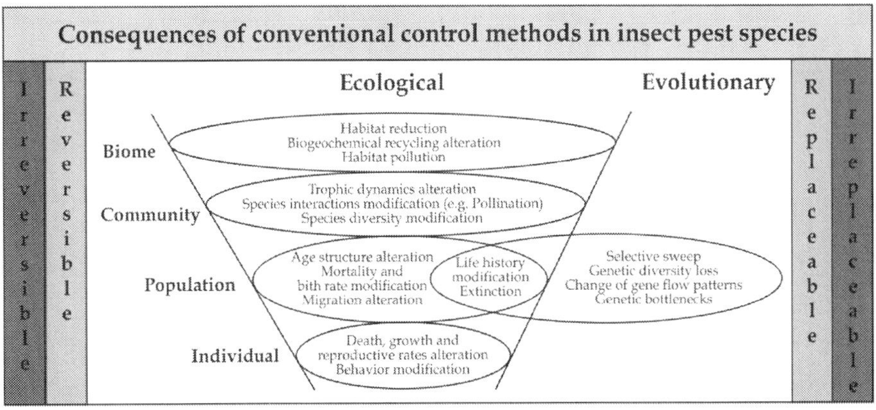

Fig. 1. Ecological and evolutionary consequences of insecticide control methods. Ecological consequences are shown at different ecosystemic levels, while evolutionary consequences occur only at the population level. Consequences can be irreversible or reversible and affect replaceable or irreplaceable environmental services.

When altered, ecological communities could suffer from an alteration of trophic dynamics and interactions, as well as a modification of species richness, diversity, and abundance; this could ultimately end interspecific relationships that could have been occurring for millions of years, affecting biodiversity irreversibly. Perhaps one of the most relevant concerns in this matter is the adverse effect of broad-spectrum insecticides on non-target organisms, especially if these contribute to an ecological service. For example, in all ecosystems, natural predators and parasites regulate the growth of herbivorous populations that could develop into potential pests. Pests can grow uncontrollably if their natural enemies are destroyed, or even if pesticides alter their predators' searching and attacking behaviors. This alteration of interactions can cause increases in pests of the same species, or even promote the appearance of new pest species (Metcalf, 1980; Van den Berg *et al.*, 1998; Mochizuki, 2003; Devine & Furlong, 2007). When this happens, additional and often more costly pesticide treatments are needed to avoid losing the crop; thus, it is possible to replace this "free" natural pest-control service, but this becomes highly expensive and environmentally hostile. Pimentel (1992) estimated that about half of the control of pest species is due to natural enemies, and that pesticides give an additional control of 10%; the main percentage is due to host-plant resistance and other limiting factors present in the agroecosystem. Losey and Vaughan (2006) estimated that if no natural predators were functioning to control native insect pests in the United States, the damages could reach $20.92 billion dollars each year. Insect pollinator species are also affected unintentionally but gravely by the use of insecticides. This is a serious matter, considering that this process is vital to plant reproduction and species preservation, and that it is a service that is not replaceable by any human technology. This problem affects not only crops, but also their wild relatives, by modifying the composition of native as well as non-native pollinators. Moreover, the lack of effective pollination reduces crop yields and the quality of the agronomical products, and can also cause the loss of the entire crop. Furthermore, it can affect the next sowing season, since seeds cannot be collected. In the United States, native pollinators may be responsible for yielding almost $3.07 billion dollars a year by allowing the successful production of fruits and vegetables (Losey & Vaughan, 2006).

Insects are not the only populations affected by the use of pesticides. Other organisms, such as birds and mammals, are directly or indirectly impacted by open exposure or the consumption of contaminated nourishment, respectively (Pimentel, 1992).

As insecticides pollute the environment, the habitat where living organisms can dwell is reduced. Insecticides may be toxic to soil microorganisms that are fundamental in maintaining the structure and function of ecosystems by enabling vital processes to take place (e.g., biogeochemical recycling, nitrogen fixation, organic matter disintegration; Atlas & Bartha, 1987; Brock & Madigan, 1988), this is why soil community alterations cannot be taken lightly. Moreover, since insecticides are often sprayed without sufficient care, they can reach neighboring crop fields and wild communities, and can even find their way into ground and surface water by soil erosion and water runoff (e.g., aldicarb, alachlor, and atrazine; Osteen & Szmedra, 1989; Relyea, 2005), finally disturbing aquatic ecosystems and contaminating potable water.

The consequences of insecticide usage are evidently costly. Simply, while the agricultural products are fixed in the markets and depend on private interests, the price of the crops freely fluctuates in the global markets. This inequity causes problems that will not be addressed in this chapter but that reveal the economic dimensions the loss of environmental services could have as a consequence of indiscriminate insecticide use; these could perhaps be considered "pest evolutionary arms race collateral effects." Studies about the environmental services jeopardized by current pest control methods and their economic costs are greatly needed. More reliable data documenting the involvement of insects in environmental services must be published to allow the generation of mathematical models and consequently the accurate estimation of the value of these services. Ideally, these efforts should be undertaken worldwide to guide each country's policy strategies regarding such practices, for example, through risk assessment studies.

2. Ecology of insect populations and life histories

More than 50 years ago, Cole (1954) published one of the first works pointing out the relevance of organisms' life history to the management of pest species. Later, almost forty seven years after Cole, Nylin (2001), indicated the necessity of studying pests' life history traits to control them, and to prevent as much as possible the development of potentially dangerous species. In this chapter, we assert that Cole and Nylin's ideas are still valid; however, in our view, trying to eliminate or at least control a species with a complex life history is almost impossible. Under current control methods, including IPM, it is unfeasible to control the population sizes of species with great reproductive potential, growth rate, and dispersal ability. It is obvious that current trends regarding pest management should be modified. To achieve this goal, the first step should be revealing life history parameters that will potentially increase the population size and distribution range of a species. Pest management strategies should be adjusted to species' life histories, so that mortality and birth rates or immigration and emigration rates can be estimated and projected.

2.1 Life history traits of insect pest species

Let us imagine the characteristics a crop-devastating insect should have. Females should be extremely fertile, being able to leave a large number of eggs, which would constitute the next generation population. For example, let us picture a highly fertile female who can leave 70 to 100 eggs per laying and the number of laying events in her lifetime is, on average, three. Moreover, let us imagine that this female belongs to a species presenting a number of generations higher than one per year; let us say there could be up to three generations in a year. Furthermore, let us assume that the devastation potential of this species is not limited by space because it can migrate very long distances, e.g., hundreds of kilometers in just one day. Additionally, let us say that these organisms are polyphagous, being able to consume different plant types, not just the ones that humans cultivate but their wild relatives as well. Finally, let's imagine that these super-organisms have the capacity of inhabiting places that present very variable conditions, where they can perfectly survive, develop, and reproduce. If such an organism existed, it should be named *Schistocerca gregaria* and it would belong to the order Orthoptera. This locust species is one of the most devastating worldwide, not only nowadays but throughout agricultural history. Considering the information described above, we could obtain a simple population growth projection for a few years in a scenario lacking natural predators and providing unlimited resources. After one year, just from the mating of one female with one male, 900 new individuals would join the population in the next generation. If these 900 individuals presented a 1:1 sex ratio, the next year it would be 400,000 individuals, and five years after that the number of individuals would reach 3.5 x 10^{13} individuals. Although these numbers appear to be unreal, they represent very well the potential a single couple of locusts could have in an agroecosystem when no restrictions are imposed. Even when most of the individuals can be eradicated by conventional control methods, the infestation can reemerge in just a few generations with a higher population density. Unfortunately, the control methods do not have an impact on the traits that contribute to the demographic success of insect populations. *Schistocerca gregaria* is not the only species with these life history characteristics. All pest species possess at least one trait endowing them with high reproductive rates and survival success. Regarding the number of generations per time unit, aphids are exceptionally capable of originating new individuals. The species *Aphis glycines*, for instance, can produce 18 generations per year in soybean fields. Moreover, all individuals of this species, when present in monocultures, are parthenogenic females (thus, they do not need males to reproduce).

Some pest species display great dispersal ability. Some control measures are focused on avoiding or at least revealing the migratory routes of these species (Riley & Reynolds, 1983; Farrow & Daly, 1987; Riley & Reynolds, 1990; Pedgley, 1993; Chapman *et al.*, 2010). *Oedalus senegalensis* is a pest species that can migrate up to 350 km in just one night (Cheke, 1990). This insect can infest and destroy crops in huge areas of western Africa in just a few days, given its outstanding dispersal skill. Orthoptera species and other insect groups that present incomplete metamorphosis can immediately devastate large crop extensions. In the case of insects with complete metamorphosis, the damage normally occurs in the next season, since individuals at the adult stage are the ones with dispersal ability, while those at the larval stage—i.e., the next generation—eat the crops. It has been observed that recent African infestations have been related to the migration routes of *S. exempta*. Moreover, the diamondback moth, *Plutella xylostella,* can perform transoceanic migrations, thus being able to continuously travel up to 3000 km over a course of days (Talekar & Shelton, 1993). During its larval stage, this insect feeds nearly exclusively on cruciferous plants and due to its dispersal potential, it is possible to find it in almost every cruciferous crop field around the world. Most of the migrations carried by insects are aided by wind currents (convergent winds in Africa), which allow the insects to invest a minimum of energy in flight (Chapman *et al.*, 2010).

Other species of insects, although they lack flight capacity, are able to colonize new agroecosystems. The orthopteran *Sphenarium purpurascens* can disperse more than 10 km annually throughout crop fields and wild environments, mostly due to its capacity to feed on greater than 50 different plant species (Cano-Santana & Oyama, 1994). Two decades ago, this grasshopper had a narrow distribution in central Mexico; at present, however, it inhabits locations hundreds of kilometers beyond its original range. *S. purpurascens*, like many other insect species, displays pronounced diversity regarding its life history traits when inhabiting crop fields as opposed to wild ecosystems. Its populations can actually change their life history traits in just a few generations (Cerritos, 2002).

2.2 Adjustable life histories: Phenotypic plasticity or swift changes in allele frequencies?

Let us now imagine a potentially crop-devastating species able to regulate its life history traits depending on environmental conditions, the resources available, and its own population density. If this species, when inhabiting locations with suitable environmental conditions and unlimited resources, could give birth to more than one generation per year and produce a huge amount of eggs each breeding, as well as being able to migrate long distances, it could be an extremely serious pest. *Locusta migratoria* is one locust species able to exhibit polyphenism, which is a biological mechanism characterized by the ability to adjust life history traits according to environmental or demographic factors (Simpson *et al.*, 2005). This kind of phenotypic plasticity has been documented in orthopterans, specifically from the Acrididae family. More than 15 locust species that are known to damage agroecosystems display polyphenism in their morphological, physiological, and behavioral traits (Song, 2005). Crop-devastating insects like *S. gregaria*, *L. migratoria*, and *L. pardalina*, for instance, possess the ability to assemble huge congregations of individuals that can migrate long distances and therefore have the potential to cause global infestations. These swarms can generally reach more than 250 million individuals (Simpson *et al.*, 2005).

The origin of this kind of plasticity was first explained, at least for *L. migratoria*, by the existence of two genotypes within their populations: one that favored the establishment of congregations and consequently infestations, and another that promoted solitary behaviors. Recently, however, it has been demonstrated that regardless of the genotype, this species has the ability to form huge groups of individuals anywhere in the world (Chapuis *et al.*, 2008). Locusts are not the only insects with phenotypic plasticity. Lepidopterans like *Polygonia c-album* and *Pararge aegeria* can modify their diapause in response to latitudinal variation and photoperiod and the heteropteran *Eurygaster integriceps*, a serious wheat and barley pest in Iran, can modify its generation time and fecundity as a result of temperature changes (Iranipour *et al.*, 2010).

Given the previous examples, it has become evident that to achieve successful pest management strategies, it is urgent to understand and unveil the genetic bases that underlie life history traits, especially those exhibited by insect pests. These traits appear to be subjected to strong selective pressures, such that in only a few generations, new genotypes increase their frequencies and eventually become genetically distinct from the ancestral populations. Few studies have been performed in this context; nevertheless, a pioneering work comparing pest and non-pest populations of the beetle *Epilachna nipponica* offers some insight. The results show that pest populations exhibit a continuous ovoposition rate, shorter immature stages, and bigger female body size. Let us now picture a species that is not only very plastic phenotypically, but also extremely diverse genotypically (Shirai & Morimoto, 1997). What could we do to control such a species? What can we do to control *S. gregaria* or *L. migratoria*?

3. Evolution of insect pest populations

One or more genes determine almost all life history traits. The modification of these genes would probably involve the modification of one of such life history characteristics. The frequency change of the different alleles of these genes and the evolutionary forces that shape their distribution form the subject matter of a field called Population Genetics. For pest species, knowledge about their genetic structure is relevant to learning how certain traits become fixed within their populations. The genetic diversity of insect populations is a result of the huge population sizes that increase the probability of mutational events. All the genetic variants (i.e., genotypes) stored in populations are consequential when selective pressures, like insecticides, occur; for instance, one genotype might be resistant to a given insecticide and will therefore increase its frequency in a few generations, thereby performing a process called "resistance evolution."

Unfortunately, for chronological reasons, Darwin never observed the effects caused by insecticides on inheritable traits; however, if he could have seen them, what he may have concluded is that strong selective pressures could modify populations in just a few generations. Darwin thought of evolution as a gradual process, but perhaps by observing the insect species subjected to insecticides, even he would have concluded that evolution could be very fast, occurring in sudden "jumps."

3.1 Effective population sizes in a pest management context

Quantification of past and present population size can provide insight into the success of an invasive population, the amount of effort required to eradicate or suppress that population, and the effectiveness of a control strategy. Habitat structure, geographic extent, mobility, size of the individual, the cryptic or elusive nature of the species, and population distribution, however, often hamper quantifying population size by direct census (Rollins *et al.*, 2006). Genetic data can be used to calculate current effective population size (the number of individuals in a population that contribute offspring to the next generation, or *Ne*; Wright, 1931), estimate minimum population size, and detect evidence of population expansion or decline (Rollins *et al.*, 2006). Due to the importance of performing conservation efforts focused on endangered species, we currently have access to a lot of examples that apply this concept. From them, we could ask: How difficult is it to achieve the local or global extinction of a species? Frankly, when humans have managed to drive a species to extinction, it has not been easy even when this species, in contrast to insects, was several orders of magnitude smaller and thus simpler to extinguish, for instance because its population size or effective population size was comparatively much lower. Pest control methods as applied today appear to be low-success practices when viewed through a population genetics, phylogeographic, and conservation genetics perspective.

Let us think of a hypothetical pest species that is affecting a given crop field. We then decide to locally eliminate it using direct methods, which will kill most of the individuals, producing a genetic bottleneck. The first alleles that will disappear from the population are the ones present at low frequencies (Hauser *et al.* 2002). After 4*Ne* generations (*Ne* being the effective population size), more alleles will be lost, which means that the loss of alleles will depend on the effective population size. However, is this the right path to effective pest control? To answer this question, we need to know: 1) how the loss of genetic diversity increases the susceptibility of a population toward extinction, and 2) how much genetic diversity is needed for a species to maintain its adaptability in response to environmental changes. These problems can both be addressed through estimation of the effective population size and the genetic diversity of the species.

In the case of insect species, they are a good model for understanding the evolutionary processes influenced by natural selection. On a neutral theory scenario, we can have an elephant population with few individuals and an insect population with numerous individuals, and both will have the same mutation rate. With the passage of time, it is evident that some evolutionary forces will act in this comparison. In insect populations, the generation time is smaller, the recombination rate is faster, and the selective pressures are bigger; since the genetic drift is dependent on the effective population size in elephant populations, genetic drift is the most important evolutionary force because of the low number of individuals, whereas in insect populations, other evolutionary forces are stronger. In the context of pest management, molecular techniques that estimate genetic diversity and identify sudden population contractions (i.e., bottlenecks), due for example to the survival of resistant individuals after the selection pressure imposed by insecticide application, can provide feedback on the effectiveness of control programs and are especially useful in situations where direct population size assessment is difficult (e.g., Hampton *et al.* 2004). If we are able to analyze these characteristics in a pest species, then we will be closer to designing better long-term control strategies.

3.2 Genetic structure and gene flow in pest populations

Local elimination of pest populations is not a solution. It is like thinking that removing the cockroaches from one apartment of an infested building would be a long-term eradication solution. Thus, the control strategy should be directed to the whole population. Information concerning the number of populations present in a given place (i.e., a building), along with the degree of connectivity between them, is vital to constructing sound management and control policies for pest populations (Rollins *et al.*, 2006). Genetic structure can be described as the distribution of genetic variation resulting from migration, selection, mutation, genetic drift, and related factors. In other words, it is a measure that will reveal the level of connectivity between populations. If this measure is significantly high, then the populations are evolving together (and thus are highly connected); inversely, if it is low, each population could be considered as an independent evolutionary unit (and the populations are poorly connected). In situations where population subdivision is unclear or boundaries are cryptic, incorrect estimation of the number of populations may bias assessment of population dynamics (Taylor, 1997). For example, Robertson and Gemmell (2004), in a study on eradicating rat pests from the Guadeloupe archipelago, concluded that populations were sufficiently isolated to be sequentially eradicated without a high risk of reintroduction; however, in a later work using genetic data for the same species, Abdelkrim *et al.* (2005) identified groups of islands that would require simultaneous eradication due to high levels of gene flow.

What would happen if we applied any control method to a population that exhibits a constant migration rate with neighboring populations? Since the population is not confined, new individuals from other populations could arrive, colonizing the area once more. If this is the case, then the overall genetic diversity of the species might be preserved in populations that are not being directly subjected to the control method. Thus, no matter how strenuous the effort to control a pest population in a particular locality is, new individuals will colonize it if their migration potential allows it. Then, it is important to determine not only the effective population size, but also the geographic area that a single population inhabits, as well as the overall species. This knowledge will aid in making more effective decisions when applying a control method, as well as contributing to a more fruitful investment in pest management.

Effective control of invasive populations may largely depend on the ability to identify their source. In many situations, the point of origin is unclear, or there may be multiple sources of an invasive population. Simple models assume that rates of movement are independent of landscape structure and use constant movement rates whatever the landscape mosaic in question (Goodwin & Fahrig, 2002), assuming that dispersal is random (Conradt *et al.*, 2001; Hunter, 2002). Because direct measurements of dispersal are typically difficult to obtain, indirect measures using population genetics may be employed (Pritchard *et al.*, 2000; Piry *et al.*, 2004). The study of gene flow can be even more informative: such an approach can be employed as a method of delimiting dispersal potential in species in which males are more likely to disperse than females (Hunter, 2002). Traditionally, sex-biased dispersal has been detected by comparing the level of population structure of bi-parentally inherited genes to those inherited from one parent only (e.g., mitochondrial genes).

Assessment of the dispersal potential may influence decisions on how to manage invasive populations. Species that experience restricted dispersal may be better candidates for control than those that disperse widely (Rollins *et al.*, 2006). A variety of methods have been developed to assign an individual to a population of origin or to exclude it from putative source populations (Wilson & Rannala, 2003; Piry *et al.*, 2004; Guillot *et al.* 2005; Rollins *et al*, 2006).

3.3 Landscape genetics as an approach to understanding pest genetic diversity

The recent improvements in molecular genetic tools, combined with existing or new statistical tools (e.g., geo-statistics, maximum likelihood, and Bayesian approaches) and powerful computers has led to the emergence of the field of landscape genetics, which is an amalgamation of molecular population genetics and landscape ecology (Turner *et al.*, 2001). This discipline aims to provide information about the interaction between landscape features and microevolutionary processes such as gene flow, genetic drift, and selection. Landscape genetics can resolve population substructure across different geographical scales at fine taxonomic levels (Smouse & Peakall, 1999). Understanding gene flow is also fundamental for ascertaining factors that enable or prevent local adaptation, and for describing dynamics that facilitate the spread of new, beneficial mutations (Sork *et al.*, 1999; Reed & Frankham, 2001). However, the aim of managers is to determine what constitutes a natural break within or between populations, the ratio of habitat (i.e., edge to interior; Chen *et al.*, 1995; Radeloff *et al.*, 2000), the isolation of habitat fragments (Collinge, 2000), subpopulation area (Kruess & Tscharntke, 2000), subpopulation quality (Hunter *et al.*, 1996; Kuussaari *et al.*, 2000; Hanski & Singer, 2001), subpopulation diversity (Gathmann *et al.*, 1994; Varchola & Dunn, 2001), and microclimate or ecological niche (Braman *et al.*, 2000). All of these phenomena contribute to determining the abundance and richness of insects on particular landscapes (Hunter, 2002). Lenormand *et al.* (1999) found a decrease in pesticide resistance with increasing distance from the treated zone by studying pesticide resistance in the mosquito *Culex pipiens*. This cline can be interpreted as a consequence of local adaptation when migration and selection act as antagonistic forces (Manel *et al.*, 2003). Landscape genetics is uniquely suited to exploring mechanisms of speciation in a complex resistance landscape, where parts of a population may experience sufficiently reduced gene flow such that drift or selection along locally steep selection gradients could lead to new species (Balkenhol *et al.*, 2009). Finally, adaptive landscape genetics explicitly deals with spatial genetic variation under selection, and can be used to study the adaptive and evolutionary potential of populations (Holderegger *et al.*, 2006, 2008; Balkenhol *et al.*, 2009).

Recently some scholars incorporated temporal changes in landscape structure (Solbreck, 1995; Onstad *et al.*, 2001), genetic change in insect populations (Singer & Thomas, 1996; Ronce & Kirkpatrick, 2001), and differential responses of predators and prey (Kruess & Tscharntke, 1994; With *et al.*, 2002) into their understanding of the spatial ecology of insects (Hunter, 2002). Roderick and Navajas (2003) suggested that identifying the origin of specific genotypes in an invasive pest population might assist in the identification of natural enemies in the native range, thus facilitating the design of effective biological control programs (Rollins *et al.*, 2006).

In the case of genetically modified crops that present insect protection features, to the extent that greater host availability increases pest adaptation to a particular host plant (Kelly & Southwood, 1999), widespread planting of transgenic insecticidal crops should favor resistance evolution (Gassmann *et al.*, 2009). Certainly, the selection pressure placed on pest populations to evolve resistance is more intense in this kind of crops because the pressure they impose is persistent instead of dependent on manual application. Resistance management of pests in insecticidal cropping systems has relied on the high dose/refuge strategy (Taylor & Georghiou, 1979; Gould, 1998). The refuge consists of growing non-transgenic host plants in close proximity to insecticidal crops. The refuge plants are expected to harbor and enable the reproduction of a large number of toxin-susceptible individuals, which will mate with any resistant individuals that emerge from the insecticidal crop, diminishing the resistance to the transgenic crop in the next generation (Gassmann *et al.*, 2009). However, the available data suggest that, in at least some cases, genetic variation serving to evolve resistance is present in the field. Numerous insect strains have responded to laboratory selection by evolving greater resistance to *Bacillus thuringiensis* Berliner (Bt) toxins (Tabashnik, 1994; Ferré & van Rie, 2002; Gassmann *et al.*, 2009), and this is suggestive of the evolutionary potential of pests to evolve resistance to transgenic Bt crops (Gassmann *et al.*, 2009). More importantly, analysis of field populations has revealed the presence of major resistance alleles for resistance to Bt crops in populations of pink bollworm

Pectinophora gossypiella (Tabashnik *et al.*, 2005), tobacco budworm *Heliothis virescens* (Gould *et al.*, 1997), the corn earworm *Helicoverpa zea* (Burd *et al.*, 2003), and the old-world bollworm *Helicoverpa armigera* (Downes, 2007; Wu *et al.*, 2006; Gassmann *et al.*, 2009).

3.4 Phylogenetic patterns

Phenotypic traits are influenced by their evolutionary history and the evolutionary forces in their actual environment. Phylogenetic patterns can reveal the effects of history in character evolution, which is relevant to understanding pest species and the ability to control them. On one hand, the history of pest relatives is essential, since many aspects that we may not know about the species in question may be shared with some of its relatives; thus, valuable information could be obtained through their study. Another key aspect that can help us determine how to fight a pest depends on the plant being cultivated, because the location inhabited by its wild relatives can provide a lot of information. These sites contain the natural enemies of the pests that attack plants, but also are home to a greater number of organisms adapted to plant defenses, and therefore monoculture in these places could be counterproductive, since the density and diversity of plants often support the defense mechanism. Additionally, most cultivars are genetically uniform; thus, natural predators can be much fiercer with these plants and the mechanisms to combat them much less effective. In many cases, gene flow between cultivated plants and their wild relatives can put their genetic diversity at risk and make wild plants more susceptible to herbivores, causing irreversible environmental damage (Ellstrand, 2003; Andersson & de Vicente, 2010).

4. Understanding the ecology and evolution of insect pest species and their crops as a basis for pest control

Which are the best methods of reducing insect pest populations in agroecosystems? For a long time, this question has been answered in terms of economic gain, and therefore control methods have been applied following this inclination. However, it is evident now that this question needs answers, and consequently actions, based on ecological and evolutionary knowledge of pest species.

Chemical and biological control methods, beyond doubt, have temporally diminished the damages caused by insect pests. The beneficial outcomes that these kinds of methods have accomplished in terms of agricultural production are considerable, but unfortunately, transient and extremely unaffordable from a biological perspective. These methods do not take into consideration the costs of locally or generally extinguishing a predatory species, of polluting the water or soil or of changing some life history traits of a pest species. The most suitable control methods should be those that, before being put into action, consider the pest species' ecological and evolutionary traits. Ecological attributes such as migration, reproductive or mortality rates, dispersion, or population growth regulatory processes (e.g., by predators or parasites) are key components in the understanding of species' short-term dynamics. Likewise, revealing the evolutionary attributes of a species such as the genetic diversity within and among populations, the gene flow between populations, the genetic structuration, and the effective population size could illuminate how fast the resistance to a given control method could evolve and how broadly, in a geographic sense, this method should be applied.

Figure 2 presents the ecological and evolutionary trait thresholds that a pest species could hypothetically display in contrast to a non-pest species. We propose that from them, it is possible to formulate better strategies to reduce, control, or eradicate pest species. This representation suggests a pest species regulation model through the application of evolutionary and ecological knowledge. Each species may have a certain surface within a multiple axis system. Each axis represents a trait, and by connecting the axes, an area is generated inside a multivariate space. Hypothetically, with colored circles, risk thresholds are represented. By quantifying each trait for a determined species, we could draw a corresponding area inside the model in a way that we could observe its shape and compare it to the proposed thresholds. The species with more devastating potential would occupy a maximum area inside the vectorial system (dotted blue line), while a species that is placed below the green circle probably will not represent a pest problem at that moment. Depending on the size and shape of the surface formed inside the system, different long-term control strategies could be proposed.

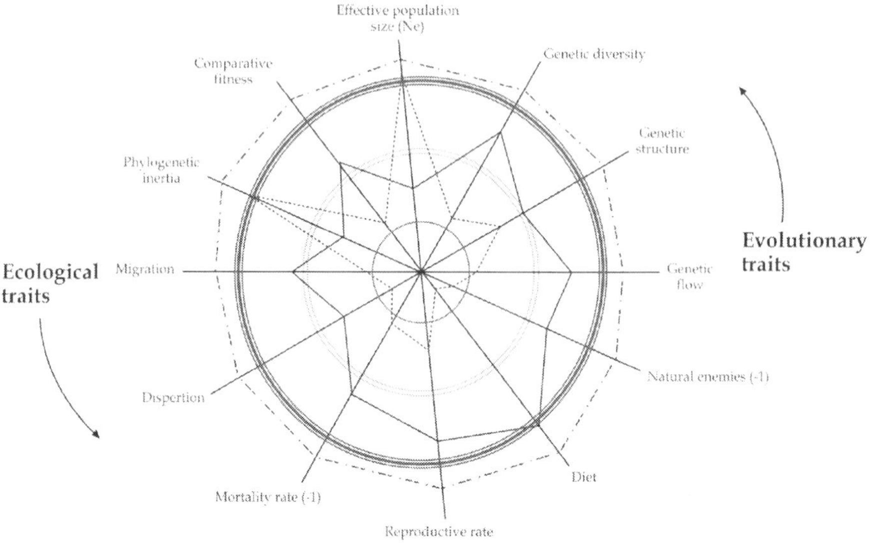

Fig. 2. Hypothetical representation of the ecological and evolutionary assessment suggested for pest control management. Each edge of the graphic represents a trait; when all of the edge's values are connected, a polygon is formed which will represent the potential of a species to become a pest. Hypothetical pest species display very different life histories and population genetics traits than non-pest species. Each of the solid line circles represents a hypothetical threshold of the damage that a species could cause; thus, the red circle illustrates a maximal threshold expected for a pest species, while the green circle will conform to a non-pest species. Polygons represent the trends for three different species: a lethal pest species (blue polygon), a moderate pest species (solid line black polygon), and an incipient pest species (dotted line).

4.1 Short-term pest control strategies: From insecticides to transgenic plants

Let us imagine an insect species that possesses all the ecological and evolutionary characteristics to become a pest, not on a short-term basis but over thousands of generations. Let us picture, then, that systematically, for decades, the same control method has been applied that has, immediately but temporarily, minimized the problems caused by the pest. Within a few generations, these insect populations will resurge even more devastatingly than before, in such a way that more complex and stronger control methods will be needed. Even though the methods used so far have proven to be inadequate, let us assume their application is continued, maybe in different presentations like a liquid or a bacterial gene that codes for a crystal-forming protein inserted into a crop plant, but leading to the same consequences. Over time, the selection pressures imposed by these methods could prompt the species' rapid adaptation and resistance evolution, thus hindering a long-term solution by these means.

The control methods predominantly used today are chemical and biological, followed by the growing development of genetically modified plants that, in some instances, could be regarded as chemical control. Numerous works have revealed that after the application of chemical agents, pest species can rapidly develop resistance and reemerge. Actually, over the last two decades, several insect populations have evolved resistance to some insecticides (Farnham, 1973; Robb, 1989; Liu & Scott, 1995; Jensen, 2000; Siqueira *et al.*, 2000). Georghiou (1990), in his classic work, reported that by 1990, around 500 species were resistant to at least one insecticide. Diverse locust species exhibit insecticide resistance through different biochemical mechanisms that reduce the lethal effect (He & Zhu, 2004; MacCuaig, 2008; Yang *et al.*, 2008). As for the utilization of genetically modified plants, some insects have acquired resistance to them, specifically to the protein codified by the transgene CryIA isolated from *Bacillus thuringiensis* (Kaiser-Alexnat *et al.*, 2005; Moser, 2007; Wolfgang, 2007). In Europe, at least four potentially devastating insects have acquired resistance: the pink bollworm *Pectinophora gossypiella* (Tabashnik *et al.*, 2005), the tobacco budworm *Heliothis virescens* (Gould *et al.*, 1997), the corn earworm *Helicoverpa zea* (Burd, 2003), and the old-world bollworm *Helicoverpa armigera* (Downes, 2007). All of the chemical agents provoke an intense selective process in the populations of insects, subsequently diminishing their population size. Nevertheless, some of the individuals of the population can be resistant and tolerate the above-mentioned compounds, thereby increasing the insect population size and diminishing the plant population size (Elrich & Raven, 1964).

Chemical control methods increase the selective pressures without being able to successfully eradicate the pest. An insect pest species with high genetic diversity and a large population size will have the potential to swiftly change its gene frequencies, thus increasing the number of resistant genotypes in just a few generations. Some chemical agents are particularly prone to promoting resistance, since with a single point mutation, the insect can block the pathway of the molecule used as insecticide.

Insecticides, Bt crops, and biological control methods, unfortunately, are not specific to the insect they are meant to control. Consequently, several non-target species are eliminated when these practices are applied, including natural enemies and endemic species (Lockwood & DeBrey, 1990; IUCN, 1996; Hoekstra, 1998; Lockwood, 1998; Stewart, 1998; Lockwood & Sergeev, 2000; Fashing *et al.*, 2010). For instance, Stewart (1998) revealed that when applying insecticides to control *Locusta pardalina* in crop fields, diverse endemic grasshopper species from South Africa were eliminated.

Besides their direct and immediate effects, insecticides also produce indirect and lasting effects. Lahr (1998) commented that organophosphate residues from insecticides applied to control *Schistocerca gregaria* were stored in water bodies, thus affecting numerous species. Recently, Fashing *et al.* (2010) analyzed the impact of insecticides used to control *S. gregaria* on African mammal species, since they feed on this locust species routinely. They analyzed the ecological implications of this particular mammal diet, because locust assimilated and stored the insecticide, which in turn affected the abundance of mammal species.

Let us assume that each of the non-target species could be represented in Figure 2, inside the group of non-pest species. The ecological and evolutionary trends that these species could display would be very different from the ones belonging to a pest species that has all the traits to demographically counteract the pressure imposed by control methods. Several non-target species have probably already been eliminated, locally or definitively, and regrettably without any available record.

Several works have analyzed the effects that biological control methods could have at an ecological level, particularly because of the lack of knowledge regarding the consequences of the artificial interactions that they impose (Louda, 1990; Thomas & Willis, 1993; Godfray, 1994; Jhonson & Stiling, 1998; McEvoy, 1999), for instance, by jeopardizing established natural interactions (e.g., competition or predation). In some cases, biological control methods fail to eradicate pest species, while in others, they promote the emergence of new pest species that parasite or prey on native species (Louda, 1990; Thomas & Willis, 1993; Godfray, 1994; Jhonson & Stiling, 1998; McEvoy, 1999). Regarding genetically modified crops with *Cry* transgenes, they are supposed to be species specific; however, there is a lack of scientific studies that support this notion, since they have not been carried out with species from the same genus or even with populations from the same pest species.

Finally, let us imagine for a moment that for many decades, there have been alternative methods able to minimize pest damages without producing so many collateral effects. And let us think these methods have been applied worldwide with favorable results. Now, what we cannot imagine is why these methods have not been used extensively in agroecosystems. Is it only an issue of lack of imagination?

4.2 Species-specific control methods that do not generate resistance

One method that does not cause resistance and could be species specific is mechanical elimination. This method refers to the removal of insects manually or by means of different kinds of tools (Faure, 1944; Van Huis, 1996; Abate *et al.*, 2000; Cerritos & Cano-Santana, 2008; Cerritos, 2009; Cerritos, 2011). This method has proven to be the most effective in controlling conspicuous insects such as locust, hemipterous and larvae. However, its practice in agricultural fields is very limited. Lockwood (1998) discusses that the main disadvantages of this method are its cost, since it could be very expensive, and that it is time-consuming, especially when the removal is undertaken manually. Nevertheless, it is a strategy that does not cause collateral effects on the environment. Lockwood (1998) mentions that because it is an extremely selective method, it does not represent a menace to non-target species. Furthermore, from an evolutionary point of view, it would be very hard for insects to develop resistance to manual removal.

The mechanical capture of pest species in some places is a common practice, especially in third world countries where insect consumption still prevails. However, although this strategy can reduce the population sizes of insect species, it is not considered a technological control system.

Although costly and labor intensive, mechanical removal is a potentially profitable method from the ecological, economic, and social perspectives. If we could compare between the chemical, biological, and mechanical methods in terms of resource investment and outcomes, we would probably find contrasting results. For instance, *S. gregaria,* a widely distributed and abundant species, caused economic losses of over 2.5 billion dollars in only the past 5 years in Africa, while around 400 million dollars were invested in chemical and biological control without success (FAO, 2008). The ecological and social damages of insecticide deployment were not evaluated in this case. Now, let us evaluate the input and profit for this example if the mechanical method could have been used. Considering that 10 million tons of edible insects could be harvested in each African country that this species inhabits, an investment of 10 million dollars would be needed, but the economic, ecological, and health gain would be immeasurable (Cerritos *et al.*, in preparation). Another example worth mentioning occurs in America, specifically in the United States, where several native grasshopper species ingest 25% of crop foliage in 17 states each year (Hewitt & Onsager, 1983). The cost of these losses has been estimated at 1 billion dollars per year (Pfadt & Hardy, 1987). To control these species, broad-spectrum insecticides have been the method of choice; in only two years (1986–88), around 5 million liters of Malathion worth 75 million dollars were used (NGMB, 1995). By implementing the mechanical method, the potential gain of the harvest could be 1 to 5 million tons annually with an investment not higher than 5 million dollars. Although in the United States, culturally, insect consumption has not been a common practice like in other countries, all that biomass could be used as livestock food. These two examples confirm the need for new and more suitable control methods to eradicate exotic pest species and maintain local or endemic pest species at low population sizes.

In Mexico, a country where insect consumption (i.e., entomophagy) is a very common practice, several grasshopper species are consumed as food. Not only is the grasshopper *Sphenarium purpurascens* one of the most significant pest species, but it is also the insect most consumed in Mexico (Cerritos & Cano-Santana, 2008). Nowadays, in just one village, hundreds of tons of this insect are produced yearly in alfalfa and corn crops, where it is especially considered a pest. The economic profit of grasshopper harvesting would be, on average, 5,000 dollars per family per year, without adding the savings related to not purchasing insecticides or investing in biological control (Cerritos & Cano-Santana 2008). The *S. purpurascens* biomass estimate that could be obtained by mechanical removal maintaining its population sizes at a level that minimizes crop damage is in the thousands of tons. The distribution and abundance of this grasshopper was estimated through demographic models, rendering the potential biomass quantity calculated per year to be almost 700,000 tons of grasshoppers in around 1 million hectares. The economic profit of this biomass extraction would exceed 50 million dollars, while the nutrimental contribution would yield 50 million rations, each with 20 grams of protein content (Cerritos *et al.*, in preparation). Beyond doubt, even when they may seem unconventional, these kinds of strategies could be applied around the world, taking advantage of native locust and grasshopper species and even considering them as a sort of mini-livestock.

If we place species like *S. gregaria* or *S. purpurascens* in our model presented in Figure 2, it is evident that both insects would be outside the higher threshold (solid line red circle), since their ecological and evolutionary traits are consistent with the characteristics displayed by highly devastating pest species. Let us review the case of *S. purpurascens* in more depth. This grasshopper, in spite of its lack of wings, has an exceptional dispersal capacity (Castellanos, 2001). It can feed on a broad diversity of plant species, from weeds to crops (Cano-Santana, 1992). The amount of natural predators and parasitoids in the environment do not seem to regulate its population size (Cerritos, 2002). Finally, the number of eggs per female and the survival rate at each stage are very high (Cerritos & Cano-Santana, 2008). From an evolutionary perspective, recent analyses have demonstrated that, at least in Central Mexico where this grasshopper dwells, several populations exhibit genetic structuration, with private genotypes and high genetic diversity (Cueva del Castillo, in preparation). When all these properties are considered, it becomes evident why chemical and biological control methods extensively applied in crop fields have been unsuccessful in controlling this pest and have caused diverse collateral ecological and social problems. Some reports reveal an increase in genetic diseases linked with insecticide application, mainly Malathion (Cerritos & Cano-Santana, 2008). Additionally, there is evidence of soil and water pollution and of the local elimination of several species (Cerritos, 2002). The strategy that could be most adequate for the control of *S. purpurascens* populations would be mechanical removal combined with other practices that do not cause collateral effects.

What happens to all pest species that are not suitable for human consumption? Up until now, we have directly linked the mechanical method to the use of insects as food, and it would seem that only edible insects could be subjected to this control method. We think this is hardly the case. We are aware that nowadays, there are not a lot of human practices that involve the exploitation of insect resources; however, since mechanical removal provides a huge amount of biomass, surely something useful and beneficial could be done with it. For instance, in Mexico, since pre-Hispanic times, several insect species have been used; perhaps the most well-known example is the hemipterous larvae *Dactylopius coccus*, an *Opuntia* spp. parasite that is the primary source of a red pigment used in the textile industry. In other instances, if not suitable for human consumption, they could perhaps be used to feed livestock.

4.3 Crop-oriented alternatives

4.3.1 Small-scale polyculture. The *milpa, chacra, nainu,* and *conuco,* among others, are traditional agroecological systems implemented by indigenous peoples from many different cultures, climates, and places in the world. These systems can be regarded as "small or medium scale polycultures" (Chávez-Servia *et al.*, 2004). While their methods vary depending to the agroecosystem of each place, all of them have active strategies for insect control, and these strategies do not allow harmful organisms to reproduce immensely, but tolerate some level of infection to avoid losing the entire crop affected (Morales & Perfecto 2000). These strategies also allow for the long-term use of soil, for example, by crop rotation and by letting the land rest. In these circumstances, insect pests simply do not find resources and cannot grow in that area or stay in it permanently, so no resistance is generated, nor is there an accumulation of chemical products in the soil (Blanco & García, 2006). In addition, in these systems, one immediate control measure can be growing different crops at the same time in the same field to help reduce infections and enhance economic effects through the profit of attaining food from both crops (Muñoz, 2003).

Some strategies use agrochemicals on seeds before planting them to prevent initial infection; others involve acting collectively in applying insecticides or known enemies of the pest in a large area, thereby preventing the pest from passing from one field to another, unprotected, parcel. Different strategies are used for controlling insect pests that infest seeds during storage, for example placement of the containers near smoke, use of powdered lime (Calcium carbonate) or application of commercial insecticides (Moreno *et al.*, 2005).

4.3.2 Large-scale polyculture. The structure and forms of large-scale mixed farming schemes are quite variable. Many rely on the same strategies described above and are carried out under the same structures, but with a modified scale. The nutrient recycling between different crops requires a little more involvement of applied sciences, and the management of the synchronic cultivation of fruit, vegetables, woody species, and fungi requires more knowledge (Altieri, 1995). In many cases, the issues regarding pest management are based on geographical and chemical barriers that impede the movement of pests, often with the help of local biological control (e.g., insects that are beneficial to crops because they defend them, like some ants species in legume crops, where the latter defend the plant against predators while nitrogen-fixing bacteria in the root system help to conserve the fertility of the soil).

5. Conclusions and future research

This chapter highlighted the need to develop new management strategies to permanently control various insect pests that attack agricultural systems. It is imperative that such proposals take into account the ecological and evolutionary properties of each insect species that can potentially become an agricultural pest. By understanding a species' genetic structure, we can assess its long-term potential to adapt and become a resistant, more devastating, and more invasive pest. On the other hand, by identifying certain life history traits, we can predict the abundance and potential distribution area of a species. From this knowledge, better control methods could be designed. An efficient long- and short-term method would be one that could avoid or minimize side effects in individuals, populations, communities, and biomes, including: 1) the evolution of pest resistance; 2) eradication of non-target species, including the pest species' natural predators; 3) elimination of relevant ecological interactions through the modification of the species' distribution and abundance; 4) modifications to the biogeochemical cycles; 5) environmental pollution; and 6) impacts on human health.

Unfortunately, at present, most commonly used methods have a high cost and high impact, not only from an economical perspective, but also from an ecological, evolutionary, and even social point of view. The deployment of chemical insecticides, including their endogenous production in genetically modified plants, as well as biological control methods, is definitely not fully compatible with our proposal. To expect an insecticide to work for the long-term is to go against the whole theory of evolution and some ecological precepts. Insecticides and biological control act as selective pressures on insect populations, causing the genotypes that can withstand these selective forces to eventually increase their frequencies in populations.

Based on several case studies, we think that the mechanical control method can be employed in relation to some insect pest species, especially those that may have an added economic value like most of the Orthoptera. Grasshoppers are the most devastating insects, not only at present but throughout history; in some countries, however, entomophagy of this group is customary. For these insects as well as some others (Coleoptera, Hemiptera, Lepidoptera), our conclusion would be not to provoke ecological disequilibria by eliminating them with insecticides or biological control; rather, it would be to mechanically remove them and use their biomass. It is clear that in some instances, the method proposed here cannot be fully functional. In a hypothetical species that is native, emergent, and non-edible, inhabiting within the same range of endemic and specific predators, with life history traits nothing like the ones that characterize a devastating species, perhaps strategies like biological control and insecticides can work as control methods. Ultimately, the fundamental step that should be taken before applying any control method, whether mechanical, chemical, or biological, is to take into consideration the ecological and evolutionary trends that a pest exhibits; only at this point can an appropriate and informed strategy can be put in place.

Our work has underlined the consequences of pest control methods when the ecological and evolutionary traits of the species are not subjected to prior analysis. In the worst scenarios, they could alter environmental processes irreversibly. No pest control methods have been applied so far that have taken into consideration the short- and long-term effects they may pose over environmental services. These practices ultimately generate economic costs that are neither easily affordable nor quantifiable. Understanding how much an environmental service costs could set the path for better decisions regarding suitable and informed pest management. It is imperative to evaluate environmental services such as those arising as a result of ecological interactions such as predation and mutualism (e.g. pollination), the cost of environmental pollution, and the costs for biogeochemical changes. If the cost of eradicating a pest species using insecticides is several million dollars per year, what would be the cost for the environmental service provided by a pest-specific predator, parasite, or parasitoid species?

At present, one of the most controversial methods is the application of chemicals using genetically modified crops as a vehicle for the endogenous production of insecticides. Besides the above-mentioned consequences of the use of conventional chemical control methods, this new method presents major problems at the genetic level, for instance, the gene flow of transgenes to conventional crop populations. This effect can be irreversible and affect the evolution and viability of plant species, especially in regions where transgenic crops are in contact with their wild relatives. Several studies have shown that gene flow from transgenic to wild plants has already occurred. For example, Wegier *et al.* (2011) confirmed the presence of transgenes in wild cotton plants in Mexico, which is the center of origin of this species. In this case, the evolutionary costs of this introgression should be evaluated, while a general question should be addressed: What would be the economic cost of losing the populations that gave rise to different crops used in agriculture today?

Right now, our team is developing a computational platform to formalize our proposal regarding the use of ecological and evolutionary properties to control pest species. This formalization requires an extensive database of each of the pest species and an efficient statistical methodology that enables us to correlate all the variables. In an upcoming study, our team will propose some strategies for some of the most devastating Orthoptera species in certain regions of Mexico. For the moment, we are reviewing all the knowledge available for *S. purpurascens* (a species with a high potential for local crop devastation) and *S. gregaria* (a species with a global distribution); with these data, we will perform a multivariate analysis. From a graphical perspective (see Figure 2), a given area will be generated with a specific form within a multivectorial system. The analytic model we propose here can lead to future long-term pest management strategies, based on ecological and evolutionary knowledge, thus preventing or at least minimizing the negative repercussions of current pest-management strategies.

6. Acknowledgements

We thank Alma Piñeyro-Nelson, Leticia Moyers, Rebeca Velázquez, Valeria Zermeño, Adriana Uscanga, Sandra Petrone, Adriana López-Villalobos and Daniel Piñero for valuable comments on early versions of this manuscript. We also thank Adriana Caballero for figure editing.

7. References

Abate, T. A., Van Huis, A. and Ampofo, J. K. O. (2000). Pest management strategies in traditional agriculture: an African perspective. *Annu. Rev. Entomol*, 45, 631-659.

Altieri, M. A. (1987). *Agroecology. The scientific basis of alternative agriculture*. Westview Press, Boulder, CO.

Altieri, M. A. (1995). *Agroecology: The science of sustainable agriculture*. 2nd edition, Westview Press, Boulder, CO.

Andersson, M. S. and de Vicente C. M. (2010). *Gene flow between crops and their wild relatives*. Baltimore, Johns Hopkins University. 564 pp.

Atlas, R. M. and Bartha, R. (1987). Microbial Biology: Fundamentals and Applications, 2nd edn., Menlo

Balkenhol, N., Gugerli, F., Cushman, S. A., Waits, L. P., Coulon, A., Arntzen, J. W., Holderegger, R., Wagner, H. H., Arens, P., Campagne, P., Dale, V. H., Nicieza, A. G., Smulders, M. J. M., Tedesco, E., Wang, H. and Wasserman, T. N. (2009) Identifying future research needs in landscape genetics: where to from here? *Landscape. Ecol.*, 24, 455-463. DOI 10.1007/s10980-009-9334-z

Begon, M., Mortimer, M. and Thompson, D. (1996) *Population Ecology, A unified study of animals and plants*. Blackwell Science, Oxford.

Berryman, A. (1973). Populations dynamics of the fir engraver, Scolytus ventralis (coleoptera: Scolytidae). Analysis of population behavior and survival from 1964 to 1971. *Can. Entomologist*, 105, 1465-88.

Blanco, R. J. L. and García, H. (2006). La agrodiversidad de la milpa. Proyecto Sierra de Santa Marta, A. C. 364 p.

Borror, D. J., Triplehorn, C. A. and Johnson, N. F. (1992). *An Introduction to the study of insects*. Saunders College, Fort Worth.

Brock, T. and Madigan M. (1988). *Biology of Microorganisms*. Prentice Hall, London.

Burd, A. D., Gould, F., Bradley, J. R., Van Duyn, J. W. and Moar, W. J. (2003). Estimated frequency of nonrecessive Bt resistance genes in bollworm, *Helicoverpa zea* (Boddie) (Lepidoptera: Noctuidae), in eastern North Carolina. *J. Econ. Entomol.*, 96, 137-142.

Cano-Santana, Z. and Oyama, K. (1994). Ámbito de hospederos de tres especies de insectos herbivoros de *Wigandia urens* (Hydrophyllaceae). Southwest. Entomol., 19, 167-172.

Castellanos, V. I. (2001). Ecología de la oviposición de Sphenarium purpurascens (Orthoptera: Pyrgomorphidae) en la reserva del Pedregal de San Angel,. Tesis. Universidad Nacional Autónoma de México, Facultad de Ciencias, Mexico D.F.

Cerritos, R. (2002). Método mecánico como alternativa en el control de plagas: el caso de *Sphenarium purpurascens* (Orthoptera: Pyrgomorphidae) en el valle de Puebla-

Tlaxcala. Tesis de Maestría, Universidad Nacional Autónoma de México, Facultad de Ciencias, Mexico D.F.

Cerritos, R. (2009). Insects as food: an ecological, social and economical approach. *CAB Reviews: Persp. Agric. Veter. Sci. Nutr. Nat. Res.* 4. No. 027.

Cerritos, R. (2011). Grasshoppers in agrosystems: pest or food. *CAB Reviews: Persp. Agric. Veter. Sci. Nutr. Nat. Res.*, 4. No. 027.

Cerritos, R. and Cano-Santana, Z. (2008). Harvesting grasshoppers *Sphenarium purpurascens* in Mexico for human consumption: A comparison with insecticidal control for managing pest outbreaks. *Crop Prot.*, 27, 473-80.

Chapman, J. W., Nesbit, R. L., Burgin, L. E., Reynolds, D. R., Smith, A. D., Middleton, D. R. and Hill J. K. (2010). Flight orientation Behaviors promote optimal migration trajectories in high-flying insect. *Science*, 327, 682-685.

Chapuis, M. P., Lecoq, M., Michalakis, Y., Loiseau, A., Sword, G. A., Piry, S. and Estoup, A. (2008). Worldwide survey in *Locusta migratoria*, a pest plagued by microsatellite null alleles. *Mol. Ecol.*, 17, 3640–3653.

Chávez-Servia, J. L., Tuxill J. and Jarvis, D. I. (2004). *Manejo de la diversidad de los cultivos en los agroecosistemas tradicionales*. Instituto Internacional de Recursos Fitogenéticos. Cali, Colombia. 286 p.

Cheke, R. A. (1990). A migrant pest in the Sahel: the Senegalese grasshopper *Oedaleus senegalensis*. *Phil. Trans. R. Soc. Lond.* B328, 539-553.

Chen, J., Franklin, J. F. and Spies, T. A. (1995). Growing-season microclimatic gradients from clearcut edges into old-growth *Old*-Growth Douglas-Fir Forests. *Ecol. Appl.* 5, 74–86.

Cole, L. C. (1954). The population consecuences of life history phenomena. *Q. Rev. Biol.*, 29, 103-137.

Collinge, S. K. (2000). Effects of grassland fragmentation on insect species loss, colonization, and movement patterns. *Ecology*, 81, 2211-2226.

Conradt, L., Roper, T. J. and Thomas, C. D. (2001). Dispersal behavior of individuals in metapopulations of two British butterflies. *Oikos*, 95, 416-424.

Dale, P. J., Clarke, B., and Fontes, E. M. (2002). Potential for the environmental impact of transgenic crops. *Nature Biotech.*, 20, 567-575.

Daly, V. H., Doyen, J. T. and Ehrlich, P. R. (1978). *Introduction to insect biology and diversity*. MaGraw-Hill, Tokio.

Devine, G. J. and Furlong, M. J. (2007). Insecticide use: Contexts and ecological consequences. *Agr. Hum. Val.*, 24 (3), 281–306.

Dominguez, R. R. (1992). *Plagas agrícolas*. Universidad Autónoma de Chapingo, Chapingo, México.

Dout, R. L. and Smith, R. F. (1971). The pesticide syndrome—diagnosis and prophylaxis. In *Biological Control*, ed. CB Huffaker, pp. 3–15. New York.

Dyer, G. A., Serratos-Hernández, J. A., Perales, H. R., Gepts, P., Piñeyro-Nelson, A., Chávez, A., Salinas-Arreortua, N., Yúnez-Naude, A., Taylor, J. E. and Alvarez- Buylla E. R. (2009). Dispersal of Transgenes through Maize Seed Systems in Mexico. *PLoS One*, 4(5), 1-9

Ehrlich, P. R. and Raven, P. H. (1964). Butterflies and plants: a study in coevolution. *Evolution*, 18, 586-608.

Ellstrand, N. C. (2003). *Dangerous liaisons? When cultivated plants mate with their wild relatives*. Baltimore: Johns Hopkins University.

FAO. (2008). Newsroom, Food and Agriculture organization of the United Nations (FAO), http://www.fao.org/newsroom/

Farnham, A. W. (1973). Genetics of resistance of Pyrethroid-selected houseflies *Musca domestica* L. *Pest. Sciences*. 4: 513.

Farrow, R. A. and Daly, J. C. (1987). Long-range movements as an adaptive strategy in the genus *Heliothis* (Lepidoptera, Noctuidae) a review of its occurrence and detection in 4 pest species. *Aust. J. Zool.* 35, 1-24.

Fashing, P. J., Nguyen, N. and Fashing, N. J. (2010). Behavior of geladas and other endemic wildlife during a desert locust outbreak at Guassa, Ethiopia: ecological and conservation implications. *Primates*. On line.

Faure, F. C. (1944). Pentatomid bugs as human food. *J. Entomol. Soc. South Afr.*, 7, 111–12.

Ferré, J. and van Rie, J. (2002). Biochemistry & genetics of insect resistance to *Bacillus thuringiensis*. *Annu. Rev. Entomol.*, 47, 501 - 533.

Finamore, A., Roselli, M., Britti, S., Monastra, G., Ambra, R., Turrini, A. and Mengheri, E. (2008). Intestinal and peripheral immune response to MON810 maize ingestión in weaning and old mice. *J. Agric. Food Chem.*, 56, 11533-11539.

Gaines, J. C. (1957). Cotton insects and their control in the United States. *Annu. Rev. Entomol.*, 2,319–38.

Gassmann A. J., Onstad, D. W. and Pittendrigh, B. R. (2006). Evolutionary analysis of herbivorous insects in natural and agricultural environments. *Wiley interscience*. DOI 10.1002/ps.1844.

Gassmann, A. J., Carrière, Y. and Tabashnik, B. E. (2009) Fitness costs of insect resistance to *Bacillus thuringiensis*. *Annu. Rev. Entomol.*, 54, 147–163.

Geier, P. W. (1966). Management of insect pests. *Annu. Rev. Entomol.*, 11,471–90.

Georghiou, G. P. (1990). *Overview of insecticide resistance, managing resistance to agrochemicals: from fundamental research to practical strategies*. American Chemical Society, Washington, D.C.

Gibbs, W. W. (2001). On the termination of species. *Scientific American*. 34, 28-37.

Gliessman, S. R. (1990). *Agroecology: Research in the Ecological Basis for Sustainable Agriculture*. Springer. New York.

Godfray, H. C. J. (1994). *Parasitoide. Behavioural and Evolutionary*. Princeton University Press, Princeton.

Goodwin, B. J. and Fahrig, L. (2002). Effect of landscape structure on the movement behavior of a specialized goldenrod beetle, Trirhabda borealis. *Can. J. Zool.*,80, 24-35.

Gould, F. (1998). Sustainability of transgenic insecticidal cultivars: integrating pest genetics and ecology. *Annu Rev Entomol.*, 43, 701–726.

Gould, F., Anderson, A., Jones, A., Sumerford, D., Heckel, D. G. and Lopez, J. (1997). Initial frequency of alleles for resistance to *Bacillus thuringiensis* toxins in field populations of *Heliothis virescens*. *Proc Natl Acad Sci* USA, 94, 3519–3523.

Guillot, G., Estoup, A., Mortier, F., and Cosson, J. F. (2005). A spatial statistical model for landscape genetics. *Genetics*, 170, 1261–1280.

Hampton, J. O., Spencer, P. B. S., Alpers, D. L., Twigg, L. E., Woolnough, A. P., Doust, J., Higgs, T., and Pluske, J. (2004). Molecular techniques, wildlife management and the importance of genetic population structure and dispersal: a case study with feral pigs. *J. Appl. Ecol.*, 41, 735–743.

Hauser, L., Adcock, G. J., Smith, P. J., Ramirez, J. H. and Carvalho, G. R.(2002). Loss of microsatellite diversity and low effective population size in an overexploited population of New Zealand snapper (*Pagrus auratus*). *Proc Natl Acad Sci USA.*, 99, 11742–11747.

He, Y. P. and Zhu, K. Y. (2004). Comparative studies of acetylcholinesterases purified from two field populations of the oriental migratory locust (*Locusta migratoria manilensis*): implications of insecticide resistance. *Pest. Biochem. Physiol.*, 78, 67-77.

Hewitt, G. and Onsager, J. (1983). Control of grasshoppers on rangeland in the United States: a perspective. *J. Ran. Manag.* 36:202–207.

Hilbeck, A. and Schmith, J. E. U. (2006). Another View on Bt Proteins - How specific are they and what else might they do? *Biopestic. Int.*, 2 (1), 1-50

Hoekstra, J. (1998). Conserving Orthoptera in the wild: lessons from *Trimerotropis infantilis* (Oedipodinae). *J. Ins Conserv.*, 2, 179–185.

Holderegger, R., Hermann, D. and Poncet, B. (2008). Land ahead: using genome scans to identify molecular markers of adaptive relevance. *Plant Ecol Div.*, 1, 273–283.

Holderegger, R., Kamm, U. and Gugerli, F. (2006). Adaptive versus neutral genetic diversity: implications for landscape. *Landscape Ecol.*, 24, 455–463.

Huffaker, C. B., Berryman, A. A. and Laing, E. J. (1984). Natural control of insects populations, In: *Ecological Entomology*, Huffaker, C. B. & Rabb, L. R., pp. 359-389, John Wiley & Sons, North Carolina.

Hunter, M. D. (2002). Landscape structure, habitat fragmentation and the ecology of insects. *Agric. For. Entomol.*, 4, 159-166.

Hunter, M. D. Malcolm, S.B. and Hartley, S.E. (1996). Population-level variation in plant secondary chemistry and the population biology of herbivores. *Chemoecology*. 45, 56.

Iranipour, S., Kharrazi, A. P. and Radjabi, G. (2010). Life history parameters of the sunn pest, *Eurigaster integriceps*, held at four constant temperatures. *J. Insect. Sci.* 10, 106.

IUCN. (1996). IUCN Red List of Threatened Animals, Gland, Switzerland.

Jensen, S. T. (2000). Insecticide resistance in the western flower thrips, *Franklinella occidentalis. Integrated Pest Manag. Rev.*, 5: 131-146.

Jhonson, D. M. and Stiling, P. D. (1998). Distribution and dispersal of *Cactoblastis cactorum* (Lepidoptera: Puralidae), an exotic opuntia-feeding moth in Florida. *Folia Entomol.*, 8:,12-22.

Kaiser-Alexnat, R., Wagner, W., Langenbruch, G., Kleespies, R., Keller, B., Meise, T. and Hommel, T. (2005). Selection of resistant European Corn Borer (*Ostrinia nubilalis*) to *Bt*-corn and preliminary studies for the biochemical characterization. *IOBC/wprs Bulletin*. 28, 115-118.

Kelly, C. K and Southwood, T. R. E. (1999). Species richness and resource availability: a phylogenetic analysis of insects associated with trees. *Proc Natl Acad Sci USA*, 96, 8013–8016.

Kenmore P. E. (1996). Integrated pest management in rice. See Ref. 133, pp. 76–97

Kogan, M. (1998). Integrated pest management: Historical Perspectives and Contemporary Developments. *Annu. Rev. Entomol.*, 43,243–70.

Kruess, A. and Tscharntke, T. (1994). Habitat fragmentation, species loss, and biological control. *Science*. PTR, 1581-1584.

Kruess, A. and Tscharntke, T. (2000). Species richness and parasitism in a fragmented landscape: experiments and field studies with insects on *Vicia sepium*. *Oecologia*. 129-137. (falta numero)

Lahr, J. (1998). An ecological assessment of the hazard of eight insecticides used in desert locusts control, to invertebrates in temporary ponds in the Sahel. *Aquat. Ecol.*, 32, 153–162.

Lenormand, T., Bourguet, D., Guillemaud, T. and Raymond, M. (1999). Tracking the evolution of insecticide resistance in the mosquito *Culex pipiens*. *Nature*, 400, 861–864.

Leyva, V. J. and Ibarra, R. J. (1992). II Curso de control biológico. Carrera de ingeniería agricola. Facultad de Estudios Superiores, Cuautitlán, UNAM.

Liu, N. and Scott, G. J. (1995). Genetics of Resistance to Pyrethroid Insecticides in the house fly, *Musca domestica*. *Pest. Biochem. Physiol.*, 52, 116-124.

Lockwood, J. A. (1998). Management of orthopteran pest: a conservation perspective. *J. Insect. Conserv.*, 2, 253–261.

Lockwood, J. A. and DeBrey, L. D. (1990). A solution for the sudden and unexplained extinction of the Rocky Mountain locust, *Melanoplus spretus* (Walsh). *Environ. Entomol.*, 19, 1194–1205.

Lockwood, J. A. and Sergeev, G. M. (2000). Comparative biogeography of grasshoppers (Orthoptera: Acrididae) in North America and Siberia: applications to the conservation of biodiversity, *J. Insect Conserv.*, 4, 161–172.

Losey, J. E. and Vaughan, M. (2006). The economic value of ecological services provided by Insects. *Bioscience*, 56, 311–323.

Losey, J. E. and Vaughan, M. (2008). Conserving the ecological services provided by insects. *Am. Entomol.* 54, 113–115. Revista fitotecnia Mexicana, enero-marzo Vol. 28-001

Louda, J. M. (1997). Ecological effect of an insect introduced for the biological control of weeds. *Science*, 271, 1088-1990.

MacCuaig, R. D. (2008). Determination of the resistance of locust to DNC in relation to their weight, age and sex. *Ann. Appl. Biol.*, 44, 634-642.

MacPhee, A. W. and MacLellan, C. R. (1971). Cases of naturally occurring biological control in Canada. In *Biological Control*, ed. CB Huffaker, pp. 312–28. New York.

Manel, S., Schwartz, K., Luikart, G. and Taberlet, P. (2003). Landscape genetics: Combining landscape ecology and population genetics. *Trends Ecol. Evol.*, 18: 189–197.

McEvoy, P. B. (1999). Biological control of plants invaders: regional patterns, field experiments, and structured populations models. *Ecol. Appl.*, 9, 387-461.

Metcalf, R. L. (1980). Changing role of insecticides. *Ann. Rev. Entomol.*, 219–256.

Mochizuki, M. (2003). Effectiveness and pesticide susceptibility of the pyrethroid-resistant predatory mite *Amblyseius womersleyi* in the integrated pest management of tea pests. *Biocontrol*, 48(2), 207–221.

Morales, H. And Perfecto, I. (2000) Traditional knowledge and pest management in the Guatemalan highlands. *Agriculture and Human Values* 17: 49–63.

Moreno, L. L. Yupit Moo, E. C. Tuxill, J. Mendoza Elos, M. Arias Reyes, L. M. Castañon Najera, G. and Chavez Servia, J. L. (2005) Sistema tradicional de almacenamiento de semilla de frijol y calabaza en Yaxcabá, Yucatán.

Moser, J. (2007). Development of resistance to Bt maize among Western corn rootworm. (2005-2008). University of Göttingen, Institute of Plant Pathology and Plant Protection, Göttingen.

Muñoz, A. (2003). *Centli-Maíz. Prehistoria, historia, diversidad, potencial, origen genético y geográfico Colegio de Posgraduados – SAGARPA*. México.

National Grasshopper Management Board (NGMB). (1995). In Proceedings and Resolutions of the 1995 meeting of the National Grasshopper Management Board, Rapid City, South Dakota;.

Newsom, L. D. (1980). The next rung up the integrated pest management ladder. *Bull. Entomol. Soc. Am.*, 26,369–74.

Nylin, S. (2001). Life history perspectives on pest insects: What's the use? *Austral Ecol.*, 26, 507-517.

Odiyo, P. O. (1990). Progress and developments in forecasting outbreaks of the African armyworm, a migrant moth. *Phil. Trns. R. Soc. Lond.* B328, 555-569.

Onstad, D. W., Spencer, J. L., Guse, C. A., Levine, E. and Isard, S. A. (2001). Modeling evolution of behavioral resistance by an insect to crop rotation. *Entomol. Exp. Appl.*, 100, 195-201.

Osteen, C. D. and Szmedra, P. I. (1989). Agricultural Pesticide Use Trends and Policy Issues. Agricultural Economics report 622, US Department of Agriculture, Economic Research Service, Washington, DC.

Pedgley, D. E. (1993). Managing migratory insect pests: a review. *Int. J. Pest Manag.*, 39, 3-12.

Pfadt, R. E. and Hardy, D. M. (1987). A historical look at rangeland grasshoppers and the value of grasshopper control programs. In J. Capinera, editor, *Integrated pest management on rangeland*, Westview Press, Boulder, CO, USA.

Pimentel, D. (1992). Ecological effects of pesticides on non- target species in terrestrial ecosystems. In R. G. Tardiff (ed.), *Methods to Assess Adverse Effects of Pesticides on Non-target Organisms*, (pp. 171–190). Toronto, Canada: John Wiley & Sons.

Pimentel, D. (2005). Environmental and economic costs of the application of pesticides primarily in the United States. Environment, Development and Sustainability, 7, 229–252.

Piry, S., Alapetite, A., Cornuet, J.-M., Paetkau, D., Baudouin, L., and Estoup, A. (2004). GENECLASS2: a software for genetic assign- ment and first-generation migrant detection. *J. Hered.* 95, 536–539. doi:10.1093/jhered/esh074

Price, P. (1984). *Insect Ecology*. Wiley-Interscience, New York.

Pritchard, J. K., Stephens, M., and Donnelly, P. (2000). Inference of population structure using multilocus genotype data. *Genetics* .155, 945–959.

Purvis, A. and Hector, A. (2000). Getting the measure of biodiversity. *Nature*, 405, 212-219.

Radeloff, V. C., Mladenoff, D. J. and Boyce, M. S. (2000). The changing relation of landscape patterns and jack pine budworm populations during an outbreak. *Oikos*, 90, 417-430.

Reed, D. H. and Frankham, R. (2001). How closely correlated are molecular and quantitative measures of genetic variation? A meta-analysis. *Evolution*. 55, 1095–1103.

Relyea, R. A. (2005). The impact of insecticides and herbicides on the biodiversity and productivity of aquatic communities. *Ecol. Appl.*, 15, 618–627.

Riley, J. R. and Reynolds, D. R. (1983). A long-range migration of grasshoppers observed in the sahelian zone of Mali by two radars. *J. Anim. Ecol.*, 52, 167-183.

Riley, J. R. and Reynolds, D. R. (1990). Nocturnal grasshopper migration in west Africa: transport and concentration by the wind, and the implications for air-to-air control. *Phil. Trans. R. Soc. Lond.*, B 328, 655-672.

Risser, P. G. (1986). Agroecosystem: structure, analysis, and modeling. See Ref. 93, pp. 321–43

Robb, K. L. (1989). *Analysis of* Franklinella occidentalis *(Pergande) as a pest of floricultural crops in California greenhouses.* University of California, Riverside.

Robertson, B. C. and Gemmell, N. J. (2004). Defining eradication units to control invasive pests. *J. Appl. Ecol.*, 41, 1042–1048. doi:10.1111/j.0021-8901.2004.00984.x

Roderick, G. K. and Navajas, M. (2003). Genes in new environments: genetics and evolution in biological control. *Nat. Rev. Gen.*, 4, 889–899. doi:10.1038/nrg1201

Rollins, L. A., Woolnough, A. P. and Sherwin, W. B. (2006). Population genetic tools for pest management: a review. *Wildlife Res.*, 33, 251–261.

Ronce, O. and Kirkpatrick, M. (2001). When sources become sinks: migrational meltdown in heterogeneous habitats. *Evolution*, 55, 1520-1531.

Schubert, D. (2002). A different perspective on GM food. *Nat. Biotech.* 20, 969.

Shirai, Y. and Morimoto, N. (1997). Life history of pest and non-pest populations in the phytophagous ladybird beetle, *Epilachna niponica* (Coleoptera, Coccinellidae). *Res. Popul. Ecol.*, 39, 163-171.

Simpson, S. J., Sword, G. A and De Loof, A. (2005). Advances, controversies and consensus in locust phase poliphenism research. *J. Orthopt. Res.*, 14, 213-222.

Singer, M. C. and Thomas, C. D. (1996). Evolutionary responses of a butterfly metapopulation to human- and climate-caused environmental variation. *Amer. Nat.*, 148, S9-S39.

Siqueira, A. A. H., Narciso, R. C. and Picanco, C. M. (2000). Insecticide resistance in populations of *Futa absoluta* (Lepidoptera:Gelechiidae). *Agric. Forest Entomol.*, 2, 147-153.

Slatkin, M. (1987). Gene flow and the geographic structure of natural populations. *Science* 236,787-92.

Slatkin, M. and Hudson, D. (1991). Pairwise comparisons of mitochondrial DNA quences in stable and exponentially growing populations. *Genetics*, 129, 555-62.

Smouse, P. E. and Peakall, R. (1999). Spatial autocorrelation analysis of individual multiallele and multilocus genetic structure. *Heredity*, 82, 561 - 573.

Solbreck, C. (1995). Variable fortunes in a patchy landscape: the habitat templet of an insect migration. *Res. Popul. Ecol.*, 37, 129-134.

Song, H. (2005). Phylogenetic perspectives on the evolution of locust phase poliphenism. *J. Orthopt. Res.*, 14, 235-245.

Sork, V. L., Nason, J., Campbell, D. R. and Fernandez, J. F. (1999). Landscape approaches to historical and contemporary gene flow in plants. *Trends Ecol. Evol.*, 14, 219–224.

Speight, M., Hunter, M. and Watt, A. (1999). *Ecology of insects, concept and applications.* Blackwell Science, Oxford.

Spiroux de Vendomois, J., Roullier, F., Cellier, D. and Séralini, G-E. (2009). A Comparison of the Effects of Three GM Corn Varieties on Mammalian Health. *Int. J. Biol. Sci.*, 5(7), 706-726.

Stewart, D. (1998). Non-targed grasshoppers as indicators of the side-effect of chemical locust control in the Karoo, South Africa. *J. Insect Conserv.*, 2, 263-276.

Strong, D. R., Lawton, J. H. and Southwood, S. R. (1984). *Insect on plants, community patterns and mechanisms.* Harvard university press, Cambridge, Massachusetts.

Tabashnik, B. E., Biggs, R. W., Higginson, D. M., Henderson, S., Unnithan, D. C., Unnithan, G. C., *et al.* (2005). Association between resistance to Bt cotton and cadherin genotype in pink bollworm. *J. Econ. Entomol.*, 98, 635-644.

Tabashnik, B. E. (1994). Evolution of resistance to *Bacillus thuringiensis*. *Annu. Rev. Entomol.* 39, 47-79.

Talekar, N. S. and Shelton, A. M. (1993). Biology, ecology and management of the diamondback moth. *Annu. Rev. Entomol.*, 38, 275-301.

Taylor, B. L. (1997). Defining 'population' to meet management objectives for marine mammals. *Mol. Gen. Mar. Mam.* 3, 49-65.

Taylor, C. E. and Georghiou, G. P. (1979). Suppression of insecticide resistance by alteration of gene dominance and migration. *J. Econ. Entomol.* 72, 105-109.

Thomas, M. B. and Willis, A. J. (1993). Biocontrol-risky but necessary. *Trends Ecol. Evol.*, 13, 325-329.

Turner, M., Gardner, R. H. & O'Neill, R. V. (2001). Landscape *Ecology in Theory and Practice*, Springer-Verlag.

Uvarov, B. P. (1964). *Grasshoppers and locust*. Centre for overseas pest research, University press, Cambridge.

Van den Berg, H., Hassan, K. and Marzuki, M. (1998). Evaluation of pesticide effects on arthropod predator populations in soya bean in farmers fields. *Biocontrol Sci. Techn.* 8(1): 125-137.

Van Huis, A. (1996). The traditional use of arthropods in sub-Saharan Africa. *Proc. Exp. Appl. Entomol.*, N.E.V. Amsterdam, 7, 3-20.

Varley, G., Grandwell, G. & Hassell, M. (1973). *Insects population ecology*. Blackwell Science, Oxford.

Wildlife Research, 2006, 33, 251-261.

Wegier, A., Piñeyro-Nelson, A., Alarcón, J., Gálvez-Mariscal, A., Álvarez-Buylla, E. R. and Piñero, D. (2011). Recent long-distance transgene flow into wild populations conforms to historical patterns of gene flow in cotton (*Gossypium hirsutum*) at its center of origin. *Mol. Ecol.* 20 (19): 4182-4194.. doi: 10.1111/j.1365-294X.2011.05258.x

Wilson, G. A. and Rannala, B. (2003). Bayesian inference of recent migration rates using multilocus genotypes. *Genetics*, 163, 1177-1191.

With, K. A., Pavuk, D. M., Worchuck, J. L., Oates, R. K. and Fisher, J. L. (2002). Threshold effects of landscape structure on biological control in agroecosystems. *Ecol. Appl.*, 12, 52-65.

Wolfgang, B. (2007). Research into the impact of Bt maize (Cry 3Bb1) on non-target organisms living in the soil. Federal Biological Research Centre for Agriculture and Forestry (BBA), Institute for Plant Protection in Field Crops and Grassland, Braunschweig.

Wright, S. (1931). Evolution in Mendelian populations. *Genetics*, 16, 97-159.

Wu, K., Guo, Y. and Head, G. (2006). Resistance monitoring of *Helicoverpa armigera* (Lepidoptera: Noctuidae) to Bt insecticidal protein during 2001 - 2004 in China. J Econ Entomol 99:893 - 898.

Yang, M. L., Zhang, J. Z., Zhu, K. Y., Xuan, T., Liu, X. J. and Guo, Y. P. (2008). Mechanisms of organophosphate resistance in a field population of oriental migratory locust, *Locusta migratoria manilensis* (Meyen). *Arch. Insect Bioch. Physiol.*, 71, 3-15.

3

Agroecological Crop Protection: Concepts and a Case Study from Reunion

Jean-Philippe Deguine[1], Pascal Rousse[2] and Toulassi Atiama-Nurbel[1]
[1]*CIRAD, UMR PVBMT, Saint-Pierre, La Réunion*
[2]*Chambre d'agriculture de La Réunion, La Réunion*
France

1. Introduction

In crop protection, chemical control rapidly revealed its limitations, as well as its possibilities, and alternative solutions to pest management problems have been recommended since at least the 1960s. A new strategy was developed under the rubric 'integrated control', envisaging the employment of a range of different control measures, constrained by their compatibility and the requirement for minimizing noxious effects on the wider environment. Despite these difficulties, a biological, then ecological, orientation has underlain the development of crop protection over the last 50 years (Pimentel, 1995; Walter, 2003). This process has been marked by multiple and diverse interpretations of the concept of IPM (Kogan, 1998). Numerous technical innovations have been proposed, without, however, bringing any really significant change in the management of pests in major crops (Lewis et al., 1997), due no doubt to an unrealistic approach to the complexities of the phenomena concerned. The debate has been re-animated recently, both by the spectacular success of the recent advances in biotechnology and by genuinely taking into account the need to preserve biological diversity. As much for socio-economic as for ecological reasons, this has given rise to a reexamination of farming systems as traditionally practiced, through an innovative agroecological approach (Dalgaard et al., 2003).

This chapter questions how agroecological concepts may contribute to sustainable pest management. In a first part, a panorama of the principles of agroecological crop protection is provided. Then, the concepts of agroecology are applied to the case study of the management of fruit fly populations in Reunion Island, describing some results obtained in research studies. Finally, an illustration of the results obtained in commercial farm conditions in Reunion Island is given.

2. Agroecological crop protection: Basis and principles

Since the 1970s, the evolution of plant protection has been driven by an improved understanding of the functioning of ecosystems (Bottrell, 1980). At this time, the desire to explore these issues favored the development of computer-based simulation models for risk assessment. The approach to these problems was considerably improved; taking into consideration the development of the plants in the particular soil/ moisture/ nutrient content and insolation context and considering the suite of pests present in the same crop.

This forms the basis for the development of a concept of integrated control and then of integrated production or integrated crop management.

The UN Conference on the Environment and Development in Rio de Janeiro in 1992 drew attention to the need to preserve the biological diversity of ecosystems in general and agro-ecosystems in particular. The subsequent publication of diverse works aimed at advancing the IPM paradigm, helped in the national adoption of IPM strategies. The simultaneous elaboration of the scientific principles underlying this field of agro-ecology, rendered these calls more credible (Altieri, 1995; Dalgaard et al., 2003). It was then necessary to move to the practical stage of conceiving growing systems which capitalized on the resilience of agro-ecosystems (Clements & Shrestha, 2004). To this end, 'agro-ecosystems management' or 'agro-ecological engineering' is today recognized as one of the up and coming concepts in crop protection (Clements & Shrestha, 2004; Gurr et al., 2004; Lewis et al., 1997; Nicholls & Altieri, 2004). More generally, this development is presented in the form of an 'IPM continuum' (Jacobsen, 1997), where it is clear that much of what is necessary will be a continuous evolution of traditional concepts and understanding in crop protection (Clements & Shrestha, 2004). The principles of a bio-centered agriculture, developed during the last few decades, have led to new orientations to crop production which will require a return to utilizing knowledge and skills progressively lost over the last few decades. IPM has been the fundamental paradigm in plant protection since the late 1960s. A major contribution of IPM to agriculture, is the incorporation of ecological principles into pest management while ensuring high productivity and profitable harvests. In agreement with this conceptual foundation, agroecological pest management largely relies on IPM.

Preserving ecosystems and biodiversity, while reducing fertilizer and pesticide use, is a challenge that must now be addressed to ensure that agriculture will be both intensive and environmentally friendly. Agroecology was defined as the study of the interactions between plants, animals, humans, and the environment within an agroecosystem (Dalgaard *et al.*, 2003). The agroecology concept was thus introduced to advocate agroecosystem-wide stand management. Agroecological pest management is thus based on ecological processes occurring between the crop and its pests (Carroll et al., 1990), but also the natural enemies of these pests (Weiner, 2003), in a quest for increased beneficial interactions that keep pest populations in check (Altieri & Nicholls, 2000; Gliessman, 2007). This crop protection strategy helps maintain bio-ecological balances between animal and plant communities within agroecosystems, while also preserving and improving the "health" of soils and plant biodiversity (Ratnadass et al., 2011). Agroecological crop protection is based on prevention at broader spatiotemporal scales. It combines plant and animal community management and thus contributes to conservation and biological control.

Deguine et al. (2009) give a definition of Agroecological Crop Protection as a crop protection system based on the science of agroecology. By prioritising preventative measures, the system seeks to establish bioecological equlibria between animal and plant communities within an agroecosystem with the goal of foreseeing or reducing the risks of infestation or outbreaks of pests. To this end, the system emphasises the conservation and improvement of the "health" of soils (fertility, biological activity, structure, etc.) and the maintenance or incorporation of plant biodiversity in the agroecosystem. Beyond the classical techniques of integrated crop protection, emphasis is placed on cultural practices and plant management systems which help maintain or create habitats to attract indigenous beneficial fauna and/or repel pest fauna. Agroecological crop protection operates at larger scales in time and space, from a single crop cycle to several years, and from a single field to an agroecosystem or a landscape. It brings together the management of plant communities (crops and non-cultivated plants in areas surrounding the field and in the wider agroecosystem) with that of the animal communities of pests, beneficials and pollinators. Agroecological crop protection thus requires concerted action by stakeholders, notably farmers and land managers. As with integrated crop protection, curative practices are only a last resort to be used in the case of absolute necessity, and then only using methods compatible with the functional biological groups which ensure the provision of ecological services. According to these criteria, the future use of pesticides may only be short term, at least in their present form, given the current status of many pesticides whose use is already restricted for environmental and toxicological reasons. According to this vision, prophylaxis, habitat management, and biological control are the principal components of crop protection.

Deguine et al. (2009) also propose a five strategy to implement the agroecological crop protection approach:

Step 1. Respect international, national and regional regulatory measures.

Step 2. Prioritise the use of preventative measures through the management of plant populations (whether cultivated on not): (i) Grow healthy plants and ensure good soil health utilising prophylaxis, varietal selection, crop rotations, whole-farm crop planning, cultural practices (such as sowing under plant cover and minimum tillage), management of weeds, rational irrigation and fertilisation, use of organic fertilisers; (ii) Reduce pest populations and increase populations of beneficial organisms (at the level of the individual field, its surroundings, of the farm and of the entire agroecosystem): crops or trap crops, planting of refuge areas, plant associations and intercropping, the *push-pull* technique, establishment of field margins, planning of ecological compensation structures (corridors, hedgerows, grassy and flowering strips etc.), techniques designed to incorporate vegetative diversity; (iii) Favour concerted actions in time and in space within the agroecosystem.

Step 3. Evaluate the real socio-economic and environmental risks by using pest scouting techniques appropriate for one field, a group of fields, a farm, or the whole ecosystem, with the assistance of the regional agricultural extension services.

Step 4. Take only need-based decisions on curative measures: (i) With the aid of decision tools and in collaboration with fellow producers, accounting for local and ever-changing multiple criteria, intervention thresholds (economic, social, environmental) and of the risk of development of resistance; (ii) In the framework of whole-of-farm management and at a range of time scales (short to long term), account for the agroecological characteristics of the agroecosystem as a whole (the spatial dimension).

Step 5. Only in the case of absolute necessity, apply curative measures: (i) Give priority to alternative control measures: cultural techniques (e.g. defoliation, plant topping), biological control, physical and biotechnical control measures; (ii) Only as a last resort: use the chemical pesticides with the lowest ecological impact, chosen to avoid the emergence of resistance.

3. Application of agroecological crop protection to the case of cucurbit fruit flies on Reunion

Fruit flies (Diptera: Tephritidae) are among the most destructive and widespread pests of horticultural systems in the tropical and subtropical areas of the world (White & Elson Harris, 1992). Although they have been the subject of many studies because of their economic impact, their control is problematic in most cases and requires large amounts of pesticides. This situation is emphasized under insular and tropical conditions as is the case of Reunion Island.

In Reunion Island, three species belonging to the Dacini tribe attack Cucurbit crops: *Bactrocera cucurbitae* (Coquillett, 1899), *Dacus ciliatus* (Loew, 1901) and *Dacus demmerezi* (Bezzi, 1917) (Fig. 1).

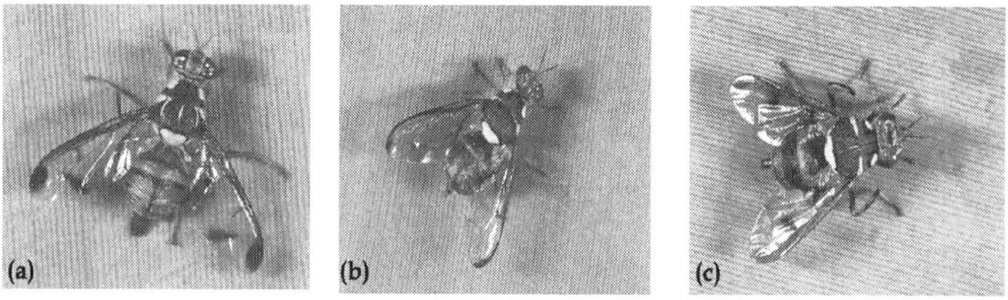

Fig. 1. The three species of fruit fly (Diptera, Tephritidae) which attack Cucurbits on Reunion (Photos: A. Franck - Cirad). (a) *Bactrocera cucurbitae;* (b) *Dacus ciliatus;* (c) *Dacus demmerezi.*

After oviposition by the females (Fig. 2), the damage caused by the larvae feeding on the fruit can reach 90% of the crop yield of zucchini, cucumber, pumpkin or chayote (Ryckewaert et al., 2010).

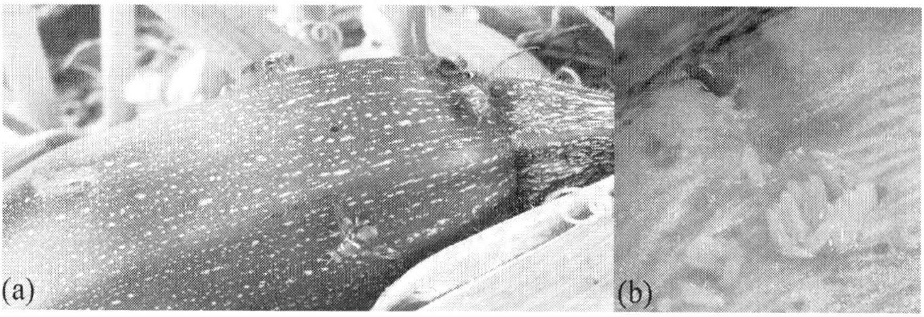

Fig. 2. Females of *Dacus demmerezi* laying eggs on a zucchini fruit (Photo: JP. Deguine - Cirad). (a) and their eggs before hatching (b) (Photo: A. Franck - Cirad).

The chemical protection currently used is not effective and also has many side effects: toxicity to natural enemies and pollinators, pollution and sanitary damage to biodiversity and to humans. There is now a demand for sustainable and agroecological crop protection (Augusseau et al., 2011).

The application of agroecological crop protection to a case study requires consideration to both the pests and to the context. It also requires knowledge and research. For the last several years, research has allowed us to obtain knowledge about the bioecology of cucurbit fruit flies, making it possible to apply the principles of agroecological crop protection. Complementary research has been developed to design and implement agroecological techniques or practices adapted to the management of Cucurbit fruit flies. Furthermore, a large-scale initiative (GAMOUR program, see Part 6.) was proposed since 2009 to assess the efficacy of agroecological cucurbit fruit fly management in agricultural pilot areas.

4. Research on bioecology of the flies

The aim of this research was to improve knowledge on the biology and ecology of the three cucurbit fruit fly species and particularly on the interactions between fly adults and host or non-host plants.

4.1 Attractiveness of non-host plants to fly adults

The study aimed to compare corn and Napier grass attractiveness on fly adults in field cages (Atiama-Nurbel et al., to be published). The two plants were established in pots and presented to adult flies in field cages. In each cage, a cohort consisting of 100 adults (50♀ and 50♂) of *B. cucurbitae* and *D. demmerezi* of known age was released. The experiment was replicated four times. The number of adult flies on the different plants as well as their location on the plant was recorded. The results showed that corn was more attractive than Napier grass to adults of the two fly species whatever their sex and their maturity.

4.2 Circadian rhythm of fly adults

The study was conducted in cucurbit crops during austral summer in a range of altitude (750 to 1,150 m) corresponding to the main areas of cucurbit cropping. The methodology consisted of recording living adults present in the cultivated field or roosting on corn planted around cucurbit fields, distinguishing species, sex and kind of activity for each adult observed. The observations were performed each hour of the day from 7:00 am to 6:00 pm. The results showed that in the fields, cucurbit fruit fly adults typically roost on corn planted around the field. The three species of FF showed circadian rhythms, and females typically move at specific times of the day from roosting sites to host fruit in order to lay eggs. Fig. 3 gives an example of the circadian rhythm of male and female adults on a zucchini crop and a corn border during the photoperiod of a day.

5. Research on agroecological techniques for fruit fly management

Taking into account the knowledge on bioecology of the Cucurbit FF, the aim of this research was to design and implement agroecological techniques. Three techniques were tested: (i) sanitation using the augmentorium technique, (ii) trap plants using corn and (iii) 'Attract and Kill' using spinosad-based bait.

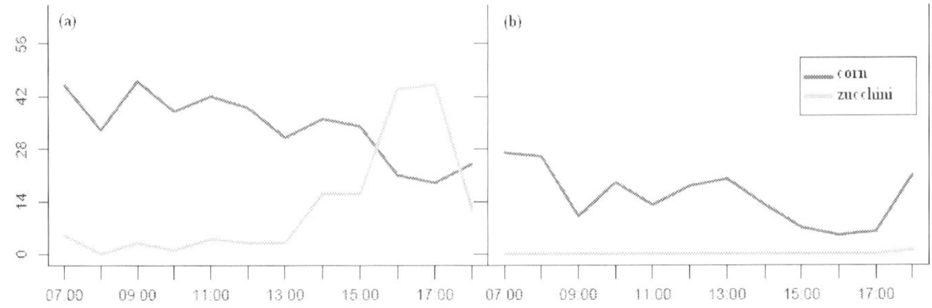

Fig. 3. Number of adult females (a) and males (b) of *Dacus demmerezi* observed on corn (56 m^2) and on zucchini (616 m^2) on the 13th February 2008 *in* Atiama-Nurbel (2008).

5.1 Sanitation using the augmentorium technique

Instead of the curative approach to reduce existing populations, the first step proposed for their management was sanitation. It is known to be an effective measure (Liquido, 1993). This method is based on an original technique originally developed by USDA in Hawaii utilizing a tent-like structure called an "augmentorium"(Jang et al., 2007; Klungness et al., 2005) which aims to sequester adult flies emerging from infested fruit while allowing the parasitoids to escape, via a net placed at the top of the structure. A prototype of augmentorium with an appropriate net mesh was developed in Reunion (Deguine et al, 2011) (Fig. 4).

(a) (b)

Fig. 4. Sanitation using the augmentorium technique (Photos: JP. Deguine - Cirad). (a) An augmentorium in a zucchini crop (see the net at the top of the augmentorium); (b) Flies adults sequestered by the net inside the augmentorium.

A first study aimed to determine the potential of numbers of flies that could be sequestred in a sanitation technique such as the augmentorium. This potential was estimated by measuring in the laboratory the emergence of several species of flies from infested fruit collected in the field from 2009 to 2010 in different sites of the island. Emergence of fly adults was measured for three species of flies: (i) *Bactrocera cucurbitae, Dacus ciliatus* and *Dacus demmerezi* attacking three species of Cucurbits (pumpkin: *Cucurbita maxima*;

cucumber: *Cucumis sativus* and zucchini: *Cucurbita pepo*). Collections of infested fruits showed the following means of emerged adults per kg of fruit: 217 for cucumber, 340 for pumpkin and 594 for zucchini (Jacquard, personal communication).

A second study focused on the performance and the efficiency of the augmentorium prototype recently tested in Reunion Island and particularly (i) the number of adult flies that can potentially be sequestred in an augmentorium in the field; (ii) the efficacy of the net mesh for fly sequestration and parasitoid escape (Deguine et al., 2011). The sequestration of three of these fly species (*B. cucurbitae, B. zonata* and *C. capitata*) and the escape of two species of their parasitoids (*Psyttalia fletcheri* and *Fopius arisanus*) were assessed on four mesh types in the Cirad laboratory in Saint-Pierre in 2008. The methodology of the experiments is described in the paper of Deguine et al. (2011). The results showed that the efficiency of the mesh chosen for the prototype of augmentorium (hole area 1.96 mm^2) proved to be perfectly effective in the laboratory with 100% of sequestration of adult flies. In the same way, 100% of the parasitoids were able to escape from the mesh if they choose to do so (Fig. 5).

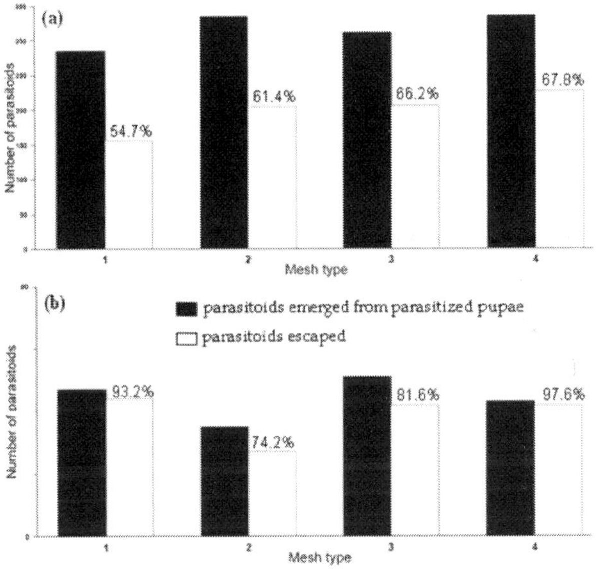

Fig. 5. Total results of the effect of four mesh sizes on *P. fletcheri* escape *(a)* (3 replications) and on *F. arisanus* escape (b) (2 replications): (i) number of parasitoids emerged from parasitized pupae of *B. cucurbitae* and (ii) number of escaped parasitoids. The percentage indicates the proportion between these two numbers (ii)/(i) *in* Deguine & Atiama-Nurbel (2010).

A third study aimed to evaluate the faisability of producing compost with infested fruit collected in the field. Preliminary tests on the feasibility of producing compost were then conducted in Saint Pierre in 2009, mixing zucchini and other components. We showed that a ratio of 50:30:20 of zucchini, sugar cane stem and chicken litter respectively was well adapted to produce compost.

These results lead us to confirm the relevance and the efficiency of the augmentorium in agroecological crop protection. As a sanitation technique against fruit flies, the augmentorium sequesters large amounts of adult flies per kg of infested fruit. As a biological control method, it may contribute to increase parasitoid populations which are often low because of the previous and significant pesticide pressure. The augmentorium can also be considered as a useful tool to produce compost in the context of sustainable agriculture. The technique of sanitation using the augmentorium is now well accepted by farmers in pilot areas in Reunion Island.

5.2 Trap plants using corn

Corn was selected on Reunion because preliminary research showed it was attractive for fly adults (see Part 4.) and it was easily available and usable by farmers. The studies were conducted in cucurbit crops during austral summer at a range of altitudes (750 to 1,150 m) corresponding to the main areas of cucurbit cropping, in four locations during three years. We recorded the living adults present in the cultivated field or roosting on corn planted around cucurbit fields, distinguishing species, sex and kind of activity for each adult observed. The observations were performed each hour of the day from 7:00 am to 6:00 pm.

The first study focused on the adult population levels on corn and zucchini. The results showed that in the field, corn concentrated more than 95 % of fly populations. Fig. 6 gives

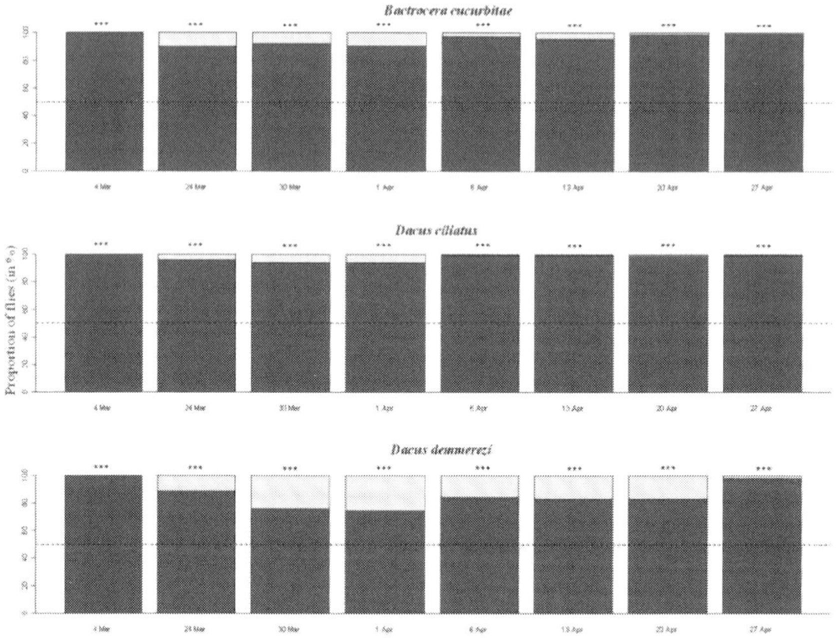

Fig. 6. Proportion of fly adults on corn (in black) and on zucchini (in grey) for the three species of Cucurbit flies at different dates (Tan Rouge, 2010). Binomial tests (5%) with H0: proportion of flies on corn (Pc) = 0.5 and H1: (Pc) ≠0.5 (***= P<0.001) *in* Bonnet (2010).

an illustration of such a concentration of the fly adults. Roosting corn plants could thus be used as trap plants and became the place to manage the populations instead of the crop, as it was showed in other parts of the world (Mc Quate et al., 2003; Mc Quate & Vargas, 2007).

The second study aimed to compare corn patches and corn strips within the field in the situation of Tan Rouge in 2010. The fly community was dominated by *B. cucurbitae,* more than 50% throughout the observation period. Corn was hosting the majority of the population (over 99% of adults observed). The sex ratio was stable on corn and only the females went to the crops. Corn was a refuge for the majority of the community during the day, where the entire population was roosting. The results of the study showed that the shape and the size of the inclusion of corn did not affect the concentration of flies. Both corn patches and corn strips were effective to trap 99 % of the fly adults. As a conclusion, different designs of trap corn plants can be recommended: borders around the cultivated field, patches or strips within the field (Fig. 7).

Fig. 7. Different designs of trap corn plants (Photos: JP. Deguine - Cirad). (a) borders of corn around the cultivated field; (b) strips of corn within the cultivated field; (c) patches of corn within the cultivated field.

5.3 Attract and kill using spinosad-based bait

Spinosad-based baits have been largely tested against fruit flies (Prokopy et al., 2004). The study was conducted to test the effectiveness of Synéis-appât® (Dow Agrosciences), a spinosad-based bait (Deguine et al., to be published). Experiments were conducted in field cages to compare efficiency of the bait on reared adult flies of the three species according ages and sexes. A total of 4.5 ml (equivalent to 5 sprays) of the bait was applied on two leaves in the upper stratum of the corn plant placed in the field cage. The bait was applied at 09.30 a.m. Adult flies were released 15 minutes after the bait application. At the beginning of each test, a release cage (30cm X 30 cm X 30cm) was placed in each field cage. At 9.45 a.m., the release cage was opened to let the adult flies (100 males and 100 females) out. At 10.00 a.m., the release cage was removed. Attractiveness and mortality were the two criteria used for measuring the efficiency of Synéis-appât®. Attractiveness and mortality were recorded from 15 minutes to 7 hours after application (a total of 29 observations for each field cage). The 15 minutes observations only continued for the 1st hour and a half, with hourly observations thereafter (for a total of ~ 13 observations). Mortality was defined as the ratio between the number of dead flies and the number of flies released in the field cage. The number of dead flies fallen on the floor of the field cage was recorded every hour from 2h to 7h after the application.

This product appeared to be effective to attract adult flies and to induce their mortality after ingestion. *B. cucurbitae* was more attracted to the bait in the first 45 minutes after application than *D. demmerezi* and *D. ciliatus*. The mortality of adult flies was significantly higher for *B. cucurbitae* than for *D. demmerezi*, and was significantly higher for the latter than for *D. ciliatus*. In conclusion, fly populations concentrated on corn trap plants could thereafter be suppressed by this food bait.

6. Evaluation of agroecological crop protection under commercial farm conditions

In Reunion Island, a pioneer project (called GAMOUR, a French abbreviation for agroecological management of cucurbit flies on Reunion), was implemented during three years from 2009 to 2011. The aim of the project was (i) to assess the efficacy of agroecological cucurbit fruit fly management under commercial farm conditions and (ii) to evaluate the economic outputs for the farmers. 26 "conventional" and 4 organic farms were contractualized to apply GAMOUR methodology during cucurbit growing season. The "conventional" farms were distributed in three pilot villages: Entre-Deux, Petite-Ile and Salazie. These pilot areas totalized about 50 ha of vegetable crops, of which 10 ha were devoted to chayote (perennially cultivated) and a variable part to other cucurbits (mainly zucchini, pumpkin, and cucumber).

The study reported below concerns the socioeconomic evaluation of the GAMOUR techniques previously tested (sanitation, traps crops, attract and kill). The economical outputs of GAMOUR application was achieved by the technical support of producers. All of them were weekly visited during two years, for material supply and for registration of yields, losses and insecticide cover sprays. These data were mostly based on farmers' declarations, even though yields were confirmed as soon as possible by cooperative's certificates. The fly damages were also assessed in the field by a counting of infested fruits on a randomly chosen 20 m cultural line. In order to compare these outputs to a classical situation, we proceeded similarly in Piton Hyacinthe with two farms experiencing similar cultural and climatic conditions than the pilot area of Petite-Ile (non perennial crops). The multiannual comparative yield production of chayote (perennial crop) could also be plotted in Salazie for the pre-GAMOUR (2007 - mid 2009) and GAMOUR application periods (since mid 2009).

From September 2009 to January 2011, we supplied the farmers with a grand total of 65 augmentoria, 636 traps baited with 2492 cue-lure blocks, 69 kg of corn seeds and 136 l of protein baits.

Table 1 shows that zucchini yields tend to be slightly higher in GAMOUR farms and losses due to fly infestations appear to be lower than in control. The more striking difference is however the mean number of insecticide sprays, which have nearly disappeared in GAMOUR farms.

Production data		Control	GAMOUR
Field surface (m^2)	mean	1980	1180
Insecticide cover sprays	Mean number per cycle	4.2	0.08
Yield (t/ha)	mean	13.1	19.3
	min	3.2	4.1
	max	20.9	31.4
Losses (%)	mean	34	13
	min	5	0
	max	70	60

Table 1. Consolidated data of the technico-economical survey of GAMOUR and control farms (sources: Vivéa, Terres Bourbon, farmers' declarations and field observations on 24 zucchini cultural cycles from 2009 to 2011).

Fig. 8 shows that chayote production was maintained at high level after the beginning of GAMOUR field application, comparable or higher during the previous years. For comparison, the reference value of the Chambre d'Agriculture for chayote yield in Reunion Island ranges from 50 to 100 t/ha/week. All along the project, no insecticide spray and a variable losses percentage of 5-25% were recorded on these crops.

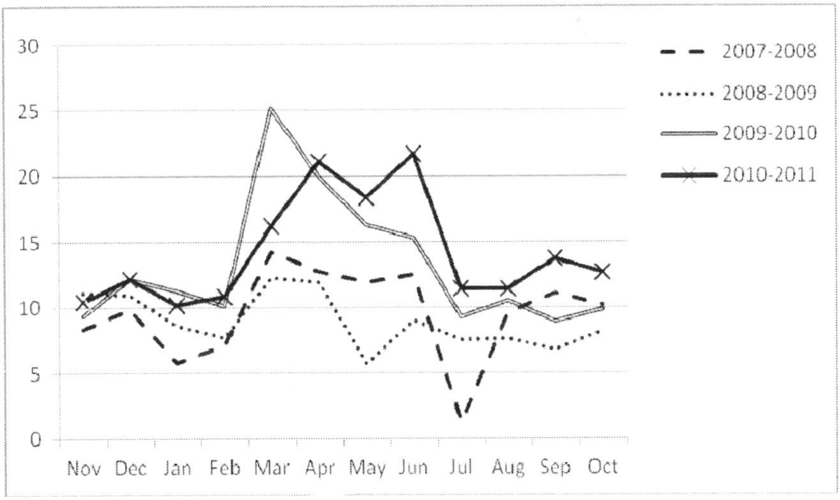

Fig. 8. Multiannual yield comparison of chayote in the Salazie pilot area. Red: before GAMOUR application; green: during GAMOUR application (source: Vivéa Réunion).

We must cautiously consider the data of yield and losses. Many of them are indeed provided by farmers' declarations, and the cross-checking with cooperatives recordings (when available) showed that they were often misevaluated despite the good faith of farmers. Moreover, the yields and losses were also influenced by additive phytosanitary, technical and climatic parameters which could not always be accurately assessed. Without any reliable comparative statistics, yield and losses are here considered as sharing the same ranges in control and GAMOUR farms.

Considering these results as a whole, we conclude on the other hand that stopping the chemical cover sprays over crops and replacing them by agroecological practices had, at worst, no negative impact on production. This answers to the main cause of concern initially expressed by the farmers in the pilot areas. A global comparative estimation of the cost of crop protection was already published (Augusseau *et al.*, 2011), combining the material and manpower costs. It assessed that GAMOUR methodology is 1.2 to 2.4 times cheaper than classical chemical protection. The farmers' are somewhat more optimistic, comparing their ancient practices, and estimate GAMOUR's protection to be at least twice cheaper. The difference between both estimations is mainly explainable by the fact that most farmers are owners of their farms and therefore not included manpower charges.Despite their lack of accuracy, all these data are another milestone for cucurbit crop protection and, further, for the evolution of agricultural practices. Agreeing with other cost-benefits analyses of similar programs (McGregor, 2005), they show that environmentally friendly agricultural practices may be profitable. This is a major step for their extension, since economical viability is fundamental for sustainable development programs: few farmers will agree to preserve their environment if they cannot "make both ends meet".

7. Conclusion

The scientific, technical and economical data presented above agree on a global objective: agroecological management of cucurbit fly populations in Reunion Island developed a sustainable methodology that farmers readily appropriate. A satisfaction survey is currently ongoing and shows that 80% of the farmers involved in the GAMOUR program are satisfied or very satisfied with the two years and half field results. Their involvement was besides recently recognized by a national award within the framework of the "Trophées de l'Agriculture Durable" (trophies for sustainable agriculture). The following step is now to extend this methodology beyond pilot areas: this will be mainly the task of education units for the next years. In parallel, the lessons of the last three years enable us now to shift the agroecological protection fundamentals on other cultures in Reunion Island.

This paper confirms that agroecology is a suitable alternative to agrochemistry for crop protection purposes.

8. Acknowledgments

The present chapter is a synthetic overview of the vast effort provided by many people who unfortunately cannot be extensively quoted here. Concerning the research studies, we acknowledge Marie-Ludders Moutoussamy, Cédric Ajaguin-Soleyen, Serge Quilici, and Elisabeth Douraguia. GAMOUR was operated by a fruitful collaboration between ASP, DAAF, Chambre d'Agriculture de La Réunion, Cirad, Farre, FDGDON, GAB, Université de

La Réunion, Takamaka Industries, SCA Terres Bourbon and Vivéa Réunion. The project was mainly funded by Europe, Conseil Régional de La Réunion, Conseil Général de La Réunion and Ministère de l'Alimentation, de l'Agriculture et de la Pêche through a CAS-DAR grant. It also received the financial support of Office de l'Eau and Crédit Agricole de La Réunion. More information is available on the website http://gamour.cirad.fr. We acknowledge all the people that have been involved in the GAMOUR project, including the farmers.

9. References

Altieri, M. A. (1995). *Agroecology. The Science of Sustainable Agriculture*, 2nd ed., Westview Press, Boulder, CO.

Altieri M.A. & Nicholls, C.I. (2000). *Applying agroecological concepts to development of ecologically based pest management strategies*, Natl Academies Press, Washington.

Atiama-Nurbel T., 2008. Interactions entre les Mouches des Cucurbitaceae et les plantes de bordures dans les systèmes horticoles à La Réunion. Master 2 Biodiversité des Ecosystèmes cultivés, Université de La Réunion, Saint-Denis, 39 p.

Augusseau, X.; Deguine, J.-P.; Douraguia, E.; Duffourc, V.; Gourlay, J.; Insa, G.; Lasne, A.; Le Roux, K.; Poulbassia, E.; Roux, E.; Suzanne, W.; Tilma, P.; Trules, E. & Rousse, P. (2011). Gamour, l'agroécologie vue de l'île de La Réunion, *Phytoma - La défense des végétaux*, No 642, pp. 33-37.

Bonnet E., 2010. Interactions entre les Mouches des Cucurbitacées (Diptera, Tephritidae), une plante hôte (courgette) et une plante piège (maïs), disposé en bandes et en patches intra-parcellaires à La Réunion. Master 2 Biodiversité des Ecosystèmes cultivés, Université de La Réunion, Saint-Denis, 35 p.

Bottrell, D. R. (1980). *Integrated Pest Management*. Council on Environmental Quality, U. S. Government Printing Office, Washington, DC.

Carroll, C.R.; Vandermeer, J.H. & Rosset, P.M. (1990). *Agroecology*. McGraw-Hill Inc., New York.

Clements, D. & Shrestha, A. (2004). *New Dimensions in Agroecology*. Food Product Press, The Haworth Press, Inc., Binghampton, NY.

Dalgaard, T.; Hutchings, N. J. & Porter., J.R. (2003). Agroecology, scaling and interdisciplinarity, *Agriculture, Ecosystems and Environment*, No 100, pp. 39-51.

Deguine J.-P., Atiama-Nurbel T., 2010. An original sanitation technique for the management of pests in organic agriculture in Reunion Island. *Proceedings of the International Conference on Organic Agriculture in Scope of Environmental Problems*, 03-07 February 2010, Famagusta (Cyprus), 139-148.

Deguine, J.P.; Atiama-Nurbel, T. & Quilici, S. (2011). Net choice is key to the augmentorium technique of fruit fly sequestration and parasitoid release, *Crop Protection*, No 30, pp. 198-202.

Deguine, J.P.; Ferron, P. & Russell, D. (2009). *Crop protection : from agrochemistry to agroecology*, Science Publisher, Enfield, NH, USA.

Gliessman, S.R. (2007). *Agroecology: the ecology of sustainable food systems*, CRC Press.

Gurr, G. M.; Wratten, S. D. & Altieri, M.A. (2004). *Ecological Engineering for Pest Management. Advances in Habitat Manipulation for Arthropods*, CSIRO and CABI Publishings, Collingwood VIC, Australia and Wallingford Oxon, UK.

Jacobsen, B. J. (1997). Role of Plant Pathology in Integrated Pest Management, *Annual Review of Phytopathology*, No 35, pp. 373-391.

Jang, E.B.; Klungness, L.M & McQuate, G.T. (2007). Extension of the use of augmentoria for sanitation in a cropping system susceptible to the alien Tephritid fruit flies (Diptera: Tephritidae) in Hawaii, *Journal of Applied Science and Environmental Management*, No 11, pp. 239-248.

Klungness, L.M.; Jang, E.B.; Ronald, F.L.; Vargas, R.I.; Sugano, J.S. & Fujitani, E. (2005). New sanitation techniques for controlling tephritid fruit flies (Diptera: Tephritidae) in Hawaii, *Journal of Applied Science and Environmental Management*, No 9, pp. 5-14.

Kogan, M. (1998). Integrated Pest Management : Historical Perspectives and Contemporary Developments, *Annual Review of Entomology*, No 43, pp. 243-270.

Lewis, W. J.; van Lenteren, J. C.; Phatak S. C. & Tumlison J.H. III. (1997). A total system approach to sustainable pest management, *Proceedings of the National Academy of Science*, No 94, pp. 12243-12248.

Liquido, N.J. (1993). Reduction of Oriental Fruit Fly (Diptera: Tephritidae) populations in papaya orchards by field sanitation, *Journal of Agricultural Entomology*, No 10, pp. 163-170.

McGregor, A. (2005). *An Interim Report on the Economic Evaluation of the Hawaii Fruit Fly Area-Wide Pest Management Program*. University of Hawaii Extension Service.

McQuate, G.T.; Jones, G.D. & Sylva, C.D. (2003). Assessment of corn pollen as a food source for two tephritid Fruit Fly Species, *Environmental Entomology*, No 32, pp. 141-150.

McQuate, G.T. & Vargas, R.I. (2007). Assessment of attractiveness of plants as roosting sites for the melon fly, *Bactrocera cucurbitae*, and oriental fruit fly, *Bactrocera dorsalis*, *Journal of Insect Science*, No 7, pp. 45-57.

Nicholls, C.I. & Altieri, M.A. (2004). Agroecological bases of ecological engineering for pest management, In: *Ecological Engineering for Pest Management. Advances in habitat manipulation for arthropods*, Gurr, G.M.; Wratten, S.D. & Altieri, M.A. (eds), pp. 33-54, CSIRO, Collingwood (Australia) & CABI, Walingford (UK).

Pimentel, D. (1995). Ecological Theory, Pest Problems, and Biologically Based Solutions, In: *Ecology and Integrating Farming Systems*, Glen, D.M.; Greaves, M.P. & Anderson, H.M. (eds), pp. 69-82, J. Wiley & Sons, Chichester.

Prokopy, R.J.; Miller, N.W.; Piñero, J.C.; Oride, L.K.; Chaney, N.; Revis, H.C. & Vargas, R.I. (2004). How effective is GF-120 fruit fly bait spray applied to border area sorghum plants for control of melon flies (Diptera: Tephritidae)?, *Florida Entomologist*, No 87, pp. 354-360.

Ratnadass, A.; Fernandes, P.; Avelino, J. & Habib, R. (2011). Plant species diversity for sustainable management of crop pests and diseases in agroecosystems: a review, *Agronomy for Sustainable Development*, DOI 10.007/s13593-011-0022-4.

Ryckewaert, P.; Deguine, J.P.; Brévault, T. & Vayssières, J.-F. (2010). Fruit flies (Diptera: Tephritidae) on vegetable crops in Reunion Island (Indian Ocean): state of knowledge, control methods and prospects for management, *Fruits*, No 65, pp. 113-130.

Walter, G. H. (2003). *Insect Pest Management and Ecological Research*. Cambridge University Press.

Weiner, J. (2003). Ecology - the science of agriculture in the 21st century, *Journal of Agricultural Science*, No 141, pp. 371-377.

White, I.M. & Elson-Harris, M.M. (1992). *Fruit flies of economic significance: their identification and bionomics*,CAB International, Wallingford (UK).

4

Quantifying the Effects of Integrated Pest Management in Terms of Pest Equilibrium Resilience

Kevin L. S. Drury
Department of Mathematics, Bethel College
USA

1. Introduction

Bistability is increasingly recognized as a mechanism underlying patterns in a wide variety of ecological time series (reviewed in Scheffer & Carpenter, 2003). One consequence of bistability is that a system can be in either of two stable states for a given set of conditions, each stable state being surrounded by a basin of attraction. Such bistability provides one explanation for pest outbreaks because pest populations can cross a threshold from a low-density, biologically-controlled equilibrium and enter into the basin of attraction of a high-density outbreak equilibrium (see e.g., Ludwig et al., 1978). In such systems, integrated pest management has the dual goal of simultaneously decreasing the magnitude of the outbreak basin of attraction and perturbing the pest population back across the threshold to the biologically controlled state. Decreases in the magnitude of the outbreak basin of attraction arise when management strategies alter basic biological parameters, such as the realized rate of population growth. For example, releasing sterile males decreases fecundity, and hence, birth rates diminish in relation to death rates (Knipling, 1970). Similarly, introduction of predators with larger half-saturation constants changes the landscape of equilibria in favor of biological control (Drury & Lodge, 2008). In the presence of such factors, the required magnitude of direct pest reductions decreases, because the state of the outbreak system moves closer to the threshold between equilibria (e.g., moving leftwards on the top line in Fig. 1). Relatively smaller pest reductions can therefore move the state of the system across the threshold and into the biologically controlled basin of attraction. Here we quantify the effects of such species manipulations in terms of equilibrium resilience to demonstrate how targets can be computed so that pest control schedules can be devised (e.g., Drury, 2007). Our definition of resilience is the distance between an equilibrium and a threshold, which is one component of classical definitions describing the characteristic return time to equilibrium following perturbations (Pimm, 1984).

2. The spruce budworm model

The Spruce budworm model has a long history in ecological pest management (Ludwig et al., 1978) and has the advantage of being a relatively simply model that nevertheless contains components of integrated pest management while generating bistable dynamics (Strogatz,

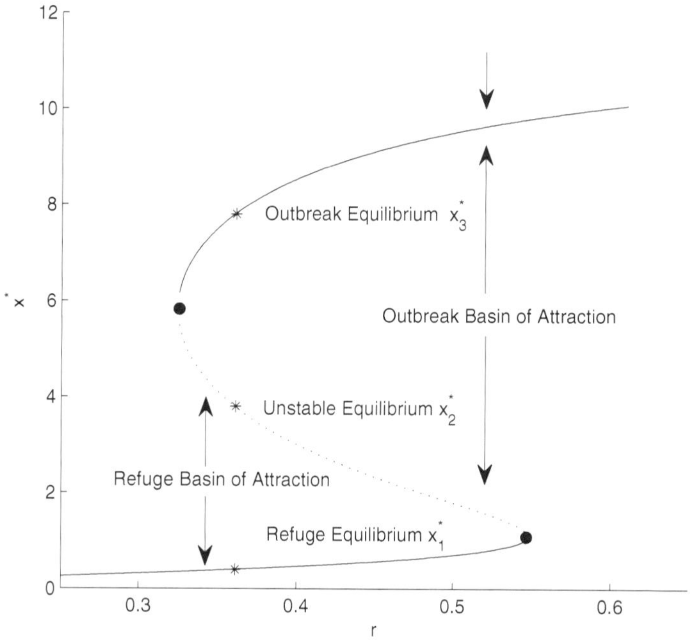

Fig. 1. As the parameter r (from Eq. (2)) is increased above 0.3, two new equilibria emerge and the potential for outbreaks is introduced. At large values of r, only the outbreak equilibrium exists.

Parameter	Meaning	Value in this study
R	population growth rate	0.25
K	carrying capacity	15
B	maximal predator attack rate	0.75
A^2	predator half-saturation constant	1

Table 1. Parameters of Eq. (1), their meanings, and representative values used in this study.

1994). Letting X represent pest population density, and using the parameters in Table 1, the model is,

$$\frac{dX}{dt} = RX\left(1 - \frac{X}{K}\right) - \frac{BX^2}{A^2 + X^2}, \tag{1}$$

where the first term represents logistic growth and the second term represents a Holling type III functional response, which means that predators switch to the pest at some intermediate density, but nevertheless become sated at higher densities thus allowing the pest to escape control.

The dynamics of Eq. (1) are often studied at equilibrium by first nondimensionalizing (see e.g., Murray, 2002). For example, with $x = \frac{X}{A}$, $r = \frac{AR}{B}$, $k = \frac{K}{A}$, and $\tau = \frac{Bt}{A}$ as nondimensional parameter groups this yields

$$r\left(1 - \frac{x}{k}\right) = \frac{x}{1+x^2}. \tag{2}$$

This strategy has the advantage of specifying the left-hand side as a linear function of x that when equal to the right-hand side, specifies equilibria. It is thus a simple matter to plot the two sides of the equation on the same axes to see where they intersect (see the solid lines in Fig. 2). Furthermore, the effects of changes to r and k can be seen by fixing one and varying the other and evaluating the effects on equilibria values. This qualitative approach has obvious intuitive appeal and provides heuristic guidance for pest management. Nevertheless, as composite parameter groupings, r and k can be difficult to interpret in practice. Additionally, direct pest control measures often require specific objectives, e.g., assessment of pest populations in relation to some economic threshold (Pedgrigo & Zeiss, 1996), which can be difficult to extract from such qualitative exercises. Thus, we revisit Eq. (1) with the goal of developing quantitative methods for assessing the effects of integrated pest management in terms of relative equilibria positions, which in turn allows quantitative predictions of necessary direct pest control measures (such as spraying insecticides).

3. Model analysis

To analyze Eq. (1), we first recognize that at equilibrium it is a cubic in X,

$$RX\left(1 - \frac{X}{K}\right) - \frac{BX^2}{A^2 + X^2} = 0 \tag{3}$$

$$-X^3 + KX^2 - X\left(A^2 + \frac{KB}{R}\right) + KA^2 = 0 \tag{4}$$

Following standard methods for solving cubic equations (Uspensky, 1948), Eq. (4) can be solved yielding three roots, X_1, X_2, and X_3, corresponding to the three possible equilibrium states of the system. As management actions change parameter values, the values of these equilibria change and at bifurcation points even emerge or vanish (see, e.g., Fig. 1-2). Our approach allows assessment of the effects of integrated pest management on pest equilibrium values. Here, we use a simplified version of Cardano's method of solving cubics to evaluate the effects of management activities that alter the intensities of R and B and the magnitudes of A and K. We then apply this strategy to the spruce budworm model to solve for the three equilibria. The method is quite general, however, and could be used to solve any cubic model, which are common in pest control, because of the ubiquity of the type III functional response.

3.1 Solving for equilibrium pest values: The general case

Consider the general monic cubic equation (i.e., with leading coefficient of 1),

$$f(x) = x^3 + ax^2 + bx + c = 0. \tag{5}$$

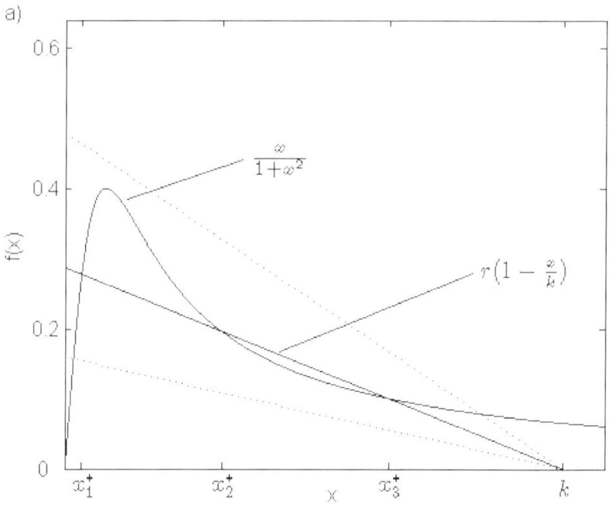

Fig. 2. The left-hand side of Eq. (2) (solid, straight line) crossing the curve described by the right-hand side in three places, correspondig to the three equilibria in Fig. (1), x_1, x_2, x_3. The two dotted straight lines signify the effects of changing r on the existence and value of equilibria. The lower dotted line, representing small values of pest intrinsic growth rate, only crosses the curve at a relatively low pest population size; pest outbreaks are not possible. In contrast, the upper dotted line only crosses the curve at a relatively high pest population size; biological control is not possible.

Cardano's technique, dating to the first half of the sixteenth century, involves simplifying Eq, (5) so that it no longer contains the second power of x. This is achieved by introducing a new variable y and setting,

$$x = y + k, \tag{6}$$

where k is as yet an arbitrary constant.

To specify k, we begin by using Taylor's formula to approximate $f(y+k)$ by

$$f(y+k) = f(k) + f'(k)y + \frac{f''(k)}{2}y^2 + \frac{f'''(k)}{6}y^3. \tag{7}$$

The terms on the right-hand side of Eq. (7) can be related to the terms of our monic cubic, Eq. (5) yielding,

$$f(k) = k^3 + ak^2 + bk + c, \tag{8}$$

$$f'(k) = 3k^2 + 2ak + b, \tag{9}$$

$$\frac{1}{2}f''(k) = 3k + a, \tag{10}$$

$$\frac{1}{6}f'''(k) = 1. \tag{11}$$

Simplification is now possible because we can choose k in Eq. (10) so that,

$$3k + a = 0, \qquad \text{or} \qquad k = -\frac{a}{3}. \tag{12}$$

With k so defined (i.e., the so-called "Tschirnhaus transformation") our solution is translated to the origin and we can substitute it into the remaining equations to arrive at,

$$f'\left(-\frac{a}{3}\right) = b - \frac{a^2}{3}, \tag{13}$$

$$f\left(-\frac{a}{3}\right) = c - \frac{ba}{3} + \frac{2a^3}{27} \tag{14}$$

After substituting $x = y - \frac{a}{3}$ and equating coefficients of like powers of y and k in Eqs. (7) and (8–11) respectively, Eq. (5) is transformed into,

$$y^3 + \alpha y + \beta = 0, \tag{15}$$

where

$$\alpha = b - \frac{a^2}{3}, \qquad \beta = c - \frac{ab}{3} + \frac{2a^3}{27}. \tag{16}$$

By a simple reverse transformation, any formula for the roots of Eq. (15) can be transformed into a formula for the roots of Eq. (5), i.e., by substituting Eq. (16) into Eq. (15) and using $x = y - \frac{a}{3}$. Thus, in the sequel, we consider only Eq. (15).

Cubic equations of the form, Eq. (15), referred to as "depressed cubics", can be solved by introducing two new variables u, v such that,

$$y = u + v. \tag{17}$$

Upon substituting this expression into Eq. (15) we see that u and v must satisfy the equation,

$$u^3 + v^3 + (\alpha + 3uv)(u + v) + \beta = 0, \tag{18}$$

which has two unknowns. Thus, the problem is indeterminant without another known relation between u and v, Eq. (17) being one. Another relation that is consistant with Eq. (18) is to take

$$\alpha + 3uv = 0, \tag{19}$$

or,

$$uv = -\frac{\alpha}{3}. \tag{20}$$

It then follows from Eq. (18) that,

$$u^3 + v^3 = -\beta, \tag{21}$$

Solving the cubic Eq. (15) can now be achieved by solving the system of two equations,

$$u^3 + v^3 = -\beta, \tag{22}$$

$$uv = -\frac{\alpha}{3}. \tag{23}$$

Because it is convenient to know the sum and product of two unknown quantities, in this case u^3 and v^3, we take the cube of the latter equation yielding,

$$u^3 v^3 = -\frac{\alpha^3}{27}. \tag{24}$$

The reason these quantities are convenient is that upon solving one in terms of the other and substituting, a single quadratic equation results. To see this rearrange Eq. (22) as follows,

$$u^3 = -(v^3 + \beta), \tag{25}$$

and substitute this expression for u^3 into Eq. (24) giving,

$$-(v^3 + \beta)v^3 = \frac{-\alpha^3}{27}. \tag{26}$$

We next expand the left-hand side of Eq. (26) giving,

$$-v^6 - \beta v^3 = \frac{-\alpha^3}{27}, \tag{27}$$

or, substituting $t = v^3$ and dividing through by (-1),

$$t^2 + \beta t - \frac{\alpha^3}{27} = 0. \tag{28}$$

Thus, apparently, u^3 and v^3 can be computed from the roots of Eq. (28). Equating v^3 to the positive root that results from applying the quadratic formula, we have,

$$v^3 = -\frac{\beta}{2} + \sqrt{\frac{\beta^2}{4} + \frac{\alpha^3}{27}}, \tag{29}$$

or

$$v = \sqrt[3]{-\frac{\beta}{2} + \sqrt{\frac{\beta^2}{4} + \frac{\alpha^3}{27}}}. \tag{30}$$

Now we take advantage of the relationship in Eq. (25) to solve for u yielding,

$$u^3 = -(v^3 + \beta), \tag{31}$$

$$= -\left(-\frac{\beta}{2} + \sqrt{\frac{\beta^2}{4} + \frac{\alpha^3}{27}} + \beta\right), \tag{32}$$

$$= \frac{\beta}{2} - \sqrt{\frac{\beta^2}{4} + \frac{\alpha^3}{27}}, \quad \text{or,} \tag{33}$$

$$u = \sqrt[3]{\frac{\beta}{2} - \sqrt{\frac{\beta^2}{4} + \frac{\alpha^3}{27}}}. \tag{34}$$

We can now write an expression for y, which was set equal to $u + v$ in Eq. (17) yielding,

$$y = u + v, \tag{35}$$

$$= \sqrt[3]{\frac{\beta}{2} - \sqrt{\frac{\beta^2}{4} + \frac{\alpha^3}{27}}} + \sqrt[3]{-\frac{\beta}{2} + \sqrt{\frac{\beta^2}{4} + \frac{\alpha^3}{27}}}. \tag{36}$$

The final step is to use the relation $x_i = y_i - \frac{a}{3}$ where $i \in \{1,2,3\}$ to return the three equilibrium solutions to our original model. Note that at certain values of α and β a bifurcation occurs and two of the real solutions disappear to be replaced by imaginary solutions. In this study, we have used parameter values that yield three real solutions. In the case of a single real equilibrium the basin of attraction of the outbreak equilibrium either does not exist (i.e., outbreaks are not possible) or consists of the entire space (i.e., biological control is not possible).

3.2 Solving for equilibrium pest values: The spruce budworm model

As we have seen, the spruce budworm model, Eq. (1), can be expanded to its cubic form yielding,

$$x^3 - Kx^2 + (A^2 + \frac{KB}{R})x - KA^2 = 0, \tag{37}$$

(after multiplying through by -1) or, more simply,

$$x^3 + px^2 + qx + r = 0, \tag{38}$$

where,

$$p = -K, \tag{39}$$

$$q = A^2 + \frac{KB}{R}, \tag{40}$$

$$r = -KA^2. \tag{41}$$

Forming the depressed cubic requires removing the quadratic term from Eq. (38). To do so, let

$$x = y - \frac{p}{3}, \tag{42}$$

and to simplify the resulting equation, let,

$$\alpha = \frac{1}{3}(3q - p^2), \tag{43}$$

$$\beta = \frac{1}{27}(2p^3 - 9pq + 27r), \tag{44}$$

yielding,

$$y^3 + \alpha y + \beta = 0. \tag{45}$$

We can now simply substitute our particular model parameters, embodied in α and β, through $p, q,$ and r into Eqs. (35–36) yielding,

$$y = u + v, \tag{46}$$

$$= \sqrt[3]{\frac{\beta}{2} - \sqrt{\frac{\beta^2}{4} + \frac{\alpha^3}{27}}} + \sqrt[3]{-\frac{\beta}{2} + \sqrt{\frac{\beta^2}{4} + \frac{\alpha^3}{27}}}. \tag{47}$$

Finally, back transforming the solution into terms of x we use the relation,

$$x_i = y_i - \frac{p}{3}, \qquad i \in \{1,2,3\}. \tag{48}$$

4. Results

Having explicitely solved the cubic equation, we can now evaluate the effects of parameter manipulation on equilibrium resilience. For example, Fig. 3 shows that the magnitude of the outbreak basin of attraction changes relatively rapidly with increasing K and R. It is not possible to evaluate the relative effects of changes to each using this graph because of the different scales and units. Nevertheless, it is apparent that these two familiar parameters both influence the magnitude of the outbreak basin of attraction.

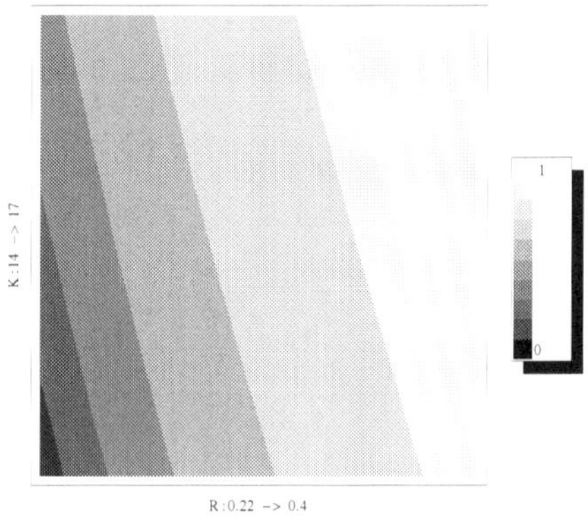

Fig. 3. Shaded contours of the magnitude of the basin of attraction of the outbreak equilibrium as intrinsic growth rate R and carrying capacity K are changed. Scale is normalized from 0 to 1 such that darker colors represent smaller, and lighter colors larger, basins of attraction. Thus, as expected, large values of R and K both yield large potential for outbreaks.

When the square root of the predator's half-saturation constant A is varied, its effects can be compared to the large changes caused by changes in K because they both have units of individuals (Fig. 4). The roughly vertical contours indicate that changes to carrying capacity have a much greater effect than changes to A, at least over ranges of each that yield bistability. Note that the slight concave down character of these contours indicates that changes in resilience following changes in A are slightly more likely at high values of K. The reason for such changes to equilibium resilience can be seen in Fig. (5). Specifically, as the predator half-saturation constant A^2 increases, the $\frac{dx}{dt}$ curve is shifted downwards, increasing the magnitude of the biologically controlled basin of attraction at the expense of the outbreak basin.

Finally, changes to the maximal predator attack rate have a large effect on the resilience of the both the biologically controlled equilibrium and the outbreak equilibrium. Specifically, increasing B, the maximal predator attack rate *increases* the magnitude of the biologically controlled equilibrium. This matches our intuition, because higher attack rates presumably favor the predator, not the pest. To analyze the effects of changing B, we used both a higher value of B and a lower value and inspect the effects on the magnitudes of each basin of attraction.

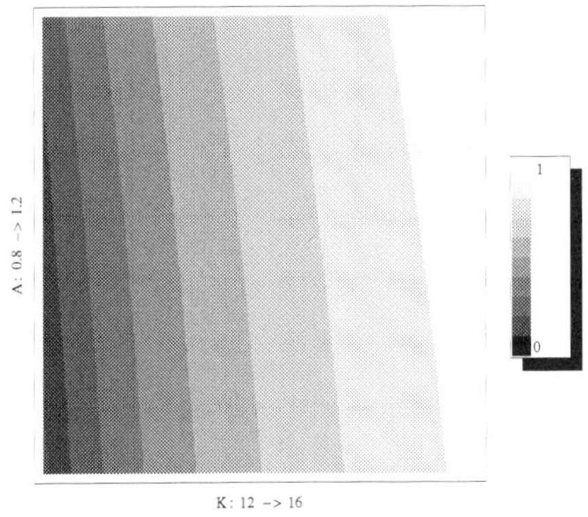

Fig. 4. Shaded contours of the magnitude of the basin of attraction of the outbreak equilibrium as carrying capacity K and the square root of half-saturation constant A are changed. Scale is normalized from 0 to 1 such that darker colors represent smaller, and lighter colors larger, basins of attraction.

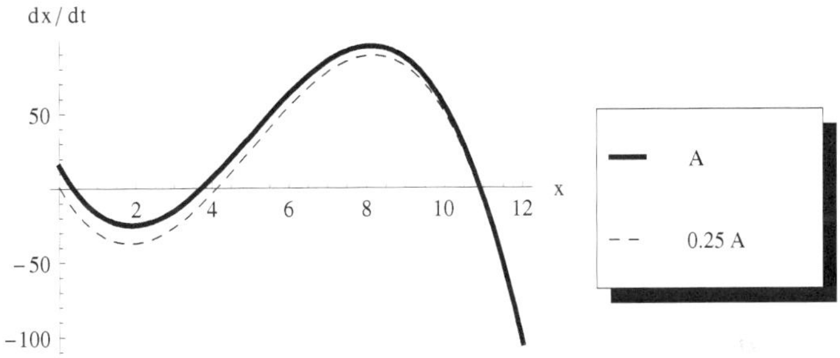

Fig. 5. Effect of changing A in the spruce budworm model. Increases in A generate increases in the magnitude of the biologically controlled basin of attraction by decreasing the value of the lower equilibrium and increasing the value of the unstable, middle equilibrium.

Figure 6 shows that increasing B lowers the entire dx/dt curve. In contrast, increasing R raises the curve, which also has an intuitive effect, because increased growth rate is expected to favor outbreaks. As Fig. 7 shows, increasing R increases the magnitude of the outbreak basin of attraction at the expense of the biologically controlled basin of attraction.

The net sum effects of changes in R and B are depicted in Fig. 8, which shows that lower attack rates and higher pest growth rates both increase the magnitude of the outbreak basin of attraction. In contrast, high attack rates and low growth rates decrease that basin and increase the biologically controlled basin of attraction.

5. Discussion

We have undertaken an analysis of a simple model of pest population dynamics with the goal of understanding the role of specific, commonly encountered model parameters on the

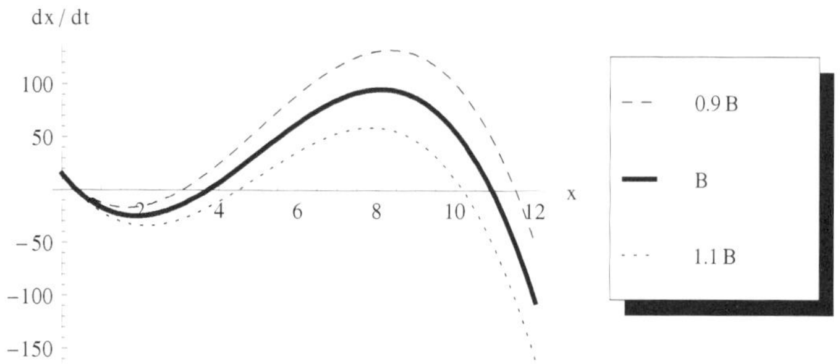

Fig. 6. The rate of change in pest population growth rate versus pest population size according to the spruce budworm model. As the maximal predator attack rate B increases, the magnitude of the biologically controlled basin of attraction increases, as evidenced by the reduced distance between the two right-most points where $dn/dt = 0$. In contrast, when B is dereased, the magnitude of the outbreak equilibrium is increased.

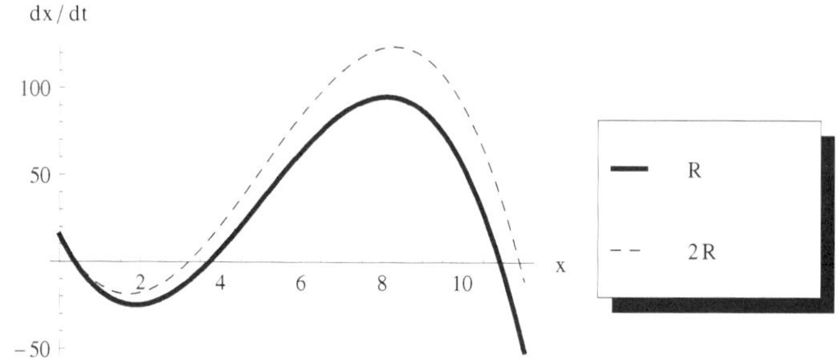

Fig. 7. The rate of change in pest population growth rate versus pest population size according to the spruce budworm model. As R is increased, the curve, dx/dt shifts upward bringing the two left-most equilibrium points where $dx/dt = 0$ closer together. Thus, the basin of attraction of the biologically controlled equilibrium decreases.

resilience of equilibria. Furthermore, the parameters we have studied are those typically affected by integrated pest management. We defined equilibrium reslience as the distance between the stable equilibria and the unstable equilibrium, which marks the threshold betweem the basins of attraction for the biologically-controlled, and outbreak equilibria.

Our results show that the pest carrying capacity has a large effect on the resilience of the outbreak equilibrium. Specifically, changes in K generate large changes in the magnitude of the outbreak basin of attraction. In terms of integrated pest management, this means that pest control strategies that decrease the pest's carrying capacity, such as intercropping to break up monocultures, move the outbreak equilibrium closer to the threshold of the biologically controlled equilibrium. Thus, smaller culling events, or smaller releases of enemies, achieve the desired shift to the biologically controlled equilibrium than otherwise.

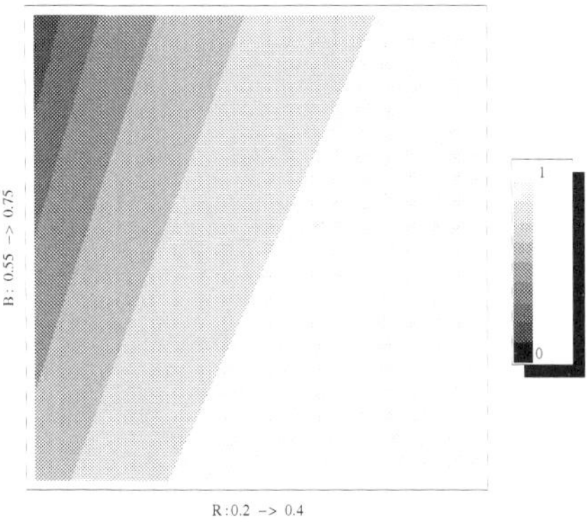

Fig. 8. Shaded contours of the magnitude of the basin of attraction of the biologically controlled equilibrium as intrinsic growth rate R and maximal attack rate B are changed. Scale is normalized from 0 to 1 such that darker colors represent smaller, and lighter colors larger, basins of attraction. At large values of R and small values of B the outbreak basin dominates, while at high values of B and low values of R, it goes to zero.

The effects of pest population growth rate are two-fold. First, increasing pest growth rate increases the outbreak basin of attraction, an intuitive result, because higher pest population growth rates are expected to favor outbreaks. At the same time, increasing r decreases the magnitude of the biologically controlled basin of attraction, making escape from predator control more feasible.

The magnitude of the outbreak basin of attraction decreases with increasing half-saturation constant. This constant is comprised of (attack rate $\times$ handling time)$^{-1}$ (Gotelli, 1995). Thus, the effects of A^2 depend on the relationship between these two biological parameters. In our analysis, for simplicity we used $A = 1$. Increases in the half-saturation constant mean that the predators require higher densities of prey in order to begin being sated. Thus, they can consume more prey before the saturating effects of the functional response take effect.

The maximum feeding rate B also has a large and intuitive effect on equilibrium resilience. Because this feeding rate is the inverse of handling time (Gotelli, 1995) introducing natural predators with shorter handling times generates meaningful decreases in the magnitude of the outbreak basin of attraction. Additionally, Kidd & Amarasekare (2011) demonstrated that predators with shorter handling times induce weak transient dynamics of short duration relative to predators with longer handling times. Thus, according to their analysis, when predators with short handling times are introduced (i.e., those with larger maximum feeding rates), large oscillations in the prey population are less likely. According to our analysis, when predators with short handling times and correspondingly higher maximal attack rates are released, outbreaks become less likely.

Using the approach we outline here, schedules for integrated pest management can be developed based on knowledge of specific equilibrium values. For example, solving the model for x_1, x_2, and x_3 allows one to compute the needed change in density to change a system from the outbreak state to the biologically controlled state. Furthermore, our approach permits evaluation of the effects of individual components of integrated pest management on distance to the threshold between equilibria. Thus, the effects of strategies such as release of sterile males and enhancement of natural predators can be understood in terms of decreased culling requirements.

Acknowledgements: Professor Adam Hammett kindly read the manuscript and made useful comments that improved clarity, especially in the final steps of the solution of the cubic.

6. References

Abramowitz, M. & Stegun, I. A. (1964). *Handbook of Mathematical Functions*, National Bureau of Standards.

Drury, K. L. S. (2007). Shot noise perturbations and mean first passage times between stable states, *Theoretical Population Biology* 72: 153–166.

Drury, K. L. S. & Lodge, D. M. (2008). Using mean first passage times to quantify equilibrium resilience in perturbed intraguild predation systems, *Theoretical Ecology* 2(1): 41–51.

Gotelli, N. (1995). *A primer of ecology*, Sinauer Associates, Inc, Sunderland, MA.

Kidd, D., P. Amarasekare. (2011). The role of transient dynamics in biological pest control: insights from a host-parasitoid community, *Journal of Animal Ecology* (1365-2656).

Knipling, E. (1970). Suppression of pest lepidoptera by releasing partially sterile males, a theoretical approach, *BioScience* 20(8): 465–470.

Ludwig, D., Jones, D. D. & Holling, C. S. (1978). Qualitative analysis of insect outbreak systems: the spruce budworm and forest, *J. Anim. Ecol.* 47: 315–332.

Murray, J. D. (2002). *Mathematical Biology*, Springer-Verlag.

Pedgrigo, L. & Zeiss, M. (1996). *Analyses in Insect Ecology and Management*, Iowa State University Press, Ames.

Pimm, S. L. (1984). The complexity and stability of ecosystems, *Nature* 307: 321–326.

Scheffer, M. & Carpenter, S. (2003)Catastrophic regime shifts in ecosystems: linking theory to observation, *Trends Ecol. Evol.* 18(12): 648–656.

Strogatz, S. H. (1994). *Nonlinear Dynamics and Chaos with Applications to Physics, Biology, Chemistry, and Engineering*, Addison-Wesley Publishing Co., Reading, MA.

Uspensky, J. (1948). *Theory of Equations*, McGraw Hill.

5

Manipulation of Natural Enemies in Agroecosystems: Habitat and Semiochemicals for Sustainable Insect Pest Control

Cesar Rodriguez-Saona[1], Brett R. Blaauw[2] and Rufus Isaacs[2]
[1]*Department of Entomology, Rutgers University, New Brunswick*
[2]*Department of Entomology, Michigan State University, East Lansing*
USA

1. Introduction

Plants are not capable of running away from their enemies, i.e., the herbivores that may eat them. However, under certain circumstances, plants can rely on the natural enemies of insect herbivores for protection. These natural enemies include other insects that are predators and parasitoids. To help protect plants from damage caused by insect herbivores, practical methods have been developed and evaluated to conserve and augment natural enemies of several agricultural pests. These strategies include improving the suitability of the crop landscape for natural enemies by manipulating the resources available for these insects, and the use of semiochemicals to attract predators and parasitoids. This chapter will review recent studies exploring the potential for manipulating the behavior of natural enemies through vegetational diversification of crop habitats and the use of semiochemicals to enhance biological control in agroecosystems, and we will discuss how these might be combined to improve crop protection.

2. Vegetational diversity

Increasing the diversity within crops is predicted to provide a greater number of opportunities for natural enemies to survive in agricultural systems (Fig. 1). Thus, pest outbreaks tend to be less common in polycultures than in monocultures (Root, 1973; Andow, 1991). Crop diversification tends to increase natural enemy abundance and diversity, providing a system more resilient to pest population increase. Overall farming diversity within the agroecosystem may also affect biological control by natural enemies, due in part to a wider range of flowering plants that provide nectar (carbohydrate) and pollen (protein) resources to insects during more times of the growing season. Vegetational diversity can also provide support for insect biological control at the local and landscape levels (Thies et al., 2003; Roschewitz et al., 2005; Bianchi et al., 2006; Gardiner et al., 2009). Farmers can make some simple changes to their crop systems to manipulate vegetational diversity, through addition of plants that provide specific functions (Landis et al., 2000; Gurr et al. 2003; Isaacs et al., 2009). Below, we provide an overview of those methods and describe situations where such changes have reduced pest infestation.

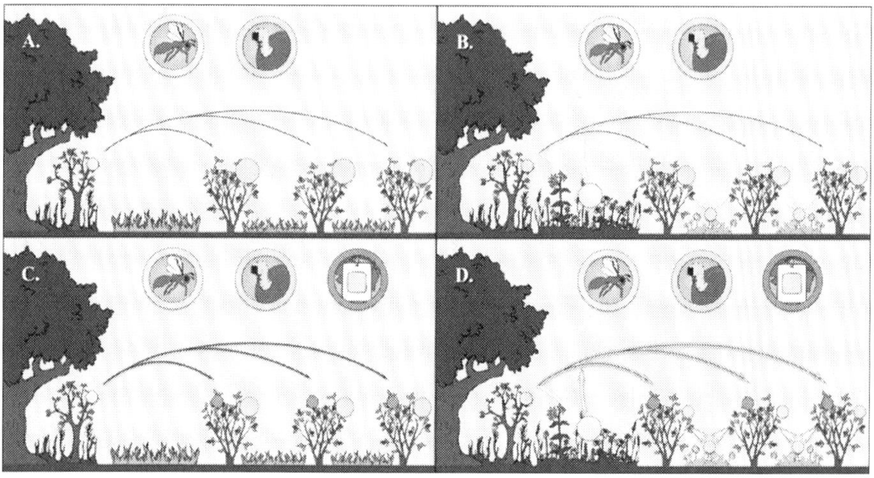

Fig. 1. Strategies for manipulating natural enemies in agroecosystems for enhanced insect pest control. A.) Conventional method with no manipulation of habitat leads to high pest numbers and few natural enemies entering the crop from the surrounding landscapes. Yellow circles represent natural enemy populations and green circles represent pest abundance. B.) The addition of inter-cropping, cover crops, or supplemental food sources to an agroecosystem may lead to an increase in natural enemy abundance and a potential decrease in insect pest abundance within field settings, but relies on the presence of insect pests to attract natural enemies into the crop. C.) The addition of semiochemical-based lures, such as herbivore-induced plant volatiles (HIPVs) may attract natural enemies into the crop to enhance biological control, but does not provide resources to directly enhance natural enemy abundance. HIPV lures are represented by pink circles. D.) Combining habitat manipulation and HIPV strategies in an agroecosystem may increase natural enemy abundance within field settings as well as directly attract natural enemies into the crop to enhance biological control.

2.1 Inter-cropping, monocultures, polycultures

The response of beneficial insect populations to habitat manipulation depends upon their ability to use or exploit one or more of the plant components of the agroecosystem (Altieri & Nicholls, 2004). Crop systems that are dominated by a single plant species only provide resources to those select organisms that can exploit that single plant species. Hence, monocultures are an example of agroecosystems with low diversity and may be more susceptible to pest or disease outbreaks (Theunissen, 1994; Altieri & Nicholls, 2004). Because of this increased susceptibility, management and external inputs are essential to support low diversity agroecosystems. On the other hand, reliance on diverse plantings, a range of natural enemies that are supported by these plants, and associated crop management strategies can in some cases help maintain pest populations below economic thresholds (Altieri & Nicholls, 2004).

Intercropping, which is the cultivation of two or more species within the same field, is a common method to increase beneficial insect diversity within agroecosystems (Fig. 1B) (Vandermeer, 1989; Theunissen, 1994). Intercropping crop plants with flowering species such as clovers can also provide a favorable habitat for a variety of beneficial insects that may not otherwise survive in a single crop environment, and hence intercropping may provide natural pest management by increasing the abundance and diversity of insect natural enemies in the agroecosystem (Theunissen, 1994). Diverse systems encourage complex food webs that involve more interactions among vegetation, pests, and natural enemies, providing resources for a diverse group of organisms and allowing for alternative resources and food sources. Thus, polycultures and natural ecosystems with higher diversity tend to be more stable and less subject to fluctuations in pest and disease populations (Altieri & Nicholls, 2004). As an example of this, Beizhou et al. (2011) recently reported that intercropping pear orchards with aromatic plants significantly reduced pest abundance and increased the ratio of natural enemies to pests when compared to orchards with only natural grass or clean tillage. They also found higher abundance of natural enemies and reduced numbers of major pests in intercropped orchards. Hence, intercropping the pear orchard with aromatic plants led to improved insect pest management by enhancing the activity of the insect natural enemy community.

2.1.1 Push-pull strategy

The 'push-pull' strategy uses a combination of stimuli to manipulate the behavior of insect pests and/or natural enemies and to alter their distribution and abundance in agroecosystems (Miller & Cowles, 1990; Khan et al., 1997). The push-pull approach works by repelling or deterring the pest insects (push) away from the main crop by using deterring chemical stimuli. Simultaneously, highly appealing stimuli are used to attract the pests (pull) from the main crop to other areas such as trap crops where the pests aggregate and are easier to control (Khan et al., 1997). While on their own each individual component (the push and the pull) of the approach may not be effective at reducing pest numbers below economic thresholds, combining the push and pull components increases the efficacy of such a strategy. Also, since the push and pull components are commonly non-toxic this strategy is compatible with supporting insect natural enemies and biological control (Khan & Pickett, 2004). A suitable push-pull strategy will be unique for each system it is used for, and hence the development of effective push-pull approaches requires an understanding of the targeted pest's biology and interactions with its hosts and natural enemies (Khan & Pickett, 2004).

Push-pull strategies are not a new idea, but one of the most successful examples was developed recently in Africa for controlling stem borers on cereal crops (Khan et al., 2001). Stem borers are moth larvae that feed on and destroy cereal crops. The adult stem borer moths are cryptic and nocturnal while the larvae feed within the crop stem making both adults and larvae difficult to see and to control. Chemical pesticides are a common method of control for stem borers, but this is not an economical or safe approach for many resource-poor, small-scale farms where these pests are common. Hence, a push-pull strategy was developed using technologies appropriate and economical for such farmers (Khan & Pickett, 2004). This strategy involves a combined use of intercropping and trap crops, and uses plants that are also locally-appropriate and that can be used in their agricultural system. While some push-pull strategies use chemical deterrents and attractants to create the push-pull effect (see Section 3), this strategy does not use any chemical deterrents or toxins, but instead uses plant species that repel the pest away from the main crop while also attracting insect natural enemies to the fields. The repelled stem borers are attracted to nearby trap plants where the moths lay their eggs, but their larvae are unable to develop, thus reducing the number of trapped pest insects (Khan and Pickett, 2004). The farms that implemented the push-pull strategy in Africa have experienced lower pest abundance, and also an overall enhancement of beneficial insect abundance (Khan et al., 2001).

While the work of Khan et al. in Africa is primarily in cereal-based farming systems, the push-pull approach may be applicable to a much wider range of agricultural pest problems in a variety of crops, if the appropriate components can be developed and implemented.

2.2 Cover crops

A major means of conserving beneficial insects and stabilizing their populations is to meet the ecological requirements of these insects within or near the cropping environment (Landis et al., 2000). To be effective, many natural enemies and pollinators need access to alternate hosts, overwintering habitats, constant food supply, and appropriate microclimates. These requirements can be fulfilled through the inclusion of a diverse assemblage of flowering plants within agricultural landscapes.

Cover crops are planted in crop fields, either in rotation with annual crops or within perennial crops. These plants have been widely used to reduce soil erosion, add or retain soil nutrients, produce organic matter, reduce soil compaction, and also aid in pest control (Bugg, 1991; Bugg & Waddington, 1994). They are used in sustainable and organic agricultural systems to enhance soil health and crop nutrition.

Flowering species such as buckwheat (*Fagopyrum esculentum* Moench) or clovers (*Trifolium* spp.) have been promoted as cover crops to provide flowering resources for insects when the crop is not in bloom. Increasingly, the use of flowering cover crops is seen as one component of an overall 'farmscaping' approach that aims to make farmland suitable for a range of beneficial organisms throughout the growing season. Ideally, cover crops will provide shelter and resources for natural enemies, enhancing their populations and hence biological control of insect pests (Bugg & Waddington, 1994). Plants should be screened for their attractiveness to not only the target biological control agent, but also to other potential competitors for floral resources (Hogg et al., 2011) (see below).

The deployment of cover crops within the row middles of perennial crops may create a conflict with insecticide-based pest management activities that would kill the natural enemies active within the field. This will depend on the type of insecticide used (since many of the more recently-developed insecticides are quite selective), on the tolerance of the natural enemies to the insecticide, and on their ability to re-colonize fields from the perimeter. The potential for killing beneficial insects inside crop fields is one reason that strip plantings of flowering plants within field margins or perimeter plantings are considered. Placement of flowering resources adjacent to crop fields may minimize exposure to insecticides while also providing the resources for natural enemies that can then re-invade the crop fields. Indeed, flowering borders adjacent to blueberry fields have enabled natural enemy populations inside the crop to rebound more rapidly after insecticide applications (Walton & Isaacs, 2011). An increase in natural enemies, particularly parasitoids, has been observed in apple orchards adjacent to flowering borders compared to orchards without flowering borders. This increase in parasitoid numbers also coincided with a doubling of parasitism of spotted tentiform leafminer (*Phyllonorycter blancardella* (Fabr.)) in apple trees adjacent to flowering borders, compared with orchards without flowering borders (Blaauw & Isaacs, unpublished data).

2.3 Selection of supplemental food sources

2.3.1 Flowering plants

Flowering field margins adjacent to crop fields can provide necessary resources for natural enemies of crop pests during periods when crop flowers are not present, thus maintaining high populations of insect predators and parasitoids, which are supported by a provision of nutrients throughout the season (Sotherton, 1984; Ahern & Brewer, 2002; Büchi, 2002; Sanchez et al., 2003; Wanner et al., 2006a, 2006b). Such natural areas were once common in most agricultural landscapes, particularly between plantings, along roadsides, or as part of woody hedgerows, but as the production of crops has increased and intensified, these non-cropped areas are becoming less common (Sotherton, 1998). Current crop production techniques shape the physical structure of agricultural landscapes (Robinson & Sutherland, 2002), and with increased reliance on mechanization and pesticides, vegetative diversity in farmlands has decreased causing negative impacts on natural enemies (Ryszkowski et al., 1993).

Flowering plant strips adjacent to fields help support beneficial insect biodiversity in agricultural landscapes (Baggen & Gurr, 1998; Dufour, 2000; Carreck & Williams, 2002; Fiedler & Landis, 2007a, 2007b; Tuell et al., 2008). Much of the testing of flowering plants has been done with non-native annual or biennial flowering species, although these often bloom in one growing season requiring annual sowing. This makes these resource plants costly to successfully maintain , whereas native perennial flowering plants are sown once, adapted to the local environment, less likely to become invasive, and may increase native beneficial insect diversity in agricultural landscapes (Stephens et al., 2006, Fiedler & Landis, 2007a, 2007b). A well-designed flowering border adjacent to a crop field will provide necessary resources and alternative food source for natural enemies during periods when crop pest and crop flower numbers are low, thus maintaining high populations of natural enemies supported by the provision of nutrients throughout the season (Landis et al., 2000; Isaacs et al., 2009; Hogg et al., 2011).

2.3.2 Nutritional requirements: Pollen and nectar resources

Sufficient flower abundance and appropriate vegetation structure are required to support diverse populations of insects (Zurbrügg & Frank, 2006), and therefore manipulation of structurally resource-poor habitats through the addition of flowering plants and grasses can increase beneficial insect populations in agricultural landscapes (Long et al., 1998; Kells et al., 2001; Rebek et al., 2005). Many beneficial insects, including natural enemies, require access to alternate hosts, overwintering habitats, a constant food supply, and appropriate microclimates in order to survive (Johnson & Triplehorn, 2005). The majority of predators and parasitoids are omnivores and require non-prey food, such as pollen and nectar, as part of their diet. Natural enemies from a broad range of orders including Hymenoptera, Diptera, Coleoptera, Heteroptera, Neuroptera, Araneae, and Acari have been observed to require and/or benefit from access to flowering resources (Wäckers et al., 2005). Access to pollen and nectar sources can significantly increase the activity, longevity, and fecundity of these predators and parasitoids (Wäckers et al., 2008; Hogg et al., 2011), and thus, the availability of flowering resources can be essential to natural enemy efficacy in biological control of pest insects (van Rijn & Sabelis, 2005). These non-prey requirements can be fulfilled with a diverse assemblage of flowering plants, which will provide necessary resources that support populations of predators and parasitoids throughout the season.

Simple addition of flowering plants to farms may not be sufficient to gain the expected increase in biological control, however, and in some cases it may be counter-productive due to supplying resources for pest insects. In recent years there has been a much greater understanding of the need to tailor resource plants for the specific natural enemies that can provide pest suppression, but further investigation is needed to tailor this to specific crop systems (Jonsson et al., 2008). In one line of investigation, the nutritional quality of plant resources has been investigated in detail, revealing the range of suitability of different plants as food for parasitoids and predators (reviewed by Wäckers, 2005). Additionally, the need to select plants that are beneficial to the natural enemies without providing resources to pest insects has driven the careful evaluation of pest and natural enemy life history traits on candidate floral resources. Baggen et al. (2003) evaluated a range of potential resource plants and found that only lacy phacelia (*Phacelia tanacetifolia* Benth.) and Nasturtium (*Tropaeolum majus* L.) provided resources for natural enemies without also enhancing moth pest performance, as the other tested plants did. In field trials of this selective planting approach, Begum et al. (2006) found higher parasitism of light brown apple moth (*Epiphyas postvittana* (Walker)) eggs in vineyard plots where three types of flowering resource plants were sown under the vines. In more recent studies, this rewarding plant strategy has been combined with attraction of natural enemies (see Fig. 1) in an attract-and-reward approach. Initial reports from combining these two tactics indicate that while this approach has potential for synergy, further work is required to realize the full potential (Simpson et al., 2011a, 2011b).

2.3.3 Alternative prey

As mentioned above, most natural enemies benefit from having access to alternate hosts/prey. Taking advantage of this knowledge, banker plant systems (or open rearing systems) can be used to augment populations of natural enemies in greenhouse and field settings on ornamental and food crops (Van Driesche et al., 2008). Although many natural enemies can be purchased and released to augment biological control, they often leave or die once the targeted pest has been controlled. By combining aspects of augmentative and conservation biological control, banker plant systems attempt to alleviate these factors (Frank, 2010). Banker plant systems generally consist of a non-crop plant that is deliberately infested with a non-pest herbivore. The non-pest herbivores serve as alternative hosts/prey for a desired parasitoid or predator of the target crop pest. A banker plant system is typically based on the use of alternative host/prey in the form of non-pest herbivores, but it can also be based around the use of surrogate food, such as pollen for generalist natural enemies (Huang et al., 2011). As a form of conservation biological control, banker plant systems provide alternative food or hosts for natural enemies so they can survive and reproduce for long periods even when no pests are present in the crop (Frank, 2010). Banker plants can also conserve released natural enemies to provide sustainable, long-term suppression of crop pests.

2.4 Shelters

Natural enemies of insect pests require shelter from environmental hazards, and a lack of shelter during periods of heat, cold, rain, or pesticide application may be highly detrimental to their survival. Availability of appropriate habitats may promote foraging, resting, overwintering, or nesting of natural enemies.

Physical environmental conditions profoundly affect natural enemy activity during the growing season. For example, excessive wind is thought to limit foraging by adult hoverflies (Beane & Bugg, 1998). Hedgerows, windbreaks, or shelter-belts can protect croplands in windy areas, and provide some protection to the windward as well as to the leeward side. Shelter can reduce soil erosion, and improve photosynthetic and water-use efficiency by crop plants, and can lead to locally elevated temperatures in the sheltered areas. Because hedgerows and windbreaks often contain flowering plants used by many natural enemies, the effects of shelter and of flowers may be difficult to separate (Beane & Bugg, 1998).

Overwintering and resting aggregations of various natural enemies are often observed in crop fields. Typical sites for such aggregations vary among species, and include herbaceous and woody plants as well as human-made structures (Beane & Bugg, 1998). Houses for lacewings have even been built and tested to provide shelter during harsh weather conditions (McEwen & Sengonca, 2001). These lacewing houses have been successfully used as a tool for augmenting biological control in crop fields by increasing the number of lacewings in the agroecosystem (McEwen & Sengonca, 2001).

At the small scale at which mite biological control operates, shelters are important for the survival of predatory species. Some plants have naturally-occurring shelters, called domatia, that predatory mites use as a protected location. They can then forage from these sites on leaves to reach pest mites, and leaves with greater domatia structures tend to have higher populations of predatory mites (Karban et al., 1995; Loughner et al., 2008). Leaves with domatia also protect predatory mites from other natural enemies (intraguild predation) (Norton et al., 2001), and lead to lower densities of leaf-feeding mites and foliar fungal pathogens (Norton et al., 2000; English-Loeb et al., 2005). These findings help explain why certain grape cultivars that possess domatia are less susceptible to mite and mildew outbreaks. Such information could be used in breeding programs to develop crop cultivars that are more likely to have lower pest mite populations, due to their harboring of predatory mites (English-Loeb et al., 2002).

2.5 Landscape influences on natural enemies

Research into landscape-level effects on biological control of insect pests has developed rapidly over the past 20 years, in concert with the expansion of the field of landscape ecology (Turner et al., 2001). New techniques and tools have become available for detailed analysis of aerial imagery or remotely sensed data of the landscapes surrounding crop fields, and combining this with measurements of pest-natural enemy interactions in crop fields has provided new insights, as reviewed by Bianchi et al. (2006). In general, increased habitat fragmentation, isolation and decreased landscape structural complexity destabilize the biotic interactions that regulate pest populations (Robinson et al., 1992; Landis et al., 2000; Tscharntke et al., 2007). In practical terms, this means that farms in more intensively managed landscapes can rely less on naturally-occurring biological control than those that are in more diverse landscapes. In more diverse landscapes that contain multiple crop types, natural habitat, perennial wooded land, and a greater availability of flowering resources, natural enemies are more likely to have their ecological requirements met near to the crop field and are less likely to disperse. This then translates into a greater abundance and diversity of natural enemy insects available during periods of pest population growth to limit the trajectory of that growth and limit pest populations (Marino & Landis, 1996; Bommarco, 1998; Thies et al., 2003; Schmidt & Tscharntke, 2005, Tscharntke et al., 2005).

Having high landscape diversity, including flowering plants, near crop fields can also interact with the management approach taken on farms to affect the natural enemy population available to control pests in crop fields. Thus, Ostman et al. (2001) found that aphid population growth was slower in organic than conventional farms, and fields set in landscapes with greater proportion of perennial crops and with more field margins received more biological control. Spider populations are also sensitive to the landscape, with spider species richness increasing with the proportion of non-crop habitat in the landscape (Schmidt et al., 2005), irrespective of whether farms were managed using organic or conventional tactics. Density of spiders was 62% higher in organically managed fields, and within the conventional fields there was a positive correlation between the proportion of non-crop habitat in the surrounding landscape and the spider density. These trends indicate that field management as well as what landscape they are set in will influence the availability of natural enemies to provide biological control services to crop fields.

The economic implications of how crop management and landscape composition affect the services that natural enemies provide have only recently been addressed. Landis et al. (2008) examined the value of biological control being provided to limit populations of soybean aphid, *Glycines max* (L.), in the context of increasing corn production for ethanol. Their analyses found a \$33/ha value of biological pest control for producers who employed integrated pest management, with this value provided largely from the surrounding landscapes. Increased planting of corn, and the associated reduction in landscape diversity led to a \$58 million/yr cost to farmers caused by reduced biological control of this insect, and this translated into lower yields and higher pesticide costs. In a more recent study, Meehan et al. (2011) have analyzed broad-scale landscape, pest, and pesticide use data across the Midwestern United States. Landscape simplification was correlated with higher pest pressure and greater dependence on pesticides, with multi-million dollar costs to farmers that can be attributed in part to the changes in availability of natural pest regulation supplied by surrounding landscapes.

This section has highlighted the aspects of agricultural habitats that can be manipulated to provide resources for natural enemies. But, as we have just seen, not all landscapes have high abundance of natural enemies. It may therefore be important to focus populations of natural enemies at crop plants where their pest-controlling services are needed, by harnessing the chemical interactions among organisms. This is an active and exciting area of research that can exploit the power of chemical signaling to manipulate natural enemies for the benefit of agriculture.

3. Semiochemical-based manipulation

The term "semiochemical" (semeion = sign or signal in Greek) is used to describe a chemical or mixture of chemicals that can act as messengers in interactions among organisms (Nordlund & Lewis, 1976; Dicke & Sabelis, 1988; Vet & Dicke, 1992). It includes chemicals that mediate interactions among individuals within the same species, i.e., intraspecific communication (= *pheromones*), and those that mediate interactions among individuals belonging to different species, i.e., interspecific communication (= *allelochemicals*). Among allelochemicals, compounds can be classified as *allomones* if their production benefits the emitter, *kairomones* if their production benefits the receiver or *synomones* if their production benefits both the emitter and receiver (Dicke & Sabelis, 1988; Vet & Dicke, 1992).

An approach to using semiochemicals for pest control is to exploit ways to chemically augment, conserve, or enhance the efficacy of natural enemies in cropping systems. Here we provide a review of semiochemical-natural enemy interactions and describe ways in which these compounds, particularly those emitted by plants, can be employed to enhance natural enemy attraction and ultimately reduce pest populations.

3.1 Herbivore-induced plant volatiles (HIPVs)

Plant volatiles play a critical role as signals in tri-trophic level interactions. These are interactions involving three trophic levels; for example, plants (1st trophic level), herbivores (2nd trophic level), and the natural enemies of herbivores (predators and parasitoids) (3rd trophic level). Peter Price and collaborators (Price et al., 1980) were the first to specifically emphasize the importance of including the third trophic level when considering plant-herbivore interactions. Their seminal contribution was of particular importance because interactions among organisms from more than two trophic levels are known to be common in nature (Hunter & Price, 1992; Ohgushi, 2005).

Plants can influence the natural enemies of herbivores by emitting behavior-modifying volatile organic compounds. Specifically, plants damaged by herbivores often produce a blend of volatiles (Paré & Tumlinson, 1999), commonly referred to as herbivore-induced plant volatiles (HIPVs) (Mumm & Dicke, 2010). These HIPVs consist of a mixture of the so-called green-leaf volatiles (C_6 aldehydes, alcohols, and acetates), terpenes (monoterpenes, sesquiterpenes, homoterpenes), and aromatic compounds, among others (Pichersky et al., 2006). The release of HIPVs may signal the presence of potential prey or hosts and, therefore, can be exploited by natural enemies to locate the prey organism (Sabelis et al., 1999; Verkerk, 2004). In the last 10 years there has been an increased interest in using these compounds to manipulate natural enemy behaviors for insect pest control in agricultural crops.

3.1.1 Brief overview of HIPVs

Vinson (1976) described five steps parasitoids and other natural enemies need to follow during host searching and selection; these are: 1) host habitat location; 2) host location; 3) host acceptance; 4) host suitability; and 5) host regulation. It is clear now that plant chemical cues are particularly important in aiding parasitoids during the first step. In order to locate the host habitat, natural enemies use long-distance cues from plants. These cues (e.g. HIPVs) originate mainly from plants damaged by the natural enemies' host or prey (e.g. pest insects). In most instances HIPVs provide natural enemies with a highly detectable and reliable signal. HIPVs are classified as synomones because they can benefit both the emitting plant as well as the responding natural enemy (Vet et al., 1991). Once the host habitat is located, natural enemies utilize compounds produced by the host or prey (kairomones), such as volatiles emitted from body scales, honeydew, or the herbivore's frass. Compared with HIPVs, kairomones are more reliable in providing information to natural enemies about the location of their host or prey; however, they are not as detectable (Vet & Dicke, 1992).

Dicke & Sabelis (1988) provided the first evidence that lima bean plants (*Phaseolus lunatus* L.) damaged by the two-spotted spider mite *Tetranychus urticae* Koch emit a blend of volatiles that attract the predatory mite *Phytoseiulus persimilis* Athias-Henriot. In these early studies, this phenomenon was referred to as a 'cry for help', because of the potential fitness benefits to the injured plants from attracting the natural enemies of herbivores (Dicke et al., 1990a; Dicke & Sabelis, 1992). A second tri-trophic system extensively studied in the early 1990s involved corn (*Zea mays* L.), the chewing herbivore *Spodoptera exigua* (Hübner), and its parasitoid *Cotesia marginiventris* (Cresson). Turlings et al. (1991, 1993) showed that *C. marginiventris* utilizes compounds emitted from corn seedlings damaged by the lepidopteran herbivore to locate its host. HIPVs can also mediate plants-aphids (sucking herbivores)-natural enemy interactions. For example, Du et al. (1998) showed that the parasitoid *Aphidius ervi* Haliday is attracted to beans, *Vicia faba* L., infested by the pea aphid *Acyrthosiphon pisum* (Harris). More recent evidence shows that egg deposition by herbivores can also induce a volatile response in plants and consequently attract egg parasitoids (Meiners & Hilker, 1997, 2000; Hilker & Meiners, 2002; Colazza et al., 2004). For example, Meiners & Hilker (1997) found that oviposition by the elm leaf beetle *Xanthogaleruca luteola* (Müller) induces volatile emissions from its host plant *Ulmus minor* Mill., that attract the egg parasitoid *Oomyzus gallerucae* (Fonscolombe).

This plant volatile response to herbivore damage often differs from artificial damage (Dicke et al., 1990a; Turlings et al., 1990; De Moraes et al., 1998), indicating that the caterpillar or other pest insect induces production of specific HIPVs in the plant. These can be induced locally, i.e. at the site of damage, as well as systemically, i.e. from distal undamaged parts of a damaged plant (Turlings & Tumlinson, 1992; Dicke et al., 1993; Röse et al., 1996).

3.1.2 Characteristics of HIPVs

The emission of HIPVs is common in plants (Dicke & Vet, 1999; Dicke & van Loon, 2000); however, the induced volatile blends are highly variable (Dicke, 2000; Turlings & Wäckers, 2004). The volatile blend often varies depending on plant cultivar (e.g. Loughrin et al., 1995; Gouinguené et al., 2001), plant age (Takabayashi et al., 1994; Turlings et al., 2002), plant part (Turlings et al., 1993), and abiotic factors (Gouinguené & Turlings, 2002). Emissions of HIPVs also vary depending on the species and age of the herbivore (Takabayashi et al., 1995; Gouinguené et al., 2003). For example, Takabayashi et al. (1995) found that corn plants emit greater quantities of volatiles when damaged by 1st and 2nd instar larvae of *Pseudaletia separata* Walker than when damaged by 5th instars. The parasitoid *Cotesia kariyai* (Watanabe) attacks young *P. separata* and is attracted to volatiles emitted by corn damaged by early instar larvae (Takabayashi et al., 1995). To cope with this variability, natural enemies can learn to associate HIPVs with the presence of their prey or host (Lewis & Tumlinson, 1988; Vet & Dicke, 1992; Vet et al., 1995; Allison & Hare, 2009). This learning capacity is thought to be more critical for generalist natural enemies than specialists (Steidle & van Loon, 2003), because the latter should have an innate response to HIPVs.

Another important characteristic of HIPVs is their specificity (Turlings & Wäckers, 2004). The specificity of certain tri-trophic systems allows natural enemies to differentiate plant volatile blends associated with their prey from those of non-prey (Dicke, 1994; Du et al., 1996; Dicke, 1999). De Moraes et al. (1998) first demonstrated the specificity of the volatile response to herbivory in plants and the effects on natural enemies. The authors found that tobacco, maize, and cotton plants produce distinct volatile blends in response to damage by larvae of two related lepidopteran herbivores: *Heliothis virescens* (Fabricius) and *Helicoverpa zea* (Boddie). The parasitoid *Cardiochiles nigriceps* Viereck exploits these differences during host location by being attracted only to HIPVs released from its host *H. virescens* (De Moraes et al., 1998). This specificity has also been reported in the tri-trophic system involving pea plants, pea aphids (*A. pisum*), and the parasitoid *A. ervi* (Du et al., 1998; Powell et al., 1998). Other studies, however, have reported a lack of specificity. For example, McCall et al. (1993) found that the parasitoid *Microplitis croceipes* (Cresson) is equally attracted to volatiles induced by its host *H. zea* and its non-host *S. exigua*. Similarly, the tri-trophic system of cabbage-caterpillars-*Cotesia* sp. lacked specificity at the herbivore level, but not at the plant level where differences in attractiveness to parasitoids were found (Geervliet et al., 1996).

3.1.3 Belowground tri-trophic interactions

There is now an abundant literature showing that tri-trophic level interactions occur aboveground (as described above). However, relatively little is known about the way organisms from different trophic levels interact belowground. This is particularly true for the roles of HIPVs in attraction of soil-inhabiting natural enemies to root volatiles. For example, Boff et al. (2001) found that the entomopathogenic nematode *Heterorhabditis megidis* Poinar, Jackson & Klein is attracted to the roots of *Thuja occidentalis* L. damaged by the weevil *Otiorhynchus sulcatus* Germar. However, the volatile responsible for this attraction was not identified. Recently, Rasmann et al. (2005) found that larvae of the corn rootworm, *Diabrotica virgifera virgifera* LeConte, feeding on corn roots induce the emission of (E)-β-caryophyllene, which in turn attracts entomopathogenic nematodes. Similar to aboveground interactions, interactions belowground can be specific at both the plant and herbivore levels (Rasmann & Turlings, 2008).

3.1.4 Plant elicitors of HIPVs

Jasmonic acid (JA) and its volatile derivative methyl jasmonate (MeJA) are phytohormones involved in plant defenses against herbivores (Karban & Baldwin, 1997), that can also play a key role in the production and emission of HIPVs (Hilker et al., 2002; Kessler et al., 2004). Plants treated topically with JA or MeJA increase their volatile emissions (Hopke et al., 1994; Gols et al., 1999; Ament et al., 2004; Hare, 2007); however, the volatiles produced often differ from those induced by herbivore damage (Dicke et al., 1999; Rodriguez-Saona et al., 2001). Other phytohormones involved in the emission of HIPVs include salicylic acid (SA) and ethylene (Schmelz et al., 2009). Salicylic acid is a phytohormone often associated with plant resistance against pathogens but also against sucking insects such as aphids and whiteflies (Walling, 2000). Exposure to exogenous (airborne) SA, or its volatile derivative methyl salicylate (MeSA), can induce a volatile response in plants (Ozawa et al., 2000). For example, activation of both the JA and SA pathways by *T. urticae* is required to attract predatory mites to damaged lima bean plants (Dicke et al., 1999). In fact, the predatory mite *P. persimilis* prefers lima beans releasing volatiles induced by *T. urticae* than those induced by JA (Dicke et al., 1999). This difference may be explained by the lack of MeSA from the blend induced by JA. In tomatoes, however, Ament et al. (2004) found that JA is necessary to induce the enzymatic conversion of SA into MeSA, and concluded that JA is essential for the emission of spider mite-induced volatiles. Despite the fact that MeSA can play an important role in predator attraction to plants (e.g. De Boer & Dicke, 2004a, 2004b; Rodriguez-Saona et al., 2011a), compared with the JA pathway, less is known on the importance of the SA pathway in the emission of HIPVs.

In addition, these phytohormones can interact synergistically or antagonistically (Walling 2000). For instance, SA can inhibit the plant's response to JA and *vice versa* (Koornneef & Pieterse, 2008). Horiuchi et al. (2001) demonstrated that the ethylene precursor, 1-aminocyclopropane-1-carboxylic acid, increases the induced volatile response to JA in lima bean and, as a result, increases the attraction of the predatory mite *P. persimilis*.

3.2 Manipulation of HIPVs to enhance biological control

This review will focus on three methods to manipulate HIPV emissions in agricultural fields: a) use of synthetic versions of HIPVs; b) increase of HIPV emissions through use of phytohormonal elicitors; and, c) increase of HIPV emissions via genetic engineering. Table 1 summarizes general physical, economical, and biological characteristics of these approaches.

3.2.1 Synthetic HIPV lures

The simplest way to manipulate natural enemy behaviors chemically in agricultural fields is to identify the natural HIPVs, produce them, and release synthetic versions of them (Fig. 1C). In this approach, natural enemies are exposed to a "supernormal" stimulus (i.e., a highly attractive HIPV), that is expected to outcompete the "normal" stimuli provided by the surrounding vegetation. Yet, compared with the large number of studies documenting the attraction of natural enemies to HIPVs under laboratory conditions, fewer studies have been conducted under field conditions (Hunter, 2002). In early studies, Flint et al. (1979) showed attraction of the common green lacewing *Chrysoperla carnea* (Stephens) to β-caryophyllene. Drukker et al. (1995) found an increased density of predatory anthocorids on pear trees near cages containing *Psylla*-infested trees compared with trees near cages containing non-infested trees. Similarly, Shimoda et al. (1997) found greater attraction of the predatory thrips *Scolothrips takahashii* Priesner to traps with *T. urticae*-infested lima bean plants compared with traps with uninfested plants. In a non-agricultural system, Kessler & Baldwin (2001) later showed that predation of *Manduca sexta* L. eggs by the generalist predator *Geocoris pallens* Stal. increases when plants of *Nicotiana attenuata* Torr. ex Wats are treated with the HIPVs (Z)-3-hexenol, linalool, and cis-α-bergamotene.

Attributes	Synthetic HIPV lures	Plant Elicitors	Genetic Engineering
I. Physical/Economics			
Longevity	Medium-long lasting. Slow-release devices (4 weeks or more)	Short-lasting - often quick activation of volatiles (likely less than a week)	Longest lasting approach. (throughout the plant's life)
Applicability/ Adoptability	Relatively easy to apply and adopt	Relatively easy to apply and adopt	May require long-time for development and adoption
Cost	Relatively cheap: will depend on cost of application, complexity of volatile blend, type of deployment device, number of point sources, etc	Can be expensive: will depend on acreage applied, cost of application and producing elicitor – e.g. JA is costly	The developmental phase can be costly
Commercial availability	Two lures commercially available to growers specifically for this purpose (see text for details)	No product commercially available for this purpose	No product commercially available
II. Biological			
Mode of action	Lures need to outcompete background volatiles May induce volatile emissions from exposed plants	A more "natural" attractant than synthetic lures; however, induced volatile blend often different from the herbivore-induced blend	The most "natural" volatile signal of the three approaches. Plants produce their own set of volatiles
Natural enemy efficacy	In the absence of host/prey, natural enemies can increase foraging time; thus, reduce their efficacy	In the absence of host/prey, natural enemies can increase foraging time	If plants are "primed" for increased induce volatile responses, it may increase natural enemy foraging efficacy
Specificity of signal	Generalized volatile signal: attract a wide range of natural enemies. Signal not specific neither at the plant nor herbivore levels	More specific blend; however, it affects natural enemies differently, some positive, negative, and neutral. Signal likely specific at the plant level but not the herbivore level	The most specific blend of the three approaches. Signal specific at both the herbivore and plant levels
Negative consequences	Natural enemy attraction likely to point source. High potential for association of HIPVs with lack of host/prey. Medium-high potential for natural enemy habituation to HIPVs	Natural enemy attraction likely to the treated habitat. Medium-high potential for association of HIPVs with lack of host/prey. High potential for habituation	Natural enemy attraction to the herbivore-damaged plant. Low potential for association of HIPVs with lack of host/prey. Low potential for habituation
Community-level effects	Medium-high potential for non-target effects, e.g. attraction of herbivores, pollinators	Highest potential for non-target effects, e.g. attraction of herbivores, negative effects on pollinators, cross-talk among defensive pathways, e.g. JA treatment can make plants more susceptible to pathogens	Reduced potential for non-target effects

Table 1. Comparative characteristics of different ways to manipulate natural enemies of herbivore by HIPVs in agriculture

The use of HIPVs to lure natural enemies to crop fields has been receiving increased attention in the last 10 years. James (2003a) evaluated the HIPVs MeSA, (Z)-3-hexenyl acetate, and (3E)-4,8-dimethyl-1,3,7-nonatriene to attract natural enemies in hop yards. The predatory mirid *Deraeocoris brevis* (Uhler), the anthocorid *Orius tristicolor* (White), and the coccinellid *Stethorus punctum picipes* (Casey) were attracted to sticky cards baited with (Z)-3-hexenyl acetate; while the geocorid *G. pallens*, hover flies, and *S. punctum picipes* were attracted to cards baited with MeSA. Synthetic MeSA also attracted green lacewing, *Chrysopa nigricornis* Burmeister (James, 2003b). In grape vineyards, sticky cards in MeSA-baited blocks captured greater number of *C. nigricornis*, *Hemerobius* sp., *D. brevis*, *S. punctum picipes*, and *O. tristicolor* (James and Price 2004). James (2005) tested 15 synthetic HIPVs and found attraction of *S. punctum picipes* to sticky traps baited with MeSA, cis-3-hexen-1-ol, and benzaldehyde. Other natural enemies were attracted to various degrees to different HIPVs (James, 2005). Similarly, Zhu & Park (2005) found attraction of the lady beetle *Coccinella septempunctata* L. to traps baited with MeSA, whereas 2-phenylethanol was more attractive to the lacewing *C. carnea* and syrphid flies. 2-Phenylethanol is also attractive to the multicolored Asian lady beetle, *Harmonia axyridis* (Pallas) (Sedlacek et al., 2009), and is currently being sold commercially by MSTRS Technologies (Ames, Iowa, USA) as the natural enemy attractant Benallure®. Phenylacetaldehyde is another plant attractant for the green lacewing *C. carnea* (Tóth et al., 2006, 2009).

To date, most studies have evaluated HIPVs individually; thus, the synergistic effects of HIPV mixtures on natural enemy attraction remain largely unknown. Yu et al. (2008) tested seven HIPVs and a mixture of nonanal and (Z)-3-hexen-1-ol in cotton fields. Interestingly, they found attraction of the syrphid fly *Paragus quadrifasciatus* Meigen to dimethyl octatriene, nonanal plus (Z)-3-hexen-1-ol, and octanal, whereas the syrphid fly *Epistrophe balteata* De Geer did not respond to any of the HIPVs tested (Yu et al., 2008), indicating differential responses of species of natural enemies to HIPVs within the same insect family. Also, most studies have used slow-release devices instead of spraying HIPVs directly onto crops. This latter approach was tested by Simpson et al. (2011c) who mixed different HIPVs (e.g. MeSA, MeJA, methyl anthranilate, benzaldehyde, (Z)-3-hexenyl acetate, and (Z)-hexen-1-ol) with the vegetable oil adjuvant Synertrol®, and sprayed them onto winegrape, broccoli, and sweet corn plants. They found greater abundance of several parasitic Hymenoptera and predatory insects near plants sprayed with the synthetic HIPVs (Simpson et al., 2011c).

3.2.1.1 MeSA – A natural enemy attractant

MeSA has received considerable attention lately for its potential to attract natural enemies in agricultural fields. This compound is a common component of the volatile blend emitted from several plant species (Pichersky & Gershenzon, 2002). MeSA is emitted from plants in response to feeding by cell-content feeders, e.g. *T. urticae* (Dicke et al., 1990b; Agrawal et al., 2002; van den Boom et al., 2004), phloem feeding, e.g. aphids (Staudt et al., 2010), and chewing herbivores, e.g. beetles (Bolter et al., 1997). In a recent meta-analysis, Rodriguez-Saona et al. (2011a) reviewed 14 publications that used MeSA to attract natural enemies in agricultural fields and found that natural enemies (i.e., coccinellids, syrphids, lacewings, predatory bugs, and parasitic Hymenoptera) are broadly attracted to MeSA. MeSA is now commercially available as PredaLure® (AgBio, Inc.; Westminster, Colorado, USA) to attract natural enemies of agricultural insect pests.

Commercial availability of PredaLure has allowed researchers a more standardized way to test natural enemy attraction to MeSA in agricultural fields, and three studies have recently done that. Lee (2010) found that PredaLures led to higher catches of lacewings and *O. tristicolor* on baited traps in strawberry fields, but the effect was found only at the point source and not at 5 or 10 m away from the lures. Ground-dwelling predators monitored using pitfall traps did not respond to the PredaLures (Lee, 2010). In soybean fields, Mallinger et al. (2011) captured greater numbers of syrphid flies and lacewings on sticky card traps adjacent to the PredaLures, but not on traps placed 1.5 m from the lures. In cranberry fields, PredaLure-baited sticky cards caught greater numbers of syrphid flies, lady beetles, and lacewings compared with unbaited traps (Rodriguez-Saona et al., 2011a). Syrphid abundance was greater on traps placed near PredaLures (0 m) than at 2.5, 5, and 10 m from the lures (Rodriguez-Saona et al., 2011a), so the spatial scale of influence over natural enemies seems to be restricted for this particular product.

3.2.1.2 Mechanism of attraction

The mechanism of natural enemy attraction to HIPVs remains unknown. Two possible mechanisms have been suggested (e.g. Khan et al., 2008): a) *Direct attraction*, where the natural enemies are attracted directly to the synthetic lure; b) *Indirect attraction*, where HIPV exposure triggers a volatile response from plants. These are not mutually exclusive mechanisms; in fact, it is likely that both mechanisms may act simultaneously. Additionally, arrestment of natural enemies near to sources of HIPVs requires further examination as a potential behavioral mechanism contributing to their location of the sources and higher abundance near to dispensers.

Direct attraction. Ample evidence exists in the literature from laboratory studies that natural enemies can respond to HIPVs (Mumm & Dicke, 2010). For example, *Anaphes iole* Girault, an egg parasitoid of *Lygus* spp., showed a strong antennal response (based on electroantennogram -EAG- analysis) to (Z)-3-hexenyl acetate and MeSA (Williams et al., 2008). Gas chromatographic-electroantennographic detection (GC-EAD) analysis showed that MeSA elicits a significant antennal response in *C. septempunctata* (Zhu & Park, 2005). Natural enemy attraction to HIPVs is often confirmed using behavioral assays (e.g. Y-tube olfactometers and wind tunnels). For example, four HIPVs: linalool, (E)-β-ocimene, (3E)-4,8-dimethyl-1,3,7-nonatriene, and MeSA attracted females of the predatory mite *P. persimilis* in Y-tube olfactometer assays (Dicke et al., 1990b; De Boer & Dicke, 2004a). In a wind tunnel, Williams et al. (2008) showed attraction of *A. iole* females to MeSA and α-farnesene. Thus, it is safe to infer that natural enemies are also being directly attracted to the synthetic lure in the field.

Indirect attraction. Less evidence exists to date on whether synthetic HIPVs can trigger a volatile response from plants under field conditions, or whether activation of this response in turn attracts the natural enemies of herbivores. In laboratory experiments, Dicke et al. (1990c) showed that undamaged lima bean plants exposed to HIPVs from *T. urticae*-damaged plants were more attractive to *P. persimilis* than unexposed plants. In the field, Simpson et al. (2011c) showed attraction of natural enemies for up to 6 days after treating plants with foliar sprays of synthetic HIPVs and, because of the extended period of activity, they concluded that plants might have been induced by exposure to the HIPVs to produce their own volatiles. Rodriguez-Saona et al. (2011a), in a greenhouse study, found that cranberry vines emit high amounts of MeSA when exposed to PredaLure dispensers, whereas unexposed vines released undetectable quantities of MeSA. In maize fields, von Mérey et al. (2011) found that plants exposed to four synthetic green leaf volatiles ((Z)-3-hexenal, (Z)-3-hexenol, (E)-2-hexenal, and (Z)-3-hexenyl acetate) emit increased quantities of sesquiterpenes compared with non-exposed plants.

It is also unclear whether synthetic HIPVs can induce the release of volatiles from exposed plants or "prime" them for an increased volatile response once they are under attack by an herbivore (Ton et al., 2007; Frost et al., 2008). For example, Peng et al. (2011) showed that cabbage plants previously exposed to HIPVs and subsequently damaged by *Pieris brassicae* L. caterpillars attracted more *Cotesia glomerata* L. parasitoids than control plants. Similar studies need to be conducted under field conditions with a range of crop plants to determine whether HIPV lures can prime volatile emissions in exposed plants.

3.2.1.3 Impact of HIPVs on pest abundance

A key question is whether HIPV deployment can ultimately increase predation or parasitism of agricultural pests, and thereby reduce their populations. So far, however, only a few studies have addressed this question. An early study by Altieri et al. (1981) found that spraying a crude extract from corn or *Amaranthus* onto plants increases parasitism rates of *H. zea* eggs by *Trichogramma* wasps. However, this study did not test for specific HIPVs. Three studies have explicitly tested the effects of HIPVs on parasitism rates in the field. Titayavan & Altieri (1990) first showed higher levels of parasitism of the aphid *Brevicoryne brassicae* (L.) by its parasitoid *Diaretiella rapae* (M'Intosh) with applications of a allylisothiocyanate emulsion in broccoli. More recently, Williams et al. (2008) reported greater parasitism of *Lygus lineolaris* (Palisot de Beauvois) eggs by *A. iole* in cotton fields when dispensers containing (Z)-3-hexenyl acetate and α-farnesene were placed near the host eggs. In field cage studies in cotton, Yu et al. (2010) found higher parasitism of *Helicoverpa armigera* (Hübner) larvae by *Microplitis mediator* Haliday in cages treated with 3-7-dimethyl-1,3,6-octatriene. This compound was also active to *M. mediator* in EAG and olfactometer assays (Yu et al., 2010). Lee (2010) found no change in pest abundance in response to deployment of MeSA in strawberry.

Two studies so far have tested the effects of HIPVs on predation rates in the field. Ferry et al. (2009) tested dimethyl disulfide to attract predators (*Aleochara bilineata* Gyllenhal)) of the cabbage root fly, *Delia radicum* (L.), in broccoli. Although they found increased predator attraction, the number of *D. radicum* eggs predated were reduced in treated compared with untreated plots. Finally, Mallinger et al. (2011) showed lower abundance of soybean aphids, *Aphis glycines* Matsumura, in field plots baited with MeSA lures (PredaLures) compared with untreated plots.

3.2.2 Phytohormonal elicitors

Alternatively to the use of HIPV lures, plants can be treated with an exogenous elicitor in the field, such as jasmonates (e.g. JA, MeJA, or cis-jasmone), to induce production and emissions of their own blend of volatiles, and as a result attract natural enemies (Rohwer & Erwin, 2008). This is a more "natural" approach for attracting predators and parasitoids of pests into crops as compared with using synthetic lures because these phytohomones often induce an attractive blend of volatiles in quantities that are more comparable with those induced by herbivore feeding. However, besides inducing volatile emissions, jasmonates induce a wide array of responses in plants including increase of defenses that can negatively affect the performance of natural enemies by reducing the quality and quantity of herbivores on plants (Thaler, 1999, 2002).

The effects of jasmonates on natural enemy attraction have been demonstrated under laboratory and field conditions. For example, in the laboratory, the predatory mite *P. persimilis* is attracted to an odor blend induced by JA from gerbera (Gols et al., 1999) and lima bean (Dicke et al., 1999) plants. Similarly, van Poecke & Dicke (2002) showed that treatment of *Arabidopsis thaliana* (L.) with JA increases attraction of *Cotesia rubecula* (Marshall) compared with untreated plants, whereas treatment with SA did not. Ozawa et al. (2004) also reported that treating maize plants with JA increases attraction for the parasitoid *Cotesia kariyai* Watanabe. However, natural enemies are sometimes less attracted to volatiles induced by JA than to those induced by herbivores (Dicke et al., 1999), indicating that there can be differences between the volatile blends induced by herbivores and JA treatment. In the field, Thaler (1999) showed that JA treatment of tomato plants increases parasitism of caterpillars near the treated plants. However, JA can affect natural enemies of herbivores differently. For instance, Thaler (2002) found that syrphid flies were negatively affected by JA treatment of tomato plants due to a decrease in herbivore abundance on JA-treated plants, but found no effects for a caterpillar parasitoid, an aphid parasitoid, or lady beetles. Also, Lou et al. (2005) demonstrated that egg parasitism of the rice brown planthopper, *Nilaparvata lugens* (Stål), by the parasitoid *Anagrus nilaparvatae* Pang et Wang on rice plants was two-fold higher when plants were surrounded by JA-treated plants than by control plants.

To our knowledge there is no commercial product currently available that uses plant elicitors (e.g. phytohormones) for the sole purpose of triggering HIPVs and attracting natural enemies in agricultural crops. This lack of commercial products may be due to the fact that phytohormones, such as JA, can activate multiple physiological responses in plants (including defenses against insects pests), but their effects on plant yield remains largely unknown. As a result, the risks of activating the JA pathway might outweigh its benefits if resistance to phytophagous insects reduces fitness of natural enemies on plants or increases plant susceptibility to pathogens (Table 1).

Practical application of HIPVs for insect pest control remains a goal that will require coordinated research by agricultural scientists and chemical ecologists. The involvement of commercial suppliers is a positive step towards development of cost-effective and efficaceous products for manipulation of natural enemies in crops.

3.2.3 Genetic engineering

Many of the risks associated with using lures or phytohormones to attract natural enemies may be avoided through genetic engineering because plants can be selected for enhanced HIPV emissions only when attacked by herbivores. Although plant breeding practices have historically ignored the effects of HIPVs on the third trophic level, this is expected to change with recent advances in molecular technologies. Two approaches can be taken: a) selective breeding, where the natural variation in the production of HIPVs among plants can be exploited in breeding programs to select for plants that enhance the foraging efficiency of natural enemies, or b) transgenic plants, where specific genes are incorporated to prime plants for an enhanced HIPV response.

3.2.3.1 Selective breeding

Plant breeding may produce crops with enhanced volatile emissions (Nordlund et al., 1981, 1988); however, to date, selective breeding for high HIPV production in plants has not been explored. Volatile emissions often differ within and among plant species (Elzen et al., 1985; Takabayashi et al., 1991), and selecting for plants that are more attractive to natural enemies may thus help biological control. For example, Elzen et al. (1985, 1986) found greater production of volatiles attractive to the parasitoid *Campoletis sonorensis* (Cameron) from glanded cotton (*Gossypium hirsutum* L.) than nonglanded cotton. However, use of highly attractive plants has the same disadvantage as synthetic lures because volatiles are not associated with the host/prey. A better approach is to select plants with greater induced volatile responses (HIPVs). For instance, HIPV emissions varied by 8-fold among maize cultivars (Gouinguené et al., 2001; Degen et al., 2004). Similarly, high variation in HIPV production among cultivars has been reported in apple (Takabayashi et al., 1991), cotton (Loughrin et al., 1995), and *Gerbera* (Krips et al., 2001). Among below-ground interactions, (E)-β-caryophyllene is a volatile induced from maize roots by herbivory that attracts entomopathogenic nematodes (Rasmann et al., 2005), and emissions of this attractant have apparently been lost in American maize varieties (Köllner et al., 2008). Thus, restoring this or other signals may enhance the effectiveness of biological control agents (e.g. Degenhardt et al., 2009). This would be particularly relevant in domesticated crops where breeding for high yielding crops might unintentionally reduce traits associated with insect resistance such as HIPV emissions (Rodriguez-Saona et al., 2011b).

3.2.3.2 Transgenic plants

There are a few literature reviews on the use of transgenic plants to augment HIPVs (Degenhardt et al., 2003; Aharoni et al., 2005, 2006; Turlings & Ton, 2006; Dudareva & Pichersky, 2008; Kos et al., 2009). Plant defense signaling pathways have been a target of genetic manipulation. For instance, mutant or genetically-modified plants with impaired JA production have been developed (Baldwin et al., 2001; Thaler et al., 2002; Ament et al., 2006), and are often less attractive to natural enemies (Thaler et al., 2002; Ament et al., 2004). Knock out of the JA pathways can also reduce direct defenses, thus making plants more susceptible to herbivory (Thaler et al., 2002; Kessler et al., 2004). Mutant plants also exist with impaired genes specifically involved in defense pathways (van Poecke & Dicke, 2002; van Poecke & Dicke, 2003; Shiojiri et al., 2006). These studies have improved our understanding on the ecological role of plant defensive pathways in tri-trophic level interactions; however, transgenic plants with modified production of HIPVs will be more useful for manipulation of natural enemy behaviors. Terpenoid biosynthesis has particularly been targeted for modification because of the dominance of terpenes in the HIPV blends of plants (Aharoni et al., 2005, 2006). For example, Kappers et al. (2005) modified the expression of a linalool/nerolidol synthase gene in *A. thaliana* to enhance constitutive emissions of the HIPV nerolidol and attraction of the predatory mite *P. persimilis* to plants. Schnee et al. (2006) transferred a sesquiterpene synthase gene that forms (E)-β-farnesene, (E)-α-bergamotene, and other herbivory-induced sesquiterpenes from maize into *Arabidopsis*, resulting in greater emissions of several sesquiterpenes and enhanced attraction of *C. marginiventris* after wasps learned to associate the presence of hosts with the emissions of these sesquiterpenes. Degenhardt et al. (2009) transformed a non-(E)-β-caryophyllene emitting maize line with a (E)-β-caryophyllene synthase gene from oregano, resulting in constitutive emissions of this sesquiterpene, less root damage and 60% fewer root herbivores than non-transformed, non-emitting lines.

An alternative to modifying plants to constitutively emit HIPVs, is to genetically "prime" plants for an enhanced HIPV response after herbivore attack (Turlings & Ton, 2006). These primed plants would thus invest less energy on potentially costly defenses such as HIPV emissions in the absence of herbivores. Although the molecular mechanisms remain largely unknown, once identified, genes involved in priming should provide a useful tool to manipulate HIPV emissions in plants.

3.3 Other sources for natural enemy attraction

The concept of using chemicals to manipulate natural enemy behavior in agricultural fields is not new (Dicke et al., 1990c). However, many of the tools currently used to isolate and identify HIPVs, such as sophisticated headspace volatile collection and gas chromatography apparatus, were not available when this research started in the 1970s-1980s. Thus, most of the early work focused on testing chemicals produced from the host/prey (kairomones), or those emitted from the natural enemies themselves (pheromones). Although researchers have so far found limited applicability for these chemicals because of their low volatility and high specificity, there is great potential for this approach and we expect significant advances in the coming years.

3.3.1 Chemicals from host/prey

Early studies to enhance the efficacy of natural enemies tested the use of kairomones under laboratory and field conditions. Lewis et al. (1975a, 1975b) showed increased egg parasitism rates of *H. zea* by *Trichogramma* spp. from 13% to 22% by spraying an extract from the host (moth) scales or synthetic kairomones onto soybean plants. The moth scales contain tricosane, which was found to be the main source of attraction (Jones et al., 1973). These field results were obtained, however, only at high host densities. At low to intermediate densities, parasitism rates were enhanced if moth scales or the synthetic kairomone (impregnated particles of diatomaceous earth) were applied around the host eggs (Lewis et al., 1979; Gross, 1981). Under these latter conditions, parasitoids apparently spent more time searching intensively in areas where the hosts were absent, resulting in lower parasitism. This problem can be overcome under unnaturally-high host densities (Lewis et al., 1975a, 1975b, 1979). This work first highlighted the potential of interfering with the natural enemies' foraging behavior by application of semiochemicals onto crops. Later studies revealed that volatiles from the ovipositor gland of female *H. zea*, which contains the moth sex pheromone, are also involved in *Trichogramma* spp. host search behaviors. Applications of the synthetic sex pheromone found in the gland increased egg parasitism in greenhouse and field experiments (Lewis et al., 1982).

Kairomones can also be used to "prime" natural enemies for enhanced searching behaviors before inundative releases. For example, Hare et al. (1997) demonstrated that laboratory-reared *Aphytis melinus* DeBach, a parasitoid of the California red scale (*Aonidiella aurantii* (Maskell)), more readily parasitized hosts when exposed to the kairomone O-caffeoyltyrosine prior to being released in the field.

3.3.2 Chemicals from natural enemies

Similar to other insects, natural enemies produce pheromones for intraspecific communication. Sex pheromones have been identified from various natural enemies since the 1970s (e.g. Robacker & Hendry, 1977; Jones, 1989; Eller et al., 1984; Swedenborg & Jones, 1992); however, so far they have been tested only to assess natural enemy activity in the field, monitor their population densities, and to predict rates of host parasitism (Lewis et al. 1971; Morse & Kulman, 1985). Because these pheromones are often produced by the females to attract males, and attraction of females instead of males is desirable in biological control, use of sex pheromones from natural enemies to manipulate their behaviors in agricultural crops has been limited. In addition, unless a stable sex ratio is known to exist in the field, trapping males does not give a reliable prediction of female abundance (Powell, 1986).

Aggregation pheromones might be useful for mass trapping and inundative releases of natural enemies into crops because these compounds attract both sexes. For example, males of the generalist predator, the spined soldier bug (*Podisus maculiventris* (Say)), produce a long-range attractant pheromone that attracts both adult sexes and immatures (Aldrich et al., 1984; Sant'Ana et al., 1997). Both sexes of adult seven-spot ladybeetle *C. septempunctata* are attracted to 2-isopropyl-3-methoxypyrazine, a compound produced by conspecifics (Al Abassi et al., 1998). Pheromones can be combined with HIPVs to enhance natural enemy attraction in agroecosystems. In fact, Jones et al. (2011) recently tested the attraction of three lacewing species to HIPVs in apple orchards and found that the combination of MeSA and iridodial, a male-produced aggregation pheromone, was a stronger attractant than each compound alone.

4. Conclusion

The idea of manipulating natural enemy behaviors to improve biological control of crop pests is an appealing concept, but research on how to best achieve this in agroecosystems is still in its infancy despite the fact that scientists have made important advances in recent years. Several factors need to be considered before these strategies are widely adopted by growers; here we discuss three of them.

4.1 Efficacy

There are many questions that remain unanswered as to how to best deploy these strategies to enhance biological control of crop pests. Habitat for natural enemies needs to be tailored to the region, crop, and management system being used to ensure the greatest potential for benefits and to minimize undesirable effects on crop yield, insect pest populations, or weed pressure. Currently, there is little information on the link between provision of habitat for beneficial insects and the economic effect on crop production, but this is a key missing piece of the puzzle and is an active area of research. Table 1 includes some risks associated with the use of HIPVs. For example, when using HIPVs to attract natural enemies, we don't know where the insects are coming from. If a fixed number of natural enemies occur in the environment, it is likely that attraction of natural enemies to one area will deprive other areas of their services. Also, we have yet to determine the optimal density and concentration of attractants to effectively manipulate biological control agents in agroecosystems. Although MeSA has proven effective as a powerful natural enemy attractant, we have not fully tested other HIPVs in single blends or mixtures. It is likely that in nature different natural enemies use information from HIPV blends differently. Specificity to attract the most important natural enemy in a particular system can be added if HIPVs are combined with other semiochemicals such as volatiles produced from the natural enemies themselves (pheromones) or the herbivores (kairomones). Natural enemy behaviors are very plastic and they can learn to associate synthetic HIPVs with the absence of prey/host. Inundating an area with HIPVs can also lead to habituation of the natural enemy's sensory system, potentially resulting in reduced foraging success. All these concerns need to be addressed before these strategies can be adopted by growers.

Not only can the plants be selected for enhanced HIPV emissions but the natural enemies themselves can also be artificially selected for a superior response to HIPVs. However, this concept has not been widely explored. In a belowground system, Hiltpold et al. (2010) selected the entomopathogenic nematode *Heterorhabditis bacteriophora* Poinar for an improved attractive response towards (E)-β-caryophyllene.

4.2 Costs

The costs of any strategy for manipulating natural enemy behaviors in agroecosystems have not been estimated. No matter how effective these strategies are, their adoption will depend on how comparable their costs are with currently available pest management practices. In fact, the benefits growers obtain from recruiting "free" natural enemies services to provide pest suppression should exceed the associated costs of deploying these strategies. For instance, habitat manipulation to create flower strips or planting alternative food-providing resources for natural enemies within a farm may use land that otherwise could be used in crop production. Even if land is used that is not appropriate for crop production, the expenses associated with preparing habitat for beneficial insects can be considerable. Cost-share programs are available in some countires for establishing beneficial insect habitat in agricultural landscapes. Despite the costs of habitat establishment or HIPV deployment, with the increasing costs of pesticides and consumer concerns about pesticide residues on fresh farm products, biological control is becoming a more attractive alternative. Although strategies to conserve and augment natural enemies described in this chapter are environmentally friendly, relying on the performance of natural enemies could, however, be risky for growers particularly when used to manage a pest in crops where the market has little to no tolerance for damage.

4.3 Combination of strategies

Individually, using strategies such as habitat diversification or the deployment of semiochemical lures to manipulating the behavior of natural enemies may enhance biological control in agroecosystems, but it is possible to further improve the efficacy of natural enemies by combining more than one strategy to manipulate their behaviors in field crops (Fig. 1D). For example, Simpson et al. (2011b) tested the concept of an "attract and reward" strategy that combines habitat manipulation with HIPVs. In their approach, several HIPVs (including methyl anthranilate, MeJA, and MeSA) were tested as attractants and buckwheat was used as a reward. They showed an increase in natural enemy abundance in fields treated with both the "attract" and "reward" strategies compared with those treated with a single strategy (Simpson et al., 2011b). This combined approach has also recently been tested in perennial crop systems, with increased predators and parasitoids (as well as herbivorous thrips) observed in response to HIPV application to vineyards, and parasitoids and thrips responding to provision of flowering plants (Simpson et al., 2011a). However, the combined treatments did not significantly affect natural enemy captures in treated plots. Although these first attempts to combine attract and reward are revealing additive rather than synergistic effects of combining strategies, we expect that further investigations of operational parameters to optimize such systems will provide a more clear view of the situations in which attract and reward can support pest management. This approach is appealing because it overcomes the concern of bringing natural enemies into areas deprived of prey/host, which may lead to association of HIPVs with lack of food, by providing them with an alternative food source to enhance their residency time in treated areas. The approach is expected to work as long as the presence of a supplementary food source (e.g. nectar or pollen) does not interfere with the natural enemy's search behavior for prey or host, which may result in the unwanted outcome of greater herbivore abundance.

Integrated Pest Management (IPM) relies on multiple strategies to maintain pest populations below an economic threshold. Biological control can be combined with chemical control to develop an integrative pest management program. IPM programs based on reduced-risk, softer chemical control tactics, i.e., those with reduced harmful effects on natural enemies, are more desirable. Behavioral manipulation of natural enemies is compatible with these pesticides to conserve or augment natural enemies. For example, semiochemical-based attractants can be used as a tool to measure the impact of insecticides on natural enemy populations. They can also be used to conserve natural enemies within farms by minimizing their exposure to pesticides such that natural enemies are removed from fields before pesticide treatments by placing these attractants in adjacent non-treated fields. HIPVs and other semiochemicals can also be used in augmentative releases of natural enemies by mass trapping natural enemies and releasing them in areas of low population and high pest pressure.

5. Acknowledgment

Work on this chapter was funded in part by a grant from the USDA Crops at Risk program of the United States Department of Agriculture (USDA) to Isaacs and Rodriguez-Saona (grant 2009-51100-20105). Brett Blaauw was supported by a USDA-SARE research grant (grant GNC09-116). The work of Isaacs and Blaauw is supported by Michigan State University's AgBio Research.

6. References

Agrawal, A.A.; Vala, F. & Sabelis, M.W. (2002). Induction of preference and performance after acclimation to novel hosts in a phytophagous spider mite: adaptive plasticity? *The American Naturalist*, Vol.159, pp. 553–565, ISSN 0003-0147.

Aharoni, A.; Jongsma, M.A.; Kim, T.Y.; Ri, M.B.; Giri, A.P.; Verstappen, F.W.A.; Schwab, W. & Bouwmeester, H.J. (2006). Metabolic engineering of terpenoid biosynthesis in plants. *Phytochemistry Reviews*, Vol.5, pp. 49-58, ISSN 1568-7767.

Aharoni, A.; Jongsma, M.A. & Bouwmeester, H.J. (2005). Volatile science? Metabolic engineering of terpenoids in plants. *Trends in Plant Science*, Vol.10, pp. 594-602, ISSN 1360-1385.

Ahern, R.G. & Brewer, M.J. (2002). Effect of different wheat production systems on the presence of two parasitoids (Hymenoptera: Aphelinidae; Braconidae) of the Russian wheat aphid in the North American Great Plains. *Agriculture, Ecosystems, and Environment*, Vol.92, pp. 201-210, ISSN 0167-8809.

Abassi, A.; Birkett, S.; Pettersson, M.A.; Pickett, J.A. & Woodcock, C.M. (1998). Ladybird beetle odour identified and attraction between adults explained. *Cellular and Molecular Life Sciences*, Vol.54, pp. 876–879. ISSN 1420-682X.

Aldrich, J.R.; Lusby, W.R.; Kochansky, J. P. & Abrams, C.R. (1984). Volatile compounds from the predatory insect *Podisus maculiventris* (Hemiptera: Pentatomidae): The male and female metathoracic scent gland and female dorsal abdominal gland secretions. *Journal of Chemical Ecology*, Vol.10, pp. 561–568, ISSN 0098-0331.

Allison, J.D. & Hare, J.D. (2009). Learned and naïve natural enemy responses and the interpretation of volatile organic compounds as cues or signals (Tansley Review). *New Phytologist*, Vol.184, pp. 768-782, ISSN 0028-646X.

Altieri, M.A. & Nicholls, C.I. (2004). *Biodiversity and Pest Management in Agroecosystems* (2nd ed), Food Products Press, ISBN 1560229255, New York.

Altieri, M.A.; Lewis, W.J.; Nordlund, D.A.; Gueldner, R.C. & Todd, J.W. (1981). Chemical interactions between plants and *Trichogramma* wasps in Georgia soybean fields. *Protection Ecology*, Vol.3, pp. 259–263.

Ament, K; Kant, M.R.; Sabelis, M.W.; Haring, M.A. & Schuurink R.C. (2004). Jasmonic acid is a key regulator of spider mite-induced volatile terpenoid and methyl salicylate emission in tomato. *Plant Physiology*, Vol.135, pp. 2025–2037, ISSN 0032-0889.

Ament, K.; Van Schie, C.C.; Bouwmeester, H.J.; Haring, M.A. & Schuurink, R.C. (2006). Induction of a leaf specific geranylgeranyl pyrophosphate synthase and emission of (E,E)-4,8,12-trimethyltrideca-1,3,7,11-tetraene in tomato are dependent on both jasmonic acid and salicylic acid signaling pathways. *Planta*, Vol.224, pp. 1197–1208, ISSN 0032-0935.

Andow, D.A. (1991). Vegetational diversity and arthropod population response. *Annual Review of Entomology*, Vol.36, pp. 561-586, ISSN 0066-4170.

Baggen, L.R. & Gurr, G.M. (1998). The influence of food on Copidosoma koehleri (Hymenoptera: Encyrtidae), and the use of flowering plants as a habitat management tool to enhance biological control of potato moth *Phthorimaea operculella* (Lepidoptera: Gelechiidae). *Biological Control*, Vol.11, pp. 9-17, ISSN 1049-9644.

Baggen, L.R.; Gurr, G.M. & Meats, A. (1999). Flowers in tri-trophic systems: mechanisms allowing selective exploitation by insect natural enemies for conservation biological control. *Entomologia Experimentalis et Applicata*, Vol.91, No.1, pp. 155–161, ISSN 0013-8703.

Baldwin, I.T.; Halitschke, R.; Kessler, A. & Schittko, U. (2001). Merging molecular and ecological approaches in plant-insect interactions. *Current Opinion in Plant Biology*, Vol.4, No.4, pp. 351-358, ISSN 1369-5266.

Beane, K.A. & Bugg, R.L. (1998). Natural and artificial shelter to enhance arthropod biologial control agents, In: *Enhancing Biological Control: Habitat Management to Promote Natural Enemies of Agricultural Pests*, eds C. H. Pickett & R. L. Bugg, pp. 239-254. University of California, ISBN 0520213629, Berkeley.

Begum, M.; Gurr, G.M.; Wratten, S.D.; Hedburg, P.R. & Nichol, H.I. (2006). Using selective food plants to maximize biological control of vineyard pests. *Journal of Applied Ecology*, Vol.43, pp. 547–554, ISSN 1365-2664.

Beizhou, S.; Jie, Z.; Jinghui, H.; Hongying, W.; Yun, K. & Yuncong, Y. (2011). Temporal dynamics of the arthropod community in pear orchards intercropped with aromatic plants. *Pest Management Science*, Vol.67, No.9, pp. 1107–1114, ISSN 1526-4998.

Bianchi, F.J.J.A.; Booij, C.J.H. & Tscharntke, T. (2006). Sustainable pest regulation in agricultural landscapes: a review on landscape composition, biodiversity and natural pest control. *Proceedings of the Royal Society B*, Vol.273, pp. 1715–1727, ISSN 0962-8452.

Biesmeijer, J.; Roberts, S.; Reemer, M.; Ohlemüller, R.; Edwards, M.; Schaffers, A.; Potts, S.; Kleukers, R.; Thomas, C.; Settele, J. & Kunin, W. (2006). Parallel declines in pollinators and insect-pollinated plants in Britain and the Netherlands. *Science*, Vol.313, pp. 351-354, ISSN 0036-8075.

Boff, M.I.C.; Zoon, F.C. & Smits, P.H. (2001). Orientation of *Heterorhabditis megidis* to insect hosts and plant roots in a Y-tube sand olfactometer. *Entomologia Experimentalis et Applicata*, *Vol.*98, pp. 329-337, ISSN 0013-8703.

Bolter, C.J.; Dicke, M.; Van Loon, J.J.A.; Visser, J.H. & Posthumus, M.A. (1997). Attraction of Colorado potato beetle to herbivore-damaged plants during herbivory and after its termination. *Journal of Chemical Ecology*, Vol.23, pp. 1003-1023, ISSN 0098-0331.

Bommarco, R. (1998). Reproduction and energy reserves of a predatory carabid beetle relative to agroecosystem complexity. *Ecological Applications*, Vol.8, pp. 846–853, ISSN 1051-0761.

Büchi, R. (2002). Mortality of pollen beetle (Meligethes spp.) larvae due to predators and parasitoids in rape fields and the effect of conservation strip. *Agriculture, Ecosystems, & Environment*, Vol.90, pp. 255-263, ISSN 0167-8809.

Bugg, R.L. & Waddington, C. (1994). Using cover crops to manage arthropod pests of orchards: a review. *Agriculture Ecosystems & Environment*, Vol.50, pp. 11-28, ISSN 0167-8809.

Bugg, R.L. (1991). Cover crops and control of arthropod pests of agriculture, *Proceedings of Cover Crops for Clean Water*, ISBN 0-935734-25-2, Jackson, TN, April 1991.

Carreck, N.L. & Williams, I.H. (2002). Food for insect pollinators on farmland: insect visits to flowers of annual seed mixtures. *Journal of Insect Conservation*, Vol.6, pp. 13-23, ISSN 1366-638X.

Carvell, C.; Roy, D.; Smart, S.; Pywell, R.; Preston, C. & Goulson, D. (2006). Declines in forage availability for bumblebees at a national scale. *Biological Conservation*, Vol.132, pp. 481-489, ISSN 0006-3207.

Colazza, S.; Fucarino, A.; Peri, E.; Salerno, G.; Conti, E. & Bin, F. (2004). Insect oviposition induces volatiles emission in herbaceous plant that attracts egg parasitoids. *Journal of Experimental Biology*, Vol.207, pp. 47-53, ISSN 0022-0949.

Cullen, R.; Warner, K.D.; Mattias, J. & Wratten, S.D. (2008). Economics and adoption of conservation biological control. *Biological Control*, Vol.45, No.2, pp. 272-280, ISSN 1049-9644.

De Boer, J.G. & Dicke, M. (2004a). The role of methyl salicylate in prey searching behavior of the predatory mite *Phytoseiulus persimilis*. *Journal of Chemical Ecology*, Vol.30, pp. 255–271, ISSN 0098-0331.

De Boer, J.G. & Dicke, M. (2004b). Experience with methyl salicylate affects behavioural responses of a predatory mite to blends of herbivore-induced plant volatiles. *Entomologia Experimentalis et Applicata*, Vol.110, pp. 181-189, ISSN 0013-8703.

De Moraes, C.M.; Lewis, W.J.; Paré, P.W.; Alborn, H.T. & Tumlinson, J.H. (1998). Herbivore-infested plants selectively attract parasitoids. *Nature,* Vol.393, pp. 570-573, ISSN 0028-0836.

Degen, T.; Dillmann, C.; Marion-Poll, F. & Turlings, T.C.J. (2004). High genetic variability of herbivore-induced volatile emission within a broad range of maize inbred lines. *Plant Physiology,* Vol.135, pp. 1928-1938, ISSN 0032-0889.

Degenhardt, J.; Hiltpold, I.; Köllner, T.G.; Frey, M.; Gierl, A.; Gershenzon, J.; Hibbard, B.E.; Ellersieck, M.R. & Turlings T.C.J. (2009). Restoring a maize root signal that attracts insect-killing nematodes to control a major pest. *Proceedings of the National Academy of Sciences of the USA,* Vol.106, pp. 13213-13218, ISSN 1091-6490.

Degenhardt, J.; Gershenzon, J.; Baldwin, I.T. & Kessler, A. (2003). Attracting friends to feast on foes: engineering terpene emission to make crop plants more attractive to herbivore enemies. *Current Opinion in Biotechnology,* Vol.14, pp. 169–176, ISSN 0958-1669.

Dicke, M. (1994). Local and systemic production of volatile herbivore-induced terpenoids: their role in plant-carnivore mutualism. *Journal of Plant Physiology,* Vol.143, pp. 465-472, ISSN 0176-1617.

Dicke, M. (1999). Are herbivore-induced plant volatiles reliable indicators of herbivore identity to foraging carnivorous arthropods. *Entomologia Experimentalis et Applicata,* Vol.91, pp. 131-142, ISSN 0013-8703.

Dicke, M. (2000). Chemical ecology of host-plant selection by herbivorous arthropods : a multitrophic perspective. *Biochemical Systematics and Ecology,* Vol.28, pp. 601-617, ISSN 0305-1978.

Dicke, M.; Gols, R.; Ludeking, D. & Posthumus, M.A. (1999). Jasmonic acid and herbivory differentially induce carnivore-attracting plant volatiles in Lima bean plants. *Journal of Chemical Ecology,* Vol.25, pp. 1907-1922, ISSN 0098-0331.

Dicke, M.; Sabelis, M.W. & Takabayashi, J. (1990a). Do plants cry for help? Evidence related to a tritrophic system of predatory mites, spider mites and their host plants. *Symposia Biologica Hungarica, Vol.*39, pp. 127-134, ISSN 0082-0695.

Dicke, M.; van Baarlen, P.; Wessels, R. & Dijkman, H. (1993). Systemic production of herbivore-induced synomones by lima bean plants helps solving a foraging problem of the herbivore's predators. *Proceedings of Experimental & Applied Entomology, N.E.V.,* Vol.4, pp. 39-44, Amsterdam, 1993.

Dicke, M. & Sabelis, M.W. (1988). How plants obtain predatory mites as bodyguards. *Netherlands Journal of Zoology,* Vol.38, pp. 148-165, ISSN 0028-2960.

Dicke, M. & van Loon, J.J.A. (2000). Multitrophic effects of herbivore-induced plant volatiles in an evolutionary context. *Entomologia Experimentalis et Applicata,* Vol.97, pp.237-249, ISSN 0013-8703.

Dicke, M. & Vet, L.E.M. (1999). Plant-carnivore interactions: evolutionary and ecological consequences for plant, herbivore and carnivore, In: *Herbivores between Plants and Predators,* eds. H. Olff, V.K. Brown, & R.H. Drent, pp. 483-520, Blackwell Science, ISBN 063205204X, London.

Dicke, M.; Sabelis, M.W.; Takabayashi, J.; Bruin, J. & Posthumus, M.A., (1990c). Plant strategies of manipulating predator-prey interactions through allelochemicals: prospects for application in pest control. *Journal of Chemical Ecology,* Vol.16, pp. 3091–3118, ISSN 0098-0331.

Dicke, M.; van Beek, T.A.; Posthumus, M.A.; Ben Dom, N.; van Bokhoven, H. & de Groot, A.E. (1990b). Isolation and identification of volatile kairomone that affects acarine predator-prey interactions. Involvement of host plant in its production. *Journal of Chemical Ecology*, Vol.16, pp. 381–396, ISSN 0098-0331.

Dicke, M. & Sabelis, M.W. (1992). Costs and benefits of chemical information conveyance: proximate and ultimate factors, In: *Insect chemical ecology, an evolutionary approach*, eds. B.D. Roitberg & M.B. Isman,B., pp. 122-155, Chapman & Hall, ISBN 0412018713, London.

Drukker, B.; Scutareanu, P. & Sabelis, M.W. (1995). Do anthocorid predators respond to synomones from *Psylla*-infested pear trees under field conditions? *Entomologia Experimentalis et Applicata*, Vol.77, pp. 193-203, ISSN 0013-8703.

Du, Y.J.; Poppy, G.M.; Powell, W.; Pickett, J.A.; Wadhams, L.J. & Woodcock, C.M. (1998). Identification of semiochemicals released during aphid feeding that attract parasitoid *Aphidius ervi*. *Journal of Chemical Ecology*, Vol.24, pp. 1355-1368, ISSN 0098-0331.

Du, Y.J.; Poppy, G.M. & Powell, W. (1996). Relative importance of semiochemicals from first and second trophic levels in host foraging behavior of *Aphidius ervi*. *Journal of Chemical Ecology*, Vol.22, pp. 1591-1605, ISSN 0098-0331.

Dudareva, N., Pichersky, E., 2008. Metabolic engineering of plant volatiles. *Current Opinion in Biotechnology*, Vol.19, pp. 181–189, ISSN 0958-1669.

Dufour, R. (2000). Farmscaping to enhance biological control, In: *NCAT Sustainable Agriculture Project*, Available from:
https://attra.ncat.org/attra-pub/summaries/summary.php?pub=145

Durnwald, E. (2009) Michigan Wildflower Farm. Portland, Michigan.
www.michiganwildflowerfarm.com/home.html

Eller, F.J.; Bartlet, R.J.; Jones, R.L. & Kulman, H.M. (1984). Ethyl (Z)-9-hexadecenoate, a sex pheromone of *Syndipnus rubiginosus*, a sawfly parasitoid. *Journal of Chemical Ecology*, Vol.10, pp. 291-300, ISSN 0098-0331.

Elzen, G.W.; Williams, H.J. & Vinson, S.B. (1986). Wind tunnel flight responses by hymenopterous parasitoid *Campoletis sonorensis* to cotton cultivars and lines. *Entomologia Experimentalis et Applicata*, Vol.42, pp. 285–289, ISSN 0013-8703.

Elzen, G.W.; Williams, H.J.; Bell, A.A.; Stipanovic, R.D. & Vinson, S.B. (1985). Quantification of volatile terpenes of glanded and glandless *Gossypium hirsutum L.* cultivars and lines by gas chromatography. *Journal of Agricultural and Food Chemistry*, Vol.33, pp. 1079–1082, ISSN 0021-8561.

English-Loeb, G.; Norton, A.P.; Gadoury, D.; Seem, R. & Wilcox, W. (2005). Tri-trophic interactions among grapevines, a fungal pathogen, and a mycophagous mite. *Ecological Applications*, Vol.15, pp. 1679-1688, ISSN 1051-0761.

English-Loeb, G.; Norton, A.P. & Walker, M.A. (2002). Behavioral and population consequences of acarodomatia in grapes on phytoseiid mites (Mesostigmata) and implications for plant breeding. *Entomologia Experimentalis et Applicata*, Vol.104, pp. 307-319, ISSN 0013-8703.

Ferry, A.; Le Tron, S.; Dugravot, S. & Cortesero, A.M., (2009). Field evaluation of the combined deterrent and attractive effects of dimethyl disulfide on *Delia radicum* and its natural enemies. *Biological Control*, Vol.49, pp. 219–226, ISSN 1049-9644.

Fiedler, A.K. & Landis, D.A. (2007a). Attractiveness of Michigan native plants to arthropod natural enemies and herbivores. *Environmental Entomology*, Vol.36, pp. 751-765, ISSN 0046-225X.

Fiedler, A.K. & Landis, D.A. (2007b). Plant characteristics associated with natural enemy abundance at Michigan native plants. *Environmental Entomology*, Vol.36, pp. 878-886. ISSN 0046-225X.

Fiedler, A.K. & Landis, D.A. (2008). Maximizing ecosystem services from conservation biological control: the role of habitat management. *Biological Control*, Vol.45, pp. 254-271, ISSN 1049-9644.

Fiedler, A.K. (2006). *Evaluation of Michigan native plants to provide resources for natural enemy arthropods*. MS, Michigan State University.

Flint, H.M.; Salter, S.S. & Walters, S. (1979). Caryophyllene: an attractant for the green lacewing. *Environmental Entomology*, Vol.8, pp. 1123–1125, ISSN 0046-225X.

Frank, S.D. (2010). Biological control of arthropod pests using banker plant systems: Past progress and future directions. *Biological Control*, Vol.52, pp.8-16, ISSN 1049-9644.

Frost, C.J.; Mescher, M.C.; Dervinis, C.; Davis, J.M.; Carlson, J.E. & De Moraes, C.M. (2008). Priming defense genes and metabolites in hybrid poplar by the green leaf volatile cis-3-hexenyl acetate. *New Phytologist*, Vol.180, pp. 722-734, ISSN 1469-8137.

Frost, C.J. (2011b). Tracing the history of plant traits under domestication in cranberries: potential consequences on anti-herbivore defences. *Journal of Experimental Botany*, Vol.62, pp. 2633–2644, ISSN 0022-0957.

Gardiner, M.M.; Landis, D.L.; Gratton, C.; DiFonzo, C.D.; O'Neal, M.; Chacon, J.M.; Wayo, M.T.; Schmidt, N.P.; Mueller, E. E. & Heimpel, G.E. (2009). Landscape diversity enhances biological control of an introduced crop pest in the North-Central USA. *Ecological Applications*, Vol.19, pp. 143–154, ISSN 1051-0761.

Geervliet, J.B.F.; Vet, L.E.M. & Dicke, M. (1996). Innate responses of the parasitoids *Cotesia glomerata* and *C. rubecula* (Hymenoptera: Braconidae) to volatiles from different plant-herbivore complexes. *Journal of Insect Behavior, Vol.*9, No.4, pp. 525-538, ISSN: 0892-7553.

Gols, R.; Posthumus, M.A. & Dicke, M. (1999). Jasmonic acid induces the production of gerbera volatiles that attract the biological control agent *Phytoseiulus persimilis*. *Entomologia Experimentalis et Applicata*, Vol.93, pp. 77–86, ISSN 0013-8703.

Gouiguené, S.; Degen, T. & Turlings T.C.J. (2001). Genotypic variation in induced odour emissions among maize cultivars and wild relatives. *Chemoecology*, Vol.11, pp. 9-16, ISSN 0937-7409.

Gouinguené, S.; Alborn, H. & Turlings T.C.J. (2003). Induction of volatile emissions in maize by different larval instars of *Spodoptera littoralis*. *Journal of Chemical Ecology*, Vol.29, pp. 145-162, ISSN 0098-0331.

Gouinguené, S.P. & Turlings T.C.J. (2002). The effects of abiotic factors on induced volatile emissions in corn plants. *Plant Physiology*, Vol.129, pp. 1296-1307, ISSN 0032-0889.

Goverde, M.; Schweizer, K.; Baur, B. & Erhardt, A. (2002). Small-scale habitat fragmentation effects on pollinator behaviour: experimental evidence from the bumblebee Bombus veteranus on calcareous grasslands. *Biological Conservation*. Vol.104, pp. 293-299, ISSN 0006-3207.

Gross, H.R.; Harrell, E.A.; Lewis, W.J. & Nordlund, D.A. (1981). *Trichogramma* Spp.: Concurrent Ground Application of Parasitized Eggs, Supplemental *Heliothis zea*

Host Eggs, and Host-Seeking Stimuli. *Journal of Economic Entomology*, Vol.74, pp. 227-229, ISSN 0022-0493.

Gurr, G.M.; Wratten, S.D. & Luna, J.M. (2003). Multi-function agricultural biodiversity: pest management and other benefits. *Basic and Applied Ecology*, Vol.4, pp. 107-116. ISSN 1439-1791.

Hare, J.D.; Morgan, D.J.W. & Nguyun, T. (1997). Increased parasitization of California red scale in the field after exposing its parasitoid, *Aphytis melinus* to a synthetic kairomone. *Entomologia Experimentalis et Applicata*, Vol.82, pp. 73–81, ISSN 0013-8703.

Hare, J.D. (2007). Variation in herbivore and methyl jasmonate-induced volatiles among genetic lines of *Datura wrightii*. *Journal of Chemical Ecology*, Vol.33, pp. 2028-2043, ISSN 0098-0331.

Hilker, M.; Rohfritsch, O. & Meiners, T. (2002). The plant's response towards insect egg deposition. In: *Chemoecology of Insect Eggs and Egg Deposition*, eds. M. Hilker & T. Meiners, pp. 205-233, Blackwell Verlag, ISBN 9781405100083, Berlin.

Hilker, M. & Meiners, T. (2002). Induction of plant responses towards oviposition and feeding of herbivorous arthropods: a comparison. *Entomologia Experimentalis et Applicata*, Vol.104, pp. 181-192, ISSN 0013-8703.

Hiltpold, I.; Baroni, M.; Toepfer, S.; Kuhlmann, U. & Turlings, T.C.J. (2010). Selection of entomopathogenic nematodes for enhanced responsiveness to a volatile root signal can help to control a major root pest. *Journal of Experimental Biology*, Vol.213, pp. 2417-2423, ISSN 0022-0949.

Hogg, B.N.; Bugg, R.L. & Daane, K.M. (2011). Attractiveness of common insectary and harvestable floral resources to beneficial insects. *Biological Control*, Vol.56, pp. 76-84, ISSN 1049-9644.

Hopke, J.; Donath, J.; Blechert, S. & Boland, W. (1994). Herbivore-induced volatiles: The emission of acyclic homoterpenes from leaves of *Phaseolus lunatus* and *Zea mays* can be triggered by a β-glucosidase and jasmonic acid. *FEBS Letters*, Vol.352, pp. 146-150, ISSN 0014-5793.

Horiuchi, J.; Arimura, G.; Ozawa, R.; Shimoda, T.; Takabayashi, J. & Nishioka, T. (2001). Exogenous ACC enhances volatiles production mediated by jasmonic acid in Lima bean leaves. *FEBS Letters*, Vol.509, pp. 332–336, ISSN 0014-5793.

Huang, N.; Enkegaard, A.; Osborne, L.S.; Ramakers, P.M.J.; Messelink, G.J.; Pijnakker, J. & Murphy, G. (2011). The banker plant method in biological control. *Critical Reviews in Plant Sciences*, Vol.30, pp. 259-278. ISSN 0735-2689.

Hunter, M.D. & Price, P.W. (1992). Playing chutes and ladders: heterogeneity and the relative roles of bottom-up and top-down forces in natural communities. *Ecology*, Vol.73, pp. 724-732, ISSN 0012-9658.

Hunter, M.D. (2002). A breath of fresh air: beyond laboratory studies of plant volatile-natural enemy interactions. *Agricultural and Forest Entomology*, Vol.4, pp. 81-86, ISSN 1461-9563.

Isaacs, R.; Tuell, J.; Fiedler, A.; Gardiner, M. & Landis, D. (2009). Maximizing arthropod-mediated ecosystem services in agricultural landscapes: the role of native plants. *Frontiers in Ecology and the Environment*, Vol.7, pp. 196-203, ISSN 1540-9295.

James, D.G. (2003a). Synthetic herbivore-induced plant volatiles as attractants for beneficial insects. *Environmental Entomology*, Vol.32, pp. 977-982. ISSN 0046-225X.

James, D.G. (2003b). Field evaluation of herbivore-induced plant volatiles as attractants for beneficial insects: methyl salicylate and the green lacewing, *Chrysopa nigricornis*. *Journal of Chemical Ecology*, Vol.29, pp. 1601-1609, ISSN 0098-0331

James, D.G. (2005). Further field evaluation of synthetic herbivore-induced plant volatiles as attractants for beneficial insects. *Journal of Chemical Ecology*, Vol.31, pp. 481-495, ISSN 0098-0331

James, D.G. & Price, T.S. (2004). Field-testing of methyl salicylate for recruitment and retention of beneficial insects in grapes and hops. *Journal of Chemical Ecology*, Vol.30, pp. 1613-1628, ISSN 0098-0331

Johnson, N.F. & Triplehorn, C.A. (2005). *Borror and DeLong's Introduction to the Study of Insects* (7th edition), Brooks Cole, ISBN 0030968356, Belmont.

Jones, R.L.; Lewis, W.J.; Beroza, M.; Bierl, B.A. & Sparks, A.N. (1973). Host seeking stimulants (kairomones) for the egg parasite Trichogramma evanescens. *Environmental Entomology*, Vol.2, pp. 593–596, ISSN 0046-225X.

Jones, R.L. (1989). Semiochemicals mediating *Microplitis croceipes* habitat, host, and mate finding behavior. *Southwestern Entomologist*, Vol.12, pp. 53-57, ISSN 0147-1724.

Jones, V.P.; Steffan, S.A.; Wiman, N.G.; Horton, D.R.; Miliczky, E.; Zhang, Q.H. & Baker, C.C. (2011). Evaluation of herbivore-induced plant volatiles for monitoring green lacewings in Washington apple orchards. *Biological Control*, Vol.56, pp. 98-105, ISSN 1049-9644.

Jonsson, M.; Wratten, S.D.; Landis, D. & Gurr, G.M. (2008). Recent advances in conservation biological control of arthropods. *Biological Control*, Vol.45, pp. 172-175, ISSN 1049-9644.

Kappers, I.F.; Aharoni, A.; van Herpen, T.W.J.M.; Luckerhoff, L.L.P.; Dicke, M. & Bouwmeester, H.J. (2005). Genetic engineering of terpenoid metabolism attracts bodyguards to *Arabidopsis*. *Science*, Vol.309, pp. 2070-2072, ISSN 0036-8075.

Karban, R. & Baldwin I.T. (1997). *Induced responses to herbivory*. The University of Chicago Press, ISBN 9780226424972, Chicago.

Karban, R.; English-Loeb, G.; Walker, M.A. & Thaler, J. (1995) Abundance of phytoseiid mites on *Vitis* species: effects of leaf hairs, domatia, prey abundance and plant phylogeny. *Experimental & Applied Acarology*, Vol.19, pp. 189-197, ISSN 0168-8162.

Kells, A., Holland, J. & Goulson, D. (2001) The value of uncropped field margins for foraging bumblebees. *Journal of Insect Conservation*, Vol.5, pp. 283-291, ISSN 1366-638X.

Kessler, A.; Halitschke, R. & Baldwin, I.T. (2004). Silencing the jasmonate cascade: induced plant defenses and insect populations. *Science*, Vol.305, pp. 665-668, ISSN 0036-8075.

Kessler, A. & Baldwin, I.T. (2001). Defensive function of herbivore-induced plant volatile emissions in nature. *Science, Vol.*291, pp. 2141-2144, ISSN 0036-8075.

Khan, Z.R.; James, D.G.; Midega, C.A.O. & Pickett, J.A. (2008). Chemical ecology and conservation biological control. *Biological Control*, Vol.45, pp. 210-224, ISSN 1049-9644.

Khan, Z.R. & Pickett, J.A. (2004). The 'push-pull' strategy for stemborer management: a case study in exploiting biodiversity and chemical ecology, In: *Ecological engineering for pest management: advances in habitat manipulations for arthropods*, eds. G.M. Gurr, S.D.

Wratten & M.A. Altieri, pp. 155-164. CSIRO and CABI Publishing, ISBN 9780851999036, UK.

Khan, Z.R.; Ampong-Nyarko, K.; Chiliswa, P.; Hassanali, A.; Kimani, S.; Lwande, W.; Overholt, W.A.; Picketta, J.A.; Smart, L.E. & Woodcock, C.M. (1997). Intercropping increases parasitism of pests. *Nature*, Vol.388, pp. 631-632, ISSN 0028-0836.

Khan, Z.R.; Pickett, J.A.; Wadhams, L. & Muyekho, F. (2001). Habitat management for the control of cereal stem borers in maize in Kenya. *Insect Science and its Application*, Vol.21, pp. 375-380, ISSN 0191-9040.

Kleijn, D. & Sutherland, W. (2003). How effective are European agri-environment schemes in conserving and promoting biodiversity? *Journal of Applied Ecology*, Vol.40, pp. 947-969, ISSN 1365-2664.

Köllner, T.G.; Held, M.; Lenk, C.; Hiltpold, I.; Turlings, T.C.J.; Gershenzon, J. & Degenhardt, J. (2008). A maize (*E*)-β-caryophyllene synthase implicated in indirect defense responses against herbivores is not expressed in most American maize varieties. *The Plant Cell*, Vol.20, pp. 482-494, ISSN 1040-4651.

Koornneef, A. & Pieterse, C.M.J. (2008). Cross talk in defense signaling. *Plant Physiology*, Vol.146, pp. 839-844, ISSN 0032-0889.

Kos, M.; van Loon, J.J.A.; Dicke, M. & Vet, L.E.M. (2009). Transgenic plants as vital components of integrated pest management. *Trends in Biotechnology*, Vol.27, pp. 621-627, ISSN 0167-7799.

Krips, O.E.; Willems, P.E.L.; Gols, R.; Posthumus, M.A.; Gort, G. & Dicke, M. (2001). Comparison of cultivars of ornamental crop *Gerbera jamesonii* on production of spider mite-induced volatiles, and their attractiveness to the predator *Phytoseiulus persimilis*. *Journal of Chemical Ecology*, Vol.27, pp. 1355-1372, ISSN 0098-0331.

Landis, D.A.; Gardiner, M.M.; van der Werf, W. & Swinton, S.M. (2008). Increasing corn for biofuel production reduces biocontrol services in agricultural landscapes. *Proceedings of the Natural Academy of Sciences of the USA, Vol.*105, pp. 20552-20557, ISSN 1091-6490.

Landis, D.A.; Wratten, S.D. & Gurr, G.M. (2000). Habitat management to conserve natural enemies of arthropod pests in agriculture. *Annual Review of Entomology*, Vol.45, pp. 175-201, ISSN 0066-4170.

Lee, J.C. (2010). Effect of methyl salicylate-based lures on beneficial and pest arthropods in strawberry. *Environmental Entomology*, Vol.39, pp. 635-660, ISSN 0046-225X.

Lewis, W.J. & Tumlinson, J.H. (1988). Host detection by chemically mediated associative learning in a parasitic wasp. *Nature* Vol.331, pp. 257-259, ISSN 0028-0836.

Lewis, W.J.; Jones, R.L.; Nordlund, D.A. & Sparks, A.N. (1975a). Kairomones and their use for management of entomophagous insects. I. Evaluation for increasing rates of parasitization by *Trichogramma* spp. in the field. *Journal of Chemical Ecology*, Vol.1, pp. 343-347, ISSN 0098-0331.

Lewis, W.J.; Nordlund, D.A.; Gueldner, R.C.; Teal, P.E.A. & Tumlinson, J.H. (1982). Kairomones and their use for management of entomophagous insects. XIII. Kairomonal activity for *Trichogramma* spp. of abdominal tips, excretion, and a synthetic sex pheromone blend of *Heliothis zea* (Boddie) moths. *Journal of Chemical Ecology*, Vol.8, pp. 1323-1331, ISSN 0098-0331.

Lewis, W.J.; Beevers, M.; Nordlund, D.A.; Gross, H.R. Jr. & Hagen, K.S. (1979). Kairomones and their use for management of entomophagous insects. IX. Investigations of various kairomone-treatment patterns for *Trichogramma* spp. *Journal of Chemical Ecology*, Vol.5, pp. 673-680, ISSN 0098-0331.

Lewis, W.J.; Jones, R.L.; Nordlund, D.A.; Gross, H.R. (1975b). Kairomones and their use for management of entomophagous insects. II. Mechanisms causing increase in rate of parasitization by *Trichogramma* spp. *Journal of Chemical Ecology*, Vol.1, pp. 349-360, ISSN 0098-0331.

Lewis, W.J.; Snow, J.W. & Jones, R.L. (1971). A pheromone trap for studying populations of *Cardiochiles nigriceps*, a parasite of *Heliothis virescens*. *Journal of Economic Entomology*, Vol.64, pp. 1417-1421, ISSN 0022-0493.

Long, R.F.; Corbett, A.; Lamb, C.; Reberg-Horton, C.; Chandler, J. & Stimmann, M. (1998). Beneficial insects move from flowering plants to nearby crops. *California Agriculture*, Vol.52, pp. 23-26. ISSN 0008-0845.

Losey, J.E. & Vaughn, M. (2006). The economic value of ecological services provided by insects. *BioScience*, Vol.56, pp. 311-323. ISSN 0006-3568.

Lou, Y.G.; Du, M.H.; Turlings, T.C.; Cheng, J.A. & Shan, W.F. (2005). Exogenous Application of Jasmonic Acid Induces Volatile Emissions in Rice and Enhances Parasitism of *Nilaparvata lugens* Eggs by theParasitoid *Anagrus nilaparvatae*. *Journal of Chemical Ecology*, Vol.31, pp. 1985-2002, ISSN 0098-0331.

Loughner, R.; Goldman, K.; Loeb, G. & Nyrop, J. (2008). Influence of leaf trichomes on predatory mite (*Typhlodromus pyri*) abundance in grape varieties. *Experimental and Applied Acarology*, Vol.45, pp. 111-122, ISSN 0168-8162.

Loughrin, J.H.; Manukian, A.; Heath R.R. & Tumlinson J.H. (1995). Volatiles emitted by different cotton varieties damaged by feeding beet armyworm larvae. *Journal of Chemical Ecology*, Vol.21, pp. 1217-1227, ISSN 0098-0331.

Mallinger, R.E.; Hogg, D.B. & Gratton, C. (2011). Methyl salicylate attracts natural enemies and reduces populations of soybean aphids (Hemiptera: Aphididae) in soybean agroecosystems. *Journal of Economic Entomology*, Vol.104, pp. 115-124, ISSN 0022-0493.

Marino, P.C. & Landis, D.A. (1996). Effect of landscape structure on parasitoid diversity and parasitism in agroecosystems. *Ecological Applications*, Vol.6, pp. 276–284, ISSN 1051-0761.

McCall, P.J.; Turlings, T.C.J.; Lewis, W.J. & Tumlinson, J.H. (1993). Role of plant volatiles in host location by the specialist parasitoid *Microplitis croceipes* Cresson (Braconidae: Hymenoptera). *Journal of Insect Behavior*, Vol.6, pp. 625-639, ISSN 0892-7553.

McEwen, P.K. & Sengonca, C. (2001). Artificial overwintering chambers for Chrysoperla carnea and their application in pest control, In: *Lacewings in the Crop Environment*, eds. P.K. McEwen & T.R. New, pp. 487-496, Cambridge University Press, ISBN 9780521772174, Cambridge.

Meehan, T.; Werling, B.P.; Landis, D.A. & Gratton, C. (2011). Agricultural landscape simplification and insecticide use in the Midwestern United States. *Proceedings of the Natural Academy of Sciences of the USA*, Vol.108, pp. 11500-11505, ISSN 1091-6490.

Meiners, T. & Hilker, M. (2000). Induction of plant synomones by oviposition of a phytophagus insect. *Journal of Chemical Ecology*, Vol.26, pp. 221-232, ISSN 0098-0331.

Meiners, T. & Hilker, M. (1997). Host location in *Oomyzus gallerucae* (Hymenoptera: Eulophidae), an egg parasitoid of the elm leaf beetle *Xanthogaleruca luteola* (Coleoptera: Chrysomelidae). *Oecologia*, Vol.112, pp. 87-93, ISSN 0029-8549.

Miller, J.R. & Cowles, R.S. (1990). Stimulo-deterrent diversion : a concept and its possible application to onion maggot control. *Journal of Chemical Ecology*, Vol.16, pp. 3197-3212, ISSN 0098-0331.

Morse, B.W. & Kulman, H.M. (1985). Monitoring damage by yellowheaded spruce sawflies with sawfly and parasitoid pheromones. *Environmental Entomology*, Vol.14, pp. 131-133, ISSN 0046-225X.

Mumm, R. & Dicke, M. (2010). Variation in natural plant products and the attraction of bodyguards involved in indirect plant defense. *Canadian Journal of Zoology*,Vol.88, pp. 628-667, ISSN 0008-4301.

Nordlund, D.A.; Lewis, W.J. & Altieri, M.A. (1988). Influences of plantproduced allelochemicals on the host-prey selection behavior of entomophagous insects, In: *Novel Aspects of Insect–Plant Interactions*, eds. P. Barbosa & D.K. Letourneau, pp. 65–90, Wiley, ISBN 9780471832768, New York.

Nordlund, D.A. (1981). Semiochemicals: a review of the terminology, In: *Semiochemicals - Their Role in Pest Control*, eds. D.A. Nordlund , R.L. Jones & W.J. Lewis, pp.13-28, John Wiley & Sons, ISBN 0471058033, New York.

Nordlund, D.A. & Lewis, W.J. (1976). Terminology of chemical releasing stimuli in intraspecific and interspecific interactions. *Journal of Chemical Ecology*, Vol.2, No.2, pp. 211-220, ISSN 0098-0331.

Norton, A.P.; English-Loeb, G. & Belden, A. (2001). Host plant manipulation of natural enemies: leaf domatia protect beneficial mites from insect predators. *Oecologia*, Vol.126, pp. 535-542, ISSN 0029-8549.

Norton, A.P.; English-Loeb, G.; Gadoury, D. & Seem, R.C. (2000). Mycophagous mites and foliar pathogens: leaf domatia mediate tritrophic interactions in grapes. *Ecology*, Vol.81, pp. 490–499, ISSN 0012-9658.

Ohgushi, T. (2005). Indirect interaction webs: herbivore-induced effects through trait change in plants. *Annual Review of Ecology, Evolution, and Systematics*, Vol.36, pp. 81-105, ISSN 1543-592X.

Östman, Ö.; Ekbom, B. & Bengtsson, J. (2001). Landscape heterogeneity and farming practice influence biological control. *Basic and Applied Ecology*, Vol.2, pp. 365-371, ISSN 1439-1791.

Ozawa, R.; Shiojiri, K.; Sabelis, M.W.; Arimura, G.; Nishioka, T. & Takabayashi, J. (2004). Corn plants treated with jasmonic acid attract more specialist parasitoids, thereby increasing parasitization of the common armyworm. *Journal of Chemical Ecology*, Vol.30, pp. 1797–1808, ISSN 0098-0331.

Ozawa, R.; Arimura, G.; Takabayashi, J.; Shimoda, T. & Nishioka, T. (2000). Involvement of jasmonate- and salicylate-related signaling pathways for the production of specific herbivore-induced volatiles in plants. *Plant and Cell Physiology*, Vol.41, pp. 391-398, ISSN 0032-0781.

Paré, P.W. & Tumlinson, J.H. (1999). Plant volatiles as a defense against insect herbivores. *Plant Physiology*, Vol.121, pp. 325-331, ISSN 0032-0889.

Peng, J.; Van Loon, J.J.A.; Zheng, S. & Dicke, M. (2011). Herbivore-induced volatiles of cabbage (*Brassica oleracea*) prime defence responses in neighbouring intact plants. *Plant biology*, Vol.13, No.2, pp. 276-284, ISSN 1438-8677.

Pichersky, E. & Gershenzon, J. (2002). The formation and function of plant volatiles: perfumes for pollinator attraction and defense. *Current Opinion in Plant Biology*, Vol.5, pp. 237-243, ISSN 1369-5266.

Pichersky, E.; Noel, J.P. & Dudareva, N. (2006). Biosynthesis of plant volatiles: nature's diversity and ingenuity. *Science*, Vol.311, pp. 808-811, ISSN 0036-8075.

Pickett, C.H. & Bugg, R.L. (1998). *Enhancing Biological Control: Habitat Management to Promote Natural Enemies of Agricultural Pests* (1st ed), University of California, ISBN 9780520213623, Berkeley.

Powell, W.; Pennacchio, F.; Poppy, G.M. & Tremblay, E. (1998). Strategies involved in the location of hosts by the parasitoid *Aphidius ervi* Haliday (Hymenoptera: Braconidae: Aphidiinae). *Biological Control*, Vol.11, pp. 104-112, ISSN 1049-9644.

Powell, W. (1986). Enhancing parasitoid activity in crops. In: *Insect parasitoids : 13th symposium of the Royal Entomological Society of London*, eds. J. Waage & D. Greathead, pp. 319-340, Academic Press, ISBN 0127289003, San Diego, CA.

Price, P.W.; Bouton, C.E.; Gross, P.; McPheron, B.A.; Thompson, J.N. & Weis, A.E. (1980). Interactions among three trophic levels: influence of plants on interactions between insect herbivores and natural enemies. *Annual Review of Ecology, Evolution, and Systematics*, Vol.11, pp. 41-65, ISSN 1543-592X.

Rasmann, S.; Kollner, T.G.; Degenhardt, J.; Hiltpold, I.; Toepfer, S.; Kuhlmann, U.; Gershenzon, J. & Turlings, T.C.J. (2005). Recruitment of entomopathogenic nematodes by insect-damaged maize roots. *Nature*, Vol.434, pp. 732-737, ISSN 0028-0836.

Rasmann, S. & Turlings, T.C.J. (2008). First insights into specificity of belowground tritrophic interactions. *Oikos*, Vol.117, pp. 362-369, ISSN 0030-1299.

Rebek, E.J.; Sadof, C.S. & Hanks, L.M. (2005). Manipulating the abundance of natural enemies in ornamental landscapes with floral resource plants. *Biological Control*, Vol.33, pp. 203-216, ISSN 1049-9644.

Robacker, D.C. & Hendry, L.B. (1977). Neral and geranial: components of the sex pheromone of the parasitic wasp *Itoplectis conquisitor*. *Journal of Chemical Ecology*, Vol.3, pp. 563-577, ISSN 0098-0331.

Robinson, G.R.; Holt, R.D.; Gaines, M.S.; Hamburg, S.P.; Johnson, M.L.; Fitch, H.S. & Martinko, E.A. (1992). Diverse and contrasting effects of habitat fragmentation. *Science*, Vol.257, pp. 524-526, ISSN 0036-8075.

Robinson, R.A. & Sutherland, W.J. (2002). Post-war changes in arable farming and biodiversity in Great Britain. *Journal of Applied Ecology*, Vol.39, pp. 157-176, ISSN 1365-2664.

Rodriguez-Saona, C.; Crafts-Brandner, S.J.; Paré, P.W. & Henneberry, T.J. (2001). Exogenous methyl jasmonate induces volatile emissions in cotton plants. *Journal of Chemical Ecology*, Vol.27, pp. 679-695, ISSN 0098-0331.

Rodriguez-Saona, C.; Kaplan, I.; Braasch, J.; Chinnasamy, D. & Williams, L. (2011a). Field responses of predaceous arthropods to methyl salicylate: a meta-analysis and case study in cranberries. *Biological Control*, Vol. 59, pp. 294-303, ISSN 1049-9644.

Rodriguez-Saona, C.; Vorsa, N.; Singh, A.P.; Johnson-Cicalese, J.; Szendrei, Z.; Mescher, M.C. & Frost, C.J. (2011b). Tracing the history of plant traits under domestication in cranberries: potential consequences on anti-herbivore defences. *Journal of Experimental Botany*, Vol.62, pp. 2633–2644, ISSN 0022-0957.

Rohwer, C.L. & Erwin, J.E. (2008). Horticultural applications of jasmonates: a review. *Journal of Horticultural Science & Biotechnology*, Vol.83, pp. 283–304, ISSN 1462 0316.

Root, R.B. (1973). Organization of a plant-arthropod association in simple and diverse habitats: the fauna of collards (*Brassica oleracea*). *Ecological Monographs*, Vol.43, pp. 95-124, ISSN 0012-9615.

Roschewitz, I.; Hucker, M.; Tscharntke, T. & Thies, C. (2005). The influence of landscape context and farming practices on parasitism of cereal aphids. *Agriculture Ecosystems and Environment*, Vol.108, pp. 218–227, ISSN 0167-8809.

Röse, U.S.R.; Manukian, A.; Heath, R.R. & Tumlinson, J.H. (1996). Volatile semiochemicals released from undamaged cotton leaves: A systemic response of living plants to caterpillar damage. *Plant Physiology*, Vol.8, pp. 487-495, ISSN 0032-0889.

Ryszkowski, L.; Karg, J.; Margarit, G.; Paoletti, M.G. & Zlotin, R. (1993). Above-ground insect biomass in agricultural landscapes of Europe, In: *Landscape ecology and agroecosystems*, eds. R.G.H. Bunce, L. Ryszkowski & M.G. Paoletti, pp. 71-82, Lewis Publishing, ISBN 9780873719186, Boca Raton.

Sabelis, M.W.; Janssen, A.; Pallini, A.; Venzon, M.; Bruin, J.; Drukker, B. & Scutareanu, P. (1999). Behavioral responses of predatory and herbivorous arthropods to induced plant volatiles: from evolutionary ecology to agricultural applications, In: *Induced Plant Defenses Against Pathogens and Herbivores*, eds. A.A. Agrawal, S. Tuzun & E. Bent, pp. 269-296, APS Press, ISBN 9780890542422, St Paul, MN.

Sanchez, J.A.; Gillespir, D.R. & McGregor, R.R. (2003). The effects of mullein plants (Verbascum thapsus) on the population dynamics of Dicyphus hesperus (Heteroptera:Miridae) in tomato greenhouses. *Biological Control*, Vol.28, pp. 313-319, ISSN 1049-9644.

Sant'Ana, J.; Bruni, R.; Abdul-Baki, A.A. & Aldrich, J.R. (1997). Pheromone-induced movement of nymphs of the predator, *Podisus maculiventris* (Heteroptera: Pentatomidae). *Biological Control*, Vol.10, pp. 123–128, ISSN 1049-9644.

Schmelz, E.A.; Engelberth, J.; Alborn, H.; Tumlinson, J.H. & Teal, P.E.A. (2009). Phytohormone-based activity mapping of insect herbivore-produced elicitors. *Proceedings of the National Academy of Sciences of the USA*, Vol.106, pp. 653-657, ISSN 1091-6490.

Schmidt, M.H.; Roschewitz, I.; Thies, C. & Tscharntke, T. (2005). Differential effects of landscape and management on diversity and density of ground-dwelling farmland spiders. *Journal of Applied Ecology*, Vol.42, pp. 281–287, ISSN 1365-2664.

Schmidt, M.H. & Tscharntke, T. (2005). Landscape context of sheetweb spider (Araneae: Linyphiidae) abundance in cereal fields. *Journal of Biogeography*, Vol.32, pp. 467–473, ISSN 0305-0270.

Schnee, C.; Köllner, T.G.; Held, M.; Turlings, T.C.J.; Gershenzon, J. & Degenhardt, J. (2006). The products of a single maize sesquiterpene synthase form a volatile defense signal that attracts natural enemies of maize herbivores. *Proceedings of the National Academy of Sciences of the USA*, Vol.103, pp. 1129–1134, ISSN 1091-6490.

Sedlacek, J.D.; Friley, K.L. & Hillman, S.L. (2009). Populations of Lady Beetles and Lacewings in Sweet Corn Using 2-Phenylethanol Based Benallure® Beneficial Insect Lures. *Journal of the Kentucky Academy of Science*, Vol.70, No.2, pp. 127-132, ISSN 1098-7096.

Shimoda, T.; Takabayashi, J.; Ashihara, W. & Takafuji, A. (1997). Response of predatory insect *Scolothrips takahashii* toward herbivore-induced plant volatiles under laboratory and field conditions. *Journal of Chemical Ecology*, Vol.23, pp. 2033-2048, ISSN 0098-0331.

Shiojiri, K.; Kishimoto, K.; Ozawa, R.; Kugimiya, S.; Urashimo, S.; Arimura, G.; Horiuchi, J.; Nishioka, T.; Matsui, K. & Takabayashi, J. (2006a). Changing green leaf volatile biosynthesis in plants: An approach for improving plant resistance against both herbivores and pathogens. *Proceedings of the National Academy of Sciences of the USA*, Vol.103, pp. 16672-16676, ISSN 1091-6490.

Simpson, M.; Gurr, G.M.; Simmons, A.T.; Wratten, S.D.; James, D.G.; Leeson, G.; Nicol, H.I. & Orre-Gordon, G.U.S. (2011b). Attract and reward: combining chemical ecology and habitat manipulation to enhance biological control in field crops. *Journal of Applied Ecology*, Vol.48, pp. 580-590, ISSN 1365-2664.

Simpson, M.; Gurr, G.M.; Simmons, A.T.; Wratten, S.D.; James, D.G.; Leeson, G. & Nicol, H.I. (2011c). Insect attraction to synthetic herbivore-induced plant volatile treated field crops. *Agriculture and Forest Entomology*, Vol.13, pp. 45-57, ISSN 1461-9555.

Simpson, M.; Gurr, G.M.; Simmons, A.T.; Wratten, S.D.; James, D.G.; Leeson, G.; Nicol, H.I. & Orre, G.U.S. (2011a). Field evaluation of the 'attract and reward' biological control approach in vineyards. *Annals of Applied Biology*, Vol.159, pp. 69-78, ISSN 0003-4746.

Sotherton, N.W. (1984). The distribution and abundance of predatory arthropods overwintering in farmland. *Annals of Applied Biology*, Vol.105, pp. 49-54, ISSN 0003-4746.

Sotherton, N.W. (1998). Land use changes and the decline of farmland wildlife: An appraisal of the set-aside approach. *Biological Conservation*, Vol.83, pp. 259-268, ISSN 0006-3207.

Staudt, M.; Jackson, B.; El-aouni, H.; Buatois, B.; Lacroze, J-P.; Poëssel, J-L. & Sauge, M-H. (2010). Volatile organic compound emissions induced by the aphid *Myzus persicae* differ among resistant and susceptible peach cultivars and a wild relative. *Tree Physiology*, Vol.30, pp. 1320-1334, ISSN 0829-318X.

Steidle, J.L.M. & van Loon, J.J.A. (2003). Dietary specialization and infochemical use in carnivorous arthropods: testing a concept. *Entomologia Experimentalis et Applicata*, Vol.108, pp. 133-148, ISSN 0013-8703.

Stephens, C.J.; Schellhorn, N.A.; Wood, G.M. & Austin, A.D. (2006). Parasitic wasp assemblages associated with native and weedy plant species in an agricultural landscape. *Australian Journal of Entomology*, Vol.45, No.2, pp. 176-184, ISSN 1326-6756.

Swedenborg, P.D. & Jones, R.L. (1992). Multicomponent sex pheromone in *Macrocentrus grandii* Goidanich (Hymenoptera: Braconidae). *Journal of Chemical Ecology*, Vol.18, pp. 1901-1912, ISSN 0098-0331.

Takabayashi, J.; Dicke, M. & Posthumus, M.A. (1991). Variation in composition of predator attracting allellochemicals emitted by herbivore-infested plants: relative influence of plant and herbivore. *Chemoecology*, Vol.2, pp. 1-6, ISSN 0937-7409.

Takabayashi, J.; Dicke, M.; Takahashi, S.; Posthumus, M.A. & van Beek, T.A. (1994). Leaf age affects composition of herbivore-induced synomones and attraction of predatory mites. *Journal of Chemical Ecology*, Vol.20, pp. 373-386, ISSN 0098-0331.

Takabayashi, J.; Takahashi, S.; Dicke, M. & Posthumus, M.A. (1995). Developmental stage of herbivore *Pseudaletia separata* affects production of herbivore-induced synomone by corn plants. *Journal of Chemical Ecology*, Vol.21, pp. 273-287, ISSN 0098-0331.

Thaler, J.S. (1999). Jasmonate-inducible plant defences cause increased parasitism of herbivores. *Nature*, Vol.399, pp. 686-688, ISSN 0028-0836.

Thaler, J.S.; Farag, M.A.; Paré, P.W. & Dicke, M. (2002). Jasmonate-deficient plants have reduced direct and indirect defences against herbivores. *Ecology Letters*, Vol.5, pp. 764–74, ISSN 1461-023X.

Thaler, J.S. (2002). Effect of jasmonate-induced plant responses on the natural enemies of herbivores. *Journal of Animal Ecology*, Vol.71, pp. 141–150, ISSN 1365-2656.

Theunissen, J. (1994). Intercropping in field vegetable crops: Pest management by agrosystem diversification—an overview. *Pesticide Science*, Vol.42, pp. 65-68, ISSN 0031-613X.

Thies, C.; Steffan-Dewenter, I. & Tscharntke, T. (2003). Effects of landscape context on herbivory and parasitism at different spatial scales. *Oikos*, Vol.101, pp. 18–25, ISSN 0030-1299.

Titayavan, M.; Altieri, M.A.; (1990). Synomone-mediated interactions between the parasitoid *Diaeretiella rapae* and *Brevicoryne brassicae* under field conditions. *Entomophaga*, Vol.35, pp. 499–507, ISSN 0013-8959.

Ton, J.; D'Allesandro, M.; Jourdie, V.; Jakab, G.; Karlen, D.; Held, M.; Mauch-Mani, B. & Turlings, T.C.J. (2007). Priming by airborne signals boosts direct and indirect resistance in maize. *Plant Journal*, Vol.49, pp. 16–26, ISSN 0960-7412.

Tóth, M.; Bozsik, A.; Szentkirályi, F.; Letardi, A.; Tabilio, M.R.; Verdinelli, M.; Zandigiacomo, P.; Jekisa, J. & Szarukán, I. (2006). Phenylacetaldehyde: a chemical attractant for common green lacewings (*Chrysoperla carnea* s.l.; Neuroptera: Chrysopidae). *European Journal of Entomology*, Vol.103, pp. 267–271, ISSN 1210-5759.

Tóth, M.; Szentkiralyi, F.; Vuts, J.; Letardi, A.; Tabilio, M.R.; Jaastad, G.; Knudsen, G.K. (2009). Optimization of a phenylacetaldehyde-based attractant for common green lacewings (*Chrysoperla carnea s.l.*). *Journal of Chemical Ecology*, Vol.35, pp. 449–458, ISSN 0098-0331.

Tscharntke, T.; Bommarco, R.; Clough, Y.; Crist, T.O.; Kleijn, D.; Rand, T.A.; Tylianakis, J.M.; van Nouhuys, S. & Vidal, S. (2007). Conservation biological control and enemy diversity on a landscape scale. *Biological Control*, Vol.43, pp. 294-309, ISSN 1049-9644.

Tscharntke, T.; Klein, A.M.; Kruess, A.; Steffan-Dewenter, I. & Thies, C. (2005). Landscape perspectives on agricultural intensification and biodiversity: ecosystem service management. *Ecology Letters*, Vol.8, pp. 857–874, ISSN 1461-023X.

Tuell, J.K.; Fiedler, A.K.; Landis, D.A. & Isaacs, R. (2008). Visitation by wild and managed bees (Hymenoptera: Apoidea) to Eastern U.S. native plants for use in conservation programs. *Environmental Entomology*, Vol.37, pp. 707-718, ISSN 0046-225X.

Turlings, C.J. & Tumlinson, J.H. (1992). Systemic chemical signalling by herbivore-injured corn. *Proceedings of the National Academy of Sciences of the USA*, Vol.89, pp. 8399-8402, ISSN 1091-6490

Turlings, T.C.J.; Alborn, H.T.; McCall, P.J. & Tumlinson, J.H. (1993). An elicitor in caterpillar oral secretions that induces corn seedlings to emit volatiles attractive to parasitic wasps. *Journal of Chemical Ecology*, Vol.19, pp. 411-425, ISSN 0098-0331.

Turlings, T.C.J.; Gouinguené, S.; Degen, T. & Fritzsche-Hoballah, M.E. (2002). The chemical ecology of plant-caterpillar-parasitoid interactions, In: *Multitrophic Level Interactions*, eds. T. Tscharntke & B. Hawkins, pp. 148-173, Cambridge University Press, ISBN 9780511060885, Cambridge.

Turlings, T.C.J.; Loughrin, J.H.; Röse, U.; McCall, P.J.; Lewis, W.J. & Tumlinson, J.H. (1995). How caterpillar-damaged plants protect themselves by attracting parasitic wasps.

Proceedings of the National Academy of Sciences of the USA, Vol.92, pp. 4169-4174, ISSN 1091-6490.

Turlings, T.C.J.; Tumlinson, J.H. & Lewis, W.J. (1990). Exploitation of herbivore-induced plant odors by host-seeking parasitic wasps. *Science*, Vol.250, pp. 1251-1253, ISSN 0036-8075.

Turlings, T.C.J.; Tumlinson, J.H.; Eller, F.J. & Lewis, W.J. (1991). Larval-damaged plants: source of volatile synomones that guide the parasitoid *Cotesia marginiventris* to the micro-habitat of its hosts. *Entomologia Experimentalis et Applicata*, Vol.58, pp. 75-82, ISSN 0013-8703.

Turlings, T.C.J. & Wäckers, F.L. (2004). Recruitment of predators and parasitoids by herbivore-damaged plants, In: *Advances in Insect Chemical Ecology*, eds. R.T. Cardé &J. Millar, pp. 21-75, Cambridge University Press, ISBN 9780521792752, Cambridge,UK.

Turlings, T.C.J. & Ton, J. (2006). Exploiting scents of distress: the prospect of manipulating herbivore-induced plant odours to enhance the control of agricultural pests. *Current Opinion in Plant Biology*, Vol.9, pp. 421–427, ISSN 1369-5266.

Turner, M.G.; Gardner, R.H. & O'Neill, R.V. (2001). *Landscape ecology in theory and practice: pattern and process*. Springer-Verlag, ISBN 9780387951225, New York.

van den Boom, C.; Van Beek, T.; Posthumus, M.A.; De Groot, A.E. & Dicke, M. (2004). Qualitative and quantitative variation among volatile profiles induced by *Tetranychus urticae* feeding on plants from various families. *Journal of Chemical Ecology*, Vol.30, pp. 69-89, ISSN 0098-0331.

Van Driesche, R.G.; Lyon, S.; Sanderson, J.P.; Bennett, K.C.; Stanek, E.J. III. & Zhang, R. (2008). Greenhouse trials of Aphidius colemani (hymenoptera: braconidae) banker plants for control of aphids (hemiptera: aphididae) in greenhouse spring floral crops. *Florida Entomologist*, Vol.91, pp. 583-591, ISSN 0015-4040.

Van Poecke, R.M.P. & Dicke, M. (2003). Signal transduction downstream of salicylic and jasmonic acid in herbivory-induced parasitoid attraction by Arabidopsis is independent of JAR1 and NPR1. *Plant, Cell & Environment*, Vol.26, pp. 1541–1548, ISSN 0140-7791.

Van Poecke, R.M.P. & Dicke, M. (2002). Induced parasitoid attraction by *Arabidopsis thaliana*: involvement of the octadecanoid and the salicylic acid pathway. *Journal of Experimental Botany*, Vol.53, pp. 1793–1799, ISSN 0022-0957.

van Rijn, P.C.J. & Sabelis, M.W. (2005). The impact of plant-provided food on herbivore-carnivore dynamics, In: *Plant-Provided Food for Carnivorous Insects: A Protective Mutualism and its Applications*, eds. F.L. Wäckers, P.C.J. van Rijn & J. Bruin, pp. 223-266, Cambridge University Press, ISBN 9780521819411, Cambridge.

Vandermeer, J.H. (1989). *The ecology of intercropping*. Cambridge University Press, ISBN 9780521346894, Cambridge.

Verkerk, R.H.J. (2004). Manipulation of tritrophic interactions for IPM, In: *Intregrated Pest Management: Potential, Constraints and Challenges*, eds. O. Koul, G.S. Dhaliwal & G.W. Cuperus, pp. 55-71, CABI, ISBN 0851996868, Oxfordshire, UK.

Vet, L.E.M.; Lewis, W.J. & Cardé, R.T. (1995). Parasitoid foraging and learning, In: *Chemical ecology of insects 2*, eds. R.T. Cardé & W.J. Bell, pp. 65-101, Chapman and Hall, ISBN 9780412039515, London.

Vet, L.E.M.; Wäckers, F.L. & Dicke, M. (1991). How to hunt for hiding hosts: the reliability-detectability problem in foraging parasitoids. *Netherlands Journal of Zoology*, Vol.41, pp. 202-213, ISSN 0028-2960.

Vet, L.E.M. & Dicke, M. (1992). Ecology of infochemical use by natural enemies in a tritrophic context. *Annual Review of Entomology*, Vol.37, pp. 141-172, ISSN 0066-4170.

Vinson, S.B. (1976). Host selection by insect parasitoids. *Annual Review of Entomology*, Vol.21, pp. 109-133, ISSN 0066-4170.

von Mérey, G.; Veyrat, N.; Mahuku, G.; Valdez, R.L.; Turlings, T.C.J. & D'alessandro, M. (2011). Dispensing synthetic green leaf volatiles in maize fields increases the release of sesquiterpenes by the plants, but has little effect on the attraction of pest and beneficial insects. *Phytochemistry*, Vol.72, pp. 1838-1847, ISSN 0031-9422.

Wäckers, F.L. (2005). Suitability of (extra-) floral nectar, pollen, and honeydew as insect food sources. In: *Plant-Provided Food for Carnivorous Insects: A Protective Mutualism and its Applications*, eds. F.L. Wäckers, P.C.J. van Rijn & J. Bruin, pp. 17-55, Cambridge University Press, ISBN 9780521819411, Cambridge.

Wäckers, F.L.; Rijn, P.C.J. & Bruin, J. (2005). *Plant-provided food for carnivorous insects: a protective mutualism and its applications*, Cambridge Univsersity Press, ISBN 9780521819411, Cambridge.

Wäckers, F.L.; van Rijn, P.C.J. & Heimpel, G.E. (2008). Honeydew as a food source for natural enemies: Making the best of a bad meal? *Biological Control*, Vol.45, pp. 176-184, ISSN 1049-9644.

Walling, L.L. (2000). The myriad plant responses to herbivores. *Journal of Plant Growth Regulation*, Vol.19, pp. 195-216, ISSN 0721-7595.

Walton, N. & Isaacs, R. (2011). Evaluation of flowering plant strips for support of beneficial insects in blueberry. *Environmental Entomology*, Vol.40, pp. 697-705, ISSN 0046-225X.

Wanner, H.; Gu, H. & Dorn, S. (2006a). Nutritional value of floral nectar sources for flight in the parasitoid wasp, Cotesia glomerata. *Physiological Entomology*, Vol.31, pp. 127-133, ISSN 0307-6962.

Wanner, H.; Gu, H.; Gunther, D.; Hein, S. & Dorn, S. (2006b). Tracing spatial distribution of parasitism in fields with flowering plant strips using stable isotope marking. *Biological Control*, Vol.39, pp. 240-247, ISSN 1049-9644.

Williams, L. III; Rodriguez-Saona, C.; Castle, S.C. & Zhu, S. (2008). EAG-active herbivore-induced plant volatiles modify behavioral responses and host attack by an egg parasitoid. *Journal of Chemical Ecology*, Vol.34, pp. 1190-1201, ISSN 0098-0331.

Yu, H.; Zhang, Y.; Wu, K.; Gao, X.W. & Guo, Y.Y. (2008). Field-testing of synthetic herbivore-induced plant volatiles as attractants for beneficial insects. *Environmental Entomology*, Vol.37, pp. 1410-1415, ISSN 0046-225X.

Yu, H.; Zhang, Y.; Wyckhuys, K.A.G.; Wu, K.; Gao, X. & Guo, Y. (2010). Electrophysiological and behavioral responses of *Microplitis mediator* (Hymenoptera: Braconidae) to caterpillar-induced volatiles from cotton. *Environmental Entomology*, Vol.39, pp. 600–609, ISSN 0046-225X.

Zhu, J. & Park, K.C. (2005). Methyl salicylate, a soybean aphid-induced plant volatile attractive to the predator *Coccinella septempunctata*. *Journal of Chemical Ecology*, Vol.31, pp. 1733-1746, ISSN 0098-0331.

Zurbrügg, C. & Frank, T. (2006). Factors influencing bug diversity (Insecta: Heteroptera) in semi-natural habitats. *Biodiversity and Conservation*, Vol.15, pp. 275-294, ISSN 0960-3115.

6

Grafts of Crops on Wild Relatives as Base of an Integrated Pest Management: The Tomato *Solanum lycopersicum* as Example

Hipolito Cortez-Madrigal
Centro Interdisciplinario de Investigación para el Desarrollo Integral Regional-Instituto Politécnico Nacional, Jiquilpan, Michoacán México

1. Introduction

After the potato, the most cultivated vegetable in the world is the tomato *Solanum lycopersicum* L. In 2009, the global area harvested was 4,393,045 ha, with one production of 152 956 115 ton. Mexico ranked 10 th place with 99, 088 ha and a production of 2,591,400 tons (FAO, 2011).

There is consensus that the origin of the tomato is South America, where is the greatest diversity of related species (wild relatives) (Peralta et al., 2005), but is also accepted that domestication of tomato occurred in Mexico (Rick & Holle, 1990; Hoyt, 1992, Perez et al., 1997). Consequently in this country, the tomato, also called "jitomate", is considered one of the basic components of Mexican cuisine. Additionally, the name "tomate" comes from the Nahuatl language of Mexico (Rick & Holle, 1990; Perez et al., 1997). After corn the tomato is the crop that has had greater genetic manipulation (Perez et al., 1997), but focused on the standpoint of productivity. It has been documented that there is an inverse correlation between the degree of domestication (productivity) of plants and damage by pests and diseases (Coley et al., 1985; Rosenthal & Dirzo, 1997); so that resistance to pests and diseases in wild relatives is higher than in native varieties of crops and these in turn show greater tolerance than hybrid modern varieties.

The tomato is one of the crops with the highest number of pests, with approximately 17 phytophagous insects. The whitefly *Bemisia tabaci* Gennadius, 1889 (Hemiptera-Sternorryncha: Aleyrodidae) and the psyllid *Bactericera* (= *Paratrioza*) *cockerelli* (Sulc, 1909) (Hemiptera-Sternorryncha: Psyllidae) are two of the most important pests (King & Saunders, 1984; Liu & Trumble, 2005; Morales et al., 2005). The conventional way of dealing with pest problems is basically through organo-synthetic pesticides, strategy that causes serious problems to the environment and human health. An alternative method is the plant resistance to pests and diseases (Kogan, 1990) and the main source of germplasm for crop improvement are the wild relatives (Hoyt, 1992; Perez et al., 1997). Thus, different species of Solanum that develop in the center of the origin of the tomato have been widely used in crop improvement by hybridization (Simons & Gur, 2005; Casteel et al., 2006; Restrepo et al.,

2008); however, the conventional hybridization between tomato and its wild relatives is not always possible (Perez et al., 1997; Peralta et al., 2005); then, various desirable traits of wild plants cannot be transferred by this technique. In this regard, the grafting technique is an alternative well documented in crop improvement (Lee, 1994; Kubota et al., 2008).

2. Grafts and their use in the pest and diseases management

Grafting is a technique by which two or more plants are joined, forming a single plant; the basal part is called "rootstock" and the superior "scion". This technique has been used since ancient times to transfer desirable characteristics of one plant (rootstock) to another (scion) (Yamakawa, 1982; Lee, 1994; Poincelot, 2004; Kubota et al., 2008). Exist several reasons for using grafts. Many plants are difficult to propagate by other techniques; desirable varieties with poor root development are candidates for grafted on strong rootstock (Poincelot, 2004). Furthermore, the use of grafts may also induce tolerance to adverse environmental factors such as salinity (Martinez-Ballesta et al., 2008), drought (Pire et al., 2007) and adverse temperatures (Venema et al., 2008), among others. The grafts also tend to produce stronger plants and yielding (Khah et al., 2006). In addition, grafted plants induce better quality of fruits (Martinez-Ballesta et al., 2008; Godoy et al., 2009). However, one of the principal uses of grafts is to induce resistance or tolerance to pests and diseases, such as nematodes and soil fungi (Lee, 1994; Kubota et al., 2008).

The first documented case of resistance of grafts to insects was the control of grape Phylloxera *Dactulosphaira vitifoliae* (Fitch) in the United States. The susceptible European grapes scion grafted onto resistant American wild grapes, provided the total control of the pest (Kogan, 1990). Since then, the resistance to pest continues (Granett et al., 1987), demonstrating the sustainability of that pest management strategy.

2.1 Grafts in herbaceous plants

Grafting in herbaceous plants has been known since the nineteenth century. Japan and Korea were the first countries to develop grafting vegetables. In Europe grafting is commonly practiced (Yamakawa, 1982; Lee, 1994; Kubota et al., 2008). However, in the Americas its use in plant breeding has only recently received attention (Red & Riveros, 2001; Kubota et al., 2008; Godoy et al., 2009; Garcia-Rodriguez et al., 2010). Perhaps, one reason is because the American agricultural areas are of greater extent than those of Japan, Korea and Europe; for example, the United States is the country with one of the lowest production of grafts (Kubota et al. 2008). The aim of grafts is to induce resistance to biotic and abiotic factors, including pests and diseases, but also to improve the quantity and quality of fruits (Lee, 1994; Cañizares & Goto, 1998; Dorais et al. 2008; Kubota, 2008). Protected vegetable production without crop rotation as control measure, has led the increase of pests and diseases that are a real problem for this type of agriculture. The main alternative for nematode and disease control was the use of fumigants such as methyl bromide, but with the recent ban on its use in the Montreal protocol, the graft in vegetables is seen as a major strategy in the pest and diseases management; and in general, to transfer valuable traits to the crops (Lee, 1994; Gonzalez et al., 2008; Kubota et al., 2008; Martinez-Ballesta et al., 2008).

In recent years, the grafting has aroused as a technique of great interest in vegetable crops such as cucumber, melon, watermelon, peppers, eggplant and tomato. The grafting has been used to induce resistance to fungal diseases (Alconera et al., 1988; Bletsas et al., 2003; Garcia-

Rodriguez et al., 2010) and bacterial (Nakahara et al., 2004; Coutinho et al., 2006), and to the nematodes *Meloidogyne javanica* Chitwood, 1949, *M. incognita* Kofoid and White, 1919 and *M. arenaria* Roberts and Thomason, 1989 (Heteroderidae) (Williamson, 1998; Sigüenza et al., 2005; Verdejo-Lucas & Sorribas, 2008).

Grafts have been performed on rootstock of local varieties with low productivity but high resistance to pests and diseases. Different species of Cucurbita have been used as a rootstock for melon and watermelon grafts (Yamakawa, 1982; Cohen et al., 2005; Sigüenza et al., 2005; Kubota et al., 2008) and *Capsicum* landraces for chili (Garcia -Rodriguez et al., 2010). In other cases, rootstock have been obtained from resistant hybrids, such as watermelon rootstock from hybrids of *Cucumis maxima* x *Cucumis moschata* (Lee, 1994), or hybrids of *Lycopersicon hirsutum* x *L. esculentum* for rootstock in tomato (Yamakawa, 1982). However, the main source of resistant rootstocks are wild plants, mainly so-called "crop relatives" (Yamakawa, 1982; Alconera et al., 1988; Gonzalez et al., 2008; Kubota et al., 2008; Venema et al., 2008).

3. Importance of crop wild relatives

During its evolution, the wild relatives of crops have developed many features that have enabled them to survive in extreme conditions; for example, on the shores of the Galapagos Islands there is a wild relative of tomato that has provided genes to the cultivated tomato conferring high tolerance salinity, so the plants can be irrigated with one-third seawater (Hoyt, 1992). Also, the main source of resistance it is found in wild plants, and close relatives of crops have been the most exploited in plant breeding (Hoyt, 1992; Ramanatha Rao & Hodgkin, 2002).

No wonder that the main source for grafts has been the rootstock of wild plants, which besides other characteristics have become resistant to pests and soil diseases, such as fungi and nematodes (Yamakawa, 1982; Alconera et al., 1988; Gonzalez et al., 2008; Kubota et al., 2008; Venema et al., 2008). Thus, has been common to graft watermelon on *Lagenaria siceraria* (Yamakawa, 1982; Lee, 1994; Yetis & Sari, 2003); the eggplant, on their wild relatives *Solanum integrifolium* and *Solanum turvum* (Yamakawa, 1982; Lee, 1994; Bletsas et al., 2003); cucumber, on *Cucurbita ficifolia, Sicyos angulatus* (Lee, 1994) and *Cucumis metuliferus* (Sigüenza et al., 2005); melons, on *Cucurbita* spp., *C. moschata*; tomato on *L. pimpinellifolium* and *L. hirsutum* (Lee, 1994); there are reports of tomato grafts onto the weed *Datura stramonium* L. that were practiced for many years in the Southeastern of The United States (Kubota et al., 2008).

When wild plants are used, besides to be resistant to pests and diseases or have some other desirable characteristic, it is advisable to know the effect of the rootstock on the fruit quality. For example, it has been documented that some rootstocks may influence the nutritional characteristics of fruits (Martinez-Ballesta et al., 2008) and even get translocation of toxic compounds into the scion, as happened with the first tomato grafts in wild solanum *D. stramonium* (Kubota et al., 2008). It has recently been documented that the effect of the rootstock towards the graft can even up the genetic level (Zhang et al., 2008).

3.1 Tomato wild relatives

As a native American plant, tomato has a wide diversity of wild relatives in that continent, among those mentioned: *S. cheesmaniae* (L. Riley) Fosberg, *S. pimpinellifolium* L., *S.*

chmielewskii (CM Rick, Kesicki, Fobes & M. Holle) D. M. Spooner, G. J. Anderson & R. K. Jansen, *S. neorickii* (CM Rick, Kesicki, Fobes & M. Holle) D. M. Sponner. G. J. Anderson & R. K. Jansen (= *L. parviflorum*), *S. habrochaites* S. Knapp & D. M. Spooner (= *L. hirsutum*), *S. chilense* (Dunal) Reiche, *S. peruvianum* L., *S. penelli* Correll and *S. lycopersicum* var. *cerasiforme* L. (Esquinas & Nuez, 1995; Peralta et al., 2005).

Such is the importance of wild relatives of tomato that modern varieties would not exist without the wild relatives; characteristics such as resistance to cold or extreme conditions and resistance to pests and diseases have been transferred from wild relatives to cultivated plants (Hoyt, 1992, Perez et al., 1997). Then, the knowledge and conservation of crop wild relatives is of utmost importance in global food production (Hoyt, 1992; Eigenbrode & Trumble, 1993; Perez et al., 1997).

Unfortunately, "modern" agricultural practices as the use of herbicides and other chemicals have led to a gradual loss of biological diversity and populations of wild relatives of crops (such tomatoes) have been drastically depleted (Hoyt, 1992; Vargas, 2008; Alvarez-Hernandez, 2009a).

It is accepted that the closest ancestor of cultivated tomato is *S. lycopersicum* var. *cerasiforme* D. M. Spooner, G. J. Anderson and R. K. Jansen, 1993 (Esquinas & Nuez, 1995; Peralta et al., 2005), grows in a wide variety of habitat from 0 to 3 300 meters above sea level (Sanchez-Peña et al ., 2006; Vargas, 2008; Alvarez-Hernandez et al., 2009a;), characterized by having round fruits with diameters ranging from 1 to 2.5 cm (Martinez, 1979; Rick et al., 1990). In some states of the Center-Western Mexico, the wild tomato is known as "tinguaraque" (Martinez, 1979). So in this paper frequently we use that name. Since 2005 we have developed studies about the tolerance of tinguaraque to phytophagous insects and its potential as rootstock in grafts with cultivated tomato. The research questions included:

- Which is the incidence of phytophagous insects on tinguaraque?
- Which is the preference of *Bactericera cockerelli* for tomato, tinguaraque and grafts from both?
- How is the incidence of insects' pest on tomato, tinguaraque and grafts from both under field conditions?
- Which characteristics present tomato fruits grafted on tinguaraque?
- Which is the response of tomato grafts on tinguaraque at different nutrimental handling systems?

4. The tinguaraque (*Solanum lycopersicum* var. *cerasiforme*) in Mexico

4.1 Importance and distribution

In Mexico, the tinguaraque is widely distributed in ecological reserves and associated crop fields where it eventually tends to become a weed (Perez et al., 1997; Sanchez-Peña et al., 2006). It features a high capacity for climate adaptation, it was found from 7-2 000 meters above sea level, with annual rainfall of 495-1 591 mm, annual mean minimum temperature from 7.1-21.6 °C, 22.6-38.4 °C mean annual maximum temperature, and between 15.8 and 28.1 °C mean annual temperature (Vargas, 2008). Sanchez-Peña et al. (2006) reported populations of wild tomato at altitudes from 12 to 1 104 masl on the Northest of Mexico.

In warm regions (<300 masl) populations of wild tomato are reduced and are associated with species that provide shade; in temperate regions these plants protect them from the cold (Vargas, 2008). Because of its creeping growth habit-climbing, it is common to found the wild tomato associated with different plants; for example, many plants were climbing among the thorny branches of the "acacia" (*Acacia* spp.) scattered among grass and weeds. The dispersion of its branches is a survival strategy to pests and herbivores (Alvarez-Hernandez et al., 2009a).

Partial collections in the Mexican state of Michoacan, showed that its distribution includes altitudes from 314 to 1 550 masl, maximum annual temperatures ranged from 26.9 to 35.2 at minimum of 11.7 °C to 26.9 °C; annual precipitation of 751 mm to 1 866 mm and with varying levels of soil fertility; similarly pH values ranged from 6.8 to 8.5 (Alvarez-Hernandez et al., 2009a, Table 1). The pH values obtained exceeding the normal limits for the development of cultivated plants whose optimal value is between 6.0 and 7.5 (Michel et al., 1998); by contrast, the cultivated tomato is considered tolerant to the acidity values of 5.5- 7.5 and higher values are limiting (Valadez, 1998).

This has allowed that tinguaraque have populations with different characteristics in response to biotic and abiotic factors of mortality according to the conditions where it develops. However, it also indicated that urban growth and agricultural production techniques, as use of herbicides, are the main factors influencing the loss of tinguaraque diversity; there are even regions where it is known there were populations of tinguaraque; however, nowadays farmers do not know about its existence (Vargas, 2008; Alvarez-Hernandez et al., 2009a).

Physicochemical variables	Sampling sites			
	Apatzingán	Acahuato	Los Reyes	Jiquilpan
pH	8.3	6.8	8.5	7.6
Sand (%)	19.7	26.0	24.0	15.9
Silt (%)	40.3	35.0	29.0	34.3
Clay (%)	40.0	39.0	47.0	49.8
Organic matter(%)	3.0	4.9	2.5	7.7
Total nitrogen (%)	0.12	0.2	0.10	0.3
Phosphorus mg/kg	17.1	17.4	16.6	15.7
Potassium meq/100 g	3.3	3.3	0.4	1.1

Table 1. Physical and chemical characteristics of soils obtained from sites with wild tomato populations in three regions of Michoacán, Mexico (Alvarez-Hernandez et al., 2009a).

In Michoacán state, populations of wild tomato were found restricted to habitat where agricultural impacts are minor, such as roadsides, areas with thorny plants, waterways, river banks, among others (Alvarez-Hernandez et al., 2009a).

4.2 Morphological and physiological characteristics of tinguaraque

Based on the fruit size, Alvarez-Hernandez et al. (2009a) identified two groups of tinguaraque in Center-western Mexico: Small-fruited (1.05 to 1.22 cm of polar diameter and 1.10 to 1.25 cm of equatorial diameter) and large-fruited (2.12 to 2.23 cm of polar diameter and 2.41 to 2.55 of equatorial diameter); the cultivated tomato fruit has an average of 10 cm (Valadez, 1998; Muñoz, 2009) and its weight ranges from 5 to 500 g (Chamarro, 1995). The fruit size is closely related to the number of seeds and the number of locules (Muñoz, 2009), variable that seems to be interesting to evaluate. One characteristic of wild tomatoes is to present a smaller number of locules than those grown; commercial cultivars are multilocular type (Valadez, 1998), while the wild have two locules (Rick et al., 1990; Alvarez-Hernandez et al., 2009a).

Another important feature in wild tomato species is the highest density of trichomes compared to cultivated varieties. Sanchez-Peña et al. (2006) compared the density of trichomes on *S. habrochaites* (C-360), *S. lycopersicum* var. *cerasiforme Vs* the commercial variety Rio Grande. They found that the density of trichomes was higher in the first species, followed by *S. lycopersicum* var. *cerasiforme*, and the cultivar had the lowest density of trichomes. In this regard, it is known that trichomes are one of the main factors that induce resistance to pests in tomato (Eigenbrode & Trumble, 1993; Wagner et al., 2004).

Wild plants as tinguaraque generally have a slower germination compared to cultivated varieties. In this regard, Alvarez-Hernandez et al. (2009a) found a tendency for greater speed and uniformity in germination of commercial tomato "Rio grande" compared to the germination of wild populations of tinguaraque; the time when 50% of seeds germinated ranged from 2.8 (2.5-3.0) to 10.6 (8.6-15.7) days in tinguaraque, whereas in the commercial cultivar was 4.4 (4.0-4.8) days. In general, the germination rate in large-fruited tinguaraques was similar to the cultivated tomato, suggesting a direct relationship between speeds of germination and fruit size (Table 2).

The observed differences in germination tinguaraques suggests two things: first, that the different climatic conditions where these populations grow and the time spent as wild plants could be determinants of the germination speed (Alvarez-Hernandez et al., 2009a); for example, tropical species of plants usually germinate faster than temperate species (Meletti & Bruckner, 2001); second, similar germination recorded in tinguaraques large fruited and the cultivar suggest that these tinguaraques perhaps have less time as wild plants, and even yet are handled by humans (Alvarez-Hernandez et al., 2009a). It is currently accepted the hypothesis that the var. *cerasiforme* is a wild tomato escaped from cultivation (Esquinas & Nuez, 1995; Peralta et al., 2005).

Population	**GT50 * (days)**	**Fiducial limits (days)**	**Prob. Chi. Sq.**
Little Apatzingan	8.5	7.4-10.6	0.0001
Big Apatzingan	4.9	4.6-5.2	0.0001
Acahuato	6.4	5.2-9.3	0.0001
Los Reyes	2.7	2.5-3.0	0.0001
Jiquilpan	10.6	8.6-15.6	0.0001
Tabasco (big)	4.9	4.7-5.2	0.0001
Cv. Rió Grande	4.3	4.0-4.7	0.0001

* Germination Time of 50% of seeds.

Table 2. Germination rate of six wild tomato ecotypes collected in Michoacán and Tabasco, Mex. and cv. Rio Grande (Alvarez-Hernandez et al., 2009a).

A practical use of knowledge of the germination rate could be used to improve crops by grafting. Having this base of time and germination percentage, it is possible to standardize the development stages of compatible species, but with different rates of development, as occurs in wild and cultivated tomato, the first slower in its development.

4.3 Phytophagous insects associated with tinguaraque

Few studies have been documented about the entomo-fauna of S. l. var. *cerasciforme*, but it is mentioned that wild tomato can tolerate high incidence of pests and diseases (Hoyt, 1992; Eigenbrode & Trumble, 1993; Nakahara et al., 2004; Sanchez-Peña et al., 2006).

After one year of sampling in three different climatic regions of Michoacan, Mexico (Apatzingan, Los Reyes and Jiquilpan), five groups of insects were recorded: whitefly (Hem: Aleyrodidae), aphids (Hem: Aphididae), leaf miners (Dip: Agromyzidae), psyllids (Hem: Psyllidae), horn and fruit worms (Lepidoptera), and fleahopper (Col: Chrysomelidae) (Alvarez-Hernandez et al., 2009a; Table 3). In general, those groups include some of the main pests of cultivated tomato (King & Saunders, 1984).

The incidence of phytophagous insects observed in tinguaraque was low and consequently damage to plants was also low; for example, only few specimens of hornworm *Manduca* spp were registered. Similarly, about three larval specimens of chrysomelids (Chrysomelinae) were recorded. In the three collection sites, the bug *Cyrtopeltis notata* (Distant) (Hemiptera: Myridae) was the most abundant phytophagous insect recorded on tinguaraque; Due the frequency and damage of this species, it could be considered a potential pest of tinguaraque (Table 3). Moreover, not all pests were equally distributed in the regions; so, the tomato psyllid *B.* (=*Paratrioza*) *cockerelli* was only registered in one región (Jiquilpan). *B. cockerelli* is considered a major pest of the cultivated tomato (Liu & Trumble, 2005). Therefore, it is important to consider populations of tinguaraque with longer coevolution with the pest, could be probably more resistant to it.

Order: Family	**Species**
Hemiptera: Aleyrodidae	*Bemisia tabaci y Trialeurodes vaporariorum*
Hemiptera: Aphididae	Species complex
Hemiptera: Myridae	*Cyrtopeltis notata* Distant
Hemiptera: Psyllidae	*Bactericera cockerelli* Sulc.
Diptera: Agromyzidae	*Lyriomiza sativae* Blanchard y *L. trifoli* Burgess
Lepidoptera: Sphingidae	*Manduca sp.*
Lepidoptera: Noctuidae	*Heliothis sp.*
Coleoptera: Chrysomelidae	*Epitrix sp.*
Coleoptera: Chrysomelidae	*Chrysomelinae*

Table 3. Major groups of phytophagous insects registered in wild populations of tinguaraque collected in Michoacán, Mex. (Alvarez-Hernandez et al., 2009a).

Diversity in that wild populations of *S. lycopersicum* develops, marks its importance as a resource adaptable to different climatic conditions prevailing in Mexico (Vargas, 2008). The wide variability of wild ecotypes of *S. lycopersicum* var. *cerasiforme* (Dunal), presumably with resistance to certain pests and diseases is an aspect useful for crop improvement. Previous reports have pointed out resistance of wild tomato to various tomato pests, including: *Liriomyza* sp., armyworm *Spodoptera exigua* (Hiibner), bugs complex (Hemiptera) (Eigenbrode & Trumble, 1993) and whitefly *B. tabaci* (Sanchez-Peña et al., 2006); resistance to early blight *Rhyzoctonia solani*, late blight *Phythophthora infestans* (Pérez et al., 1997) and potato rot *Ralstonia* (= *Pseudomonas*) *solanacearum* (Nakaho et al., 2004) has been documented. However, genetic improvement through hybridization is usually slow, expensive and eventually there are barriers to conventional hybridization (Perez et al., 1997; Poincelot, 2004). Grafts on wild relatives or plants resistant to pests and diseases have proven to be an important tool for crop improvement (Poincelot, 2004; Kubota et al., 2008). Therefore, it was interesting to know the response of tinguaraque and its grafts with cultivated tomato to the incidence of the insect pests.

5. Incidence of pests in grafts of tomato with tinguaraque

5.1 The tomato psyllid *Bactericera cockerelli*

Few are the documented studies about grafting in vegetables with native species in Mexico (Garcia-Rodriguez et al., 2010), therefore the wealth of germplasm has been wasted, and in some cases at risk of disappearing. Therefore, the study was aimed to evaluate the resistance of grafting of tomato in its wild relative *S. lycopersicum* var. *cerasiforme* of the region of Jiquilpan, with emphasis on the tomato psyllid *B.* (= *Paratrioza*) *cockerelli* (Hem: Psyllidae). This insect is one of the major pest of tomato, with losses of up to 85%. Although often ineffective, its control is based on the chemical method; however, other control strategies have been suggested, including plant resistance (Liu & Trumble, 2005; Casteel et al., 2006).

In field conditions we evaluated the incidence of phytophagous insects on *S. lycopersicum* var. *cerasiforme*, ecotype Jiquilpan. Results showed low incidence of insect pests on tinguaraque and particularly *B. cockerelli* was one of the species with lower incidence. In order to confirm this observation, we established an experiment including tomato, tinguaraque, and graft of both. In laboratory conditions, plants were confronted with a known number of adults of *B. cockerelli* and its preference for each plant was registered. The incidence of pests was also considered in field conditions.

Consistently, the insect preferred tomato, graft and tinguaraque in that order. When treatments were exposed individually, the highest incidence occurred in tomato psyllid (16.0 ± 10.1) and lowest in tinguaraque (7.5 ± 3.0) and graft (8.3 ± 6.8), in that order. When the three treatments were presented simultaneously, the preference of adult psyllids was 22.8 times higher in tomato than tinguaraque, and three times higher than for grafts (Table 4). This was confirmed in field trials where the largest number of adults, nymphs and oviposition was recorded in the cultivated tomato, and the lower number in tinguaraque. The graft showed intermediate number, but without differences with the tinguaraque (Table 5).

Treatment	Incidence (%)	
	Individual bioassay	Multiple bioassay
	Mean (%) ± DS[1]	Media (%) ± DS[1]
Tomato	16.00 ± 10.1 a	15.03 ± 9.26 a
Graft	8.33 ± 6.89 b	4.99 ± 2.57 b
Tinguaraque	7.50 ± 3.03 b	0.66 ± 0.71 c
N	6	8

[1] Means ± standard deviation, with the same letter into column, are not statistically different (Tukey, 0,05).

Table 4. Incidence of *Bactericera cockerelli* (adults) on tomato, tinguaraque and graft of both when they were exposed in individual and multiples bioassays (Cortez-Madrigal, 2010).

5.2 Incidence of other pests

The main groups of phytophagous insects recorded were: aphid species complex (Hemiptera: Aphididae), *Bemisia tabaci* and *Trialeurodes vaporariorum* (Hemiptera: Aleyrodidae); complex bugs (Hemiptera), highlighting the species *C. notata* (Myridae) and the leaf miners *Liriomyza* spp. (Diptera: Agromyzidae). Although was observed a trend towards a higher incidence of insects in cultivated tomato, statistically differences only were registered for miners and aphids, where the highest and lowest incidence was for tomato (3.9±3.18) and tinguaraque (0.68±0.79). The graft showed an intermediate incidence (2.18±2.16). The highest and lowest incidence of aphids was in tomato and tinguaraque in that order (0.758 ± 0.98 y 0.237 ± 0.36). The graft showed an intermediate relation respect to tomato and tinguaraque, but there were no statistical differences between them (Table 6).

Treatment	Adults	Eggs	Nymphs	N
Tomato	1.04 ± 1.01 a	0.46 ± 0.43 a	1.13 ± 0.97 a	12
Graft	0.27 ± 0.46 b	0.12 ± 0.09 b	0.27 ± 0.24 b	12
Tinguaraque	0.35 ± 0.71 b	0.10 ± 0.10 b	0.21 ± 0.19 b	12

Mean ± standard deviation after log (x+1) transformation followed by the same letter within columns do not differ statistically (Tukey, 0.05). N= number of repetitions.

Table 5. Incidence of *Bactericerca cockerelli* on tomato, tinguaraque and graft of both in field conditions from Jiquilpan, Michoacan, Mexico (Cortez-Madrigal, 2010).

Treatment	Leaf miner	Aphids	N
Tomato	3.9 ± 3.18 a	0.758 ± 0.98 a	12
Graft	2.18 ± 2.16 b	0.316 ± 0.35 ab	12
Tinguaraque	0.68 ± 0.79 c	0.237 ± 0.36 b	12

Means ± standard deviation after log (x+1) transformation followed by the same letter within columns do not differ statistically (Tukey, 0.05).

Table 6. Average incidence per plant of leaf miner and aphids on tomato, tinguaraque and graft of both under field conditions in Jiquilpan, Michoacan, Mexico. Year 2007 Cortez-Madrigal, 2010.

Although there were no statistical differences in the incidence of whitefly, graphically shows the trend of lower incidence in tinguaraque; contrary, tomato, followed by graft showed the highest incidence of the pest. Only in Hemiptera complex the incidence was similar in tomato, tinguaraque and grafting (Fig. 1).

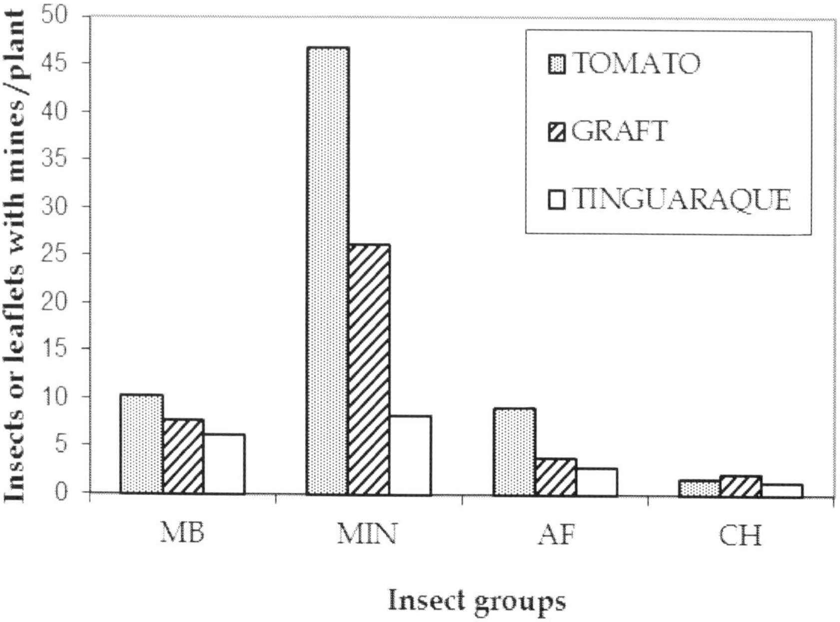

Fig. 1. Incidence of whitefly (MB), leaf miner (MIN), aphids (AF) and bugs (CH) in tomato, grafting and tinguaraque (Cortez-Madrigal, 2010).

5.3 Incidence of pests in tomato grafted with different ecotypes of tinguaraque

Given the wide variability of conditions where tinguaraque grows in Mexico, we considered convenient to evaluate different ecotypes, from temperate regions to warm places. In accordance with the above-mentioned, the study aimed to evaluate the incidence of phytophagous insects in tomato grafting on various ecotypes of *S. lycopersicum* var. *cerasiforme* native from Michoacan, Mexico. The experiment was established in the region of Apatzingan Valley at an altitude of 300 masl. The climate is a Bs1 (h ') w (W) corresponding to the semi-dry warm climate with summer rains (Garcia, 1988). The mean temperature, annual minimum and maximum are 28, 20 and 37.7 °C, respectively. The average rainfall, minimum and maximum is 834, 500 and 972.8 mm, in that order. The type of soil was a vertisol pelico (INEGI, 1983).

Thirteen treatments were established: five wild ecotypes of *S. lycopersicum* var. *cerasiforme* natives from Michoacan (GAp, ChAp, Ac, LR and Jiq) and one from Tabasco (Tab); six grafts of tomato cv. Toro onto tinguaraque (I-GAP ... I-Tab), and the cv. Toro as control (Tom).

From November 17, 2007 to February 16, 2008, weekly samplings were implemented in Ciudad Morelos, Municipality of Paracuaro, Michoacan, Mexico.

The main species of insects registered were: whitefly *B. tabaci*, psyllid *B. cockerelli* and the aphid *Aphis gossypii*. Results showed a wide variability of responses from the tinguaraque ecotypes and its grafts, generally with lower incidence of pests compared to those registered on tomato without grafting. Again, tolerance of tinguaraque and its grafts toward diverse insect pests was registered (*B. cockerelli* and *B. tabaci*).

For Whitefly adults only one tendency to lower incidence on grafting was registered. The lowest incidence was in the graft I-GAp (13.14 ± 7.18), compared with 17.6 ± 10.4 in the ungrafted tomato. The graft with tinguaraque Tabasco (larger fruit) showed an incidence similar to that of the cultivated tomato (Table 7).

The incidence of whitefly nymphs showed significant differences ($p = 0.0001$). Grafts showed an intermediate response, where stood the treatments I-GAp and I-ChAp (which were native tinguaraque), with significant differences respect to the cultivar (Table 7). Regarding *B. cockerelli*, treatments with lower incidence of adults were tinguaraques small fruit and grafts, where I-ChAp and I-GAp were the best. Conversely, the highest incidence of the adults occurred in the tinguaraques large fruit (Tab and GAp), and the commercial variety (Tom). Regarding the incidence of nymphs of *B. cockerelli*, there were no differences between treatments.

For the aphids, the lowest incidence occurred in the graft GAp (1.52 ± 1.22) along with tinguaraques small fruit; the highest incidence occurred in the commercial cultivar (4.82 ± 5.22) without differences with tinguaraques large-fruit. Most of the grafts showed an intermediate response (Table 7).

Treatment	Insect species				
	B. tabaci		*B. cockerelli*		Aphididae
	Adults	Nymphs	Adults	leaflets with eggs	
Tom	17.6±10.4* abc	9.7±8.5 a	2.2±1.9 ab	1.6±1.2 ab	4.8±5.2 a
I-LR	15.6±8.4 abc	7.5±5.9 abcd	1.7±1.7 bcd	1.2±1.3 ab	2.3±2.5 bcd
I-Jiq	15.3±8.5 abc	7.5±5.8 abc	1.6±1.6 abcd	1.3±1.3 ab	2.3±2.3 bcd
I-ChAp	14.5±7.9 bc	6.6±5.6 bcd	1.3±1.1 d	1.0±0.9 ab	2.2±2.1 bcd
I-Ac	15.7±8 abc	7.1±5.6 abcd	1.6±1.6 abcd	1.1±1.0 ab	2.2±2.7 bcd
I-GAp	13.1±7.1 c	5.8±4.6 bcd	1.3±1.1 cd	0.9±0.7 b	1.5±1.2 d
I-Tab	16.0±9.3 abc	7.8±6.8 abc	1.9±2.2 abcd	1.3±1.6 ab	2.8±3.2 bcd

*Means ± standard deviation after log (x+1) transformation followed by the same letter within columns do not differ statistically (Tukey, 0.05).

Table 7. Incidence of phytophagous insects in a cultivated variety of tomato and their grafts with different ecotypes of tinguaraque *S. lycopersicum* var. *cerasiforme* (Alvarez-Hernandez et al., 2009b).

According to a multivariate analysis of the incidence of pests, new groups of plants were formed; in the case of *B. cockerelli* (adults and eggs) five groups were formed: one consisting of the cultivated tomato (Tom) and tinguaraque G-Ap, very close to the group formed by the tinguaraque Tabasco (Tab), corresponding all of large fruit. Another group was formed by grafting and tinguaraques Jiquilpan (Jiq) and Los Reyes (LR). The tinguaraque "chico apatzingan" (ChAp) as a single group. Finally, the graft Tabasco (I-Tab) and tinguaraque Acahuato (Ac) formed another group (Fig. 2).

The commercial variety and tinguaraques large fruit were usually the ones that had the highest incidence of pests. Tinguaraques Small-fruit showed lower incidence and in turn, the grafts showed an intermediate trend. This coincides with what is stated about the incidence of pests and the degree of domestication of plants (Coley et al. 1985; Rosenthal & Dirzo, 1997). Modern varieties of tomatoes have been genetically manipulated more than tinguaraques, and within these, there may be some that are already handled by humans, as in the case of tinguaraque Tabasco, which is marketed in their origin region.

5.4 Development studies

Recent unpublished studies on the incidence of whitefly (*B. tabaci* and *T. vaporariorim*) on grafts of tomato with tinguaraque under different nutritional levels, the results confirm previous studies (Alvarez-Hernandez et al., 2009a, b; Cortez -Madrigal, 2010;) in the sense that grafts are less affected than ungrafted tomato. Additionally, the production of grafted plants was similar to that of ungrafted plants (Table 8).

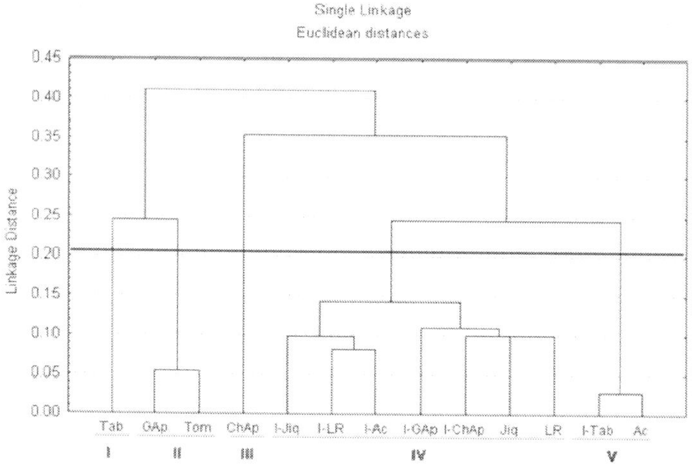

Fig. 2. Dendrogram showing the formation of groups of tomato, tinguaraques and grafts of boths based on the incidence of *B. cockerelli*. Apatzingan, Michoacan, Mexico. 2007. Tom = tomato, Tab = Tinguaraque Tabasco, Jiq = Tinguaraque Jiquilpan, LR = Tinguaraque Los Reyes, Ac = Tinguaraque Acahuato, GAp = Big tinguaraque Apatzingan, ChAp = Small tinguaraque Apatzingan, I = Graft.

Treatment	Means[1]± STD
Ungrafted with compost (U-C)	58.84±38.5 A
Ungrafted with fertilizer (T-F)	28.39±20.6 AB
Grafted withaout fertilizer (G-WF)	16.46±10.8 BC
Fertilized graft (G-F)	7.63±5.3 C
Ungrafted or fertilized (U-WF)	6.76±3.2 C

[1] Mean ± standard deviation after log transformation (x +1) followed by the same letter do not differ statistically (Tukey, 0.05).

Table 8. Incidence of whitefly *B. tabaci* and *T. vaporariorum* on tomato grafted and ungrafted under different nutritional levels. Jiquilpan, Mich. 2010.

6. Tomato fruit quality grafted on tinguaraque

An important aspect to consider is to know the quality of fruit grafting; studies such: size and production, color, acidity, soluble solids and sugars in the fruit should be included. Alvarez-Hernandez (2009) characterized biochemically fruit quality of grafts of tomato on tinguaraque and concluded that fruits of the grafts were not different from the fruit without grafting (Table 9).

Treatment	Variable			
	pH	Soluble solids (ºBrix)	Humidity (%)	Density
ChAp	5.07*±0.05	6.0*±0.0	90.73	1.48
GAp	5.02±0.05	6.0±0.0	91.43	7.06
Ac	5.35±0.1	7.75±0.5	89.94	1.31
LR	4.77± 0.05	7.75±0.5	89.05	1.02
Jiq	4.87± 0.05	7.5±0.57	90.13	0.94
Tab	5.37± 0.05	5.25±0.5	88.39	7.17
I-ChAp	4.67± 0.05	6.25±0.5	97.37	10.67
I-GAp	4.55± 0.05	6.75±0.5	93.99	9.75
I-Ac	4.45± 0.05	6.0±0.0	97.44	7.95
I-LR	4.45± 0.05	6.5±0.57	96.52	9.80
I-Jiq	4.5±0.0	5.5±0.57	96.41	10.67
I-Tab	4.5±0.00	6.75±0.5	97.44	10.50
Tom	4.52±0.09	7.0±0.0	94.28	9.79

*Means ± standar deviation.

Table 9. Physical and chemical characteristics of tomato fruits, tinguaraque and graft of both. Parácuaro, Michoacán, Mexico (Alvarez-Hernandez, 2009). Tinguaraques: ChAp, GAp, Ac, LR, Jiq y Tab; grafts: I-ChAp...I-Tab; commercial variety: Tom.

Previous reports indicate resistance of *S. lycopersicum* var. *cerasiforme* to various pest and diseases of tomato (Eigenbrode & Trumble, 1993; Perez et al., 1997; Nakahara et al., 2004; Sanchez-Peña et al., 2006). The results of our studies agree with those mentioned by Eigenbrode & Trumble (1993) in the sense that wild tomato has resistance to leaf miner *Liryomiza* spp. more does not match the resistance indicated by these authors for the complex of Hemiptera. In our case, resistance of tinguaraque was clearer to *Liryomiza* spp., *B. cockerelli* and aphids (Aphididae), but not for the bugs complex, consisting mostly of the species *C. notata* (Hem: Myridae).

The differences in the incidence of pests found between tinguaraques small fruit and large fruit is probably related to the density of trichomes. In this regard, Sanchez-Peña et al. (2006) found higher densities of trichomes on wild tomatoes than in the cultivated variety, but there were also significant differences between populations tinguaraque. It is known that the main mechanisms of pest resistance in tomato depends on the density and type of trichomes, which have distinguished seven types, including glandular and non-glandular trichomes (Simmons & Gurr, 2005); the first are involved in production of allelochemicals as acilsugars (Mutschler et al., 1996; De Resende et al., 2008), zingiberene (Freitas et al., 2002) and decanonas (Muigai et al., 2002), substances that cause insect repellency or mortality. Similarly, non-glandular trichomes play a role as physical barriers in the establishment and development of some insects (Eigenbrode & Trumble, 1993; Wagner et al., 2004).

Trichomes, mainly glandular, are generally more abundant in wild than in cultivated species (Sanchez-Peña et al., 2006; Simmons et al., 2006), and in some cases there has been a strong correlation between incidence of phytophagous insects and density of trichomes (Simmons et al., 2004; Alba et al., 2009). However, in other cases the production of allelochemicals has not clearly correlated with the density of trichomes, suggesting that independent mechanisms of resistance are involved (Nombela et al., 2000; Muigai et al., 2002), where the pH of the leaf would be a major factor; has been documented, for example that *B. tabaci* prefers cotton sheets with a pH of 6-7.25 (Berlinger, 1983).

The fact that the grafted material have shown lower incidence of pests than the commercial cultivar, suggests that the graft favored tolerance to recorded tomato pests. The incidence of insects was three times lower in grafts than in ungrafted tomato; however, mechanisms involved in this tolerance are unknown. Might think that secondary substances anti-herbivores are synthesized in the wild rootstock and from there translocated into the susceptible scion; however, some grafts with the lower incidence of pests were formed by wild rootstock obtained from tinguaraques in which the highest incidence of insects occurred. Therefore, the tolerance of grafts to insects could be multifactorial, as has been noted by other authors (Muigai et al., 2002).

The resistance of the tomato wild relatives has been used to obtain plants with resistance to pests and disease, mainly through hybridization (Casteel et al., 2006; Restrepo et al., 2008), slower than the development of grafts. Although the use of grafts in vegetables is a common practice in much of Asia and Europe (Lee, 2003; Nakahara et al., 2004; Verdejo-Lucas and Sorribas, 2008), in American countries has been little explored and less commonly used to transfer resistance to pests and diseases (Gonzalez et al., 2008; Garcia-Rodriguez et al., 2010).

Usually, grafts have been directed to pathogen and soil pests resistance (Lee, 1994; Kubota et al., 2008) where is located the rootstock resistant and little has been documented about its effect on the aerial pests. Although some scientist written disclosure mentioned the grafts resistance to aerial pests, do not show experimental evidence that support his claim (Kubota & Viteri, 2007). The results obtained by us show that through grafts were formed new groups of plants with a lower incidence of pests than on commercial variety without grafting; even, some of the best treatments were grafts.

Insects as Paratrioza and whiteflies are major pests of cultivated tomatoes and other vegetables, so these results may be important utility in the production of these crops, initially at the greenhouse and gardens level. However, other pest as the hornworm *Manduca* spp., bollworm *Heliothis* spp. and pinworm *Keiferia lycopersicella* (Walsingham) must be included in future studies.

7. Conclusions

The grafting of cultivated tomato on the wild tomato *S. lycopersicum* var. *cerasiforme* has potential in the management of foliar pests such as *B. cockerelli*, *Liriomyza* spp. complex of aphids (Aphididae) and apparently to *B. tabaci*. The grafting technique developed by us is simple and inexpensive, so it can be implemented by any producer. Its use is primarily focused on low-income farmers who grow tomatoes in small areas, although it is feasible to use in greenhouse crops with greater use of inputs.

Although by mean of graft was not reduced completely insect damage, it is important to consider that his action was on several species, some considered key pests of tomato. We understand the use of grafting as a tool of integrated pest management. Under this view, other control strategies should be evaluated, where ecological methods should be prioritized. For example, micoinsecticides, yellow traps and even low-toxicity insecticides, among others. For countries considered origin center of crops, such as Mexico, to conserve and use wild relatives of crops as source of resistance to pests and diseases should be a priority. In Mexico grow many wild relatives of crops, including *S. lycopersicum* var. *cerasiforme*. Growing adjacent to agricultural fields and modern farming techniques, such as herbicide application, threaten its permanence. The development of grafts in wild relatives can give them more value and contribute to the conservation of these species.

The fruits of tomato grafted on tinguaraque were not modified, at least in their basic biochemical characteristics. Since the tinguaraque is edible, it is feasible to think is not necessary to develop toxicological studies of grafted fruit. However, the organoleptic quality whether it should be investigated. Some compounds of interest could be found in greater concentration in tinguaraque and be transferred by grafting to tomato. This would be a plus to the fruits of the grafts.

8. Aknowledgments

To Comisión de Fomento de Actividades Académicas (COFAA) and Secretaria de Investigación y Postgrado (SIP) of the Instituto Politécnico Nacional (IPN), Mexico for financial support granted. The Consejo de Ciencia y Tecnología del Estado de Michoacán (COECYT), Mexico for the financial support granted. To Luz Marcela Zacarias-Nuñez for their support in reviewing the English version.

9. References

Alba, J. M.; Montserrat, M. & Fernández-Muñoz, R. (2009). Resistance to the two-spotted spider mite (*Tetranychus urticae*) by acylsucroses of wild tomato (*Solanum pimpinellifolium*) trichomes studied in a recombinant imbred line population. *Experimental and Applied Acarology*, 47, 35-47. ISSN: 1572-9702

Alconero, R., Robinson, W., Dicklow, B. & Shail, J. (1988). Verticillium wilt resistance in eggplant, related solanum species, and interspecific hybrids. *HortScience,* 23, 388-390. ISSN: 0018-5345

Alvarez-Hernandez, J.C. (2009). *Injerto de Jitomate (Solanum lycopersicum* L.) *en germoplasma silvestre como fuente de resistencia a plagas y enfermedades.* Tesis de maestría en ciencias, Instituto Politecnico Nacional-CIIDIR, Jiquilpan, Mich., Mex. .

Alvarez-Hernandez, J.C., Cortez-Madrigal, H. & Garcia-Ruiz, I. (2009a). Exploracion y caracterización de poblaciones silvestres de jitomate (Solanaceae) en tres regiones de Michoacan, Mexico. *Polibotanica,* 28, 139-159. ISSN: 1405-2768

Alvarez-Hernandez, J. C., Cortez-Madrigal, H., Garcia-Ruiz, I. & Ceja-Torres, L.F., Perez-Dominguez, J. F. (2009b). Incidencia de plagas en injertos de jitomate (*Solanum lycopersicum*) sobre parientes silvestres. *Revista Colombiana de Entomología,* 35, 2, 150-155. ISSN: 0120-0488

Berlinger, M. J. (1983). The importance of pH in food selection by the tobacco whitefly, *Bemisia tabaci. Phytoparasitica,* 11, 151-160. ISSN: 1876-7184

Bletsos, F., Thanassoulopoulos, C. & Rouparias, D. (2003). Effect of grafting on growth, yield, and Verticillium wilt of eggplant. *HortScience,* 38, 2, 183-186. ISSN: 0018-5345

Cañizares, K. A. L. & Goto, R. (1998). Crescimento e produção de híbridos de pepino em função da enxertia. *Horticultura Brasileira,* 16, 110-113. ISSN 0102-0536

Casteel, C.L., Walling, L. L. & Paine, T. D. (2006). Behavior and biology of tomato psyllid, Bactericera cockerelli, in response to the Mi-1.2 gene. *Entomologia Experimentalis et Applicata,* 121, 67-72. ISSN: 1570-7458

Cohen, R., Burger, Y., Horev, C., Porat, A. & Edelstein, M. (2005). Performance of Galia-type melons grafted on to *Cucurbita* rootstock in *Monosporascus cannonballus*-infested and non-infested soils *Annals of Applied Biology,* 146, 381–387. ISSN: 1744-7348

Coley, P.D., Bryant, J.P. & Chapin, F.S. (1985). Rosource availability and plant antiherbivore defense. *Science,* 230, 4728, 895-899. ISSN: 1095-9203

Cortez-Madrigal, H. (2010). Resistencia a insectos de tomate injertado en parientes silvestres, con énfasis en *Bactericera Cockerelli* SULC. (Hemiptera: Psyllidae). *Bioagro,* 22, 1, 11-16. ISSN: 1316-3361

Coutinho, C. S., Fermino, S. A. C., Dos Santos, B. A., Araujo, D. C. L. & Da Silva, L. C. (2006). Potential of Hawaii 7996 hybrid as rootstock for tomato cultivars. *Bragantia,* 65, 89-96. ISSN: 0006-8705

Chamarro, L. J. (1995). Anatomía y fisiología de la planta, In: *El cultivo del tomate,* Nuez, V. F., pp. 45-91, Mundi-Prensa, ISBN 84-7114-549-9, España.

De Resende, J. T., Maluf, W. R., Cardoso, M. das G., Faria, M. V., Gonçalves L. D. & Do Nascimento I. R. (2008). Resistance of tomato genotypes with high level of acylsugars to *Tetranychus evansi* Baker & Pritchard. *Scientia Agricola,* 65, 31-35. ISSN: 01039016

Dorais M., Ehret, D.L. & Papadopoulos, A. P. (2008). Tomato (*Solanum lycopersicum*) health components: from the seed to the consumer. *Phytochemistry Review,* 7, 231–250. ISSN: 1572-980X

Eigenbrode, S. D. & Trumble, J. T. 1993. Resistance to beet armyworn, Hemipterans, and *Liriomyza* spp. in *Lycopersicon* accesiones. *Journal of the American Society for Horticultural Science,* 118, 442-456. ISSN: 0003-1062

Esquinas, A. J. T. & Nuez, V. F. (1995). Situación taxonómica, domesticación y difusión del tomate, In: *El cultivo del tomate,* Nuez, V. F., pp. 15-42, Mundi-Prensa, ISBN 84-7114-549-9, España.

FAOSTAT. (2011). Tomatoes production. http://faostat.fao.org/site/567/DesktopDefault.aspx?PageID=567#ancor (última revisión: 4/jul/011).

Freitas , J. A., Maluf, W. R., Cardoso, M. D., Gomes L. A. A. & Bearzoti, E. (2002). Inheritance of foliar zingiberene contents and their relationship to trichome densities and whitefly resistance in tomatoes. *Euphytica,* 127, 275–287. ISSN: 1573-5060

Garcia, E. (1988). *Modificaciones al sistema de clasificación climática de Köppen, para adaptarlo a las condiciones de la Republica Mexicana* (4ta. Ed). UNAM, ISBN, México.

Garcia-Rodríguez, M. R., Chiquito-Almanza, E., Loeza-Lara,P. D., Godoy-Hernández, H., Villordo Pineda, E., Pons-Hernández, J. L., González-Chavira, M. M. & Anaya-López, J. L. (2010). Producción de chile ancho injertado sobre criollo de Morelos 334 para el control de *Phytophthora capsici. Agrociencia,* 44, 701-709. ISSN: 1405-3195

Godoy H., H. & Castellanos, J. Z. (2009). El injerto en tomate, In: *Manual de producción de tomate en invernadero,* Castellanos, J.Z., pp. 93-104. INTAGRI, ISBN 978-607-95302-0-4, México.

Gonzalez, F. M., Hernandez, A., Casanova, A., Depestre, T., Gomez, L. & Rodriguez, M. G. (2008). El injerto herbáceo: alternativa para el manejo de plagas del suelo. *Revista de Protección Vegetal,* 23, 2, 69-74. ISSN: 1010-2752

Granett, J., Goheen, A.C. & Lider, L.A. (1987). Grape phylloxera in California. *California Agriculture,* 41, 1-2, 10-12. ISSN: 0008-0845

Hoyt, E. (1992). *Conservando los parientes silvestres de las plantas cultivadas,* Addison-Wesley Iberoamericana, ISBN 0-201-51830-3, Wilmington, Delawre, E.U.A.

Khah, E.M., Kakava, E., Mavromatis, A., Chachalis, D. & Goulas, C. (2006). Effect of grafting on growth and yield of tomato (*Lycopersicon esculentum* Mill.) in greenhouse and open-field. *Journal of Applied Horticulture,* 8, 1, 3-7. ISSN: 0972-1045

King, A. B. S. & Saunders, J. L. (1984). *Las plagas invertebradas de cultivos anuales alimenticios en América Central,* Administración de Desarrollo Extranjero, ISBN 0 902500 12 0, Londres.

Kogan, M. (1990). La resistencia de la planta en el manejo de plagas, In: *Introducción al manejo integrado de plagas,* Metcalf, R.L. & Luckman, W.H., pp. 123- 172. Limusa Noriega, ISBN 968-18-3275-2, México, D.F.

Kubota, Ch. & Viteri, F. (2007). Injerto en cucurbitaceas. Alternativas para luchar contra patógenos del suelo, incrementar rendimientos y reducir costos de producción. *Productores de Hortalizas,* 22-23.

Kubota, Ch., McClure, M.A., Kokalis-Burelle, N., Bausher, M.E. & Rosskopf, N. (2008). Vegetable grafting: History, use, and current technology status in North America. *HortScience,* 43, 6, 1664-1669. ISSN: 0018-5345

Lee, J.M. (1994). Cultivation of grafted vegetables 1. Current status, grafting methods, and benefits. *HortScience,* 29, 235-239. ISSN: 0018-5345

Liu, D. & Trumble, J. T. (2005). Interactions of plant resistance and insecticides on the development and survival of *Bactericera cockerelli* (Sulc.) (Homoptera: Psyllidae). *Crop Protection,* 24, 111-117. ISSN: 0261-2194

Martinez-Ballesta, M. C., Lopez-Perez, L., Hernandez, M., Lopez-Berenguer, C., Fernandez-Garcia, N. & Carvajal, M. (2008). Agricultural practices for enhanced human health. *Phytochemistry Review*, 7, 251–260. ISSN: 1572-980

Martinez, M. (1979). *Catálogo de nombres vulgares y científicos de plantas mexicanas* (1ra. Ed), Fondo de Cultura Económica, México.

Meletti, L.M.M. & Bruckner, C.H. 2001. Melhoramento genético, In: *Melhoramento de fruteiras tropicais*, Bruckner, C.H. & Picanço, M.C., pp. 345-385. Cinco continentes, Porto Alegre (Brasil).

Michel F., R., Chirinos U., H. & Lagos B., A. G. (1998). *Manual de agronomía*, Laboratorios A-L de México S. A. de C. V, Guadalajara, Jalisco, México.

Morales, F.J., Rivera-Bustamante, R., Salinas, P. R., Torres-Pacheco, I., Dias, P. R., Aviles, B. W. & Ramirez, J. G. (2005). Whiteflies as vectors of viruses in legume and vegetable mixed cropping systems in the tropical lowlands of México, In: *Whitefly-borne viruses in the tropics: building a knowledge base for global action*, Anderson, P. K. & Morales, F. J., pp. 177-196. Centro Internacional de Agricultura Tropical. ISBN 958-694-074-8, Cali, Colombia.

Muigai, S. G., Schuster, D. J., Snyder, J. C., Scott, J. W., Bassett, M. J. & Mcauslane, H. J. (2002). Mechanisms of Resistance in *Lycopersicon* Germplasm to the Whitefly *Bemisia argentifolii*. *Phytoparasitica*, 30, 347-360. ISSN: 1876-7184

Muñoz R., J. de J. (2009). Manejo del cultivo de tomate en invernadero, In: *Manual de producción de tomate en invernadero*, Castellanos, J.Z., pp. 45-92. INTAGRI, ISBN 978-607-95302-0-4, México.

Mutschler, M. A., Doerge, R. W., Liu, S. C., Kuai, J. P., Liedl, B. E. & Shapiro, J. A. (1996). QTL analysis of pest resistance in the wild tomato *Lycopersicon pennellii*: QTLs controlling acylsugar level and composition. *Theoretical and Applied Genetics*, 92, 709-718. ISSN: 1432-2242

Nakaho, K., Inoue, H., Takayama, T. & Miyagawa, H. (2004). Distribution and multiplication of *Ralstonia solanacearum* in tomato plants with resistance derived from different origins. *Journal of General Plant Pathology*, 70, 115-119. ISSN: 1610-739X

Nombela, G., Beitia, F. & Muniz, M. (2000). Variation in tomato host response to *Bemisia tabaci* (Hemiptera: Aleyrodidae) in relation to acyl sugar content and presence of the nematode and potato aphid resistance gene *Mi*. *Bulletin of Entomological Research*, 90, 161-167. ISSN: 0007-4853

Peralta, I. E., Knapp, S. & Spooner, D. M. (2005). New species of wild tomatoes (*Solanum* section *Lycopersicon*: Solanaceae) from Northern Peru. *Systematic Botany*, 30, 424-434.

Perez, G. M., Marquez, S. F. & Peña, L. A. (1997). *Mejoramiento genético de hortalizas*, Universidad Autónoma Chapingo, Chapingo, México.

Pire, R., Pereira, A., Diez, J. & Fereres, E. (2007). Evaluación de la tolerancia a la sequía de un portainjerto venezolano de vid y posibles mecanismos condicionantes. *Agrociencia*, 41(4): 435-446. ISSN: 1405-3195

Poincelot, R.P. (2004). *Sustainable horticulture today and tomorrow*. Prentice Hall, ISBN 0-13-618554-1, New Jersey.

Ramanatha Rao, V. & Hodgkin, T. (2002). Genetic diversity and conservation and utilization of plant genetic resources. *Plant Cell, Tissue and Organ Culture*, 68, 1–19. ISSN: 1573-5044

Restrepo S., E. F., Vallejo C., F. A. & Lobo A., M. (2008). Evaluación de poblaciones segregantes producidas a partir de cruzamientos entre tomate cultivado y la accesión silvestre PI134418 de *Solanum habrochaites* var. *glabratum* resistente al pasador del fruto. *Acta Agronomica,* 57, 1-8. ISSN: 0496-3490

Rick, C. M. & Holle, M. (1990). Andean *Lycopersicon esculentum* var. *cerasiforme*. Genetic variation and its evolutionary significance. *Economic Botany,* 44, 69-78. ISSN: 0013-0001

Rick, C. M., Laterrot, H & Philouze, J. (1990). A revised key for the *Lycopersicon* species. *Tomato Genetics Cooperative Report,* 40, 31. ISSN: 0495-8306

Rojas P., L. & Riveros F., B. (2001). Efecto del método de edad de plántulas sobre el prendimiento y desarrollo de injertos en melón (*Cucumis melo*). *Agricultura Técnica,* 61, 3, 1-16. ISSN: 0718-5839

Rosenthal, J.P. & Dirzo, R. (1997). Effects of life history, domestication and agronomic selection on plant defense against insects: Evidence from maizes and wild relatives *Evolutionary Ecology,* 11, 337-355. ISSN: 1573-8477

Sanchez-Peña, P., Oyama, K., Nuñez-Farfan, J., Fornoni, J., Hernandez-Verdugo, S., Marquez-Guzman, J. & Garzon-Tiznado, J. A. (2006). Sources of resistance to whitefly (*Bemisia* spp.) in wild populations of *Solanum lycopersicum* var. *cerasiforme* (Dunal) Spooner G. J. Anderson et R. K. Jansen, in Northwestern México. *Genetic Resources and Crop Evolution,* 53, 711-719. ISSN: 1573-5109

Sigüenza, C., Schochow, M., Turini, T. & Ploeg, A. (2005). Use of *Cucumis metuliferus* as a Rootstock for melon to manage *Meloidogyne incognita. Journal of Nematology* 37(3):276-280. ISSN: 0022-300X

Simons, A. T.; Gur, G. M. 2005. Trichomes of *Lycopersicon* species and their hybrids: effects on pests and natural enemies. *Agricultural and Forest Entomology,* 7, 265–276. ISSN: 1461-9563

Simons, A. T., Gur, G. M., McGrath , D., Martin, P. M. & Nicol, H. I. (2004). Entrapment of *Helicoverpa armigera* (Hübner). *Helicoverpa armigera* (Hübner) (Lepidoptera: Noctuidae) on glandular trichomes of *Lycopersicon* species. *Australian Journal of Entomology,* 43, 196-200. ISSN: 1440-6055

Simons, A. T., Nicol, H. I. & Gur, G. M. (2006). Resistance of wild *Lycopersicon* species to the potato moth, *Phthorimaea operculella* (Zeller) (Lepidoptera: Gelechiidae). *Australian Journal of Entomology,* 45, 81-86. ISSN: 1440-6055

Valadez, L. A. (1998). *Producción de hortalizas (*1ra. Ed), Limusa. México.

Vargas C., D. (2008). *Caracterización ecogeográfica y etnobotánica y distribución geográfica de Solanum lycopersicum var. cerasiforme (Solanaceae) en el occidente de México.* Tesis doctorado en ciencias, Centro Universitario de Ciencias agropecuarias, Universidad de Guadalajara. Las agujas, Zapopan, Jal. México.

Venema, J.H., Dijk, B.E., Bax, J.M., Van Hasselt, P.R. & Elzenga, T.M. (2008). Grafting tomato (*Solanum lycopersicum*) onto the rootstock of a high-altitude accession of *Solanum habrochaites* improves suboptimal-temperature tolerance. *Environmental and Experimental Botany,* 63, 359-367. ISSN: 0098-8472

Verdejo-Lucas, S. & Sorribas, F. J. (2008). Resistance response of tomato rootstock SC 6301 to *Meloidogyne javanica* in plastic house. *European Journal of Plant Pathology,* 121, 103-107. ISSN: 0929-1873

Wagner, G. J., Wang, E. & Shepherd, R. W. (2004). New Approaches for Studying and Exploiting an old protuberance, the plant trichome. *Annals of Botany*, 93, 3-11. ISSN: 1095-8290

Williamson, V. (1998). Root-knot nematode resistance genes in tomato and their potential for future use. *Annual Review of Phytopathology*, 17, 277-293. ISSN:0066-4286

Yamakawa, K. (1982). Use of Rootstocks in Solanaceous Fruit-Vegetable Production in Japan. *Plant Breeding Division, Vegetable and Ornamental Crops Research Station JARQ*, 15, 3, 175-179. ISSN: 0021-3551

Yetisir, H. & Sari, N. (2003). Effect of different rootstock on plant growth, yield and quality of watermelon. *Australian Journal of Experimental Agriculture*, 43, 10, 1269–1274. ISSN: 0816-1089

Zhang, Z.J., Wang, Y.M., Long, L.K., Lin, Y., Pang, J.s., & Liu, B. (2008). Tomato rootstock effects on gene expression patterns in eggplant scions. *Russian Journal of Plant Physiology, 55, 1, 93–100.* ISSN: 1608-3407

7

Techniques to Estimate Abundance and Monitoring Rodent Pests in Urban Environments

Regino Cavia, Gerardo Rubén Cueto and Olga Virginia Suárez
Departamento de Ecología, Genética y Evolución, Universidad de Buenos Aires and Consejo Nacional de Investigaciones Científicas y Técnicas
Argentina

1. Introduction

Different techniques have been developed in order to study the ecology of animals. The application of each technique depends on the studied animal species, on the type of habitats where they live and the objectives of the study. Most of the ecological studies focus on a unique population, which is defined as a group of organisms of the same species that coexist at the same time and in the same area (Krebs, 1978); or on a community, which is defined as a group of populations that exist at the same time and in the same area (Begon et al., 1987). One of the most important characteristics of a population is its size or abundance. This is determined by the number of individuals born, the number of individuals that dead and the number of individuals going into or out of the area that the population occupies per unit of time (Begon, 1979; Krebs, 1978). On the other hand, some of the characteristics of the animal community are the species composition, its absolute abundances and relative abundances, the richness, dominance, diversity, equitability, trophic structure and the niche structure (Krebs, 1978). Except for the trophic and niche structure, the other characteristics mentioned above are inferred from the abundances of the individual species that make up the community.

Studying the factors which control the species' abundance is one of the main topics of ecology. This topic has been explained by studying the natural variations in abundance according to space and time, or due to experimental manipulation (Aplin et al., 2003). In relation to pest species, knowing the factors which limit the populations' growth allows to take decisions to control them.

In the population studies the sampling method used by the researcher must be the most appropriate to study the particular species and to allow answering the question that has been posed. Regarding the community studies, for practical purposes, it is necessary to limit the community studied to a group of populations that share a determined characteristic and are adequately sampled according to the selected method. In general, this subgroup inside a community is related phylogenetically (e.g. community of insects, birds, rodents, etc) or they are exploiting the same resources in the same way. According to Magurran (1988) the diversity is more informative and easy to understand when it is applied to a limited and well-defined taxonomic group.

The presence or absence at the same time and site is the minimum information that we can have of the populations in a community. The complete list of all the individuals or census is the maximum information that we can have of the size of the population and the diversity of a community (Aplin et al., 2003; Krebs, 1978; Magurran, 1988; Southwood, 1978). A census is rarely conducted in natural ecosystems due to limitations of time, money, personal and/or the difficulty to reach all the individuals in the study area. On the other hand, a census can cause interferences and destroy populations or cause damage or destruction of their habitats; therefore, sampling is generally used instead of a census (Magurran, 1988; Rabinovich, 1980; Southwood, 1978).

According to the study objectives, the absolute or relative sizes of the populations can be estimated (Krebs, 1989). The absolute size estimate allows assessing the density, i.e. the number of individuals present in a determined area or volume (Krebs, 1978). Different methods have been developed to estimate the absolute size of populations, which can be used when certain requirements are met (Brownie et al., 1986; Hayne, 1949; Krebs, 1966, 1989). It is possible to estimate the density with the population size and the size of the studied area. Although it is sometimes easy to calculate the size of the studied area; it is difficult in samplings with traps or another system which depends on the attraction of animals. The studied area would depends on several factors, such as the influence area of the trap, its bait (what distance is necessary for a baited trap to attract an animal) and the animals' mobility. At the same time, the individuals' mobility may depend on species, season, habitat conditions, age, sex and the reproductive condition of the individuals, among other factors.

Sometimes it is not necessary to know the absolute size, but it is wanted to know the spatial and/or temporal variations of abundance; thus, trend indicators or relative abundance indices can be used. The relative abundance is defined as an abundance measurement that is relative to the sampling effort, showing the number of individuals with regard to a measurement different from the surface or volume (Seber, 1973). For example, the number of individuals trapped is used with regard to the number of traps or set nets; the number of animals observed during a period of time, etc. The use of relative abundance estimators allows the comparison of the abundance between sites or of the same site at different times, even if the absolute abundance values are unknown. Two of the relative abundance estimators frequently used are the trap success (Seber, 1973) and the relative density index (Begon, et al., 1987), both of them are calculated as the number of different animals captured / number of active trapping elements * the time that the elements have been active. The trapping elements can be traps, nets, etc.

Another way to estimate relative abundances is with the record of animals signs or related elements that can infer the presence or absence of an animal species in the studied area, and they can also estimate the size of the population calibrating the signs quantity with the abundance (Krebs, 1978; Rabinovich, 1980; Southwood, 1978). The main advantage that these methods or "population indices" present (sensu Southwood, 1978) is that they require in general less effort and expenditure than other methods. For a lot of species it is possible to count footprints, nests, burrows or other habitat alterations, while for birds it is possible to use the record of their songs or calls as signs of their presence (Aplin et al., 2003; Krebs, 1978; Rabinovich, 1980; Southwood, 1978). Finally, it is possible to estimate the relative abundance of an animal by means of surveys or questionnaires (Krebs, 1978). This methodology uses the experience of other people to determine the presence or absence of a species, or to estimate the relative abundance (Filion, 1987).

The most common techniques to estimate the rodents abundance are those based on the use of capture traps or on the record of signs, due to the fact that most of them have crepuscular habits and its direct count can only be used in special cases (Aplin et al., 2003). There is a wide variety of designs for kill or live capture traps, been their designs creative and/or old, some of them were described by Chani (1980) and Hawthorne (1987). Among the relative abundance estimators for rodents, Aplin et al. (2003) pointed out three methods that involve the use of signs that are widely employed: the use of footprint traps, the record of food consumption and the count of burrows. Yo et al. (1987) proposed particularly for *Rattus norvegicus* the count of gnawed wood pegs as a method to estimate abundance and the use of space. Since these mammals gnaw different materials in order to limit the length of their incisors, the record of the marks left on the pegs are good to estimate abundance, independently from the food availability in the environment.

Maybe urban environments are the least studied in relation to the ecology of rodents, probably due to the methodology problems they present: 1) some trap designs could be dangerous for people. For example, snap traps can hurt a person or pet if they activate it accidentally, 2) the difficulty of reaching some sites such as inside houses, shops or industries and 3) there is a big risk of losing the material used to sample, especially the traps which are valuable elements. Another difficulty to study rodents in urban ecosystems is its environmental heterogeneity; so if it is necessary to compare the results of the different environments, the method used to estimate the abundance should be the same. The selected sampling technique should be able to sample for example: inside a house, a shop, an industry, and also open areas such as gardens, parks, lawns, public spaces, etc.

As a brief summary we will mention some of the experiences carried out in cities and the sampling techniques employed. In the 50s. a study was conducted in Baltimore City laying the foundations for the biology of *R. norvegicus* in urban environments (Davis, 1951a, 1951b, 1951c). In that study, trap sampling was conducted using live capture traps, which is the technique also used in various more recent studies (Battersby et al., 2002; Castillo et al., 2003; Cavia et al., 2009; Ceruti et al., 2002; Glass et al., 1988; Traweger & Slotta-Bachmayr, 2005). There is a growing number of studies that estimate abundance with the record of signs. For example, bait stations were used in drains to estimate the population abundance of *R. norvegicus* in Enfield City, England (Channon et al., 2000) and the record of rat bites in patients in the hospitals of New York were used to determine areas with different outbreak risk (Childs et al., 1998). In another study the possible causes of rats and mice infestation in dwellings (Langton et al., 2001) were determined by recording the signs of rodent activity obtained during an inspection of 17100 dwellings in different regions of England. Among the studies conducted to estimate the rodents abundance using surveys, it can be pointed out one conducted to householders in Manchester, the United Kingdom, which determined that 44% of dwellings were infested with *Mus musculus* and 49% with *Rattus* spp. When these results were compared with samplings using footprint traps in dwellings, they were consistent for *M. musculus*, while the abundance for *Rattus* spp. was apparently overestimated (Marshall & Murphy, 2003).

In this chapter different methods to estimate rodent abundance in urban environments are evaluated. For this purpose samples were carried out in a coastal area, in a cars warehouse, in an urban reserve, in a shantytown and in a residential neighborhood. The different methods to estimate abundance that were tested are: record of activity of rodent burrows, visual record of animals, glue traps, wood pegs, bait stations, bait stations with hair-hunting traps, Sherman live traps and cage traps, and the use of surveys.

2. Evaluation of methods to estimate rodents abundance, preliminary survey

Study area

The samplings described in this section were conducted in a coastal area in the city of Buenos Aires where waste materials and soil were deposited in order to gain land from the river and subsequently covered by spontaneous vegetation, figures 1 and 2. The objective of these preliminary surveys were to prove the methods to estimate the rodent relative abundance by counting burrow entrances, active individuals, kill capture with glue traps, live capture with cage traps and by recording consumption in bait stations.

2.1 Materials and methods

Count of burrow entrances

After detecting a colony of *R. norvegicus*, an inspection of the place was carried out in order to find burrow entrances. The number of entrances was recorded in an area of 60 by 30 m, table 1.

Glue traps

Glue traps consisted of pieces of cardboard of 30 by 30 cm covered with a thin layer of commercial glue (Pega-Rat) and baited with peanut butter, figure 3. The glue was placed according to the manufacturer's instructions. Twelve glue traps were placed at burrow entrances in the afternoon and checked in the morning of the following day during four consecutive days, table 1.

Count of active rodents

Observations were performed during four days beginning on 19 June 2001, at three different times: in the morning from 11:00 to 12:00hs, at midday from 13:30 to 15:00hs and in the afternoon from 16:00 to 17:00hs, table 1. In each period three records of five minutes were registered, separated by breaks of five minutes. Two observers stood in the middle of the studied area (60 by 30 meters). The total surface was divided and each observer registered the records in a quadrant of 30 by 30 meters. The quadrant limits were marked with paint over the waste materials. At each interval of five minutes each observer recorded the number of individuals of *R. norvegicus* observed in their area. The average number of individuals observed in five minutes was calculated for the three different periods of time.

Fig. 1. Aerial view showing the coastal area where the samplings were conducted (Source: ©2006 Google Earth, imagery date, April 21, 2000).

Fig. 2. View of a section of the coastal area where some of the samplings described in this chapter were conducted. The de la Plata river can be seen on the right side of the image.

Fig. 3. Photograph of a glue trap placed at a *R. norvegicus* burrow entrance.

Live traps

A capture-mark-recapture sampling was conducted using cage traps, figure 4. Between 14 and 24 August 2001 eight cage traps were set in six opportunities, working one hour between 11:00 and 13:00hs, and they were checked every 15 minutes, table 1. The captured animals were marked with synthetic paint on the back. A different color was used for each day. This way of marking animals was used to prove if it was possible to identify them in subsequent counts. The trap success (TS) was assessed to estimate the relative abundance:

$$TS = {}^{I}\!\big/_{(T * t - 1/2 * ST)} \quad (1),$$

where I is the number of captured individuals, T is the number of set traps, t is the number of intervals of time in hours, nights, etc. that the traps were set active and ST is the number of traps that were sprung without captures during an interval t. Half of the sprung traps without captures is subtracted from $T*t$ since it is not possible to know if the traps were inactive from the beginning, during or at the end of the considered interval. Thus, it is assumed that an average of these traps were inactive half of the interval.

Use of nontoxic bait stations

Twelve bait stations were placed 10m apart of each other on a transect. The bait stations consisted of transparent two liters plastic bottles, containing two grams of a mixture of fat, peanut butter and paraffin. The bait stations were set for three nights, after this period of time it was recorded whether the bait had been consumed, table 1. The proportion of bait stations with rodent activity was calculated (PropABS) to estimate the relative abundance:

$$\text{PropABS} = {}^{ABS}\!\big/_{(TBS - MBS)} \quad (2),$$

where ABS is the number of bait stations with rodent activity, TBS is the total number of bait stations and MBS is the number of missing bait stations. A bait station was considered with rodent activity when bait consumption was recorded.

2.2 Results

A total of 41 active burrow entrances were recorded (figure 5) in 1800 m^2. Only two animals were captured using glue traps. The remains that were found did not allow the identification of the sex nor the size class, because they were eaten by other animals. Up to 17 different individuals were observed at a five-minute interval by direct observations. The number of observed individuals per period of time was (mean ± standard deviation): 6.75 ± 4.33 individuals in the morning, 5 ± 3.30 individuals at midday and 7.91 ± 3.14 individuals in the afternoon. Twelve *R. norvegicus* were captured using cage traps and none individual was recaptured. The trap success was 0.25 individuals/trap per hour, and the highest trapping frequency (5/12) was recorded during the first 15 minutes of the sampling. The bait was completely consumed after three nights in 100% of the set bait stations (PropABS=1).

Fig. 4. Photograph showing a styrofoam bait station of 250 cm^3 prepared with sticky tape on the front (on the left), a cage trap (in the centre) and a Sherman trap (on the right). Observe a one Argentinean peso coin on the bottom left corner to determine the size of the elements.

3. Comparison between the active rodents count and the use of bait stations to detect changes in abundance

A sampling was conducted in the same coastal area as in the previous sampling with the objective of assessing if, when modifying experimentally the size of the population the changes were detected by the methods for estimating the abundance by counting active animals and the use of bait stations.

Fig. 5. Photograph showing a *R. norvegicus* burrow entrance.

3.1 Materials and methods

Rodent abundance was modified using rodenticides in a part of the coastal area. Two or three blocks of five grams of a commercial rodenticide were placed into the burrow entrances found on 1 October 2001, table 1. This area covered a coastline of approximately 150m long, which is called the treated area. The rest of the place is the control area (not treated).

In the treated area the activity of *R. norvegicus* was recorded by means of the active individuals count in the same way as it had been performed in the preliminary sampling. Two samplings were conducted: one before the use of the rodenticide between 27/08 and 10/09/01, and another after its use between 10 and 18/10/01, table 1. The daily mean animals observed at five-minute intervals between the samplings before and after the use of rodenticide were compared by means of the Mann Whitney test (Daniel, 1978).

The relative abundance was also recorded, before and after the use of rodenticide, by means of bait stations with nontoxic bait, which were placed on transects 100m apart: one in the control area and the other one in the treated area on 28/9 and 12/10/2001, respectively (table 1.). Each transect consisted of 12 bait stations placed every 10m. The bait stations were set active during the same time and had the same characteristics as the pilot sampling, but in this opportunity they had 18-20g of bait, since in the pilot sampling it had been consumed after three days. The proportions of bait stations with rodent activity before and after the use of rodenticide were compared statistically by means of a proportions test (Zar, 1996).

	Date	Activities
Preliminary survey	28/05/01	Burrow entrances survey
	28/05 - 01/06/01	Glue traps sampling
	19 - 22/06/01	Direct observations of individuals at three different times
	14 - 24/08/01	Live capture and marking of rodents
	14 - 15/09/01	Sampling with bait stations
Comparison of bait stations and direct observations	27/08 - 10/09/01	Direct observations of rodents at three different times
	29/09 - 2/10/01	Sampling with bait stations (control and treated)
	01/10/01	Use of rodenticide in burrow entrances
	10 - 18/10/01	Direct observations of rodents at three different times
	12 - 15/10/01	Sampling with bait stations (control and treated)

Table 1. Schedule of activities performed with the objective of testing different rodent sampling techniques.

3.2 Results

In the treated area, up to 11 different individuals were detected at a five-minute interval in the previous sampling and two in the sampling after the use of rodenticide. The mean number of individuals seen at a five-minute interval decreased from 1.32 to 0.45 between both samplings, these differences were marginally significant ($U=3.5$; $p=0.1$), figure 6.a.

The proportion of bait stations with rodent activity decreased significantly after the use of rodenticide in relation to the previous moment in the treated area ($p=0.039$). In the control area, changes in the proportion of bait stations with rodent activity were not observed between the moments before and after the use of rodenticide ($p=1.000$), figure 6.b.

4. Comparison of use of pegs and bait stations

With the objective of comparing the use of wood pegs and nontoxic bait stations as methods to estimate the rodent relative abundance, two samplings were performed: one in the coastal area of the de la Plata river and the other one in a judicial cars warehouse in an urban area in the city of Buenos Aires.

4.1 Materials and methods

In the coastal area of the de la Plata river three plots of 2000 m^2 (100 by 20 m) were selected. In each plot they were set 32 stations with eight pine pegs, eight pine pegs scented with vanilla essence, eight pine pegs scented with almond essence and eight bait stations consisting of plastic containers of 20 cm depth and 10cm diameter with a mixture of cow fat, paraffin and peanut butter. The pegs were scented putting them into water and edible essences during 72hs. The stations were placed 10m apart, making a grid that occupied all the plot. The different elements were placed in the stations alternately and systematically. Both the pegs and the bait stations were set active for 3 days and checked every day, recording the gnawing evidence on the pegs and the consumption of bait. The bait stations where consumption was detected were replaced for a new one each time they were checked.

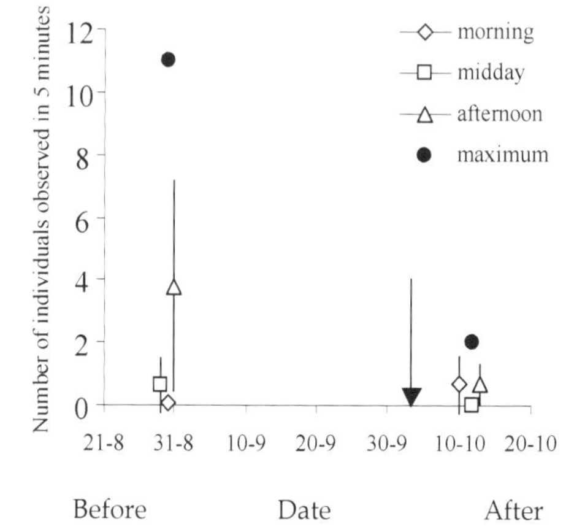

a)

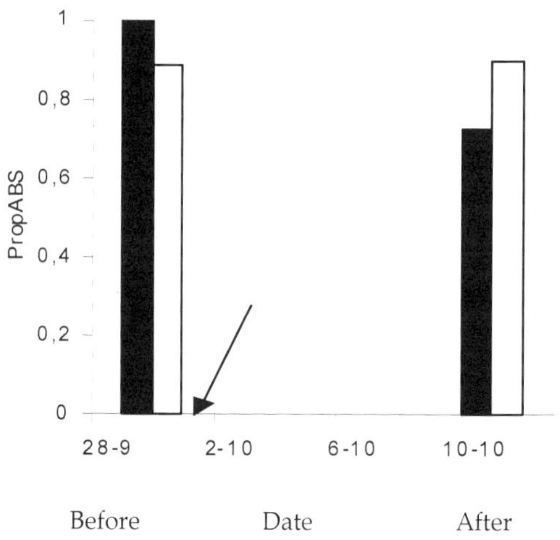

b)

Fig. 6. a) Mean number of animals observed at five-minute intervals at different times: in the morning (11:00-12:00hs), at midday (13:30-15:00hs) and in the afternoon (16:00-17:00hs) and the maximum number of animals seen at five-minute intervals (maximum) before and after the use of rodenticide in the treated area; and b) proportion of bait stations with signs of rodent activity (PropABS) in the control area (white bars) and treated area (black bars) before and after the use of rodenticide. The arrows in both figures indicate the moment at which the rodenticide was set in the treated area (1/10/2001).

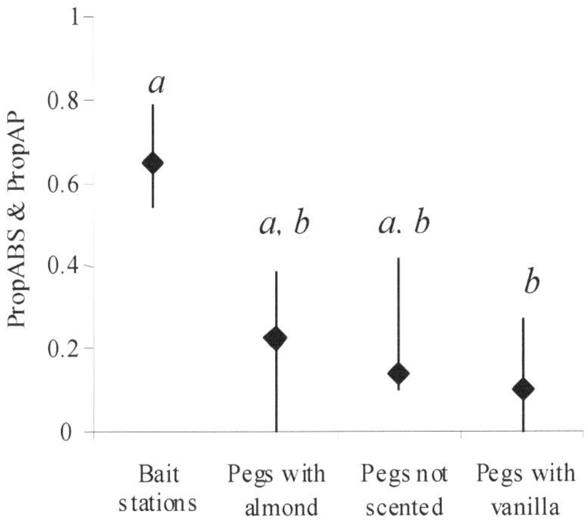

Fig. 7. Median proportion of bait stations (PropABS) and pegs (PropAP) with signs of rodent activity per grid in the coastal area. X^2= 6.517; gl= 3; n= 3; p= 0.089. The proportions that share the indices *a* and *b* do not show differences for the Wilcoxon test per pairs (X^2_c>0.05; p>0.2).

In the judicial cars warehouse, where complaints of rodent infestation have been made, five grids of 1900 m^2 (190 by 10 m) were placed. Each grid consisted of two parallel transects 10m apart, with 20 stations each one 10 m apart. The bait stations and the pegs were placed alternately in each station. In this site only pegs without essence were used. The bait stations and the pegs had the same characteristics as in the sampling performed in the coastal area and they were also set active during three nights.

For both samplings, the proportion of bait stations with rodent activity was calculated as in the previous section. The proportion of pegs with signs of rodent activity (PropAP) was calculated as follows:

$$\text{PropAP} = AP/(TP - MP) \quad (3),$$

where *AP* is the number of pegs with signs of rodent activity, *TP* is the number of set pegs and *MP* is the number of missing pegs. The gnawing evidence on the wood was considered as signs of rodents on pegs. The possible differences in the proportions of pegs and bait stations with rodent activity per plot were assessed using a Friedman test. If differences were detected, comparisons per pairs were performed using a Wilcoxon test (Daniel, 1978). In the sampling performed in the coastal area, the missing pegs and bait stations were assessed.

4.2 Results

The proportion of bait stations with signs of rodent activity was marginally higher than the proportion of pegs (X^2= 6.517; gl= 3; p< 0.089) in the coastal area and significantly higher in the cars warehouse (X^2= 5; gl= 1; p< 0.025), figures 7 and 8. The use of essences did not increase the proportion of pegs with rodent activity (p>0.29). In both environments the bait stations were more sensitive than the pegs to detect the presence of rodents; since in some plots the presence of rodents was detected with bait stations and not with pegs. The loss of sampling elements was one of the problems. There were more missing pegs than bait stations, probably because it was more difficult to find them in the field due to their small size, figure 9.

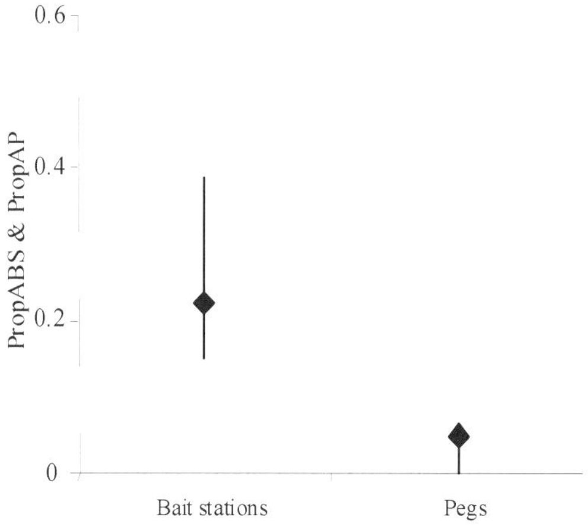

Fig. 8. Median proportion of bait stations (PropABS) and pegs (PropAP) with rodent activity per grid in a judicial cars warehouse X^2= 5; gl= 1; n= 5; p= 0.025.

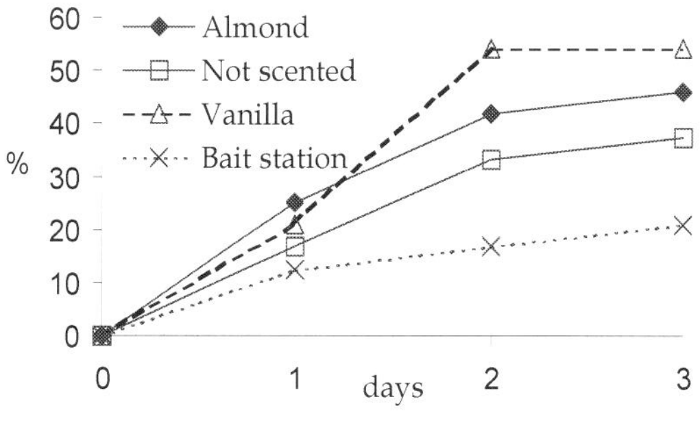

Fig. 9. Missing pegs and bait stations (in %) over the three days of sampling in the coastal area.

5. Evaluation of the use of bait stations to estimate the relative abundance in an urban reserve

According to the previous results, where it was observed that the bait stations could be used to detect the presence of rodents; in an urban reserve it was assessed if there was a good relation between the proportion of bait stations with signs of rodent activity and the relative abundance estimated using traps. For this purpose samplings were performed in an urban reserve which presents various environments with rodents of different species and with different abundances.

5.1 Materials and methods

A total of five samplings were conducted in the urban reserve, one in spring 2002 and four between autumn 2004 and summer 2005.

In spring 2002, 10 transects were placed. Thirty live capture traps, 15 Sherman traps and 15 cages, and 15 bait stations with hair-hunting traps were set on the transects. The arrangement of the traps and bait stations along the transect is shown in the figure 10.a. The traps were active during three consecutive nights and checked every day in the morning. The bait stations with hair-hunting traps consisted of styrofoam containers of $250cm^3$ containing 10g of a mixture of peanut butter, fat and vanilla essence. A strip of sticky tape was placed on the entrance of the container, so hair of the animals that entered stick on this tape. This allows to identify the individuals' species that have visited the bait stations, figure 4 and 11.

In autumn, winter and spring 2004 and summer 2005, nine transects were placed: three transects in a riparian thicket on the coast of the de la Plata river, three in an alders forest and three in a grassland dominated by *Cortaderia selloana*. Forty live capture traps (20 Sherman and 20 cage traps), and 20 bait stations with hair-hunting traps were set on the transects. The arrangement of the traps and bait stations along the transect is shown in the figure 10.b. The traps and stations were baited and checked like in spring 2002, and were active for the same period of time.

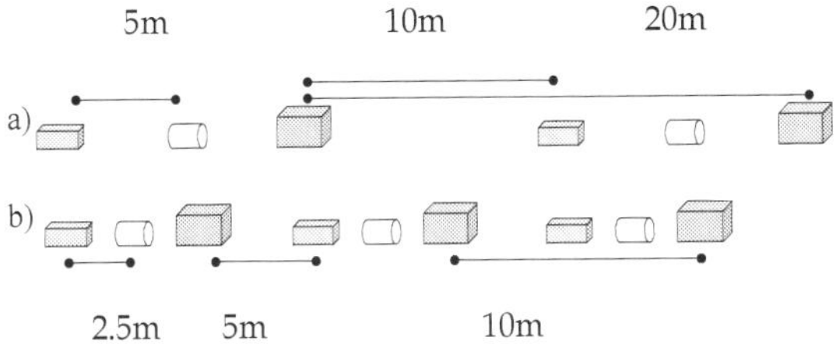

Fig. 10. Display of the different elements on the transects in a) spring 2002 and b) between autumn 2004 and summer 2005. The grey figures represent the Sherman traps, the light blue ones represent the cage traps and the white ones represent the bait stations.

Evidence of rodent activity in the bait stations was recorded third night after the setting. Incisor marks on the bait and/or rodent hair left on the sticky tape were considered as evidence of rodent activity in the bait stations (figure 11).

The trap success was calculated for each transect and this was compared with the proportion of bait stations per transect which showed signs of rodent activity by means of a correlation analysis. For this purpose the trap success was calculated and corrected as follows:

$$TS_c = \frac{(I - 1/6 * I)}{(T * N - 1/2 * ST)}, \quad (4)$$

Where *I* is the number of captured individuals, *T* is the number of set traps, *N* is the number of nights that the traps were active and *ST* is the number of traps that were sprung without captures.

Fig. 11. Photograph where gnawing evidence is observed on the surface of the nontoxic bait on the floor of the bait station as parallel marks, and hair left on the sticky tape.

It was necessary to correct the abundance the bait stations are exposed to, because it had been considered that if an animal is caught in a trap, it can not visit a bait station, so a correction factor of $1/6*I$ was subtracted from the number of captured rodents. It is not possible to know when each individual has been captured, if at dusk, in the middle of the night or at dawn. However, it can be assumed that on average all individuals have been caught in the trap half of the night. It was considered that as each period of sampling consisted of a three-night sampling, i.e. 6 half nights, the factor of correction should be 1/6 per capture.

The association between the proportion of bait stations with signs of rodent activity and the trap success was assessed by means of a simple linear correlation (Sokal & Rohlf, 1995).

5.2 Results

A total of 132 rodents (of six species) and 129 red opossums (*Lutreolina crassicaudata*) were captured, so the association between the percentage of bait stations with rodent activity and the trap success of rodents, of opossums and of both together was analyzed. The proportion of bait stations with rodent activity was correlated with the trap success of rodents (r= 0.4837; p= 0.000) and with the trap success of rodents and opossums together (r= 0.4665; p= 0.001), but it was not correlated with the trap success of opossums (r= 0.1233; p= 0.414), figure 12. These results confirm the observations performed in the field where it was determined that the marks corresponded to rodent incisors and rodent hairs, and not to marks made by opossums, figure 11.

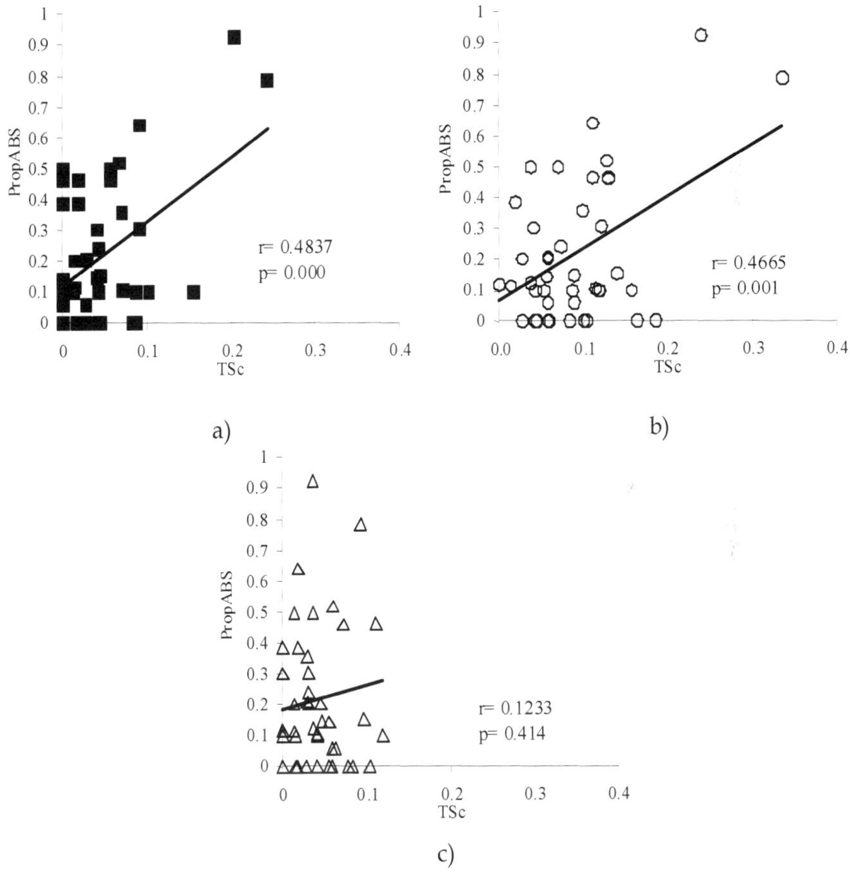

Fig. 12. Relation between the corrected trap success (TSc) and the proportion of bait stations with signs (PropABS). The lines represent the trend of association. a) TSc of rodents, b) TSc of rodents and opossums and c) TSc of opossums.

6. Evaluation of the use of bait stations and surveys about the presence of rodents to estimate the relative abundance in residential environments

With the objective of evaluate the use of bait stations and surveys about the presence of rodents to estimate the relative abundance, and evaluate their association with the trap success, samplings were conducted in different blocks of a shantytown and of a residential neighborhood in the city of Buenos Aires.

6.1 Materials and methods

A total of four samplings were conducted in the residential neighborhood: in winter, autumn, and spring 2004 and summer 2005; and three samplings were conducted in a shantytown: in summer, autumn and winter 2004. In each opportunity six different blocks in the residential neighborhood and between two and four different blocks in the shantytown were selected.

In each block, cooperation was requested to the people of the dwellings. In a residential neighborhood a dwelling was defined as a house, shop, house-shop or industry built in a lot. In the shantytown a dwelling was defined as a house, shop or house-shop that had a different number provided by the Secretary of Housing. Once "the responsible person" of the dwellings has agreed to cooperate with this study, a survey was conducted to know about the presence of rodents and their signs in their neighborhood and dwelling. The questions were the followings:

1. Have you ever had rats or mice in your house/job or have you seen gnawed objects or droppings?
2. Have you ever seen rats or mice in the neighbourhood?

For both questions if the person answered in an affirmative way, the following question was asked: when was the last time you saw one? in order to separate the recent events from the non-recent ones.

In each of the dwellings where people agreed to cooperate, bait stations with the same characteristics as in the section 5 were placed, figure 4. A fixed number of bait stations were set per dwelling. According to Aplin et al. (2003), in residential environments this way of distributing the rodent sampling elements is more adequate than the display on transects. It was decided that there would be two bait stations per dwelling, but exceptionally this number varied between one and eight when the dwelling was too small or big in size. Evidence of rodent activity was recorded in the bait stations seven days after their setting in the residential neighborhood and three and seven days after their setting in the shantytown. Bait stations were removed on the same day live Sherman and cage traps were set in each dwelling. In the same way as for the bait stations, a fixed number of each type of trap was placed, setting two Sherman and two cage traps in each dwelling. In some exceptional cases, this number varied between one and four Sherman and between one and eight cage traps and they were checked every day in the morning. The species of each captured animal was determined. The animals were sacrificed and collected.

The indices of rodent relative abundance were estimated on the base of the capture data, of the activity in bait stations and of surveys about rodents. In order to analyze the consistency of the associations between the indices, the analysis were performed at three different spatial scales: 1) taking each block as a sampling unit (analysis per block), 2) joining blocks according to their proximity defining different areas, and using these areas as a sampling unit (analysis per areas) and 3) using as different sample units the samplings performed in each period of the year (joining areas) and in each environment separately (shantytown and residential neighborhood; analysis per period).

For each sampling unit (which depended on the analyzed scale) the rodent relative abundance was estimated with the data of the surveys using the following indices: 1) the proportion of people who reported having had rodents in their dwelling during the last year (PropD365); 2) the proportion of people who reported having had rodents in their dwelling during the last semester (PropD180) and 3) the proportion of people who reported having had rodents in their dwelling during the last quarter (PropD90). The same indices were calculated to estimate the proportion of people who reported having seen rodents in their neighborhood at different time scales. The indices were calculated as follows:

$$\text{PropD365} = {PD365}/{P}, \quad (5)$$

$$\text{PropD180} = {PD180}/{P}, \quad (6)$$

$$\text{PropD90} = {PD90}/{P}, \quad (7)$$

$$\text{PropN365} = {PN365}/{P}, \quad (8)$$

$$\text{PropN180} = {PN180}/{P}, \quad (9)$$

$$\text{PropN90} = {PN90}/{P}, \quad (10)$$

where *PD365, PD180* and *PD90* is the number of people who reported having had rodents in their dwelling during the last 365, 180 and 90 days respectively; *PN365, PN180* and *PN90* is the number of people who reported having seen rodents in their neighborhood during the last 365, 180 and 90 days respectively; and *P* is the total number of people surveyed in the sampling unit.

The proportion of bait stations with signs of rodent activity (PropABS, equation 2) and the proportion of dwellings with rodent activity (PropDABS) were estimated using the data of rodent activity detected in the bait stations:

$$\text{PropDABS} = {DABS}/{D} \quad (11)$$

where *DABS* is the number of dwellings with at least one bait station with signs of rodent activity and *D* is the number of sampled dwellings in each sampling unit.

Finally, the rodent relative abundance was estimated with the data of the captures using the trap success (TS, equation 1), and the proportion of dwellings with captured rodents (PropDR) as:

$$\mathrm{PropDR} = {DR}/{D} \quad (12),$$

where *DR* is the number of dwelling with at least one captured rodent and *D* is the number of sampled dwellings in each sampling unit.

Firstly, in order to analyze the association between the different indices of relative abundance, non-parametric Spearman correlations were used due to the low number of sampling unites considered and lack of normality in the distribution of the indices (Daniel, 1978). Then, it was analyzed if there was a functional relationship between the trap success (since it is a relative abundance index widely accepted) and the other indices of relative abundance estimated using simple regression models:

$$y_i = a + b.x_i$$

being for the model y_i the trap success, x_i the other indices, a the intercept and b the slope of the line. The model was adjusted and the hypothesis of the zero slope was tested with a randomization method (Manly, 1991), 5000 randomizations were performed using the RT program (Manly, 1996).

6.2 Results

In the shantytown and in the residential neighborhood 30.0% of the people surveyed reported having had rodents in their dwellings and 41.0% reported having seen them in the neighborhood during the last 90 days (total of people surveyed = 429). Evidence of rodent activity was detected in 49 out of 805 bait stations set in 382 dwellings. A total of 25 *R. rattus*, 52 *R. norvegicus* and 28 *M. musculus* were captured with a total trapping effort of 1769 cage-nights and 1837 Sherman trap-nights, set in 347 dwellings.

All the indices showed positive associations with the other indices at the three analyzed scales. When each block was considered as a sampling unit, most of the associations were significant with a probability lowered than 0.05 (Table 2). The weaker associations were observed between the trap success and the proportion of people who reported having seen rodents in the neighborhood during the last 90, 180 and 365 days. A weak association was also observed between the proportions of people who reported having seen rodents in their neighborhood during the last 90 days and the proportion of dwellings with captured rodents. The proportion of people who reported having had rodents in their dwellings during the last 90 days showed a higher coefficient of association with trap success than the proportion of people who reported having had rodents during the last 180 or 365 days. Both the proportion of bait stations with signs of rodent activity and the proportion of dwellings with bait stations with signs of rodent activity proved to be associated with the rest of the analyzed indices, being low the coefficient of Spearman association with the trap success and with the proportion of dwellings with rodent capture.

In the analysis per area, the general patterns of associations observed at block scale were maintained; except for the proportion of people who reported having seen rodents in their neighborhood during the last 90 days, which was significantly related to the trap success ($p<0.05$) and the associations between the trap success and the proportion of bait stations with signs of rodent activity and of dwellings with signs of rodent activity, which were marginally significant ($p<0.10$), table 4.

At a larger spatial scale (per period) several associations lose their statistical significance; however, the associations between the trap success and the proportion of people who reported having had rodents in their dwellings during the last 90, 180 and 365 days continue to be significant, table 5. The proportion of bait stations with signs of rodent activity and the proportion of dwellings with signs of rodent activity showed the same association pattern with the other indices of relative abundance that had been observed at the "per area" scale. The decline of significance in the correlations could be due to the lower number of sample units, and not necessarily due to the absence of association between the indices. This is because at higher scale there are less sampling unites as a consequence of pooling the sampling units of the lower scale; and, although the correlation coefficients increased, some were not significant because the degrees of freedom decreased, tables 2, 3 and 4.

The regression analyses were performed at block scale due to the fact that: 1) there are more sampling unites, 2) it demands less sampling effort per replica making this analysis scale the most feasible to use in future works, and 3) the highest number of significant associations was observed at this scale. The regression analysis was not performed for the proportion of people who reported having seen rodents in their neighborhood during the last 180 and 365 days due to the fact that the associations were marginally significant. The proportions of people who reported having had rodents in their dwellings during the last 180 and 365 days were not analyzed either, because the information provided by these indices is redundant in relation to the proportion of people who reported having had rodents in their dwellings during the last 90 days, being this index the one which presents a higher association with the trap success.

	PropD180	PropD90	PropN365	PropN180	PropN90	TS	PropDR	PropABS	PropDABS
PropD365	0.932*	0.904*	0.588*	0.509*	0.550*	0.561*	0.540*	0.547*	0.533*
PropD180		0.964*	0.581*	0.586*	0.628*	0.573*	0.567*	0.620*	0.611*
PropD90			0.582*	0.656*	0.695*	0.579*	0.580*	0.590*	0.581*
PropN365				0.675*	0.688*	0.284+	0.338*	0.391*	0.419*
PropN180					0.953*	0.257+	0.313*	0.479*	0.479*
PropN90						0.248+	0.294+	0.492*	0.494*
TS							0.972*	0.326*	0.314*
PropDR								0.389*	0.380*
PropABS									0.993*

Table 2. Coefficient r of Spearman correlations test between the indices of relative abundance considering each block as the sampling unit (N=34). PropD365, PropD180 and PropD90: proportion of people who reported having had rodents in their dwellings during the last 365, 180 and 90 days, respectively; PropN365, PropN180 and PropN90: proportion of people who reported having seen rodents in their neighborhood during the last 365, 180 and 90 days, respectively; TS: trap success; PropDR: proportion of dwellings with rodent capture; PropABS: proportion of bait stations with signs of rodent activity; and PropDABS: proportion of dwellings with bait stations with signs of rodent activity. * $p<0.05$ y +$p<0.10$.

	PropD180	PropD90	PropN365	PropN180	PropN90	TS	PropDR	PropABS	PropDABS
PropD365	0.963*	0.945*	0.665*	0.583*	0.671*	0.668*	0.573*	0.688*	0.716*
PropD180		0.963*	0.664*	0.636*	0.722*	0.679*	0.595*	0.779*	0.798*
PropD90			0.631*	0.718*	0.778*	0.716*	0.672*	0.706*	0.731*
PropN365				0.700*	0.731*	0.359+	0.335*	0.584*	0.650*
PropN180					0.964*	0.359+	0.484*	0.729*	0.753*
PropN90						0.362+	0.431*	0.768*	0.793*
TS							0.934*	0.368+	0.381+
PropDR								0.370+	0.379+
PropABS									0.993*

Table 3. Coefficient r of Spearman correlations test between the indices of relative abundance joining blocks according to their proximity, considering these new areas as sampling units (N=17). Symbols and abbreviations idem table 2.

	PropD180	PropD90	PropN365	PropN180	PropN90	TS	PropDR	PropABS	PropDABS
PropD365	1.000*	0.929*	0.500	0.714*	0.893*	0.750*	0.679+	0.886*	0.886*
PropD180		0.929*	0.500	0.714*	0.893*	0.750*	0.679+	0.886*	0.886*
PropD90			0.464	0.679+	0.821*	0.929*	0.857*	0.829*	0.829*
PropN365				0.571+	0.679+	0.357	0.607+	0.886*	0.886*
PropN180					0.893*	0.429	0.536+	0.829*	0.829*
PropN90						0.607+	0.679+	0.886*	0.886*
TS							0.893*	0.600+	0.600+
PropDR								0.600+	0.600+
PropABS									1.000*

Table 4. Coefficient r of Spearman correlations test between the indices of relative abundance per period of the year, maintaining the residential neighborhood and shantytown samplings separately (N=7). Symbols and abbreviations idem table 2.

The four models of regression presented intercept close to zero and positive slopes. For the proportion of people who reported having had rodents in their dwelling during the last 90 days and the proportion of dwellings with rodent capture, the slopes were significantly different from zero, while the proportion of bait stations with signs of rodent activity and the proportion of dwellings with bait stations with signs of rodent activity were only marginally significant, table 5.

Regressor	Model	*t*	*p*
PropD90	*a*: -0.002		
	b: 0.076	3.99	0.0004
PropDR	*a*: 0.001		
	b: 0.145	7.65	0.0002
PropABS	*a*: 0.016		
	b: 0.082	1.75	0.0828
PropDABS	*a*: 0.017		
	b: 0.042	1.49	0.0880

Table 5. Simple linear regression models $y_i = a + b\,x_i$, where y_i is the trap success, x_i are the other indices of relative abundance, a is the intercept and b is the slope of the line or regression coefficient. $t_i = b_i / SE(b_i)$, i = 1 until n, and p = exact probability of the value t_i for the regression coefficient estimated with a randomization method. Symbols and abbreviations idem table 2.

7. Discussion

The different evaluated methods detected evidence of rodent activity; however, the count of burrow entrances and animals only allowed to detect the presence of *R. norvegicus*. The differences in body size between *R. norvegicus* and the other smaller native species, in their behavioral habits or simply because they were not present in the area could be the cause for not detecting them with these techniques. The methods of kill trapping of animals are only accepted in particular cases and the methods producing a quick death and without suffering for the captured animals are advisable (Beaver et al., 2001). Taking into account this recommendation, the glue traps should not be used under any circumstance, because the animals captured could die due to stress or simply because of tiredness when trying to get released (Kravetz, *personal comments*). In addition, they present other disadvantages such as its use is limited to closed environments, with low humidity and without environmental dust since the external environmental conditions limit the glue adherence. Another problem of the glue as a method of kill trapping is the risk of capturing and killing unwanted species. This type of trapping is frequently used by pest controllers because it is economical and allow the capture of several animals per trap, while other killing traps (e.g. snap trap) are more expensive and become inactive after the first capture. When rodenticides are used to control rodents, the animals die at the site and many times in places that are difficult to reach; thus the pest controllers prefer to use glue traps at sites where it is risky to use toxic substances and it is also necessary to remove the animals from the site, such as food warehouses, food industries, supermarkets, etc.

The use of footprint traps was not considered as it is possible to be used in closed spaces, but with some limitations in open areas, and they can be disturbed by other animals, wind, etc.

In relation to the detection of differences in the abundance, the bait stations were more sensitive than the direct observation; probably because of the low number of days the counts were performed, and because of the large variation per interval of time and between the different times of the day in the number of active animals. A correct estimate of the abundance using this last method requires a lot of intervals of observation. The count of burrow entrances can also be used to estimate the abundance, but its use would be limited by the visibility conditions of the entrances in relation to its size and by the visibility of the habitat. This technique may not be appropriate under high cover conditions or where the rodent density may not be as high as the one observed in this sampling, since even 17 individuals were recorded in 1800 m^2 in a period of five minutes (in a time of the day where *R. norvegicus* has low activity, Macdonald et al., 1999) and an average of two individuals were captured with eight traps in an hour.

The bait stations were useful to detect the presence of rodents in the coastal area; they detected changes in the abundance due to the use of rodenticide, and showed an association with the trap success both in natural environments as the urban reserve, and in residential environments as the studied neighborhoods.

The proportion of bait stations with signs of rodents seem to be an adequate variable to estimate the abundance of small mammals (Blackwell et al., 2002; Brown et al., 1996; Gurnell et al., 2001; Gurnell et al., 2004), while the quantity of consumed bait would be affected by competition, microhabitats preferences and the risk of predation (Brown, 1988; Kotler, 1997). The use of bait stations allows performing monitoring programs of pest species in big areas due to its low cost (Battersby & Greenwood, 2004). Disposable containers could be used as bait stations and then discarded after the sampling, which simplifies the post sampling activities, since they do not need to be disinfected and washed like in the case of traps. In addition, due to its low cost, they do not represent an expensive element for people, being low the risk of loss due to theft. This allows its use in a wide variety of public spaces such as lawns, parks and in the streets. On the other hand, the bait stations are easy to be prepared and set, and in relation to the wood pegs they are easier to locate and are more effective to detect rodents. However, in the same way as it happens with the methods that involve animal trapping, there are a number of factors that will affect this index of relative abundance; thus, the indices can only be comparable under similar conditions and during short intervals of time.

The use of bait stations as a method of rodent sampling has the disadvantage of not providing any information regarding the individuals; such as species, body size, sex, reproductive condition, etc. The addition of sticky tape to the bait station where samples of animal hair were left could allow the identification of rodent species that visited it, due to the fact that the hair has specific characteristics (Busch, 1986; Cavia et al., 2008; Day, 1966). Nevertheless, for this purpose it is necessary to have an identification key according to the morphological characteristics of the hairs of the species likely to be present in the study area.

The use of surveys is a methodology widely employed in the field of sociology (Galtung, 1978; Kerlinger, 1988) and there is a significant number of works where they are used to assess the condition of the population of wild species with some risk of preservation or with an economic relevance (Filion, 1978). Sometimes hunters, park rangers, naturalists, etc. are surveyed because they are considered well-qualified. In the present study non-qualified people were surveyed. Surveyed people remembered quite accurately the moment and the place where they were in contact with rodents, probably because of being afraid of them. The surveys can only be used in inhabited places (residential and/or work) where people stay most of the time. The question asked to people about whether they had rodents in their dwelling seems to be more adequate than the question about whether they had rodents in their neighborhood, since the responses of the first question were associated to the other indices of relative abundance at the three analyzed scales, while the responses to the second question showed associations only with some indices and at some scales. Besides, the proportion of people who reported having had rodents in their dwelling during the last 90 days presented a linear relation with the trap success, indicating that both indices vary proportionally.

It would be useful to compare the indices used with absolute values of rodent abundance, but for this purpose some assumptions need to be met, and they are sometimes difficult to guarantee. In order to estimate the abundance with capture-mark-recapture samplings like the ones performed in the reserve; it is necessary to have recapture rates higher than 20%, which were not reached in this environment and with the sampling design made. In the case of removal samplings such as the ones performed in the neighborhoods, in order to apply the pattern of capture per effort unit (Hayne, 1949), there should be a decrease in the number of animals captured in the following days and this did not occur in these samplings. Due to the impossibility of calculating absolute abundances with the data obtained, the indices were contrasted, and the trap success was considered as the most reliable way to estimate the relative abundance because it is associated with the absolute abundance (Bronner & Meester, 1987) and it is widely used to estimate the rodent abundance.

8. References

Aplin, K.P., Brown, P.R., Jacob, J., Krebs, C.J., & Singleton, G.R. (2003). *Field methods for rodent studies in Asia and the Indo-Pacific.* Australian Center for International Agricultural Research, Camberra.

Battersby, J.E., & Greenwood, J.J.D. (2004). Monitoring terrestrial mammals in the UK: past, present and future, using lessons from the bird world. *Mammal Review,* Vol.34, No.1-2, pp. 3-29.

Battersby, S.A., Parsons, R., & Webster, J.P. (2002). Urban rat infestations and the risk to public health. *Journal of Environmental Health Research,* Vol.1, No.2, pp. 4-12.

Beaver, B.V., Reed, W., Leary, S., McKiernan, B., Bain, F., Shultz, R., et al. (2001). Report of the AVMA panel on euthanasia. *Journal of American Veterinary Medical Association,* Vol.218, No.5, pp. 669-696.

Begon, M. (1979). *Investigating Animal Abundance: Capture-recapture for biologist.* Edward Arnold, London.

Begon, M., Harper, J.L., & Townsend, C.R. (1987). *Ecología, individuos, poblaciones y comunidades* (M. Costa, Trans.). Ediciones Omega, S.A., Barcelona.

Blackwell, G.L., Potter, M.A., & McLennan, J.A. (2002). Rodent density indices from tracking tunnels, snap-traps and Fenn traps: do they tell the same story? *New Zealand Journal of Ecology*, Vol.26, No.1, pp. 43-51.

Bronner, G., & Meester, J. (1987). Comparison of methods for estimating rodent numbers. *Sud Africa Journal of Wildlife Research*, Vol.17, No.2, pp. 59-63.

Brown, J.S. (1988). Patch use as an indicator of habitat preference, predation risk, competition. *Behavioral Ecology and Sociobiology*, Vol.22, pp. 37-47.

Brown, K.P., Moller, H., Innes, J., & Alterio, N. (1996). Calibration of tunnel tracking rates to estimate relative abundance of ship rats (*Rattus rattus*) and mice (*Mus musculus*) in a New Zealand forest. *New Zealand Journal of Ecology*, Vol.20, No.2, pp. 271-275.

Brownie, C., Hines, J.E., & Nichols, J.D. (1986). Constant - parameter capture - recapture models. *Biometrics*, Vol.42, pp. 561-574.

Busch, M. (1986). Identificación de algunas especies de pequeños mamíferos de la provincia de Buenos Aires mediante caracteristicas de sus pelos. *Physis (Buenos Aires)*, Vol.44, No.107, pp. 113-118.

Castillo, E., Priotto, J., Ambrosio, A.M., Provensal, M.C., Pini, N., Morales, M.A., et al. (2003). Commensal and wild rodents in an urban area of Argentina. *International Biodeterioration & Biodegradation*, Vol.52, No.3, pp. 135-141.

Cavia, R., Andrade, A., Zamero, M., Fernandez, S.M., Muschetto, E., Cueto, G.R., et al. (2008). Hair structure of small rodents from central Argentina: A tool for species identification. *Mammalia*, Vol.72, pp. 35-42.

Cavia, R., Cueto, G., & Suárez, O. (2009). Changes in rodent communities according to the landscape structure in an urban ecosystem. *Landscape and Urban Planning*, Vol.90, pp. 11-19.

Ceruti, R., Ghisleni, G., Ferretti, E., Cammarata, S., Sonzogni, O., & Scanziani, E. (2002). Wild rats as monitors of environmental lead contamination in the urban area of Milan, Italy. *Environmental Pollution*, Vol.117, No.2, pp. 255-259.

Chani, J.M. (1980). *Guia de métodos de captura para el estudio de los vertebrados*. Universidad de Mar del Plata, Mar del Plata.

Channon, D., Cole, M., & Cole, L. (2000). A Long-term study of *Rattus norvegicus* in the London Boroygh of Enfield using returns as an indicator of sever population levels. *Epidemiol. Infect.*, Vol.125, pp. 441-445.

Childs, J.E., McLafferty, S.L., Sadek, R., Miller, G.L., Khan, A.S., DuPree, E.R., et al. (1998). Epidemiology of rodent bites and prediction of rat infestation in New York city. *American Journal of Epidemiology*, Vol.148, No.1, pp. 78-87.

Daniel, W.W. (1978). *Applied Nonparametric Statistics*. Houghton Mifflin Company, Boston.

Davis, D.E. (1951a). A comarison of reproductive potential of two rat populations. *Ecology*, Vol.32, pp. 469-475.

Davis, D.E. (1951b). The relation between level of population and size and sex of norway rats. *Ecology*, Vol.32, pp. 459-461.

Davis, D.E. (1951c). The relation between the level of population and the prevalence of leptospira, salmonella, and capillaria in norway rats. *Ecology*, Vol.32, pp. 465-468.

Day, M.G. (1966). Identification of hair and feather remains in the gut and faeces of stoat and weasels. *Journal of Zoology (London)*, Vol.148, pp. 201-217.

Filion, F.L. (1978). Increasing the effectiveness of mail surveys. *Wildlife Society Bulletin*, Vol.6, No.3, pp. 135-141.

Filion, F.L. (1987). Encuestas humanas en la gestión de la vida silvestre (B. Orejas Miranda, Trans.). In *Manual de técnicas de gestión de vida silvestre* (4th edition ed. R. Rodriguez Tarrés (Ed.), 463-477, The Wildlife Society, Inc., Maryland.

Galtung, J. (1978). *Teoría y método de la investigación social.* (5th edition. Vol. 1). Editorial Universitaria de Buenos Aires, Buenos Aires.

Glass, A.G., Korch, G.W., & Childs, J.E. (1988). Seasonal and habitat differences in growth rates of wild *Rattus norvegicus*. *Journal of Mammalogy,* Vol.69, No.3, pp. 587-592.

Gurnell, J., Lurz, P.P.W., & Pepper, H. (2001). *Practical Techniques for Surveying and Monitoring Squirrels.* Edinburgh: Forestry Commission Practice Note 11. Forestry Commission.

Gurnell, J., Lurz, P.W.W., Shirley, M.D.F., Cartmel, S., Garson, P.J., Magris, L., et al. (2004). Monitoring red squirrels *Sciurus vulgaris* and grey squirrels *Sciurus carolinensis* in Britain. *Mammal Review,* Vol.34, No.1-2, pp. 51-74.

Hawthorne, D.W. (1987). Daños provocados por animales silvestres y técnicas de control. In *Manual de técnicas de gestión de vida silvestre* (ed., R. Rodriguez Tarrés, Vol. 431-462, The Wildlife Society, Inc., Maryland.

Hayne, D.W. (1949). Two methods for estimating population from trapping records. *Journal of Mammalogy,* Vol.30, No.4, pp. 399-411.

Kerlinger, F.N. (1988). Investigación de encuestas. In *Investigación del Comportamiento* (427-439, Mc Graw-Hill.

Kotler, B.P. (1997). Parch use by gerbils in a risky environment: manipulating food and safety to toest four models. *Oikos,* Vol.78, pp. 274-282.

Krebs, C.J. (1966). Demographic changes in fluctuating populations of *Microtus californicus. Ecological Monographs,* Vol. 36, pp. 239-273.

Krebs, C.J. (1978). *Ecology: The experimental analysis of distribution and abundance* (Second edition ed.). Harper & Row, New York.

Krebs, C.J. (1989). *Ecological Methodology.* Harper & Row, New York.

Langton, S.D., Cowan, D.P., & Meyer, A.N. (2001). The occurrence of commensal rodents in dwellings as revealed by the 1996 English House Condition Survey. *Journal of Applied Ecology,* Vol.38, No.4, pp. 699-709.

Macdonald, D.W., Mathews, F., & Berdoy, M. (1999). The behaviour and ecology of *Rattus norvegicus*: from opportunism to kamikaze tendencies. In *Ecologically-based rodent management* (ed.,G.R. Singleton, H. Leirs, L.A. Hinds & Z. Zhang (Eds.), 49-80, Australian Center for International Agricultural Research, Camberra.

Magurran, A.E. (1988). *Ecological diversity and its measurement.* Croom Helm, London.

Manly, B. (1991). *Randomization and Monte Carlo methods in biology.* Chapman and Hall, London.

Manly, B. (1996). RT: a program for randomization testing (Version 2.0). Dunedin: CASM.

Marshall, P.A., & Murphy, R.G. (2003). Investigating residents′ perceptions of urban rodents in Manchester, UK. In *Rats, mice and people: Rodent biology and management* (ed.,G.R. Singleton, L.A. Hinds, C.J. Krebs & D.M. Spratt (Eds.), 473-476, ACIAR, Camberra.

Rabinovich, J.E. (1980). *Introducción a la ecología de poblaciones animales.* (1st edition, Vol. 313). Compañia Editorial Continental, S.A., Mexico DF.

Seber, G.A.F. (1973). *The estimation of animal abundance and related parameters.* Charles Griffin & Company Limited, London.

Sokal, R.R., & Rohlf, F.J. (1995). *Biometry, the principles and practice of statistics in biological research.* (Third edition). W.H. Freeman and Company, New York.

Southwood, T.R.E. (1978). *Ecological Methods. With particular reference to the study of insect populations* (2nd edition). Chapman and Hall, New York.

Traweger, D., & Slotta-Bachmayr, L. (2005). Introducing GIS-modelling into the management of a brown rat (*Rattus norvegicus* Berk.) (Mamm. Rodentia Muridae) population in an urban habitat. *Journal of Pest Science,* Vol.78, No.1, pp. 17-24.

Yo, S., Marsh, R.E., & Salmon, T.P. (1987). Correlation of two census methods (food consumption and gnawing evidence) for assessing norway rat popoulations. In *Vertebrate pest control and management materials* (ed.,S.A. Shumake & R.W. Bullard (Eds.), Vol. 5, 81-88, American Society for Testing and Materials, Philadelphia.

Zar, J.R. (1996). *Biostatistical Analysis* (Third edition ed.). Printice Hall, New Jersey.

8

Insectigation in Vegetable Crops: The Application of Insecticides Through a Drip, or Trickle, Irrigation System

Gerald M. Ghidiu
Rutgers – The State University
New Brunswick, New Jersey
USA

1. Introduction

Drip, or trickle, irrigation can be defined as a method of uniformly delivering water to a plant's root zone through point or line sources (emitters) on or below the soil surface at a small operating pressure (Dasberg & Or, 1999). Modern drip irrigation systems use low pressure (~34.48-68.95 kPa [5-10 psi]) to force water through plastic or metal tubing with emitters spaced at regular intervals down it's length to deliver water to the plant's root zone, and can be either a surface system (tubing on top of the soil) or a subsurface system (tubing buried beneath the soil). Water savings with drip irrigation can be as high as 80% compared with other irrigation methods (Bogle & Hartz, 1986).

The basic concept of efficient irrigation using less water dates back centuries ago, but the idea of drip irrigation using tubing was used by crop producers as early as the 1860's in Europe for subsurface irrigation using perforated metal irrigation pipe (Ross et al.,1978). Modern day commercial drip irrigation was not possible until the development of plastics during World War II which enabled drip irrigation equipment and supplies to be economical for use by crop producers. During the late 1970's researchers were successful with injecting liquid fertilizers through drip irrigation systems, followed soon thereafter with other agrichemicals including insecticides and fungicides. Today, many agricultural chemicals are labeled for application through various irrigation systems, including overhead, sprinkler, and drip/trickle, in vegetable and other crops.

2. History of drip chemigation

Although researchers experimented with drip irrigation systems before the 1950's, the first use of a modern drip irrigation system was conceived by Symcha Blass (Blass, I. & S. Blass, 1969), a retired British Water Agency worker. His ideas of micro-tubing for irrigation included special low volume water emitters that overcame the clogging of drip holes by particles carried in the water. He patented his plastic emitters in 1959 with Kibbutz Harzerim in Israel under the trade name 'Netafilm Company', producing the first practical surface irrigation drip emitters.

The initial use of drip tubing in conjunction with a plastic row cover, together called 'plasticulture' (Lamont Jr., 2004) was conducted in a cucurbit field at Old Westbury Gardens, Long Island, NY in 1963 by R. Chapin of Chapin Watermatics, Inc, and N. Smith, a Nassau County Agricultural Agent (Ayars et al., 2007). Over the next decade, plastic row covers and drip tape improved, and inexpensive, consistent emitters with a constant discharge rate were quickly developed. Commercial drip irrigation rapidly expanded to >54,000 ha in the USA by 1975, being used on various crops for water management in 35 different states, and between 1982 and the late 1990's drip irrigation increased in the United States 650% (Anonymous, 2000).

As drip/trickle chemigation quickly became adopted by growers, research investigations on the injection of agricultural chemicals though the same system rapidly increased. Fertilizers were first injected into a drip irrigation system in 1979 to tomatoes and eggplant in New Jersey (Paterson, 1980), and to cucurbits and other vegetables in California (Hall, 1982). Insecticides were first injected into a drip irrigation system in bell peppers in New Jersey for the control of European corn borer (*Ostrinia nubilalis* Hubner) in 1980 (Ghidiu & Smith, 1980), and the following year in lima beans for control of the Mexican bean beetle (*Epilachna varivestis* Mulsant) (Ghidiu, 1981). However, the insecticides were not effective against these pests when applied systemically and neither trial resulted in insect pest reduction. The first successful drip application of an insecticide for insect pest reduction was conducted in 1985 (Wildman & Cone, 1986), where asparagus aphid (*Brachycorynella asparagi* [Mordvilko] numbers in asparagus were significantly reduced using disulfoton (Di-Syston 6E, Miles Inc., Elkhart, IN) as compared with the untreated. Ghidiu (1992) used a small ¼ hp electric pump to inject carbofuran (Furadan 4F, FMC Corporation, Philadelphia, PA) and methomyl (Lannate 1.8L, E. I. DuPont de Nemours & Co., Wilmington, DE) through a drip irrigation system under black plastic mulch for European corn borer control in bell peppers, but reported no reduction in borer damage and significant phytotoxicity to the pepper plants, demonstrating that injected materials must not only be efficacious but must also be safe to the plants. Successful insectigation trials were reported with entomopathogenic nematodes for the control of spotted cucumber beetles (*Diabrotica undecimpunctata howardi* [Barber]) in 1986 (Reed et al., 1986), followed by the effective control of aphid (Aphididae spp.) populations by chemigating imidacloprid in vegetables in Arizona in the mid-1990's (Kerns & Palumbo, 1995; Palumbo, 1997), and effective control of spotted cucumber beetles in melons in Virginia with drip-applied imadcloprid and thiamethoxam (Kuhar & Speese, 2002). In a 3-yr field trial starting in 2004, chlorantraniliprole was shown to be highly effective against the European corn borer in bell pepper when applied through a drip irrigation system (Ghidiu et al., 2009). Further, chlorantraniliprole applied through a drip irrigation system significantly reduced armyworms (*Spodoptera* spp.) and fruitworms (*Helicoverpa* zea [Boddie]) in tomatoes in field tests in both Virginia (Kuhar et al., 2009) and Florida (Schuster et al., 2009). And Ghidiu (2009) reported that chlorantraniliprole and thiamethoxam injected via a drip irrigation system significantly reduced damage to eggplant foliage caused by flea beetles (*Epitrix* spp.) and leafminers (*Liriomyza* spp.).

3. Effectiveness of insectigation

During the mid-1990's, researchers reported effective control of beetles, aphids, whiteflies, and several other insect pests using foliar applications of a newly-developed class of insecticides, the neonicotinoids. These new-chemistry insecticides are especially suited for application through a drip irrigation system because they are highly soluble, they are root systemic and essentially non-phytotoxic to most plants, they are highly effective against specific pests, and they are considered by the USEPA to be reduced-risk pesticides. Felsot et al. (1998) examined the distribution of imidacloprid in soil when applied through a drip irrigation system and concluded that it is a good candidate for insect control via drip irrigation systems.

More recently, another new class of insecticides, the anthranilic diamides, has been shown to be highly toxic to numerous caterpillar pests (Lahm et al., 2005). One of these insecticides, chlorantraniliprole, is xylem-mobile through root uptake and controls caterpillars and other leaf-feeding pests (Lahm et al., 2007). Like the neonicotinoid-class insecticides, chlorantraniliprole is also highly soluble, root systemic, and effective against specific insect pests, especially caterpillars, leafminers, and beetles. Because both of these materials are selective against certain insect pests, they are ideal materials for a pest management program.

Currently, the USEPA has approved and labeled numerous insecticides of different classes for application through a drip irrigation system in fruits and vegetables for the control of a wide variety of insect pests:

Common name	US Brand name	Insecticide class
azadirachtin	Aza-direct	limonoid insect growth regulator (neem)
chlorantraniliprole	Coragen	anthranilic diamide
clothianidin	Belay	neonicotinoid
dimethoate	Dimate	organic phosphate
diazinon	Diazinon	organic phosphate
dinotefuron	Venom	neonicotinoid
imidacloprid	Admire PRO	neonicotinoid
malathion	Malathion 8 Aquamul	organic phosphate
methomyl	Lannate	carbamate
oxamyl	Vydate	carbamate
rosemary+peppermint oils	Ecotec	botanical
thiamethoxam	Platinum	neonicoinoid
thiamethoxam + chlorantraniliprole	Durivo	neonicotinoid

4. Advantages and disadvantages of insectigation

There are both advantages and disadvantages to injecting agricultural chemicals, including insecticides, into a drip/trickle irrigation system. Additionally, some states in the U.S. require that the irrigation operator register with a specific State Department (such as Department of Water, Natural Resources, Agriculture, or Environmental Control, etc.) before using any chemigation with overhead or drip/trickle systems. The irrigation operator may also be required to keep records of each chemigation application including the date, type and brand name of chemical, the field area covered by the injection, and the amount of material used.

4.1 Advantages of insectigation

1. The total insecticide input for control of targeted insect pests in most crops is significantly reduced when compared with that of traditional foliar applications, while at the same time essentially 100% protection of the plant is obtained because these materials are root systemic and translocate throughout the plant, resulting in a more even distribution of the pesticide within the plant. For some vegetable crops, 1-2 drip/trickle irrigation applications of an insecticide during the season result in equivalent control, or better control, of insect pests than that of multiple foliar sprays. Kuhar et. al. (2009) reported that a single injection of the high labeled rate of chlorantraniliprole (Coragen; E.I. DuPont de Nemours Inc., Wilmington, DE) into a drip system was as effective as 4 foliar applications of the pyrethroid lambda-cyhalothrin (Warrior II; Syngenta Crop Protection, Inc., Wilmington, DE) for control of caterpillar damage in fresh market tomatoes, and Ghidiu et at. (2009) reported that 2 injections of chlorantraniliprole into a drip system was as effective as 7 applications of a standard grower foliar spray program consisting of 2 applications of acephate (Orthene 97; United Phosphorus, Inc., King of Prussia, PA) followed by 5 applications of indoxacarb (Avaunt 30WDG; E.I. DuPont de Nemours, Inc., Wilmington, DE) for control of European corn borer in bell peppers.
2. Less energy is required to transport water and insecticide solutions at the lower pressures and velocities in drip/trickle systems as compared with other irrigation systems. Also, because fewer applications are needed, less energy input is required than by tractor or other application methods, and no soil compaction occurs that results from heavy tractors or spray equipment being operated over the field. The fewer times a tractor goes over the field, the less potential for plant damage caused by the tractor operation.
3. Pathogen movement through the field via water flowing over the soil surface can be reduced through the use of a drip/trickle irrigation system if plastic mulch is used in combination with drip tubing. This is especially true for plant diseases such as *Phytophthora capsici*, a soil-borne fungus, which produces spores that are spread via water splashing up onto the foliage by rainfall, operation of field equipment and sprayers, etc.
4. Weather is not a factor during application, as injection of an insecticide into a drip/trickle irrigation system can be made in wind or rain, or when fields are too muddy or soft to operate ground equipment without getting stuck.

5. Applicator exposure, both in terms of physical contact and time of exposure, to insecticides during application is significantly reduced.
6. For many growers in areas that are experiencing urban encroachment, insectigation can be completed without spray drift, eliminating 'application visibility' that concerns these growers.
7. Plant growth may be enhanced through the use of drip irrigation systems because with frequent drip waterings, it is possible to maintain a more optimum balance between soil water, plant needs and aeration. Healthy plants are less susceptible to insect pest problems than unhealthy plants or plants under water stress.
8. And because many of these new-chemistry insecticides are selective to specific insect pests, they are generally less toxic or disruptive to non-target species and beneficial organisms, including insect predators, parasites and pollinators. The injection of insecticides through a drip/trickle irrigation system thus fits well into an integrated pest management program.

4.2 Disadvantages of insectigation

1. The initial capital expenditures for a complete drip system and additional injection equipment can oftentimes be greater than that for sprinkler, overhead or other irrigation systems. However, the more the system is used, the smaller the costs per hour of operation. And most drip systems have many re-useable components (pumps, filters, tubing, hoses, injectors, etc).
2. Drip/trickle systems generally require consistent maintenance and monitoring of all equipment for constant pressure, leaks in any part of the system, plugged emitters, etc. Specific safety equipment is required, and additional safety precautions must be followed.
3. Water carries particulates that can clog the emitters if the filters malfunction or are not maintained properly (backwashing, cleansing, etc). If plastic mulch is used in combination with the drip/trickle system, clogged emitters are sometimes difficult to locate, resulting in uneven distribution of the insecticide when pumped into the drip system. Also, if plastic drip lines and tubing are not properly rinsed after each use, it is possible that emitters may become clogged with residue.
4. Drip line repairs can sometimes be time consuming and costly. Bright, direct sunlight may affect some plastic tubing used for drip irrigation, shortening their useable life span. Also, small rodents (mice, rabbits, chipmunks, etc) and certain soil insects (crickets, wireworms, ground beetles, others) may chew on drip hose, especially in droughts, causing small leaks that subsequently result in loss of pressure and uneven distribution of the insecticide. Oftentimes, such small leaks are difficult to detect under row covers such as black plastic until the row becomes saturated and wet spots appear.
5. After the final harvest, clean-up costs of drip/trickle systems may be higher than with other irrigation systems. Costs may include removal of plastic row covers, all drip lines and tubing, and injection and safety equipment. The disposal cost of used plastics (plastic mulches, drip lines) continually increases.
6. Top-dressed fertilizers and some herbicides may need additional sprinkler application for activation, especially if a plastic mulch is not used over the row.

7. Salts can accumulate as a result of inadequate flushing of the drip/trickle irrigation system (Dasberg & Or, 1999), particularly at the perimeter of the wetted area.

5. Drip chemigation system requirements and operation

Drip chemigation systems that will have insecticides injected into them must be properly engineered, installed, and maintained over the season to ensure a uniform distribution of outflow (Ross, 2004). Chemigation requires that two separate hardware systems be joined together, operating as a single system. The components of the first system, a typical drip/irrigation system, include:

- main water source (well, irrigation pond, etc). Public water supplies cannot be used.
- main water pump
- water filter system (with sand screens, screen filters, flush valve or drain)
- backflow prevention valve/backwash controller
- pressure gauge
- low pressure shutoff valve
- low pressure sensor/shutoff switch
- pressure relief valve
- various diameter hoses and polytubing to carry water to the plant roots (main lines, lateral and drip lines, etc). Drip lines are available with a wide range of emitter spacings, and can be found with spacings of 4″, 8″, 12″, 18″, or even 24″, with flow rates of 12 to 64 gallons or more per 100′/hour.

Most growers that currently use some form of drip/trickle irrigation as a water-management tool can easily, and inexpensively, add the necessary equipment to properly inject agrichemicals. The components of the second system, in addition to the typical drip irrigation system equipment listed above, include the following pieces of equipment that are needed before injecting any agricultural chemical into the system:

- chemical mix tank, such as a 19 L [5 gallon] plastic jug, preferably with an agitator and an outlet filter to prevent clogging of emitters by the chemical solution
- containment tray or pan to catch any chemical solution leakage or spillage
- positive displacement pump or other reliable solution metering device which provides a consistent flow rate at low pressures
- backflow prevention valve to prevent backflow of solution into the mixing container
- low pressure shutoff valve to shut the injection system off should a loss of pressure occur

A basic drip/trickle irrigation system with an additional insectigation system using a positive displacement metering pump for injection of an insecticide is shown in Figure 1.

6. Injection pumps

The injection pump is a critical component of a chemigation system, and must be properly installed and maintained to ensure an even flow of the chemical solution to every emitter in the irrigation system. Injection should be on the downflow side of the main pump filters to avoid potential site contamination as a result of the filter back-flush operation. Two basic types of chemical injection pumps are available for the application of agrichemicals (insecticides, fertilizers, etc) through low-pressure drip/trickle irrigation systems: the volumetric water flow pump and the positive metering pump. Regardless of type used, it should be a pump that provides consistent flow rates at low pressures. For agricultural use, both should be made of materials that are resistant to corrosion by fertilizers, acids, chlorine, etc., and both should have adjustable injection rates at various pressures. Both types of pumps have models that may deliver a flow rate as little as 11.4 liter/h (3 gal/h) at operating pressures of 20 kPa (3 psi) or more.

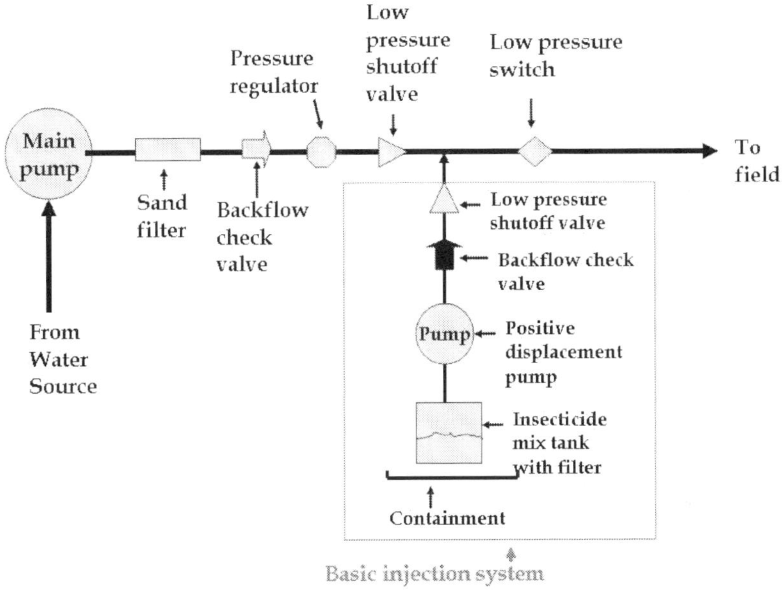

Fig. 1. A basic drip/trickle irrigation system with chemigation capabilities using a positive displacement injection metering pump.

6.1 Volumetric pumps

There are two types of volumetric water flow pumps available. Both operate on the flow of water through the pump to suction out the chemical solution in the mixing tank.

6.1.1 Venturi pumps

Venturi pumps (Fig.2) are powered by water that flows through a constriction (Fig. 3) in the pump tubing, creating a change in water flow velocity, which creates a pressure differential that forms a vacuum. The vacuum pulls the chemical solution out of the chemical solution tank through a small diameter hose and injects a precise amount of chemical solution into the irrigation system in proportion to a certain volume of water. However, the injection rate varies with the pressure differential, and an accurate regulating valve and flow meter are needed for calibrating the system if a precise metering of the chemical solution is necessary. The advantages of Venturi pumps are that these pumps are relatively inexpensive and very simplistic in that they essentially have no moving parts (except for the solution moving through it). All the suctioning activity is completed by the vacuum created within the pump. A valve in the main line between the inlet and outlet of the Venturi pump assists in the control of the volume of water flow. As with all other injection pumps, a backflow check valve is required. Filters just before the pump to remove particulates in the water are not as critical in the Venturi pump as it is with other pumps, but they are still recommended to be installed to prevent any potential clogging of the system.

Courtesy of E. Simmone, UF-IFAS.

Fig. 2. Venturi pump with a flow control valve between the water in port and the water out port.

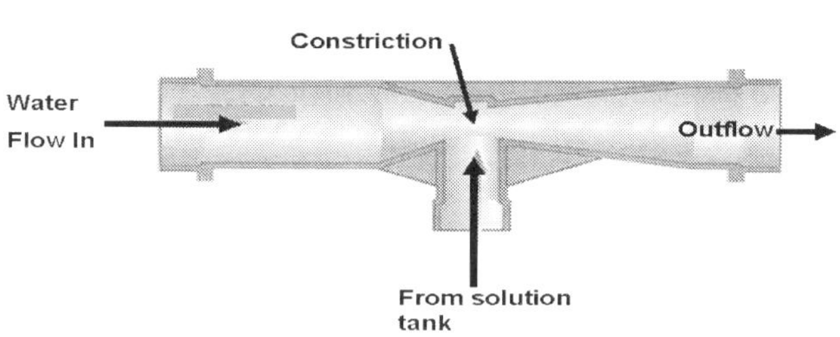

Fig. 3. Cross-sectional view of a Venturi pump. Note constriction in the center of the pump to create a suction to pull solution from chemical tank.

6.1.2 Proportional liquid injectors

A more complex water-driven pump than the Venturi pump is the proportional liquid injector (Figs. 4, 5). This pump injects a precise amount of chemical solution proportional to a volume of water, and operates with system flows of as low as 3.8 liter/hour (1 gal/hr). Water pressure is the power source (electric is not needed), as the water flow operates a piston inside the pump which takes up the required percentage of concentrate from the chemical solution tank, and in-line water pressure forces the solution downstream through the irrigation system (Fig. 6). The dose (concentration) of the chemical solution picked up is directly proportional to the volume of water entering the pump, regardless of variations in flow or pressure which may occur in the main water line. A bypass valve allows clean water to be supplied without operation of the pump, and also allows the pump to be easily dismantled while the irrigation system is operating. Proportional liquid injectors have moving parts within the pump, and these units cost significantly more than the Venturi pumps. In-line filters (300 mesh- 60 microns, depending on water quality) to remove particulates in the water supply are critical to the trouble-free operation of the pump. As with other chemigation injection units, a backflow check valve is required for proportional liquid injectors.

Fig. 4. A Chemilizer[R] volumetric injection pump.

Fig. 5. A Dosatron[R] injection pump injecting a injecting a pesticide in bell pepper, Bridgeton, NJ.

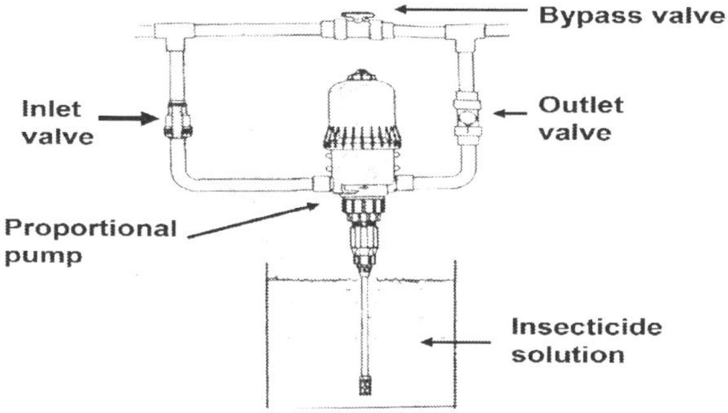

Fig. 6. Diagrammatic sketch of a proportional liquid injector pump connected to a drip/trickle irrigation system with a chemical solution mix tank.

6.2 Positive displacement metering pumps

Several types of positive displacement metering pumps are available, including both electric (Fig. 7) and gasoline-operated metering pumps (Fig. 8) and hydraulic metering pumps. The positive displacement pump has an expanding cavity on the suction side of the pump where a liquid solution is sucked in from the insecticide mix container, and a decreasing cavity on the outlet side of the pump where the solution is forced out into the irrigation system as the cavity collapses. The volume of liquid discharged is always constant for each cycle of operation by the pump, hence the term 'positive displacement'. These pumps can be easily and quickly transported between fields.

Fig. 7. A positive displacement electric (1/3 hp) metering pump with a 5-gallon jug for the concentrated insecticide solution and a containment tray beneath to catch all drips, spillage, etc. In the center of the picture is a flush valve to quickly and completely rinse all solution from within the pump.

Hydraulic drive metering pumps are positive displacement pumps that use the water pressure in the irrigation main lines from the irrigation well to power the pump instead of electric or gasoline power. Pumping cylinders can be mounted in parallel for large volume applications, or for injection of two non -compatible materials. Some models of hydraulic metering pumps have an adjustable piston stroke length to quickly change the precise flow of chemical solution to be injected, while other models change the injection rate by using a variable frequency drive on the pump which can vary the speed of the pump with the water flow rate.

Fig. 8. A John Blue[R] E-Z meter injection pump. This pump is operated by a gasoline-operated engine.

7. Insectigation and water management

- Underwatering during injection will prevent the insecticide from uniformly reaching the root zone of all plants, reducing the systemic uptake and thus its effectiveness against the targeted insect pests.
- Overwatering (excessive watering during or after injection) increases the potential of the injected agrichemical to leach or move away from the root zone. Agrichemicals applied via a drip/trickle irrigation system are highly soluble, and too much water may reduce their effectiveness. Some insecticides, such as methomyl (Lannate L; E.I. DuPont de Nemours & Co., Wilmington, DE) and oxamyl (Vydate L; E.I. DuPont de Nemours & Co., Wilmington, DE), specify that best results are obtained when the product is applied at the end of the irrigation cycle, minimizing flush time to prevent the loss of efficacy. Other insecticides, such as rynaxypyr[R] (Coragen[R]), specify that best results are obtained when the product is injected at the beginning of the irrigation cycle (without over-irrigating). Thus it is important to carefully follow directions on the pesticide label.
- Uniform applications of the insecticide solution are necessary for consistent, effective control. Uniformity of application is controlled primarily by the duration of the injection period. Too short of an injection period will result in non-uniform distribution of the agrichemical, and not all plants will receive insecticide treatment alike. For very large fields, it may be best to establish irrigation blocks to reduce the size of the irrigated field which may result in a more uniform distribution of the injected material.
- After insectigation is complete, thoroughly rinse the irrigation system with clean water for the minimum injection time to ensure clog-free operation. It may take a considerable amount of time to completely remove all of the injected chemical from the drip/trickle irrigation system. For a thorough rinsing of the chemigation system, clean water should be pumped through the entire system for approximately twice the amount of time it takes water to leave the pump and reach the most distant emitter (the minimum injection time - see below).

7.1 Timing of insectigation applications

Injecting an insecticide via a drip/trickle irrigation system offers great flexibility in application timing. Depending on the presence of insect pests, the time required between root uptake and translocation throughout the plant needs to be considered. As a general rule, pest control is usually obtained within 24 hours after injection, depending on factors such as emitter spacing, length of time of injection, selection of insecticide, and plant growth stage.

- The overall objective of insectigation is to have an equal amount of insecticide released through every emitter in the system in order to have a uniform applicaiton to the root zone of all plants.
- The minimum injection time is the time needed for water to leave the injection pump and reach the most distant emitter in the field. To determine the minimum injection time, inject approximately 4 liters (about 1 gallon) mixture of water with a few drops of a household detergent soap, or with a few drops of a soluble food dye, through the system. Record the time beginning when the injection starts until the soap bubbles or dye reaches the furtherst (very last) emitter - this is the minimum amount of time it takes for an injection to fill the system. Any injection time less than this will result in unequal application of the insecticide to the plants.
- Extending the length of time to complete the insecticide injection will improve uniformity of application delivery, especially in larger fields. Too short of an injection time will result in unequal application of the insecticide. As a general rule, the maximum injection time should last for no longer than 2 hours per irrigation block or zone (if the system is zoned).
- It is recommended that the injection of the insecticide be targeted to the middle third of an irrigation cycle. For example, if the irrigation cycle is 180 minutes, injection of the chemical should commence after the first 60 minutes.
- Run the drip/trickle irrigation system at the correct operating pressure for at least 30-60 minutes before injecting any insecticide. This will prime the system, wet the root zone of the plants, and ensure rapid, even uptake of the injected material.

7.2 Calculation of rates of insecticides (amount to inject)

To calculate the rate or amount of an insecticide to inject in a drip/trickle irrigation system, it is necessary to first determine the effective wetting zone. The wetting zone can be modified by changing the placement of the drip tape, the drip tape emitter spacing, the drip tape flow rate, or the frequency of water applications (the time the irrigation system is operating).

Example 1 (crop grown on bare ground). Crop is on beds 1.5 m (5 ft) wide planted to bare ground (drip irrigation but no plastic mulch). After applying enough water to wet the root zone of the plants, determine the width of the wet zone. The width of the wet zone X the total length of the rows under irrigation will yield the area (squared) of the wet zone. The rate should be based on this area. For example, if the total length of the rows is 8,712 row ft and the wet zone covers 2.5 ft wide, the total area to be treated is 8,712 ft X 2.5 ft = 21780 sq ft, or 0.5 acre (since there are 43,560 sq ft per acre). Refer to the insecticide label for the application rate/acre (amount of product per acre) and inject ½ of that amount to the crop, since the area to be treated is only ½ acre. In this example, if the label states 3.0 fl oz per acre (88.7 ml of product per hectare) per application, then inject 1.5 fl oz (44.4 ml) of product through the irrigation system.

Example 2 (crop grown on geds covered with plastic mulch). Crop is a single row on beds 1.5 m (5 ft) wide on black plastic mulch row cover with drip irrigation under the plastic. The mulched row (after plastic is laid) is now 0.91 m wide (3 ft) under the plastic, and this represents the wetting zone. As in Example 1 above, the width of the wet zone X the total length of the rows under irrigation will yield the total area (squared) of the wet zone. The rate should be based on this area. If the total length of the rows is 8,712 row ft and the plastic mulch covers 3.0 ft wide, the total wetting area would be 8,712 ft X 3.0 ft = 26,136 sq ft, or 0.6 acre (since there are 43,560 sq ft per acre). Refer to the insecticide label for the application rate/acre (amount of product per acre), and inject 0.6 of that amount to the crop. In this example, if the label states 3.0 fl oz per acre (88.7 ml of product per hectare) per application, then inject 1.8 fl oz (53.2 ml) of product through the injection system.

The amount of product injected for both examples remains the same whether the crop is single row per bed, double rows per bed, or more. The amount of product injected is always based on the area of the irrigation wetting zone, and not on the crop width or number of rows per bed. Many of the newer insecticide labels now have tables that list the amount of product per row foot to be injected based on different wetting zones.

Dilute the appropriate amount of insecticide as calculated for injection with water in a dedicated mix tank or poly jug. It is recommended to use a dilution rate of at least 5 parts water to one part of the insecticide. As a general rule, the greater the dilution rate, the better potential for increased uniformity of applicaton. Mix the solution thoroughly before injection (an agitator in the mix tank may be necessary for some insecticides). The insecticide solution should be injected into the irrigation system at a point before the final filters, or have a filter on the chemical solution tank, to prevent any particulate matter from reaching and clogging the emitters.

7.3 Additional safety equipment for chemigation

The U.S. Environmental Protection Agency requires that the water source be protected from contamination by chemical solutions in case of unscheduled system shut down. It is important that the agrichemical injection pump be completely interlocked with the irrigation system so that the chemical injection pump will quickly shut down if the main irrigation pump were to stop, or if there was a loss of pressure in the irrigation system. This will prevent a free flow of chemical solution if there is a pressure drop or loss (resulting from a power loss, a break or hole in the drip lines, etc), and it will also prevent the irrigation and drip lines from filling up with the chemical solution if the main water pump stops for any reason.

A flow sensor installed downflow from both the injection pump and the main pump should be interlocked with the shutoff valves of both the main pump and the injection pump to shut down both the irrigation system and the insecticide injection system if water pressure at any point in the irrigation system drops or ceases. A two-way interlock between these pumps will also shut down both systems if one of the pumps stops or malfunctions (pump breakdown, power outage, etc.).

It is important to tightly seal leaks throughout the system, especially at the end of the drip tape where leaks often form puddles. Sealing these leaks will reduce or eliminate exposure of the injected insecticides to pollinators and other beneficial organisms and will result in a more uniform distribution of the insecticide to the plant roots. It is also important to seal all hose joints and connections to prevent leaks which may contaminate the environment. And a containment tray or pan under the injection pump will catch any insecticide solution leakage that may occur during injection. This material can be re-injected at the end of the injection period.

Pay particular attention to the directions and restrictions on the pesticide label, as many products are permitted for use only in overhead or sprinkler irrigation systems but cannot be applied via a drip/trickle irrigation system. Only products specifically labeled for application through a drip/trickle irrigation system can be applied in this manner.

8. Conclusion

Insectigation offers growers a sound option in place of traditional foliar sprays of insecticides for control of specific insect pests of vegetables produced using a drip/trickle irrigation system. Use of the drip/trickle irrigation system for application of insecticides allows for precise placement of systemic insecticides into the root zone of vegetable crops, eliminating the need for multiple foliar sprays of insecticides. Many growers currently use drip irrigation systems for water management, and the addition of an agrichemical injection system is a cost-effective method of pesticide application. It enables growers to apply an insecticide under virtually any weather condition for control of a wide range of insect pests, including aphids, whiteflies, leafhoppers, leafminers, beetles, caterpillars, and others while at the same time reducing the total insecticide inputs as compared with foliar sprays. The overall benefits of using chemigation include less application labor, less energy inputs, less time needed for application, less pesticide inputs, less worker and applicator exposure to the pesticide, less potential of soil-borne disease problems, a more even distribution of the pesticide, and less soil compaction. It suits a pest management program well because many of the new-chemistry insecticides labeled for drip/trickle irrigation system application are selective to specific insect pests and, because they are applied to the plant root zone, are generally less toxic to beneficial and non-target organisms. And in an urban state such as New Jersey, where urban populations border rural populations, insectigation can be conducted with no spray drift or misapplications, eliminating the ever important 'application visibility' that concerns both growers and the public.

Many University fact sheets are available on the internet that include information and instructions on how to inject agricultural chemicals into irrigation systems, including:

University of Florida IFAS Extension publication #BUL250, *Injection of Chemicals Into Irrigation Systems: Rates, Volumes, and Injection Periods* (http://edis.ufl.edu/ae116)

University of Florida Publication #HS980, *How to Conduct an On-Farm Dye Test and Use the Results to Improve Drip Irrigation Management in Vegetable Production* (http://edis.ifas.ufl.edu/HS222)

South Dakota State University Fact Sheet 862, *Chemigation Management* (http://agbiopubs.sdstate.edu/articles/FS862.pdf)

Washington State University Fact Sheet FS035E, *Calculating Chemigation Injection Rates* (http://cru.cahe.wsu/CEPublications/FS035E/FS035E.pdf)

Oregon State Unversity Bulletin *Pacific Northwest Insect Management Handbook: Guidelines – Chemigation.* http://insects.ippc.orst.edu/pnw/insects?31ADJV09.dat)

In addition, there are several commercially-produced technical brochures currently available that thoroughly describe drip/trickle chemigation system requirements, equipment set-up, injection pumps, calibration, safety equipment, application timing, water use and maintenance, etc., including *Drip Chemigation: Best Management Practices* (2008 Technical Update K-14954 from DuPont Crop Protection, E.I. DuPont de Nemours and Company, Wilmington, DE) and *Best Use Guidelines for Drip Application of Crop Protection Products* (2009 Technical Bulletin, Syngenta Crop Protection, Greensboro, NC 27419). Although these brochures are oriented towards the injection of insecticides into a drip/trickle irrigation system, the information is applicable to other agrichemicals applied through a drip/trickle system.

Also, it is important to refer to the current manufacturer labels of specific pesticides that can be used for chemigation. These labels list pests controlled, pesticide rates, restrictions, use directions, suggested application timing, required safety equipment, and other information necessary for successful chemigation. The pesticide label is a legal and binding document, and the pesticide user/applicator must carefully read, fully understand and adhere to all directions, instructions and restrictions on the label.

9. Glossary

Backflow check valve - a safety device that prevents the flow of water backwards from the irrigation delivery system to the water source (main pump or the injection pump). Operation is automatic and quick closing to prevent contamination.

Chemigation - the application of agrichemicals (fertilizers, insecticides, herbicides, fungicides, etc.) to crops through an irrigation system (overhead, drip, etc.

Chemical tank agitator - a device within the chemical tank that maintains constant mixture throughout the chemical injection process.

Containment device - a pan, tray, or dike that will contain any chemical leaks or drips from the chemical injection pump.

Drip irrigation - also called trickle irrigation, the application of water to the soil using low pressure and low volumes through emitters in tubing or piping. Drip irrigation generally uses flow rates of >3 gallons per hour at >10 psi.

Emitter - delivers water from a pipe or tube to the plant root zone. Also called a "dripper", flow rates are generally between 0.6 - 16 L/h (0.16 - 4.0 gal/h). A pressure compensating emitter discharges water at a constant rate over a wide range of drip line pressures.

Insectigation - the application of soluble insecticides through a drip irrigation system. A type of chemigation.

Interlocking controls - device that interlocks the chemigation pump with the well pump so if one pump fails the other is automatically shut off.

Laterals - pipes or tubing that go from the control valves to the drip emitter tubes.

Low pressure shutoff valve - a device that shuts off the pump when pressure in the delivery system suddenly drops. Can be installed in the irrigation pipeline or at the pump.

Low pressure sensor - a device that detects sudden drop in pressure in the delivery system and relays the signal to the shutoff valve.

Main lines - pipes or tubing from the water source to the control valves of an irrigation system.

Plasticulture - the application of plastics in agriculture for plant production, including row covers, drip irrigation, plastic tunnels, etc.

Positive displacement pump - a water pump that assures proper rate of injection of a liquid.

Pressure regulator - a device that maintains constant pressure downflow. It cannot increase pressure.

Subsurface irrigation (SDI) - drip systems that are buried beneath the soil surface, not recovered between cropping cycles. Some SDI systems are semi-permanent.

Vacuum relief valve - a device that prevents back siphoning, usually installed upflow of the mainline backflow check valve.

10. References

Anonymous. 2000. Annual irrigation survey 1999. *In*: Irrigation Journal 50(1): 8-15.

Ayars, J.E., D.A. Bucks, F.R. Lamm & F.S. Nakayama. 2007. Micro-irrigation for crop production. F.R. Lamm, J.E. Ayars, F.S. Nakayama {ed.}, Elsevier Publ. Co., Amsterdam, the Netherlands. 13: 4.

Blass, I. & S. Blass. 1969. Irrigation dripper unit and pipe system. U.S. Patent Office, Patent No. 3,420,064 issued January 7, Washington, D.C., United States of America.

Bogle, O. & T.K. Hartz. 1986. Comparison of drip and furrow irrigation for melon production. Hort. Science 21: 242-244.

Dasberg, S. & D. Or. 1999. Drip irrigation. Springer-Verlag Berlin and Heidelberg GmbH & Co.K. 171 pp.

Felsot, A.S., W. Cone, J. Yu & J.R. Ruppert. 1998. Distribution of imidacloprid in soil following subsurface drip irrigation. Bulletin Environmental Contamination and Toxicology 60: 363-370.

Ghidiu, G.M. 1981. Vydate injected through a trickle irrigation system to control Mexican bean beetle. In: *Rutgers University Vegetable Entomology Research Results - 1981*. New Jersey Agricultural Experiment Station Report No. 2: 4.

Ghidiu, G.M. 1992. Chemigation with carbofuran for insect control in bell peppers. In: *Rutgers University Vegetable Entomology Research Results - 1992*. New Jersey Agricultural Experiment Station, NJ Cooperative Extension Bulletin 104A: 12-13.

Ghidiu, G.M. 2009. Control of insect pests of eggplant with insecticides applied through a drip irrigation system under black plastic. In: *Rutgers University Vegetable Entomology Research Results - 2009*. New Jersey Agricultural Experiment Station. NJ Cooperative Extension Bulletin 104R: 8-11.

Ghidiu, G.M. & N.L. Smith. 1980. Trickle irrigation system injected insecticides to control the European corn borer in bell pepper. In: *Rutgers University Vegetable Entomology*

Research Results - 1980, New Jersey Agricultural Experiment Station, NJ Cooperative Extension Report No. 1: 5-6.

Ghidiu, G.M., D.L. Ward & G.S. Rogers. 2009. Control of European corn borer in bell peppers with chlorantraniliprole applied through a drip irrigation system. J. Vegetable Science 15: 193-201.

Hall, B.J. 1982. Row crop fertigation. American Vegetable Grower, April 1982. 30(4): 72-73.

Kerns, D.L. & J.C. Palumbo. 1995. Using Admire on desert vegetable crops. IPM Series No. 5, University of Arizona Cooperative Extension Publication No. 195017.

Kuhar, T.P., H.B. Doughty, M. Cassell, A. Wallingford & H. Andrews. 2009. Control of Lepidopteran larvae in fall tomatoes through drip irrigation systems. In: *Arthropod Pest Management Research on Vegetables in Virginia for 2009.* VPI&SU Eastern Shore AREC Report #308: 27-28.

Kuhar, T.P. & J. Speese. 2002. Evaluation of drip line injected and foliar insecticides for controlling cucumber in melons, 2001. Entomological Society of America. Arthropod Management Tests 27: E46.

Lahm, G.P., T.M. Stevenson, T.P. Selby, J.H. Freudenberger, D. Cordova, L. Flexner, C.A. Belilin, C.M. Dubas, B.K. Smith, K.A. Hughes, J.G. Hollingshaus, C.E. Clark & E.A. Benner. 2007. Rynaxypyr: a new insecticidal anthranilic diamide that acts as a potent and selective ryanodine receptor activator. Bioorganic and Medicinal Chemistry Letters 17: 6274-6279.

Lahm, G.P., T.P. Selby, J.H. Freudenberger, T.M. Stevenson, B.J. Myers, G. Seburyamo, B.K. Smith, L. Flexner, C.E. Clark & D. Cordova. 2005. Insecticidal anthranilic diamides: a new class of potent ryanodine receptor activators. Bioorganic and Medicinal Chemistry Letters 15: 4898-4906.

Lamont, W.J. Jr. 2004. Plasticulture - An Overview. In: *Production of Vegetables, Strawberries and Cut Flowers Using Pplasticulture.* [ed.] W.J. Lamont, Jr. pp. 1-8. Natural Resource, Agriculture and Engineering Service Cooperative Extension Publication NRAES-133.

Palumbo, J.C. 1997. Evaluation of aphid control in lettuce with Admire applied through drip irrigation. Entomological Society of America. Arthropod Management Tests 22: E61.

Paterson, J.W. 1980. Fertilizing vegetables via drip/trickle irrigation. Special Research Report, Rutgers - the State University of New Jersey. 6 pp.

Reed, D.K., G.L. Reed & C.S. Creighton. 1986. Introduction of entomogenous nematodes into trickle irrigation systems to control striped cucumber beetle (Coleoptera: Chrysomelidae). J. Economic Entomolgy 79: 1330-1333.

Ross, D.S. 2004. Drip irrigation and water management. In: *Production of Vegetables, Strawberries, and Cut Flowers Using Plasticulture.* [ed.] W.J. Lamont, Jr. pp. 14-35. Natural Resource, Agriculture and Engineering Service Cooperative Extension Publication NRAES-133.

Ross, D.S., R.C. Funt, C.W. Reynolds, D.R. Coston, H.H. Fries & J.N. Smith. 1978, Trickle irrigation - an introduction. Northeast Regional Agricultural Engineering Service (NRAES) Bulletin No. 4. 24 pp.

Schuster, D.J., A. Shurtleff & S. Kalb. 2009. Management of armyworms and leafminers on fresh market tomatoes, Fall 2007. Arthropod Management Tests 34: E79.

Wildman, T.E. & W.W. Cone. 1986. Drip chemigation of asparagus with disulfoton: *Brachycorynella asparagi* (Homoptera: Aphididae) control and disulfoton degradation. J. Economic Entomology 76: 1617-1620.

9

Generalist Predators, Food Web Complexities and Biological Pest Control in Greenhouse Crops

Gerben J. Messelink[1], Maurice W. Sabelis[2] and Arne Janssen[2]
[1]*Wageningen UR Greenhouse Horticulture*
[2]*IBED, Section Population Biology, University of Amsterdam*
The Netherlands

1. Introduction

Biological control of pest species has traditionally mainly focused on specific natural enemies for each pest (Huffaker & Messenger, 1976; Hokkanen & Pimentel, 1984; Van Lenteren & Woets, 1988; Hoy, 1994). However, pest-enemy interactions are often embedded in rich communities of multiple interacting pests and natural enemies and the interactions among these species affect the efficacy of biological pest control (Sih et al, 1985; Janssen et al, 1998; Prasad & Snyder, 2006; Evans, 2008). The effect of interactions among various species of predators and parasitoids on biological control of a shared pest species has received ample attention (see Letourneau et al., 2009), showing that it can range from larger to smaller than the effect of each enemy species separately (Rosenheim et al., 1995; Rosenheim et al., 1998; Losey and Denno, 1998; Colfer & Rosenheim, 2001; Venzon et al., 2001; Cardinale et al., 2003; Snyder & Ives, 2001, 2003; Finke & Denno, 2004; Cakmak et al., 2009). However, it is not only predator diversity, but also the diversity of herbivorous prey that may affect the suppression of a particular pest species through competition or indirect interactions mediated by host plant or shared predators (Holt,1977; Karban & Carey, 1984). Hence, designing effective biological control programs for more than one pest species requires an understanding of all interactions occurring among species within biocontrol communities, not just those among pests and their natural enemies or among different species of natural enemies.

Greenhouse crops are often considered as simple ecosystems with low biodiversity (Enkegaard & Brødsgaard, 2006). Especially modern greenhouses appear sterile compared to outdoor crops, as plants are grown on hydroponic systems in greenhouses that are closed from the environment because of modern energy saving techniques (Bakker, 2008). However, the general experience is that infestations by several small pest species cannot be avoided, and the release of natural enemies against these pests adds to the diversity (van Lenteren et al., 2000; Cock et al., 2010). Thus, apparently "clean" greenhouse crops often accommodate complex artificial communities of multiple pests and natural enemies. Furthermore, there seems to be a tendency that these communities increase in food web complexity during the last decades (Enkegaard & Brødsgaard, 2006). One reason for this increased diversity is the invasion of exotic pest species (global trade, global warming) (Roques et al., 2009). Second, more species than before develop into pests as a result of the reduced use of pesticides and the use of more selective pesticides (van der Blom et al., 2009). A third reason is that biological control programs increasingly include generalist predators (Gerson & Weintraub, 2007; Sabelis et al., 2008), and such generalists potentially interfere more with other natural enemies than specialists. Thus, recent developments further increase food web complexity in biological control programs and emphasize that such complexities need to be considered when designing biological control programs.

Here, we review the ecological theory relevant to interactions in food webs occurring within arthropod communities and we discuss the possible implications for biological control in greenhouses. This review is restricted to the most important greenhouse pests, namely aphids (Ramakers, 1989; Blümel, 2004), thrips (Lewis, 1997; Shipp and Ramakers, 2004), spider mites (Helle & Sabelis, 1995; Gillespie & Raworth, 2004) and whiteflies (Byrne & Bellows, 1991; Avilla et al., 2004), and their natural enemies.

2. Food web theory and effects in greenhouse crops

Consumption (i.e. herbivory, predation and parasitism) and competition are considered the two most important interactions determining the structure of communities (Chase et al., 2002). Within communities of natural enemies and pests, species may interact through exploitative competition, induced plant defences, apparent competition or apparent mutualism via shared natural enemies, or through predation and parasitism, which includes omnivory, intraguild predation and hyperpredation or hyperparasitism (Fig. 1). Besides these density-mediated interactions, species interactions can be modified through trait changes of the interacting individuals (which includes changes in behaviour and induced plant responses). In the following, we summarize the current theory on these interactions and their relevance for biological control.

2.1 Exploitative competition and induced plant responses

Herbivores can interact through exploitative competition for the plant (Fig. 1), but this is undesirable for biological control, because it occurs at high pest densities, which may exceed the economic damage threshold. We will therefore refrain from discussing resource competition among herbivores here. Herbivores can also interact via the plant when the presence of one species induces a defence response in the plant that also affects a second species (Karban & Carey, 1984). These plant responses can both result in increased resistance or increased susceptibility (e.g. Karban & Baldwin, 1997; Sarmento et al., 2011). Induced plant resistance against insects consists of direct defences, such as the production of toxins and feeding deterrents that reduce survival, fecundity or reduce developmental rate (Kessler & Baldwin, 2002), and indirect defences such as the production of plant volatiles

that attract carnivorous enemies of the herbivores (Dicke and Sabelis, 1988; Schaller, 2008). Several biochemical pathways are involved in these processes (Walling, 2000). Recent studies have shown that plant-mediated interactions between herbivores are very common and could be important in structuring herbivore communities (Kessler et al., 2007). Models of interactions that are mediated by inducible changes in plant quality predict a range of outcomes including coexistence, multiple equilibria, dependence on initial conditions and competitive exclusion of some herbivore species (Anderson et al., 2009). However, these models assume that herbivore populations are well mixed and possible variation in induction caused by variation in population densities is ignored.

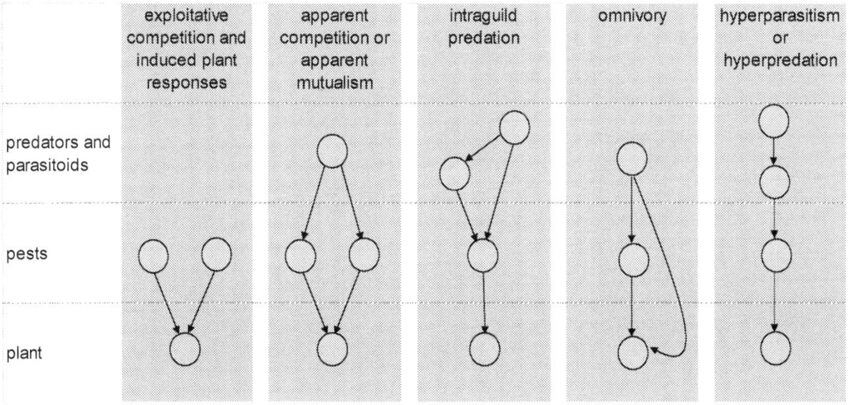

Fig. 1. Schematic diagrams of the direct and indirect interactions among plants, pests and natural enemies that will be treated in this chapter. Arrows indicate consumption. From left to right: *exploitative competition*: two pest species compete for the same plant, but also affect each other's densities through induced plant defences; *apparent competition or apparent mutualism*: indirect interactions between two prey species mediated by a shared natural enemy (with pests on the same plants this automatically includes exploitative competition and induced plant defences); *intraguild predation*: predators consume another natural enemy with whom they also compete for the same pest species; *omnivory*: consumption of species from more than one trophic level, "true" omnivores are predators that feed on both pests and plants; *hyperparasitism or hyperpredation*: the consumption of natural enemies by other natural enemies with whom they do not compete for shared prey, but they differ by the fact that hyperpredators can develop on alternative prey, whereas true hyperparasitoids are obligate. Except for induced plant responses, these interactions are density mediated.

Several studies documented indirect interactions between herbivores through induced changes in plant quality (Karban & Baldwin 1997), but studies on greenhouse crops are limited. In tomato, it has been demonstrated that infestations by caterpillars of a noctuid moth increased resistance to spider mites, aphids and another lepidopteran pest (Stout et al., 1998). Likewise, infestations by whiteflies induced resistance against leafminers (Inbar et al., 1999). Similar results were found on cucumber (Zhang et al., 2005). Induced susceptibility may also occur, for example, infestations of tomato plants by whiteflies increased susceptibility to aphids (Nombela et al., 2008). On lima bean, similar results were found for whiteflies and spider mites (Zhang et al., 2009). The spider mite *Tetranychus evansi* Baker & Pritchard was found to down-regulate plant defences (Sarmento et al., 2011), and the closely related species *Tetranychus urticae* Koch can profit from this induced susceptibility (Sarmento et al. in press). Induced resistance may also affect the behaviour of omnivores that facultatively feed on plants. The omnivorous western flower thrips switched from feeding on the host plant to feeding on spider mite eggs when defences of the plants were induced (Agrawal et al., 1999). Moreover, they performed worse on eggs of spider mites from induced plants than on eggs from spider mites on non-induced plants (Agrawal & Klein, 2000). In conclusion, plant-mediated interactions among pest species are probably a common phenomenon in greenhouse crops, where they may influence the biological control of multiple pests.

2.2 Apparent competition and apparent mutualism

Generalist predators can mediate indirect interactions among prey species that might otherwise not interact (Holt & Lawton 1994; Janssen et al., 1998; Harmon & Andow 2004; van Veen et al., 2006) (Figure 1). If, for example, the density of one prey species increases, the density of the shared predator subsequently increases and ultimately, the second prey species decreases in abundance. Holt (1977) suggested the term "apparent competition" for this interaction between prey, because the dynamics of the two species resemble that of species competing for resources, whereas in fact it is mediated by the shared predator. Apparent competition is usually defined as a reciprocal negative interaction between prey, but most empirical studies show non-reciprocal indirect interactions (Chaneton & Bonsall, 2000). Hence, only one of the two prey species is negatively affected by the predator-mediated prey interaction. Originally, the theory of apparent competition considered equilibrium densities. However, generalist predators can also cause "short-term" apparent competition between prey species when predators aggregate in habitat patches containing both prey, or when their feeding rate on one prey is enhanced by the presence of another prey (Holt & Kotler 1987, Müller and Godfray 1997).

The opposite effect may also occur between two prey that share a natural enemy, i.e. a positive indirect effect of one prey population on densities of the other (apparent mutualism). This occurs when increases in the density of one prey species result in satiation of the shared predator or in predator switching (when a predator eats disproportionately more of the most common type of prey), consequently reducing the consumption of the second prey species (Murdoch 1969; Abrams & Matsuda 1996). This effect is apparent in the short-term, when the densities have not yet reached an equilibrium (transient dynamics), because eventually, the predator populations will increase because of the higher densities of prey (Abrams & Matsuda 1996) and result in apparent competition. Apparent mutualism may also occur in the long term when population densities do not reach equilibria, but show cycles, resulting in repeated satiation of the shared predators and repeated reduced predation on the other prey (Abrams et al., 1998). Hence, depending on the time scale and on the type of dynamics, theory predicts that a shared natural enemy can generate positive or negative indirect effects between prey species.

Apparent competition and apparent mutualism are inherently related to diet choice and switching of the predators from feeding on one prey to feeding on the other or both prey, but effects of mixed diets on predator performance are also relevant. Mixed diets are known to have positive effects on reproduction in some predator species (Wallin et al., 1992; Toft 1999; Evans et al., 1999).

When generalist predators are released in greenhouse crops, pest species such as thrips, whiteflies, spider mites and aphids can be involved in apparent competition or apparent mutualism. Examples of such generalist predators are anthocorid and mirid bugs and several species of predatory mites. For example, the predatory mite *Amblyseius swirskii* Athias-Henriot is able to control both whiteflies and thrips effectively (Nomikou et al., 2002; Messelink et al., 2006). On greenhouse cucumber, it has indeed been shown this predator mediates apparent competition between the two pests: whitefly control was substantially better in the presence of thrips (Messelink et al., 2008). Moreover, better pest control was also achieved by positive effects of a mixed diet of thrips and whiteflies on juvenile survival and developmental rate (Messelink et al., 2008). So far, this aspect of mixed diets has been ignored in theoretical models about apparent competition. Not only whiteflies, but also spider mites were controlled better by the presence of thrips through apparent competition (Messelink et al., 2010). Although *A. swirskii* is not an effective predator of spider mites because it is strongly hindered by the webbing, it can prevent the formation of new colonies of spider mites when there are other prey, such as thrips, available. Thus, generalist predators can even have significant effects on prey species which they cannot suppress successfully on their own.

Although the theory of predator-mediated interactions has long been neglected in biological control, there has been a long-standing interest in the use of alternative hosts for enhancing biological control (Stacy, 1977). The method by which these alternative hosts are facilitated is based on the introduction of a non-crop plant harbouring the alternative hosts. It is often referred to as the "banker plant method" (Frank, 2010; Huang et al., 2011). A widely applied system in greenhouse crops is the use of monocotyledonous plants with grain aphids that serve as alternative hosts for parasitoids of aphids that attack the crop (Huang et al., 2011). The elegance of this system is that the grain aphids are host-specific and pose no threat to the crop. Another method is based on banker plants that provide pollen to generalist predators (Ramakers & Voet, 1995). For example, pollen can serve as food for generalist predatory mites and enhance the biological control of thrips and whiteflies (van Rijn et al., 2002; Nomikou et al., 2010). In fact, all kinds of "open rearing" systems of natural enemies in greenhouse crops (e.g. rearing sachets containing small cultures of predatory mites, bran and an astigmatic mites) are based on the principles of apparent competition, but there is little awareness that apparent mutualism may also occur.

2.3 Intraguild predation

Natural enemies can compete for the same prey species, but this is frequently combined with predation by one species of natural enemy on another (Rosenheim et al., 1995), which is called intraguild predation (IGP, Figure 1). The predator that kills and eats the other natural enemy is called the intraguild predator and the other natural enemy is the intraguild prey (Polis et al., 1989; Holt and Polis 1997). General theory predicts that IGP can only result in stable coexistence of the species when the intraguild prey is the superior competitor for the shared prey, and only in systems with intermediate levels of productivity (Holt and Polis 1997). These conditions are very restrictive and thus predict that IGP is not common in nature. However, it has become clear that IGP generally occurs in many ecosystems, including in biological control systems (Polis et al., 1989; Rosenheim et al., 1995, Janssen et al. 2006, 2007). There may be several reasons for this discrepancy between theory, predicting that systems with strong IGP will be rare, and reality, where IGP is common. Factors that can contribute to the coexistence of intraguild predators and intraguild prey are now increasingly included in theoretical models. Examples of such factors are structured populations with intraguild prey stages that are invulnerable or intraguild predator stages that do not prey on the other predator (Mylius et al., 2001), anti-predator behaviour (Heithaus, 2001), switching intraguild predators (Krivan, 2000) or alternative prey (Daugherty et al., 2007; Holt & Huxel, 2007). Based on theory, intraguild predation is expected not to benefit biological control (Rosenheim et al., 1995), but in practice, results are mixed (Janssen et al., 2006; 2007; Vance-Chalcraft et al., 2007).

Intraguild predation has been described for many natural enemies that are used for biological control in greenhouse crops (Rosenheim et al. 1995; Janssen et al., 2006). Here, we summarize the results for natural enemies of thrips, whiteflies, aphids and spider mites. The omnivorous predator *Macrolophus pygmaeus* (Rambur) (formely identified as *Macrolophus caliginosus* Wagner) is an intraguild predator of natural enemies of aphids; it consumes the eggs of the syrphid *Episyrphus balteatus* de Geer (Frechette et al., 2007) and parasitized aphids (Martinou, 2005). This predator did not prey on nymphal stages of *Orius majusculus* (Reuter), but in turn, the nymphal stages of *M. pygmaeus* were vulnerable for predation by *O. majusculus* (Jakobsen et al., 2004). Predatory bugs of the genus *Orius* act as intraguild predators of phytoseiid mites (Gillespie & Quiring, 1992; Venzon et al., 2001; Brødsgaard & Enkegaard, 2005; Chow et al., 2008), the aphidophagous predatory midge *Aphidoletes aphidimyza* (Rondani) (Hosseini et al., 2010) and aphid parasitoids (Snyder & Ives, 2003). Many generalist predatory mites are intraguild predators of other predatory mites (Schausberger & Walzer, 2001; Buitenhuis et al., 2010; Montserrat et al., 2008; Van der Hammen et al., 2010) or juvenile stages of predatory bugs (Madali et al., 2008). Finally, a number of studies show intraguild predation among specialist natural enemies of aphids. The syrphid *E. balteatus* feeds on freshly parasitized as well as unparasitized aphids (Brodeur & Rosenheim, 2000). Syrphid larvae may also consume the aphidophageous gall midge *A. aphidimyza*, but predation rates are low in the presence of aphids (Hindayana et al., 2001). In turn, this midge does not prey on *E. balteatus* (Hindayana et al., 2001), but may consume parasitized aphids (Brodeur & Rosenheim, 2000).

None of these studies demonstrates a negative effect of intraguild predation on biological control in greenhouse crops. Although the potential risk of intraguild predation disrupting biological control appears to be low in many cases (Janssen et al., 2006), there are also examples of negative effects of intraguild predation on biological control.

2.4 Omnivory

Omnivory in its broadest sense can be defined as the consumption of species of more than one trophic level. Under this definition, intraguild predators are also omnivores. Predators that feed on both animals and plants are a particular case of trophic omnivory, also referred to as "true omnivory" (Coll & Guershon, 2002). The first theoretical models on its dynamical consequences showed that omnivory destabilizes food webs (Pimm & Lawton, 1978), which is remarkable, considering the fact that omnivory is a common feature of food webs (Coll & Guershon, 2002, Polis & Strong, 1996). More specific theory for plant-feeding omnivores shows that omnivores can stabilize the dynamics and persistence of populations by switching between consuming plants and prey, especially when the searching efficiency of the predator for prey is low relative to that for plant tissue (Lalonde et al., 1999). Hence, this theory suggests that biological control with plant-feeding omnivores may stabilize pest population dynamics. The question is, whether these equilibrium densities are acceptable for pest control (Lalonde et al., 1999). Other aspects of plant-feeding omnivory, such as the persistence of predators in the absence of prey, or the nutritional benefits for predators of feeding on plants may also result in positive contributions to biological control.

Many predators that are used for biological control are true omnivores, feeding on pests and plant-provided food such as pollen, nectar and plant saps. For example, many generalist predatory mites and bugs can complete their life cycle feeding on pollen. However, not all greenhouse crops produce pollen (e.g. male-sterile cucumber) or edible pollen, but some omnivores, such as the mirid bug *M. pygmaeus*, can also live and reproduce on plant saps. Although considered as a pest species, western flower thrips, *Frankliniella occidentalis* (Pergande) are in fact omnivorous predators that feed on spider mites, predatory mites, whiteflies and plants (Trichilo & Leigh, 1986; Faraji et al., 2001; Janssen et al., 2003, van Maanen et al., in prep.). The consumption of prey in addition to plant material by mirid bugs and thrips can increase reproduction rates (Janssen et al., 2003; Perdikis & Lykouressis, 2004). The quality of the host plant can affect the predation rates of omnivores on pests (Agrawal et al., 1999; Agrawal & Klein, 2000; Magalhães et al., 2005; Hatherly et al., 2009) or the extent to which intraguild predation occurs (Janssen et al., 2003, Shakiya et al., 2009). Thus for biological control with predators that can also feed on the plant, it is important to known that the dynamics will be affected by plant quality.

2.5 Hyperpredation and hyperparasitism

In contrast to intraguild predation, natural enemies can also be consumed by other predators or parasitoids without sharing a prey with these enemies. Thus there is no competition for prey between the natural enemies. This consumption is well known for parasitoids, so-called hyperparasitism. Hyperparasitism is well-studied for its dynamical consequences, both theoretically (Beddington & Hammond, 1977; May & Hassell, 1981) and empirically (Sullivan & Völkl, 1999). These studies indicate that obligate hyperparasitoids (secondary parasitoids that can develop only in or on a primary parasitoid) always lead to an increase of the pest equilibria, which might be detrimental to biological control. In case the hyperpredator is a true predator, there is no agreement in the literature on the name of this type of interaction. Some prefer to use the term "secondary predation" (Rosenheim et al., 1995), or "higher-order predation" (Rosenheim, 1998; Symondson, 2002) for predators consuming other predators, which includes both hyperpredation and intraguild predation. Even more confusing is that some interactions are described as hyperpredation, whereas it would be more consistent to typify them as apparent competition (e.g. Courchamp et al., 2000; Roemer et al., 2001) or intraguild predation (e.g. Roemer et al., 2002). We suggest to use the term hyperpredation in cases where predators eat other predators without sharing a prey, because of its similarity to hyperparasitism. However, an important difference is that hyperpredators can develop on alternative prey or food, whereas most hyperparasitoids specifically reproduce on or in other parasitoids. In the presence of alternative prey, hyperpredation can be classified as apparent competition between the alternative prey and the specialist natural enemy. To our knowledge, no specific theory has been formulated on the effects of hyperpredation on prey populations in the presence of alternative prey. Theory on apparent competition predicts that the presence of one prey lowers the equilibrium densities of the second prey. For hyperpredation, this would mean that increases in the densities of the alternative prey will results in lower equilibrium densities of the specialist natural enemy, which would consequently release the prey of the specialist from control. In the short-term, satiation effects of the hyperpredator might result in apparent mutualism between the alternative prey and the specialist natural enemy, hence, a reduced negative effect on pest control by the specialist natural enemy.

In greenhouse crops, predatory mites that are used for control of thrips and whiteflies have been observed to be hyperpredators. They feed on eggs of predatory midge *A. aphidimyza*, but not on aphids, the pest that is controlled by predatory larvae of midges (Messelink et al., 2011). In sweet pepper, the biological control of aphids by *A. aphidimyza* was seriously disrupted through this hyperpredation by the predatory mite *A. swirskii* (Messelink et al., 2011). Hyperparasitism is common in the biological control of aphids in greenhouses and can also disrupt biological control (Messelink, personal observations).

2.6 Effect of flexible behaviour

The interactions in food webs described above all concern density-mediated interactions among species. However, it is generally recognized that traits of individuals, such as behaviour or defence levels, can change in response to the presence of individuals of other species (so-called trait-mediated interactions, Werner & Peacor, 2003). For example, anti-predator behaviour, can strengthen or weaken density-mediated effects (Prasad & Snyder, 2006; Janssen et al., 2007). Many of these behavioural changes are mediated by chemical cues, which are released or left behind by both natural enemies and prey (Dicke & Grostal, 2001). Theoretical models of community dynamics now increasingly try to study the consequences of these behavioural-mediated interactions (e.g. Holt & Kotler, 1987; Abrams, 2008). These models show that the effects of such interactions may change the dynamics substantially.

Many interactions among natural enemies and pests in greenhouses can be affected by changes in the behaviour of pest and natural enemy. First of all, it is known that pest species can avoid their enemies. For example, whiteflies can learn to avoid plants with generalist predatory mites (Nomikou et al., 2003) and spider mites avoid plants with the predator *Phytoseiulus persimilis* Athias-Henriot (Pallini et al., 1999) or with thrips, which is a competitor and intraguild predator (Pallini et al., 1997). Aphids are well-known for their antipredator responses, for example, they kick at natural enemies, or they walk away or drop off the plants when perceiving a natural enemy (Villagra et al., 2006). Aphids as well as thrips release alarm pheromones that alert conspecifics (Bowers et al., 1972; Teerling et al., 1993; de Bruijn et al. 2006). Thrips can avoid predation by predatory bugs and predatory mites by using spider mite webbing as a refuge (Pallini et al. 1998; Venzon et al. 2000). They can defend themselves against predators by swinging with their abdomen and producing defensive droplets (Bakker & Sabelis, 1989), or even by counter-attacking the vulnerable egg stages of their phytoseiid predators (Faraji et al., 2001, Janssen et al. 2002). Natural enemies also respond to threats of other (intraguild) predators or counter-attacking prey. Predatory mites avoid ovipositing near counter-attacking thrips (Faraji et al., 2001) or intraguild predators (Choh et al., 2010, van der Hammen et al., 2010), or retain eggs in the presence of intraguild predators (Montserrat et al., 2007). Aphid parasitoids are known to avoid intraguild predation once they detect the chemical cues of their predators (Nakashima et al., 2006). The effects of intraguild predation can also be changed by the prey preference of the intraguild predator. For example, the syrphid *E. balteatus* is an intraguild predator of aphid parasitoids because it consumes parasitized aphids, but when given a choice, it prefers to oviposit in aphid colonies without parasitized aphids (Pineda et al., 2007), thus weakening the effects of intraguild predation.

Interactions among species may change over time through learning or experience (Nomikou et al., 2003). For example, the predatory bug *O. majusculus* was more successful at preying on aphids after learning how to avoid the prey's kicking response (Henaut et al., 2000). Furthermore, predation rates on a specific pest might change through the presence of alternative food: the predatory bug *O. laevigatus* increased the predation rates on thrips in the presence of pollen (Hulshof & Linnamäki, 2002). Thus somehow, the pollen seemed to stimulate the feeding behaviour of these predators. In contrast, the presence of unsuitable prey may reduce the efficacy of a natural enemy for the target pest. For example, studies with parasitoids demonstrated that spending foraging time or eggs on less-suitable hosts will decrease parasitoid foraging success and ultimately decrease parasitoid population size (Meisner et al., 2007). Such "distraction" effects may also occur in greenhouses when mixtures of aphid species are present in a crop. The reason why parasitoids attack unsuitable or marginal hosts in the study by Meisner et al. (2007) is not clear, perhaps the parasitoids and marginal hosts have not coevolved and there has been no selection on the parasitoid to discriminate between the marginal host and other host species. It is also possible that the parasitoids cannot assess host suitability as this may vary through the presence of symbiotic bacteria that induce resistance to parasitoids (Oliver et al., 2003). The examples presented above show that multiple prey effects can change the behaviour of shared natural enemies and may determine the outcomes of biological control.

Summarizing, changes in interactions or interaction strengths through flexible behaviour are common among the pests and natural enemies in greenhouse crops. Thus, when designing and interpreting results of multi-species experiments, it should be realized that both density-mediated interactions and behavioural mediated interactions affect biological control. The potential diversity and complexity of an artificial food web in a greenhouse vegetable crop is presented in the next section.

3. A case study: Food web complexity in sweet pepper

The complexities of arthropod communities associated with biocontrol systems vary among crops, because crops differ in susceptibility to pests species and suitability for natural enemies. Sweet pepper is one of the crops where the release of natural enemies for biological control has resulted a complex system of multiple pests and natural enemies, including several different species of generalist predators. The most important pests in sweet pepper in greenhouses in temperate regions are western flower thrips, *F. occidentalis*, two-spotted spider mites, *T. urticae* and aphids, mostly the green peach aphid, *Myzus persicae* (Sulzer) and the foxglove aphid *Aulacorthum solani* (Kaltenbach) (Ramakers, 2004), whereas in Mediterranean countries, one of the major pest species is the tobacco whitefly, *Bemisia tabaci* Gennadius (Calvo et al., 2009). Many other pest species can attack sweet pepper, such as caterpillars of noctuid moths, broad mites, leaf miners and mirid bugs, but they are less important (Ramakers, 2004).

Anthocorid bugs are commonly used as generalist predators in sweet pepper. *Orius laevigatus* (Fieber) is most used in Europe, *O. insidiosus* (Reuter) in Northern America (Brødsgaard, 2004; Shipp and Ramakers, 2004). Although anthocorid bugs are mainly released for thrips control, they can also contribute to the control of whiteflies (Arnó et al., 2008), aphids (Alvarado et al., 1997), and spider mites (Venzon et al., 2002). The omnivorous predator *M. pygmaeus* is also released often, and is know to suppress whiteflies (Gerling et al., 2001), aphids (Alvarado et al., 1997), thrips (Riudavets & Castañé, 1998) and spider mites (Hansen et al., 1999). Finally, generalist predatory mites are commonly released in sweet pepper. The first releases started with the phytoseiid *Neoseiulus barkeri* (Hughes) (= *Amblyseius mckenziei*) for the control of thrips (Ramakers, 1980). Since then, several other phytoseiids, such as *Neoseiulus cucumeris* (Oudemans) or *Iphiseius degenerans* (Berlese), are released in sweet pepper (Ramakers, 2004). Nowadays, *A. swirskii* is a very popular species, because this predatory mite not only controls thrips (Messelink et al., 2006), but also whiteflies (Nomikou et al., 2002; Calvo et al., 2009), broad mites (van Maanen et al. 2010) and it can contribute to the control of spider mites (Messelink et al., 2010). Populations of both generalist predatory bugs and predatory mites can establish in sweet pepper crops even when prey is scarce, because of the continuous presence of flowers that produce pollen (Ramakers, 1980; Van den Meiracker & Ramakers, 1991).

Specialist predators released in sweet pepper crops are the predatory mite *P. persimilis* against spider-mites (Gillespie & Raworth, 2004), and the aphidophagous predators *A. aphidimyza* and *E. balteatus* (Ramakers, 1989; Blümel, 2004) against aphids. Furthermore, several specialist parasitoids are released: for aphids mainly *Aphidius colemani* Viereck, *Aphidius ervi* Haliday or *Aphelinus abdominalis* Dalman and for whiteflies mainly *Eretmocerus mundus* Mercet and *Er. eremicus* Rose & Zolnerowich (Cock et al., 2010).

The simultaneous occurrence and need to control several pest species in sweet pepper results in a complex food web of interacting species (Fig. 2). The presence of western flower thrips in this food web contributes strongly to the complexity. Although *F. occidentalis* is primarily considered a phytophagous species that feeds on plant tissue, plant nectar or pollen, it is actually an omnivore, feeding facultatively on spider mite eggs (Trichilo & Leigh, 1986), predatory mite eggs (Faraji et al., 2001; Janssen et al., 2003), or on whitefly crawlers (van Maanen et al., in prep.).

The food web presented in Figure 2 shows that the interactions between a certain pest and its natural enemy are often embedded in a complex web of interactions. For example, intraguild predation is often accompanied by apparent competition between the intraguild prey and several other alternative prey species. Furthermore, the intraguild predators or hyperpredators can also feed on plant-provided food, with the result that plant quality may affect intraguild predation or hyperpredation (Agrawal & Klein, 2000; Janssen et al., 2003). This emphasizes the complexity of biological control, where effects of some interactions may override the effects other interactions (Polis & Holt, 1989). Thus, the study of particular species interactions, such as those between a pest and its natural enemy, should be embedded in empirical studies and models that capture the essence of realistic food webs. Although it may be difficult to disentangle all possible interactions and their importance for biological control, the understanding of such interactions will help in designing effective communities of natural enemies for the suppression of multiple pests.

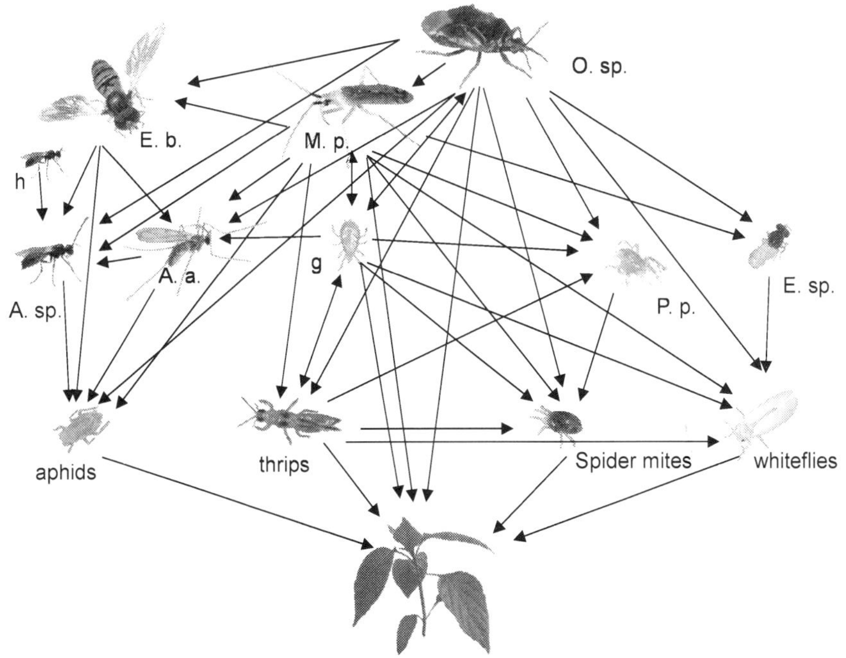

Fig. 2. A food web of pest species and their most commonly used natural enemies in sweet pepper crops. The generalist predators are bugs from the genus *Orius* (O. sp.), the mirid bug *Macrolophus pygmaeus* (M. p.) and generalist phytoseiid predatory mites (g). Specialist enemies of aphids are parasitoids from the genus *Aphidius* (A. sp.), the predatory midge *Aphidoletes aphidimyza* (A.a.) and the syrphid *Episyrphus balteatus* (E.b.). Parasitoids are commonly attacked by several species of hyperparasitoids (h). The specialist predator of spider mites is *Phytoseiulus persimilis* (P.p.). The main whitefly species in sweet pepper is *Bemisia tabaci*, which can be controlled by specialist whitefly parasitoids from the genus *Eretmocerus* (E. sp.).

4. Conclusions

Food web theory can provide insight into how various interactions between species might affect species dynamics and their possible effects on biological control. However, since models are necessarily based on simplifying assumptions, theoretical predictions are bound to differ from empirical studies (e.g. Janssen et al., 2006; Rosenheim & Harmon, 2006). For example, theory often predicts equilibrium dynamics, whereas biological control systems often concern short-term (transient) dynamics, which might differ from long-term dynamics (Bolker et al., 2003; Briggs & Borer, 2005). A second reason is that real food webs are much more complex than theoretical models assume (Rosenheim et al, 1995; May, 1999, Coll & Guershon, 2002; Bolker et al., 2003; Cardinale et al., 2003; Janssen et al., 2006, 2007; Letourneau et al., 2009). The presence of multiple pests and natural enemies will result in joint effects of several types of interactions, and there is limited theory that takes such complexity into account. Although theory is increasingly extended with aspects such as anti-predator behaviour (Heithaus, 2001), predator switching (Krivian (2000) or alternative prey (Holt & Huxel, 2007), there is still a large gap between theory and practice. Theory might be further extended by connecting different types of interactions, such as omnivory and apparent competition between prey. We further recommend to implement effects of mixed diets in the theory of apparent competition. Furthermore, no specific theory exists on hyperpredation in the presence of alternative prey. Greenhouse crops are ideally suited to test theoretical predictions, because artificially created communities in biocontrol systems can easily be manipulated. Similarly, such greenhouse experiments could give insight into short-term dynamics of interactions for which more theory is needed since current theory focuses on what happens in or near equilibrium states (Briggs & Borer, 2005).

This review shows that both density-mediated interactions and behaviour-mediated interactions are common in greenhouse crops and affect the results of biological control. Especially the use of generalist predators may give rise to several types of interactions and food web complexity. Generalist predators were long considered as less effective than specialist natural enemies (Huffaker & Messenger, 1976; Hokkanen & Pimentel, 1984; Van Lenteren & Woets, 1988; Hoy, 1994). Moreover, recent criteria for risk assessments of natural enemies consider generalist predators as less desirable than specialist natural enemies (van Lenteren et al., 2006). However, several studies show that generalists can be effective control agents, especially because populations of generalists can be established easily (Messelink et al., 2010). The role of generalist predators was recognized earlier by Murdoch et al. (1985), who argued that the biggest advantage of generalist predators is their persistence in systems (see also Chang & Kareiva, 1999; Symondson et al., 2002). In contrast, augmentative releases of specialist natural enemies often involve problems with timing, costs and quality (Bloemhard & Ramakers, 2008). Generalist predators can establish into crops prior to pest infestations, which makes the system resilient to pest invasions. Moreover, growers need to respond less to infestations with pest species. In the near future, we expect that biological control systems in greenhouses will increasingly shift from augmentative releases of specialist natural enemies to inoculative releases of generalist predators. For example, whitefly control was mainly based on releases of specialist parasitoids for decades (van Lenteren & Woets, 1988; Avilla et al., 2004). This has changed since the introduction of generalist predatory bugs and predatory mites that also feed on whiteflies. This has been so successful in some crops that most, if not all, biological control is done by means of generalist predators (Messelink, personal observations). Thrips control has a long tradition of using generalist predators, and in crops such as sweet pepper, these predators are very effective (Ramakers, 2004). So far, biological control of aphids is mainly based on frequent releases of specialist natural enemies such as parasitoids and predatory midges (Ramakers, 1989; Blümel, 2004), which is expensive and often not successful (Bloemhard & Ramakers, 2008). Recent experiments showed that inoculative releases of the generalist predator *M. pygmaeus* can also effectively control aphids in sweet pepper (Messelink, 2011). Hence, we expect that future control of aphids and other pests will increasingly be based on generalist predators. In general, we suggest that generalist predators deserve more attention in biological control programs for greenhouse crops.

An interesting aspect of using generalist predators is that pest control strongly depends on the diversity of pests in the crop (see paragraph 2.2). The fact that a mixture of two pests can increase the survival and developmental rate of a generalist predator offers new opportunities to enhance pest control by optimizing the diet for predators. Because many crops do not or hardly provide food for generalist predators, it may be possible to add food that is supplemental to the diet of a certain natural enemy species. Research should furthermore focus on ways to enhance establishment of generalist predators by offering alternative prey in open rearing systems or banker plant systems (Huang et al., 2011), by food sprays (Wade et al., 2008), or by selecting plants that provide food or shelter in the crop (Wäckers et al., 2005). Finally, it is desirable that future research focuses on selecting predators that are adapted to important crops and perform well on the pests and food sources present in these crops, rather than selecting natural enemies for any particular pest species.

Finally, we conclude that it is important to consider all possible interactions among species in arthropod food webs in order to detect interactions that are potentially detrimental or beneficial for biological control. Detrimental effects can mainly be expected from hyperpredators or hyperparasitoids, and, in theory, IGP can also disrupt biological control. Furthermore, apparent mutualism may be negative for pest control. Hence, it is clear that the results of biological control of a particular pest species may be negatively affected by the presence of other pests or natural enemies. However, this review also showed many examples of plant-mediated and predator-mediated interactions that are beneficial for pest control. Future research should focus on more complementarity and synergy among natural enemies. The literature provides interesting examples of such interactions based on predator facilitation (Losey & Denno, 1998), pest stage complementarity (Calvo et al., 2009) or microhabitat complementarity (Onzo et al., 2004).

Nowadays, there are unique possibilities to manipulate communities of natural enemies by choosing from several species that are commercially available (van Lenteren, 2000; Enkegaard & Brødsgaard, 2006). Thus, biodiversity can be created and manipulated to maximise sustainable pest control. At the same time, such systems can be used to study the manipulation of biodiversity on the dynamics of communities of plant-inhabiting arthropods under relatively controlled conditions and at larger spatial scales than can usually be realized with communities under field conditions. Based on the abundance, diversity and potential risk of pest species, it is possible to adapt the strategies of natural enemy releases. In conclusion, greenhouse experiments that evaluate multiple pest control with diverse assemblages of natural enemies are not only needed to further develop biological control strategies, but also offer excellent opportunities to test and extend theories on multispecies interactions.

5. Acknowledgment

We thank all our colleagues for fruitful discussions, particularly Roos van Maanen and Pierre Ramakers. Chantal Bloemhard is thanked for creating and editing the insect pictures. Support from the Dutch Ministry of Economic Affairs, Agriculture and Innovation made this review possible.

6. References

Abrams, P. A. & Matsuda, H. (1996). Positive indirect effects between prey species that share predators. *Ecology*, Vol.77, No.2, pp. 610-616, ISSN 0012-9658

Abrams, P.A., Holt, R.D. & Roth, J.D. (1998). Apparent competition or apparent mutualism? Shared predation when populations cycle. *Ecology*, Vol.79, No.1, pp. 201-212, ISSN 0012-9658

Abrams, P.A. (2008). Measuring the impact of dynamic antipredator traits on predator-prey-resource interactions. *Ecology*, Vol.89, No.6, pp. 1640-1649, ISSN 0012-9658

Agrawal, A.A., Kobayashi, C. & Thaler, J.S. (1999). Influence of prey availability and induced host-plant resistance on omnivory by western flower thrips. *Ecology*, Vol.80, No.2, pp. 518-523, ISSN 0012-9658

Agrawal, A.A. & Klein, C.N. (2000). What omnivores eat: direct effects of induced plant resistance on herbivores and indirect consequences for diet selection by omnivores. *Journal of Animal Ecology*, Vol.69, No.3, pp. 525-535, ISSN 0021-8790

Alvarado, P., Balta, O. & Alomar, O. (1997). Efficiency of four heteroptera as predators of *Aphis gossypii* and *Macrosiphum euphorbiae* (Hom.: Aphididae). *Entomophaga*, Vol.42, No.1-2, pp. 215-226, ISSN 0013-8959

Anderson, K.E., Inouye, B.D. & Underwood, N. (2009). Modeling herbivore competition mediated by inducible changes in plant quality. *Oikos*, Vol.118, No.11, pp. 1633-1646, ISSN 0030-1299

Arnó, J., Roig, J. & Riudavets, J. (2008). Evaluation of *Orius majusculus* and *O. laevigatus* as predators of *Bemisa tabaci* and estimation of their prey preference. *Biological control*, Vol.44, No.1, pp. 1-6

Avilla, J., Albajes, R., Alomar, O., Castane, C. & Gabarra, R. (2004). Biological control of whiteflies on vegetable crops, In: *Biocontrol in protected culture*, K. M. Heinz, R. G. Van Driesche and M. P. Parrella (Eds.), pp. 171-184, Ball Publishing, ISBN 1-883052-39-4, Batavia, Illinois

Bakker, J.C. (2008). Developments in greenhouse horticultural production systems. *IOBC/wprs*, Vol.32, pp. 5-12, ISBN 978-92-9067-206-7

Bakker, F.M. & Sabelis, M.W. (1989). How larvae of *Thrips tabaci* reduce the attack success of phytoseiid predators. *Entomologia experimentalis et applicata*, Vol.50, No.1, pp. 47-51

Beddington, J.R. & Hammond, P.S. (1977). On the dynamics of host-parasite-hyperparasite Interactions. *Journal of Animal Ecology*, Vol.46, No.3, pp. 811-821, ISSN 0021-8790

Bloemhard, C. & Ramakers, P. (2008). Strategies for aphid control in organically grown sweet pepper in the Netherlands. *IOBC/wprs*, Vol.32, pp. 25-28, ISBN 978-92-9067-206-7

Blümel, S. (2004). Biological control of aphids on vegetable crops, In: *Biocontrol in protected culture*, K. M. Heinz, R. G. Van Driesche and M. P. Parrella (Eds.), pp. 297-312, Ball Publishing, ISBN 1-883052-39-4, Batavia, Illinois

Bolker, B., Holyoak, M., Krivan, V., Rowe, L. & Schmitz, O. (2003). Connecting theoretical and empirical studies of trait-mediated interactions. *Ecology*, Vol.84, No.5, pp. 51101-1114, ISSN 0012-9658

Bonsall, M.B. & Hassell, M.P. (1997). Apparent competition structures ecological assemblages. *Nature*, Vol.388, No.6640, pp. 371-373, ISSN 0028-0836

Bowers, W.S., Webb, R.E., Nault, L.R. & Dutky, S.R. (1972). Aphid alarm pheromone: isolation, identification, synthesis. *Science*, Vol.177, No.4054, pp. 1121-1122, ISSN 0036-8075

Briggs, C.J. & Borer, E.T. (2005). Why short-term experiments may not allow long-term predictions about intraguild predation. *Ecological Applications*, Vol.15, No.4, pp. 1111-1117, ISSN 1051-0761

Brodeur, J. & Rosenheim, J.A. (2000). Intraguild interactions in aphid parasitoids. *Entomologia Experimentalis Et Applicata*, Vol.97, No.1, pp. 93-108, ISSN 0013-8703

Brødsgaard, H.F. & Enkegaard, A. (2005). Intraguild predation between *Orius majusculus* (Reuter) (Hemiptera: Anthocoridae) and *Iphiseius degenerans* Berlese (Acarina: Phytoseiidae). IOBC/wprs, Vol.Bulletin 28, pp. 19-22

Buitenhuis, R., Shipp, L. & Scott-Dupree, C. (2010). Intra-guild vs extra-guild prey: effect on predator fitness and preference of *Amblyseius swirskii* (Athias-Henriot) and *Neoseiulus cucumeris* (Oudemans) (Acari: Phytoseiidae). *Bulletin of Entomological Research*, Vol.100, No.2, pp. 167-173, ISSN 0007-4853

Byrne, D.N. & Bellows, T.S. (1991). Whitefly biology. *Annual Review of Entomology*, Vol.36, pp. 431-457, ISSN 0066-4170

Cakmak, I., Janssen, A., Sabelis, M.W. & Baspinar, H. (2009). Biological control of an acarine pest by single and multiple natural enemies. *Biological Control*, Vol.50, No.1, pp. 60-65, ISSN 1049-9644

Calvo, F.J., Bolckmans, K. & Belda, J.E. (2009). Development of a biological control-based Integrated Pest Management method for *Bemisia tabaci* for protected sweet pepper crops. *Entomologia Experimentalis Et Applicata*, Vol.133, No.1, pp. 9-18, ISSN 0013-8703

Cardinale, B.J., Harvey, C.T., Gross, K. & Ives, A.R. (2003). Biodiversity and biocontrol: emergent impacts of a multi-enemy assemblage on pest suppression and crop yield in an agroecosystem. *Ecology Letters*, Vol.6, No.9, pp. 857-865, ISSN 1461-023X

Casula, P., Wilby, A. & Thomas, M.B. (2006). Understanding biodiversity effects on prey in multi-enemy systems. *Ecology Letters*, Vol.9, No.9, pp. 995-1004, ISSN 1461-023X

Chaneton, E.J. & Bonsall, M.B. (2000). Enemy-mediated apparent competition: empirical patterns and the evidence. *Oikos*, Vol.88, No.2, pp. 380-394, ISSN 0030-1299

Chang, G.C. & Kareiva, P. (1999). The case for indigenous generalists in biological control, In: *Theoretical approaches to biological control*, B. A. Hawkins and H. V. Cornell, pp. 103-115, Cambridge University Press, ISBN 0-521-57283-5, Cambridge

Choh, Y., van der Hammen, T., Sabelis, M.W. & Janssen, A. (2010). Cues of intraguild predators affect the distribution of intraguild prey. *Oecologia*, Vol.163, No.2, pp. 335-340, ISSN 0029-8549

Chow, A., Chau, A. & Heinz, K.M. (2008). Compatibility of *Orius insidiosus* (Hemiptera : Anthocoridae) with *Amblyseius* (*Iphiseius*) *degenerans* (Acari : Phytoseiidae) for control of *Frankliniella occidentalis* (Thysanoptera : Thripidae) on greenhouse roses. *Biological Control*, Vol.44, No.2, pp. 259-270, ISSN 1049-9644

Cock, M.J.W., van Lenteren, J.C., Brodeur, J., Barratt, B.I.P., Bigler, F., Bolckmans, K., Consoli, F.L., Haas, F., Mason, P.G. & Parra, J.R.P. (2010). Do new access and benefit sharing procedures under the convention on biological diversity threaten the future of biological control? *Biocontrol*, Vol.55, No.2, pp. 199-218, ISSN 1386-6141

Colfer, R.G. & Rosenheim, J.A. (2001). Predation on immature parasitoids and its impact on aphid suppression. *Oecologia*, Vol.126, No.2, pp. 292-304, ISSN 0029-8549

Coll, M. & Guershon, M. (2002). Omnivory in terrestrial arthropods: Mixing plant and prey diets. *Annual Review of Entomology*, Vol.47, pp. 267-297, ISSN 0066-4170

Courchamp, F., Langlais, M. & Sugihara, G. (2000). Rabbits killing birds: modelling the hyperpredation process. *Journal of Animal Ecology*, Vol.69, No.1, pp. 154-164, ISSN 0021-8790

Daugherty, M.P., Harmon, J.P. & Briggs, C.J. (2007). Trophic supplements to intraguild predation. *Oikos*, Vol.116, No.4, pp. 662-677, ISSN 0030-1299

de Bruijn, P.J.A., Egas, M., Janssen, A. & Sabelis, M.W. (2006). Pheromone-induced priming of a defensive response in western flower thrips. *Journal of Chemical Ecology*, Vol.32, No.7, pp. 1599-1603, ISSN 0098-0331

Dicke, M. & Sabelis, M.W. (1988). How plants obtain predatory mites as bodyguards. *Netherlands Journal of Zoology*, Vol.38, No.2-4, pp. 148-165, ISSN 0028-2960

Dicke, M. & Grostal, P. (2001). Chemical detection of natural enemies by arthropods: An ecological perspective. *Annual Review of Ecology and Systematics*, Vol.32, pp. 1-23, ISSN 0066-4162

Enkegaard, A. & Brødsgaard, H.F. (2006). Biocontrol in protected crops: Is lack of biodiversity a limiting factor?, In: *Ecological and Societal Approach to Biological Control*, J. Eilenberg and H. M. T. Hokkanen, pp. 91-122, Springer, ISBN 1-4020-4320-1, Dordrecht

Evans, E.W., Stevenson, A.T. & Richards, D.R. (1999). Essential versus alternative foods of insect predators: benefits of a mixed diet. *Oecologia*, Vol.121, No.1, pp. 107-112, ISSN 0029-8549

Evans, E.W. (2008). Multitrophic interactions among plants, aphids, alternate prey and shared natural enemies - a review. *European Journal of Entomology*, Vol.105, No.3, pp. 369-380, ISSN 1210-5759

Faraji, F., Janssen, A. & Sabelis, M.W. (2001). Predatory mites avoid ovipositing near counterattacking prey. *Experimental and Applied Acarology*, Vol.25, No.8, pp. 613-623, ISSN 0168-8162

Finke, D.L. & Denno, R.F. (2004). Predator diversity dampens trophic cascades. *Nature*, Vol.429, No.6990, pp. 407-410, ISSN 0028-0836

Finke, D.L. & Denno, R.F. (2005). Predator diversity and the functioning of ecosystems: the role of intraguild predation in dampening trophic cascades. *Ecology Letters*, Vol.8, No.12, pp. 1299-1306, ISSN 1461-023X

Frank, S.D. (2010). Biological control of arthropod pests using banker plant systems: Past progress and future directions. *Biological Control*, Vol.52, No.1, pp. 8-16, ISSN 1049-9644

Frechette, B., Rojo, S., Alomar, O. & Lucas, E. (2007). Intraguild predation between syrphids and mirids: who is the prey? Who is the predator? *Biocontrol*, Vol.52, No.2, pp. 175-191, ISSN 1386-6141

Gerson, U. & Weintraub, P.G. (2007). Mites for the control of pests in protected cultivation. *Pest Management Science*, Vol.63, No.7, pp. 658-676, ISSN 1526-498X

Gillespie, D.R. & Quiring, D.J.M. (1992). Competition between *Orius tristicolor* (White) (Hemiptera: Anthocoridae) and *Amblyseius cucumeris* (Oudemans) (Acari: Phytoseiidae) feeding on *Frankliniella occidentalis* (Pergande) (Thysanoptera, Thripidae). *Canadian Entomologist*, Vol.124, No.6, pp. 1123-1128, ISSN 0008-347X

Gillespie, D.R. & Raworth, D.A. (2004). Biological control of two-spotted spider mites on greenhouse vegetable crops, In: *Biocontrol in protected culture*, K.M. Heinz, R.G. Van Driesche and M.P. Parrella (Eds.), pp. 201-220, Ball Publishing, ISBN 1-883052-39-4, Batavia, Illinois

Hansen, D.L., Brødsgaard, H.F. & Enkegaard, A. (1999). Life table characteristics of *Macrolophus caliginosus* preying upon *Tetranychus urticae*. *Entomologia Experimentalis Et Applicata*, Vol.93, No.3, pp. 269-275, ISSN 0013-8703

Harmon, J.P. & Andow, D.A. (2004). Indirect effects between shared prey: Predictions for biological control. *Biocontrol*, Vol.49, No.6, pp. 605-626, ISSN 1386-6141

Hatherly, I.S., Pedersen, B.P. & Bale, J.S. (2009). Effect of host plant, prey species and intergenerational changes on the prey preferences of the predatory mired *Macrolophus caliginosus*. *Biocontrol*, Vol.54, No.1, pp. 35-45, 1386-6141

Heithaus, M.R. (2001). Habitat selection by predators and prey in communities with asymmetrical intraguild predation. *Oikos*, Vol.92, No.3, pp. 542-554, ISSN 0030-1299

Helle, W. & Sabelis, M.W. (1985). *Spider mites, their biology, natural enemies and control, World crop pests, Volume 1A*, Elsevier, ISBN 0-444-42372-9, Amsterdam

Henaut, Y., Alauzet, C., Ferran, A. & Williams, T. (2000). Effect of nymphal diet on adult predation behavior in *Orius majusculus* (Heteroptera : Anthocoridae). *Journal of Economic Entomology*, Vol.93, No.2, pp. 252-255, ISSN 0022-0493

Hindayana, D., Meyhofer, R., Scholz, D. & Poehling, H.M. (2001). Intraguild predation among the hoverfly *Episyrphus balteatus* de Geer (Diptera : Syrphidae) and other aphidophagous predators. *Biological Control*, Vol.20, No.3, pp. 236-246, ISSN 1049-9644

Hokkanen, H. & Pimentel, D. (1984). New approach for selecting biological-control agents. *Canadian Entomologist*, Vol.116, No.8, pp. 1109-1121, ISSN 0008-347X

Holt, R.D. (1977). Predation, apparent competition and structure of prey communities. *Theoretical Population Biology*, Vol.12, No.2, pp. 197-229, ISSN 0040-5809

Holt, R.D. & Kotler, B.P. (1987). Short-term apparent competition. *American Naturalist*, Vol.130, No.3, pp. 412-430, ISSN 0003-0147

Holt, R.D. & Lawton, J.H. (1994). The ecological consequences of shared natural enemies. Annual Review of Ecology and Systematics, Vol.25, pp. 495-520, ISSN 0066-4162

Holt, R.D. & Polis, G.A. (1997). A theoretical framework for intraguild predation. *American Naturalist*, Vol.149, No.4, pp. 745-764, ISSN 0003-0147

Holt, R.D. & Huxel, G.R. (2007). Alternative prey and the dynamics of intraguild predation: Theoretical perspectives. *Ecology*, Vol.88, No.11, pp. 2706-2712, ISSN 0012-9658

Hosseini, M., Ashouri, A., Enkegaard, A., Weisser, W.W., Goldansaz, S.H., Mahalati, M.N. & Moayeri, H.R.S. (2010). Plant quality effects on intraguild predation between *Orius laevigatus* and *Aphidoletes aphidimyza*. *Entomologia Experimentalis Et Applicata*, Vol.135, No.2, pp. 208-216, ISSN 0013-8703

Hoy, M.A. (1994). Parasitoids and predators in management of arthropod pests, In: *Introduction to insect pest management*, R.L. Metcalf and W.H. Luckmann, pp. 129-198, John Wiley & Sons, ISBN 978-0-471-58957-0, New York

Huang, N.X., Enkegaard, A., Osborne, L.S., Ramakers, P.M.J., Messelink, G.J., Pijnakker, J. & Murphy, G. (2011). The banker plant method in biological control. *Critical Reviews in Plant Sciences*, Vol.30, No.3, pp. 259-278, ISSN 0735-2689

Huffaker, C.B. & Messenger, P.S. (1976). *Theory and practice of biological control*, Academic Press, ISBN 0-12-360350-1, New York

Hulshof, J. & Linnamäki, M. (2002). Predation and oviposition rate of the predatory bug *Orius laevigatus* in the presence of alternative food. *IOBC/wprs*, Vol.25, pp. 107-110, ISBN 92-9067-137-4

Inbar, M., Doostdar, H., Leibee, G.L. & Mayer, R.T. (1999). The role of plant rapidly induced responses in asymmetric interspecific interactions among insect herbivores. *Journal of Chemical Ecology*, Vol.25, No.8, pp. 1961-1979, ISSN 0098-0331

Jakobsen, L., Enkegaard, A. & Brodsgaard, H.F. (2004). Interactions between two polyphagous predators, *Orius majusculus* (Hemiptera : Anthocoridae) and *Macrolophus caliginosus* (Heteroptera : Miridae). *Biocontrol Science and Technology*, Vol.14, No.1, pp. 17-24, ISSN 0958-3157

Janssen, A., Pallini, A., Venzon, M. & Sabelis, M.W. (1998). Behaviour and indirect interactions in food webs of plant-inhabiting arthropods. *Experimental and Applied Acarology*, Vol.22, No.9, pp. 497-521, ISSN 0168-8162

Janssen, A., Faraji, F., van der Hammen, T., Magalhaes, S. & Sabelis, M.W. (2002). Interspecific infanticide deters predators. *Ecology Letters*, Vol.5, No.4, pp. 490-494, ISSN 1461-023X

Janssen, A., Willemse, E. & van der Hammen, T. (2003). Poor host plant quality causes omnivore to consume predator eggs. *Journal of Animal Ecology*, Vol.72, No.3, pp. 478-483, ISSN 0021-8790

Janssen, A., Montserrat, M., HilleRisLambers, R., Roos, A.M.d., Pallini, A. & Sabelis, M.W. (2006). Intraguild Predation Usually does not Disrupt Biological Control, In: *Trophic and Guild Interactions in Biological Control*, J. Brodeur and G. Boivin (Eds.), pp. 21-44, Springer Netherlands, ISBN 1-4020-4766-5, Dordrecht

Janssen, A., Sabelis, M.W., Magalhaes, S., Montserrat, M. & Van der Hammen, T. (2007). Habitat structure affects intraguild predation. *Ecology*, Vol.88, No.11, pp. 2713-2719, ISSN 0012-9658

Kaplan, I. & Eubanks, M.D. (2002). Disruption of cotton aphid (Homoptera: Aphididae) - natural enemy dynamics by red imported fire ants (Hymenoptera: Formicidae). *Environmental Entomology*, Vol.31, No.6, pp. 1175-1183, ISSN 0046-225X

Karban, R. & Carey, J.R. (1984). Induced resistance of cotton seedlings to mites. *Science*, Vol.225, No.4657, pp. 53-54, ISSN 0036-8075

Karban, R. & Baldwin, I.T. (1997). *Induced responses to herbivory*, University of Chicago Press, ISBN 0226424952, Chicago

Kessler, A. & Baldwin, I.T. (2002). Plant responses to insect herbivory: The emerging molecular analysis. *Annual Review of Plant Biology*, Vol.53, pp. 299-328, ISSN 1543-5008

Kessler, A. & Halitschke, R. (2007). Specificity and complexity: the impact of herbivore-induced plant responses on arthropod community structure. *Current Opinion in Biology*, Vol.10, No.4, pp. 409-414, ISSN 1369-5266.

Krivan, V. (2000). Optimal intraguild foraging and population stability. *Theoretical Population Biology*, Vol.58, No.2, pp. 79-94, ISSN 0040-5809

Lalonde, R.G., McGregor, R.R., Gillespie, D.R., Roitberg, B.D. & Fraser, S. (1999). Plant-feeding by arthropod predators contributes to the stability of predator-prey population dynamics. *Oikos*, Vol.87, No.3, pp. 603-608, ISSN 0030-1299

Letourneau, D.K., Jedlicka, J.A., Bothwell, S.G. & Moreno, C.R. (2009). Effects of Natural Enemy Biodiversity on the Suppression of Arthropod Herbivores in Terrestrial Ecosystems. *Annual Review of Ecology Evolution and Systematics*, Vol.40, pp. 573-592, ISSN 1543-592X

Losey, J.E. & Denno, R.F. (1998). Positive predator-predator interactions: Enhanced predation rates and synergistic suppression of aphid populations. *Ecology*, Vol.79, No.6, pp. 2143-2152, ISSN 0012-9658

Magalhães, S., Janssen, A., Montserrat, M. & Sabelis, M.W. (2005). Host-plant species modifies the diet of an omnivore feeding on three trophic levels. *Oikos*, Vol.111, No.1, pp. 47-56, ISSN 0030-1299

Martinou, A.F., Milonas, P.G. & Wright, D.J. (2009). Patch residence decisions made by *Aphidius colemani* in the presence of a facultative predator. *Biological Control*, Vol.49, No.3, pp. 234-238, ISSN 1049-9644

May, R.M. & Hassell, M.P. (1981). The dynamics of multiparasitoid-host interactions. *American Naturalist*, Vol.117, No.3, pp. 234-261, ISSN 0003-0147

May, R. (1999). Unanswered questions in ecology. *Philosophical Transactions of the Royal Society of London Series B-Biological Sciences*, Vol.354, No.1392, pp. 1951-1959, ISSN 0962-8436

Meisner, M., Harmon, J.P. & Ives, A.R. (2007). Presence of an unsuitable host diminishes the competitive superiority of an insect parasitoid: a distraction effect. *Population Ecology*, Vol.49, No.4, pp. 347-355, ISSN 1438-3896.

Messelink, G.J., Van Steenpaal, S.E.F. & Ramakers, P.M.J. (2006). Evaluation of phytoseiid predators for control of western flower thrips on greenhouse cucumber. *Biocontrol*, Vol.51, No.6, pp. 753-768, ISSN 1386-6141

Messelink, G.J., van Maanen, R., van Steenpaal, S.E.F. & Janssen, A. (2008). Biological control of thrips and whiteflies by a shared predator: Two pests are better than one. *Biological Control*, Vol.44, No.3, pp. 372-379, ISSN 1049-9644

Messelink, G.J., Van Maanen, R., Van Holstein-Saj, R., Sabelis, M.W. & Janssen, A. (2010). Pest species diversity enhances control of spider mites and whiteflies by a generalist phytoseiid predator. *BioControl*, Vol.55, No.3, pp. 387-398, ISSN 1386-6141

Messelink, G.J., Bloemhard, C.M.J., Cortes, J.A., Sabelis, M.W. & Janssen, A. (2011). Hyperpredation by generalist predatory mites disrupts biological control of aphids by the aphidophagous gall midge *Aphidoletes aphidimyza*. *Biological Control*, Vol.57, No.3, pp. 246-252, ISSN 1049-9644

Messelink, G.J., Bloemhard, C.M.J., Kok, L. & Janssen A. (2011). Generalist predatory bugs control aphids in sweet pepper, *IOBC/wprs Bulletin*, Vol.68, pp. 115-118

Montserrat, M., Bas, C., Magalhaes, S., Sabelis, M.W., de Roos, A.M. & Janssen, A. (2007). Predators induce egg retention in prey. *Oecologia*, Vol.150, No.4, pp. 699-705, ISSN 0029-8549

Montserrat, M., Magalhaes, S., Sabelis, M.W., de Roos, A.M. & Janssen, A. (2008). Patterns of exclusion in an intraguild predator-prey system depend on initial conditions. *Journal of Animal Ecology*, Vol.77, No.3, pp. 624-630, ISSN 0021-8790

Müller, C.B. & Godfray, H.C.J. (1997). Apparent competition between two aphid species. *Journal of Animal Ecology*, Vol.66, No.1, pp. 57-64, ISSN 0021-8790

Murdoch, W.W. (1969). Switching in general predators: experiments on predator specificity and the stability of prey populations. *Ecological Monographs*, Vol.39, No.4, pp. 335-354, ISSN 0012-9615

Murdoch, W.W., Chesson, J. & Chesson, P.L. (1985). Biological control in theory and practice. *American Naturalist*, Vol.125, No.3, pp. 344-366, ISSN 0003-0147

Mylius, S.D., Klumpers, K., de Roos, A.M. & Persson, L. (2001). Impact of intraguild predation and stage structure on simple communities along a productivity gradient. *American Naturalist*, Vol.158, No.3, pp. 259-276, ISSN 0003-0147

Nakashima, Y., Birkett, M.A., Pye, B.J. & Powell, W. (2006). Chemically mediated intraguild predator avoidance by aphid parasitoids: Interspecific variability in sensitivity to semiochemical trails of ladybird predators. *Journal of Chemical Ecology*, Vol.32, No.9, pp. 1989-1998, ISSN 0098-0331

Nombela, G., Garzo, E., Duque, M. & Muniz, M. (2009). Preinfestations of tomato plants by whiteflies (*Bemisia tabaci*) or aphids (*Macrosiphum euphorbiae*) induce variable resistance or susceptibility responses. *Bulletin of Entomological Research*, Vol.99, No.2, pp. 183-191, ISSN 0007-4853

Nomikou, M., Janssen, A., Schraag, R. & Sabelis, M.W. (2002). Phytoseiid predators suppress populations of *Bemisia tabaci* on cucumber plants with alternative food. *Experimental and Applied Acarology*, Vol.27, No.1-2, pp. 57-68, ISSN 0168-8162

Nomikou, M., Janssen, A. & Sabelis, M.W. (2003). Herbivore host plant selection: whitefly learns to avoid host plants that harbour predators of her offspring. *Oecologia*, Vol.136, No.3, pp. 484-488, ISSN 0029-8549

Nomikou, M., Sabelis, M.W. & Janssen, A. (2010). Pollen subsidies promote whitefly control through the numerical response of predatory mites. *Biocontrol*, Vol.55, No.2, pp. 253-260, ISSN 1386-6141

Oliver, K.M., Russell, J.A., Moran, N.A. & Hunter, M.S. (2003). Facultative bacterial symbionts in aphids confer resistance to parasitic wasps. *Proceedings of the National Academy of Sciences of the United States of America*, Vol.100, No.4, pp. 1803-1807, ISSN 0027-8424

Onzo, A., Hanna, R., Janssen, A. & Sabelis, M.W. (2004). Interactions between two neotropical phytoseiid predators on cassava plants and consequences for biological control of a shared spider mite prey: a screenhouse evaluation. *Biocontrol Science and Technology*, Vol.14, No.1, pp. 63-76, ISSN 0958-3157

Pallini, A., Janssen, A. & Sabelis, M.W. (1997). Odour-mediated responses of phytophagous mites to conspecific and heterospecific competitors. *Oecologia*, Vol.110, No.2, pp. 179-185, ISSN 0029-8549

Pallini, A., Janssen, A. & Sabelis, M.W. (1998). Predators induce interspecific herbivore competition for food in refuge space. *Ecology Letters*, Vol.1, No.3, pp. 171-177, ISSN 1461-023X

Pallini, A., Janssen, A. & Sabelis, M.W. (1999). Spider mites avoid plants with predators. *Experimental and Applied Acarology*, Vol.23, No.10, pp. 803-815, ISSN 0168-8162

Perdikis, D.C. & Lykouressis, D.P. (2004). *Myzus persicae* (Homoptera : Aphididae) as suitable prey for *Macrolophus pygmaeus* (Hemiptera : Miridae) population increase on pepper plants. *Environmental Entomology*, Vol.33, No.3, pp. 499-505, ISSN 0046-225X

Pimm, S.L. & Lawton, J.H. (1978). Feeding on more than one trophic level. *Nature*, Vol.275, No.5680, pp. 542-544, ISSN 0028-0836

Pineda, A., Morales, I., Marcos-Garcia, M.A. & Fereres, A. (2007). Oviposition avoidance of parasitized aphid colonies by the syrphid predator *Episyrphus balteatus* mediated by different cues. *Biological Control*, Vol.42, No.3, pp. 274-280, ISSN1049-9644

Polis, G.A., Myers, C.A. & Holt, R.D. (1989). The ecology and evolution of intraguild predation: potential competitors that eat each other. *Annual Review of Ecology and Systematics*, Vol.20, pp. 297-330, ISSN 0066-4162

Polis, G.A. & Strong, D.R. (1996). Food web complexity and community dynamics. *American Naturalist*, Vol.147, No.5, pp. 813-846, ISSN 0003-0147

Prasad, R.P. & Snyder, W.E. (2006). Diverse trait-mediated interactions in a multi-predator, multi-prey community. *Ecology*, Vol.87, No.5, pp. 1131-1137, ISSN 0012-9658

Ramakers, P.M.J. (1980). Biological control of *Thrips tabaci* (Thysanoptera: Thripidae) with *Amblyseius* spp. (Acari: Phytoseiidae) *IOBC/wprs*, Vol.3, No.3, pp. 203-208

Ramakers, P.M.J. (1989). Biological control in greenhouses, In: *Aphids, their biology, natural enemies and control*, A.K. Minks and P. Harrewijn (Eds.), pp. 199-208, Elsevier, ISBN 0-444-42799-6, Amsterdam

Ramakers, P.M.J. & Voet, S.J.P. (1995). Use of castor bean, Ricinus communis, for the introduction of the thrips predator *Amblyseius degenerans* on glasshouse-grown sweet peppers. *Mededelingen Faculteit Landbouwkundige en Toegepaste Biologische Wetenschappen*, Universiteit Gent., Vol.60, No.3a, pp. 885-891

Ramakers, P.M.J. (2004). IPM program for sweet pepper, In: *Biocontrol in protected culture*, K.M. Heinz, R.G. Van Driesche and M.P. Parrella (Eds.), pp. 439-455, Ball Publishing, ISBN 1-883052-39-4, Batavia, Illinois

Riudavets, J. & Castañé, C. (1998). Identification and evaluation of native predators of *Frankliniella occidentalis* (Thysanoptera : Thripidae) in the Mediterranean. *Environmental Entomology*, Vol.27, No.1, pp. 86-93, ISSN 0046-225X

Roemer, G.W., Coonan, T.J., Garcelon, D.K., Bascompte, J. & Laughrin, L. (2001). Feral pigs facilitate hyperpredation by golden eagles and indirectly cause the decline of the island fox. *Animal Conservation*, Vol.4, pp. 307-318, ISSN 1367-9430

Roemer, G.W., Donlan, C.J. & Courchamp, F. (2002). Golden eagles, feral pigs, and insular carnivores: How exotic species turn native predators into prey. *Proceedings of the National Academy of Sciences of the United States of America*, Vol.99, No.2, pp. 791-796, ISSN 0027-8424

Roques, A., Rabitsch, W., Rasplus, J.-Y., Lopez-Vaamonde, C., Nentwig, W. & Kenis, M. (2009). Alien Terrestrial Invertebrates of Europe, In: *Handbook of Alien Species in Europe*, pp. 63-79, Springer Netherlands, ISBN 978-1-4020-8280-1, Dordrecht

Rosenheim, J.A. & Harmon, J.P. (2006). The influence of intraguild predation on the suppression of a shared prey population: an emperical reassessment, In: *Trophic and Guild Interactions in Biological Control*, J. Brodeur and G. Boivin (Eds.), pp. 1-20, Springer Netherlands, ISBN 1-4020-4766-5, Dordrecht

Rosenheim, J.A., Kaya, H.K., Ehler, L.E., Marois, J.J. & Jaffee, B.A. (1995). Intraguild predation among biological control agents: theory and evidence. *Biological Control*, Vol.5, No.3, pp. 303-335, ISSN 1049-9644

Rosenheim, J.A. (1998). Higher-order predators and the regulation of insect herbivore populations. *Annual Review of Entomology*, Vol.43, pp. 421-447, ISSN 0066-4170

Sabelis, M.W., Janssen, A., Lesna, I., Aratchige, N.S., Nomikou, M. & van Rijn, P.C.J. (2008). Developments in the use of predatory mites for biological pest control. *IOBC/wprs*, Vol.32, pp. 187-199, ISBN 978-92-9067-206-7

Sarmento, R.A., Lemos, F., Bleeker, P.M., Schuurink, R.C., Pallini, A., Oliveira, M.G.A., Lima, E.R., Kant, M., Sabelis, M.W. & Janssen, A. (2011). A herbivore that manipulates plant defence. *Ecology Letters*, Vol.14, No.3, pp. 229-236, ISBN 1461-023X

Sarmento , R.A., Lemos, F., Dias, C.R., Kikuchi, W.T., Rodrigues, J.C.P., Pallini, A., Sabelis, M.W. & Janssen, A. 2011. A herbivorous mite down-regulates plant defence and produces web to exclude competitors. *PLoS ONE* (in press).

Schaller, A. (2008). *Induced plant resistance to herbivory*, Springer, ISBN 978-1-4020-8181-1, Dordrecht

Shipp, J.L. & Ramakers, P.M.J. (2004). Biological control of thrips on vegetable crops, In: *Biocontrol in protected culture*, K.M. Heinz, R.G. Van Driesche and M.P. Parrella (Eds.), pp. 265–276., Ball Publishing, ISBN 1-883052-39-4, Batavia, Illinois

Sih, A., Crowley, P., McPeek, M., Petranka, J. & Strohmeier, K. (1985). Predation, competition, and prey communities: a review of field experiments. *Annual Review of Ecology and Systematics*, Vol.16, pp. 269-311, ISSN 0066-4162

Snyder, W.E. & Ives, A.R. (2001). Generalist predators disrupt biological control by a specialist parasitoid. *Ecology*, Vol.82, No.3, pp. 705-716, ISSN 0012-9658

Snyder, W.E. & Ives, A.R. (2003). Interactions between specialist and generalist natural enemies: Parasitoids, predators, and pea aphid biocontrol. *Ecology*, Vol.84, No.1, pp. 91-107, ISSN 0012-9658

Stacey, D.L. (1977). 'Banker' plant production of *Encarsia formosa* Gahan and its use in the control of glasshouse whitefly on tomatoes. *Plant Pathology*, Vol.26, No.2, pp. 63-66, ISSN 0032-0862

Stout, M.J., Workman, K.V., Bostock, R.M. & Duffey, S.S. (1998). Specificity of induced resistance in the tomato, *Lycopersicon esculentum*. *Oecologia*, Vol.113, No.1, pp. 74-81, ISSN 0029-8549

Sullivan, D.J. & Völkl, W. (1999). Hyperparasitism: Multitrophic ecology and behavior. *Annual Review of Entomology*, Vol.44, pp. 291-315, ISSN 0066-4170

Symondson, W.O.C., Sunderland, K.D. & Greenstone, M.H. (2002). Can generalist predators be effective biocontrol agents? *Annual Review of Entomology*, Vol.47, pp. 561-594, ISSN 0066-4170

Teerling, C.R., Pierce, H.D., Borden, J.H. & Gillespie, D.R. (1993). Identification and bioactivity of alarm pheromone in the western flower thrips, *Frankliniella occidentalis*. *Journal of Chemical Ecology*, Vol.19, No.4, pp. 681-697, ISSN 0098-0331

Trichilo, P.J. & Leigh, T.F. (1986). Predation on spider mite eggs by the western flower thrips, *Frankliniella occidentalis* (Thysanoptera: Thripidae), an opportunist in a cotton agroecosystem. *Environmental Entomology*, Vol.15, No.4, pp. 821-825, ISSN 0046-225X

Vance-Chalcraft, H.D., Rosenheim, J.A., Vonesh, J.R., Osenberg, C.W. & Sih, A. (2007). The influence of intraguild predation on prey suppression and prey release: A meta-analysis. *Ecology*, Vol.88, No.11, pp. 2689-2696, ISSN 0012-9658

van den Meiracker, R.A.F. & Ramakers, P.M.J. (1991). Biological control of the western flower thrips *Frankliniella occidentalis*, in sweet pepper, with the anthocorid predator. *Mededelingen van de Faculteit Landbouwwetenschappen, Rijksuniversiteit Gent*, Vol.56, No.2a, pp. 241-249

van der Blom, J., Robledo, A., Torres, S. & Sanchez, J.A. (2009). Consequences of the wide scale implementation of biological control in greenhouse horticulture in Almeria, Spain. *IOBC/wprs*, Vol.49, pp. 9-13, ISBN 978-92-9067-223-4

van der Hammen, T., de Roos, A.M., Sabelis, M.W. & Janssen, A. (2010). Order of invasion affects the spatial distribution of a reciprocal intraguild predator. *Oecologia*, Vol.163, No.1, pp. 79-89, ISSN 0029-8549

van Lenteren, J.C. & Woets, J. (1988). Biological and integrated pest control in greenhouses. *Annual Review of Entomology*, Vol.33, pp. 239-269, ISSN 0066-4170

van Lenteren, J.C. (2000). A greenhouse without pesticides: fact or fantasy? Crop Protection, Vol.19, No.6, pp. 375-384, ISSN 0261-2194

van Lenteren, J.C., Bale, J., Bigler, E., Hokkanen, H.M.T. & Loomans, A.M. (2006). Assessing risks of releasing exotic biological control agents of arthropod pests. *Annual Review of Entomology*, Vol.51, pp. 609-634, ISSN 0066-4170

van Maanen, R., Vila, E., Sabelis, M.W. & Janssen, A. (2010). Biological control of broad mites (*Polyphagotarsonemus latus*) with the generalist predator *Amblyseius swirskii*. *Experimental and Applied Acarology*, Vol.52, No.1, pp. 29-34, ISSN 0168-8162

van Maanen, R., Broufas, G., Oveja, M.F., Sabelis, M.W. & Janssen, A. Intraguild predation among herbivorous insects: western flower thrips larvae feed on whitefly crawlers. In prep.

van Rijn, P.C.J., van Houten, Y.M. & Sabelis, M.W. (2002). How plants benefit from providing food to predators even when it is also edible to herbivores. *Ecology*, Vol.83, No.10, pp. 2664-2679, ISSN 0012-9658

van Veen, F.J.F., Memmott, J. & Godfray, H.C.J. (2006). Indirect Effects, Apparent Competition and Biological Control., In: *Trophic and Guild Interactions in Biological Control*, J. Brodeur and G. Boivin (Eds.), pp. 145-169, Springer, ISBN 1-4020-4766-5, Dordrecht

Venzon, M., Janssen, A., Pallini, A. & Sabelis, M.W. (2000). Diet of a polyphagous arthropod predator affects refuge seeking of its thrips prey. *Animal Behaviour*, Vol.60, pp. 369-375, ISSN 0003-3472

Venzon, M., Janssen, A. & Sabelis, M.W. (2001). Prey preference, intraguild predation and population dynamics of an arthropod food web on plants. *Experimental and Applied Acarology*, Vol.25, No.10-11, pp. 785-808, ISSN 0168-8162

Venzon, M., Janssen, A. & Sabelis, M.W. (2002). Prey preference and reproductive success of the generalist predator *Orius laevigatus*. *Oikos*, Vol.97, No.1, pp. 116-124, ISSN 0030-1299

Villagra, C.A., Ramirez, C.C. & Niemeyer, H.M. (2002). Antipredator responses of aphids to parasitoids change as a function of aphid physiological state. *Animal Behaviour*, Vol.64, pp. 677-683, ISSN 0003-3472

Wäckers, F.L., van Rijn, P.C.J. & Bruin, J. (2005). *Plant-provided Food for Carnivorous Insects: A Protective Mutualism and its Applications*, Cambridge University Press, ISBN-13 978-0-521-81941-1, Cambridge

Wade, M.R., Zalucki, M.P., Wratten, S.D. & Robinson, K.A. (2008). Conservation biological control of arthropods using artificial food sprays: Current status and future challenges. *Biological Control*, Vol.45, No.2, pp. 185-199, ISSN 1049-9644

Walling, L.L. (2000). The myriad plant responses to herbivores. *Journal of Plant Growth Regulation*, Vol.19, No.2, pp. 195-216, ISSN 0721-7595

Werner, E.E. & Peacor, S.D. (2003). A review of trait-mediated indirect interactions in ecological communities. *Ecology*, Vol.84, No.5, pp. 1083-1100, ISSN 0012-9658

Zhang, L.P., Zhang, G.Y., Zhang, Y.J., Zhang, W.J. & Liu, Z. (2005). Interspecific interactions between *Bemisia tabaci* (Hem., Aleyrodidae) and *Liriomyza sativae* (Dipt., Agromyzidae). *Journal of Applied Entomology*, Vol.129, No.8, pp. 443-446, ISSN 0931-2048

Zhang, P.J., Zheng, S.J., van Loon, J.J.A., Boland, W., David, A., Mumm, R. & Dicke, M. (2009). Whiteflies interfere with indirect plant defense against spider mites in Lima bean. *Proceedings of the National Academy of Sciences of the United States of America*, Vol.106, No.50, pp. 21202-21207, ISSN 0027-8424

10

Feral Pigeons: Problems, Dynamics and Control Methods

Dimitri Giunchi[1], Yuri V. Albores-Barajas[2], N. Emilio Baldaccini[1], Lorenzo Vanni[1] and Cecilia Soldatini[2]
[1]*University of Pisa*
[2]*University of Venice*
Italy

1. Introduction

Feral pigeons are birds now largely present with naturalized populations all around the world (Lever, 1987). The Rock Dove (*Columba livia*), which is their ultimate ancestor, was originally present in coastal and inland cliffs of central and western Palearctic and in the northern Ethiopian regions, as well as in those of the Indian subcontinent (Goodwin, 1983). These wild populations gave rise to domestic breeds as a result of artificial selection, having been the pigeons one of the first birds subjected to domestication (Sossinka, 1982). Domestics readily go feral, they have done so widely and in different times and locations', both in their natural range and in all continents where they were transported as captive birds, and subsequently introduced (Johnston & Janiga, 1995; Lever, 1987). Pigeons are granivorous birds tightly linked to arid and rocky habitats, so that feral populations remain linked to human settlements both as a consequence of their domestic origin and by these biological characteristics, that act in synergy (Baldaccini, 1996a). According to Goodwin (1978) the synanthropism of ferals is mainly a consequence of the food resources becoming available with the development of agriculture or otherwise mainly depends on the presence of buildings that constitute a vicariant habitat with respect to the natural one, as suggested by Hoffmann (1982). Food resources and human buildings are the key ecological factors that bring ferals into most cities and towns worldwide (Haag-Wackernagel, 1995), extensively in agricultural habitats and wherever man has constructed suitable recoveries to dwell in, forming stable or increasing populations of millions of individuals as stated by BirdLife International (2004) for Europe or Sauer et al. (2008) for the USA. The way by which feral pigeons established in urban habitats has been illustrated from a historical point of view by Ghigi (1950) and van der Linden (1950) and recently reviewed by Johnston & Janiga (1995), Haag-Wackernagel (1998) and Baldaccini & Giunchi (2006). Even in the Old World, synanthropic wild Rock Doves have a very marginal contribution to the constitution of feral populations (Ballarini et al., 1989; Johnston & Janiga, 1995).

2. Problem overview

The presence of feral pigeons in urban habitat and their degree of interactions with human life and activities can be perceived in many ways, ranging from harmless and tame birds to harmful pests, depending on the personal cultural background (Jerolmack, 2008; Johnston & Janiga, 1995). Nevertheless, feral pigeons have a formidable capacity to become pest by any standard. Factors that have been identified as important in becoming a pest include the main characteristics of pigeons, such as being a granivore, having an alimentary storage crop, high reproductive rate, colonial habits and group foraging (Johnston & Janiga, 1995).

2.1 Public health risks

Feral pigeons are of considerable epidemiological importance, being reservoirs and potential vectors of a large number of microorganisms and source of antigens of zoonotic concern, causing both infections and allergic diseases, that can be lethal (Haag-Wackernagel, 2006; Haag-Wackernagel & Bircher, 2009; Haag-Wackernagel & Moch, 2004; Magnino et al., 2009; Rosický, 1978). Pathogens can be transmitted to humans mainly via excreta, secretions, or dust from feathers spread into the environment, thus a direct contact with pigeons can be unimportant (Curtis et al., 2002; Geigenfeind & Haag-Wackernagel, 2010). Pigeons breeding and roosting sites host an endless number of arthropods that may infest humans as bugs, fleas, mites and ticks. The latter are of particular human concern, as the soft tick *Argas reflexus* (Haag-Wackernagel & Bircher, 2009; Mumcuoglu et al., 2005). Lists of the different pathogenic organisms and of the most common parasitic arthropods identified in feral pigeons are reported by Johnston & Janiga (1995), Haag-Wackernagel & Moch (2004) and Haag-Wackernagel (2006). *Chlamydophila psittaci* is one of the most common pathogenic bacteria affecting at least European population of ferals (Magnino et al., 2009 and references therein); infection by different serotypes of *Salmonella* is on the contrary low (e.g. Pedersen et al., 2006). Regarding disease-producing fungi, Gallo et al. (1989) reported a percentage of pigeons infected by yeasts ranging from 7% (rural habitat) up to 22% (urban centre). According to these data, the most common pathogens transmitted to humans are *Chlamydophila psittaci* and the yeast *Cryptococcus neoformans*, while infections caused by *Salmonella* are very rare (Haag-Wackernagel & Moch, 2004), thus confirming a relationship between host population density and pathogenic transmission rate (Grenfell & Bolker, 1998). According to Haag-Wackernagel & Moch (2004), the risk of transmission of pathogens from pigeons to healthy humans is low, even for people in close contact with pigeons or their nests. On the contrary, immuno-depressed patients have a greater risk of infection in comparison to healthy people (Haag-Wackernagel & Moch, 2004). Feral pigeons, both in urban areas and in countryside, came in contact with different, often closely related, animal species thus enlarging their potential role as vectors of pathogens and parasites (Bevan 1990; Pedersen et al., 2006; Rosický, 1978). Pigeons have apparently introduced many avian pathogens into wild populations wherever they have been naturalized, infecting taxa as seabirds, penguins, raptors, other columbids and passerines (Phillips et al., 2003 and references therein).

Feral pigeons can also be the source of accidents of various nature, from the trivial slipping on surfaces littered by pigeon droppings, to the most serious problem of hazards to aircraft (bird-strike). As open habitats, in many cases not far from cities, airports attract selectively flocks of pigeons that are listed as one of the species more commonly involved in bird-strike events (Cleary et al., 2006; Dolbeer et al., 2000).

2.2 Infrastructural damages

Urban architectural problems constitute another factor of the negative relationship of humans and pigeons. Litter that accumulates under and on the surfaces used to roost or to nest are not only problematic from hygienic and urban deface reasons, they also cause structural and aesthetic damages to man-made structures accelerating their deterioration and increasing the costs of maintenance (Haag-Wackernagel, 1995; Pimentel et al., 2000). Damages are of particular relevance in the case of historic cities and towns, where buildings constitute ideal sites for nesting and roosting, contributing in a direct way to the growth of feral pigeon populations (Ballarini et al., 1989). Medieval buildings, for instance, whose external walls are plentiful of holes due to the building methods, constitute an ideal place for nesting (Ragni et al., 1996). Fowling of churches, architectural treasures and sculptures constitutes a serious problem for their conservation (Ballarini et al., 1989; Mendez-Tovar et al., 1995). Marbles and other calcareous stones are particularly damaged by the acidity of pigeon droppings that soil their surface. Indeed Bassi & Chiatante (1976) demonstrated that droppings from pigeons constitute a highly favourable substrate for fungal growth, that contributes to damaging the marble's surface both mechanically and by the excretion of acidic metabolities.

Pigeons do not only soil buildings but also foul foodstuffs; problems are relevant in particular places as grain elevators or food industries, all sites where scaring pigeons is of paramount importance for hygienic purposes related to food preparation (Gingrich & Osterberg, 2003).

2.3 Pigeons and agriculture

Agricultural landscape represents for pigeons an important and well exploited source of food that can influence in a direct way the population size of a given city. According to Hetmanski et al. (2010), the number of pigeons is significantly higher in towns located in agricultural landscape than in those surrounded by forests, at least in Poland. Countryside can host colonies in a variety of locations such as bridges, ruins or otherwise it can be visited by pigeons for feeding purposes with fast commuting foraging flights. This is a character that ferals largely share with Mediterranean Rock Doves (Baldaccini et al., 2000), whose occurrence may differ from town to town depending on a number of variables influencing pigeons' habits and needs; in fact in some cases foraging flights can be extremely rare (e.g. Sol & Senar, 1995). The distribution of food resources and the annual trend of reproductive attempts appear to exert a leading role in shaping the characteristics of these flights, as previously suggested both for feral pigeons (Soldatini et al., 2006) and for wild rock doves (Baldaccini et al., 2000 and references therein). The distances covered in such commuting flights vary between 3 and 20 km (see Rose et al., 2006 for a review), mainly depending on the landscape and distribution of food resources (Hetmanski et al., 2010; Soldatini et al., 2006). These foraging flights can be a significant source of damage for agriculture which adds to the damages done by colonies resident in the countryside. Pigeons can take seeds at the moment of sowing, destroy the just sprouted cotyledon leaves or feed widely on mature crops (Johnston & Janiga 1995).

The size of damage can vary according to main cultivations present in the area. For instance, in countries were wheat and maize are intensely cultivated, most of the damage occurs during crops storage (Saini & Toor, 1991) because pigeons cannot feed actively on spikes. In other cases, such as sunflower fields, the damage can be greater, occurring both at sowing time and before harvesting, as pigeons are able to eat seeds directly from the flowers (van Niekerk & van Ginkel, 2004). Very little is known about the details of habitat selection by feral pigeons during their feeding flights towards croplands. Data collected during the springs 2010-2011 in the Pisa Province (central Italy), showed a preference for harvested fields of *Brassica* sp. and sowed fields of legumes (soybean *Glicine max* and chickpea *Cicer arietinum*) and sunflowers (*Helianthus annuus*) while other kinds of crops showed a strong negative selection (Fig. 1).

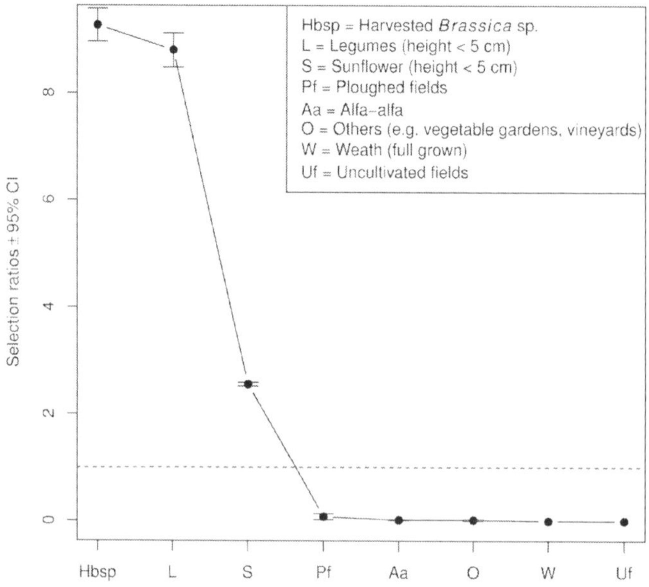

Fig. 1. Selection ratios (± 95% CI) according to design I (Manly et al., 2002) calculated on feeding feral pigeons (*n* = 12846 observations) in the agricultural landscape study area of Pisa Province (central Italy). The horizontal line indicates the threshold = 1 for positive selection.

2.4 Costs

While problems posed by pigeons have been largely assessed, only a few studies have quantified the direct costs and economic losses related to the species both in urban or countryside habitats (Bevan, 1990; Haag-Wackernagel, 1995; Phillips et al., 2003; Pimentel et al., 2000; Zucconi et al., 2003). From an economic point of view, all the different negative interactions causing damages or risks and all the actions to counteract or to evaluate the presence of ferals represent a cost. An example of the various sources of costs is presented in Fig. 2; as it can be noticed, some costs are independent on the number of pigeons present in a given site, while others are not.

Fig. 2. Chart of the source of costs due to feral pigeons' presence (from Zucconi et al. 2003, modified).

Published papers sometimes report total estimates for large areas: for instance Pimentel et al. (2000) estimated in 1.1 billion dollars per year the cost of pigeons in the USA, only for direct damages. In other cases, data are only incidental, giving examples of the costs relative to single cases or cities. The only available paper specifically focused on an analysis of the costs directly linked to pigeon presence is that by Zucconi et al. (2003), where the economic estimates were based on damage costs by private individuals and Municipalities in some sample cities from Italy. According to this paper, if we consider the costs of cleaning streets and squares, the percentage attributable to the presence of pigeons is in the range of 2.5-3.5% of the total cleaning costs in a single city. The cost was therefore estimated at 7-9 euros/pigeon/year. But if we consider the cleaning costs of historical buildings and artworks then the percentage increases to 10-15% of the total cleaning costs, with an individual cost ranging between 16 and 23 euros/pigeon/year, even though it should be noted that the costs related to damage to artworks are very difficult to estimate. According to Zucconi et al. (2003) it is impossible to make a reliable estimate of the sanitary and birdstrike costs in Italy. Costs of 2669 million dollars have been estimated in damages to civil aviation aircrafts in a period of seven years for the USA (Dolbeer et al., 2000).

In farmlands, the loss due to pigeon presence in Italy was estimated between 20-43 million euros/year, considering an estimated of crop loss of about 0.5-1% of the total yield (Zucconi et al., 2003). A more recent assessment suggested that the loss of sunflower seeds for South Africa caused by four species of *Columbiformes* amounts to 8.4% (van Niekerk, 2009). In a pilot study we conducted in farmlands surrounding the city of Pisa (ca. 200 ha), the daily average feral pigeons' density found in various types of crops was 5.7 ind/ha for sunflower and 19.1 ind/ha for legumes fields (soybean). Assuming each pigeon feeds only on farmland and has a food daily requirement of around 70 g of seeds (Johnston & Janiga, 1995), these values will be equivalent to a maximum damage of 400 and 1337 g of seeds/day respectively for sunflowers and soybean fields. If we consider that the number of individuals/ha reached peak values of 38 pigeons/ha, it is easy to understand how damages may be high; in these cases farmers are often forced to seed the fields again.

Given the amount of damage and the costs linked to pigeon presence it is often necessary to carry out several actions to reduce the number of pigeons present in the cities and, as a consequence, also the number of pigeons foraging in farmland. Actually, according to Haag-Wackernagel (1995), the damages caused by feral pigeons are reduced proportionally to the reduction of their number. An important component of the active costs related to pigeon control is the use of deterring systems on buildings, that can be easily estimated based on known prices of the components of the system. The costs of proofing with deterring systems was estimated by Zucconi et al. (2003) in 30,000-40,000 euros for 1 km^2 in an Italian city centre. In many European cities pharmacological sterilization methods are used to control pigeon population. For this method, costs range from 18-19 euros/pigeon/year for 800 ppm (Ovistop™, Acme Drugs, Italy) or 5000 ppm nicarbazin (OvoContol P™, Innolytics LLC, USA) up to 30 euros/pigeon/year for progesterone based products.

3. Population dynamics

Any properly designed control protocol involving lowering the number of an avian pest needs a thorough understanding of the population processes of the considered species (Feare, 1991). Estimates of the demographic parameters and of their variability are indeed crucial when selecting the control strategy as they provide sensible hints regarding the feasibility of attaining the objective of the control itself. Moreover, the same data collected during the control period could give useful information for adjusting the programme to the new characteristics of the population, especially when the likelihood of compensatory mechanisms (e.g. density-dependent variations in mortality or immigration rate) is not negligible, as for feral pigeons. The aim of this section is not to provide a thorough review of the available data on the demography of feral pigeons; instead, we discuss some data which are important in light of population control. The first thing to consider is that pigeons belonging to the same city constitute a single management unit. This is true for foraging behaviour, as suggested by the data collected from downtown area of Montreal which indicate that pigeons behave as a single population of consumers (Morand-Ferron et al., 2009), but it is particularly evident on the demographic point of view. Indeed, while breeding dispersal is almost absent (Hetmanski, 2007; Johnston & Janiga, 1995), juvenile dispersal within a given city is significant and, as estimated by data collected in Poland, approximately 30% of fledglings disperse each year on average (Hetmanski, 2007). As expected, the degree of dispersal is higher for high-density colonies (Fig. 3a) and most juveniles tend to move toward colonies with low density of breeding pairs. This implies that any local population reduction within a city would likely be compensated by the natural pattern of dispersal of young birds. On the other hand, the available data suggest that the rate of exchange among cities is almost absent (Hetmanski, 2007; Johnston and Janiga 1995).

As most bird pests, feral pigeons are *r*-selected organisms (Newton, 1998). Indeed, pigeon life-span is relatively short and rarely exceeds three years (Haag, 1990; Johnston & Janiga, 1995). This value is rather low considering the bird's size, as, according to the allometric equation reported in Atanasov (2008), the maximum life span of pigeon should be about 15 year. Mortality rates are thus high and this implies a high turnover rate. On the other hand, feral pigeons have a high breeding potential. They become sexually mature when six months old (Johnston & Janiga, 1995), although one-year-old birds usually represent a small fraction of the breeding segment of the population (Hetmanski, 2004; Johnston & Janiga,

1995). Moreover, while clutch size is small (only two eggs), the breeding season is long and could be regarded as lasting almost all year, with a spring-summer peak (Giunchi et al., 2007a; Hetmanski, 2004; Johnston & Janiga, 1995). Interestingly, the contribution of winter breeding attempts to the yearly number of fledglings is rarely negligible (Hetmanski, 2004; Johnston & Janiga, 1995) and in some cases absolutely relevant (e.g. 41% in Lucca, Italy; Giunchi et al., 2007a). This means that any action aimed at reducing the population size of feral pigeons, not only should be targeted to the whole or at least a significant part of the city, but also, especially if aimed at controlling the breeding output, should be continuous throughout the year. Other important things to consider are that replacement clutches are common and also the time needed for completing a single clutch is relatively short. Both parents share incubation and chick development is quite fast, given the use of the energy-rich cropmilk (Shetty et al., 1992). Moreover, pigeons can overlap clutches (Hetmanski & Wolk, 2005; Johnston & Janiga, 1995), which enables the clutch interval to be shortened, thereby increasing the number of clutches within a season. All these features indicate that feral pigeons are characterized by a high intrinsic demographic rate of increase (Neal, 2004).

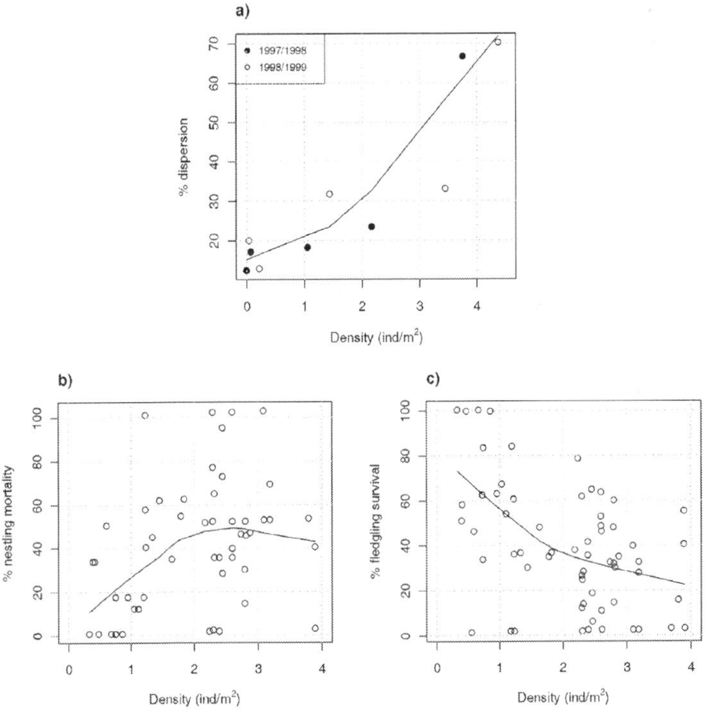

Fig. 3. Examples of density-dependence of three demographic parameters in feral pigeons (smoothing calculated by means of LOWESS; span=0,70). a) Percentage dispersion of juveniles (from Hetmanski 2007a, modified). b-c) Percentage of nestling mortality and fledging survival as a function of population density (from Haag 1988, modified).

A third important common feature of pigeon populations is the density-dependence of demographic parameters. Indeed, reproductive success, inter-clutch interval, adult mortality, immigration and recruitment rate show various degrees of density-dependence (references in Hetmanski, 2007; Hetmanski & Barkowska, 2007; Hetmanski & Wolk, 2005; Johnston & Janiga, 1995), being high at low density and low at higher density of birds (see Fig. 3). This means that populations of feral pigeons have a high compensatory potential, which is particularly evident when considering the rapid recovery of populations subjected to considerable harvesting during pest control activities (Johnston & Janiga, 1995; Kautz & Malecki, 1991; Senar et al., 2009; Sol & Senar, 1992).

All the above mentioned characteristics, associated to the mild climate and the high levels of food availability and productivity typical of most temperate and boreal urban ecosystems (Müller & Werner, 2010), leads to hypothesize that most populations of feral pigeons have reached the limit of the carrying capacity of the urban environment, after the substantial increase occurred during the second half of the last century (1940-1970), following changes in agricultural practices and the human demographic explosion after World War II (Johnston & Janiga, 1995). This implies that, excluding recent colonized cities or newly built outskirts of cities (e.g. Haag, 1988; Senar et al., 2009), most of the historical (and largest) populations of feral pigeons should be almost stable, provided that the environmental conditions which affect population abundance (e.g. human population density, prevalent structural characteristics of buildings, habitat features of the surrounding landscape; Buijs & Van Wijnen, 2001; Hetmanski et al., 2010; Johnston & Janiga, 1995; Jokimäki & Suhonen, 1998; Sacchi et al., 2002) did not change significantly. This pattern is clearly confirmed for Hamburg, where four censuses conducted during the second half of the last century indicates that feral pigeon population increased markedly from 1953 to 1966 and remained at a high level thereafter (Rutz, 2008; Fig. 4a). Moreover, periodic censuses performed during the last decades of the 20th century in a small number of cities (e.g. Barcelona, Bratislava) revealed a noticeable intra-annual, but a very low inter-annual variability of counts of resident pigeons (Johnston & Janiga, 1995). This low inter-annual variability is confirmed for two Italian cities, characterized by very different environmental conditions: Venice and Pisa. Venice (urban area: ca. 7 km^2, inhabitants: ca. 70,000) is located in Northern Italy and it is an island in a large wetland, while Pisa (urban area: ca. 10.3 km^2, inhabitants: ca. 90,000) is located in central Italy and it is surrounded by large agricultural areas where pigeons could find plenty of food. Given these conditions, the number of pigeons in Venice foraging in the mainland is rather small (e.g. < 900 pigeons/day recorded in October 2004; Baldaccini et al., unpubl. data) and birds rely on food resources within the city, favoured by the extremely high tourist presence during spring-summer months (Soldatini et al., 2006). On the other hand, the number of commuting pigeons we observed in Pisa is quite high (e.g. > 6500 pigeons/day recorded from two observation points in October 1995; Baldaccini et al. unpubl. data) and pigeons make extensive use of farmland for feeding. In spite of these differences, data indicate that in absence of significant control measures, both populations did not show any positive trend in recent years (Fig. 4b, c). Obviously, these two case studies do not represent the whole variability of abundance of pigeons, but clearly show that at least in those cities which have a sufficiently long history of presence of feral pigeons, the local populations do no show any significant inter-annual trend, at least over short-mid periods.

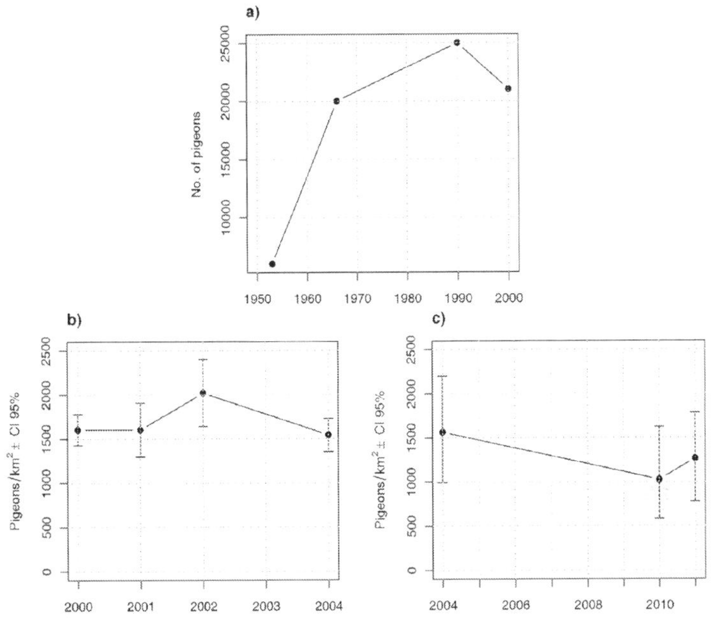

Fig. 4. a) Number of pigeons counted in the city of Hamburg in the period 1953-2000 (from Rutz 2008, modified). b) Venice: density of feral pigeons estimated by means of late-autumn uncorrected quadrat counts in the period 2000-2004 (from Giunchi et al. 2007b, modified). c) Pisa: density of feral pigeons estimated in February by means of distance sampling in the period 2004-2011 (Vanni et al., unpublished data).

4. Monitoring and control methods

4.1 Monitoring

The definition of pest should incorporate the requirement that the species actually cause economic damage (Hone, 1994) and it is the damage that justifies any control programme. However, most control programmes regarding feral pigeons lack of appropriate damage estimations, simply relying on pigeon numbers as a surrogate of the impact of the species. This approach clearly shows how monitoring and control are intimately related. More generally, estimates of pest abundance are essential not only for the assessment of pest population size to justify control, but also for the choice of appropriate control methods, with a plausible estimate of their costs and effectiveness. Unfortunately, while the development of pest control techniques for feral pigeons have involved a significant amount of research (see below), in comparison, research aimed to develop unbiased methods for estimating pigeon population size has aroused far less interest. Pigeons counts are intrinsically difficult both because of the characteristics of urban environments (complex structure and poor visibility) and of the pigeons themselves (clustered distribution and high density; Buijs & Van Wijnen, 2001; Giunchi et al., 2007b; Johnston & Janiga, 1995; Jokimäki & Suhonen, 1998). Probably for these difficulties, several authors adopted *ad hoc* and uncalibrated indexes of population abundance, such as: (1) counts of naturally occurring flocks (e.g., Buijs & Van Wijnen, 2001; Haag-Wackernagel, 1995); (2) counts of birds attracted with food (Dobeic et al., 2011; Sacchi et al., 2002); (3) counts carried out by walking along a random sample of square, non-overlapping sampling units ('quadrat counts'; Senar, 1996; Sol & Senar, 1992). While still widely used in wildlife management due to their relatively low costs, population indexes are however highly criticized because their critical assumption (proportionality between index and true population density) is usually violated in real situation (see Sutherland, 1996; Williams et al., 2002 and references therein). In the case of feral pigeons, this often led to the impossibility of an objective evaluation and quantification of the actual effects of most pest control programmes (see Giunchi et al., 2007b for further details). More reliable population estimates have been obtained by combining the quadrat counts with the use of 'correction factors', which take into account the imperfect bird detectability and are estimated by using a mark-resight procedure on a subsample of the study area (Sacchi et al., 2002; Senar, 1996). In fact, this method can produce accurate results, but it is costly as it requires catching a significant number of birds, and entails that the correction factor is estimated for each condition, as the number of birds that will pass undetected in different surveys is variable, depending on the characteristics of the study area and on the density and behaviour of pigeons themselves (Giunchi et al., 2007b). Recently, Giunchi et al. (2007b) proposed the use of distance sampling as a valuable alternative for estimating pigeons abundances. The method consists in counting pigeons on line-transects randomly distributed over the urban area and then adapted to the urban road network. During censuses the position of detected birds is accurately determined and then used for estimating detection probability according to the procedures of distance sampling (Buckland et al., 2001). The main problem of the method is that, contrary to the recommendations of Buckland et al. (2001), as transects followed the urban road network, (1) they do not represent a random sample of various habitats of the city, and (2) they are located on roadways where pigeon density is low, since birds are usually disturbed by road traffic. These conditions, intrinsically related to the structure of urban habitats, could lead to a significant underestimate of population density which can be reduced by left-truncating the data in order to exclude the low-density area near each transect. In spite of the possible biases due to the not rigorously random distribution of transects and to the spiked nature of collected distances, distance sampling in urban environment turned out to be highly repeatable, as suggested by the estimates collected in two consecutive year (2010 and 2011) in Pisa, with the same methodologies (see Fig. 4c), even though the high variability of the estimates has to be acknowledged. Provided that censuses were performed when pigeons are not at their annual population peak (i.e. late summer-autumn), the methods turned out to be consistent in different cities, with different architectural characteristics, as exemplified by Pisa, Bolzano and especially Venice, where the urban road network is not used by motor vehicle and thus roads and squares constitute available habitat for pigeons, which, on the contrary, could find a lot of food there (e.g. wastes, or food provided by the citizen or tourists) (Fig. 5). Moreover, it should be noted that the above-mentioned theoretical problems mostly affect the accuracy of distance sampling, but not its repeatability, given their dependence on the structural characteristics of the urban environment, which should be roughly the same in different years. This means that even a systematically biased distance sampling should be an unbiased tool for detecting population trends. On the contrary, the repeatability of other *ad hoc* methods (e.g. quadrat counts) probably depend also on the density of pigeons, as commonly observed for several indexes of abundance (Sutherland, 1996). This means that any control programme aimed at significantly reducing pigeon population size has to calibrate the adopted index of abundance, in order to estimate correctly the population trend and thus to evaluate the effect of the control. Given the above consideration, distance sampling should be regarded as a rather promising approach for monitoring feral pigeons, also considering its relatively low operative costs (Giunchi et al., 2007b). Actually, it should be noted that in recent years distance sampling has been increasingly used for estimating bird population size in urban habitat (Fuller et al., 2009) and that the method has been included in the guidelines for managing feral pigeons by some Italian local administrations (e.g. Piedmont Regional Authority, www.regione.piemonte.it/sanita/sanpub/animale/dwd/colombi.pdf). We believe that techniques aimed at giving reasonable estimates of pigeon populations size, such as distance sampling, have to be considered as a critical component of any effective management programme, because they help to assess both the costs for control and its effectiveness, by objectively quantifying their effects on pigeons abundance.

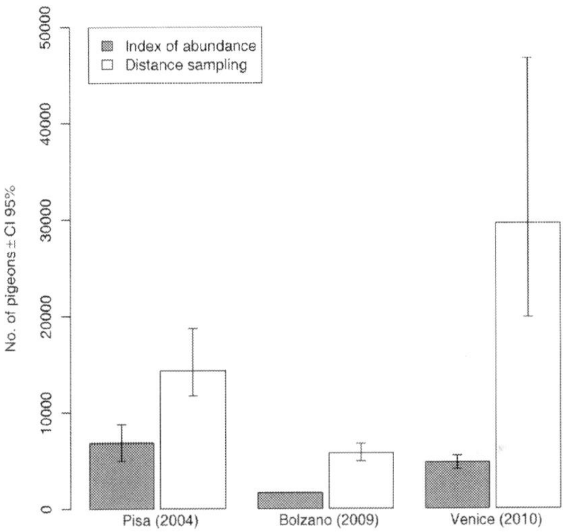

Fig. 5. Comparison of late-winter counting results between distance sampling and indexes of abundance in three Italian cities. **Pisa.** Index of abundance = quadrat counts carried out on 6.25-ha sample units (n = 40) proportionally allocated to two strata ; distance sampling carried out on 40 line transect proportionally allocated to the same two strata (more details in Giunchi et al., 2007b). **Bolzano.** Index of abundance = counts of feeding flocks at 11 traditional feeding sites used by public authorities to control pigeon food, to facilitate captures for epidemiological investigation and for distributing chemosterilants (Baldaccini & Mongini, 1991; Carsaniga, 1996); distance sampling carried out on 40 transects proportionally allocated to three strata. **Venice.** Index of abundance = quadrat counts carried out on 6-ha sample units (n = 36); distance sampling = data collected on 36 line transects put in the very centre of each quadrat and crossing almost all the length of the unit itself. All distance sampling analyses closely followed the approach detailed in Giunchi et al. (2007b), except for Venice, where data were not left-truncated. It is important to note that, given its peculiar urban structure, with a high densities of narrow streets, detection probability is particularly reduced in Venice.

4.2 Control

Given the above-mentioned peculiar interactions between humans and feral pigeons, it is important that control actions should be calibrated on the approach that the inhabitants of a given city have towards the pigeons, so that the control actions are accepted and will have an increased chance of success (Conover, 2002). Methods used to control pigeon populations could be essentially clustered in three main categories: 1) culling; 2) decrease of reproductive success; 3) reduction of habitat carrying capacity.

4.2.1 Culling

Several models indicate that for monogamous species with high mortality rates and high productivity, such as feral pigeons, culling is likely less effective than the reduction of reproductive potential for controlling population (Barlow et al., 1997; Dolbeer, 1998). Actually, even though culling has been widely applied to feral pigeon populations in several cities in the past (see e.g. Feare, 1991; Johnston & Janiga, 1995; Murton et al., 1972; Sol & Senar, 1992) and it is still used in several contexts (see e.g. Senar et al., 2009), no scientific study has demonstrated the efficacy of this approach in significantly affecting population size. As indicated above, the high intrinsic demographic rates of pigeons and the strong density dependence of several demographic parameters determine that pest control mortality is often compensatory (Feare, 1991; Johnston & Janiga, 1995) up to a relatively high threshold level estimated to be over 30% of the population/year by Kautz & Malecki (1991). Given the size of most pigeon populations, especially those producing significant damages, these figures could be high (thousands of individuals), which poses several technical problems. Moreover massive killing of pigeons is difficult to accept by many citizens, which determines further problems of ethical nature.

4.2.2 Decrease of reproductive success

Egg removal, egg puncturing or dummy eggs have been used in several cities, especially from public urban dovecotes set up with the aim to limit reproductive success (Baldaccini & Giunchi, 2006; Jacquin et al., 2010; Johnston & Janiga, 1995). This kind of method is almost inapplicable in 'natural' colonies, which are often difficult to reach, and it is costly in urban dovecotes, requiring cleaning and maintenance. Moreover this practice could affect egg laying cycles of birds, suggesting that feral pigeons respond to egg-removal by multiplying reproduction attempts (Jacquin et al., 2010). Furthermore egg quality is negatively affected by egg removal, suggesting that such management procedures can lead to an increase of reproductive physiological costs and to a decrease of female condition, raising issues about its potential consequences on parasite resistance and health status of urban populations (Jacquin et al., 2010). In any case, we are not aware of any quantitative estimation of the efficacy of this kind of approach for pest control.

The use of chemosterilants (e.g. cytostatic agents, synthetic progestinic and estrogenic drugs or drugs that interfere with the birds' metabolic activities) has received much more attention (see Ballarini et al., 1989; Giunchi et al., 2007a and references therein). Some results in terms of reduction of the population size and improvement in the health status of the birds have been reported (e.g. Baldaccini, 1996b; Dobeic et al., 2011), even though there are no evidences of significant long-term effects. The recent development of new reproductive inhibitors based on nicarbazin (e.g. Ovistop™, OvoContol P™) provided new interest for this kind of approach (Avery et al., 2008; Giunchi et al., 2007a; Yoder et al., 2006). While some authors report significant but sometimes puzzling effects of drug distribution on usually small populations (e.g. Bursi et al., 2001), no well controlled data on the long-term effects of these chemosterilants are available. More generally, as the effects are only partial (a maximum of 59% reduction of productivity under controlled conditions; Avery et al., 2008) and temporary (Yoder et al., 2005), drugs such as nicarbazin are likely to produce only short lasting reductions of pigeon abundance in the field, with a rapid recovery as soon as the treatment is stopped (Giunchi et al., 2007a).

4.2.3 Reduction of carrying capacity

Carrying capacity reduction through habitat modification is at present the most reliable way to obtain long-lasting effects on pigeon populations (Haag, 1993); moreover this method is usually well accepted (and sometimes requested) by citizens. Carrying capacity reduction should act on two main factors: nest/roost sites and food.

The limitation of nest and roost sites may be achieved by applying exclusions or scare techniques (Johnston & Janiga, 1995). Different kinds of tactile or mechanical repellents had been used to deter pigeons (Haag-Wackernagel, 2000; Seamans et al., 2007; Williams & Corrigan, 1994). Chemical, acoustic, and visual repellents are known to be effective only for short time periods as pigeons habituate to them within a few days (Johnston & Janiga, 1995), while no deterring effect was observed when using an ultrasonic or repellent odour system (Haag-Wackernagel, 2000). On the other hand, mechanical devices, such as porcupine wires, can be surmounted if bird motivation to access a given site is high enough (Haag-Wackernagel, 2000). Buildings and structures can be also designed to reduce the attractiveness to pigeons (Haag-Wackernagel & Geigenfeind, 2008; Williams & Corrigan, 1994). While applied in midtown areas, train stations, airports and historical buildings, exclusion methods are rarely integrated into a systematic pest control program, as wrongly thought to be ineffective (Magnino et al., 2009). However, they proved to be highly effective in Perugia (Italy) resulting in a reduction of 23% of the population of feral pigeons in one year (Ragni et al., 1996).

As suggested by Haag (1991, 1993), control of food supply is the basis for a successful control programme, also determining a general improvement of the population quality and resistance to parasites and pathogens. Food resources management may be particularly effective when feral pigeon populations mostly depend on food resources located within the urban environment (see Murton et al., 1972; Rose et al., 2006; Sol & Senar, 1995). In this case it can be possible to manage food availability, although both theoretical considerations and field data indicates that this may be difficult (see Giunchi et al. 2007a and references therein). Besides published data (see e.g. Haag 1993), as a successful example we may report the case of Venice, where, until a few years ago it was allowed the distribution of corn for feeding the pigeons as a touristic attraction. It was estimated that pigeons were fed 350 tons of corn per year and the number of pigeons present in St Mark's Square was critically high, reaching concentrations of >10,000 individuals in 1.3 ha. In May 2008, the local Authorities decided to ban the distribution of corn and since then the number of pigeons has decreased dramatically, down to a maximum of 1000 individuals at one time in St Mark' Square.

Quadrat count estimations of birds density, obtained in late autumn (November, $n = 9$ years) and in late winter (February-March, $n = 7$ years) from 1996 confirmed the decreasing trend in both census periods (Pearson correlations, late autumn: $r = -0.81$, $P = 0.008$; late winter: $r = -0.86$; $P = 0.013$). But more in detail, considering densities recorded in the city before and after 2008 we can assess that differences are significant both in late autumn (ANOVA: $F_{1,8} = 6.82$, $P= 0.035$) and in late winter ($F_{1,6} = 8.89$, $P = 0.031$). Substantial differences were recorded also in foraging flights. Indeed the number of commuting birds recorded before (2004) and after (2009) the ban occurred had dramatically decreased all over the year (t-test: departing flock sizes $t_{11} = 7.44$, $P << 0.001$; returning flock sizes $t_{11} = 5.36$, $P << 0.001$; number of departing flocks in 2004 N=818 *vs.* N=213 in 2009 and of retuning flocks in 2004 N= 590 *vs.* N= 170 in 2009). Thus, the reduction of food resources within the city had not been compensated for by any increase in foraging flights towards the countryside. This is probably due to the fact that Venice is an island in a wetland that pigeons must fly over to reach mainland foraging sites and experimental data by Wagner (1972) reported the avoidance by pigeons in crossing a body of water. On the contrary, the management of food resources should be less effective in cities where most birds fly for food to adjacent agricultural areas (see e.g. Soldatini et al., 2006). In this last case, bird scaring devices and reflecting strips as well as gas cannons are extensively used by farmers, but with a very low long term effectiveness. The use of culling of limited numbers of individuals as scaring method linked together with scarecrows and gas cannons is applied in some Italian provinces but the results of these methodologies are still under considerations (Baldaccini et al., unpublished data).

4.3 A population model

All the above considered control methods have their own drawbacks, depending on the characteristics of feral pigeon populations (e.g. size), on the features of the urban habitat (e.g. age of buildings), and on the characteristics of the surrounding landscape (e.g. distribution of food resources). This means that the different techniques could be more effective/easy to apply in different context and suggest the usefulness of a combination of methods in order to reach better results in shorter time. To evaluate the possible effects of the use of some combination of control methods on feral pigeon populations, we simulated a number of scenarios by means of the software VORTEX 9.50 (Miller & Lacy, 2005). The aim of these simulations was not to provide a precise demographic forecast of a given population subjected to pest control, instead to give some hints regarding the choice of a proper pest control programme.

4.3.1 Methods

The values used as initial input for simulations are reported in Table 1. On the whole, the approach we followed was roughly the same adopted in Giunchi et al. (2007a) and we do not report all the details here. The main differences, with respect to the above-mentioned paper were:

1. In order to extend the considered scenarios, we modelled two populations, which we called 'Murton' and 'Haag', as demographic parameters were partly derived from papers published by Murton et al. (1972, 1974) and by Haag (1988, 1990). The 'Murton' population was characterized by a comparable mortality rate between adults and

juveniles (values derived from Murton et al., 1972), while the 'Haag' population had a rather high juvenile mortality and low adult mortality (values obtained as the average of those reported in Haag, 1988).

2. Density dependence was modelled not by varying the percentage of breeding females in the population, instead the number of fledglings (NF) per female. This latter parameter is indeed more frequently reported than the former one, which is only a matter of speculation in a few papers (Johnston & Janiga, 1995). The equation we used was of the same type of that adopted for the percentage of breeding females in Giunchi et al. (2007a):

$$NF(N) = NF(0) - \left[\left(NF(0) - NF(K) \right) \left(NK \right)^{B} \right] \quad (1)$$

where NF(N), NF(K) and NF(0) are the number of fledglings per females that breed when the population size is N, at carrying capacity (K), and at extremely low density (near 0), respectively, while the exponent B is a constant which determines the form of the curve. To simplify calculations, we considered only the case of B = 2. This appears a reasonable assumption, given that, as suggested by Fowler (1981), density dependence in reproductive success can often be modelled with a quadratic function (see also Fig. 3). NF(K) was chosen by trial and error as the values which determined a fundamental stability of the population defined by the other demographic parameters listed in Table 1 in the absence of density-dependent reproduction and with a carrying capacity much higher than the initial population size (10,000 birds). Interestingly, at least for the 'Haag' population, this value was quite comparable to that reported for a numerically stable colony in the city centre of Basel (Haag, 1988). Given this comparability, NF(0) for the Haag population was set to the value reported in Haag (1988) for a recently settled colony in the periphery of Basel, where density of pigeons was rather low. We then assumed that the 'Murton' population behaves in the same way, and thus we hypothesized the same proportional increase.

3. We considered two types of scenarios. In one scenario both populations were near carrying capacity (K = 5,000), while in the other K was set to 10,000. In this way we modelled two different situations: old populations, with relatively stable numbers, and relatively recent populations with increasing size.
4. To simplify calculations, we did not consider any environmental variability, also because no data in this regard could be found in the literature.

A series of simple simulations was performed to investigate the effects of different degrees of reduction of fertility with a reduction of K. We considered three scenarios for the reduction of fertility (-15%, -30% and -60% of the fertility of the whole population) with a maximum set to the maximum effect obtained with the recently proposed chemosterilants based on nicarbazin (see Avery et al., 2008) and four scenarios for the reduction of K (no reduction, -1%/year, -2%/year, -4% year). We did not simulate an abrupt reduction of carrying capacity, because this is often difficult to obtain in the field.

All pest control programme lasted 10 years. We did not consider culling in our simulation because of the lack of evidence regarding its efficacy and its above-mentioned technical problems. In order to simplify calculations and in absence of detailed information useful for modelling, mortality rate was considered density-independent, although some data regarding American populations indicated an increased survival of pigeons following an experimentally induced decrease of population density (Kautz & Malecki, 1991). In this regard, it is important to notice that Haag (1988, 1990) did not report any remarkable difference in mortality and in age distribution of pigeons in colonies characterized by significantly different densities. Obviously, it is important to emphasize that this choice had the consequence of increasing the theoretical effect of the simulated pest control, because it cut down the recovery potential of the modelled population when density was low (see also Newton 1998).

Variable	'Murton'	'Haag'
Number of simulations	100	100
Period	10 years	10 years
Initial population size (N)	5,000	5,000
Start at stable age distribution	Yes	Yes
Carrying capacity (K)	5,000, 10,000	5,000, 10,000
Demographic closure	Yes	Yes
Inbreeding	0 lethals	0 lethals
Catastrophes	0	0
Mortality at age 0	43	82
Mortality at age 1	34	10
Breeding system	Long term monogamy	Long term monogamy
Age of first breeding	1	1
Maximum Age of Reproduction	7	7
Sex ratio at birth	0.5	0.5
% females breeding	100	100
Density dependence	Yes	Yes
Number of fledglings when N=K	1.3	2.2
Number of fledglings when N=0	2.4	4.0
B	2	2
% males in the breeding pool	100	100

Table 1. Summary of the input parameters used in the simulations.

4.3.2 Results

When population size was very near to K, we observed a rather similar outcome for both 'Murton' and 'Haag' populations regarding the fertility control (Fig. 6). In both cases, when the fertility control was high (-60%) the impact of the reduction of K was not significant. Less strong reduction of fertility had rather less impact on the populations and a rather poor additive effect with respect to the decrease of K. When the population was increasing, the impact of the reduction of K was obviously lower than in the former cases, and it was only evident after a few years, when the populations began to level off (Fig. 7). For both populations, the final outcome of the simulations depended on the reduction of K only when the fertility control rates were low; otherwise the differences were slight or absent. It should be noted that only the strongest controls (last scenarios) inverted the positive trend of population size.

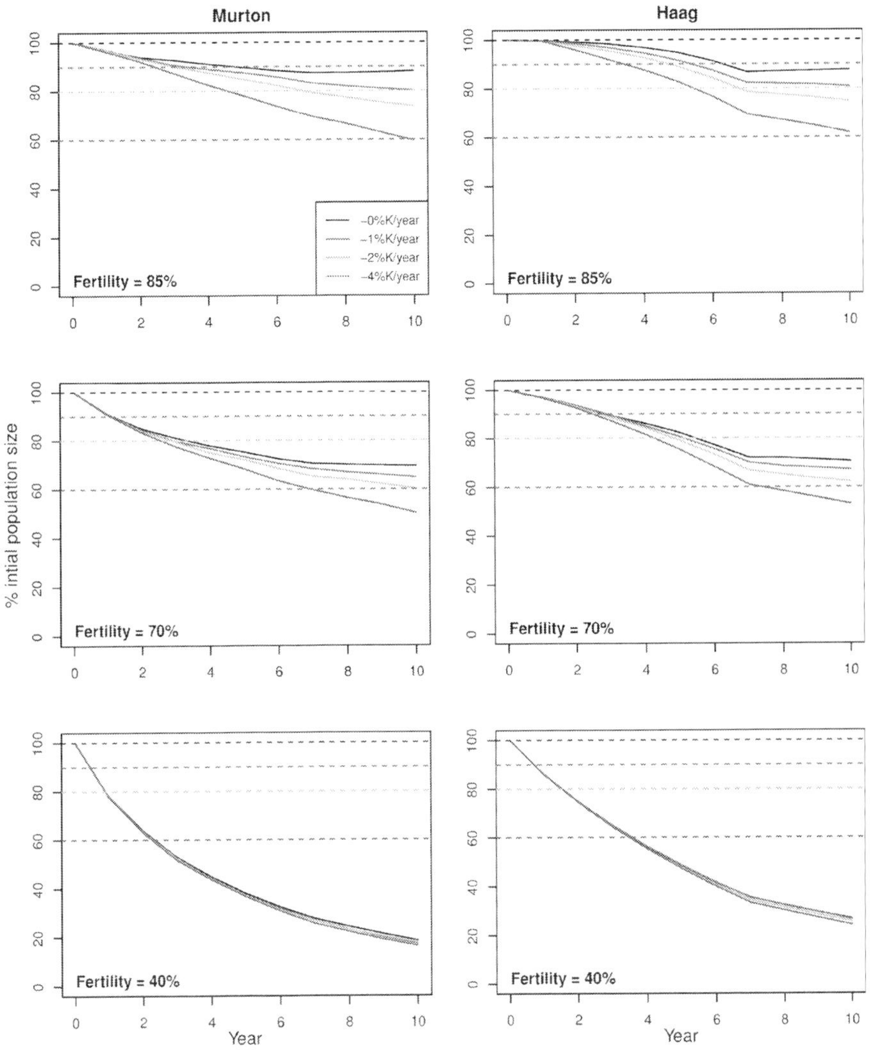

Fig. 6. Feral pigeons population size near the carrying capacity (K) of the urban habitat. Trends (continuous lines) of the 'Murton' and 'Haag' populations predicted after 10 years of various degrees of fertility control (-15, -30, -60%) combined with different degrees of reduction of carrying capacity. Broken lines refer to the cumulative population reduction which could be obtained by only reducing K [e.g. red line: -0.1%/year * 10 years = -10%)].

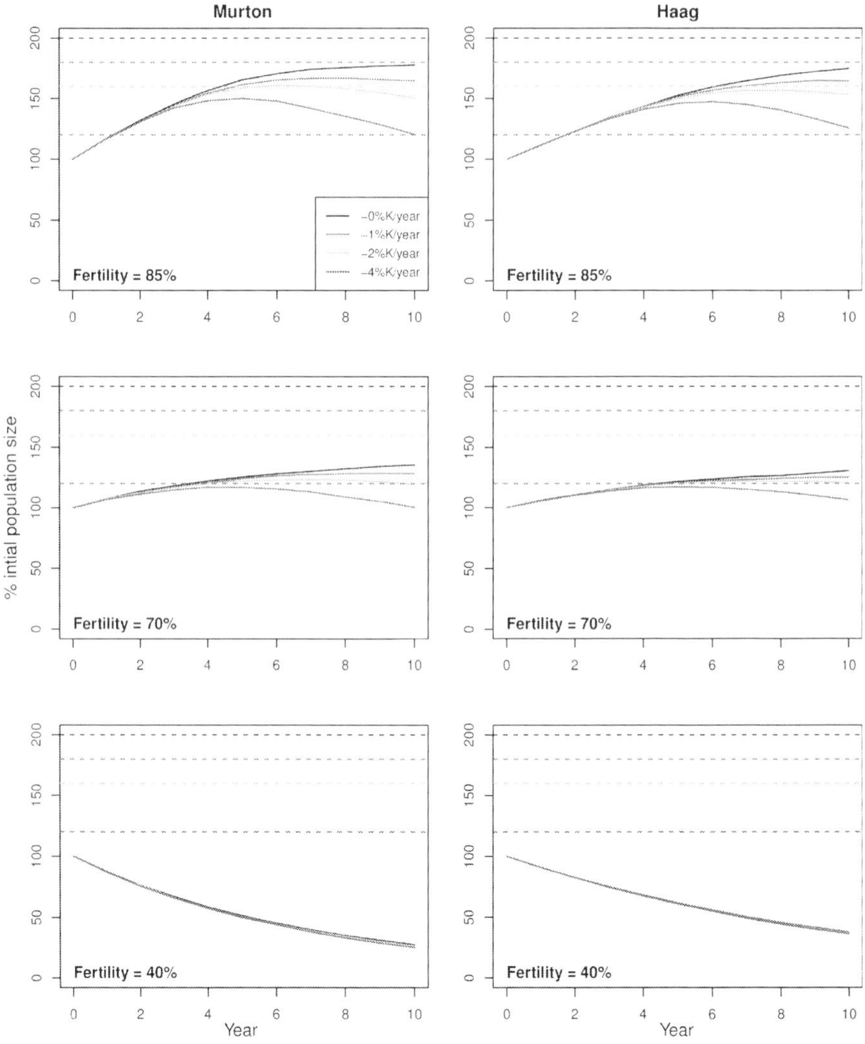

Fig. 7. Feral pigeons population size at 50% of the carrying capacity (K) of the urban habitat. Trends (continuous lines) of the 'Murton' and 'Haag' populations predicted after 10 years of various degrees of fertility control (-15, -30, -60%) combined with different degrees of reduction of carrying capacity. See Fig. 6 for other details.

5. Concluding remarks

The potential for compensation is one of the most important features which has to be taken into account when starting a pest control programme on feral pigeons. For this reason, an effective management policy should focus on the reduction of carrying capacity of the urban habitat, possibly integrating other approaches according to the characteristics of both the city and pigeon population. Carrying capacity reduction through habitat modification is indeed the most effective way for obtaining long-lasting effects on pigeon abundance. This could be obtained by focusing on all the ecological resources, and not only on foods, as, in some situations it is not easy to control all birds' feeding sites. Among the other above-considered control methods, culling is probably the less reliable, as in most cases it revealed to be not effective and often problematic both under the practical and ethical point of view, especially for large populations. Our simulations also indicate that, under certain circumstances, fertility control could be profitably combined with the reduction of carrying capacity. When it is difficult to distribute chemosterilants to the population (e.g. because the population is very large or food resources are abundant and widespread), the additive effects are only marginal. On the other hand, when the fertility control could be strong (e.g. small populations), the need for a reduction of carrying capacity is less stringent. It is however important to underline that: (1) as mentioned above, our simulations overestimated the effects of the fertility control; (2) the reduction of the carrying capacity is the only way for capitalizing the results obtained by any other method, whose effects are often fleeting and reversible. For this reason, this last method should be considered even in those cases where it is not the best for obtaining strong and rapid results (e.g. when it is not possible to have a strong impact on carrying capacity at least on the short term or when the population is far from the carrying capacity), also because several actions aimed at reducing the ecological resources for pigeons (e.g. exclusion from the most profitable nest/roost sites) could sum up over several years. In this context, budget availability is an important factor to consider in choosing the management policy, as fertility control may result expensive compared to e.g. food reduction especially if applied for long periods. Therefore, it is extremely important for the city council to carefully evaluate its capacity to afford the application of different control methods for several years.

In general terms an effective management policy needs a strong local background knowledge in order to be calibrated to the characteristics of the considered population. This implies a carefully balanced integration of control methods, proper monitoring and reliable modelling, in order to forecast the effects of control actions (Chee & Wintle, 2010). It is therefore important to understand the behavioural and ecological characteristics of the pigeon population before starting a control programme, and to analyse the capability of covering the costs and the participation and awareness of the municipal Authorities and city dwellers.

6. Acknowledgments

We are grateful to the Venice City Council and to the Pisa Province Council for funding and the long term collaboration and availability. Thanks are also due to: Chiara Caccamo, Riccardo Fiorin, Gioia Garavini, Ilaria Gemelli, Vyron Georgalas, Andrea Ragazzoni, Federico Riccato, Davide Righetti, and Giuseppe Vecchio, for their help in the field.

7. References

Atanasov, A. T. (2008). The near to linear allometric relationship between the total metabolic energy per life span and the body mass of aves. *Journal of Animal and Veterinary Advances,* Vol. 7, No. 4, pp. 425–432, ISSN 1993-601X

Avery, M. L.; Keacher, K. L. & Tillman, E. A. (2008). Nicarbazin bait reduces reproduction by pigeons (*Columba livia*). *Wildlife Research,* Vol. 35, No. 1, pp. 80–85, ISSN 1035-3712

Baldaccini, N. E. (1996a). Inurbamento: processo attivo alla ricerca di spazi da colonizzare. *Controllo delle popolazioni ornitiche sinantropiche: "problemi e prospettive",* ISS/WHO/FAO-CC/IZSTe/96.27, Roma, October 1993, pp. 2-13

Baldaccini, N. E. (1996b). La diminuzione della capacità riproduttiva come mezzo di controllo dei colombi urbani. *Controllo delle popolazioni ornitiche sinantropiche: "problemi e prospettive",* ISS/WHO/FAO-CC/IZSTe/96.27, Roma, October 1993, pp. 103-105

Baldaccini, N. E. & Giunchi, D. (2006). Le popolazioni urbane di colombo: considerazioni sulla loro genesi e sulle metodologie di gestione. *Biologia Ambientale,* Vol. 20, pp. 125-141

Baldaccini, N. E.; Giunchi, D.; Mongini, E. & Ragionieri, L. (2000). Foraging flights of wild rock doves (*Columba l. livia*): a spatio-temporal analysis. *Italian Journal of Zoology,* Vol. 67, No. 4, pp. 371–377, ISSN 1748-5851

Baldaccini, N. E. & Mongini, E. (1991). Diminuzione del numero di colombi di città a Bolzano in seguito a interventi di controllo. *Supplemento alle Ricerche di Biologia della Selvaggina* Vol. 17, pp. 215-217, ISSN 1121-3973

Ballarini, G.; Baldaccini, N. E. & Pezza, F. (1989). *Colombi in città. Aspetti biologici, sanitari e giuridici. Metodologie di controllo.* Istituto Nazionale di Biologia della Selvaggina, Bologna

Barlow, N. D.; Kean, J. M. & Briggs, C. J. (1997). Modelling the relative efficacy of culling and sterilisation for controlling populations. *Wildlife Research,* Vol. 24, No. 2, pp. 129-141, ISSN 1035-3712

Bassi, M. & Chiatante, D. (1976). The role of pigeon excrement in stone biodeterioration. *International Biodeterioration Bulletin,* Vol. 12, No. 3, pp. 73-79, ISSN 0020-6164

Bevan, R.D.R. (1990) The costs of feral pigeons. In *British Ornithological Union, Proceedings of a Symposium,* October 1990, ADAS/British Ornithologists' Union, London, pp. 10–11.

BirdLife International (2004). *Birds in Europe,* BirdLife International, ISBN 0946888531, Wageningen, The Netherlands.

Buckland, S. T.; Anderson, D. R.; Burnham, K. P.; Laake, J. L.; Borchers, D. L. & Thomas, L. (2001). *Introduction to Distance Sampling: Estimating Abundance of Biological Populations,* Oxford University Press, ISBN 0198509278, New York

Buijs, J. A. & Van Wijnen, J. H. (2001). Survey of feral rock doves (*Columba livia*) in Amsterdam, a bird-human association. *Urban Ecosystems,* Vol. 5, No. 4, pp. 235-241, ISSN 1573-1642

Bursi, E.; Gelati, A.; Ferraresi, M. & Zannetti, G. (2001). Impiego della nicarbazina nel controllo della riproduzione del colombo randagio di città. *Annali della Facoltà di Medicina Veterinaria di Parma,* Vol. 21 , pp. 97-115, ISSN 0393-4802

Carsaniga, G. (1996). Controllo dei colombi di città a Bolzano: analisi di un'esperienza pilota. *Controllo delle popolazioni ornitiche sinantropiche: "problemi e prospettive"*, ISS/WHO/FAO-CC/IZSTe/96.27, Roma, October 1993, pp. 74-78.

Chee, Y. E.; Wintle B. A. (2010). Linking modelling, monitoring and management: An integrated approach to controlling overabundant wildlife. *Journal of Applied Ecology*, Vol. 47, No. 6, pp. 1169-1178, ISSN 0021-8901

Cleary, E. C., Dolbeer, R. A. & Wright, S. E. (2006). Wildlife strikes to civil aircraft in the United States 1990-2005, In: DigitalCommons@University of Nebraska - Lincoln, Available from http://digitalcommons.unl.edu/birdstrikeother/7/

Conover, M. R. (2002). *Resolving Human-Wildlife Conflicts: The Science of Wildlife Damage Management,* Lewis Publishers, ISBN 156670538X, Boca Raton, Fla

Curtis, L.; Lee, B. -S; Cai, D.; Morozova, I.; Fan, J. -L; Scheff, P.; Persky, V.; Einoder, C. & Diblee, S. (2002). Pigeon allergens in indoor environments: a preliminary study. *Allergy*, Vol. 57, No. 7, pp. 627-631, ISSN 0105-4538

Dobeic, M.; Pintarič, Š.; Vlahović, K. & Dovč, A. (2011). Feral pigeon (*Columba livia*) population management in Ljubljana. *Veterinarski arhiv., Vol.* 81, No. 2, pp. 285–298, ISSN 0372-5480

Dolbeer, R. A. (1998). Population dynamics: the foundation of wildlife damage management for the 21st century. In: DigitalCommons@University of Nebraska - Lincoln, Available from http://digitalcommons.unl.edu/vpc18/9/

Dolbeer, R. A.; Wright, S. E. & Cleary, E. C. (2000). Ranking the hazard level of wildlife species to aviation. *Wildlife Society Bulletin*, Vol. 28, No. 2, pp. 372-378, ISSN 0091-7648

Feare, C. J. (1991). Control of bird pest populations. In: *Bird population studies: relevance to conservation and management,* C.M. Perrins, J.D. Lebreton & G.J.M. Hirons (Eds), 463-478, Oxford University Press, ISBN 0198577303, Oxford, UK

Fowler, C. W. (1981). Density dependence as related to life history strategy. *Ecology*, Vol. 62, No. 3, pp. 602-610, ISSN 0012-9658

Fuller, R. A.; Tratalos, J., & Gaston, K. J. (2009). How many birds are there in a city of half a million people? *Diversity and Distributions*, Vol. 15, No., pp. 328–337, ISSN 1366-9516

Gallo, M. G.; Cabeli, P. & Vidotto, V. (1989). Sulla presenza di lieviti patogeni nelle feci di colombo torraiuolo (*Columba livia*, Gmelin 1789, forma urbana) della città di Torino. *Parassitologia* Vol. 31, No. 2-3, pp. 207-212, ISSN 0048-2951

Geigenfeind, I. & Haag-Wackernagel, D. (2010). Detection of *Chlamydophila psittaci* from feral pigeons in environmental samples: problems with currently available techniques. *Integrative Zoology* , Vol. 5, No. 1, pp. 63–69, ISSN 1749-4869

Ghigi, A. (1950). *Piccioni Domestici e Colombicoltura*, REDA, Roma

Gingrich, J.B.; Osterberg, T.E. (2003). Pest birds: Biology and Management at Food Processing Facilities. In: *Food plant sanitation*, Y.H. Hui, B.L. Bruinsma, J.R. Gorham, W.K. Nip, P.S. Tong & P. Ventresca (Eds), 317–339, Marcel Dekker, ISBN 0824707931, New York

Giunchi, D.; Baldaccini, N. E.; Sbragia, G. & Soldatini, C. (2007a). On the use of pharmacological sterilisation to control feral pigeon populations. *Wildlife Research*, Vol. 34, No. 4, pp. 306-318, ISSN 1035-3712

Giunchi, D.; Gaggini, V. & Baldaccini, N. E. (2007b). Distance sampling as an effective method for monitoring feral pigeon (*Columba livia* f. *domestica*) urban populations. *Urban Ecosystems*, Vol. 10, No. 4, pp. 397-412, ISSN 1573-1642

Goodwin, D. (1978). *Birds of man's world*, Cornell University Press, ISBN 080141167X, Ithaca, NY

Goodwin, D. (1983). *Pigeons and Doves of the World*, 3rd edition. Cornell University Press, ISBN 0801414342, New York

Grenfell, B. T. & Bolker, B. M. (1998). Cities and villages: infection hierarchies in a measles metapopulation. *Ecology Letters*, Vol. 1, No. 1, pp. 63–70, ISSN 1461-023X

Haag, D. (1988). Die dichteabhängige Regulation im Brutschwarm der. Strassentaube *Columba livia* forma *domestica*. *Der Ornithologische Beobachter* , Vol. 85, pp. 209-224.

Haag, D. (1990). Lebenserwartung und Altersstruktur der Strassentaube *Columba livia* forma *domestica*. *Der Ornithologische Beobachter*, Vol. 87, pp. 147-151

Haag, D. (1991). Population density as a regulator of mortality among eggs and nestlings of feral pigeons (*Columba livia domestica*) in Basel, Switzerland. In: *Nestling mortality of granivorous Birds due to Microorganisms and Toxic Substances*, J. Pinowski, B.P. Kavanagh & W Górski (Eds.), 21-31. Polish Scientific Publishers, ISBN 8301104767, Warsaw

Haag, D. (1993). Street Pigeons in Basel. *Nature*, Vol. 361, No. 6409, p. 200, ISSN 0028-0836

Haag-Wackernagel, D. (1995). Regulation of the street pigeon in Basel. *Wildlife Society Bulletin*, Vol. 23, No. 2, pp. 256-260, ISSN 0091-7648

Haag-Wackernagel, D. (1998). *Die Taube: Vom heiligen Vogel der Liebesgöttin zur Strassentaube*, Schwabe, ISBN 3796510167, Basel

Haag-Wackernagel, D. (2000). Behavioural responses of the feral pigeon (Columbidae) to deterring systems. *Folia Zoologica*, Vol. 49, No. 2, pp. 101-114, ISSN 0139-7893

Haag-Wackernagel, D. (2006). Human diseases casued by feral pigeons. In: *Advances in Vertebrate Pest Management Vol. 4*, C. Feare & D. P. Cowan (eds.), 31-58, Filander Verlag, ISBN 3930831643, Furth

Haag-Wackernagel, D. & Bircher, A. J. (2009). Ectoparasites from feral pigeons affecting humans. *Dermatology*, Vol. 220, No. 1, pp. 89-92, ISSN 1018-8665

Haag-Wackernagel, D. & Geigenfeind, I. (2008). Protecting buildings against feral pigeons. *European Journal of Wildlife Research, Vol.* 54, No. 4, pp. 715-721, ISSN 1439-0574

Haag-Wackernagel, D. & Moch, H. (2004). Health hazards posed by feral pigeons. *Journal of Infection*, Vol. 48, No. 4, pp. 307-313, ISSN 0163-4453

Hetmanski, T. (2004). Timing of breeding in the Feral Pigeon Columba livia f. domestica in Slupsk (NW Poland). *Acta Ornithologica*, Vol. 39, No. 2, pp. 105-110.

Hetmanski, T. (2007). Dispersion asymmetry within a feral pigeon *Columba livia* population. *Acta Ornithologica*, Vol. 42, No. 1, pp. 23-31

Hetmanski, T. & Barkowska, M. (2007). Density and age of breeding pairs influence feral pigeon, *Columba livia*, reproduction. *Folia Zoologica*, Vol. 56, No. 1, pp. 71–83.

Hetmanski, T.; Bochenski, M.; Tryjanowski, P. & Skórka, P. (2010). The effect of habitat and number of inhabitants on the population sizes of feral pigeons around towns in northern Poland. *European Journal of Wildlife Research*, Vol. 57, No. 3, pp. 421-428, ISSN 1439-0574

Hetmanski, T. & Wolk, E. (2005). The effect of environmental factors and nesting conditions on clutch overlap in the feral pigeon *Columba livia* f. *urbana* (Gm.). *Polish Journal of Ecology,* Vol. 53, No. 4, pp. 523-534, ISSN 1505-2249

Hoffmann, H. (1982). *Das Taubenbuch,* Krüger, ISBN 3810507113, Frankfurt

Hone, J. (1994). *Analysis of Vertebrate Pest Control.* Cambridge University Press, ISBN 0521415284, Cambridge.

Jacquin, L., Cazelles, B., Prevot-Julliard, A.-C., Leboucher, G. & Gasparini, J. (2010). Reproduction management affects breeding ecology and reproduction costs in feral urban Pigeons (*Columba livia*). *Canadian Journal of Zoology, Vol.* 88, No. 8, pp. 781-787, ISSN 0008-4301

Jerolmack, C. (2008). How pigeons became rats: The cultural-spatial logic of problem animals. *Social Problems,* Vol. 55, No. 2, pp. 72–94, ISBN 0037-7791

Johnston, R. F. & Janiga, M. (1995). *The Feral Pigeons,* Oxford University Press, ISBN 0195084098, London

Jokimäki, J. & Suhonen, J. (1998). Distribution and habitat selection on wintering birds in urban environments. *Landscape and Urban Planning,* Vol. 39, No. 4, pp. 253-263, IISSN 0169-2046

Kautz, J. E. & Malecki, R. A. (1991). *Effects of Harvest on Feral Rock Dove Survival, Nest Success and Population Size.* U.S. Dept. of the Interior, Fish and Wildlife Service, Washington, D.C

Lever, C. (1987). *Naturalized Birds of the World,* Wiley, ISBN 978-0582460553, New York.

van der Linden, C. G. (1950). *Le Pigeon Voyageur,* Payot, Paris

Magnino, S.; Haag-Wackernagel, D.; Geigenfeind, I.; Helmecke, S.; Dovc, A.; Prukner-Radovcic, E.; Residbegovic, E.; Ilieski, V.; Laroucau, K.; Donati, M.; Martinov, S. & Kaleta, E. F. (2009). Chlamydial infections in feral pigeons in Europe: Review of data and focus on public health implications. *Veterinary Microbiology,* Vol. 135, No. 1-2, pp. 54-67, ISSN 0378-1135

Manly, B. F.; McDonald, L.; Thomas, D. L.; McDonald, T. L. & Erickson, W. P. (2002). *Resource Selection by Animals: Statistical Design and Analysis for Field Studies,* 2nd edition, Kluwer Academic Publishers, ISBN 1402006772, Dordrecht, The Netherlands

Mendez-Tovar, L. J.; Mainou, L. M.; Pizarro, S. A.; Fortoul-Vandergoes, T. & Lopez-Martinez, R. (1995). Fungal biodeterioration of colonial facades in Mexico City. *Revista Mexicana de Micologia,* Vol. 11, pp. 133–144

Miller, P. S. & Lacy, R. C. (2005). *A Stochastic Simulation of the Extinction Process. Version 9.50 User's Manual.* Conservation Breeding Specialist Group (SSC/IUCN), Apple Valley, MN, Available from http://www.vortex9.org/vortex.html

Morand-Ferron, J.; Lalande, É. & Giraldeau, L. (2009). Large scale input matching by urban feral pigeons (*Columba livia*). *Ethology, Vol.* 115, No. 7, pp. 707-712, ISSN 0179-1613

Müller, N. & Werner, P. (2010). Urban biodiversity and the case for implementing the convention on biological diversity in towns and cities. In: *Urban Biodiversity and Design,* N. Müller, P. Werner, & J.G. Kelcey (Eds), 1–33, Wiley, ISBN 9781444332667, Chichester, UK

Mumcuoglu, K. Y.; Banet-Noach, C.; Malkinson, M.; Shalom, U. & Galun, R. (2005). Argasid ticks as possible vectors of West Nile virus in Israel. *Vector-Borne & Zoonotic Diseases*, Vol. 5, No. 1, pp. 65–71, ISSN 1530-3667

Murton, R. K.; Thearle, R. J. P. & Coombs, C. F. B. (1974). Ecological studies of the feral pigeon *Columba livia* var. III. reproduction and plumage polymorphism. *Journal of Applied Ecology*, Vol. 11, No. 3, pp. 841-854, ISSN 0021-8901

Murton, R. K.; Thearle, R. J. P. & Thompson, J. (1972). Ecological studies of the feral pigeon *Columba livia* var. I. Population, breeding biology and methods of control. *Journal of Applied Ecology*, Vol. 9, No. 3, pp. 835-874, ISSN 0021-8901

Neal, D. (2004). *Introduction to Population Biology*. Cambridge University Press, ISBN 0511078692, Cambridge

Newton, I. (1998). *Population Limitation in Birds*. Academic Press, ISBN 0125173652, San Diego

van Niekerk, J. H. (2009). Loss of sunflower seeds to columbids in South Africa: economic implications and control measures. *Ostrich, Vol.* 80, No. 1, pp. 47-52, ISSN 698572454

van Niekerk, J. H. & van Ginkel, C. M. (2004). The feeding behaviour of pigeons and doves on sown grain crops on the South African Highveld. *Ostrich*, Vol. 75, No. 1-2, pp. 39-43, ISSN 698572454

Pedersen, K.; Clark, L.; Andelt, W. F. & Salman, M. D. (2006). Prevalence of shiga toxin-producing *Escherichia coli* and *Salmonella enterica* in rock pigeons captured in Fort Collins, Colorado. Journal of Wildlife Diseases, Vol. 42, No. 1, pp. 46-55, ISSN 0090-3558

Phillips, R. B., Snell, H. L. & Vargas, H. (2003). Feral rock doves in the Galápagos Islands: Biological and economic threats. *Noticias de Galápagos*, Vol. 62, (December 2003), pp. 6-11

Pimentel, D.; Lach, L.; Zuniga, R. & Morrison, D. (2000). Environmental and economic costs of nonindigenous species in the United States. *BioScience*, Vol. 50, No. 1, pp. 53-65, ISSN 0006-3568

Ragni, B.; Velatta, F. & Montefameglio, M. (1996). Restrizione dell'habitat per il controllo della popolazione urbana di Columba livia. *Controllo delle popolazioni ornitiche sinantropiche: "problemi e prospettive"*, ISS/WHO/FAO-CC/IZSTe/96.27, Roma, October 1993, pp. 106-110

Rose, E., Nagel, P. & Haag-Wackernagel, D. (2006). Spatio-temporal use of the urban habitat by feral pigeons (*Columba livia*). *Behavioral Ecology and Sociobiology*, Vol. 60, No. 2, pp. 1-13, ISSN 0340-5443.

Rosický, B. (1978). Healt risks associated with animals in different types of urban areas: Present status and new ecological conditions due to urbanization. *Annali dell'Istituto Superiore di Sanità*, Vol. 14, No. 2, pp. 273-286, ISSN 0021-2571

Rutz, C. (2008). The establishment of an urban bird population. *The Journal of Animal Ecology*, Vol. 77, No. 5, pp. 1008–1019, ISSN 1365-2656

Sacchi, R.; Gentilli, A.; Razzetti, E. & Barbieri, F. (2002). Effects of building features on density and flock distribution of feral pigeons *Columba livia* var. *domestica* in an urban environment. *Canadian Journal of Zoology*, Vol. 80, No. 1, pp. 48-54, ISSN 0008-4301

Saini, H. K. & Toor, H. S. (1991). Feeding ecology and damage potential of feral pigeons, Columba livia, in an agricoltural habitat. *Le Gerfaut*, Vol. 81, pp. 195-206, ISSN 0016-9757

Sauer, J. R.; Hines, J. E. & Fallon, J. (2008). The North American Breeding Bird Survey, Results and Analysis 1966-2007. Version 5.15. 2008. USGS Patuxent Wildlife Research Center, Laurel, MD.

Seamans, T. W.; Barras, S. C. & Bernhardt, G. E. (2007). Evaluation of two perch deterrents for starlings, blackbirds and pigeons. *International Journal of Pest Management*, Vol. 53, No. 1, pp. 45–51, ISSN 0967-0874

Senar, J. C. (1996). Bird census techniques for the urban habitat: a review. *Controllo delle popolazioni ornitiche sinantropiche: "problemi e prospettive"*, ISS/WHO/FAO-CC/IZSTe/96.27, Roma, October 1993, pp. 36-44

Senar, J. C.; Carrillo, J.; Arroyo, L.; Montalvo, T. & Peracho, V. (2009). Estima de la abundancia de palomas (*Columba livia* var.) de la ciudad de Barcelona y valoración de la efectividad del control por eliminación de individuos. *Arxius de Miscel-lània Zoològica, Vol.* 7, No. 1, pp. 62–71, ISSN 1698-0476

Shetty, S.; Bharathi, L.; Shenoy, K. B. & Hegde, S. N. (1992). Biochemical properties of pigeon milk and its effect on growth. *Journal of Comparative Physiology B.*, Vol. 162, No. 7, pp. 632-636, ISSN 0174-1578

Soldatini, C.; Mainardi, D.; Baldaccini, N. E. & Giunchi, D. (2006). A temporal analysis of the foraging flights of feral pigeons (*Columba livia* f. *domestica*) from three Italian cities. *Italalian Journal of Zoology*, Vol. 73, No. 1, pp. 83-92, ISSN 1748-5851.

Sol, D. & Senar, J. C. (1992). Comparison between two censuses of Feral Pigeon Columba livia var. from Barcelona: an evaluation of seven years of control by killing. *Butlleti del Grup Catala d'Anellament*, Vol. 9, pp. 29-32, ISSN 1130-2070.

Sol, D. & Senar, J. C. (1995). Urban pigeon populations: stability, home range, and the effect of removing individuals. *Canadian Journal of Zoology*, Vol. 73, No. 6, pp. 1154-1160, ISSN 0008-4301

Sossinka, R. (1982). Domestication in birds. *Avian biology* Vol. 6, , pp. 373–403, ISBN 0122494016

Sutherland, W. J. (1996). *Ecological Census Techniques: a Handbook.* Cambridge University Press, ISBN 052147244X, Cambridge.

Wagner, G. (1972). Topography and pigeon orientation. In: *Animal Orientation and Navigation*, S.R. Galler, K. Schmidt-Koenig, G.J. Jacobs & R.E. Belleville (Eds.), 259-273, NASA, ISBN , Washington DC.

Williams, D. E. & Corrigan, R. M. (1994). Pigeons (rock doves). In: *The Handbook: Prevention and Control of Wildlife Damage*, S.E. Hygnstrom, R.M. Timm & E.G. Larson (Eds.), E87-E96, University of Nebraska, Lincoln, Available at http://tinyurl.com/3nwtx6x

Williams, B. K.; Nichols, J. D. & Conroy, M. J. (2002). *Analysis and Management of Animal Populations: Modeling, Estimation, and Decision Making*. Academic Press, San Diego, ISBN 0127544062

Yoder, C., Avery, M. & Wolf, E. (2006). Nicarbazin: an avian reproductive inhibitor for pigeons and geese. In: DigitalCommons@University of Nebraska - Lincoln, Available from http://tinyurl.com/3r6vxk6

Yoder, C. A.; Miller, L. A. & Bynum, K. S. (2005). Comparison of nicarbazin absorption in chickens, mallards, and Canada geese. *Poultry Science, Vol.* 84, No. 9, pp. 1491-1494, ISSN 0032-5791

Zucconi, S., Galavotti, S. & Deserti, R. (2003). I colombi in ambiente urbano - Sintesi del progetto di ricerca Nomisma. *Disinfestazione,* Novembre/Dicembre 2003, pp. 9-21.

11

Biological Control of Dengue Vectors

Mario A Rodríguez-Pérez, Annabel FV Howard
and Filiberto Reyes-Villanueva
*Centro de Biotecnología Genómica, Instituto
Politécnico Nacional, Ciudad Reynosa, Tamaulipas
México*

1. Introduction

Biological control is the deliberate use of natural enemies to reduce the number of pest organisms. It comprises methods that have gained acceptance for controlling nuisance arthropods partly due to the emergence of insecticide resistance and also because people have become more aware about the need to limit environmental pollution. In the case of arthropod-borne disease vectors, biological control is a potentially effective strategy for regulating and preventing transmission of diseases such as dengue, malaria and lymphatic filariasis, amongst others. Dengue is an arbovirus transmitted by species of *Aedes* mosquitoes. *Aedes aegypti* and *Aedes albopictus* are the primary and secondary worldwide vectors; they breed in peridomestic man-made water containers, and their control is the most effective way to reduce the viral transmission.

In this chapter we present an outline of the conceptual development of biological control since it was proposed by Harry S. Smith in 1919, to the current understanding of applied biological control involving basically autecology of insects that has led to Integrated Pest Management (IPM) principles. The potential of a natural enemy to regulate vector abundance will provide quantitative insight into Paul DeBach´s principles, *ie.* an optimal biological agent according to its search capacity, host specificity, and tolerance to environmental factors. We will also explain Pavlovsky´s theory to understand the origin of dengue as a human disease evolving from enzootic cycles. Likewise, we will introduce the reader to population regulation and transmission control describing the concepts of "functional response" and "numeric response" according to the classical Holling modelling.

We then explain the evolution of vector synanthropism and outline why dengue transmission can only be reduced by controlling the *Aedes* mosquito vector. In this regard we introduce the reader to the parasites, predators and pathogen complexes of the dengue vectors so that the he understands the present situation in terms of the biological control of the *Aedes* mosquito. We will use classic and recent papers and reviews to describe novel lines of research, and pros and cons of the use of natural enemies for dengue vector control. We hope that the chapter will work as a source of key literature references for students and researchers. Finally, the need for an integrated vector management (IVM) strategy aimed at controlling dengue will be put forward, and the potential for the deployment of biological control tools in future programmes will be made.

2. Biological control: Basic concepts

Biological control is part of a larger phenomenon called natural control, because the environment is always exerting selection pressure on populations. The selection is expressed in terms of mortality rates inflicted by all environmental factors and living beings as natural control. In this context, Harry S. Smith in 1919 proposed the term "natural control". In reality, this idea is closely linked to the Darwinian concept of "struggle for the existence or survival of the fittest" because a given species interacts within its ecosystem, and coexists surviving the attack of microorganisms, animals and plants that depend directly or indirectly on it. All these consumers who share resources (competitors) in the context of Smith, make up the complex known as "biotic" factors, which are constantly adapting and optimizing the manner of obtaining energy from the species in question. Conversely, all those factors from the environment such as temperature, humidity, pH, chemicals, substrates, etc., which also cause mortality on populations, constitute the complex of "abiotic" factors, the other component of natural control.

Four decades later in 1964, Paul DeBach, a student of Dr. Harry S. Smith, emphasized the term "regulation" as synonymous of "control" to specify the total mortality exerted by the biotic and abiotic factors on populations. He explained that since the mortality is dynamic, the population size will also be fluctuating in time and space. In theory, if a graph is depicted using the population changes against a reasonable period of time, such as a year, we might determine the average density (equilibrium) around which decreases and increases may occurred in the population. DeBach used this criterion to define the concept of biological control: If we may remove all biotic factors acting on a particular species, obviously its average density would be higher than normal. He pointed out that the difference between both average densities (with and without biotic factors) is the "effect" of biological control. In addition to the large number of studies published by DeBach on biocontrol, he also has the merit of having edited and published in 1964, together with Evert L. Schlinger, the book entitled "Biological Control of Insect Pests and Weeds", which is the classic in this discipline.

Now then, biotic factors include the "natural enemies", a term used by DeBach and by Huffaker and Messenger (1976) (another classic book as well). Natural enemies are the predators, parasites, parasitoids, and pathogens of each species. Predators kill rapidly, and require several preys, usually smaller in size, to complete their life cycle. Parasites are generally much smaller than the host, live on or inside it, and may or may not kill it. Pathogens are smaller still; they are the microorganisms that consume nutrients from a host which may be killed or not (Price, 1970). A parasitoid is an insect similar in size to its host, which is parasitized in immature stage and always dies. Whilst all the other types can be used against *Aedes* mosquitoes, parasitoids cannot. Biological control is divided into two types: natural and applied. The former is geared to the regulation of the populations by natural enemies without any human intervention, while the latter is obviously artificial. Applied biological control in turn can be divided into three types: 1) Classic biological control includes cases where there is an introduction of a foreign or exotic natural enemy to a region or country where it does not exist, 2) The increase of local natural enemies which can be performed by inoculative (if the natural enemy establishes itself in the habitat and in the target organism using a single release) and inundative releases (where the natural enemy regulates the target population temporarily, or only while he remains alive and therefore many releases may be required), and 3) Applied biological control by conservation of natural enemies, which include all those agricultural practices (plowing, planting dates management, irrigation, etc...) or other activities (manipulation of weeds, with alternate preys and hosts or maintaining the nectar source for the natural enemies as adults) aimed at increasing the level of regulation of the pest population.

3. Quantitative expression of the regulation capacity of a natural enemy

Populations change in time and space. Changes can occur in the density of a species whether host or prey, affecting the biology of its parasites, parasitoids and predators. Scientists have been modelling functional responses since the 1920s although the term "functional response" was introduced in 1949 by Solomon. Holling (1959) explained the concept of functional response as the population rate of a host or prey consumed by a carnivore per time unit, and the concept of numerical response as those changes induced by the host or prey on survival rate, emigration, and mostly fecundity of the carnivore, which in turn depends on the food amount eaten. Therefore, functional response determines the numerical response. The former was considered as a response of the carnivores at individual level while the later as response at population level.

Since the beginning both concepts called the attention of ecologists. Functional response is crucial because taking the number of prey eaten per individual at the end of a period of time, the product of this estimator by the density of the predator will allow prediction of the host or prey population consumed after that time interval (Royama, 1971). In other words, if the predation rate increases, also the reproduction of the carnivores will increase and this will be reflected in an increment of its population; its density will follow the one of the host or prey resulting in a plot where oscillations of both will be very close. This is a top-down regulation system with negative feedbacks because when the host or prey is scarce it diminishes the carnivore population and this permits the host or prey to recover its original density (Holling, 1961). However these models rely on theoretical assumptions that must be accomplished, for instance, host searching has to be a random process, its populations should have a stable age structure, spatial distribution without clusters, no emigration, etc., and these requirements do not exist in nature. Despite these drawbacks the concept is applied to have an idea of how much a predator could diminish the prey population.

Holling (1966) proposed three basic types of functional response: Type I is a linear relationship where the predator consumes the same rate along the prey density until it reaches a "satiation state" which is a plateau in the graph. Type II describes a situation in which the number of prey consumed per predator initially rises quickly as the density of prey increases but then levels off with further increase in prey density. Finally, Type III resembles Type II in having an upper limit to prey consumption, but differs in that the response of predators to prey is depressed at low prey density (Figure 1). Explanation of the three models is not in the objectives of this chapter, but we will describe only the characteristics of the Holling Type II model also known as the "disc equation" because it has been the most widely used and accepted by researchers to describe numerous prey-carnivore systems (Tully, et al. 2005). We recommend to the reader the review of Jeschke et al. (2002) in which the authors carry out a detailed analysis of 47 models of functional response, 32 of them were for parasites and predators.

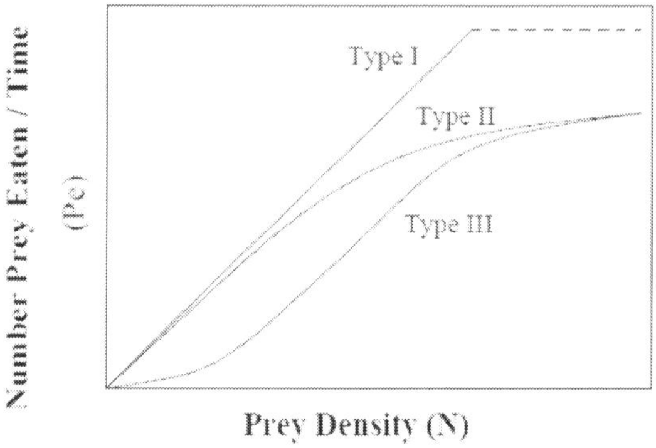

Fig. 1. Three types of functional response relating prey density (N) and the number of prey eaten by one predator (Pe).

Type II response incorporates predator handling time, which refers to the act of subduing, killing, and eating a prey, including cleaning and resting before moving on to search for more prey (Juliano and Williams, 1985). The number of prey attacked increases at a constant initial rate under this model, but then increases at an ever decreasing rate as satiation is approached. The result is the hyperbolic curve of the "disc equation" which is mathematically equivalent to the Michaelis-Menten model of enzyme kinetics and the Monod formula for bacterial growth (Abrams, 2000). This name derives from the way Holling conducted his experiment; he distributed randomly paper discs on a table (available preys), and a blindfolded person (the predator) with his fingertips "searched" around on the table the discs. Each disc was removed but replaced immediately after each encounter, and using the number of "consumed" preys from the exposed ones per time unit, he found his deterministic disc equation:

$$Na = \frac{aTNo}{(1 + aThNo)}$$

Where Na is the number of prey killed, No is the initial density of prey, T is the time available for searching during the experiment, a is the instantaneous rate of discovery or also known as attack rate varying with No, and Th is the total amount of time (constant) the predator handles each prey killed. So, the first step is to plot the numbers of eaten prey per density per unit time. Once detected the tendency of the hyperbolic curve, the computation of a and Th proceeds in the second step, which is the simplification of the equation by reciprocal linear transformation (Livdahl and Stiven, 1983) which is as follows:

$$\frac{1}{Na} = \frac{1}{aTNo} + \frac{Th}{T}$$

This equation is the linear regression model Y = α+ βX, where the intersection value α is the reciprocal of the attack rate $1/a$ and the slope β is the factor Th/T, however T is constant and the factor stays just as Th and is equal to $-\frac{\beta}{a}$; then both values are used in the disc equation as follows:

$$Na = \frac{\alpha * No}{\left(1 + \alpha * (-\frac{\beta}{a}) * No\right)}$$

Therefore, after conducting an experiment, coefficients α and β are computed regressing Na/No as the response variable on No. The maximum predation rate is $1/Th$ and is the maximum value that Na/No can have. Attack rate, a determines how steeply the curve rises with increasing prey density (No). Once the expected values are plotted together with the observed ones, the degree by which the model explains the functional response is computed by the determination coefficient R^2 and a χ^2 goodness of fit test.

Another similar option to analyzing data of the same experiment is to calculate the reciprocals $1/Na$ and $1/N$ and conducting a regression of the former as the response variable on the later (Williams and Juliano, 1985); this method also produces the next linear equation:

$$\frac{1}{Na} = \frac{1}{aN} + Th$$

Here, the attack rate a is $1/\beta$, and Th is the intercept α, both coefficients are used in the disc equation to have the hyperbolic curve. To know which method is better the determination coefficient R^2 for each regression is calculated; the higher value will indicate the better fitting. Nevertheless, regardless of the method, linearization is polemic because it may produce bias in parameter estimates. To directly fit the disc equation with data generated from experimental protocols, or to use nonlinear procedures (logistic regression), the reader could examine the excellent review of Juliano (2001).

4. Origin and ecology of dengue

4.1 How diseases adapt from zoonoses to anthropozoonoses

Pavlovsky in 1962 published his book in Russian about the theory of natural nidality of transmissible diseases and the book was available in English in 1964. The *nidus* is a nest or focus of infection, *i.e.*, a place where a disease occurs in the wild and is then transmitted to humans by arthropod vectors when they invade the *nidus*. These diseases are zoonoses. It is the triangle host-pathogen-vector interactions that functions on a permanent and tridimensional space within a "pathobiocenosis" which is the community where the three species converge. The conceptual framework includes what Pavlovsky pointed out as "polivectorial focus" where several vectors and pathogens interact in the same three-dimensional space. For example, a polivectorial focus is the nest of the cactus rat *Neotoma spp.* in North America where the *Lutzomyia spp.* sand flies with the protozoan *Leishmania mexicana,* and the *Triatoma infestans* kissing bug with the flagellate *Trypanosoma cruzii* coexist infecting the rodent. Humans that inhabit areas near to the rat nests could potentially be bitten by those vectors seeking a blood meal and so the parasites that produce the leishmaniasis or Chagas disease will be spread to the humans. If the parasite transmission cycle occurs among the rats only it is known as enzootic cycle or zoonose. The rat is the primary host, but if the human is infected by the vector's bite, then it becomes a secondary host and the disease shifts from zoonose to anthropozoonose. Given that the human invaded the *nidus,* he is said to be infected by tangential transmission, and the rodent is the reservoir or amplifier host. The reservoir is the host where the pathogen survives in the wild as the mosquito *A. aegypti* for dengue viruses. Usually, the primary hosts or reservoirs have co-evolved with the pathogens and developed immune defenses that will cause them to become asymptomatic. On the other hand, the secondary hosts have evolved recently, so there is still some susceptibility to the infection with the pathogen that produces symptoms ranging from minor to quite severe ones which may even cause death.

4.2 The evolution of dengue (Pavlovsky´s principle)

According to Pavlovsky a dengue *nidus* is an enzootic cycle in Asia with *Aedes (Finlaya) niveus* mosquitoes as vector and several primates of the genus *Macaca* as hosts. Or in Africa with *Erythrocebus patas, Cercopithicus aethiopica,* or *Papio anubis* monkeys as reservoir hosts (Roche 1983) and as vectors, the mosquitoes *Aedes* (*Stegomyia*) *africanus, Aedes* (*S.*) *luteocephalus, Aedes* (*S.*) *opok, Aedes* (*Diceromyia*) *taylori, Aedes* (*D.*) *furcifer* or *Aedes polynesiensis* in the South Pacific islands (Moncayo, et al. 2004) (Figure 2).

Fig. 2. Zoonotic cycle of dengue viruses circulating between *Aedes* mosquitoes and monkeys in the jungle. *A. aegypti* is adapted to live near to the ground, while *A. albopictus* tends to be catholic, and the other vectors are canopy residents.

In Malaysia, blood samples were taken from over 2,300 domestic (in urban areas) and wild (in the forest) animals belonging to 55 species and 28 genera to detect anti-dengue antibodies. In addition, 25,000 adult mosquitoes were tested for dengue viruses but all were negative despite high levels of antibodies in the majority of wild monkeys leading to the first evidence of dengue zoonoses (Rudnick, 1965). However, these zoonoses could become epidemics if the viruses are spread by vectors at high rates in monkey populations as in Sri Lanka where the exposure rate of dengue epizootic, based on antibody detection in asymptomatic monkeys, had been reported to be up to 94% in an area of 3 km^2 (Peiris, et al. 1993; de Silva, et al. 1999). Humans could become tangentially infected when invading this dengue *nidus*, permitting the transfer of arboviruses from zoonoses to anthropozoonoses.

In theory, dengue hemorrhagic fever (DHF) occurred after the dissemination and co-adaptation of the dengue viruses in the vector (reservoir) and in the human (host). The first DHF cases were reported in Manila, Asia, in 1954 (Quinlos, et al. 1954), and in 1981 in Cuba, America (Guzman, et al. 1995). Thus, according to Pavlovsky, the benign form of dengue fever is associated with those from the original *nidus*. The circulating serotypes in the urban areas are thought to be the same as those in enzootic cycles, though they differ genetically due to independent evolution (Cordellier, et al. 1983; Roche, et al. 1983). Five, six, four, and two genotypes of the serotypes 1, 2, 3, and 4, respectively have been defined, hence, dengue serotypes contains a total of 17 dengue genotypes; two (belonging to serotype 2 and 4) out of which are circulating in the *nidus* amongst monkey populations in the forest (Holmes and Twiddy, 2003).

According to Pavlovsky principle, the older a serotype is that is circulating in a particular area the less pathogenic it will tend to be to humans. It has been documented that *A. aegypti* and *A. albopictus* are less susceptible to a wild genotype strain than to an epidemics genotype of the serotype 2 (Moncayo, et al. 2004). Similarly, the American genotype (AM) of serotype 2, the oldest strain thought to cause the first dengue fever epidemics in the Americas, has been, in appearance, supplanted by the Southeast Asia (SA) genotype, a more pathogenic strain to humans that is efficiently transmitted by *A. aegypti* mosquitoes (Armstrong and Rico-Hesse, 2003). In addition, an emerging genotype of the serotype 3 had evolved in Sri Lanka which is more pathogenic to humans and transmissible by *A. aegypti* mosquitoes. This apparently has displaced the native genotype which is less pathogenic to humans, and there is some evidence that *A. aegypti* is less competent at acquiring and transmitting the native strain (Lambrechts, et al. 2009). Thus, the incidence of DHF cases seems to be normal during dengue fever outbreaks because of the serotype and genotype diversity circulating which leads to multiple infections in the human populations.

4.3 Co-evolution between viruses and *Aedes* mosquitoes

Although the terms vector and transmitter are used interchangeably in the literature, the term vector involves a more intimate evolutionary co-existence among blood-sucking arthropods and the pathogens they transmit; this has led to the concept of biological transmission, which is a basic concept in medical entomology. Extended co-existence allows adjustment in the arthropod bionomics to acquire more tolerance (vector competence) towards a virulent pathogen. This gradual adaptation of the viruses to the vectors could lead to a successful evasion of their immune defenses resulting in minimal damage. Consequently, a pathogen that has achieved such adaptation to survive optimally in the vector populations could be successfully transmitted to healthy individuals from an exposed host population. This phenomenon may be observed in the case of vectors-viruses interactions.

The dengue virus was isolated by first time at the end of World War II by Dr. Susumu Hotta´s group (Kimura and Hotta, 1944). They found the viruses in blood samples taken from Japanese soldiers. Two decades later, in Singapore the presence of dengue viruses in field-collected *A. aegypti* and *A. albopictus* mosquitoes was shown (Rudnick and Chan, 1965). *A. aegypti* females are easily infected with the serotype 2 which is transmitted successfully; conversely, this is not the case when using the other three serotypes. If mosquitoes were fed with blood contaminated with similar viral titres, the amount of viruses which could reach salivary glands of those infected females would be higher with serotype 2, than that of those mosquitoes fed with other serotypes. In addition, the viruses will infect the salivary glands of mosquitoes in a shorter period of time (a shorter extrinsic incubation period) than that when using the other dengue serotypes. This phenomenon has been documented for the SA genotype of the serotype 2 which showed a higher replication rate in both the vector and definitive hosts than that of the AM genotype in America (Cologna, et al. 2005).

4.4 Evolution of vector synanthropism

Vectors for dengue fever including the primary *A. aegypti* and secondary *A. albopictus* mosquitoes are species whose origin was in a forest habitat; this could be also alike for other vector species associated with dengue zoonoses in Africa and Asia (Figure 1). The wild vectors in the forest prefer tree heights and foliage-canopy, depositing eggs in breeding sites of rainwater accumulated in hollow trees and in the axils of epiphytic plants such as the family Bromeliaceae. The forest habitat is a permanent shaded site; hence, these mosquitoes exhibit negative phototaxis during visual flight orientation and show specific preferences for resting or moving towards dark sites. This innate behavior facilitated their adaptation to survive in shacks or huts of the primitive man. Today it is common for people in underdeveloped tropical areas around the world to store drinking water in containers of clay or other material that remain in or close to their houses. These containers are dark and relatively cold, and serve as the perfect replacement of the typical larval breeding site, as mentioned, that was in a hollow tree in the forest. These artificial containers would represent the transition between the forest and urban habitats.

A. albopictus feeds less on humans than *A. aegypti* which is highly anthropophylic. *A. aegypti* shows evident endophilic and endophagic behaviour. It appears that females tended to stay indoors because there was availability of blood from people and oviposition sites in a form of artificial containers, which allows the survival of successive generations within the same household. When more large human settlements appeared, the vector adapted to standing water in flowerpots, buckets, old tires, etc. which are abundant in the exterior and interior of houses in villages and modern urban areas. A feature of the *Aedes* lifecycle that lent itself to the utilization of these small and artificial larval habitats is that the eggs can survive desiccation. This allowed them to utilize small habitats that potentially dry out.

Aedes mosquitoes are active during the day, and as such they cannot be controlled using insecticide-treated bednets in the same way the malaria vectors can be controlled. And yet it is becoming increasily important that these mosquitoes are effectively controlled. Dengue is reported to be the most rapidly spreading mosquito-borne disease in the world (World Health Organisation, 2009). Recent estimates are that 50 million dengue infections occur each year, with 2.5 billion people at risk of infection in dengue endemic countries. Dengue distribution is spread across the tropics but also reaches sub-tropical areas too. Given the high mutation rate of dengue viruses it is difficult to develop a dengue vaccine. Nevertheless, vaccine development is ongoing but dengue is rapidly becoming a public health problem in the Americas, Africa and Asia, and at present the only feasible way to control it is by controlling the *Aedes* mosquito vector (World Health Organisation, 2009).

There are many ways in which *Aedes* mosquitoes can be controlled. But as with other mosquito vectors, *Aedes* mosquitoes are becoming resistant to the insecticides used, and in any case, people are becoming more sensitive to environmental pollution. This is especially true because a major source of mosquito larvae are people's drinking water storage jars. Thus in the rest of this chapter we outline the different types of biological control available to use against *Aedes* mosquitoes, and we will explain how these biological tools are playing a role in IVM programmes.

5. Biological control tools

5.1 Single celled organisms

Aedes aegypti is host of entomopathogenic microorganisms but historically just a few species have been reported and isolated from the dengue vector as natural host (Hembree, 1979). There is a lot of information about entomopathogens evaluated against *A. aegypti* but most are at experimental level. We think that the spore-forming bacteria *Bacillus sphaericus* (*Bs*) and *Bacillus thuringiensis israelensis* (*Bti*) are regarded as the most promising microbial control agent against the dengue vector.

As soon as *Bs* and *Bti* appeared, they demonstrated their usefulness as control tools particularly when the dengue vector began to show signs of resistance to chemical insecticides (Sun, et al. 1980). *Bs* was discovered in 1964 (Kellen and Meyers, 1964) while *Bti* was isolated in 1977 (Goldberg and Margalit, 1977). Both are highly effective not only at killing larvae of Culicidae and Simuliidae (Federici, 1995) but also at killing adults of *Aedes*, *Culex* and *Anopheles* mosquitoes (Klowden and Bulla Jr., 1984). However *Bs* is more selective than *Bti* because it is specially toxic to *Culex* and *Anopheles* larvae, and tolerant to high levels of organic pollution (Regis, et al. 2000); however some mosquitoes already are resistant to *Bs* (Rodcharoen and Mulla, 1994). *Bti* acts when its spore-crystal containing toxic proteins (protoxins) is ingested by larvae. Then the pro-toxins are solubilized in the alkaline pH of the gut and activated into toxins which cause a detergent-like rearrangement of lipids in the epithelial membrane, leading to its disruption and cytolysis (Gill, et al. 1992). The mode of action of *Bs* is similar but less known. Since their isolation, both bacteria have been intensively investigated and virtually thousands of papers have been published. Most papers have been focused on enhancing the toxicity of the proteins associated to the crystals, and currently hundreds of bio-formulates have been produced biotechnologically. We will mention as an example the Programme for Eradication of *Aedes aegypti*, launched in 1997 in Brazil to fight dengue fever transmission. Although today still in Brazil the use of biological agents to control mosquitoes has been restricted to experimental and operational research, they have discovered new technologies to improve the efficacy of these bacteria. For example, the *Bti* tablet experimental formulation C4P1-T, shows good persistence, killing more than 70% of *A. aegypti* larvae within 40 days after treatment of tanks in shade, and 25 days in tanks exposed to sunlight. In addition, the *Bs* formulations showed up to 100 days persistence against *Culex quinquefasciatus* larvae after the third application in shaded tanks, as did a *Bti* formulation, Inpalbac, after the 4th treatment. Tested under identical conditions, some of the experimental formulations revealed performances almost or equally as good as the best commercial products tested, VectoBac and VectoLex (Regis, et al. 2000).

For the rest of entomopathogens there are only evaluations at laboratory or semi-field conditions, for instance, some iridoviruses have been explored in relation to their sublethal effects on *A. aegypti* (Marina, et al. 2003), the protozoan *Ascogregarina culicis* has been explored against larvae in cemeteries (Vezzani and Albicocco, 2009), and the microsporidian *Edhazardia aedis* examined at semi-natural and laboratory conditions (Becnel and Johnson, 2000; Barnard, et al. 2007). In conclusion, the only promising entomopathogen in this group is *Bti* especially those new formulates with better efficacy than the traditional formulates. However, *Bti* does not always persist for a long time under field conditions.

5.2 Fungi (Ascomycetes: Hypocreales)

Entomopathogenic Ascomycetes could be a promising biological control tool. The conidia of these fungi, once germinated, directly penetrate the adult mosquito cuticle then produce a blend of organic compounds, causing internal mechanical damage, nutrient depletion and death (Gillespie and Clayton, 1989). These fungi have been successfully used under field conditions to kill malaria vectors (Scholte, et al. 2005), and to modify wild mosquito blood feeding behaviour (Howard, et al. 2010). While a wide range of these fungi have been used in experiments with dengue vectors (Scholte, et al. 2004), there are two main species that are currently being used by many laboratories worldwide: *Metarhizium anisopliae* and *Beauveria bassiana.*

Most work has been carried out against adult mosquitoes. Scholte, et al. (2007) found that *M. anisopliae* caused significant mortality to *A. albopictus,* and found high levels of infection. Studies have showed that *A. aegypti* had significantly increased mortality after exposure to *M. anisopliae* (Scholte, et al. 2007; de Paula, et al. 2008, Reyes-Villanueva, et al. 2011) and *B. bassiana* (de Paula, et al. 2008, Garcia-Munguia, et al. 2011). Worryingly, susceptibility to fungal infection is significantly reduced following a blood meal, but after digestion fungal susceptibility returned to pre-feeding levels (de Paula, et al. 2011a). Also wild strains may be less susceptible to fungal infection than some colony strains used (de Paula, et al. 2011b). Nevertheless fungal virulence can be increased by the co-exposure to an insecticide (de Paula et al. 2011a).

In addition to direct effects on mortality, an interesting pre-lethal effect has been seen. Recently our group has shown that fungal infections can affect fecundity of female *A. aegypti.* When infected with *M. anisopliae,* fecundity was reduced by up to 99% (Reyes-Villanueva, et al. 2011) and after infection with *B. bassiana,* fecundity was reduced by 95% (Garcia-Munguia, et al. 2011). Whilst no field trial data has been published using entomopathogenic fungi against dengue vectors, field trials are reported to be underway (de Paula, et al. 2011a) and hopefully the results will soon be known.

Many different methods have been put forward for the deployment of entomopathogenic fungi, these include black cotton sheets attached the ceilings (Scholte, et al. 2005), direct application onto mud walls (Mnyone, et al. 2010) and treated window covers (Howard, et al. 2010). Our team in Mexico have shown that both *M. anisopliae* (Reyes-Villanueva, et al. 2011) and *B. bassiana* (Garcia-Munguia, et al. 2011) can be transferred from infected *A. aegypti* males to females during mating, and this could be utilised in field applications in the future.

Although much of the recent work has focussed on the adults, fungi can be used to kill *A. aegypti* larvae as well. For mosquito larvae the fungal conidia are either ingested through the mouth or enter the siphon. Here they can cause a physical blockage by vegetative growth of the fungi, and the release of midgut toxins causes death. *M. anisopliae* is effective at killing *A. aegypti* larvae (Ramoska, et al. 1981) but there are conflicting reports with one study stating that *B. bassiana* is pathogenic (Miranpuri and Khachatourians, 1990), while other studies found that it is not (Clark, et al. 1968; Geetha and Balaraman, 1999). Work targeting *A. aegypti* eggs has also been carried out. Both *M. anisopliae* and *B. bassiana* have proved to be ovicidal (Luz, et al. 2007), but high levels of humidity were required (Luz, et al. 2008). Further work using *M. anisopliae* found that oil-based formulations can enhance the ovicidal effect (Albernaz, et al. 2009).

Questions still remain about fungal longevity and viability under tropical conditions. In Tanzania, *M. anisopliae* in suspension did not lose viability whereas when the fungus was impregnated onto black cotton cloths and exposed to the ambient heat and humidity, the viability had reduced to 63% three weeks after application (Scholte, et al. 2005). Similarly, in Benin, *B. bassiana* conidia in suspension did not lose viability, but after 20 days exposure to field conditions viability of *B. bassiana* on polyester netting was reduced to 30% (Howard, et al. 2011b). This inability of entomopathogenic fungi to withstand tropical temperatures has also been found in several laboratory studies (Rangel, et al. 2005, Lekimme, et al. 2008, Darbro and Thomas, 2009) and could pose an obstacle for the deployment of these entomopathogenic fungi for dengue vector control. Work focussing on the formulation needs to be carried out to ensure that this promising control tool can withstand the tropical climates in which it is likely to be used.

5.3 Invertebrates

5.3.1 *Toxorhynchites*

Toxorhynchites is the largest genus (52 species out of 90) of four in the subfamily Culicinae of Culicidae (Diptera) (Harbach, 2011). *Toxorhynchites* mosquitoes are diurnal and carnivorous in its larval stages but not haematophagous as adults (Steffan and Evenhuis, 1981). Larvae are generalist predators and their range of prey is so wide that they exhibit a strong cannibalism with the biggest larvae easily consuming the small ones of their own species. All these species have a precocious vitelogenesis since the pupal stage (Watts and Smith, 1978), which allows them to oviposit a short time after mating. This is because they are inhabitants of tropical and subtropical forests where the availability of temporary breeding sites for oviposition is unpredictable. *Toxorhynchites* usually lay eggs in rock depressions and tree holes, axils of bromeliads, cut bamboo canes, and so on, where they find live mosquito larvae as prey (Clark-Gil and Darsie, 1983). For species living in suburban or urban habitats egg-laying occurs in man-made water containers, such as discarded tires, buckets, cans, and graveyard flowerpots, where there are larvae of domestic mosquitoes such as *Culex spp.* and *A. aegypti* and *A. albopictus* (Rubio and Ayesta, 1984).

Both sexes of the Neotropical *Toxorhynchites theobaldi* (currently *T. moctezuma*) stay close to the breeding sites from where they emerged waiting for females to mate. Most females mate during the first ten days in the laboratory, kept in a cage without a cup with water, but after introducing the cup into the cage they oviposit within the following four days. Eggs are white or yellowish and hydrophobic floating individually on the water surface (Rodriguez and Reyes-Villanueva, 1992). The pattern of oviposition of all species have a bimodal activity with two peaks; the lower in the morning and the higher one in the afternoon (Arredondo-Bernal and Reyes-Villanueva, 1989; Bonnet and Hu, 1951). The preferences for oviposition in flowerpots in a Mexican cemetery were described by Reyes-Villanueva, et al. (1987) for *T. theobaldi* (today known as *T. moctezuma*). They examined 584 containers and found 1,009 eggs in 204 flowerpots (35%). Most eggs (66%) were found in shady flower containers, which indicate the preference of females to stay and oviposit in the shady microhabitats.

There are few field studies evaluating the impact of *Toxorhynchites* adults released in areas with man-made containers harboring *A. aegypti* larvae. One 4th instar of *T. brevipalpis* at 22-25°C living in tires consumes around 12 *A. aegypti* larvae during 24 hours, while in the laboratory they eat on average 16 prey larvae at 26°C (Trpis, 1972). In a survey in Africa nine tires and nine tins were sampled weekly between April 1969 and March 1970 registering the number of *T. brevipalpis* and *A. aegypti* larvae per container. So, to estimate the larval population of both predator and prey per hectare which was the surface of the tire dump, an extrapolation was done based on the mean number of larvae per container and the percent of containers with water in the dumping (Trpis, 1973). By this way, he was able to obtain the numerical response of *T. brevipalpis* to *A. aegypti*, existing a lag time of a month between both, and with the predator population always following the prey one.

The same author estimated the absolute population of *T. brevipalpis*. Two hundred wild specimens of both sexes of *T. brevipalpis* were collected by hand-nets from a 1-ha tire dump at Dar es Salaam, Tanzania; each specimen was marked after anaesthetization by applying a spot of enamel paint to the front of the mesonotum. However, only 195 marked mosquitoes (140 males and 55 females) were released in the middle of the habitat. Then, 24 h after the release a new capture was carried out. From the number of mosquitoes marked and released (M) and the number marked (m) in the total recapture sample (T), the size of the population (N) was estimated according to the formula $N=MT/m$, which is the Lincoln index. Of the 337 mosquitoes in the second collection, 19 (15 male and 4 female) had been marked. The size of the *T. brevipalpis* population of the 1-ha habitat was therefore estimated as 3,459 mosquitoes. The author also calculated the *A. aegypti* population in the area and this was of 570 females, which was around 33% compared to the predator population (Trpis, 1973).

In New Orleans, USA, *T. amboinensis* was examined by Focks, et al. (1985). They did 29 weekly releases of 6-8 day old females; releases comprised 100, 200 and 300 females per block of a neighborhood formed by 16 blocks (4x4 area) during March-July 1982. The *A. aegypti* population was monitored by using two ovitraps per block. The response variables measured weekly were three: the average number of *A. aegypti* eggs per ovitrap, number of exhuviae of *Aedes spp.* and *Culex spp.* per container in treated and control blocks, and the proportion of containers in treated area positive for predators. There was a reduction of 45% in the *A. aegypti* population compared to the one of control blocks after the release of 100 *T. amboinensis* females, while no significant increase in control was achieved at 200 and 300 females. Likewise, the *C. quinquefasciatus* population also was diminished by around 40% with 100 females; while ovitraps placed around the experimental areas demonstrated that the females released had little dispersion between blocks. This study showed the potential of

T. amboinensis as biocontrol agent used in inundative releases at urban habitats infested with the dengue vector.

T. moctezuma oviposition rate was examined at Northeast Mexico by Alvarado-Castro and Reyes-Villanueva (1995). They performed six releases of 20, 40 and 80 inseminated, 10-day old females in the center of a pecan orchard *Carya illinoensis*, with a discarded tire filled with 3 liters of water, and fastened at the trunk of each tree. Ten trees with tires comprised the experimental area arranged in two lines of five trees each, predators were released at the center, and the egg number per tire of the predator were counted daily for 17 days after each release. Daily means were 48.23 and 35.88 eggs for both 20-female releases, 95.65 and 65.12 eggs, and 242.94 and 108.12 eggs for both 40- and both 80-female releases, respectively. There was a linear trend well defined only for the releases of 20 and 40 females with a decrease rate of oviposition of 0.25 per day, and most eggs (56-66%) were laid during the first four days after release.

Although in the above experiment there were no larvae of *A. aegypti* in tires, the high numbers of eggs deposited daily by the released females of *T. moctezuma*, suggest this predator is promising as biocontrol agent against the dengue vector larval populations. Females are able to locate larval breeding sites of *A. aegypti* and ovipisit on them, but a strong limiting factor is the expensive production of adults of this predator. At least for the experiment of Alvarado-Castro and Reyes-Villanueva (1995) to develop and produce 100 pupae of the predator required the use of around 10,000 larvae of *A. aegypti*.

5.3.2 Copepods

The most successful type of invertebrate used for mosquito larva control is the cyclopoid copepods, most notably *Mesocyclops*. These are 1-2 mm long crustaceans that are one of the most numerous multicellular organisms on earth that can be found in many geographical locations, and therefore the use of copepods for mosquito control does not require exotic introductions. Because of their size, copepods mainly kill the first instar larvae, and they prefer *Aedes* larvae over *Anopheles* and *Culex* larvae. Copepods can live for 1-2 months, are quite hardy and they self-replicate readily. Because they eat a variety of aquatic prey, they can maintain populations in water storage containers even if mosquitoes are not found (Marten and Reid, 2007). They can also be easily moved from one container to other container habitats. Therefore they offer the potential of sustainable mosquito control. Furthermore, copepods can be easily and cheaply mass produced and transported, even under field conditions where they are required. Nam, et al. (2000) used a method using plastic garbage bins in which thousands of copepods could be produced in just 3 weeks. They then transported these copepods to the various field locations using hollowed out polystyrene blocks that they were able to send using the Vietnam postal service. No seriously adverse environmental effects have been reported from the use of copepods.

The major success story for the use of *Mesocyclops* against dengue vectors comes from Vietnam. A study carried out in northern Vietnam using *Mesocyclops* as the primary control measure was able to reduce *A. aegypti* levels to 0-0.3% of baseline estimates and *A. albopictus* to 0-14.1% of baseline levels (Kay, et al. 2002). This project was then expanded into 3 provinces in central Vietnam, with similar findings. The authors report that *Aedes* mosquitoes were eliminated from several study communes and several years into the programme no dengue was detectable in the three treated rural communes (covering a population of 27,167), but dengue transmission was still evident in the control areas (Nam, et al. 2005). Following country-wide programme expansion it was reported that *A. aegypti* had been eradicated from 32 of 37 communes, covering a human population of 309,730. Dengue has not been reported in the treated areas for years, where the authors estimate that 386,544 people have been protected, but dengue transmission remained in the untreated areas, (Kay and Nam, 2005). *Mesocyclops* use is also proving to be sustainable; 7 years after official involvement ceased, *Mesocyclops* are still being used by community members to keep *Aedes* populations at bay and local transmission of dengue has been eliminated in areas where they are being used (Kay, et al. 2010).

It is not just in Asia that copepods have been successful in field trials. A field trial in Mexico used copepods in water tanks, tires and vases to control *A. aegypti*. It was found that the most effective control was in the cemetery vases, with 67.5% reduction over the 3 month study period (Gorrochotegui-Escalante, et al. 1998). Copepods have also been used to successfully control *A. aegypti* populations in Argentina (Marti, et al. 2004) and *A. albopictus* populations in Japan were effectively controlled by *Mesocyclops* and *Macrocyclops* copepods (Dieng, et al. 2002).

A disadvantage of *Mesocyclops* is that they are the intermediate hosts for the Guinea worm *Dracunculus medinensis*. This is a helminth human parasite that infects people when they ingest infected *Mesocyclops* in drinking water. Therefore, *Mesocyclops* cannot be used to control dengue vectors in areas where Guinea worm transmission still takes place. There is a global Guinea worm eradication programme that has made great progress, however, Sudan, Ethiopia, Ghana and Mali all remain endemic for Guinea worm. Chad is the only other country that reported cases in 2010. Another disadvantage is that as with some other biological control options, *Mesocyclops* are susceptible to insecticides like Temephos (Kaul, et al. 1990), but they are unaffected by *Bti*. In addition they are sensitive to chlorine in the water (Marten and Reid, 2007). Nevertheless, despite these disadvantages *Mesocyclops* have been sustainably used to almost eradicate dengue from areas of Vietnam (Kay, et al. 2010), and along with fish (see below) are probably the best biological control tool of dengue vector mosquitoes that is currently available for operational use.

5.4 Fish

Another biological control method that has been used to control mosquitoes is the deployment of fish that will eat the mosquito larvae and pupae. Many different types of fish are used, but to avoid damaging ecosystems the World Health Organisation (WHO) advocates the use of native larvivorous fish (World Health Organisation, 2002).

Fish can be incredibly effective at reducing *Aedes* mosquito numbers under field conditions. In Mexico, the mean container index (CI) (percentage of water-holding containers infested with *Aedes* larvae or pupae) in cement tanks was around 87% before indigenous fish species were introduced, and mosquito numbers were recorded for a year. The results show that each of the 5 fish species eliminated mosquito breeding in the tanks, while the CI in the control remained at 86% (Martinez-Ibarra, et al. 2002). Similarly, the Chinese cat fish *Clarias fuscus* reduced the Breteau Index (BI) (the number of positive containers per 100 houses) from 50 (before fish introduction) to 0 just 15 days after fish introduction (Neng, et al. 1987).

In Northeastern Brazil, before the deployment of *Betta splendens* fish, 70.4% of the tanks were infested with *A. aegypti*, one year later the infestation rate was just 7.4%, dropping to 0.2% 11 months later (Pamplona, et al. 2004). Furthermore, a study in Thailand found that in rural areas 43.7% of containers without fish had *A. aegypti* larvae, compared to just 7.0% of containers that had fish; this effect was also seen in an urban area (40.6% vs 8.3%) (Phuanukoonnon, et al. 2005). This study in Thailand compared a range of control methods and found that keeping fish was the most effective (Phuanukoonnon, et al. 2005).

Larvivorous fish have further advantages. Unlike some of the invertebrate predators, people feel familiar with fish (Martinez-Ibarra, et al. 2002), and this means that they are able to apply this control tool themselves. This happened in Brazil where the successful use of *B. splendens* fish was broadcast in the media, resulting in the people placing these fish in their water storage containers of their own accord (Lima, et al. 2010). In addition, the success of a trial in Mexico was attributed in part to the adoption of the larvivorous fish as pets by the local children (Martinez-Ibarra, et al. 2002). As well as being pets, some fish can be farmed and eaten by local communities (Howard, et al. 2007). Several indigenous Mexican species used to control *A. aegypti* can be eaten (Martinez-Ibarra, et al. 2002) and the Chinese cat fish *C. fuscus* is not only edible but also highly larvivorous and tolerant of harsh environmental conditions (Neng, et al. 1987). Keeping fish can also be more cost-effective than other control methods like insecticide spraying (Neng, et al. 1987) and larvicide application (Seng, et al. 2008). Furthermore, fish have not only been found to be more cost-effective and long lasting than *Bti*, but they were also found to be much more effective as a control method (Lima, et al. 2010). Further advantages of larvivorous fish are that they are self-sustaining, so in general water bodies only have to be treated once, or at least less frequently than for other control tools. This can lead to sustainable mosquito control. In addition, fish survival does not depend on the presence of mosquito larvae whereas other biological control agents often depend on the mosquito population not being entirely eliminated (Wright, et al. 1972). Fish are effective at controlling the older larval stages of *Aedes*, something that is not readily achieved by the copepod predators (Russell, et al. 2001). Also, unlike for chemical larvicides, mosquito larvae cannot build up a physiological resistance to fish.

As with all mosquito control tools, there are some disadvantages of using fish. Larvivorous fish can only be used in certain water bodies conducive to their survival (Lima, et al. 2010), and they will only thrive and reproduce under certain conditions that can be specific to the different fish species. In addition, not all containers that allow *Aedes* breeding are suitable for fish. Fish obviously cannot be used in habitats that are prone to drying out. They are also not well suited to the smaller containers where the water may become too hot during the day, and where oxygen levels may not be high enough. The ability to withstand chlorine can be an important characteristic because in many countries chlorine is added to the drinking water, and that is then stored in large tanks by householders. A study comparing the chlorine tolerance of two larvivorous fish, *B. splendens* and *Poecilia reticulata*, found that *P. reticulata* was unable to withstand chlorine concentrations within the limits for human consumption in Brazil (Cavalcanti, et al. 2009). Not always but, somehow there are reluctance of certain individuals to use fish in tanks because their presence stinks drinking water.

Under laboratory conditions, *B. splendens* repelled *A. aegypti* females from laying eggs in the water where the fish were, but *P. reticulata* (Pamplona, et al. 2009) and *Gambusia affinis* (Van Dam and Walton, 2008) did not. This repellency can be a problem because these fish cannot be very effective at controlling successive generations, especially when untreated oviposition sites are available (although in an integrated approach, those sites could be removed or treated with another control tool). Fish can also have an effect on non-target organisms. A study comparing *P. reticulata* with a native Australian fish found that the Australian fish outperformed in terms of the larvivorous potential, but this fish species also ate native tadpoles, and as such should only be utilised in water containers where the tadpoles would not be found (Russell, et al. 2001). Some fish can also reduce their larval intake in the presence of commercial fish food (Ekanayake, et al. 2007). Whilst the use of fish has proved popular in certain trials, and shows great promise for sustainable control of dengue vectors, the implementation of larvivorous fish should be accompanied by adequate participatory education to make it more acceptable for communities, and therefore potentially more sustainable.

5.5 Plants

As entomopathogenic fungi seem promising for adult control, plants could be a promising biological control tool for aquatic stage mosquitoes. Plants produce compounds to protect themselves from insects, and these compounds can effect insect development in many ways. Hundreds of plant species have been tested for their effects against mosquitoes (Shaalan, et al. 2005) with a recent review published by Fallatah and Khater (2010). Much of the research against *A. aegypti* mosquitoes has focussed on by-products of plants already utilised for economic gain, or on already recognised medicinal plants. In the former bracket, avocado seed extracts were found to be able to kill *A. aegypti* larvae (Leite, et al. 2009). Similarly, unripe black pepper extracts were found to be effective at killing pyrethroid-resistant *A. aegypti* (Simas, et al. 2007). Ethanolic extracts also fall into this category, since ethanol is a by-product from sugar cane refinement (Wandscheer, et al. 2004). In the latter bracket, 14 Mexican medicinal plants were tested and a range of toxicity was found, with some being highly toxic and others showing very little larvicidal effect (Reyes-Villanueva, et al. 2008). The neem tree (*Azadirachta indica*) is a well known medicinal plant that has been widely tested against mosquitoes (Howard, et al. 2009; Fallatah and Khater, 2010). When tested against dengue vectors neem was found to be effective at relatively low doses (Wandscheer, et al. 2004) but oviposition was inhibited (Coria, et al. 2008). It is important that oviposition is not affected, because if mosquitoes do not expose their progeny to the neem then control cannot be sustainable (Howard, et al. 2011a).

Plants have not yet been used to control dengue vectors in field trials, and are not currently under consideration for inclusion into IVM trials, but many laboratory trials have been conducted with a view to identifying promising candidates. However, as well as testing whether plant extracts can kill mosquitoes, it is important that the effect on non-target organisms is evaluated. These could be native aquatic fauna, other biological control tools, or mammals that have access to the water into which the botanical larvicides are to be placed. A recent laboratory study tested the bioefficacy of two plants against *A. aegypti* mosquitoes and the larvivorous fish *P. reticulata* (Patil, et al. 2011). Both plants were found to be highly effective as larvicide but *Plumbago zeylanica* was found to have a slight toxic effect against the fish, although the authors concluded that these plant species could be used alongside this larvivorous fish in IVM programmes (Patil, et al. 2011). Sodium anacardate from cashew nut shell liquid was evaluated against *A. aegypti* eggs, larvae and pupae and found to be highly toxic to all life stages, although the dose required to kill the pupae was much higher than that needed to kill larvae (Farias, et al. 2009). This is not uncommon for botanical products (Howard, et al. 2009). The authors also tested the effect against mice. They used a dose much higher than the dose required to kill the mosquitoes, and found that even at 0.3 g/kg there was no apparent damage to the mice. They concluded that this botanical mosquitocidal compound was safe for mammals (Farias, et al. 2009).

There are several advantages that plants offer. Plants could be used in water sources that are too small to house larvivorous fish or that have a tendency to dry up completely for long periods of time. These small habitats are more prone to fluctuating temperatures, and evidence has shown that some plants can be effective at a range of temperatures, with increasing toxicity at the higher temperatures (Wandscheer, et al. 2004; Patil, et al. 2011). In addition, many plants are widely available where they are required, and they can be grown by rural communities which could provide sustainable and relatively cheap mosquito control. Plants are biodegradable, relatively safe for the environment and communities are familiar with many of the plants that have proven insecticidal. As with the other biological control tools discussed, they can be used to manage insecticide-resistant mosquito populations.

Despite these advantages, there are several reasons why plants are not being used in IVM programmes. Most plants toxic to larvae of *A. aegypti* are wild species, and therefore not cultivated. In addition to the fact that they are not available in practical amounts, phytochemicals can display heat and UV instability which can reduce the applied dose to levels that are no longer effective. Some plant parts are more effective than others; for example root infusions of *Solanum nigrescens* were toxic but leaf infusions were not (Reyes-Villanueva, et al. 2008), and variation can occur between the same plant products produced in different geographical areas (Schmutterer, 1995). Further disadvantages of using botanical products to control *A. aegypti* include the pronounced taste of some of the plants. For example the use of black pepper may not be acceptable in drinking water (Simas, et al. 2007). Whilst aqueous extracts are normally less effective than organic chemical extracts (Simas, et al. 2007), they could be more applicable for use by rural communities. Thus effectiveness in the laboratory may not immediately translate to field success, especially when community-based control tools are required. Work should be carried out looking at the most ubiquitous and larvicidal plants with a view to community deployment in future IVM trials. In addition, work should continue towards the commercialisation of botanical products for dengue vector control.

6. Integrated vector management

Integrated vector management (IVM) is a comprehensive strategy which aims to achieve a maximum impact on vector borne diseases like dengue. IVM was adopted by the WHO in 2004 (World Health Organisation, 2004) as a strategy to improve the cost-effectiveness, efficacy, ecological soundness and sustainability of vector control. The emphasis of IVM is on examining and analyzing the local situation, making decisions at decentralized levels, and utilising the appropriate mosquito control tools (World Health Organisation, 2009). One of the features of IVM is the use of a range of interventions, often in combination and simultaneously, that work together to reduce dengue transmission.

For dengue control, there are three main categories of intervention. These are biological, as described in detail above, the use of chemicals to kill the adult and immature mosquito stages, and the physical removal, periodic cleaning or covering of container habitats. These categories can be used to target all life stages of the *Aedes* mosquito, as shown in (Table 1). The use of education is also an important component, because communities need to know how and why to control dengue vectors. Community-participation in these methods is not only crucial for sustainability (Wang, et al. 2000), but also leads to more effective *Aedes* control, as shown in an IVM trial in Guantanamo, Cuba (Valerberghe, et al. 2009).

There is no silver bullet for dengue vector control, and each of the intervention categories has their disadvantages. Biological tools are not always feasible in certain small container habitats. Chemicals can pollute the environment, be expensive, and insecticide resistance has developed (World Health Organisation, 2009), and not all water storage containers can be removed/cleaned/covered. One of the benefits of IVM is that it overcomes the disadvantages of using individual methods, and a combination of mosquito control tools can be more effective that any tool used in isolation. Authors of a study in Taiwan concluded that integrated pest control was the best and most effective method for dengue control (Chen, et al. 1994). A study in Thailand that looked at the effectiveness of individual methods also concluded that a combination of the control methods increased effectiveness (Phuanukoonnon, et al. 2005). Furthermore, a systematic review and meta analysis of 56 publications detailing the results from field studies found IVM to be the most effective method of reducing entomological indices like the BI and CI (Erlanger, et al. 2008).

Life stage	Intervention	Measure	Sample reference
Egg	Chemical	Insecticide-impregnated ovitraps	(Perich, et al. 2003)
	Physical	Autocidal ovitraps	(Cheng, et al. 1982)
		Removing, cleaning of containers	(Chen, et al. 1994)
Larvae and pupae	Biological	Fish	(Martinez-Ibarra, et al. 2002)
		Mesocyclops	(Nam, et al. 2005)
		Bacillus thuringiensis israelensis(*Bti*)	(Lima, et al. 2010)
		Spinosad	(Darriet, et al. 2010)
	Chemical	Temephos	(Phuanukoonnon, et al. 2005)
		Pyriproxyfen	(Darriet, et al. 2010)
	Physical	Removing, cleaning of containers	(Chen, et al. 1994)
Adults	Biological	Entomopathogenic fungi	(Reyes-Villanueva, et al. 2011)
	Chemical	ULV fogging	(Osaka, et al. 1999)
		Aerosol cans	(Osaka, et al. 1999)
		Repellents	(Jahn, et al. 2010)
		Lethal ovitraps	(Kittayapong, et al. 2008)
	Physical	House modification	(Vanlerberghe, et al. 2011)
		Sticky ovitraps	(Ordonez-Gonzalez, et al. 2001)

Table 1. Some methods that can be used in IVM programmes to control dengue vector mosquitoes, for a full list of possible control tools see World Health Organisation (2009).

Operational large-scale IVM programmes are already being carried out in a range of countries. IVM has been carried out in Singapore since the mid 1970s, in China since the early 1980s (Neng, et al. 1987), and Taiwan since the late 1980s (Chen, et al. 1994, Wang, et al. 2000). A successful regional IVM campaign focussed on the use of predacious copepods was expanded to a national campaign in Vietnam in the mid 1990s (Nam, et al. 2000). Dengue control programmes in Brazil (Lima, et al. 2010) and Thailand (Phuanukoonnon, et al. 2005) are centred around community participation, health education, larval control (including biological control), chemical control of adult mosquitoes and physically removing/covering containers. In Cuba routine *Aedes* control comprises physically removing container habitats and chemical control of adult and larval mosquitoes, backed up by health education (Vanlerberghe, et al. 2009). Not only have these IVM control programmes been carried out for many years in some countries, but the notion of enforcement has been adopted in a few countries. A study from China describes how fines were handed out for non-compliance, with incentives given to those households adequately maintaining the dengue control methods (Neng, et al. 1987). Specific laws aimed at ensuring that householders carryout dengue control measures have been in effect since 1968 in Singapore, and 1988 in Taiwan (Chen, et al. 1994). As in these other countries, mosquito control legislation is enforced by handing out fines in Cuba (Vanlerberghe, et al. 2009).

Not only do IVM programmes show that *Aedes* mosquitoes can be successfully controlled, but more importantly, IVM can be effective at reducing dengue disease burden. An IVM trial that was targeted at high-transmission areas in Thailand used a combination of biological larval control, chemical adult control and physically preventing oviposition. Not only did they report a dramatic reduction in the number of *Aedes* positive containers, but there was also a significant reduction in adult *Aedes* mosquitoes. Crucially, there were no dengue cases reported in the treated area, whilst in the control there were 322.2 cases per 100,000 people; baseline data was similar for the two areas at around 230 cases per 100,000 (Kittayapong, et al. 2008). Similarly, a programme utilising all the major categories of intervention was carried out in Taiwan between 1987 and 1993 (Chen, et al. 1994). The authors reported that in 1988, there were 10,420 dengue cases however, between 1990 and 1993, no dengue cases were reported (Chen, et al. 1994). In a later report from Liu-Chiu island (off the coast of Taiwan) IVM was able to nearly eradicate *A. aegypti* mosquitoes and there were no dengue cases reported by the end of the study, even though mosquito habitats were still present (Wang, et al. 2000).

7. Potential of biological control methods in the future

At present, there is no vaccine for dengue, and vector control remains the cornerstone of any dengue control effort. The future of dengue vector control must involve IVM programmes, ideally with a combination of governmental top-down and community-based bottom-up approaches. Attention must be paid to the WHO guidelines on dengue control (World Health Organisation, 2009), as well as to new research that may also be effective. Ultimately, sustainable mosquito control requires behavioural change at both individual and community levels so that the number of larval habitats is reduced and remains low. Because the main dengue vector has a preference for breeding in domestic water containers, the potential of the community to sustainably control mosquito populations is probably higher than for malaria programmes, where the malaria vector breeds in natural habitats that are not always easy to find.

The WHO says that IVM should be composed of an integration of non-chemical (biological) and chemical vector control methods. Furthermore, they say "productive larval habitats should be treated with chemicals only if environmental management [physical] methods and other non-chemical [biological] methods cannot be easily applied or are too costly" (World Health Organisation, 2009). Thus, biological larval control tools appear to be given more emphasis than chemical tools. Perhaps because of this, there has been a shift towards using more biological control methods, with chemical control trials becoming less frequent (Erlanger, et al. 2008). A review of 21 studies comparing biological, chemical and educational dengue prevention programs found that biological interventions were the most effective; nearly all the biological interventions eliminated mosquito larval populations, whereas the chemical interventions were judged to be the least effective, and were not thought to offer a long-term solution (Ballenger-Browning and Elder, 2009). A separate review of 56 field studies found that the relative effectiveness of biological control was better than chemical or environmental/physical control measures, but that an integrated approach was best (Erlanger, et al. 2008).

This switch from chemical to biological control tools is in part due to raising insecticide resistance in mosquito populations. Another reason is that chemical control tools are usually associated with top-down campaigns, where the government was solely in charge of implementing mosquito control, like insecticide spraying. In these cases the insecticide and equipment used was rarely available to the communities themselves. Top-down campaigns usually relied on the mass-production of one product that was easy to store. Recently there has been a shift to more bottom-up campaigns because it has been recognised that these are more likely to be cost-effective and sustainable. The same characteristic that makes biological control unattractive from a commercial point of view (namely the difficulty in making money from organisms that cannot be mass produced, stored and shipped from cost-effective industrial plants) is especially appealing to the resource-poor community members affected by dengue, because many of the biological tools can be produced on a small-scale without the need for expensive and complicated infrastructure.

Utilising indigenous biological control tools is appropriate in under-resourced countries because biological control tools are *in situ* in many areas where they will be required. In addition, they can be easy to reproduce under field conditions. An important point is whether the control tool can be produced in large enough quantities to be used in control programmes. For example a simple and effective way of increasing and transporting copepod populations has been devised in Vietnam using polystyrene blocks (Nam, et al. 2000). Fish can be farmed where needed and locally-produced *Bti* was used in a trial in Vietnam where dengue transmission was successfully suppressed (Kittayapong et al. 2008). However, there are some tools whose biological characteristics do not lend themselves to intentional deployment such as the corixid bug *Micronecta quadristrigata*. Attempts to culture this invertebrate predator in the laboratory were unsuccessful because it readily flew from one container to the other (Nam, et al. 2000). In addition, some fish are not easily transported (Russell, et al. 2001). By their nature biological control tools are natural, living organisms and as such there are certain considerations to be made before deciding which should and can be used in certain settings. Their ability to survive in the intended control area is of course important, and for this reason some water bodies are more suited to invertebrate predators, and some to the vertebrate ones.

Biological control has an advantage over physical control due to the "egg trap effect". In essence, if you remove containers then the reduction in mosquitoes is generally proportional, because there will still be some that *Aedes* mosquitoes can lay eggs in, and from which they can emerge. But with biological control, mosquitoes that emerge from untreated containers waste most of their eggs on containers treated with the biological agent and this can cause a population collapse (Marten and Reid, 2007). This can be of particular importance in terms of dengue control because unlike for malaria, dengue can be transmitted vertically from infected adult mosquitoes to their eggs, and adult mosquitoes can emerge from water bodies already infected with and infectious for dengue.

The main risks with biological control are the safety of the biological control tool to non-target organisms, and the consequence of permanent establishment of the tool into areas where it may not naturally be found (Various, 1995). For this reason, WHO says that only native organisms should be used (World Health Organisation, 2002), and many native *Mesocyclops* species and fish types exist that can be used. Formulated and registered biopesticides such as *Bs* and *Bti* are being produced that could overcome the risks of classical biological control. These biopesticides are usually mass-produced and could complement the use of classical biological control tools in IVM programmes.

Being able to produce control tools where they are needed can lead to more cost-effective and sustainable control. Crucially, local production and trading of biological control tools could lead to an increase in the socio-economic status of communities. Control programmes incorporating biological control tools that lead to successful mosquito suppression, along with an increase in the socioeconomic status of the community, not only have the potential to be more sustainable than some top-down insecticide-based control programmes but they can also lead to an increased sense of understanding, ownership and empowerment among the community. This is important because eventually communities will be charged with monitoring and implementing mosquito control. This process will be made easier if the control tools used are already familiar to the communities and are readily available, like some of the biological control tools discussed above.

The successful future of dengue control lies in engaging, empowering and entrusting affected communities with mosquito control in their environment using many methods in an IVM approach. For this to occur, cheap, readily accessible and effective mosquito control tools need to be researched and developed. Biological control tools certainly have the potential to fulfil these criteria.

8. Conclusions

Traditionally, IVM control programmes have been based on two components: chemical control (temephos as larvicide and organophosphates and pyrethroids as adulticides applied by ultra-low volume space spraying), and the community contribution to remove the water in artificial containers. However, dengue is associated to the lowest socio-economical strata of the endemic (developing) countries worldwide where the community lacks a culture of participation. Therefore, although there are reports of resistance of *A. aegypti* to chemicals, nowadays their application is still a major tool that health agencies have against the vector, whose populations are invading new habitats due to the global warming. Nevertheless, the persistent use of chemicals conveys a high risk for a serious and real trouble of resistance; if their application continues it is not far from the day in which no chemical reduces sufficiently the vector densities to below their transmission threshold.

Despite the vast number of technical reports and scientific papers published yearly about *A. aegypti* biocontrol, most of the natural enemies of the mosquitoes *Aedes* incriminated in suburban and urban transmission are at experimental level. Based on our review, we think that the efforts of Brazil will produce at short term, good formulates of *Bacillus sphaericus* and *B. thuringienis* subsp. *israelensis* to control larval populations at accessible costs in developing countries and with no risk of pollution as threat to human and his environment.

So far, there no low-cost production of viruses with practical potential to be used against *A. aegypti* in developing countries; neither are there artificial cultures to produce the protozoans and microsporidians evaluated as parasites of the dengue vector. The *Toxorhynchites* mosquitoes are good larval predators but high numbers of *Aedes* larvae are required to produce sufficient adults that need to be used in inundative releases. Their production although easy is impractical; likewise the strong cannibalism tendency is an obstacle for their mass production. A similar case is the huge complex of plants reported as toxic to larvae or adults. It is difficult to cultivate them to use their crude extracts as bioinsecticides; what proceeds is to carry on research to identify the chemical structure of the active compounds to produce them synthetically and use them as bioinsecticides; but this needs of a lot of time and great investments in biotechnology, which is prohibitive for the economy of endemic countries. Also somehow impractical is the use of larvivorous fishes; most *A. aegypti* populations are produced in small man-made containers as tires, buckets, cans, bottles and so on, located at the backyards of houses. Nevertheless in the tropical Central and South American countries, it is common to have large cement-built deposits to store water in houses; the use of fishes in those structures is effective.

Copepods are the group with the most potential; they are very cheap to yield them as biolarvicides as today they are being produced and used in Vietnam. Actually copepods show a great potential to be used in the IVM control programmes against *A. aegypti* worldwide. Finally another promising group is formed by the fungus Ascomycetes *Metarhizium anisopliae* and *Beauveria bassiana*. They are effective to control immatures and adults, although we think they are more effective as adulticides by indirect exposure of mosquitoes to surfaces impregnated with conidia at doses superior to 10^8 spores ml^{-1}. Fungal dissemination among female populations by releasing conidia-contaminated virgin males of *A. aegypti* deserves further research. These fungi are cheaply produced by using natural substrates like rice, sorghum, etc. in plastic bags in laboratory, to have a low cost production in an IVM control programme for dengue in any developing country.

9. Acknowledgments

Carolina Briceño-Dávila helped with drafting the zoonotic cycle of dengue viruses and is thanked. This study was funded (article processing charges were covered by grant of SIP No. 20111028) by the Secretaría de Investigación y Posgrado (SIP)-Instituto Politécnico Nacional (IPN)- Megaproyecto II sobre Dengue de la Red de Biotecnología through the Fondo de Investigación Científica y Desarrollo Tecnológico del IPN. Mario A. Rodríguez-Pérez holds a scholarship from Comisión de Operación y Fomento de Actividades Académicas (COFAA)/IPN. This study was also funded by Consejo Nacional de Ciencia y Tecnología (CONACYT)-México (Grant No. 168394).

10. References

Abrams, P. A. 2000. The evolution of predator-prey interactions: theory and evidence. Annu. Rev. Ecol. Syst. 31: 79-105.

Albernaz, D. A., M. H. Tai, and C. Luz. 2009. Enhanced ovicidal activity of an oil formulation of the fungus *Metarhizium anisopliae* on the mosquito *Aedes aegypti*. Medical and Veterinary Entomology 23: 141-147.

Alvarado-Castro, J. A. y F. Reyes-Villanueva. 1995. Tasa de oviposicion en llantas de hembras de *Toxorhynchites theobaldi* liberadas en una huerta de nogal en el Noreste de Mexico. Southwest. Entomol. 20: 215-221.

Armstrong, P. M. and R. Rico-Hesse. 2003. Efficiency of dengue serotype 2 virus strains to infect and disseminate in *Aedes aegypti*. Am. J. Trop. Med. Hyg. 68:539-544.

Arredondo-Bernal, H. C. and F. Reyes-Villanueva. 1989. Diurnal pattern of oviposition of *Toxorhynchites theobaldi* in the field. J. Am. Mosq. Control Assoc. 5: 25-28.

Barnard, D. R., R. D. De Xue, M. A. Rotstein and J. J. Becnel. 2007. Microsporidiosis (Microsporidia: Culicosporidae) alters blood-feeding responses and DEET repellency in *Aedes aegypti* (Diptera: Culicidae). J. Med. Entomol. 44: 1040-1046.

Becnel, J. J. and M. A. Johnson. 2000. Impact of *Edhazardia aedis* (Microsporidia: Culicosporidae) on a Seminatural Population of *Aedes aegypti* (Diptera: Culicidae). Biological Control 18: 39-48.

Ballenger-Browning, K. K., and J. P. Elder. 2009. Multi-modal *Aedes aegypti* mosquito reduction interventions and dengue fever prevention. Tropical Medicine and International Health 14: 1542-1551.

Bonnet, D. D. and S. M. K. Hu. 1951. The introduction of *Toxorhynchites brevipalpis* Theobald into the territory of Hawaii. Proc. Hawaii. Entomol. Soc. 14: 237-242.

Cavalcanti, L. P. d. G., F. J. d. P. Junior, R. J. S. Pontes, J. Heukelbach, and J. W. d. O. Lima. 2009. Survival of larvivorous fish used for biological control of *Aedes aegypti* larvae in domestic containers with different chlorine concentrations. Journal of Medical Entomology 46: 841-844.

Chen, Y.-R., J.-S. Hwang, and Y.-J. Guo. 1994. Ecology and Control of Dengue Vector Mosquitoes in Taiwan. Kaohsiung J Med Sci 10: S78-S87.

Cheng, M.-L., B.-C. Ho, R. E. Bartnett, and N. Goodwin. 1982. Role of a modified ovitrap in the control of *Aedes aegypti* in Houston, Texas, USA. Bulletin of the World Health Organisation 60: 291-296.

Clark, T. B., W. R. Kellen, T. Fukuda, and J. E. Lindegren. 1968. Field and laboratory studies on the pathogenicity of the fungus *Beauveria bassiana* to three genera of mosquitoes. Journal of Invertebrate Patholology 11: 1-7.

Clark-Gil, S. and R. F. Darsie, Jr. 1983. The mosquitoes of Guatemala, their identification, distribution and bionomics, with keys to adult females and larvae in English and Spanish. Mosq. Syst. 15: 151-284.

Coria, C., W. Almiron, G. Valladares, C. Carpinella, F. Luduena, M. Defago, and S. Palacios. 2008. Larvicide and oviposition deterrent effects of fruit and leaf extracts from *Melia azedarach* L. on *Aedes aegypti* (L.) (Diptera: Culicidae). Bioresour Technol 99: 3066-70.

Cordellier, R., B. Bouchite, J. C. Roche, N. Monteny, B. Diaco and P. Akoliba 1983. Circulation silvatique du virus dengue 2 en 1980, dans les savannes sub-soudaniennes du Cote d'Ivoire. Cah ORSTOM Ser. Entomol. Med. Parasitol. 21: 165–179.

Cologna, R., P. M. Armstrong, and R. Rico-Hesse. 2005. Selection for virulent dengue viruses occurs in humans and mosquitoes. J. Virol. 79: 853-859.

De Silva A. M., W. P. J. Dittus, P. H. Amerasinghe and F. P. Amerasinghe. 1999. Serologic evidence for an epizootic dengue virus infecting toque macaques (*Macaca sinica*) at Polonnaruwa, Sri Lanka. Am. J. Trop. Med. Hyg. 60: 300–306.

Darbro, J. M., and M. B. Thomas. 2009. Spore persistence and likelihood of aeroallergenicity of entomopathogenic fungi used for mosquito control. Am J Trop Med Hyg 80: 992-997.

Darriet, F., S. Marcombe, M. Etienne, A. Yebakima, P. Agnew, M.-M. Yp-Tcha, and V. Corbel. 2010. Field evaluation of pyriproxyfen and spinosad mixture for the control of insecticide resistant *Aedes aegypti* in Martinique (French West Indies). Parasites & Vectors 3: 88.

DeBach, P. and E. L. Schlinger (eds.). 1964. Biological Control of Insect Pests and Weeds. London, Chapman and Hall, 844 p.

De Paula, A. R., E. S. Brito, C. R. Pereira, M. P. Carrera, and R. I. Samuels. 2008. Susceptibility of adult *Aedes aegypti* (Diptera: Culicidae) to infection by *Metarhizium anisopliae* and *Beauveria bassiana*: prospects for Dengue vector control. Biocontrol Science and Technology 18: 1017-1025.

De Paula, A. R., A. T. Carolino, C. O. Paula, and R. I. Samuels. 2011a. The combination of the entomopathogenic fungus *Metarhizium anisopliae* with the insecticide Imidacloprid increases virulence against the dengue vector *Aedes aegypti* (Diptera: Culicidae). Parasites & Vectors 4: 8.

De Paula, A. R., A. T. Carolino, C. P. Silva, and R. I. Samuels. 2011b. Susceptibility of adult female *Aedes aegypti* (Diptera: Culicidae) to the entomopathogenic fungus *Metarhizium anisopliae* is modified following blood feeding. Parasites & Vectors 4: 91.

Dieng, H., M. Boots, N. Tuno, Y. Tsuda, and M. Takagi. 2002. A laboratory and field evaluation of *Macrocyclops distinctus, Megacyclops viridis* and *Mesocyclops pehpeiensis* as control agents of the dengue vector *Aedes albopictus* in a peridomestic area in Nagasaki, Japan. Medical and Veterinary Entomology 16: 285-291.

Ekanayake, D. H., T. C. Weeraratne, W. A. P. P. de Silva, and S. H. P. P. Karunaratne. 2007. Potential of some selected larvivorous fish species in *Aedes* mosquito control. Proceedings of the Peradeniya University Research Sessions, Sri Lanka 12: 98-100.

Erlanger, T. E., J. Keiser, and J. Utzinger. 2008. Effect of dengue vector control interventions on entomological parameters in developing countries: a systematic review and meta-analysis. Medical and Veterinary Entomology 22: 203-221.

Fallatah, S. A. B., and E. I. M. Khater. 2010. Potential of medicinal plants in mosquito control. Journal of the Egyptian Society of Parasitology 40: 1-26.

Farias, D. F., M. G. Cavalheiro, S. M. Viana, G. P. G. de Lima, L. C. B. da Rocha-Bezerra, N. Ricardo, M. P. S, and A. F. U. Carvalho. 2009. Insecticidal action of sodium anacardate from Braziliam cashew nut shell liquid against *Aedes aegypti*. Journal of the American Mosquito Control Association 25: 386-389.

Federici, B. A. 1995. The future of microbial insecticides as vector control agents. J. Amer. Mosq. Contr. Assoc. 11: 260-265.

Focks, D. A., S. A. Sackett, D. A. Dame and D. L. Bailey. 1985. Effect of weekly releases of *Toxorhynchites amboinensis* (Doleschall) on *Aedes aegypti* (L.) (Diptera: Culicidae) in New Orleans, Louisiana. J. Econ. Entomol.78: 622-626.

Garcia-Munguia, A. M., J. A. Garza-Hernandez, E. A. Rebollar-Tellez, M. A. Rodriguez-Perez, and F. Reyes-Villanueva. 2011. Transmission of *Beauveria bassiana* from male to female *Aedes aegypti* mosquitoes. Parasites & Vectors 4: 24.

Geetha, I., and K. Balaraman. 1999. Effect of entomopathogenic fungus, *Beauverai bassiana* on larvae of three species of mosquitoes. Indian Journal of Experimental Biology 37: 1148-1150.

Gill, S. S., E. A. Cowles and P. V. Pictrantonio. 1992. The mode of action of *Bacillus thuringiensis* endotoxins. Annu. Rev. Entomol. 37: 615–36.

Gillespie, A. T., and N. Clayton. 1989. The use of entomopathogenic fungi for pest control and the role of toxins in pathogenesis. Pestic Sci 27: 203-215.

Goldberg, L. J. and J. Margalit. 1977. Bacterial spore demonstrate rapid larvicidal activity against *Anopheles sergentii, Uranotaenia unguiculate, Aedes aegypti, Culex pipiens, Culex unititatius.* Mosq. News 37: 355–8.

Gorrochotegui-Escalante, N., I. Fernanez-Salas, and H. Gomez-Dantes. 1998. Field evaluation of *Mesocyclops longisetus* (Copepoda: Cyclopoidea) for the control of larval *Aedes aegypti* (Diptera: Culicidae) in Northeastern Mexico. Journal of Medical Entomology 35: 699-703.

Guzman, M. G., V. Deubel, J. L. Pelegrino, D. Rosario, M. Marrero, C. Sariol and G. Kouri. 1995. Partial nucleotide and amino acid sequences of the envelope and the envelope/nonstructural protein-1 gene junction of four dengue-2 virus strains isolated during the 1981 Cuban epidemic. Am. J. Trop. Med. Hyg. 52: 241-246.

Harbach, R. E. 2011. Mosquito Taxonomic Inventory, http://mosquito-taxonomic-inventory.info/ accessed on August 9, 2011.

Hembree, S. C. 1979. Preliminary report on some mosquito pathogens from Thailand. Mosq. News 39:575–582.

Holling, C. S. 1959. Some characteristics of simple types of predation and parasitism. Can. Entomol. 91: 385-398.

Holling, C. S. 1961. Principles of insect predation. Annu. Rev. Entomol. 6: 163-182.

Holling, C. S. 1966. The functional response of invertebrate predators to prey density. Mem. Entomol. Soc. Can. 48: 1-86.

Holmes, E. C. and S. S. Twiddy. 2003. The origin, emergence and evolutionary genetics of dengue virus. Infect. Genet. Evol. 3: 19-28.

Howard, A. F. V., E. A. Adongo, A. Hassanali, F. X. Omlin, A. Wanjoya, G. Zhou, and J. Vulule. 2009. Laboratory evaluation of the aqueous extract of *Azadirachta indica* (neem) wood chippings on *Anopheles gambiae* s.s. (Diptera: Culicidae) mosquitoes. J Med Entomol 46: 107-114.

Howard, A. F. V., E. A. Adongo, J. Vulule, and J. Githure. 2011a. Effects of a botanical larvicide derived from *Azadirachta indica* (the neem tree) on oviposition behaviour in *Anopheles gambiae s.s.* mosquitoes. Journal of Medicinal Plants Research 5: 1948-1954.

Howard, A. F. V., G. Zhou, and F. X. Omlin. 2007. Malaria mosquito control using edible fish in western Kenya: preliminary findings of a controlled study. BMC Publ Health 7: 199.

Howard, A. F. V., R. N'Guessan, C. J. M. Koenraadt, A. Asidi, M. Farenhorst, M. Akogbeto, B. G. J. Knols, and W. Takken. 2011b. First report of the infection of insecticide-resistant malaria vector mosquitoes with an entomopathogenic fungus under field conditions. Malar J 10: 24.

Howard, A. F. V., R. N'Guessan, C. J. M. Koenraadt, A. Asidi, M. Farenhorst, M. Akogbeto, M. B. Thomas, B. G. J. Knols, and W. Takken. 2010. The entomopathogenic fungus *Beauveria bassiana* reduces instantaneous blood feeding in wild multi-insecticide-resistant mosquitoes in Benin, West Africa. Parasit Vectors 3: 87.

Huffaker, C. B. and P.S. Messenger (eds.). 1976. Theory and Practice of Biological Control. Academic Press, New York.

Jahn, A., S. Y. Kim, J. H. Choi, D. D. Kim, Y. J. Ahn, C. S. Yong, and J. S. Kim. 2010. A bioassay for mosquito repellency against *Aedes aegypti:* method validation and bioactivities of DEET analogues. J Pharm Pharmacol 62: 91-7.

Jeschke, J. M., M. Kopp and R. Tollrian. 2002. Predator functional responses: discriminating between handling and digesting prey. Ecol. Monogr. 72: 95-112.

Juliano, S. A. and F. M. Williams.1985. On the evolution of handling time. Evolution 39:212-215.

Juliano, S. A. 2001. Non-linear curve fitting: Predation and functional response curves. In: Scheiner, S. M. and Gurevitch, J., editors. *Design and analysis of ecological experiments*. 2nd edition, 178–196. New York: Chapman and Hall.

Kaul, S. M., V. K. Saxena, R. S. Sharma, V. K. Raina, B. Mohanty, and A. Kumar. 1990. Monitoring of temephos (abate) application as a cyclopicide under the guineaworm eradication programme in India. J Commun Dis 22: 72-6.

Kay, B. H., and V. S. Nam. 2005. New strategy against *Aedes aegypti* in Vietnam. Lancet 365: 613-617.

Kay, B. H., T. T. T. Hanh, N. H. Le, T. M. Quy, V. S. Nam, P. V. D. Hang, N. T. Yen, P. S. Hill, T. Vos, and P. A. Ryan. 2010. Sustainability and cost of a community-based strategy against *Aedes aegypti* in northern and central Vietnam. American Journal of Tropical Medicine and Hygiene 82: 822-830.

Kay, B. H., V. S. Nam, T. V. Tien, N. T. Yen, T. V. Phong, V. T. Diep, T. U. Ninh, A. Bektas, and J. G. Aaskov. 2002. Control of *Aedes* vectors of dengue in three provinces of Vietnam by use of *Mesocyclops* (copepoda) and community-based methods validated by entomologic, clinical and serological surveillance. Am J Trop Med Hyg 66: 40-48.

Kellen, W. R. and C. M. MeyerS. 1964. *Bacillus sphaericus* Neide as a pathogen of mosquitoes. *J Invertebr Pathol 7*: 442-448.

Kittayapong, P., S. Yoksan, U. Chansang, C. Chansang, and A. Bhumiratana. 2008. Suppression of dengue transmission by application of integrated vector control strategies at sero-positive GIS-based foci. American Journal of Tropical Medicine and Hygiene 78: 70-76.

Kimura, R. and S. Hotta. 1944. On the inoculation of dengue virus into mice. Nippon lgakku 3379: 629-633.

Klowden, M. J. and M. A. Bulla Jr. 1984. Oral toxicity of *Bacillus thuringiensis* subsp. *israelensis* to adult mosquitoes. Appl. Environ. Microbiol. 48: 665-667.

Lambrechts, L., C. Chevillon, R. G. Albright, B. Thaisomboonsuk, J. H. Richardson, R. G. Jarman and T. W. Scott. 2009. Genetic specificity and potential for local adaptation between dengue viruses and mosquito vectors. BMC Evolutionary Biology 9:160.

Leite, J. J. G., E. H. S. Brito, R. A. Cordeiro, R. S. N. Brilhante, J. J. C. Sidrim, L. M. Bertini, S. M. de Morais, and M. F. G. Rocha. 2009. Chemical composition, toxicity and larvicidal and antifungal activities of *Persea americana* (avocado) seed extracts. Revista da Sociedade Brasileira de Medicina Tropical 42: 110-113.

Lekimme, M., C. Focant, F. Farnir, B. Mignon, and B. Losson. 2008. Pathogenicity and thermotolerance of entomopathogenic fungi for the control of the scab mite, *Psoroptes ovis*. Exp. Appl. Acarol. 46: 95-104.

Lima, J. W. d. O., L. P. d. G. Cavalcanti, R. J. S. Pontes, and J. Heukelbach. 2010. Survival of *Betta splendens* fish (Regan, 1910) in domestic watre containers and its effectiveness in controlling *Aedes aegypti* larvae (Linnaeus, 1762) in Northeast Brazil. Tropical Medicine and International Health 15: 1525-1532.

Livdahl, T. P. and A. E. Stiven. 1983. Statistical difficulties in the analysis of predator functional response data. Can. Entomol. 115: 1365-1370.

Luz, C., M. H. Tai, A. H. Santos, L. F. Rocha, D. A. Albernaz, and H. H. Silva. 2007. Ovicidal activity of entomopathogenic hyphomycetes on *Aedes aegypti* (Diptera: Culicidae) under laboratory conditions. J Med Entomol 44: 799-804.

Luz, C., M. H. Tai, A. H. Santos, and H. H. Silva. 2008. Impact of moisture on survival of *Aedes aegypti* eggs and ovicidal activity of *Metarhizium anisopliae* under laboratory conditions. Mem Inst Oswaldo Cruz 103: 214-5.

Marina, C. F., J. E. Ibarra, J. I. Arredondo-Jiménez, I. Fernandez-Salas, P. Liedo and T. Williams. 2003. Adverse effects of covert iridovirus infection on life history and demographic parameters of *Aedes aegypti*. Entomol. Exp. Appl. 106: 53–61.

Marten, G. G., and J. W. Reid. 2007. Cyclopoid copepods. Journal of the American Mosquito Control Association 23: 65-92.

Marti, G. A., M. V. Micieli, A. C. Scorsetti, and G. Liljesthrom. 2004. Evaluation of *Mesocyclops annulatus* (Copepoda: Cyclopoidea) as a control agnet of *Aedes aegypti* (Diptera: Culicidae) in Argentina. Mem. Inst. Oswaldo Cruz 99: 535-540.

Martinez-Ibarra, J. A., Y. G. Guillen, J. I. Arredondo-Jimenez, and M. H. Rodriguez-Lopez. 2002. Indigenous fish species for the control of *Aedes aegypti* in water storage tanks in Southern Mexico. BioControl 47: 481-486.

Miranpuri, G. S., and G. G. Khachatourians. 1990. Larvicidal activity of blastospores and conidiospores of *Beauveria bassiana* (strain GK 2016) against age groups of *Aedes aegypti*. Vet Parasitol. 37: 155-62.

Mnyone, L. L., M. J. Kirby, D. W. Lwetoijera, M. W. Mpingwa, E. T. Simfukwe, B. G. J. Knols, W. Takken, and T. L. Russell. 2010. Tools for delivering entomopathogenic fungi to malaria mosquitoes: effects of delivery surfaces on fungal efficacy and persistence. Malar J 9: 246.

Moncayo, A. C., Z. Fernandez, D. Ortiz, M. Diallo, A. Sall, S. Hartman, C. T. Davis, L. Coffey, C. C. Mathiot, R. B. Tesh and S. C. Weaver. 2004. Dengue emergence and adaptation to peridomestic mosquitoes. Emerg. Infect. Dis. 10: 1790-1796.

Nam, V. S., N. T. Yen, M. Holynska, R. J. W, and B. H. Kay. 2000. National progress in dengue vector control in Vietnam: survey for *Mesocyclops* (Copepoda), *Micronecta* (Corixidae), and fish as biological control agents. American Journal of Tropical Medicine and Hygiene 62: 5-10.

Nam, V. S., N. T. Yen, T. V. Phong, T. U. Ninh, L. Q. Mai, L. V. Lo, L. T. Nghia, A. Bektas, A. Briscombe, J. G. Aaskov, P. A. Ryan, and B. H. Kay. 2005. Elimination of dengue by community programs using *Mesocyclops* (Copepoda) against *Aedes aegypti* in Central Vietnam. American Journal of Tropical Medicine and Hygiene 72: 67-73.

Neng, W., W. Shusen, H. Guangxin, X. Rongman, T. Guangkun, and Q. Chen. 1987. Control of *Aedes aegypti* larvae in hosehold water containers by Chinese cat fish. Bulletin of the World Health Organisation 65: 503-506.

Ordonez-Gonzalez, J. G., R. Mercado-Hernadez, F.-S. A. E, and I. Fernanez-Salas. 2001. The use of sticky ovitraps to estimate dispersal of *Aedes aegypti* in northeastern Mexico. Journal of the American Mosquito Control Association 17: 93-97.

Osaka, K., D. Q. Ha, Y. Sakakihara, H. B. Khiem, and T. Umenai. 1999. Control of dengue fever with active surveillance and the use of insecticidal aerosol cans. Southeast Asian J Trop Med Public Health 30: 484-8.

Pamplona, L. d. G. C., J. W. d. L. Lima, J. C. d. L. Cunha, and E. W. d. P. Santana. 2004. Evaluation of the impact on *Aedes aegypti* infestation in cement tanks of the Municipal District of Caninde, Ceara, Brazil after using *Betta splendens* fish as alternative biological control. Revista da Sociedade Brasileira de Medicina Tropical 37: 400-404.

Pamplona, L. d. G. C., C. H. Alencar, J. W. d. L. Lima, and J. Heukelbach. 2009. Reduced oviposition of *Aedes aegypti* gravid females in domestic containers with predatory fish. Tropical Medicine and International Health 14: 1347-1350.

Patil, C. D., S. V. Patil, B. K. Salunke, and R. B. Salunkhe. 2011. Bioefficacy of *Plumbago zeylanica* (Plumbaginaceae) and *Cestrum nocturnum* (Solanaceae) plant extracts against *Aedes aegypti* (Diptera: Culicidae) and nontarget fish *Poecilia reticulata*. Parasitolgy Research 108: 1253-1263.

Pavlovsky, E. N. 1962. Natural Nidality of Transmissible Diseases. Translation from the Russian edition by Plous F. K., Jr. English translation edited by Levine N. D., (editor). University of Illinois Press, Urbana and London.

Peiris JS, W. P. J. Dittus and C. B. Ratnayake. 1993. Seroepidemiology of dengue and other arboviruses in a natural population of Toque macaques (*Macaca sinica*) at Polonnaruwa, Sri Lanka. J Med Primatol *22:* 240–245.

Perich, M. J., A. Kardec, I. A. Braga, I. F. Portal, R. Burge, B. C. Zeichner, W. A. Brogdon, and R. A. Wirtz. 2003. Field evaluation of a lethal ovitrap against dengue vectors in Brazil. Medical and Veterinary Entomology 17: 205-10.

Phuanukoonnon, S., I. Mueller, and J. H. Bryan. 2005. Effectiveness of dengue control practices in household water containers in Northeast Thailand. Tropical Medicine and International Health 10: 755-763.

Price, P. W. 1970. Insect Ecology. 1st ed. John Wiley and Sons.

Quinlos F. N., L. E. Lim, A. Juliano, A. Reyes and P. Lacson. 1954. Haemorrhagic fever observed among children in the Philippines. Phillip J. Pediatrica 3:1-19.

Ramoska, W. A., S. Watts, and H. A. Watts. 1981. Effects of sand formulated *Metarhizium anisopliae* spores on larvae of three mosquito species. Mosquito News 41: 725-728.

Rangel, D. E. N., G. Braga, U. L, A. J. Anderson, and D. W. Roberts. 2005. Variability in conidial thermotolerance of *Metarhizium anisopliae* isolates from different geographical origins. J Invertebr Pathol 88: 116-125.

Regis, L., S. B. Silva and M. A. B. Melo-Santos, 2000. The Use of bacterial larvicides in mosquito and black fly control programmes in Brazil. Mem. Inst. Oswaldo Cruz 95, Suppl. I: 207-210.

Reyes-Villanueva, F., M. H. Badii, M. L. Rodriguez and M. Villarreal-Leal. 1987. Oviposition of *Toxorhynchites theobaldi* in different types of artificial containers in Mexico. J. Am. Mosq. Control Assoc. 3: 651-654.

Reyes-Villanueva, F., O. J. Gonzalez-Gaona, and M. A. Rodriguez-Perez. 2008. Larvicidal effect of medicinal plants against *Aedes aegypti* (L.) (Diptera: Culicidae) in Mexico. BioAssay 3: 7.

Reyes-Villanueva, F., J. A. Garza-Hernandez, A. M. Garcia-Munguia, P. Tamez-Guerra, A. F. V. Howard, and M. A. Rodriguez-Perez. 2011. Dissemination of *Metarhizium anisopliae* of low and high virulence by mating behavior in *Aedes aegypti*. Parasites & Vectors 4: 171.

Roche, J. C., R. Cordellier, J. P. Hervy, J. P. Digoutte and N. Monteny. 1983. Isolement de 96 souches de virus dengue 2 a partir de moustiques captures en Cote d'Ivoire et Haute-Volta. Ann. Virol. 134: 233–244.

Rodcharoen, J. and M. S. Mulla. 1994. Resistance development in *Culex quinquefasciatus* (Diptera: Culicidae) to *Bacillus sphaericus*. J. Econ. Entomol. 87: 1113-1140.

Rodriguez, A. D. y F. Reyes-Villanueva. 1992. Comportamiento sexual de *Toxorhynchites theobaldi* bajo condiciones de laboratorio. Southwest. Entomol. 17: 255-260.

Royama, T. 1971. A comparative study of models for predation and parasitism. Researches on Population Ecology S1:1–90.

Rubio, Y. and C. Ayesta. 1984. Laboratory observations on the biology of *Toxorhynchites theobaldi*. Mosq. News 44: 86-90.

Rudnick, A. and Y. C. Chan. 1965. Dengue Type 2 virus in naturally infected *Aedes albopictus* mosquitoes in Singapore. Science, 149, 638-639.

Rudnick R. A. 1965. Studies of the ecology of dengue in Malaysia: a preliminary report. J. Med. Entomol. 2: 203-208.

Russell, B. M., J. Wang, Y. Williams, M. N. Hearnden, and B. H. Kay. 2001. Laboratory evaluation of two native fishes from tropical north Queensland as biological control agents of subterranean *Aedes aegypti* Journal of the American Mosquito Control Association 17: 124-126.

Schmutterer, H. 1995. The Neem Tree. VCH, Weinheim, Germany.

Scholte, E.-J., B. G. J. Knols, R. A. Samson, and W. Takken. 2004. Entomopathogenic fungi for mosquito control: a review. J Insect Sci 4: 19.

Scholte, E.-J., K. Ng'habi, J. Kihonda, W. Takken, K. P. Paaijmans, S. Abdulla, G. F. Killeen, and B. G. J. Knols. 2005. An entomopathogenic fungus for control of adult African malaria mosquitoes. Science 308: 1641-1642.

Scholte, E.-J., W. Takken, and B. G. J. Knols. 2007. Infection of adult *Aedes aegypti* and *Ae. albopictus* mosquitoes with the entomopathogenic fungus *Metarhizium anisopliae*. Acta Tropica 102: 151-158.

Seng, C. M., T. Setha, J. Nealon, D. Socheat, N. Chantha, and M. B. Nathan. 2008. Community-based use of the larvivorous fish *Poecilia reticulata* to control the dengue vector *Aedes aegypti* in domestic water storage containers in rural Cambodia. Journal of Vector Ecology 33: 139-144.

Shaalan, E. A.-S., D. Canyon, M. W. F. Younes, H. Abdel-Waheb, and A.-H. Mansour. 2005. A review of botanical phytochemicals with mosquitocidal potential. Environ Int 31: 1149-1166.

Simas, N. K., E. d. C. Lima, R. M. Kuster, C. L. S. Lage, and A. M. d. O. Filho. 2007. Potential use of *Piper nigrum* ethanol extract against pyrethroid-resistant *Aedes aegypti* larvae. Revista da Sociedade Brasileira de Medicina Tropical 40: 405-407.

Smith, H.S., 1919. On some phases of insect control by the biological method. J. Econ. Entomol. 12: 288–292.

Solomon, M. 1949. The natural control of animal populations. Jour. Anim. Ecol. 18:1–35.

Sun, C. N., G. P. Georghiou and K. Weiss. 1980. Toxicity of *Bacillus thuringiensis* subsp. *israelensis* to mosquito larvae variously resistant to conventional insecticides. Mosq. News 40:614-618.

Steffan, W. A. and N. L. Evenhuis. 1981. Biology of Toxorhycnhites. Annu. Rev. Entomol. 26: 159-181.

Trpis, M. 1972. Development and predatory behavior of *Toxorhynchites brevipalpis* (Diptera, Culicidae) in relation to temperature. Environ. Entomol. 1: 537-546.

Trpis, M. 1973. Interaction between the predator *Toxorhynchites brevipalpis* and its prey *Aedes aegypti*. Bull. Wld. Hlth. Org. 49: 359-365.

Tully, T. P. Cassey and R. Ferriere. 2005. Functional response: rigorous estimation and sensitivity to genetic variation in prey. Oikos 111: 479-487.

Van Dam, A. R., and W. E. Walton. 2008. The effect of predatory fish exudates on the ovipositional behavior of three mosquito species: *Cules quinquefasciatus, Aedes aegypti* and *Culex tarsalis*. Medical and Veterinary Entomology 22: 399-404.

Vanlerberghe, V., M. E. Toledo, M. Rodriguez, D. Gomez, A. Baly, J. R. Benitez, and P. Van der Stuyft. 2009. Community involvement in dengue vector control: cluster randomised trial. British Medical Journal 338: b1959.

Vanlerberghe, V., E. Villegas, M. Oviedo, A. Baly, A. Lenhart, P. J. McCall, and P. Van der Stuyft. 2011. Evaluation of the effectiveness of insecticide treated materials for household level dengue vector control. PLoS Negl Trop Dis 5: e994.

Various. 1995. Biological Control Benefits and Risks. Cambridge University press.

Vezzani, D. and A. P. Albicocco. 2009. The effect of shade on the container index and pupal productivity of the mosquitoes *Aedes aegypti* and *Culex pipiens* breeding in artificial containers. Medical and Veterinary Entomology, 23: 78–84.

Wandscheer, C. B., J. E. Duque, M. A. N. da Silva, Y. Fukuyama, J. L. Wohlke, J. Adelmann, and J. D. Fontana. 2004. Larvicidal action of ethanolic extracts from fruit endocarps of *Melia azedarach* and *Azadirachta indica* against the dengue mosquito *Aedes aegypti*. Toxicon 44: 829-835.

Wang, C. H., N. T. Chang, H. H. Wu, and C. M. Ho. 2000. Integrated control of the dengue vector *Aedes aegypti* in Lui-Chui village, Ping-Tung country, Taiwan. Journal of the American Mosquito Control Association 16: 93-99.

Williams, F. M. and S. A. Juliano. 1985. Further difficulties in the analysis of functional response experiments and a resolution. Can. Entomol. 117: 631-640.

Watts, R. B. and S. M. Smith. 1978. Oogenesis in *Toxorhynchites rutilus* (Diptera: Culicidae). Can. J. Zool. 56: 136-139.

World Health Organisation. 2002. Malaria entomology and vector control. Learner's guide. WHO/CDS/CPE/SMT/2002.18.

World Health Organisation. 2004. Global strategic framework for integrated vector management. WHO/CDS/CPE/PVC/2004.10.

World Health Organisation. 2009. Dengue guidelines for diagnosis, treatment, prevention and control WHO/HTM/NTD/DEN/2009.1.

Wright, J. W., R. F. Fritz, and J. Haworth. 1972. Changing concepts of vector control in malaria eradication. Ann. Rev. Entomol. 17: 75-102.

12

Fruit Flies (Diptera: Tephritoidea): Biology, Host Plants, Natural Enemies, and the Implications to Their Natural Control

M. A. Uchôa
Laboratório de Insetos Frugívoros,
Universidade Federal da Grande Dourados
Brazil

1. Introduction

Brazil is the third world largest producer of fruits, surpassed only by China (94.4 millions of tons) and India (51.14 million tons) (Vitti, 2009). The fruit growing area in Brazil currently takes up 2.3 millions of hectares, with an annual production superior to 36.8 millions of tones. The horticulture generates six millions of direct jobs, totalizing about 27% of total labor force employed in agriculture in the Country, and makes a gross domestic product (GDP) of about US$ 11 billion. In the farms of fruit growing, in general, there are a demand for intensive and qualified labor, creating jobs and ensuring a rural Well-being of the farmers and their employees, both on small farms as on large farms. However, Brazil occupies the 17th position among world exporters of fruits (Ibraf, 2009; Vitti, 2009).

Part of Brazilian fruit production is lost in the field due the attack by larvae of different species of fruit flies (Diptera: Tephritoidea). Herein, fruit flies are referred as the guild of all specialized species with frugivorous larvae, that in South America, especially in Brazil, belong to two families: Tephritidae and Lonchaeidae (Diptera: Tephritoidea) (Uchôa & Nicácio, 2010). On the other hand, the fruit flies are interesting animals of the scientific point of view, because they have polytene chromosomes like those found in species of *Drosophila* (Drosophilidae), which are very important for genetics studies. Fruit Flies also can be easily reared in the laboratory to serve as experimental animals for research in several areas of the biological and environmental sciences (Uchôa et al., 2004).

The fruit flies belong to two families: Tephritidae and Lonchaeidae (Tephritoidea). They have great economic importance because they are considered the key pests that most adversely affect the production and marketing of fruits and vegetables around the world. The tephritids are able of inserting the ovipositor to drop their eggs into the living tissues of host plants, such as green fruit, fruit in process of maturation or ripe fruits. If females of Lonchaeidae lay their eggs inside or over the fruits, flowers, or inside terminal shoots of Euphorbiaceae is still unknown. According Lourenção et al. (1996), *Neosilba perezi* (Romero & Ruppel) is a key pest in shoots of cassava clones. Both families of fruit flies cause direct and indirect damages. The direct ones are because their eggs hatch and the larvae eat the underlying flesh of the fruits. The indirect damage is due to depreciation of the fruits in the market retailers; opening holes through which can penetrate pathogenic microorganisms or decomposers, or yet, causing the early fall of fruits attacked in the field. Some species of fruit flies are also the major bottleneck in the exports of fresh fruits and vegetables between nations. This is because the importing countries generally impose stringent quarantine barriers to the producing and exporting Countries where fruit flies do occur, fearing the entry exotic species inside the imported products in their territories (Uchôa & Nicácio, 2010; White & Elson-Harris, 1992).

Tephritidae is the most species rich family of fruit flies, with around 5,000 described species, in six subfamilies (Tachiniscinae, Blepharoneurinae, Phytalmyiinae, Trypetinae, Dacinae, and Tephritinae); about 500 genera, and probably many undescribed species worldwide. Tephritids are peculiars because they are among the few groups of dipterans strictly phytophagous, except the Tachiniscinae, which are thought be parasitoids of Lepidoptera, and at least, some species of Phytalmyiinae that feed on live or dead bamboos (Poaceae) or on trees recently fallen of other plant families. Blepharoneurinae feed in flowers, fruits, and make galls in Cucurbitaceae; Trypetinae and Dacinae feed in fruits or in seeds of a wide range of plant families, and Tephritinae eat in flowers, make gall, or are leaf-miners in a wide array of plant taxa: Aquifoliaceae, Scrophulariaceae, Verbenaceae, but mainly in flowerheads of Asteraceae (Norrbom, 2010; Uchôa & Nicácio, 2010).

The Lonchaeidae fruit flies have about 500 described species worldwide, in two subfamilies, and nine genera. Dasiopinae is represented only by *Dasiops* Rondani, and the Lonchaeinae, with the other eight remaining genera, being *Neosilba* the most studied and economically important genus in Neotropics, with 20 described species, from which 16 are reported in Brazil. The genus *Dasiops*, with about 120 described species worldwide, have few species reported in Brazil. The lonchaeids eat in flowers or fruits from different plant taxa (e. g. Asteraceae) or feed on organic matter, especially decaying plants (Macgowan & Freidberg, 2008; Uchôa & Nicácio, 2010).

The fruit fly species economically important in Brazil belong to six genera: *Anastrepha* Schiner, *Bactrocera* Macquart, *Ceratitis* McLeay, *Rhagoletis* (Loew) (Tephritidae), *Dasiops* Rondani, and *Neosilba* McAlpine (Lonchaeidae). The genera *Bactrocera* and *Ceratitis* in Brazil are represented by only one species each: *B. carambolae* Drew & Hancock, and the Mediterranean-Fruit fly, *C. capitata* (Wiedemann), both introduced in Brazil (Nicácio & Uchôa, 2011). The species of *Rhagoletis* have some economic importance in South of Brazil.

2. Fruit flies species with economic importance in South America

The genus *Anastrepha* is originally from the Neotropical Region, with a total of 252 species described worldwide to date, being 112 recorded in Brazil (Nicácio & Uchôa, 2011; Norrbom & Uchôa, 2011), where about 14 species of *Anastrepha* (Tab. 1), along with *Bactrecera carambolae*, *Ceratitis capitata* (Wiedemann) (Tephritidae), and some species of *Dasiops* and *Neosilba* (Lonchaeidae) are the main species of fruit flies with actual or potential economic importance to the Brazilian crop fruits or vegetables (Nicácio & Uchôa, 2011).

Bactrecera carambolae is native to the Indo-Australian region. It attacks at least 26 species of host fruits worldwide, most of them of commercial interest (e.g., Star Fruit, mango, sapodilla, cherry, guava, jabuticaba, rose apple, jackfruit, breadfruit, orange, tangerine, tomato, etc.). It was introduced in Northern Brazil (Oiapoque, Amapá) in 1996 from French Guiana, carried probably by airplane flights (aircraft) between Indonesia and Suriname (Oliveira et al., 2006). *B. carambolae* is a species in process of eradication from the Region North of Brazil.

The genus *Ceratitis* has 89 described species worldwide, occurring mainly in tropical Africa. In Brazil occurs only *Ceratitis capitata* which is distributed in almost all tropical and warm temperate areas in the world (Virgilio et al., 2008). *C. capitata* is originally from Africa, with abundant populations in the Mediterranean region which borders with Europe. It has been found in Brazil for the first time in 1901, in the state of São Paulo (Uchôa & Zucchi, 1999).

The genus *Rhagoletis*, with 70 described species occurs mainly in the Holarctic and Neotropical regions, being reported 21 species in the last one. *Rhagoletis* species infest mostly fruits of Juglandaceae, Rosaceae, Rutaceae, and Solanaceae. In the Brazilian territory are reported three species (*Ragoletis adusta* Foote, from the state of São Paulo, *R. ferruginea* Hendel, in Bahia, Paraná, and Santa Catarina, and *R. macquarti* (Loew), in Goiás, and Minas Gerais (Foote, 1981; Ramírez et al., 2008), but the species of *Rhagoletis* have not been considered as key pests in Brazil. On the other hand, some species in this genus are pest of fruits in Peru and Chile (Salazar et al., 2002).

Lonchaeidae is the second family of fruit flies with economic importance in South America, where some species of the genera *Dasiops* and *Neosilba* are primary pests in crop fruits. The species of *Dasiops* attack cultivated or wild passion fruit species: green or ripe fruits, or floral buds (Passifloraceae), depending on the *Dasiops* species (Norrbom & Mcalpine, 1997; Uchôa et al., 2002; Uchôa & Nicácio, 2010). The *Neosilba* species are generally polyphagous, attacking many species of fruit, native or exotic, cultivated or wild ones. The *Neosilba* species most commonly involved in the infestation of fruits and vegetables are: *N. zadolicha* Steyskal & McAlpine, *N. pendula* (Bezzi), *N. glaberrima* (Wiedemann), and *N. inesperata* Strikis & Prado. These four *Neosilba* species, plus *N. perezi*, are considered of greatest economic importance in South America because of their damage in crop fruits, vegetables, or in cassava plantations (Lourenção et al., 1996; Nicácio & Uchôa, 2011).

From the species of fruit flies pests that occurs in Central and South America, *Anastrepha obliqua* (Macquart), *Anastrepha fraterculus* (Wiedemann), and *Ceratitis capitata*, are the most polyphagous and with greater distribution in Brazil (Uchôa & Nicácio, 2010), Argentina Guillén & Sánchez (2007), Bolivia, Ovruski et al. (2009), Colombia, Canal (2010), Venezuela, Katiyar et al. (2000), and Peru, Harris & Olalquaiga (1991). Similar pattern is reported in Central America (Reyes et al., 2007), where *Anastrepha ludens* also occurs. Consequently, that that three first species are the most often involved in the colonization of fruits and vegetables sold in the market retailers. The status of these three species as pests of horticulture is motivated by three main factors: the existence of several host species, their wide distribution in the Neotropics (from Mexico to Argentina), and the direct damage that they can cause to fruits and vegetables (Uchôa & Nicácio, 2010). Populations of the Mexican fruit fly *Anastrepha ludens* occurs in North America: Mexico and USA (Florida); in Central America: Belize, Costa Rica, El Salvador, Guatemala, Honduras and Nicaragua, but it is not recorded in South America (Oliveira et al., 2006).

3. Why the control of the fruit flies is so difficult?

The control of fruit flies (including lance flies) in the South American orchards is still done mainly through of spray chemical pesticide. However, worldwide, the widespread use of chemical pesticides to protect agricultural products against insects and other arthropod pests is of increasing concern (Cancino et al., 2009), especially because of consequent environmental pollutants, and human food contamination by pesticides residues with disastrous consequences on our health and environments.

The adult female of the tephritid fruit flies (e.g. *Anastrepha* spp., *Bactrocera* spp., *Rhagoletis* spp., and *Ceratitis capitata*) are able to lay their eggs inside the fruit tissue, pouncing the skin and fruit pulp with their aculeus (ovipositor). After oviposition the wounds over the fruit surface become healed, and the eggs can mature and hatch inside the fruit tissue. The newly emerged larvae are now sheltered from the external environment, making difficult any effort with pesticides to control them.

4. Life history of *Anastrepha* species (Trypetinae: Tephritidae)

The complete life cycle of *Anastrepha fraterculus* in the field is still unknown, but under laboratory conditions (25ºC, and 70-80% RH), the life cycle from egg to the first female oviposition, occurred in about 80 days. The adult longevity in that condition was 161 days to both males and females. The eggs hatch in about 3 days, larvae is completed around 13 days, pupae emerged in about 14 days, and the female gained sexual maturation and started oviposition after 7 days from emergence (Salles, 2000). Differently from other phytophagous groups of Diptera, the adult females of several *Anastrepha* species need to feed on proteinaceous materials to maturing their eggs.

In nature or in laboratory, when the third-instar larvae of *Anastrepha* spp. are fully mature, they fall off from the fruit and dig in the soil to pupation, that occurs at depths between 2 and 5 cm (Hodgson et al. 1998). Nicácio & Uchôa (2011) found that depending on the climatic conditions (between 15-30ºC, and 60-90% RH) the emergence is faster. Under this condition, the adults can emerge, depending on the species, between 14 and 22 days after they have buried themselves in the soil to pupation.

The sexual behavior of *Anastrepha sororcula* Zucchi was studied in laboratory. This species is a key pest of guava (*Psidium guajava* L.) in Brazil. The age of sexual maturation to the males of *A. sororcula* in laboratory was completed between 7 and 18 days, at an average, 12 days after emergence. The males exhibited signaling behavior to the females, characterized by the distension of the pleural area of the abdomen, forming a small pouch on each side, and by the protrusion of a tiny membranous pouch of rectal cuticle that surrounds the anal area. During this display, the males produced rapid movements of wing vibrations, producing an audible sound. A droplet was liberated from the anal area during wing vibration movements. After attracting the females, the males accomplished a series of elaborated movements of courtship behavior (Fig. 1). On the other hand, females became sexually mature between 14 and 24 days, on average, at 19 days after emergence. The daily exhibition of sexual activities was confined almost exclusively to the period from 16:00 to 17:30h. *A. sororcula* presented a sharp protandry pattern (Facholi & Uchôa, 2006). These asynchronous developments between males and females of fruit flies may play an important evolutionary role. If males and females of the same progeny (offspring) reach sexual maturity at different times in nature, the chance of inbred mating decreases, which increases the genetic variability of the species (Nicácio & Uchôa, 2011).

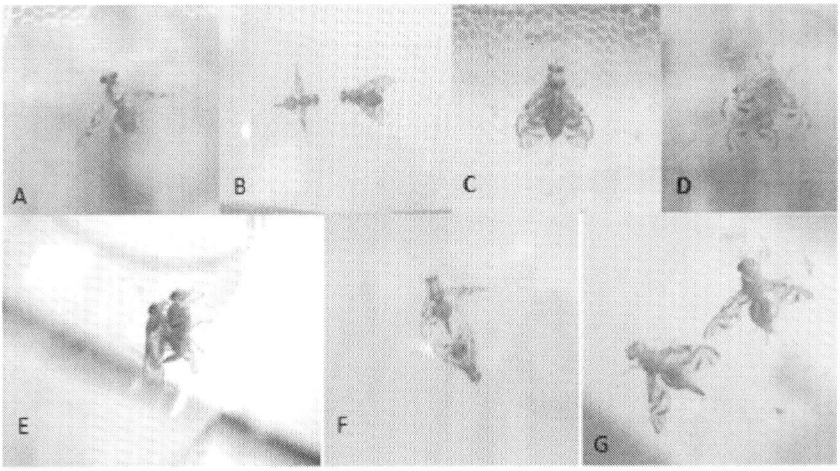

Fig. 1. Ethogram of the typical sequence of the mating behavior of *Anastrepha sororcula*: (A) Male signaling to the female with wing vibration, abdominal tip distension, and protrusion of their anal pouch; (B) the female attracted to the male approaches, and goes running to that chosen one, making alternating movements of rotation with their wings; (C) the male fly forward to mount the female, trying the copulation, or sometimes, he rises by the head of the female trying the copulation; (D) male with hind legs, raises the ovipositor of the female to connect their genitals for coupling; (E) regularly the male vibrates their body over the female's body; (F) the male goes down from female dorsum and both walk with their heads diametrically opposed for the separation of their genitals, and (G) after decoupling, both start rubbing hind legs on their terminalia (Facholi & Uchôa, 2006).

The longest fase on life cycle of *Anastrepha* species is, probably, adult. For some studied species (e. g. *A. fraterculus* and *A. sororcula*) in laboratory conditions (around 25-27 °C, 60-80% RH) they are able to live for about 180 days. Probably this trait enables the survival some species of *Anastrepha* in natural environment, enabling them to wait for the adequate stage of development of their host fruit in nature.

5. Host plants to fruit flies pests in South America

Although *Anastrepha* is the most biodiverse genus of Neotropical fruit flies, only 14 species are polyphagous, they are with a wide distribution in South America, and able to attack grown fruit and/or vegetables of commercial value. *Anastrepha pickeli* Lima has been recorded as polyphagous, because it is reported breeding in two species of different families (Uchôa et al., 2002; Zucchi, 2008). But, taking in account that the fruits of *Manihot esculenta* Crantz (Euphorbiaceae), and that of *Quararibea turbinata* (Swartz) (Bombacaceae), are not edible, *A. pickeli* is not considered a key pest (Tab. 1).

Ceratitis capitata is cosmopolitan, one of the most important key pest of fruit and vegetable crops worldwide, and certainly, the most widespread species of frugivorous tephritid around the world. This species feeds in more than 400 fruit species from 75 plant families. In Brazil, *C. capitata* is recorded in 60 species of host fruits from 22 families, of which 22 are native (Uchôa et al., 2002; Uchôa & Nicácio, 2010) (Tab. 1).

Species	Host Fruits	Plant Family	Distribution	References
**Anastrepha antunesi* Lima	*Spondias cf. macrocarpa* Engl. *Eugenia stipitata* McVaugh *Psidium guajava* L. *Spondias purpurea* L.	Anacardiaceae Myrtaceae Anacardiaceae	Brazil Peru Venezuela	Uramoto et al., 2008 Zucchi, 2008 White & Elson-Harris, 1994
**A. bahiensis* Lima	*Psidium guajava* L. *Myrciaria cauliflora* (Mart.) *Brosimum potabile* Ducke *Helicostylis tomentosa* (Poep. et Endl.) *Rollinia* aff. *sericea* (Fries) *Ampelocera edentula* Kuhlm.	Myrtaceae Moraceae Annonaceae Ulmaceae	Brazil Colombia Brazil	Zucchi, 2008 White & Elson-Harris, 1994 Uramoto et al., 2008 Costa et al., 2009
**A. bistrigata* Bezzi	*Pouteria gardneriana* (D.C.) *Psidium australe* Cambess. *Psidium guajava* L.	Sapotaceae Myrtaceae	Brazil	Zucchi, 2008
***A. fraterculus* (Wiedemann)	*Rollinia laurifolia* Schltdl. *Myrcianthes pungens* (Berg.) *Psidium guajava* L. *P. kenedianum* Morong *Syzygium jambos* (L.) + 81 Host fruits in Zucchi (2008)	Annonaceae Myrtaceae +18 Plant Families in Zucchi (2008)	Brazil Argentina Bolivia Colombia Ecuador Guyana Paraguay Peru Suriname Uruguay Venezuela	Uramoto et al., 2008 Ovruski et al., 2003 White & Elson-Harris, 1994 Zucchi, 2008 Uchôa & Nicácio, 2010 Castañeda et al., 2010
***A. grandis* (Mcquart)	*Citrullus lanatus* (Thunb.) *Cucumis sativus* L. *Cucurbita maxima* Duchesne *Cucurbita moschata* Duchesne *Cucurbita pepo* L.	Cucurbitaceae	Argentina Bolivia Brazil Colombia Ecuador Paraguay Peru Venezuela	White & Elson-Harris, 1994 Uchôa et al., 2002 Zucchi, 2008 Castañeda et al., 2010
**A. leptozona* Hendel	*Anacardium occidentale* L. *Alibertia* sp. *Pouteria torta* (Martius) *Pouteria cainito* Radlk.	Anacardiaceae Rubiaceae Sapotaceae	Bolivia Brazil Guyana Venezuela	White & Elson-Harris, 1994 Zucchi, 2008 Uchôa & Nicácio, 2010 Silva et al., 2010
**A. macrura* Hendel	*Ficus organensis* (Miq.) *Schoepfia* sp. *Pouteria lactescens* (Vell.)	Moraceae Olacaceae Sapotaceae	Argentina Brazil Ecuador Paraguay Peru Venezuela	White & Elson-Harris, 1994 Norrbom, 1998 Uchôa & Nicácio, 2010

Species	Host Fruits	Plant Family	Distribution	References
***A. obliqua* (Macquart)	*Anacardium humile* St.Hil. *Anacardium othonianum* Rizzini *Spondias cytherea* Sonn. *Psidium kennedianum* + 37 Host fruits in Zucchi (2008)	Anacardiaceae Myrtaceae + 5 Plant families in Zucchi (2008)	Argentina Brazil Bolivia Colombia Ecuador Paraguay Peru Venezuela	Zucchi, 2008 Uchôa & Nicácio, 2010 Silva et al., 2010 Castañeda et al., 2010 Katiyar et al., 2000
***A. pseudoparallela* (Loew)	*Mangifera indica* L. *Psidium guajava* *Passiflora alata* Curtis *Passiflora edulis* Sims. *Passiflora quadrangularis*	Anacardiaceae Myrtaceae Passifloraceae	Argentina Brazil Ecuador Peru	Zucchi, 2008 White & Elson-Harris, 1994
***A. serpentina* (Wiedemann)	*Spondias purpurea* L. *Mammea americana* L. *Salacia campestris* Walp. *Alibertia* sp. *Coffea canephora* L. *Ficus gomelleira* Kunth & Bouché *Achras sapota* L. *Chrysophyllum cainito* L. *Cotia* sp. *Manikara* spp. *Pouteria* spp. *Pouteria torta* *Pouteria ramiflora* (Martius) *Mimusops coriacea* (A. DC.) *Mimusopsis commersonii* (G. Don.)	Anacardiaceae Clusiaceae Hippocrateaceae Rubiaceae Moraceae Sapotaceae	Argentina Brazil Colombia Ecuador Guyana Peru Suriname Venezuel	Zucchi, 2008 White & Elson-Harris, 1994 Uramoto et al., 2008 Silva et al., 2010 Uchôa & Nicácio, 2010 Uchôa, M. A. – unpubl.
***A. sororcula* Zucchi	*Spondias purpurea* L. *Licania tomentosa* Fritsch *Terminalia catappa* L. *Casearia sylvestris* Swartz *Byrsonima orbignyana* A.Jussieu *Mouriri elliptica* Martius *Psidium cattleianum* Sabine *Psidium kennedyanum* Morong *Schoepfia* sp. *Physalis angulata* L. + 21 Host Fruits in Zucchi (2008)	Anacardiaceae Chrysobalanaceae Combretaceae Fabaceae Flacourtiaceae Oxalidaceae Malpighiaceae Melastomataceae Myrtaceae Olacaceae Oxalidaceae Rosaceae Rubiaceae Solanaceae	Brazil Colombia Ecuador Paraguay	Zucchi, 2008 Uchôa et al., 2002 Uchôa & Nicácio, 2010 Castañeda et al., 2010

Species	Host Fruits	Plant Family	Distribution	References
***A. striata* Schiner	*Spondias mombin* L. *Spondias purpurea* L. *Rolinia mucosa* Jacq. *Attalea excelsa* Martius *Chrysobalanacus icaco* *Persea americana* L. *Byrsonima crassifolia* L. Rich. *Artocarpus heterophyllus* Lam. *Campomanesia cambessedeana* O. Berg. *Eugenia stipitata* McVaugh *Psidium acutangulum* DC *Psidium australe* Cambess. *Psidium guajava* L. *Psidium guineense* SW *Citrus sinensis* L. *Passiflora edulis* *Pouteria cainito* L.	Anacardiaceae Annonaceae Araceae Chrysobalanaceae Lauraceae Malpighiaceae Moraceae Myrtaceae Rutaceae Passifloraceae Sapotaceae	Bolivia Brazil Colombia Ecuador Guyana Peru Suriname Venezuela	White & Elson-Harris, 1994 Uchôa et al., 2002 Zucchi, 2008 Uchôa & Nicácio, 2010
***A. turpiniae* Stone	*Andira cuyabensis* Benthan *Andira humilis* Martius *Psidium kennedyanum* *Psidium guajava* *Psidium guineense* *Eugenia dodoneifolia* Cambess. Syzygium jambos L. *Jacaratia heptaphylla* (Vell.) *Terminalia catappa* L. *Mangifera indica* L. *Spondias purpurea* L. *Prunus persicae* L. *Citrus sinensis*	Fabaceae Myrtaceae Caricacea Combretaceae Anacardiaceae Rosaceae Rutaceae	Brazil	Uchôa & Nicácio, 2010 Uchôa et al., 2002 Zucchi, 2008
***A. zenildae* Zucchi	*Licania tomentosa* *Terminalia catappa* *Andira cuyabensis* *Banara arguta* Briquel *Mouriri elleptica* *Sorocea sprucei saxicola* (Hassler) + 20 Host fruits in Zucchi (2008)	Chrysobalanaceae Combretaceae Fabaceae Flacourtiaceae Melastomataceae Moraceae + 6 Plant Families in Zucchi (2008)	Brazil	Uchôa & Nicácio, 2010 Uchôa et al., 2002 Zucchi, 2008

Species	Host Fruits	Plant Family	Distribution	References
***Bactrocera carambolae* Drew & Hancock	*Benincasa hispida* (Thunb.) *Cucumis sativus* L. *Cucurbita pepo* L. *Lagenaria siceraria* (Molina) *Luffa acutangula* (L.) *Luffa aegyptiaca* (Mill.) *Momordica charantia* L. *Trichosanthes cucumerina* L. *Psidium guajava* *Syzygium samarangense* (Blume) *Prunus persica* (L.) *Citrus aurantium* L. *Citrus maxima* Merr. *Manilkara zapota* (L.) *Capsicum annuum* L. *Lycopersicon esculentum* Mill.	Cucurbitaceae Myrtaceae Rosaceae Rutaceae Sapotaceae Solanaceae	Brazil Guyana Suriname	Oliveira et al., 2006
***Ceratitis capitata* (Wiedemann)	*Juglans australis* Grisebach *Hancornia speciosa* Gomez *Licania tomentosa* *Terminalia catappa* *Mouriri elliptica* *Inga laurina* *Syzygium jambos* *Chrysophyllum gonocarpum* Engler *Pouteria ramiflora* > 400 Host species worldwide (Uchôa & Nicácio 2010)	Juglandaceae Apocynaceae Chrysobalanaceae Combretaceae Melastomataceae Mimosaceae Myrtaceae Sapotaceae + 68 Plant families worldwide (Uchôa & Nicácio 2010)	Argentina Brazil Bolivia Chile Colombia Ecuador Paraguay Peru Uruguay Venezuela	Ovruski et al., 2003 White & Elson-Harris, 1994 Uchôa et al., 2002 Uchôa & Nicácio, 2010

Table 1. Species of Fruit Flies (Diptera: Tephritoidea: Tephritidae) with *potential or **real economic importance in South America.

Herein are considered species with **real economical importance those that have been historically reared from cultivated fruit species with economic value and, with *potential economical importance those that the adults are polyphagous and were reared from some genera of fruit trees in which occur species of fruit with commercial value.

The knowledge of trophic interactions between frugivorous Tephritoidea and their host plants is absolutely necessary to guide strategies for integrated management of fruit fly pests (polyphagous or oligophagous), and for the conservation of stenophagous and monophagous species in their natural environments. Currently in Brazil, from the total of 112 species of *Anastrepha* reported in our territory, are known the host plants for only 61 species (54.46%), being unknown where 51 *Anastrepha* species (45.54%) are breeding neither whom are their natural enemies (Nicácio & Uchôa, 2011).

6. Native parasitoids of *Anastrepha* species and *Ceratitis capitata*

Hymenoptera parasitoids are the most important natural enemies of pest tephritoid larvae throughout both the Neotropical and Nearctic Regions. These entomophagous insects help reduce naturally, sometimes substantially, populations of Tephritidae and Lonchaeidae pests in the field (Ovruski et al., 2009; Uchôa et al., 2003). Mass-rearing and augmentative releases of braconid parasitoids have been considered an important component of area-wide management programs for some species of fruit flies, including widespread polyphagous species of *Anastrepha* and *Ceratitis capitata* (Marinho et al., 2009; Palenchar et al., 2009).

Biological control of frugivorous tephritoid larvae with native parasitoids is a promising component of integrated pest management programs (IPM), because it is environmentally safe and works in synergy with sterile insect technique. Braconidae is the most abundant and species rich parasitoid family of fruit flies in the Neotropical Region. Species of this group also serve as bioindicators of the presence and absence of populations of their host insects (Nicácio et al. 2011).

Tritrophic interactions among wild tephritoids, their host plants and parasitoids, have been a largely neglected field of study in some regions. It could suggest possible applications for native parasitoid species upon frugivorous tephritoid key pests (Cancino et al., 2009). The autochthonous parasitoids are particularly interesting, because of their evolved interactions over extensive periods of time with their hosts (Nicácio et al., 2011), they can be effective in lowering pest populations in orchards (Cancino et al., 2009), keeping tephritoids outbreak in check without diminishing the local biodiversity, as may occur with the use of exotic natural enemies (Nicácio et al., 2011; Uchôa et al., 2003).

Nicácio et al. (2011) evaluated the incidence of parasitoids in larvae of fruit flies that infest several species of native and exotic fruit trees in the South Pantanal Region, Mato Grosso do Sul, Brazil. Ninety-two species of fruits from 36 families and 22 orders were sampled. From 11 species of host fruits, we obtained 11,197 larvae of fruit flies; being Braconidae and Figitidae the main recovered parasitoids. The Braconidae totaled 99.45%, represented by three species: *Doryctobracon areolatus* (Szépligeti), *Utetes anastrephae* (Viereck), and *Opius bellus* Gahan. The Figitidae were represented by *Lopheucoila anastrephae* (Rohwer) from puparia of *Neosilba* spp. (Lonchaeidae), infesting pods of *Inga laurina* (Swartz). *D. areolatus* was associated with two species of *Anastrepha*: *A. rhedia* Stone in *Rheedia brasilensis* Planchon & Triana, and *A. zenildae* Zucchi in *Sorocea sprucei saxicola* (Hassler) C.C. Berg. In *Ximenia americana* L., 14% of the larvae of *Anastrepha* spp. were parasitized and, *D. areolatus* reached more than 96% of total parasitism in this host fruit. The braconids were specific to Tephritidae (Tab. 2), and the Figitidae species were associated only with larvae of *Neosilba* spp. (Lonchaeidae) (Tab. 4).

Parasitism rates found in surveys in which the fruits were removed from the field and carried to laboratory condition, certainly are unreal, because the fruits were picked up from the natural environments, with possibly, some eggs, and larvae of first and second instars of the fruit flies. So, when this immature tephritoids have left the field and have arrived in the laboratory, they have had no more chance to be parasitized (Uchôa *et al.*, 2003). Another mortality factor related of parasitoid attack that is not measured by percentage of parasitism is the damage caused by the scars left by the ovipositor of parasitoid, even when ovipositions failed, and the possibility of subsequent infections by viruses, bacteria, fungi, protozoa and nematodes (Nicácio et al., 2011) on the frugivorous larvae of tephritoids. There are still no methodologies available, however, to unambiguously to evaluate these causes of mortality to immature frugivorous flies, and this is an area that will require further research. In the future is important to look for oviposition scars by parasitoids upon the third-instar larvae or puparium of dead tephritoids to establish if they are correlated or not to death of flies (Nicácio et al., 2011).

Species of Parasitoids	Species of Fruit Flies	Species of Host Fruits	Host Family	Country	References
Alysiinae *Asobara anastrephae* (Muesebek)	*Anastrepha obliqua* (Macquart) *Anastrepha bahiensis* Lima	*Spondias lutea* L.	Anacardiacaee	Brazil	Uchôa et al., 2003 Silva et al., 2010 Costa et al., 2009
Idiasta delicata Papp	*Anastrepha* sp.	*Duckeodendron cestroides* Kuhlm.	Duckeodendraceae	Brazil	Costa et al., 2009
Phaenocarpa pericarpa Wharton & Carrejo	*A. distincta* Greene	*Inga* sp.	Fabaceae	Venezuela	Trostle et al., 1999
Opiinae *Doryctobracon areolatus* (Szépligeti)	*Anastrepha amita* Zucchi *Anastrepha fraterculus* (Wiedemann) *Anastrepha leptozona* Hendel *Anastrepha serpentina* (Wiedemann) *Anastrepha obliqua* (Macquart) *Anastrepha rheedia* Stone *Anastrepha zenildae* Zucchi *Ceratitis capitata* (Wiedemann)	*Citharexylum myrianthum* Cham. *Psidium guajava* L. *Pouteria ramiflora* (Martius) *Puoteria torta* (Martius) *Spondias purpurea* *Rheedia brasiliensis* Planchon & Triana *Sorocea sprucei saxicola* (Hassler) *Mouriri elliptica* Martius	Verbenaceae Myrtaceae Sapotaceae Anacardiaceae Clusiaceae Moraceae Melastomataceae	Brazil Argentina Bolivia Brazil Brazil Brazil Brazil Brazil	Marinho et al., 2009 Ovruski et al., 2009 Nicácio et al., 2011 Nicácio et al., 2011 Alvarenga et al., 2009 Nicácio et al., 2011 Nicácio et al., 2011

Species of Parasitoids	Species of Fruit Flies	Species of Host Fruits	Host Family	Country	References
Doryctobracon brasiliensis (Szépligeti)	*Anastrepha fraterculus*	*Psidium guajava* *Eugenia uniflora* L. *Feijoa sellowiana* O. Berg. *Prunus persicae* *Prunus salicina* Lindl.	Myrtaceae Rosaceae	Argentina Brazil Bolivia Brazil	Ovruski et al., 2009 Ovruski *et al.*, 2009 Marinho et al., 2009
Doryctobracon crawfordi (Viereck)	*Anastrepha fraterculus*	*Psidium guajava* *Prunus persicae* (L.)	Myrtaceae Rosaceae	Bolivia	Ovruski et al., 2009
Doryctobracon fluminensis (Lima)	*Anastrepha pickeli* Lima 1934 *Anastrepha montei* Lima	*Manihot esculenta* Crantz	Euphorbiaceae	Brazil	Uchôa et al., 2003 Alvarenga et al., 2009
Opius bellus Gahan	*Anastrepha alveatoides* Blanchard *Anastrepha pickeli* *A. fraterculus*	*Ximenia americana* L. *Manihot esculenta* *Psidium guajava* *Prunus persicae*	Olacaceae Euphorbiaceae Myrtaceae Rosaceae	Brazil Brazil Bolivia	Nicácio *et al.*, 2011 Alvarenga et al., 2009 Ovruski et al., 2009
Utetes anastrephae (Viereck)	*Anastrepha fraterculus* *Anastrepha obliqua*	*Eugenia uniflora* *Psidium guajava* *Spondias lutea* L. *Spondias purpurea* L. *Prunus persicae* *Manihot esculenta*	Myrtaceae Anacardiaceae Rosaceae Euphorbiaceae	Argentina Bolivia Brazil Bolivia Brazil	Ovruski et al., 2009 Uchôa et al., 2003 Ovruski et al., 2009 Alvarenga et al., 2009

Table 2. Trophic interactions between koinobiont braconid parasitoids, tephritid fruit flies, and host plants in South America.

Nine native species of braconid parasitoids have been recorded in several states of Brazil, and in other South American Countries. The most promising species to study with the view to apply in biocontrol programs against fruit fly pests are *Doryctobracon areolatus*, *Utetes anastrephae* and *Opius bellus* (Tab. 2), because they are ubiquitous, frequent and abundant in several regions of South America. Going forward is important to focus in studies on their biology and behavior, in order to multiply them in laboratory for use in programs of integrated pest management in horticulture.

7. Insect predators on *Anastrepha* species and *Ceratitis capitata*

The main predators for frugivorous larvae of tephritids worldwide has been the ants: *Solenopsis geminata* (Fabricius), *Solenopsis* spp., and *Pheidole* sp. (Hymenoptera: Formicidae) (Aluja et al., 2005); the myrmeleontid *Myrmeleon brasileiensis* (Navás) (Neuroptera) (Missirian *et al.*, 2006); some species rove beetles, probably *Belonuchus* Nordmann (Coleoptera: Staphylinidae), and Carabidae (Coleoptera) (Uchôa, M. A., unpubl.). Galli & Rampazo (1996) listed the carabids *Calosoma granulatum* Perty, *Calleida* sp., and *Scarites* sp., and the staphylinids: *Belonuchus haemorrhoidalis* (Fabricius), and *Belonuchus rufipennis* (Fabricius), among the predators of *Anastrepha* spp. larvae in Brazil. Because all these predators are generalist upon larvae of *Anastrepha* species, they probably are also able of preying upon *Ceratitis capitata* larvae. Therefore, when these insects are present, it is important conserve their populations in the orchards to help in natural control of fruit flies.

8. Food attractants, parapheromones and pheromones to fruit flies

Three kinds of attractants have been proposed to catch fruit flies in traps: food lures, parapheromones, and sex pheromones. Although the McPhail traps baited with food lures are the most usually employed in the field to catch tephritids worldwide, they have low attractiveness to fruit flies, normally attracting adults only from a short distance, about 10 m far from the source, depending if the wind is blowing continuously. The most usual baits are hydrolyzed proteinaceous from soybean, corn or torula yeast. According to Aluja et al. (1989) only 30% of the flies that are attracted to near the traps with food baits are actually captured.

Some blends of synthetic dry food lures (ammonium acetate + trimethylamine hydrochloride + putrescine) have been prepared to catch *Ceratitis capitata*, *Anastrepha* and *Bactrocera* species (Leblanc et al., 2010), but like the hydrolysate proteinaceous baits, it has the inconvenient of catching nontarget insects from several Orders, such as Diptera (e.g. Calliphoridae, Tachinidae), Lepidoptera, Hymenoptera, Neuroptera, Orthoptera, and in some places, till small vertebrates such as amphibians (Uchôa, M. A., unpubl.).

The compounds called parapheromones, such as trimedlure, cuelure and methyl eugenol are efficient on capturing fruit flies. They have been applied in traps to capture species of *Ceratitis*, *Dacus* and *Bactrocera* in the field. Differently from the common food baits, like hydrolyzed proteinaceous (corn, soybean) or torula yeast, the parapheromones are considered more selective for catching fruit flies. This is an interesting trait of these chemicals due to avoid the capture of non-target insects. But, on other hand, due the fact they capture almost exclusively male specimens, they are a problem in cases when the aim of the research is to survey the diversity of fruit flies species. Because, in some taxa, the accurate identification is based mainly in females. Furthermore, they are comparatively more expensive and harder to find in the local markets than the food baits.

The pheromones are considered biochemically ideals to control fruit flies, because generally they are species-specific, environmentally safe, being non-toxic till to the target species. However, unlike other insects such as moths, beetles, and the true bugs; Tephritidae have a complex communication system, involving short range vision and acoustic signaling, beyond the chemical language (see **life history of *Anastrepha* species**). Although in Mexico has been reported the capture of *A. suspensa* females in traps baited with virgin males (Perdomo et al., 1975, 1976), in Brazil, Felix et al. (2009) found that Jackson and McPhail traps baited with food bait were significantly more attractive to females of *Anastrepha sororcula* that traps baited with fruit fly sexually mature conspecific males. The last authors did not found significant capture of *A. sororcula* females in the traps baited with conspecific virgin males releasing sex pheromone; conspecific female neither conspecific couples. So, probably, sex pheromone of *Anastrepha* fruit flies did not show high potential to be applied in field to control this group of horticultural pests. For Lonchaeidae, only food baits based on protein hydrolysates have been used. Lonchaeids are well captured into the same McPhail traps used for sampling of tephritids.

9. Life history of *Dasiops* and *Neosilba* species (Lonchaeidae)

The species of *Dasiops* (Dasiopinae) are probably stenophagous (see Aluja & Mangan, 2008), feeding mainly on flowers or fruits *Passiflora* spp. (Malpighiales: Passifloraceae) (Nicácio & Uchôa, 2011; Uchôa et al., 2002). On other hand, *Neosilba* species (Lonchaeinae) are mainly polyphagous, attacking a broad array of host plant groups in South America (Tab. 3). *Neosilba perezi* attacks the terminal buds of cassava (Euphorbiaceae), but this behavior of feeding on tissue different of fruits and flowers is uncommon for other Lonchaeidae species in South America, where the lance flies colonize fruits of both, native or exotic species (Tab. 3). Caires et al. (2009) found five species of *Neosilba* [*Neosilba bifida* Strikis & Prado, *N. certa* (Walker), *N. pendula* (Bezzi), *N. zadolicha* McAlpine & Steyskal, and *Neosilba* morphotype MSP1] feeding in fruits of a mistletoe plant, *Psittacanthus acinarius* (Martius) (as *Psittacanthus plagiophyllus* Eichler) (Santalales: Loranthaceae) in the Brazilian Pantanal.

10. Pest status of *Dasiops* and *Neosilba*

Up to date at least 34 species of Lonchaeidae that feed on live tissue of plants are reported in Americas. *Dasiops* species are probably stenophagous (Aluja & Mangan, 2008), feeding in flowers or fruits of *Passiflora* (Passifloraceae). Some of them (e.g. *D. inedulis*), are important pest in flower buds of passion fruits in South America (Peñaranda et al., 1986; Uchôa et al., 2002). By other hand, some species of the same genus have been proposed to be biocontrol agents for weed *Passiflora* introduced in Hawaii (Norrbom & McAlpine, 1997). In Brazil four *Dasiops* species are reported (*D. frieseni* Norrbom & McAlpine *D. inedulis* Stayskal, *D. longulus* Norrbom & McAlpine, and *D. ypezi* Norrbom & McAlpine). *D. inedulis* and *D. longulus* were reared from flower buds, but *D. frieseni* and *D. ypezi* were recovered from fruits (Tab. 3).

Currently 21 species of *Neosilba* McAlpine are recorded in the Neotropical Region. From this total, interestingly, only five species [*Neosilba dimidiata* (Curran) from Colombia and Trinidad, *N. fuscipennis* (Curran) from Panama, *N. longicerata* (Hennig) from Peru, *N. major* (Malloch) from Colombia, Peru and Mexico, and *N. oaxacana* McAlpine & Steyskal from Mexico], are not yet reported in Brazil. As far as we know the species of the genus *Neosilba* are highly polyphagous, attacking plant tissues, especially fruit (Tab. 3).

Species	Host's Floral Buds (FLB), Apical Buds (AB), Fruits (FRU), or Pods (PO)	Plant Family	Country	References
Dasiopinae *Dasiops alveofrons* McAlpine	*Prunus armeniaca* L. (FRU)	Rosaceae	USA	McAlpine, 1961
Dasiops brevicornis (Williston)	?	?	Jamaica	Norrbom & McAlpine, 1997
Dasiops caustonae Norrbom & McAlpine	*Passiflora molissima* (H.B.K.) (FRU)	Passifloraceae	Venezuela	Norrbom & McAlpine, 1997
Dasiops curubae Steyskal	*Passiflora molissima* (H.B.K.) (FLB)	Passifloraceae	Colombia	Steyskal, 1980
Dasiops dentatus Norrbom & McAlpine	*Passiflora ligularis* Juss. (FRU)	Passifloraceae	Peru	Norrbom & Mcalpine, 1997
Dasiops frieseni Norrbom & McAlpine	*P. alata* W. Curtis (FRU)	Passifloraceae	Brazil	Aguiar-Menezes et al., 2004
Dasiops gracilis Norrbom & McAlpine	*P. edulis* Sims (FLB and FRU)	Passifloraceae	Venezuela	Norrbom & Mcalpine, 1997
	P. ligularis Juss. (FRU)	Passifloraceae	Colombia	Norrbom & Mcalpine, 1997
	P. ligularis Juss. (FRU)	Passifloraceae	Costa Rica	Norrbom & Mcalpine, 1997
	P. pinannatistipula (Cav.) (FRU)	Passifloraceae	Colombia	Norrbom & Mcalpine, 1997
Dasiops inedulis Steyskal	*Passiflora edulis* Sims (FLB)	Passifloraceae	Brazil	Uchôa et al., 2002
	P. edulis (FLB)		Brazil	Aguiar-Menezes et al., 2004
	P. edulis (FLB)		Colombia	Chacon & Rojas, 1984
	P. edulis (FLB)		Colombia	Peñaranda et al., 1986
	P. edulis (FLB)		Panama	Steyskal, 1980
	P. lindeniana Planch. (FRU)		Venezuela	Norrbom & Mcalpine, 1997
	P. rubra L. (FRU)		Venezuela	Norrbom & Mcalpine, 1997

Species	Host's Floral Buds (FLB), Apical Buds (AB), Fruits (FRU), or Pods (PO)	Plant Family	Country	References
Dasiops longulus Norrbom & McAlpine	*Passiflora alata* (FLB)	Passifloraceae	Brazil	Aguiar-Menezes et al., 2004
	P. edulis (FRU)	Passifloraceae	Brazil	Norrbom & Mcalpine, 1997
Dasiops passifloris McAlpine	*Passiflora suberosa* L. (FRU)	Passifloraceae	USA	Steyskal, 1980
Dasiops rugifrons Hennig	*Passiflora alata* (FRU)	Passifloraceae	Venezuela	Norrbom & Mcalpine, 1997
	?	?	Peru	Korytkowski & Ojeda, 1971
Dasiops rugulosus Norrbom & McAlpine	?	?	Trinidad	Norrbom & Mcalpine, 1997
Dasiops ypezi Norrbom & McAlpine	*Passiflora ligularis* (FRU)	Passifloraceae	Colombia	Norrbom & Mcalpine, 1997
	P. edulis (FRU)	Passifloraceae	Brazil	Uchôa, M. A.- Unpubl.
Lonchaeinae *Neosilba batesi* (Curran)	*Mangifera indica* L. (FRU) *Carica papaya* L. (FRU) *Persea americana* Mill. (FRU) *Citrus sinensis* (L.) (FRU)	Anacardiaceae Caricaceae Lauraceae Rutaceae	Mexico Guatemala Colombia	McAlpine & Steyskal, 1982 Ahlmark & Steck, 1997
Neosilba bella Strikis & Prado	*Inga edulis* Martius (PO) *Inga velutina* Willd. (PO)	Fabaceae	Brazil	Strikis et al., 2011
Neosilba bifida Strikis & Prado	*Sorocea sprucei saxicola* (Hassler) (FRU)	Moraceae	Brazil	Uchôa & Nicácio, 2010
	Psittacanthus acinarius (Martius) (FRU)	Loranthaceae	Brazil	Caires et al., 2009

Species	Host's Floral Buds (FLB), Apical Buds (AB), Fruits (FRU), or Pods (PO)	Plant Family	Country	References
Neosilba certa (Walker)	*Opercunina alata* (Hamilton) (FRU)	Convovulaceae	Brazil	Uchôa & Nicácio, 2010
	Terminalia catappa L. (FRU)	Combretaceae	Brazil	Uchôa & Nicácio, 2010
	Ficus insipida Willdenow (FRU)	Moraceae	Brazil	Uchôa & Nicácio, 2010
	Syzygium jambos L. (FRU)	Myrtaceae	Brazil	Uchôa & Nicácio, 2010
	Pouteria glomerata (Miquel) (FRU)	Sapotaceae	Brazil	Uchôa & Nicácio, 2010
	Pouteria torta (Martius) (FRU)	Sapotaceae	Brazil	Uchôa & Nicácio, 2010
	Physalis angualata L. (FRU)	Solanaceae	Brazil	Uchôa & Nicácio, 2010
	Psittacanthus acinarius (Martius) (FRU)	Loranthaceae	Brazil	Caires et al., 2009
	Inga velutina Willd. (PO)	Fabaceae	Brazil	Strikis et al., 2011
	Pouteria caimito (Ruiz & Pav.) (FRU)	Sapotaceae	Brazil	Strikis et al., 2011
	Coffea arabica L. (FRU)	Rubiaceae	Brazil	Souza et al., 2005
Neosilba dimidiata (Curran)	*Annona* spp. (FRU)	Annonaceae	Colombia Trinidad	Peña & Bennett, 1995 McAlpine & Steyskal, 1982
Neosilba flavipennis (Morge)	*Brassica rapa* L. (Roots)	Brassicaceae	Peru	Urrutia & Korytkowski, unpublished
Neosilba fuscipennis (Curran)	Unknown	Unknown	Panama	McAlpine & Steyskal, 1982

Species	Host's Floral Buds (FLB), Apical Buds (AB), Fruits (FRU), or Pods (PO)	Plant Family	Country	References
Neosilba glaberrima (Wiedemann)	*Spondia dulcis* Parkinson (FRU)	Anacardiaceae	Brazil	Uchôa & Nicácio, 2010
	Annona crassiflora Martius (FRU)	Annonaceae	Brazil	Uchôa & Nicácio, 2010
	T. catappa (FRU)	Combretaceae	Brazil	Uchôa & Nicácio, 2010
	Ficus insipida (FRU)	Moraceae	Brazil	Uchôa & Nicácio, 2010
	Syzygium jambos (FRU)	Myrtaceae	Brazil	Uchôa & Nicácio, 2010
	Ximenia americana L. (FRU)	Olacaceae	Brazil	Uchôa & Nicácio, 2010
	Alibertia edulis A. Richard (FRU)	Rubiaceae	Brazil	Uchôa & Nicácio, 2010
	Genipa americana L. (FRU)	Rubiaceae	Brazil	Uchôa & Nicácio, 2010
	Coffea arabica L. (FRU)	Rubiaceae	Brazil	Souza *et al.*, 2005
	Pouteria ramiflora (Martius) (FRU)	Sapotaceae	Brazil	Uchôa & Nicácio, 2010
	Pouteria torta (Martius) (FRU)	Sapotaceae	Brazil	Uchôa & Nicácio, 2010
Neosilba inesperata Strikis & Prado	*T. catappa* (FRU)	Combretaceae	Brazil	Uchôa & Nicácio, 2010
	Opercunina alata (Hamilton) (FRU)	Convovulaceae	Brazil	Uchôa & Nicácio, 2010
	Strychnos pseudoquina St.Hilarie (FRU)	Loganiaceae	Brazil	Uchôa & Nicácio, 2010

Species	Host's Floral Buds (FLB), Apical Buds (AB), Fruits (FRU), or Pods (PO)	Plant Family	Country	References
Neosilba inesperata Strikis & Prado	*Inga laurina* (Swartz) (PO)	Fabaceae	Brazil	Uchôa & Nicácio, 2010
	Psidium cattleianum Sabine (FRU)	Myrtaceae	Brazil	Uchôa & Nicácio, 2010
	Schoepfia sp. (FRU)	Olacaceae	Brazil	Uchôa & Nicácio, 2010
	Eryobotria japonica (Thunb.) (FRU)	Rosaceae	Brazil	Strikis & Prado, 2009
	Citrus jambhiri Lush (FRU)	Rutaceae	Brazil	Uchôa & Nicácio, 2010
	Pouteria ramiflora (Martius) (FRU)	Sapotaceae	Brazil	Uchôa & Nicácio, 2010
	Physalis angulata L. (FRU)	Solanaceae	Brazil	Uchôa & Nicácio, 2010
	Solanum sisymbriifolium Lamarck (FRU)	Solanaceae	Brazil	Uchôa & Nicácio, 2010
Neosilba longicerata (Hennig)	Unknown	Unknown	Peru	McAlpine & Steyskal, 1982
Neosilba major (Malloch)	*Capsicum annuum* L. (FRU)	Solanaceae	Colombia Peru Mexico	McAlpine & Steyskal, 1982
Neosilba morphotype MSP1	*Allogoptera leucocalyx* (Drude) (FRU)	Arecaceae	Brazil	Uchôa & Nicácio, 2010
Neosilba nicrocaeruela (Malloch)	*Carica papaya* L. (FRU) *Pouteria* sp. (FRU)	Caricaceae Sapotaceae	Brazil	McAlpine & Steyskal, 1982 Strikis et al., 2011
Neosilba oaxacana McAlpine & Steyskal	?	?	Mexico	McAlpine & Steyskal, 1982
Neosilba peltae McAlpine & Steyskal	? *Passiflora edulis* Sims	? Passifloraceae	Mexico Brazil	McAlpine & Steyskal, 1982 Strikis et al., 2011

Species	Host's Floral Buds (FLB), Apical Buds (AB), Fruits (FRU), or Pods (PO)	Plant Family	Country	References
Neosilba parva (Hennig)	Unknown	Unknown	Brazil	Bittencourt et al., 2006
Neosilba pendula (Bezzi)	*Anacardium humile* Saint Hilaire (FRU) *Annona* spp. (FRU)	Anacardiaceae Annonaceae	Brazil Brazil Colombia Venezuela	Uchôa & Nicácio, 2010 Peña & Bennett , 1995
	T. catappa (FRU)	Combretaceae	Brazil	Uchôa & Nicácio, 2010
	Opercunina alata (FRU)	Convovulaceae	Brazil	Uchôa & Nicácio, 2010
	Andira cuyabensis Benthan (FRU)	Fabaceae	Brazil	Uchôa & Nicácio, 2010
	Banara arguta Briquel (FRU)	Flacourtiaceae	Brazil	Uchôa & Nicácio, 2010
	Inga laurina (Swartz) (PO)	Fabaceae	Brazil	Uchôa & Nicácio, 2010
	Ficus insipida (FRU)	Moraceae	Brazil	Uchôa & Nicácio, 2010
	Psidium cattleianum (FRU)	Myrtaceae	Brazil	Uchôa & Nicácio, 2010
	Schoepfia sp. (FRU)	Olacaceae	Brazil	Uchôa & Nicácio, 2010
	Citrus jambhiri (FRU)	Rutaceae	Brazil	Uchôa & Nicácio, 2010
	Chrysophyllum soboliferum Rizzini (FRU)	Sapotaceae	Brazil	Uchôa & Nicácio, 2010
	Pouteria ramiflora (FRU)	Sapotaceae	Brazil	Uchôa & Nicácio, 2010
	Pouteria torta (FRU)	Sapotaceae	Brazil	Uchôa & Nicácio, 2010
	Coffea arabica L. (FRU)	Rubiaceae	Brazil	Souza et al., 2005
	Psittacanthus acinarius (Martius) (FRU)	Loranthaceae	Brazil	Caires et al., 2009

Species	Host's Floral Buds (FLB), Apical Buds (AB), Fruits (FRU), or Pods (PO)	Plant Family	Country	References
Neosilba pseudopendula (Korytkowski & Ojeda)	*Coffea arabica* L. (FRU)	Rubiaceae	Brazil	Souza et al., 2005
Neosilba perezi (Romero & Ruppel)	*Manohot esculenta* Crantz (Apical Buds)	Euphorbiaceae	Brazil	Lourenção et al., 1996
Neosilba pradoi Strikis & Lerena	*Inga laurina* (Swartz) (PO)	Fabaceae	Brazil	Uchôa & Nicácio, 2010
Neosilba zadolicha McAlpine & Steyskal	*Anacardium humile* Saint Hilaire (FRU)	Anacardiaceae	Brazil Colombia	Uchôa & Nicácio, 2010 McAlpine & Steyskal, 1982
	Anacardium othonianum Rizzini (FRU)	Anacardiaceae	Brazil	Uchôa & Nicácio, 2010
	Spondia dulcis Parkinson (FRU)	Anacardiaceae	Brazil	Uchôa & Nicácio, 2010
	Annona crassiflora Martius (FRU)	Annonaceae	Brazil	Strikis et al., 2011
	Annona muricata L. (FRU)	Annonaceae	Brazil	Strikis et al., 2011
	Rollinia mucosa (Jacq.) (FRU)	Annonaceae	Brazil	Strikis et al., 2011
	Hancornia speciosa Gomez (FRU)	Apocynaceae	Brazil	Uchôa & Nicácio, 2010
	Licania tomentosa Fritsch (FRU)	Chrysobalanaceae	Brazil	Uchôa & Nicácio, 2010
	Buchenavia sp. (FRU)	Combretaceae	Brazil	Uchôa & Nicácio, 2010

Species	Host's Floral Buds (FLB), Apical Buds (AB), Fruits (FRU), or Pods (PO)	Plant Family	Country	References
Neosilba zadolicha McAlpine & Steyskal	*T. catappa* (FRU)	Combretaceae	Brazil	Uchôa & Nicácio, 2010
	Operculina alata (FRU)	Convovulaceae	Brazil	Uchôa & Nicácio, 2010
	Strychnos pseudoquina (FRU)	Loganiaceae	Brazil	Uchôa & Nicácio, 2010
	Byrsonima orbignyana A. Jussieu (FRU)	Malpighiaceae	Brazil	Uchôa & Nicácio, 2010
	Mouriri elliptica Martius (FRU)	Melastomataceae	Brazil	Uchôa & Nicácio, 2010
	Inga laurina (PO)	Fabaceae	Brazil	Uchôa & Nicácio, 2010
	Ficus insipida (FRU)	Moraceae	Brazil	Uchôa & Nicácio, 2010
	Syzygium jambos (FRU)	Myrtaceae	Brazil	Uchôa & Nicácio, 2010
	Psidium kennedyanum Morong (FRU)	Myrtaceae	Brazil	Uchôa & Nicácio, 2010
	Schoepfia sp. (FRU)	Olacaceae	Brazil	Uchôa & Nicácio, 2010
	Ximenia americana (FRU)	Olacaceae	Brazil	Uchôa & Nicácio, 2010
	Passiflora coccinea Aublet (FRU)	Passifloraceae	Brazil	Uchôa & Nicácio, 2010
	Pasiflora edulis (FRU)	Passifloraceae	Brazil	Uchôa & Nicácio, 2010
	Alibertia edulis (FRU)	Rubiaceae	Brazil	Uchôa & Nicácio, 2010
	Genipa americana (FRU)	Rubiaceae	Brazil	Uchôa & Nicácio, 2010

Species	Host's Floral Buds (FLB), Apical Buds (AB), Fruits (FRU), or Pods (PO)	Plant Family	Country	References
Neosilba zadolicha McAlpine & Steyskal	*Tocoyena formosa* (Cham. & Schlencht.) (FRU)	Rubiaceae	Brazil	Uchôa & Nicácio, 2010
	Citrus jambhiri (FRU)	Rutaceae	Brazil	Uchôa & Nicácio, 2010
	Pouteria glomerata (FRU)	Sapotaceae	Brazil	Uchôa & Nicácio, 2010
	Pouteria ramiflora (FRU)	Sapotaceae	Brazil	Uchôa & Nicácio, 2010
	Pouteria torta (FRU)	Sapotaceae	Brazil	Uchôa & Nicácio, 2010
	Physalis angulata (FRU)	Solanaceae	Brazil	Uchôa & Nicácio, 2010
	Psittacanthus acinarius (Martius) (FRU)	Loranthaceae	Brazil	Caires et al., 2009
	Quararibea quianensis Aubl. (FRU)	Bombacaceae	Brazil	Strikis et al., 2011
	Duckeodendron cestroides Kuhlm. (FRU)	Duckeodendraceae	Brazil	Strikis et al., 2011

Table 3. Species list of Lance Flies (Diptera: Tephritoidea: Lonchaeidae) with economic importance, and their host plants in the Neotropical Region.

11. Native parasitoids of Lonchaeidae species

Eight species of Eucoilinae parasitoids (Figitidae: Cynipoidea) have been associated to frugivorous larvae of *Neosilba* in Brazil. However, up to date, only four of these parasitoid species were associated to their host larvae and host plant. *Aganaspis nordlanderi* Wharton was recovered from pupae of *N. pendula* (Bezzi) whose larvae were feeding in fruits of tangerine, *Citrus reticulata* Blanco (Rutaceae). *Lopheucoila anastrephae* (Rhower) was reared from pupae of *N. batesi* (Curran), obtained as larvae in *Passiflora* fruits (Passifloraceae), and from *N. pendula* attacking orange, *Citrus sinensis* (L.) (Rutaceae). *Odontosema anastrephae* Borgmeier was recovered from larvae of *N. pendula* in fruits of *Caryocar brasiliense* Camb. (Caryocaraceae), and *Trybliographa infuscata* Gallardo, Díaz & Uchôa was recovered from *N. pendula* in orange, *Citrus sinensis* and *Caryocar brasiliense*. In all the cases the species of *Neosilba* were collected in the larval third-instars, and only one specimen of Eucoilinae emerged from each pupa (Tab. 4).

Species of Parasitoids	Species of Lonchaeids	Species of Host Fruits	Family	Country	References
Aganaspis nordlanderi Wharton	*Neosilba pendula* (Bezzi)	*Citrus reticulata* Blanco	Rutaceae	Brazil	Gallardo et al., 2000
Aganaspis pelleranoi (Bréthes)	Not associated	Not associated	Not associated	Brazil	Guimarães et al., 2003
Lopheucoila anastrephae (Rhower)	*Neosilba batesi* (Curran)	*Passiflora* sp. *Citrus sinensis* (L.).	Passifloraceae Rutaceae	Argentina Brazil Peru Venezuela	Guimarães *et al.*, 2003 Uchôa et al., 2003
Odontosema albinerve Kieffer	Not associated	Not associated	Not associated	Brazil	Guimarães & Zucchi, 2011
Odontosema anastrephae Borgmeier	*Neosilba pendula*	*Caryocar brasiliense* Camb.	Caryocaraceae	Brazil	Uchôa, M. A. - unpublished Guimarães et al., 2003
Tropideucoila rufipes Ashmead	Not associated	Not associated	Not associated	Brazil	Guimarães & Zucchi, 2011
Tropideucoila weldi Lima	Not associated	Not associated	Not associated	Brazil	Guimarães et al., 2003
Trybliographa infuscata Gallardo, Díaz & Uchôa	*Neosilba pendula*	*Caryocar brasiliense* Camb. *Citrus sinensis*	Caryocaraceae Rutaceae	Brazil	Uchôa et al., 2003 Guimarães et al., 2003

Table 4. Trophic interactions between parasitoids, lonchaeid fruit flies, and host plants in South America.

12. Current status and future perspectives on the control of fruit flies

Currently the control of fruit fly is made with chemical pesticide spraying, a concerning reality because most tropical fruits are eaten raw, making the residue over them an environmental and human health problem. In Brazil, some farmers have reduced the impact of pesticides in orchards, spraying sugar solution on certain rows of fruit trees in the orchards, where fruit flies are attracted to the food source. So, they spray insecticides in this crowd of tephritids. This practice reduces the amount of insecticides in the environment, decreasing the risk of residues in the fruits.

Several researchers in the Americas (e.g. in Brazil) are looking for powerful and specific attractants to catch fruit flies in traps. These natural chemicals can be present in the host fruits of the fruit flies. If isolated, identified and synthesized these natural attractants can be important in both cases: surveys on species diversity in natural environments, and for the management of pest species in orchards, enabling the reduction in the use of chemical insecticides. This technique in association with biological control with native parasitoids, probably, will be possible in the near future. *Doryctobracon areolatus* and *Utetes anastrephae* are good candidates for keeping population of *Anastrepha* species and *Ceratitis capitata* in low levels, making possible to produce clean fruits and vegetables.

13. Conclusions

Anastrepha is the most biodiverse and economically important genus of Tephritidae in Brazil, but from the total of 112 species reported in the Country to date, only 14 species can be considered as pest or potential pests. In Brazil two very economically important tropical species of fruit flies: *Anastrepha ludens* (Loew) and *Anastrepha suspensa* (Loew) do not occur.

In South America occur at least eight species of Braconidae parasitoids. *Doryctobracon areolatus, Utetes anastrephae,* and *Opius bellus* are the most ubiquitous and with wide distribution, being *D. areolatus* the best candidate for biological control programs of *Anastrepha* species, and maybe also, for *Ceratitis capitata*. There are not enough studies to know how *Neosilba,* and *Dasiops* species lay their eggs in the host plants: if endophytic, like the tephritids, or if the eggs are scattered in the target part of the host plants and the newly hatched larvae are able to penetrate in the plant tissue by them. The Lonchaeidae can occupy the same ecological niche occupied by the tephritids. In some host plants, the lonchaeids can be more abundant and important as pest that the tephritids, including some fruit species with economic importance, such as *Citrus* spp. (Rutaceae), *Spondias dulcis* Parkison (Anacardiaceae), and species of *Passiflora* (Passifloraceae). The Lonchaeidae have, at least, eight species of Eucoilinae (Figitidae) parasitoids in Brazil, but the biology of both groups (lonchaeids and its parasitoids) is unknown. *Lopheucoila anastrephae, Trybliographa infuscata* and *Aganaspis nordlanderi*, have been the most abundant and frequent parasitoids in larvae of third-instars of *Neosilba* species in *Citrus* orchards in Brazil.

14. Research needs

For solving some bottlenecks to enable the monitoring and control of fruit flies with non-polluting methods, the following topics are specially in need of researches: regional surveys about species diversity; prospecting for more specific attractants to use in traps; developing of artificial diets to rearing larvae of Tephritoidea to multiply their parasitoids; improvement of mass rearing methods to both: fruit flies and their parasitoids; studies on tritrophic relationship with their host plants and parasitoids; basic biology, and behavior.

15. References

Aguiar-Menezes, E.L.; Nascimento, R.J. & Menezes, E.B. 2004. Diversity of fly species (Diptera: Tephritoidea) from *Passiflora* spp. and their Hymenopterous parasitoids in two municipalities of the Southeastern Brazil. Neotropical Entomology 33 (1): 113-116.

Ahlmark, K. & Steck, G.J. 1997. A new U.S. record for a secondary fruit infester, *Neosilba batesi* (Curran) (Diptera: Lonchaeidae). Insecta Mundi 11 (2): 116.

Aluja, M.; Sivinski, J.; Rull, J. & Hodgson, P.J. 2005. Behavior and predation of fruit fly larvae (*Anastrepha* spp.) (Diptera: Tephritidae) after exiting fruit in four types of habitats in tropical Veracruz, Mexico. Environmental Entomology 34 (6): 1507-1516.

Aluja, M.; Cabrera, M.; Guillén, J.; Celedonio, H. & Ayora, F. 1989. Behavior of *Anastreopha ludens, A. obliqua* and *A. serpentina* (Diptera: Tephritidae) on a wild mango tree (*Mangifera indica*) harbouring three McPhail traps. Insect Science and its Application 10 (3): 309-318.

Aluaja, M. & Mangan, R.L. 2008. Fruit fly (Diptera: Tephritidae) host status determination: critical conceptual, methodological, and regulatory considerations. Annual Review of Entomology 53: 473-502

Alvarenga, C.D.; Matrangolo, C.A.R.; Lopes, G.N.; Silva, M.A.; Lopes, E.N.; Alves, D.A.; Nascimento, A.S. & Zucchi, R.A. 2009. Moscas-das-frutas (Diptera: Tephritidae) e seus parasitóides em plantas hospedeiras de três municípios do norte do estado de Minas Gerais. Arquivos do Instituto Biológico 76 (2): 195-204.

Bittencourt, M.A.L.; Silva, A.C.M. Bomfim, Z.V. Silva, V.E.S.; Araújo, E.L. & Strikis, P.C. 2006. Novos registros de espécies de *Neosilba* (Diptera: Lonchaeidae) na Bahia. Neotropical Entomology 35 (2): 282-283.

Caires, C.S. ; Uchôa, M.A.; Nicácio, J. & Strikis, P.C. 2009. Frugivoria de larvas de *Neosilba* McAlpine (Diptera, Lonchaeidae) sobre *Psittacanthus plagiophyllus* (Santalales, Loranthaceae) no sudoeste de Mato Grosso do Sul. Revista Brasileira de Entomologia 53 (2): 272-277.

Canal, N.A. 2010. New species and records of *Anastrepha* Schiner (Diptera: Tephritidae) from Colombia. Zootaxa 2425: 32-44.

Cancino, J., Ruiz, L., Sivinski, J., Galvez, F.O. & Aluja, M. 2009. Rearing of five hymenopterous larval-prepupal (Braconidae, Figitidae) and three pupal (Diapriidae, Chalcidoidea, Eurytomidae) native parasitoids of the genus *Anastrepha* (Diptera: Tephritidae) on irradiated *A. ludens* larvae and pupae. Biocontrol Science and Technolology 19 (S1): 193-209.

Castañeda, M.R.; Osorio, A.; Canal, N.A. & Galeano, P.E. 2010. Species, distribution and hosts of the genus *Anastrepha* Schiner in the Department of Tolima, Colombia. Agronomia Colombiana 28 (2): 265-271.

Chacon, P. & Rojas, M. 1984. Entomofuna asociada a *Passiflora mollissima, P. edulis* y *P. quadrangularis* en el Departamento del Valle del Cauca. Turrialba 34 (3): 297-311.

Costa, S.G.M.; Querino, R.B.; Ronchi-Teles, B.; Penteado-Dias, A.M.M . & Zucchi, R.A. 2009. Parasitoid diversity (Hymenoptera: Braconidae and Figitidae) on frugivorous larvae (Diptera: Tephritidae and Lonchaeidae) at AdolphoDucke Forest Reserve, Central Amazon Region, Manaus, Brazil. Brazilian Journal of Biology 69 (2): 363-37

IBRAF-Instituto Brasileiro de Frutas. 2009. www.ibreaf.org.br/estatisticas/ Exportação/ Comparativo_das_Exportações_Brasileiras_de_Frutas_frescas_2010-2009.pdf <Accessed on June 30, 2011).

Facholi, M.C.N. & Uchôa, M.A. 2006. Comportamento sexual de *Anastrepha sororcula* Zucchi (Diptera, Tephritidae) em laboratório. Revista Brasileira de Entomologia 50 (3): 406-412.

Felix, C.S.; Uchôa, M.A. & Faccenda, O. 2009. Capture of *Anastrepha sororcula* (Diptera: Tephritidae) in McPhail and Jackson traps with food attractant and virgin adults. Brazilian Archives of Biology and Technology 52 (1): 99-104.

Foote, R.H. 1981. The genus *Rhagoletis* Loew south of the United States (Diptera: Tephritidae). U. S. Department of Agriculture, Technical Bulletin, No.1607, 75p.

Gallardo, F.; Díaz, N. & Uchôa, M.A. 2000. A new species of *Trybliographa* (Hymenoptera: Figitidae, Eucoilinae) from Brazil associated with fruit infesting Lonchaeidae (Diptera). Revista de la Sociedad Entomológica Argentina 59 (1-4): 21-24.

Galli, J.C. & Rampazzo, E.F. 1996. Enemigos naturales predadores de *Anastrepha* (Diptera, Tephritidae) capturados con trampas de suelo en huertos de *Psidium guajava* L. Boletín de Sanidad Vegetal: Plagas 22 (2): 297-300.

Guillén, D. & Sánchez, R. 2007. Expansion of the national fruit fly control programme in Argentina. pp. 653-660. *In*: Vreysen, M.J.B; Robinson, A.S. & Hendrichs, J. (eds.). *Area-Wide Control of Insect Pests*. IEA. Vienna, Austria. 792p. ISBN 978-1-4020-6059-5.

Guimarães, J.A.; Gallardo, F.E.; Díaz, N.B. & Zucchi, R.A. 2003. Eucoilinae species (Hymenoptera: Cynipoidea: Figitidae) parasitoids of fruit-infesting dipterous larvae in Brazil: identity, geographical distribution and host associations. Zootaxa 278: 1-23.

Guimarães, J.A. & Zucchi, R.A. 2011. Chave de identificação de Figitidae (Eucoilinae) parasitóides de larvas frugívoras na região Amazônica. pp.104-109.*In* Silva, R.A.; Lemos, W. P. & Zucchi, R. A. (eds.). *Moscas-das-frutas na Amazônia Brasileira: Diversidade, hospedeiros e inimigos naturais*. Embrapa, Macapá. 299p. ISBN 978-85-61366-02-5

Harris, E.J. & Olalquaiga, G. 1991. Occurrence and distribution patterns of Mediterranean fruit fly (Diptera: Tephritidae) in desert areas in Chile and Peru. Environmental Entomology 20 (1): 174-178.

Hodgson, P.J.; Sivinski, J.; Quintero, G. & Aluja, M. 1998. Depth of pupation and survival of fruit fly (*Anastrepha* spp.: Tephritidae) pupae in a range of agricultural habitats. Environmental Entomology 27 (6): 1310-1314.

Katiyar, K.P., Molina, J.C. & Matheus, R. 2000. Fruit flies (Diptera: Tephritidae) infesting fruits of the genus *Psidium* (Myrtaceae) and their altitudinal distribution in Western Venezuela. Florida Entomologist 83 (4): 480-486.

Korytkowski, C.A. & Ojeda P.,D. 1971. Revision de las especies de la Familia Lonchaeidae en el Peru (Diptera: Acalyptratae). Revista Peruana de Entomologia 14 (1): 87-116.

Leblanc, L.; Vargas, R.I. & Rubinoff, D. 2010. Capture of pest fruit flies (Diptera: Tephritidae and nontarget insects in biolure and torula yeast traps in Hawaii. Envoronmental Entomology 39 (5): 1626-1630.

Lourenção, A.L.; Lorenzi, J.O. & Ambrosano, G.M.B. 1996. Comprtamento de clones de mandioca em relação a infestação por *Neosilba perezi* (Romero & Ruppell) (Diptera: Lonchaeidae). Sciencia Agricola 53 (2-3): 304-308.

Macgowan, I. & Freidberg, A. 2008.The Lonchaeidae (Diptera) of Israel, with descriptions of three new species. Israel Journal of Entomology 38 (1): 61-92.

Marinho, C.F.; Souza-Filho, M.F.; Raga, A. & Zucchi, R.A. 2009. Parasitóides (Hymenoptera: Braconidae) de moscas-das-frutas (Diptera: Tephritidae) no estado de São Paulo: plantas associadas e parasitismo. Neotropical Entomology 38 (3): 321-326.

McAlpine, J.F. 1961. A new species of *Dasiops* (Diptera: Lonchaeidae) injurious to apricots. Canadian Entomologist 93 (7): 539-544.

McAlpine, J.F. & Steyskal, G.C. 1982. A revision of *Neosilba* McAlpine with a key to the world genera of Lonchaeidae (Diptera). Canadian Entomologist 114 (2): 105-137.

Missirian, G.L.B.; Uchôa, M.A. & Fischer, E. 2006. Development of *Myrmeleon brasiliensis* (Navás) (Neuroptera, Myrmeleontidae) in laboratory with different natural diets. Revista Brasileira de Zoologia 23 (4): 1044-1050.

Nicácio, J.N. & Uchôa, M.A. 2011. Diversity of frugivorous flies (Diptera: Tephritidae and Lonchaeidae) and their relationship with host plants (Angiospermae) in environments of South Pantanal Region, Brazil. Florida Entomologist 94 (3): 443-466.

Nicácio, J.N.: Uchôa, M.A.; Faccenda, O.; Guimarães, J.A. & Marinho. C.F. 2011. Native larval parasitoids (Hymenoptera) of frugivorous Tephritoidea (Diptera) in South Pantanal Region, Brazil. Florida Entomologist 94 (3): 407-419.

Norrbom, A.L. 2010. Tephritidae (Fruit Flies, Moscas de Frutas). pp. 909-954. *In* Brown, B.V.; Borkent, A.; Cumming, J.M.; Wood, D.M.; Woodley, N.E. & Zumbado, M.A. (eds.). *Manual of Central American Diptera*. NRC-CNRC Research Press. Ottawa. 1,442p. ISSB 978-0-660-19958-0

Norrbom, A.L. & McAlpine, J.F.1997. A review of Neotropical species of *Dasiops* Rondani (Diptera: Lonchaeidae) attacking *Passiflora* (Passifloraceae). Memories of the Entomological Society of Washington 18 (1): 189-211.

Norrbom, A.L. & Uchôa, M.A. 2011. New species and records of *Anastrepha* (Diptera: Tephritidae) from Brazil. Zootaxa 2835: 61-67.

Oliveira, M.R.V., Paula-Moraes, S.V. & Lopes, F.P.P. 2006. Moscas-das-frutas (Diptera: Tephritidae) com potencial quarentenário para o Brasil. EMBRAPA, Brasília. 261p. ISBN 978-85-87697-39-4

Ovruski, S.M., Schliserman, P., & Aluja, M. 2003. Native and introduced host plants of *Anastrepha fraterculus* and *Ceratitis capitata* (Diptera: Tephritidae) in Northwestern Argentina. Journal of Economic Entomology 96 (4): 1108-1118.

Ovruski, S.M., Schliserman, P., Nunez-Campero, S.R., Orono, L.E., Bezdjian, L.B., Albornoz-Medina, P. & Van Nieuwenhove, G.A. 2009. A survey of Hymenoptereous larval-pupal parasitoids associated with *Anastrepha fraterculus* and *Ceratitis capitata* (Diptera: Tephritidae) infesting wild guava (*Psidium guajava*) and peach (*Prunus persicae*) in the Southernmost section of the Bolivian Yungas forest. Florida Entomologist 92 (2): 269-275.

Palenchar, J.; Holler, T.; Moses-Rowley, A.; Mcgovern, R. & Sivinski, J. 2009. Evaluation of irradiated caribean fruit fly (Diptera: Tephritidae) larvae for laboratory rearing of *Doryctobracon areolatus* (Hymenoptera: Braconidae). Florida Entomologist 92 (4): 535-537.

Peña, J.E. & Bennett, F.D. 1995. Arthropods associated with *Annona* spp. in the Tropics. Florida Entomologist 78 (2): 329-349.

Peñaranda, A.; Ulloa, P.C. & Hernández, M.R. 1986. Biología de la mosca de los botones florales del maracuyá *Dasiops inedulis* (Diptera: Lonchaeidae) en el Valle del Cauca. Revista Colombiana de Entomologia 12 (1): 16-22.

Perdomo, A.J.; Baranowski, R.M. & Nation, J.L. 1975. Recapture of virgin female caribbean fruit flies from traps baited with males. Florida Entomologist 58 (4): 291-295.

Perdomo, A.J.; Nation, J.L. & Baranowski, R.M. 1976. Attraction of Female and Male Caribben Fruit Flies to food-baited and male-baited traps under field conditions. Environmental Entomology 5 (6): 1208-1210.

Ramírez, C.C.; Salazar, M.; Palma, R.E.; Cordeiro, C. & Meza-Basso, L. 2008. Phylogeographical analysis of Neotropical *Rhagoletis* (Diptera: Tephritidae): did the Andes uplift contribute to current morphological differences? Neotropica Entomology, 37 (6): 651-661.

Reyes, J.; Carro, X.; Hernández, J.; Mendez, W.; Campos, C.; Esquivel, H.; Salgado, E. & Enkerlin, W. 2007. A multi-institutional approach to create fruit fly-low prevalence and fly-free areas in Central America. pp. 627-640. *In*: Vreysen, M.J. B; Robinson, A. S. & Hendrichs, J. (eds.). *Area-Wide Control of Insect Pests*. IEA. Vienna, Austria. 792p. ISBN 978-90481-7521-5.

Salazar, M. Theoduloz, C., Vega A., Poblete, F., González, E., Badilla, R.& Meza-Basso, L. 2002. PCR–RFLP identification of endemic Chilean species of *Rhagoletis* (Diptera: Tephritidae) attacking Solanaceae. Bulletin of Entomological Research 92 (4): 337-341.

Salles, L.A. 2000. Biologia e ciclo de vida de *Anastrepha fraterculus* (Wied.), pp.81-86. *In*: Malavasi, A. & Zucchi, R. A. (eds). *Moscas-das-frutas de importância econômica no Brasil: Conhecimento básico e aplicado*. Holos. Ribeirão Preto. 327p. ISNB 85-86699-13-6

Souza, S.A.S; Resende, A.L.S; Strikis, P.C; Costa, J.R; Ricci, M.S.F; Aguiar-Menezes, E.L. 2005. Infestação natural de moscas frugívoras (Diptera: Tephritoidea) em café arábica sob cultivo orgânico arborizado e a pleno sol, em Valença, RJ. Neotropical Entomology 34 (4): 639-648

Steyskal, G.C. 1980. Two-winged flies of the genus *Dasiops* (Diptera: Lonchaeidae) attacking flowers or fruits of species of *Passiflora* (passion fruit, granadilla, curuba, etc.). Proceedings of the Entomological Society of Washington 82 (2): 166-170.

Silva, J.G. , Dutra, V.S., Santos, M.S., Silva, N.M.O., Vidal, D.B., Nink, R.A., Guimarães, J.A. & Araújo, E.L. 2010. Diversity of *Anastrepha* spp. (Diptera: Tephritidae) and associated braconid parasitoids from native and exotic hosts in Southern Bahia, Brazil. Environmental Entomology 39 (5): 1457-1465.

Strikis, P.C. & Lerena, M.L.M. 2009. A new species of *Neosilba* (Diptera: Lonchaeidae) from Brazil. Iheringia Série Zoologia 99 (3): 273-275.

Strikis, P.C. & Prado, A.P. 2009. Lonchaeidae associados a frutos de nêspera, *Eryobotria japonica* (Thunb.) Lindley (Rosaceae), com a descrição de uma espécie nova de *Neosilba* (Diptera: Tephritoidea). Arquivos do Instituto Biológico 76 (1): 49-54.

Strikis, P.C., De Deus, E.G., Silva, R.A., Pereira, J.D.B., Jesus, C.R. & Massaro-Júnior, A.L. 2011. Conhecimento sobre Lonchaeidae na Amazônia brasileira. *In*: Silva, R.A., Lemos, W.P. & Zucchi, R. A. (eds.). *Moscas-das-frutas na Amazônia Brasileira: diversidade, hospedeiros e inimigos naturais*. Embrapa. Macapá, 299p. ISBN 978-85-61366-02-5

Trostle, M., Carrejo, N.S., Mercado, I. & Wharton, R.A. 1999. Two new species of *Phaenocarpa* Foerst (Hymenoptera: Alysiinae) from South America. Proceedings of the Entomological Society of Washington 101 (1): 197-207.

Uchôa, M.A.; Nicácio, J.N. & Bomfim, D.A. 2004. Biodiversidade de moscas-das-frutas (Diptera: Tephritidae) e seus hospedeiros no Brasil Central. *In*: XX Congresso Brasileiro de Entomologia, Resumos, p.155, 2004, Sociedade Entomológica do Brasil, Gramado-RS, Brazil.

Uchôa, M.A. & Nicácio, J.N. 2010. New records of Neotropical fruit flies (Tephritidae), lance flies (Lonchaeidae) (Diptera: Tephritoidea), and their host plants in the South Pantanal and adjacent areas, Brazil. Annals of the Entomological Society of America 103 (5): 723-733.

Uchôa, M.A. & Zucchi, R.A. 1999. Metodología de colecta de Tephritidade y Lonchaeidae frugívoros (Diptera: Tephritoidea) y sus parasitoides (Hymenoptera). Anais da Sociedade Entomológica do Brasil 28 (4): 601-610.

Uchôa, M.A.; Oliveira, I.; Molina, R.M.S. & Zucchi, R.A. 2002. Species diversity of frugivorous flies (Diptera: Tephritoidea) from hosts in the cerrado of the state of Mato Grosso do Sul, Brazil. Neotropical Entomology 31 (4): 515-524.

Uchôa, M.A.; Molina, R.M.S.; Oliveira, I.; Zucchi, R.A.; Canal N.A. & Díaz, N.B. 2003. Larval endoparasitoids (Hyemnoptera) of frugivorous flies (Diptera, Tephritoidea) reared from fruits of the cerrado of the State of Mato Grosso do Sul, Brazil. Revista Brasilieira de Entomologia 47 (2): 181-186.

Uramoto, K.; Martins, D.S. & Zucchi, R.A. 2008. Fruit flies (Diptera, Tephritidae) and their associations with native host plants in a remnant area of the highly endangered Atlantic Rain Forest in the state of Espírito Santo, Brazil. Bulletin of Entomological Research 98 (5): 457-466.

Virgilio, M.; Backeljau, T.; Barr, N. & De Meyer, M. 2008. Molecular evaluation of nominal species in the *Ceratitis fasciventris*, *C. anonae*, *C. rosa* complex (Diptera: Tephritidae). Molecular Phylogenetics and Evolution 48 (1): 270-280.

Vitti, A. 2009. Analise da competitividade das exportações brasileiras de frutas selecionadas no mercado internacional. Economia aplicada, ESALQ-Universidade de São Paulo, Piracicaba. MSc Thesis, 105p.

Zucchi, R.A. 2008. Fruit flies in Brazil - *Anastrepha* species their host plants and parasitoids. Available in: www.lea.esalq.usp.br/anastrepha/, updated on May 04, 2011. <Accessed on May 26, 2011>.

White, I. M. & Elson-Harris, M.M.1994. Fruit Flies of economic significance: their Identification and Bionomics. CAB International-ACIAR. Wallingford, UK. 601p. ISBN 0-851198 -790-7

13

From Chemicals to IPM Against the Mediterranean Fruit Fly *Ceratitis capitata* (Diptera, Tephritidae)

Synda Boulahia Kheder, Imen Trabelsi and Nawel Aouadi
*National Agronomic Institute of Tunisia,
Entomology Ecology Lab, Cité Mahrajène, Tunis
Tunisia*

1. Introduction

In Tunisia, the *Citrus* culture is important especially in the region of Cap-bon that is located in the North-eastern tip of Tunisia, and which is the main production area with about 15300 ha. In this region, the main source of income of approximately 25 000 rural households comes from *Citrus* farms, although most farmers have small orchards less than 5 ha (Zekri and Laajimi, 2000). The *Citrus* production reached during the campaign 2010/2011 an average of 354 000 T. The production of the oranges Maltaise that is the main variety, is about 50% of the total production. The other varieties most cultivated are the oranges Thomson, the clementines, the lemons, then the oranges Meski and Valencia late. Most of the production (80-90%) is sold in the local market, providing it in fresh fruits for up to 6 months per year. Some varieties and particularly the oranges Maltaise, are annually exported mainly to European Union with an average of about 23 000 T (Jemmazi, 2011 pers. com.).

However, the productivity of *Citrus* orchards is still below the desired level because the sub-sector is exposed to several constraints, such as the aging plantations, the climatic conditions, the availability of water and the problems of diseases and pests. Among these, the Mediterranean fruit fly is the most important pest, as *Citrus* leafminer, aphids, scales, mites and thrips recently (Jerraya, 2003; Trabelsi and Boulahia Kheder, 2010).

2. Current situation of the control of the medfly

The mediterraneean fruit fly *Ceratitis capitata* is a harmful pest of many summer fruits and *Citrus*. The control of this pest is mainly chemical by terrestrial or airlift ways. These treatments using particularly Malathion, concern an area of about 10 000 ha in the region of Cap-bon. The treatments are made by the national company "SONAPROV" specialized in treatments, following the instructions of the Ministry of Agriculture. When the medfly populations tend upwards and their level reached the thresholds of 2-3 medflies/trap/day, the treatments were made. Thus, the treatments begin from September and are repeated until November totalizing an average of 6 passages. The farmers also made several chemical treatments on *Citrus*, up to 10 times, with very toxic Organophosphates, especially Malathion + Lysatex (food attractant) and Dimethoate. The biological products such as

Spinosad, which is homologated in Tunisia against the medfly, isn't frequently used because the farmers don't master its technique of application and consider it as not very effective.

On the other hand, there is an effort to develop biological control in *Citrus* orchards, especially against the *Citrus* leafminer for which the parasitoid *Semielacher petiolatus* (Hymenoptera: Eulophidae) was imported, mass-reared and released in the orchards. The ladybugs *Novius cardinalis* and *Cryptolaemus montrouzieri* (Coleoptera: Coccinellidae) are also reared and released against the Australian mealybug *Icerya purchasi* (Hemiptera: Margarodidae).

Because of all the inconveniences of Organophosphates products on the environment, human health, auxiliaries, and resistance recently reported in Spain (Magana and al., 2008), an effort has been deployed during these last years to reduce their use and to find alternative methods to control the medfly. Among these, the mass-trapping technique used in many countries such as Spain and Greece (Miranda and al., 2001; Ros and al., 2002), was tried from 2006 in Tunisia.

3. Use of mass-trapping against the medfly

3.1 First experience: On summer fruits

With the aim of substituting the chemical applications by alternative methods without side effects, the mass trapping was first tried on summer fruits in the region of Raf Raf (North-East of Tunisia). This region is very favorable to the medfly multiplication because of its proximity to the sea and the presence of summer fruits with overlapping maturities (medlar, apricots, peaches, pears, figs etc). The idea was: if mass-trapping succeeds to control medfly in this area, it can be transposed to other cultures and regions. The principle of mass-trapping is to capture the maximum of flies in a region by means of traps baited by food attractants. The attractants commonly used were hydrolysates of proteins, as females of medfly and especially the immature, are attracted by such component to mature their eggs (Placido-Silva and al., 2005; Quilici and al., 2004). For this first experience the attractant used was the Diammonium Phosphate (DAP) well known as a fertilizer by the farmers. This product was used at a concentration of 30 g/l and was renewed each week. The traps manufactured in Tunisia, were of the type Mac Phail but entirely transparent (Fig.1). They were installed from May 17th until December 12th 2006 at a density of 40 traps/ ha in an area of approximately 2 ha composed by family orchards.

The captures were monitored by sampling 36 traps whose contents were analyzed to separate non target insects from the medflies whose were counted each week.

The total number of medflies captured reached 90 000 during all the period. The density of medflies was the most important in August with an average of 330 medflies/trap/week (Fig. 2). At the end of this month, the captures reached a maximum of 1400 medflies/trap/week, which indicates a notable catch capacity of traps baited by DAP. Moreover, the percentage of medflies females captured was approximately 70%.

On the other hand, we noticed through this first experience that high proportions of non target insects were captured. Much more, the auxiliaries such lacewings (Nevroptera: Chrysopidae) and hoverflies (Diptera: Syrphidae), were rather numerous. Indeed the numbers of these insects captured during May until November, were respectively 35 and 804 that's equivalent to 1,5 and 33,5 individuals/trap/week.

Fig. 1. Food trap manufactured in Tunisia used for the mass-trapping of the medfly.

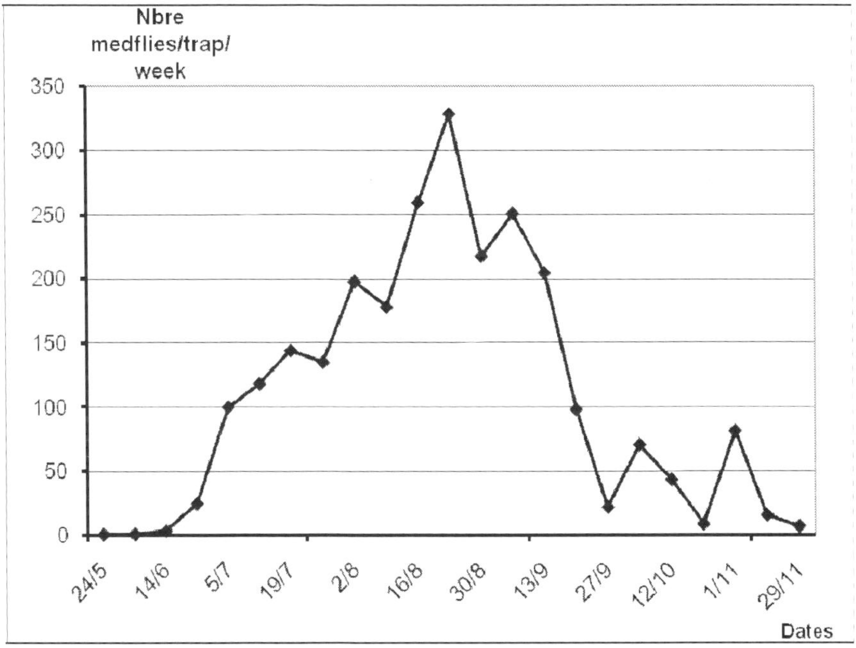

Fig. 2. Evolution of captures of medfly on summer fruits in the region of Raf Raf.

Regarding the mass-trapping effect on fruits, a survey conducted targeting the farmers (without statistical data), showed that this technique allowed a reduction of damages on figs, peaches and pears of approximately 30% compared to the previous years.

Although the disadvantage that the DAP must be renewed each week and attracts many non target insects, we considered that the results obtained with this attractant were promising; that's why we took again the essays on *Citrus* (Boulahia Kheder and Jerraya, 2010).

3.2 Next step: On citrus fruits

We chose for this second essay several orchards of clementines Cassar and oranges Thomson extending on a surface of 15 ha in the region of Takelsa in Cap-bon. For this essay the traps used were plastic Mac Phail traps baited with DAP. These traps were considered by Gazit and al., (1998), the best to capture females of *C. capitata* with food-based synthetic attractants. The traps were installed with a density of 40 traps/ha, on the 2sd of October 2006 until the 10th of January 2007. They were hung on the southeast side of the tree canopy, at 1,5-2 m above the ground.

In the middle of the "block" a plot of 1 ha containing 40 traps was considered as a sample. The traps were checked and their contents (medflies and non target insects) were counted and recorded every week.

The effectiveness of mass-trapping on the fruits was estimated by monitoring the punctures rate of clementines and oranges on 5 trees per species, 20 fruits per each tree. The fruits were marked when they were still healthy on October 17th and were monitored until the harvest to estimate the final punctures rate. At the harvest, the damages in the plot protected by mass-trapping were compared to those obtained in a control plot (without any treatments) and to those of an orchard treated according to a predetermined schedule, located few kms from experimental orchard.

The monitoring of the captures shows that the medfly has done 3 generations from October 9th to January10th, the first one in late October and the others in early and late December. The most important one was the second, with a peak of 28 medflies/trap/week, coinciding with the maturity of clementines and oranges Thomson (Fig. 3).

Regarding the impact on fruits, on clementines the punctures started at late November and reached 7% at the harvest (Fig. 3). This percentage is comparable to that obtained in an orchard received 5 chemical treatments against the medfly (8,5%) and is significantly different from the control plot with 30% of damage.

As for oranges Thomson, at mid-December their puncture rate was 20% (Fig. 3), while in the treated orchard it was 1%. In the control, it reached about 33%. Further observations showed that oranges are attacked even when they are ripe; the puncture rate with mass-trapping reached 35% at early January (Fig. 3). This behavior was probably favored by the fact that when the clementines were harvested at early December, the only host available was the oranges Thomson. This phenomenon was described by Segura et al. (2002) as the return on the host explained by the strong trend of females of the medfly to lay their eggs in the host-plant in which they made their larval development.

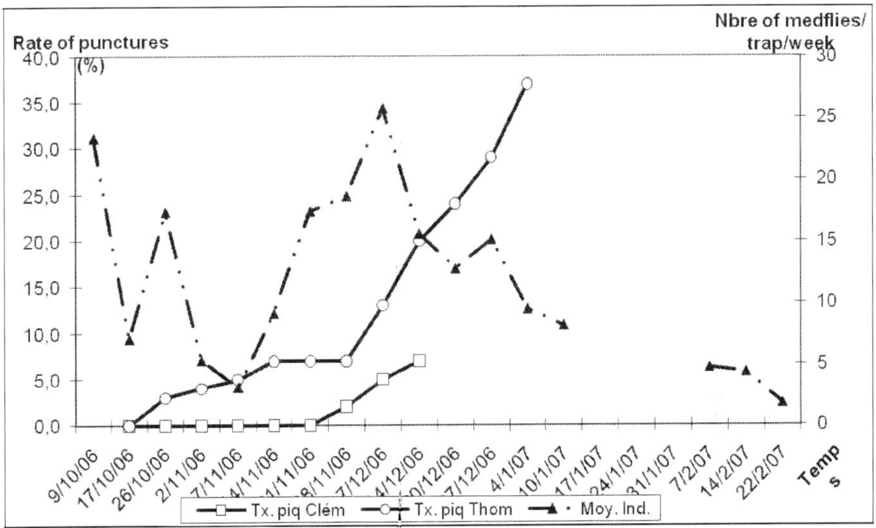

Fig. 3. Impact of mass-trapping on the evolution of catches and medfly damage on oranges Thomson and clementines in an orchard of the region of Takelsa.

In summary we can say that the mass-trapping could protect the clementines as effectively as chemical treatments, but it was insufficient for oranges Thomson whose receptive period is very long. The result obtained for clementines is close to that of Tison and al. (2003) and could be improved by an early installation of traps, before the ripening of fruits.

Comments on the selectivity of traps to auxiliaries have shown that the Mac Phail traps were more selective than transparent one, even when they are baited with the DAP, since the captures of lacewings and hoverflies were low with respectively 19 and 10 captured during 19 weeks with an average of 1 and 0,5 per week.

At the end of the second experiment we concluded that mass-trapping using DAP was able to protect adequately the clementines; but for oranges Thomson the level of medflies should be lowered further either by completing the mass-trapping by cultural practices or reasoned chemical treatments or by increasing the density of traps/ha (Boulahia Kheder and Jerraya, 2010). However this attractant has several disadvantages: low selectivity to auxiliaries particularly to the hoverflies, short-acting (7 days) and manipulation not practical. For further essays we added the first alternative in order not to increase the cost of trapping.

The following autumn, another trial was conducted in the same orchards in Takelsa region, but by installing the traps much more earlier and by using the synthetic food-based attractants Trimethylamine (TMA), Ammonium acetate (AA) and Putrescine (P) instead of DAP. These substances were the more appropriate for mass-trapping of the medfly (Miranda and al., 2001; Heath and al., 2004). These attractants formulated in separate patches, are known to be very female *C. capitata* selective, and to have a long duration of action (about 8 weeks). The traps used were plastic Mac Phail traps. They were hung at the same conditions: 1,5-2m on the south side, and clearly visible in the canopy. Three plots were considered: the first one (I) was the control, in the second (II) which surface is 1 ha, 40 traps were baited by DAP and in the third (III) in the same way, 40 traps were baited by the 3 attractants with DDVP to compare the performances of the 2 types of attractants. The DDVP was placed on a cotton dental attached to the lid of trap and the cotton was moistened every 4 to 6 weeks. The attractants patchs were changed every 8 weeks. The traps were installed on June 28th 2007 in plot III and from October 2006 in plot II.

In plots II and III the contents of traps were checked weekly for analyzis and counting. All arthropods captured were identified to assess the selectivity degree of traps and attractants. The identification of families of Diptera was made in collaboration with Dr Martinez (INRA Montpellier).

The impact of mass-trapping on fruits was estimated by monitoring changes of puncture rate of 2 clementine and 2 orange trees per plot. On each tree, 100 fruits (25/orientation) were marked when they were still free of damage and followed from October 25th each week to detect punctures until the harvest ie November 28th and end of December respectively for clementines and oranges.

While during the previous trial conducted from October to December 2006, the medfly has developed 3 overlapping generations, it has developed 2 clearly individualized generations in the same period in 2007. Indeed the first one occurred in early October and the second in December (Fig. 4). The comparison of fluctuations and population number of medflies during the fall in 2006 and 2007 that are respectively relatively warm and dry, cool and rainy, suggests that the medfly is much more influenced by the availability of host plant, than by the annual variation of temperature, these whatever their seasonal fluctuations allow the development of the medfly (Boulahia Kheder and al., 2008).

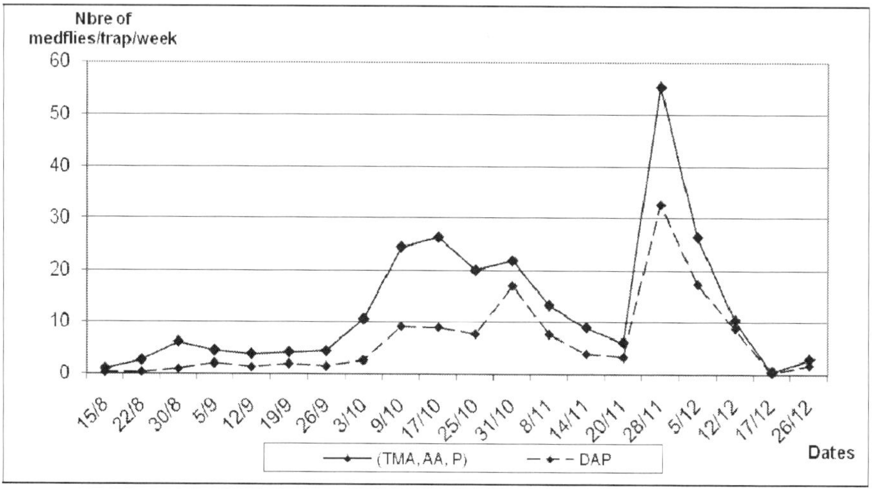

Fig. 4. Evolution of captures of medfly on oranges Thomson in an orchard of the region of Takelsa (August-December, 2007).

Regarding the protective effect of fruits by mass-trapping, it was very satisfaying for clementines with no damaged fruits at the harvest for the 2 attractants, while the percentage of punctured fruits in control was about 5%.

By cons, on oranges Thomson although the mass-trapping protected the fruits comparing to control where the damage reached about 60%, the rate of punctured fruits was rather high with respectively 28 and 35% for the (TMA, AA, P) and DAP (Fig. 5). So, the 3 attractants are more efficient than DAP. This product has captured 2-3 times less flies than (TMA, AA, P) and the proportion of females was 70 and 90% respectively for DAP and (TMA, AA, P).

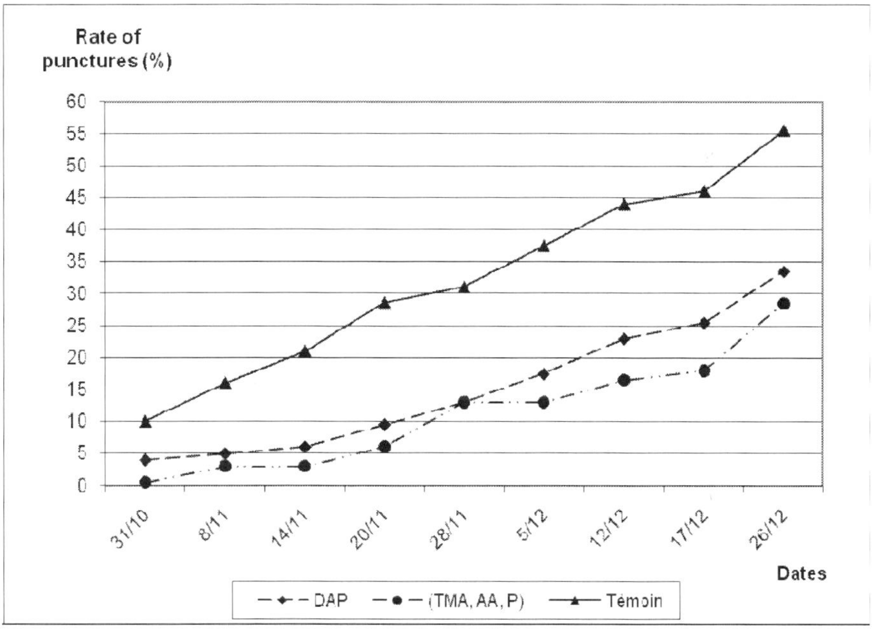

Fig. 5. Evolution of damage of medfly until the harvest with mass-trapping on the oranges Thomson (Takelsa, 2007).

Through this study we also tried to determine the selectivity of (TMA, AA, P) comparing to DAP. The identification of arthropods captured by the 2 attractants showed that the Diptera were the most represented order with 17 and 18 families respectively for DAP and (TMA, AA, P) (Table 1). In these families there are some insects that are beneficials such as Tachinidae, Syrphidae but this qualitative study dit not allow to determine precisely the selectivity of attractants (Boulahia Kheder and al., 2008). A subsequent study conducted on figs, showed that the 3 attractants were very selective towards non target insects as their captures didn't exceed 3% of total insects (Boulahia Kheder and al., 2011).

In conclusion of the 2 first years of trials we could say that mass-trapping provide a real alternative to control the medfly in *Citrus* orchards but its efficiency is variable and depends on the varieties. Indeed, for the oranges Thomson whose receptivity to medfly is very long (from early October to January) the mass-trapping alone can't protect them sufficiently and needs to be completed by other measures. Similar result was obtained by Médiouni and al. (2010) on Washington Navel oranges with 25-32% of damaged fruits at the harvest. That's why the insertion of the mass-trapping in IPM program constituted the next step of the work.

Fauna collected	Attractants	
	DAP	(TMA, AA, P)
Insects		
Diptera	Brachycera : Muscidae, Sarcophagidae, Lauxaniidae, Chloropidae, Lonchaeidae, Phoridae, Syrphidae, Trixoscelicidae, Odiniidae, Anthomyzidae, Tachinidae, Calliphoridae, Ephydridae, Sciomyzidae. Nematocera :Scatopsidae, Mycetophilidae, Psychodidae.	Brachycera : Muscidae, Sarcophagidae, Lauxaniidae, Chloropidae, Lonchaeidae, Phoridae, Syrphidae, Trixoscelicidae, Calliphoridae, Ephydridae, Pipunculidae, Drosophilidae, Stratiomyidae, Sphaeroceridae, Heleomyzidae, Therevidae. Nematocera : Scatopsidae, Bibionidae.
Coleoptera	Coccinellidae (1 specie) Staphylinidae	Coccinellidae (4 species) Staphylinidae
Hymenoptera	Especially wasps Vespidae and others Formicidae, rare parasitoïds : Braconidae, Ichneumonidae	Vespidae and others Formicidae Braconidae Chalcidoïdae
Hemiptera	Leafhoppers, bugs	Leafhoppers, bugs, aphids
Nevroptera	Chrysopidae Hemerobidae	Chrysopidae Hemerobidae
Thysanoptera	Aeolothripidae	Aeolothripidae
Orthoptera	Ensifera	Ensifera
Spiders	Several species	Several species

Table 1. Insects and spiders captured by Mc Phail traps baited with 2 types of attractants (Takelsa, Sept.-Nov., 2007).

4. IPM based on mass-trapping against the medfly

Several measures were tried between 2008 and 2010 to increase the effectiveness of mass-trapping such as the cultural practices, the chemosterilization, the applications of gibberillic acid, and chemical treatments if necessary with spinosad or other products.

To improve the performance of mass-trapping we have tried to use traps with better capacities of captures than the Mac Phail traps. So we compared these one to Moskisan, an improved version of Mac Phail: these traps have 4 inlets instead of one in Mac Phail traps. The attractants used were a new formulation of the 3 synthetic food attractants ammonium acetate (AA) (29,8%), trimethylamine (TMA) (12,4%) and putrescine (P) (0,2%) formulated in a unique patch (Unipack®). These lures have an improved duration of action that is of 4 months.

To compare Moskisan and Mac phail traps, a trial was conducted on figs from July 1st to August 12th 2009 in the region of Sidi-Thabet to compare their capacities of capture to medfly. The results obtained showed that in average the Moskisan traps captured significantly more medfly than Mac Phail ones (Table 2). Moreover, the details of captures per date shows that the number of medflies caught varies according to the population level. Indeed when this is very high, the Moskisan traps are more efficiency than Mac Phail, but when the level is lower the catching capacities of the two traps are similar (Figure 6).

These results allowed us to conclude that Moskisan traps are more efficient than Mac Phail for mass-trapping use and to choose these traps for the next experiments (Boulahia Kheder and al., 2011).

Traps	Number of meflies/trap/day
MacPhail	2.41 b*
Moskisan	4.74 a

*The means followed by different letters differ significantly.

Table 2. Average number of medflies captured by Moskisan and Mac Phail traps from July 1st to August 12th 2009.

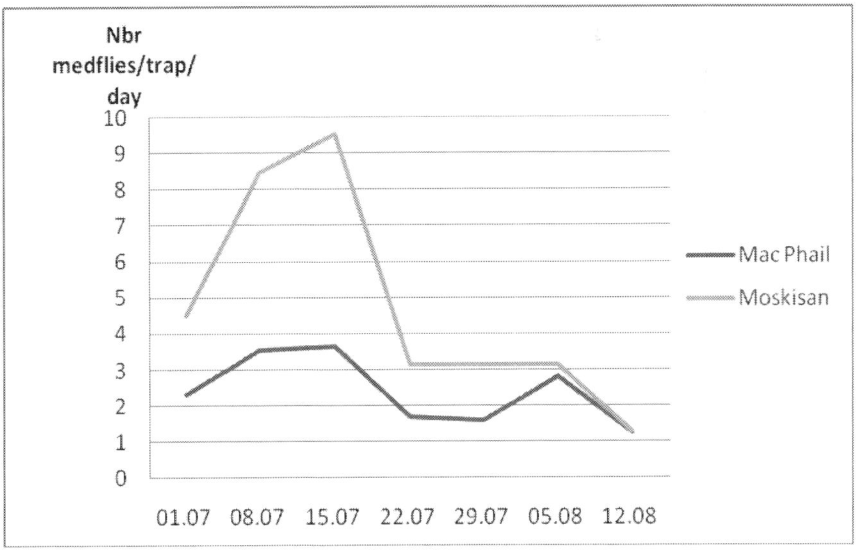

Fig. 6. Evolution of captured medflies in mass-trapping conditions according to traps baited with (AA, TMA, P).

The first combination tried was on oranges Valencia Late that's a variety also susceptible to medfly. It was the mass-trapping associated to cultural practices and to one insecticide application. This was made with deltamethrine at late February to maintain the medfly population at a low level. Cultural practices consisted of the collect of the dropped fruits each week. Mass-trapping used Moskisan traps baited with (AA, TMA, P) at a density of 40 traps/ha. The traps were installed from late January when the fruits were at early ripening. The plot conducted with this IPM program was compared to another conducted by mass-trapping alone. The effectiveness on production was evaluated by the weekly check for punctures of 200 fruits chosen randomly from 20 trees.

The application of IPM, combining mass-trapping, one chemical spraying and farming practices against the medfly, allowed a protection of harvest twice better than mass-trapping alone with a rate of punctured fruits of about 15% (Fig. 7). And we think that farming method rather than the deltamethrine spray, was the key factor in increasing the efficiency of the mass-trapping because on the totality of dropped fruits, about 52% were damaged by *C. capitata* (Table 3).

So this measure must be included in an IPM program against *C. capitata* and since, it was made in all programs. Fruit protection could also be improved by an earlier installation of the traps and by a chemical treatment rather at the end of March to prevent the medfly's populations increase (Trabelsi and Boulahia Kheder, 2011).

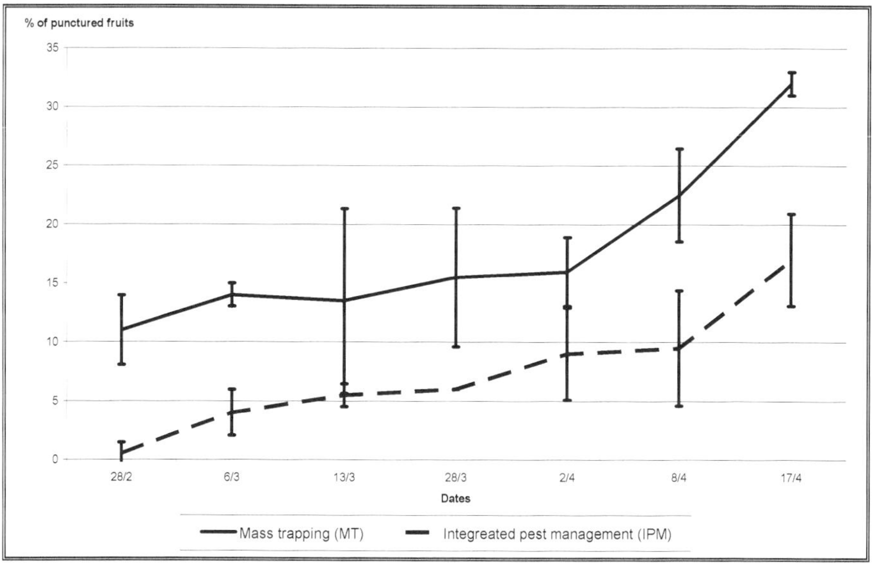

Fig. 7. Evolution of puncture rates on oranges Valencia late with mass-trapping and IPM.

Dates	Total number of dropped fruits	Damaged fruits		Healthy fruits	
		Nbre	%	Nbre	%
20/02	277	144	51,98	133	48,01
28/02	124	62	50	62	50
06/03	293	79	26,96	214	73,03
13/03	102	75	73,52	27	26,47
19/03	50	32	64	18	36
28/03	109	77	70,64	32	29,35
02/04	34	30	88,23	4	11,76
08/04	52	39	75	13	25
17/04	61	36	59,01	25	40,98
Total	1102	574	52,08	528	47,91

Table 3. Proportion of fruits damaged by the mdefly versus the total of dropped fruits.

The second program applied was the combination of mass-trapping, cultural practices, chemosterilization and 2 chemical treatments using Organophosphates products when the medfly population exceeds the threshold. The chemosterilization was used by several authors and was not sufficient to allow a good protection of *Citrus* fruits and must be applied several years successively (Bachrouch at al., 2008; Navarro-Llopis and al., 2004, 2008). The idea was to combine the mass-trapping and the chemosterilization to improve their efficiency.

The orchard chosen for this trial was the same that for the previous work, located in the region of Takelsa. The trial began very early to ensure an efficiency maximum, from September 8th 2009 when the oranges Thomson were still dark green and small size (5-6 cm in diameter) and observations were made until January 6th 2010. The trial was made in an area of about 2 ha. The chemosterilant traps (Adress ®) placed at a density of 24 traps/ha, were baited by Trimedlure, Ammonium acetate, Trimethylamine and Putrescine and contained the Lufenuron gel sterilizing the females and males of *C. capitata*. The mass-trapping used always the same traps and attractants, but half of the traps contained a disk insecticide of cypermethrine (killdisc) and the other was filled by water. The aim was to compare the efficiency of dry traps and those with water that are less expensive, to make available for the farmers the less costly system; although it is known that water traps have the disadvantage to capture more non target species including beneficial insects (Miranda and al., 2001). Twenty chemosterilant traps were placed in the center of the plot and 48 traps for mass-trapping around the perimeter. The mass-trapping was reinforced at the periphery of the plot as a barrier to prevent the intrusion of medflies as recommended by Cohen and Yuval (2000). So our hypothesis was that the chemosterilant traps should sterilize the few of medflies that succeed to penetrate in the center of the plot, leading to low damage on fruits.

The impact of control methods (mass-trapping and chemosterilization) on the harvest was assessed by the rate of punctured and dropped fruits. Five trees per treatment and in a control plot were considered in which 80 fruits were marked from October 14th and checked each 2 weeks for punctures or drop. At the harvest the number of fruits checked per treatment was increased to 800. In addition a sample of 30 punctured fruits was collected 3 times, on late November then on early and late December to compare the evolution of eggs laid in fruits belonging to mass-trapping, chemosterilization and control.

The weekly monitoring of medflies caught in traps placed in all the experimental plot, shows that the captures were the lowest in the center of the plot where the chemosterilization was applied but with no significant difference with mass-trapping.

On the production the result was similar as the lowest damage was obtained on fruits from trees treated by chemosterilization: 14,6% versus 25%, 28 and 45% respectively in mass-trapping with water, with killdisc and control (Fig. 8).

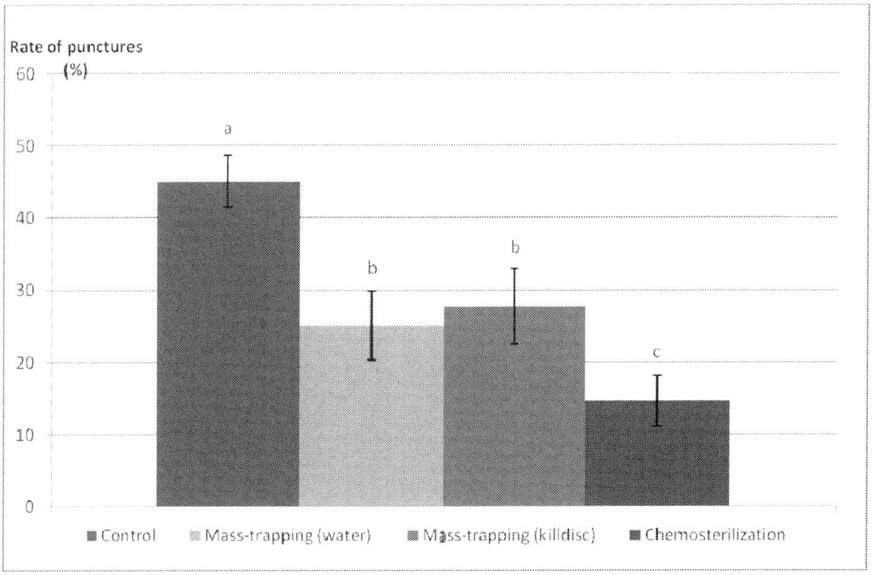

Fig. 8. Estimation of the medfly damage on oranges Thomson at the harvest with IPM based on mass-trapping and chemosterilization.

The rate of dropped fruits was very low (0,75%), compared to the other treatments (Table 4). Finally the examination of punctured fruits shows that the fruits collected from trees treated by chemosterilant traps, contained 10 to 40 larvae of medfly times lower than in mass-trapping and control respectively (Table 5). This result is consistent with that of Navarro-Llopis and al., (2004) and could be improved if this technique is maintained in the orchard for at least 4 consecutive years.

Plots	Number of fruits	%*
Control	25	6,25
Mass-trapping	8	2
Mass-trapping + killdisc	9	2,25
Chemosterilization	3	0,75

*These percentages were calculated on 400 marked fruits/plot.

Table 4. Percentages of punctured and fallen fruits per plot (Takelsa, September 2009-January 2010).

Plots / Dates	Control		Mass-trapping		Chemosterilization	
	Number of fruits	Number of pupae	Number of fruits	Number of pupae	Number of fruits	Number of pupae
November 25th	30	26	30	6	30	0
December 9th	30	15	30	3	30	0
December 23rd	30	29	30	10	30	2
Total	90	70	90	19	90	2

Table 5. Number of pupae collected from punctured fruits per plot.

Based on the results obtained, we can conclude that the program combining mass-trapping, chemosterilization, cultural practices allowed a good protection of oranges Thomson at the harvest with about 15% of punctured fruits, that represents a gain of 5-10% versus to mass-trapping alone. Moreover, most of the punctures were sterile explaining the low rate of dropped fruits.

Another IPM program tested was mass-trapping, cultural practices and 2 applications of gibberillic acid. The applications of gibberellic acid on oranges when they had the size of a golf ball, by delaying the ripening period, allowed a good protection of them from the medfly in Brazil (Malavasi and al., 2004).

Inspired by this result, we have tried to apply it in combination with mass-trapping. So, always on oranges Thomson, mass-trapping was associated with farming practices, reasoned chemical control and 2 gibberellic acid applications in an IPM program.

The traps used for mass-trapping were always of the type Moskisan® with a density of 40/ha, baited by the 3 synthetic attractants Ammonium acetate, Trimethylamine and Putrescine (Unipack®). In addition to the application made usually by farmers in the spring to improve fruit set, 2 gibberellic acid applications were made in early august and late September on small size fruits (3 and 6 cm of diameter) in order to delay their colour-break, critical stage for the attack of medfly. The dose used was 1g of gibberellic acid/hl. Chemical control was reasoned according to the medfly population level. Farming practices consisted in regular collect of dropped fruits. This program was compared to mass-trapping with traps filled with water or containing killdisc and the chemical control with 3 applications of Organophosphates products. In all plots the fallen fruits were regularly collected except in the control one.

The efficiency of the IPM program based on mass-trapping and gibberellic acid applications was evaluated by the weekly monitoring of the medfly populations level and rate of punctured fruits until the harvest.

In the plot with IPM program, the medfly populations level was the lowest, but with no significant difference between the others modalities and significantly different compared to the control plot. Moreover, since there is no difference between the 2 types of mass-trapping we considered that using traps with water is the most economically advantageous technique.

Otherwise, this IPM program reduced significantly the damage on oranges Thomson with approximately 11 % of punctured fruits at harvest against 33 % in the control field (Fig. 9).

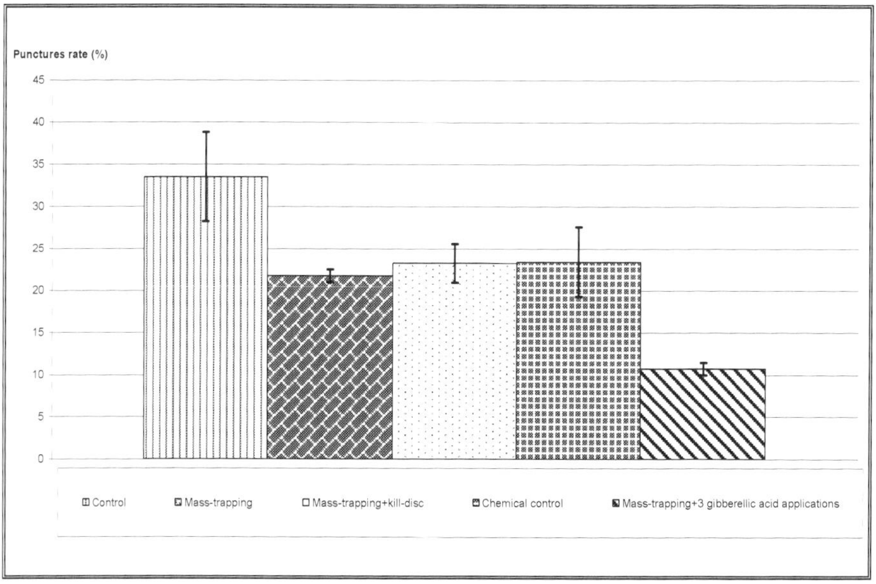

Fig. 9. Estimation of the medfly damage on oranges Thomson at the harvest with several IPM programs based on mass-trapping (Sidi Thabet, February 2010).

This program was significantly better than mass-trapping and chemical treatments in protecting fruits. The control of a sample of 400 fruits received gibberellic acid applications in the IPM plot, showed that on November 17th 60% of them were dark green while none of the fruits in the control was at this stage, but most of them were already ripe (Table 6) (Fig. 10). Thus, delaying the maturity of fruits at late December that's a cool period (average temperature < 9°C), the gibberellic acid allowed the fruits to escape the infestation of medfly because of the very low population level at this period. Indeed, from the beginning of the fruit ripening to the harvest, the punctured rate of fruits sprayed by gibberellic acid increased only about 6% against an increase of respectively 19, 21 and 32% in plots treated respectively by mass-trapping, chemical treatments and control.

Phenological stages / Treatments	Dark green fruits		Light green fruits		Ripe fruits	
	Number	%	Number	%	Number	%
IPM	241	60,25 %	159	39,75%	0	0
Control	0	0	43	10,75%	357	89,25%

* These percentages were calculated on 400 randomly selected fruits/treatment.

Table 6. Fruits phenological stages in the IPM (mass-trapping + 2 gibberellic acid applications + farming practices + reasoned chemical control) and control plots 2 months before harvest (Sidi Thabet, November 17th 2009).

a

b

Fig. 10. Effect of gibberellic acid applications on oranges Thomson (a: untreated oranges; b: oranges treated 3 times by gibberellic acid) (Sidi Thabet, November 17th).

So we can conclude that IPM based on mass-trapping and 2 gibberellic acid sprays improve the resistance of oranges Thomson to medfly by delaying 3 weeks their ripening. This is a promising result as the gibberellic acid is a natural substance without risk for human. Moreover, this substance did not affect the technological characteristics of juice (Ben Amor, 2009). The question is whether this substance does not cause adverse effects on the future production of the tree.

In conclusion of these trials, we can say that when the mass-trapping is inserted in IPM programs, its efficiency to protect fruits until the harvest, increase significantly provided that the farmers participation are involved in the operation. This is very important because the farmers must collect at least 2 times / week the fall fruits and must monitor the population level of medfly by traps to spray chemicals when the threshold is reached. So there is a need for training and supervision of the farmers to acquire the basics of IPM practices.

5. Prospects of extention of IPM in citrus orchards

Following these successful trials carried out on small areas, and the promising results obtained with IPM programs based on mass-trapping in protecting the oranges Thomson, until the harvest, the Tunisian Ministry of Agriculture decided to extend this alternative method to a larger surface. This project was conducted in the region of Takelsa in the Cap-bon (Tunisia) on about 300 ha of *Citrus*. The program applied has combined the mass-trapping to cultural practices and aerial sprays. All these operations, except the collect of fruits fall, were supported by the Ministry of Agriculture.

Moskisan traps were distributed to farmers depending on the size of their orchards. To begin trapping before the receptivity of fruits, the farmers installed the traps at a density of 40/ha around mid-August. The traps were baited by the 3 synthetic food attractants ammonium acetate (AA) (29,8%), trimethylamine (TMA) (12,4%) and putrescine (P) (0,2%) formulated in a unique patch (Biolure Unipack Suterra LLC U.S.A). These lures have a duration of action of 4 months. A killdisc of pyrethrine was added in each trap to make it easier for farmers.

The treatments were applied by aerial way using the bioinsecticide spinosad at the dose of 1 L/Ha mixed in 6 L of water. Four treatments were carried out (on August 19th, September 17th, October 7th and November 1st) following the instructions of the Ministry of Agriculture when the medfly populations tend upwards and when their level reached the thresholds of 2-3 medflies/trap/day. The treatments were carried out by means of helicopters of the SONAPROV by alternate bands of 20 meters. In some orchards inaccessible by aerial way, the farmers treated by Organophosphates such as malathion and dimethoate.

To evaluate this experiment, we considered 2 orchards, one received aerial treatments, and the other terrestrial applications, as samples.

The IPM program based on aerial treatments combined with mass-trapping was more effective to protect fruits, than that based on terrestrial treatments combined with mass-trapping. This result was confirmed by the data obtained at the harvest. Indeed, the final rate of punctured fruits obtained with aerial treatments was 2,05% on the 10th of March (24 punctured fruits over 1168 fruits observed). In the orchard with terrestrial applications, it was 8,51% (317 punctured fruits over 3341 fruits observed) on February 1st (Fig. 11).

So we can conclude that the mass-trapping supplemented by 4 aerial treatments with spinosad and the systematic collect of fallen fruits was very effective to protect the oranges Thomson against the medfly (Boulahia Kheder and al., 2011).

This result is so important that it allowed the adhesion of the farmers to mass-trapping and improve their "confidence" to the spinosad. Although homologated for several years in

Tunisia, this product is under used by the farmers against medfly because they consider it inefficient and costly.

Thus, with this first experiment where the mass-trapping was used at a relatively large scale, the surfaces treated against medfly by chemical products can be gradually reduced to the profit of the IPM programs based on mass-trapping.

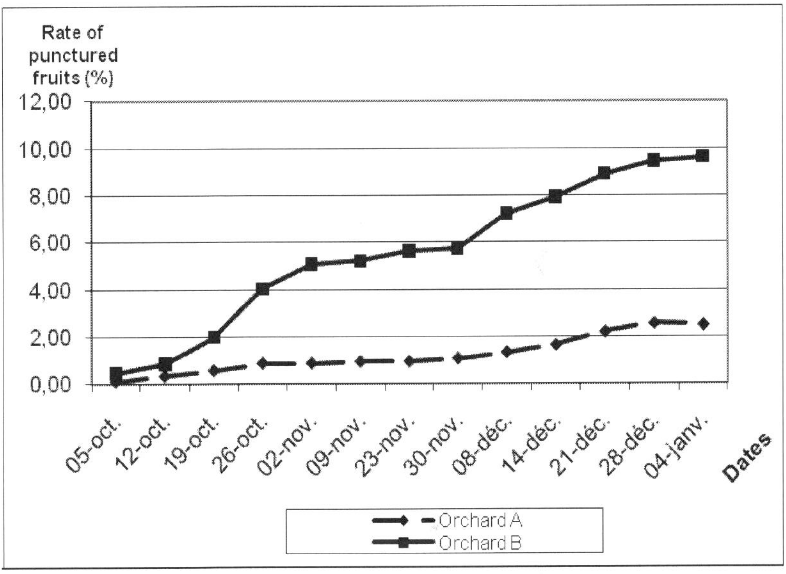

Fig. 11. Effect of the IPM programs applied on the production in 2 orchards: A (mass-trapping + aerial treatments + cultural practices) and B (mass-trapping + terrestrial treatments + cultural practices).

6. Conclusion

Several trials conducted during 5 years to promote the use of mass-trapping to replace the chemical sprays to control the medfly on *Citrus*, obtained good results when this technique was included in an IPM program. Supplemented by cultural practices, and other measures the mass-trapping protected the oranges Thomson and clementines until the harvest with a percentage of damage around 2-10%. This allowed the acceptation of this technique by the farmers and has resulted to an increase demand for traps.

The bases of IPM against the medfly in *Citrus* orchards are thus initiated; it remains to consolidate them by the training of farmers to involve them in the monitoring of population level of the medfly.

7. Acknowledgments

The authors thank all the persons who contributed in this research particularly Prof. A. Jerraya who is the proponent of mass-trapping project in Tunisia, the Ministry of

Agriculture (Direction of Control and Protection of the Quality of Agricultural Products) partner in this project for its support, and especially Mr F. Loussaïef for his collaboration, Mr. S. Rezgui for doing some statistical analysis, the students Ms L. Ben Amor and W. Salleh for their participation in the research on mass-trapping of the medfly; Mrs F. Jrad and M. Fezzani the technicians of the Entomology Ecology lab in the Agronomic Institute of Tunisia for their significant help in the orchards and in the lab; the Directors of INPFCA Sidi-Thabet, SMVDA Mraïssa, and the farmers Mrs Gabtni and Mr Fourati for allowing to conduct the experiments on their orchards.

The author also thank very much Dr. M. Martinez from the LNPV (Montpellier) for his help to identify the Diptera collected by traps in the regions of Raf Raf and Tekelsa.

This work was supported in part by the Research Unit on IPM of cultures of the National Agronomic Institute of Tunisia.

8. References

Bachrouch O.; Mediouni-Ben Jemâa J.; Alimi E.; Skillman S.; Kabadou T. & Kerber E. (2008). Efficacy of the lufenuron bait station Technique to control Mediterranean fruit fly (Medfly) *Ceratitis capitata* in *Citrus* orchards in Northern Tunisia. *Tunisian Journal of Plant Protection*, 3, 1, pp. 35-46

Ben Amor L. (2009). Effet de l'acide gibbéréllique sur l'amélioration de la qualité des oranges Thomson et leur résistance à la cératite. Master thesis, National Institute Agronomic of Tunisia, 77 p.

Boulahia Kheder S. ; Jerraya A. ; Fezzani M. & Jrad F. (2008). Le piégeage de masse : une méthode alternative de lutte contre la mouche méditerranéenne des fruits *Ceratitis capitata* (Diptera : Tephritidae) sur Agrumes. *8th International Conference on agricultural pests* CIRA organized by AFPP (Montpellier France, October, 2008)

Boulahia Kheder S. ; Jerraya A. ; Fezzani M. & Jrad F. (2010). Premiers résultats en Tunisie sur la capture de masse, moyen alternatif de lutte contre la mouche méditerranéenne des fruits *Ceratitis capitata* (Diptera, Tephritidae). *Annales de l'INRAT*, Vol. 82, pp. 168-180.

Boulahia Kheder S.; Salleh W.; Awadi N.; Fezzani M. & Jrad F. (2011). Efficiency of different traps and baits used in mass-trapping of the meditrraneean fly Ceratitis capitata Wied. (Diptera ; Tephritidae) ; *Integrated Control in Citrus Fruit Crops, IOBC/wprs Bulletin* Vol. 62, 2011, pp. 215-219.

Boulahia Kheder S. ; Loussaïef F.; Ben Hmidène A.; Trabelsi I.; Jrad F. ; Akkari Y. & Fezzani M., (2011). Evaluation of IPM programs based on mass-trapping against mediterranean fruit fly *Ceratitis capitata* (Diptera, Tephritidae) on *Citrus* in Tunisia, (to submit)

Cohen H. & Yuval B. (2000). Perimeter trapping strategy to reduce Mediterranean fruit fly (Diptera : Tephritidae) damage on different host species in Israel. *J. Econ. Entomol.*, 93,3, pp. 721-725

Gazit Y.; Rössler Y.; Epsky N.D. & Heath R.R. (1998). Trapping females of the Mediterranean fruit fly (Diptera: Tephritidae) in Isräel: Comparaison of lures and trap type. *J. Econ. Entomol.*, 91, 6, pp. 1355-1359.

Heath R.R.; Epsky N.D.; Midgarden D. & Kasoyannos B.I. (2004). Efficacy of 1, 4-Diaminobutane (putrescine) in a food-based synthetic attractant for capture of Mediterranean and Mexican fruit flies (Diptera: Tephritidae). *J. Econ. Entomol.*, 97, 3, pp. 1126-1131

Jerraya A. (2003). *Principaux nuisibles des plantes cultivées et des denrées stockées en Afrique du Nord. Leur biologie, leurs ennemis naturels, leurs dégâts et leur contrôle.* Edition Climat Pub. 415 p.

Magana C.; Hernandez-Crespo P.; Brun-Barak A.; Conso-Ferrer F.; Bride J-M.; Castanera P.; Feyereisen R. & Ortego F.(2008). Mechanisms of resistance to malathion in the medfly *Ceratitis capitata, Insect Biochemistry and Molecular Biology.* 38, pp. 756-762.

Malavasi A. ; Duarte A. L. A. ; Silva J. A. ; Greany P. D. & McDonald R. E. (2004). Effect of exogenous application of Gibberelic acid on 'Pera' oranges in Sao Paulo, Brazil: Entomological and Horticulutural aspects. *Proceedings of the International Society of Citriculture*, 3, pp. 988-990.

Mediouni-Ben Jemâa J.; Bachrouch O.; Allimi E. & Dhouibi M.H. (2010). Mass trapping based on the use of female food-attractant tri-pack as alternative for the control of the medfly *Ceratitis capitata* in *Citrus* orchards in Tunisia. *Tunisian journal of plant protection*, 5, 1, pp. 71-81

Miranda M.A. ; Alonso R. & Alemany A. (2001). Field evaluation of Medfly (Dipt., Tephritidae) female attractants in Mediterranean agrosystem (Balearic Islands, Spain). *J. Appl. Ent.*, 125, pp. 333-339

Navarro-Llopis V. ; Sanchis-Cabanas J. ; Ayala I. ; Casaña-Giner V. & Primo-Yufera E. (2004). Efficacy of lufenuron as chemosterilisant against *Ceratitis capitata* in field trials. *Pest Manag. Sci.*, 60, pp.914-920

Navarro-Llopis V.; Alfaro F. ; Dominguez J. ; Sanchis J. & Primo J. (2008). Evaluation of traps and lures for mass trapping of Mediterranean fruit fly in *Citrus* Grove. *J. Econ. Entomol.*, 101, 1, pp. 126-131

Placido-Silva M.D.C.; Zucolotto F.S. & Joachim-Bravo I. S. (2005). Influence of protein on feeding behavior of Ceratitis capitata Wiedemann (Diptera: Tephritidae): Comparison between males and females. *Neotropical Entomology* 34(4), pp. 539-545.

Quilici S. ; Nergel L. & Franck A. (2004). Influence of some physiological parameters on the response to visual stimuli and olfactory stimuli in *Ceratitis capitata* females. *Proc. Int. Soc. Citriculture*, 2004: 966-969

Ros J. P.; Gomlia J.; Reurer M.; Pons P. & Castillo E. (2002). The use of mass-trapping against Medfly (Ceratitis capitata (Wied.)) in a sustainable agriculture system on Minorca Island, Spain. *Proceedings of 6th International Fruit Fly Symposium* 6-10 May, Stellenbosch, South Africa, pp. 361-364.

Segura D. F., Vera M. T. & Cladera J. L. (2002). Host utilization by the Mediterranean fruit fly, Ceratitis capitata (Diptera : Tephritidae). *Proceedings of 6th International Fruit Fly Symposium*, 6-10 May Stellenbosch. South Africa, pp. 83-90.

Tison G. ; Paolacci G. & Martin P. (2003). Evaluation de systèmes de piégeage pour les mouches des fruits (Ceratitis capitata Wied). Rapport d'expérimentation, 9 p.

Trabelsi I. & Boulahia Kheder S. (2010). Sur la présence en Tunisie du thrips des agrumes Pezothrips kellyanus (Thysanoptera : Thripidae). *Annales de l'INRAT*, Vol. 82, pp. 181-186.

Trabelsi I. & Boulahia Kheder S. (2011). The use of mass-trapping technique in an integrated pest management program against the Mediterranean fruit fly Ceratitis capitata (Diptera: Tephritidae). *Integrated Control in Citrus Fruit Crops, IOBC/wprs Bulletin* Vol. 62, 2011, pp. 183-188.

Zekri S. & Laajimi A. (2000). Etude de la compétitivité du sous-secteur agrumicole en Tunisie. Available from Internet <http://ressources.ciheam.org/om/pdf/C57/01600238. pdf>

Part 2

Integrated Pest Management – Current Applications

14

Bark Beetles Control in Forests of Northern Spain

Arturo Goldazarena[1], Pedro Romón[2] and Sergio López[1]
[1]Neiker-Basque Institute of Agricultural Research and Development
[2]FABI- Forestry and Agricultural Biotechnology Institute
[1]Spain
[2]South Africa

1. Introduction

Bark beetles (Coleoptera: Curculionidae: Scolytinae) are an insect group that contains at least 6000 species from 181 genera around all the world (Wood, 1982). It is well known that some species are among the most destructive insects of coniferous forests representing a continuous threat (Ayres & Lombardero, 2000). The knowledge of these insects in Spain is very far away of desirable. Some studies have advanced in the knowledge of the taxonomic and faunistic composition of diverse forest types (Plaza & Gil, 1982; Plaza, 1983; Gil & Pajares, 1986; Pajares, 1987; González, 1990; Lombardero, 1994; Lombardero & Novoa, 1994; Riba, 1994; Fernández, 1996; López et al., 2007a), but the understanding of their dynamics and interrelationship biotic factors is very low. The main objective of the present work is to present a synthesis of the research works that have been performed so far in Spanish mainland, particularly in its northern plateau. First, we will introduce the situation of the conifer species *Pinus radiata* D.Don, the most widely species planted in this geographic zone, and its relationship with both bark beetles and associated phytopathogenic fungi (mainly ophiostomatoid fungi and *Fusarium circinatum*). Secondly, a brief and overall approach to the chemical ecology of bark beetles is given and finally there will be exposed some particular cases of species considered as serious threats for spanish forestry, focusing on studies aimed to the development of integrated pest management strategies.

2. The status of *Pinus radiata* D.Don in Northern Spain and its association with bark beetles (Coleoptera: Scolytinae) and ophiostomatoid fungi (Sordariomycetes: Ophiostomatales) and *Fusarium circinatum* (Sordariomycetes: Hypocreales)

The commonly known as Monterey pine (*Pinus radiata*), arising from Mexico, was introduced in Spain during the first half of the XIX century, initiating its plantation in the Basque Country. The first introduction was carried out by Mr. Carlos Adán de Yarza in his botanical garden of Zubieta (Lekeitio, Vizcaya), planting the first stands in 1840 near Amoroto (Vizcaya). The common name "insignis" became popular, both in Spanish and Basque language, due to be more frequent at that time the scientific synonymy *Pinus insignis*. Monterey pine has been planted in Spain in the North Atlantic orle between the parallels 42 and 44 ° N in altitudes below 800 m (Autonomic Communities of Galicia, Asturias, Cantabria, Basque Country and Navarra), showing up also in the Mediterranean province of Gerona (Cataluña), with similar latitude, and promptly in the botanic grounds of the laurisilva and the fayal-brezal (*Myrica-Erica*) of the Canary islands (28-29 ° N). The worldwide and national areas occupied with commercial plantations of *P. radiata* are indicated in **Table 1**.

Country	Area	Autonomic Community	Area
Chile	1,400,000 ha.	Basque Country	160,000 ha.
New Zealand	1,200,000 ha.	Galicia	60,000 ha.
Australia	650,000 ha.	Asturias	26,000 ha.
Spain	270,000 ha.	Cantabria	15,000 ha.
South Africa	55,000 ha.	Navarra	5,500 ha.
Others	30,000 ha.	Canarias	3,000 ha.
		Gerona	500 ha.
Worldwide total	3,605,000 ha.	National total	270,000 ha.

Table 1. Worldwide and national areas (hectares) intended to the lumber exploitation of Monterey pine (adapted from Michel 2004).

This area of artificial plantation gathers in its whole the characteristics that favour the appearing of pests and diseases. Thus, all of them form (1) pure masses of more or less extension where the nutrient abundance and the facility of dispersion support the population increase of the harmful species; (2) contemporary masses that favour the appearance of pests and diseases related to different ages intervals; and (3) exotic masses which are submitted to intercontinental transports with high risk of introduction of exotic pests (Alonso Zarazaga & Goldarazena, 2005; López et al., 2007b) and diseases (Landeras et al., 2005) of greater potential virulence due to be free of natural enemies or under favourable environmental conditions. Besides, the economic profitability of Monterey pine, at first like cellulose pulp and now also as saw wood (50% for each intention nowadays in Northern Spain), has sponsored its plantation in geographic areas that are far from the optimum conditions. This has caused physiologically weak forestry masses, favouring the attack of pests and diseases due to a higher incidence of abiotic noxious agents such as frosts and droughts.

Bark beetles can cause damage both by the direct boring action and by the inoculation of phytopathogenic fungi. Once the host tree has been localized, pioneer specimens produce aggregation pheromones, which joined with the volatile kairomones of the recently attacked tree, attract other members of the same species. While they feed and build their galleries under the bark, some species cut the vascular flow causing the dead of the tree. At the same time, they inoculate spores and fragments of mycelium into the phloem, playing an important role in the transmission of some phytopathogenic fungi such as the causal agents of Dutch elm disease and blue-staining fungi that alter the structure of the wood and reduce its worth (Pajares, 1987).

Some species of bark beetles are able to cause significant damage to forests. For example, during the 1970's, outbreaks of *Ips typographus* destroyed 2 million m^3 of timber in

Scandinavia (Bakke, 1983). Regarding to United States, *Dendroctonus ponderosae* has caused an average annual loss of about 1.5 billion board feet particularly in *Pinus contorta* in western North America since 1895 (Wood, 1982). According to Carter et al. (1991), outbreaks of *Dendroctonus frontalis* in southeastern Texas covered 3,200,000 hectares with heavy economic losses from 1974 to 1980. Furthermore, from 1999 to 2003, *Dendroctonus frontalis* caused losses of about USD $1.5 billion in the southeastern U.S.A. (Nowak, 2005). The current outbreak of *Dendroctonus ponderosae* in Canada has impacted about 12 of the 14.3 million ha of *Pinus contorta* in the British Columbia (Westfall, 2006). In Spain, the most damaging species is *Tomicus piniperda,* which can cause annual losses of up to 72 million € in the Basque Country region (northern Spain) (Amezaga, 1993). Among the scolytid species present in the Iberian Peninsula, *Tomicus piniperda* and *Ips sexdentatus* represent the most relevant directly damaging species, as it will be explained later (see section 4).

Many sapstain fungi, especially ophiostomatoid fungi (Sordariomycetes: Ophiostomatales), are associated with phloeophagous bark beetles and might help to overcome the defences of attacked trees (Kirisits, 2004). They are commonly called "blue-stain fungi" because the discoloration they cause, namely blue, gray, brown or even black on the sapwood of trees, mostly on conifers (Kirisits, 2004). This sap stain is caused by the growth of the hyphae in the ray parenchima cells and resin ducts, disrupting the sap flow, and tracheids are also colonized in later stages of infection. As a consequence of this discoloration, the lumber defect is largely cosmetic. Most ophiostomatoid fungi that cause sapstain are moderately to weakly virulent pathogens, but some species can display relative high virulent levels and cause the death of infected trees when they are inoculated in high doses (Kirisits, 2004). Although some blue-staining fungi can cause strong damage on the strength properties of the wood assigned to furniture industry (Seifert, 1993) and these losses can amount to a 50% price reduction in the Basque Country (Maderas Elorriaga Company, Muxika, Vizcaya, personal communication, from about 180 €/m^3 to 85 €/m^3), the knowledge of conifer bark beetle-associated blue-staining fungi in the Iberian Peninsula is very limited (Fernández et al., 2004; Villareal et al., 2005). Romón et al. (2007a) isolated 16 species of *Ophiostoma* sensu lato or their asexual states from 13 bark beetle species and the root weevil *Brachyderes incanus* (Coleoptera: Curculionidae: Entiminae). Among the isolated taxa, species such as *Ophiostoma ips, O. minus, O. piceae* and *O. pluriannulatum* are important agents of blue-staining (Seifert, 1993), whereas *O. ips, O. minus* and *L. wingfieldii* pathogenicity is well recognized (Raffa & Smalley, 1988; Lieutier et al., 1989; Yamaoka et al., 1990; Fernández et al., 2004). Sixty-nine of the reported associations by Romón et al. (2007a) had not been previously recorded until them. Studies based on fungus-conifer scolytids association have not been only carried out in *P. radiata* inhabiting bark beetle populations, but also in *P. pinaster* (Bueno et al., 2010). They identified twenty-five taxa belonging to the Eumycotina group from isolations of *I. sexdentatus* adults and naturally colonized tissues (sapwood and phloem).

On the other side, pitch canker disease, caused by the fungus *Fusarium circinatum,* is one of the most important pathogens of *Pinus radiata.* This fungus species is endemic to southeastern United States (Dwinell et al., 1985). Pitch canker disease was recently identified and reported in California, predominately in planted urban *P. radiata* and in native Monterey pine forest (Correll et al., 1991). Since being first reported in the United States, it has been also found in different countries around the world: Japan (Muramoto et al., 1989),

Mexico (Rodriquez, 1989), South Africa (Viljoen et al., 1994) and Spain (Landeras et al., 2005). Its accidental introduction in these areas is probably due to softwood lumber, seedling and seed exports. In California, wounding has not resulted. The transmission of the pathogen to cones seems to be not (Correll et al., 1991) or branches (Fox et al., 1991), despite the presence of significant airborne inoculum. In contrast, a complex of insects have been demonstrated to be able to transmit *F. circinatum* (Fox et al., 1991; Hoover et al., 1995; Storer et al., 2004). Within this species-complex there are included some bark beetles species, like *Ips paraconfusus*, *I. mexicanus*, *I. plastographus*, *Pityophthorus carmeli*, *P. nitidulus*, *P. setosus* and *Conophthorus radiatae* and the anobid beetle *Ernobius punctulatus* (Coleoptera: Anobiidae). All of them appear to be phoretically associated with the fungus, and are known to visit and infest non-diseased trees. Romón et al. (2007b) isolated *F. circinatum* from adult bark beetles collected from baiting logs in two stands of *P. radiata* located in two plots of Biscay province (Morga and Muxika). Five conifer bark beetles were found to carry the inoculum: *Pityophthorus pubescens* (25.00%), *Hylurgops palliatus* (11.96%), *Ips sexdentatus* (8.57%), *Hypothenemus eruditus* (7.89%), *Hylastes attenuatus* (7.40%) and *Orthotomicus erosus* (2.73%) (Table 2). Frecuency of occurrence of *F. circinatum* is given within parentheses. In addition, *Brachyderes incanus* (14.28%) had the second highest frequency of occurrence of the fungus. Frequencies of occurrence of fungi were computed using the formula of Yamaoka where F = (NF / NT) 100 (%) and F represents the frequency of occurrence (%) of the fungus, NT represents the total number of samples from which isolation attempts were made and NF represents the number of samples from which the fungus was isolated.

Insect species	Total no. samples	NF	F (%)*
Pityophthorus pubescens	32	8	25.00
Brachyderes incanus	42	6	14.28
Hylurgops palliatus	117	14	11.96
Ips sexdentatus	35	3	8.57
Hypothenemus eruditus	38	3	7.89
Hylastes attenuatus	54	4	7.40
Orthotomicus erosus	73	2	2.73
Dryocoetes autographus	45	-	-
Hylastes ater	32	-	-
Tomicus piniperda	18	-	-
Xyleborus dryographus	9	-	-
Hylurgus ligniperda	5	-	-

* Frequency of occurrence F = (NF / NT) 100 (%), where NT represents the total number of samples from which isolations attempts were made, and NF represents the number of samples from which *F. circinatum* was isolated.

Table 2. Variation about frecuency of occurrence of *Fusarium circinatum* of several bark beetles (Coleoptera: Curculionidae: Scolytinae) and weevils (Coleoptera: Curculionidae: Entiminae) species in northern Spain.

Table 3 includes the relationship of different fungi species, belonging to the Orders Ophiostomatales, Sphaeropsidales and Hypocreales associated with conifer inhabiting bark beetles (and the weevil _Brachyderes incanus_), with special emphasis to isolations detected in the Basque Country (northern Spain).

Bark beetles species	Order Ophiostomatales	Order Sphaeropsidales	Order Hypocreales
Hylastes attenuatus	_Ophiostoma ips_** _O. olivaceum_ _O. piceae_** _O. piliferum-like_ _O. quercus_ _O. stenoceras_ _Leptographium guttulatum_** _L. truncatum-like_ _L. wingfieldii_*** _Pesotum fragans_	_Diplodia pinea_**	_Fusarium circinatum_*** _F. oxysporum_** _F. lateritium_*
Hylastes ater	_Ophiostoma ips_** _O. minus_ _O. penicillatum_ _O. piceae_** _O. piliferum_ _O. floccosum_* _O. olivaceum_* _O. piliferum-like_* _O. pluriannulatum_** _O. quercus_* _O. rectangulosporium-like_* _O. stenoceras_* _Leptographium stenoceras_ _L. guttulatum_** _L. lundbergii_ _L. serpens_ _L. wingfieldii_***		_Fusarium moniliformis_**
Hylurgops palliatus	_Ophiostoma ainoae_ _O. bicolor_ _O. cucullatum_ _O. galeiformis_ _O. japonicum_ _O.neglectum_ _O. penicillatum_ _O. piceae_** _O. piceaperdum_ _O. simplex_	_Diplodia pinea_**	_Fusarium circinatum_*** _F. proliferatum_**

Bark beetles species	Order Ophiostomatales	Order Sphaeropsidales	Order Hypocreales
Hylurgops palliatus	*O. stenoceras* *O. ips*** *O. olivaceum** *O. piliferum-like** *O. pluriannulatum*** *O. quercus** *O. rectangulosporium-like** *Ceratocystiopsis alba* *C. minuta* *Leptographium guttulatum*** *L. lundbergii* *L. procerum* *L. wingfieldii**** *L. truncatum-like** *Pesotum fragans**		
Ips sexdentatus	*Ophiostoma ainoae* *O. araucariae* *O. brunneo-ciliatum* *O. clavatum* *O. ips*** *O. japonicum* *O. minus** *O. obscura* *O. piceae* *O. piceaperdum* *O. olivaceum** *O. pluriannulatum*** *O. rectangulosporium-like** *O. stenoceras** *Ceratocystiopsis minuta* *Leptographium guttulatum*** *L. truncatum-like** *Pesotum fragans*		*Fusarium circinatum**** *F. moniliformis***
Dryocoetes autographus	*Ophiostoma ainoae* *O. araucariae* *O. cucullatum* *O. galeiformis* *O. japonicum* *O. obscura* *O. neglectum* *O. piceae*		

Bark beetles species	Order Ophiostomatales	Order Sphaeropsidales	Order Hypocreales
Dryocoetes autographus	*O. piceaperdum* *O. simplex* *O. ips*** *O. minus** *O. olivaceum** *O. piliferum-like** *O. rectangulosporium-like** *Ceratocystiopsis alba* *C. minuta* *Leptographium guttulatum***		
Orthotomicus erosus	*Ophiostoma ips*** *O. piceae*** *O. pluriannulatum*** *O. canum-like* *O. floccosum* *O. olivaceum* *O. stenoceras* *O. rectangulosporium-like* *Leptographium guttulatum*** *L. wingfieldii**** *Pesotum fragans*		*Fusarium circinatum**** *F. moniliformis*** *F. culmorum** *F. lateritium**
Tomicus piniperda	*Ophiostoma canum* *O. clavatum* *O. floccosum* *O. galeiformis* *O. huntii* *O. ips*** *O. minus* *O. piceae*** *O. piceaperdum* *O. piliferum* *O. pluriannulatum*** *Ceratocystiopsis minuta* *Leptographium euphyes* *L. guttulatum*** *L. lundbergii* *L. procerum* *L. wingfieldii****	*Diplodia pinea***	
Pityophthorus pubescens	*Ophiostoma ips*** *O. piliferum-like** *Leptographium guttulatum***		*Fusarium circinatum**** *F. lateritium**

Bark beetles species	Order Ophiostomatales	Order Sphaeropsidales	Order Hypocreales
Hypothenemus eruditus	*Ophiostoma pluriannulatum*** *O. quercus** *Pesotum fragans**		*Fusarium circinatum**** *F. culmorum** *F. lateritium**
Xyleborus dispar	*Ophiostoma pluriannulatum*	*Diplodia pinea*	
Xyleborus dryographus	*Ophiostoma olivaceum* *O. piceae*** *Sporothrix schenckii-like*		*Fusarium moniliformis***
Hylurgus ligniperda	*Leptographium guttulatum*		
Pityogenes calcaratus	*Ophiostoma ips* *Leptographium guttulatum***		
Brachyderes incanus	*Ophiostoma piceae*** *O. pluriannulatum*** *O. quercus* *Leptographium guttulatum*		*Fusarium circinatum**** *Fusarium moniliformis*

Table 3. Fungal species frequently associated with bark beetles colonizing conifers. Fungal taxa underlined indicate high percentage of isolation with the corresponding insect within a row. Fungal species with an upper symbol are present in the Basque Country, as follows: Saprophytic species (*), facultative pathogens (**) and strictly pathogens (***)(Adapted from Kirisits, 2004; Romón et al., 2007a, 2007b).

3. Chemical ecology of the host tree colonization by bark beetles: Basis for the development of a sustainable strategy for the protection of forestry masses

Each species of bark beetles is adapted to only one or a few host tree species probably due to natural selection driven by trees biochemicals. It is likely that each species of tree has coevolved chemicals to defend against the selection pressures of bark beetles and other insects (Berryman et al., 1985; Byers, 1995). Plant chemicals can be attractive, repellent, toxic or nutritious to bark beetles and have effects on: (1) finding and accepting the host tree (selection and suitability); (2) feeding stimulation and deterrence; (3) host resistance; (4) pheromone biosynthesis and communication; and (5) attraction of predators, parasites and competitors of bark beetles (Byers, 2004).

Bark beetles must locate a suitable host from among a relatively few that are widely scattered in the forest. During dispersal flight, insect must discriminate between potential conifer hosts and avoid any unsuitable host and non-hosts decidious trees, so the ability to detect and recognize different olfactory signals in a complex olfactory landscape represents an important cue for bark beetles colonization processes (Raffa, 2001; Zhang & Schlyter,

2004). The host tree is restricted usually to one or a few species and in most cases the insects seek weakened, less resistant trees. It is expected that insects can detect certain volatile host plant chemicals that indicate its suitability [see Fig. 1 for massive attack of *Tomicus piniperda*, the aggregation of this bark beetle species is considered to be predominantly mediated by host tree kairomonal compounds blend (Vité et al., 1986) although *trans*-verbenol has been recently suggested as its potential pheromone (Poland et al., 2003)]. Pheromones and/or kairomones involved in the semiochemical communication of bark beetle species most commonly colonizing *Pinus radiata* in Spain are indicated in Table 4.

Fig. 1. Massive attack of *Tomicus piniperda*, causing, after one year since the population outbreak, a scattered distribution pattern of nearly dead trees within a *Pinus radiata* plantation near Morga, Biscay province, Basque Country (Spain).

Bark beetle species	Pheromone	Kairomone
Hylurgops palliatus	-	Alpha-pinene (Perttunen, 1957) Beta-pinene (Volz, 1988; Byers, 1992) Terpinolene Myrtenol Ethanol 3-carene *cis*-verbenol *trans*-verbenol
Ips sexdentatus	Ipsdienol (Vité et al., 1972) Ipsenol	Myrcene (Hughes, 1974) 2-methyl-3-buten-2-ol (Serez, 1987) *cis*-verbenol
Tomicus piniperda	*trans*-verbenol (Poland et al., 2003)	Alpha-terpineol (Kangas et al., 1970) *cis*-carveol *trans*-carveol Beta-pinene (Volz, 1988) Myrtenol Ethanol (Byers, 1992) Alpha-terpinolene Delta-3-carene Alpha-pineneoxide (Czokajlo, 1998) Alpha-pinene (Song et al., 2005)

Table 4. Pheromonal and kairomonal components of most common conifer bark beetles distributed among *Pinus radiata* plantations in northern Spain, compendium revision.

Host plant's suitability to bark beetles varies with its nutritional quality and composition of anti-insect toxins (Scriber, 1984). Nonhost trees are probably less nutritional, and the beetle would not be adapted to detoxifying some of the nonhost toxins that have evolved for use against other herbivorous insects. A beetle would save much time and energy if it can discriminate the host from the nonhost and determine the intraspecific suitability of a host by olfactory means, deciding not to land due to the detection of volatile compounds called allomones (Byers, 1995). It is more difficult to isolate repellents and inhibitors used in avoidance behavior than to isolate attractants since tests of avoidance require one to first isolate the attractive host odors and then present these with and without the possibly inhibitory odor compounds. Several studies indicate that at least some species of bark beetles avoid nonhost volatiles during their search for host trees. For example, the attraction of both *T. piniperda* and *Hylurgops palliatus* to ethanol (1-6 g/day) was reduced by odors from cut logs of *Betula pendula* and *Populus tremula* (Schroeder, 1992). In this way, one monoterpenic compound called verbenone has shown relatively promising results as an anti-aggregant pheromone (Lindgren et al., 1989). Verbenone (4,6,6-trimethylbicyclo[3.1.1]-

hept-3-en-2-one) is a simple oxidation product of *trans*-verbenol, which in turn is a biological oxidation product of α-pinene (Birgersson & Leufvén, 1988), one of the most ubiquitous monoterpenes in the Pinaceae. *a*-Pinene is quite toxic to a number of coniferophagous insects, whereas *trans*-verbenol and verbenone appear to be less toxic (Lindgren et al., 1996). Thus, insects inhabiting environments high in *a*-pinene could be expected to have either a high tolerance or an effective detoxification system. Verbenone has been found in relatively large amounts (mg) in hindguts of *Dendroctonus ponderosae* (Pierce et al., 1987), *Dendroctonus frontalis* (Renwick & Vité, 1968), *Dendroctonus brevicomis* (Byers et al., 1984) and *Dendroctonus pseudotsugae* (Rudinsky et al., 1974), and in low amounts (ng) in *T. piniperda* (Lanne et al., 1987), but it appears to be absent in *I. paraconfusus, I. typographus* and *Pityogenes chalcographus* (Byers, 1983; Birgersson et al. 1984, 1990). During the last two decades, verbenone has been demonstrated as a good natural repellent for the control of damages caused by several insect species such as *Hylobius pales* (Salom et al., 1994), *Hylobius abietis* (Lindgren et al., 1996), *D. ponderosae* (Lindgren et al., 1989), *Ips pini* and *Ips latidens* (Lindgren & Miller, 2002). It has been hypothesised that bark beetles species which required relatively fresh host tissue would be more affected by the presence of verbenone, whereas species inhabiting aged tissues would have higher tolerance to verbenone and/or a more efficient detoxifying system (Lindgren, 1994; Lindgren & Miller 2002; Romón et al., 2007b).

Bark beetle pheromones used in aggregation and for avoidance of competition consist of many varied structures. Plant compounds, predominantly the monoterpenes *α*-pinene or myrcene, are used as kairomonal precursors for their pheromonal components (Hendry et al., 1980). Many of the same pheromonal compounds are used by species in throme same genus, such as ipsenol, ipsdienol, and *cis*-verbenol in the genus *Ips* (Byers, 1995) or *exo*-brevicomin, frontalin, *trans*-verbenol and verbenone in the genus *Dendroctonus* (tribe Tomicini) (Borden, 1982). Some compounds such as *cis*- and *trans*-verbenol may be found in *Ips* (Tribe Ipini) as well as *Tomicus* (Tribe Tomicini). However, *cis*-verbenol has so far only been proven as an aggregation pheromone component for species in the tribe Ipini, whereas *trans*-verbenol has semiochemical activity only in the tribe Tomicini. The base structure of ipsenol, ipsdienol, and myrcenol resembles the plant monoterpene myrcene; likewise, *cis*- and *trans*-verbenol resemble *α*-pinene. These structural similarities support the hypothesis that in many cases bark beetles use plant compounds as precursors for their pheromone components (Byers, 1995).

Aggregation pheromone components were first identified in bark beetles from males of *Ips paraconfusus* as a synergistic blend of (*S*)-(-)-ipsenol, (*S*)-(+)-ipsdienol, and (4*S*)-*cis*-verbenol (Wood et al., 1968). Several other *Ips* species were soon discovered to produce and respond to various blends of these compounds (Vité et al., 1972). The similarity of chemical structure between myrcene and ipsenol and ipsdienol led to propose, and demonstrate by gas chromatography and mass spectrometry (GC-MS), that myrcene is precursor of pheromones in *Ips* spp. (Hughes, 1974; Byers et al., 1979).

More resistant tree genotypes may have evolved through natural selection lower levels of pheromonal precursor terpenes or attractive kairomones and/or higher concentrations of other toxic monoterpenes. Evolution of plant chemicals that increase tree's resistance to colonization by bark beetles requires that (1) the plant chemicals are detrimental to the beetle; (2) the host chemistry is genetically driven; (3) population variation in genotypes of these trees exists; and (4) the bark beetle exerts selection pressure on the tree by killing or reducing fertility. The beetle population should coevolve, if possible, by shifting their genotype frequencies to those that offer more protection against the plant chemicals.

Host tree chemistry affects most aspects of bark beetle biology, moreover, bark beetles probably differentially affect survival of host trees and alter genotypic frequencies and host chemistry (Byers, 1995). Geographic and intraspecific variation in toxicity of host compounds has been little studied. Thus, more studies are needed in stands with ongoing outbreaks of bark beetles to determine if natural selection can slant trees intraspecific variation that will determine their monoterpene properties.

4. An overview to conifer-inhabiting bark beetle species with most forestry importance in spanish mainland, particularly northern Spain: Concrete cases

Alonso-Zarazaga (2002) listed 128 bark beetles species present across iberian and balearic area, including both conifer and decidious-inhabiting species. Two more alien species to Iberian Peninsula should be added to this checklist, i.e. *Gnathotrichus materiarius* and *Xylosandrus germanus* (López et al., 2007b). *Gnathotrichus materiarius* is considered as a nearctic native species, whereas the origin of *Xylosandrus germanus* is asiatic. According to Kirkendall and Faccoli (2010) and references therein, owing to commercial trading, bark beetles mainly travel in wood and in wooden packing materials such as crating, dunnage and pallets. Both alien species were found in sawmills and wood-proccesing companies of Basque Country that use imported lumber from France. Nowdays, their establishment to different *P. radiata* stands in the Basque Country is fully confirmed (Goldarazena et al., personnal observation). Although these two species are not considered as highly dangerous species, the prevention of the entry and early detection of invasive species, through different pathways including treatment of imported commodities, should be a priority task, in order to avoid potential negative environmental impacts that would be generated.

Within all of these species present in spanish mainland, few species are capable of killing healthy trees (Gil & Pajares, 1986). The register of produced damage data are very scarce, because there have not been studies focused on tracing the economical incidence of these insects in contrast with other countries. First reports were produced in 1907, related to some attacks of *Tomicus piniperda* and other decidious bark beetle species (Gil & Pajares, 1986). During mid-1950´s and mid-1970´s several attacks of *Ips acuminatus* (Fig. 2) were registered in *P. sylvestris* stands of Guadalajara province (1954-57) and Cuenca province (1972-1973) (Gil & Pajares, 1986) but there are no economic data available for these events. In addition, less important sporadic attacks of *I. sexdentatus* and *O. erosus* had occurred in north and central Spain in last decades (Gil & Pajares, 1986). According to Grégoire & Evans (2004), three species of conifer-inhabiting bark beetles species, i.e. *Ips sexdentatus, Ips acuminatus* and *Tomicus piniperda* should be considered as significant pests for Spanish mainland. Not only these species, but also *Pityophthorus pubescens* is a serious candidate to be considered as potential forest pest, taking into account its importance as vector of *Fusarium circinatum*, fact that has been previously mentioned in text (Romón et al, 2007b). Following sub-sections will deepen about different studies carried out with these species in Spanish mainland so far. Although quantitative data of economical damage produced by bark beetles are difficult to estimate, some data are provided.

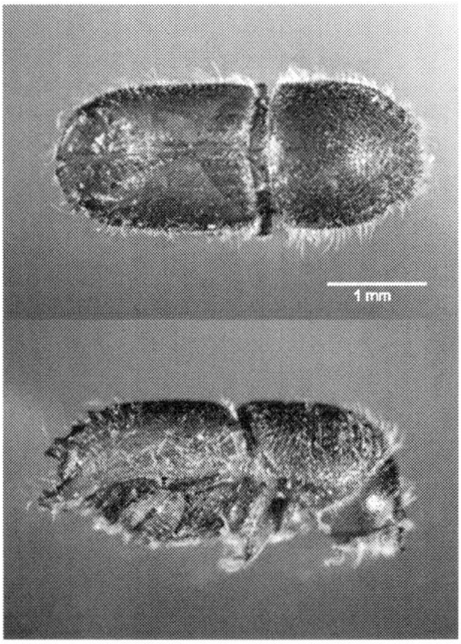

Fig. 2. *Ips acuminatus*, dorsal and lateral view. Photographs taken from López et al., 2007a.

4.1 The six-toothed bark beetle *Ips sexdentatus* (Börner) (Fig. 3)

This species is considered as one of the forest pests that higher damages cause in conifer stands of Iberian Peninsula (Gil & Pajares, 1986). Although endemic populations of *I. sexdentatus* tend to colonize weakened or dead trees, it is well reported that healthy trees can be attacked under epidemic conditions. Frecuently, improper management of logs, for example storaging them for long time, adverse abiotic and climatic conditions (storms, fires, droughts) generate breeding resources for *I. sexdentatus* and favor the generation of these population outbreaks. As a significant example, 11,997 ha of a mixed forest with predominant presence of *P. pinaster* were affected by a fire during 2005 in Gualadajara province (Sánchez et al., 2008). Due to this fact, *Ips sexdentatus* populations significantly increased causing severe damages in some zones. Thus, a massive trapping program was carried out in following years, setting 99 and 237 Theysohn traps in 2006 and 2007 respectively. This trapping methodology led to the capture of 4,928,270 beetles. Moreover, different silvicultural techniques (extraction of affected timber) were applied in parallel. All of these measurements contributed to reduce the negative effects of this population outbreak. On the other hand, 25,000 ca. trees were killed by *I. sexdentatus* in Castilla y León province in 2000 (Consejería de Medio Ambiente, Junta de Castilla y Léon 2001, as cited in Bueno et al., 2010).

Fig. 3. *Ips sexdentatus*, dorsal and lateral view. Photographs taken from López et al., 2007a.

In addition, it is remarkable that the incidence of "Klaus" named windstorm during January 2009 affected 37.9 million m^3 of maritime pine (*P. pinaster*) in Aquitanie (southern France) (Inventaire Forestier Nationale, 2009). As a consequence a great amount of timber was left as suitable breeding material for *I. sexdentatus*. The importation of those *P. pinaster* logs to different sawmills and timber-processing industries located at the Basque Country is a common commercial activity. So, it must be taken into account that the importation and storage of such infested logs for a long time would have consequences for the forest management and put into risk the adjacent *P. radiata* stands (Goldarazena et al., personnal observations).

Some studies focusing on verbenone have been carried out in order to test it as a potential component of IPM strategies for the protection of different pine species stands against this species. Two compounds, verbenone and *trans*-conophthorin, have been mainly considered as the most potential anti-aggregative semiochemicals. Biological implication of verbenone in bark beetles has been previously mentioned (see subsection 3), so it is worthwhile to remark the bioactivity of the second compound. The spyroketal conophthorin [5*S*,7*S*-(-)-7-methyl-1,6-dioxaspiro(4.5)decane] is a non-host bark volatile found in angiosperm trees, such as *Betula pubescens* and *B. pendula* (Betulaceae) in Europe (Byers et al., 1998), and *Populus tremuloides, P. trichocarpa* (Salicaceae), *B. papyrifera* and *Acer macrophyllum* (Aceraceae) in North America (Huber et al., 1999). On insects, it was first identified from the abdomina of workers of some Hymenoptera: Vespidae, i.e. *Paravespula vulgaris* (Francke et al., 1978), and *P. germanica* and *Dolichovespula saxonica*, together with males of the bark beetle species *Leperisinus fraxini* (Francke et al., 1979). Later, Kohnle et al. (1992) found it in the frass of the fir bark beetle *Cryphalus piceae*, reducing field response of the insect to attractants. The name conophthorin comes from the genus *Conophthorus*, that includes species known to produce it with an inhibitor effect to aggregation pheromones or host kairomones, that is, *Conophthorus coniperda*

(Birgersson et al. 1995; de Groot et al., 1998, Rappaport et al., 2000) and *C. resinosae* (Pierce et al., 1995; de Groot & DeBarr, 2000; Rappaport et al., 2000). In addition, there are several studies about the repellent effect of (*E*)-(-)-conophthorin and racemic conophthorin to pheromone baited traps in bark beetles species that are not known to produce it, as seen in *Xylosandrus germanus* (Kohnle et al. 1992), *Dendroctonus ponderosae* (Huber et al., 1999), *D. pseudotsugae* (Huber et al., 1999, 2000, 2001), *Dryocoetes confusus* (Huber et al., 2000) *Pityophthorus setosus* (Dallara et al., 2000), *C. cornicolens* and *C. teocotum* (Rappaport et al. 2000), *I. pini* (Huber et al., 2000, 2001), *I. duplicatus* (Zhang et al., 2001), *I. sexdentatus* (in France) (Jactel et al., 2001) and *I. typographus* (Zhang & Schlyter, 2003). In spite of this fact, there are reports referring to conophthorin as attractant as well, according to results observed in *I. mexicanus*, *Lasconotus pertenuis* (Coleoptera: Colydiidae) and *P. carmeli* (Dallara et al. 2000) and in *Epuraea thoracica* (Coleoptera: Nitidulidae) (Kohnle et al., 1992).

Romón et al. (2007b) detected a significant negative dose-dependent relationship between verbenone release rate and catches of *I. sexdentatus* in *P. radiata* stands with traps baited with an specific *Ips sexdentatus* attractant blend (Myrcene 250 mg/day + Ipsdienol 0.20 mg/day + Ipsenol 0.40 mg/day) (Fig. 4). Four verbenone release rate were used, as follows: 0.01, 0.2, 1.8 and 3.1 mg/24h (at 22-24°C). It has been previously mentioned that bark beetles species which requiere relative fresh host tissues would have a less tolerance to verbenone. So, if we asume that *I. sexdentatus* requires relatively fresh phloem and may attack healthy and live trees when endemic populations outbreak, it is feasible to show a negative response to the presence of verbenone, as these results show. In contrast with these results, bark beetle species (*Tomicus piniperda, Orthotomicus erosus, Dryocoetes autographus, Hypothenemus eruditus, Xyleborus dryographus, Hylastes ater, H. attenuatus* and *Hylurgus ligniperda*), trapped accidentally while carrying out that bioassay, have been shown to be not affected significantly by verbenone.

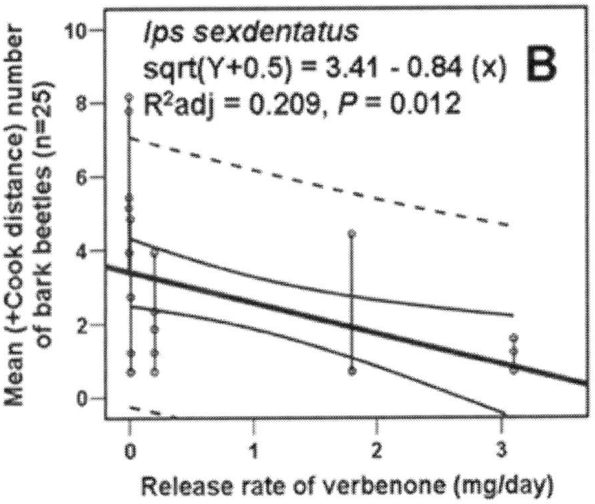

Fig. 4. Effect of four different release rates of verbenone on the attraction of *Ips sexdentatus*. Slope of regression line is significantly different from zero (*t*-test, $P < 0.001$). Confidence limits (95%) (thin solid lines) are associated with the regression line (thick solid line). Dashed lines represent confidence limits (95%) for catches in control trap.

Moreover, Etxebeste and Pajares (2011) tested verbenone and *trans*-conophthorin against *I. sexdentatus* populations present in mixed pine stands (*P. sylvestris, P. nigra* and *P. pinaster*). Two verbenone release rate (2, 40 and 60 mg/day) and one *trans*-conophthorin release rate (0.3 mg/day) were tested in two different field trapping bioassays with Ipsdienol 2.35 mg/day as attractant. Both verbenone and conophthorin and their combination significantly elicited a reduction of trap catches. Verbenone at 2 and 40 mg/day reduced the catches in a similar percentage (73% and 82% respectively) whereas *trans*-conophthorin reduced them by 45-49%. The strongest effect was showed by the blend of both compounds (verbenone at 40 mg/day plus *trans*-conophthorin at 0.3 mg/day) with a trap reduction rate of 90%. Another experiment was conducted to determine the potential of verbenone (at 60 mg/day) as a tool for tree protection. All control considered trees, that is, with no verbenone releasing device, were attacked by *I. sexdentatus*, whereas verbenone treated trees were less attacked.

4.2 The pine shoot beetle *Tomicus piniperda* L. (Fig. 5A) and Mediterranean pine shoot beetle *Tomicus destruens* (Wollaston) (Fig. 5B)

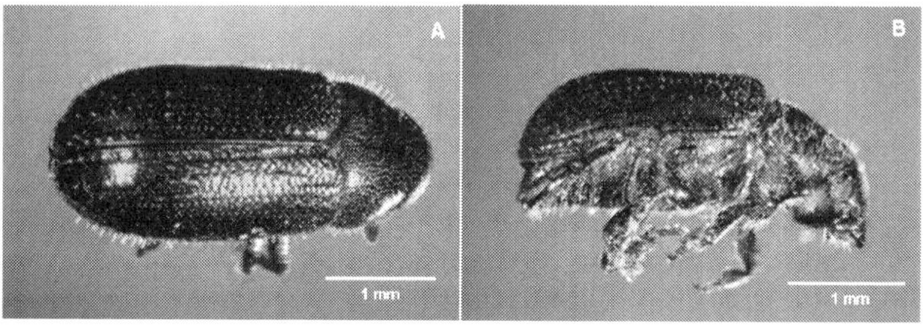

Fig. 5. (A) *Tomicus piniperda*, dorsal view; (B) *Tomicus destruens*, lateral view. Photographs taken from López et al., 2007a.

Genus *Tomicus* is represented by *Tomicus piniperda, Tomicus minor* and *Tomicus destruens* (Mediterranean pine shoot beetle) in Spain, since the latter taxa was definitively distinguished from *T. piniperda* and considered as a distinct species by the use of molecular techniques (Gallego & Galián, 2001; Kerdelhué et al., 2002). Morphologically, the colour of the elytra, colour of the antennal club, distribution of the antennal setae and distribution of the punctures along the elytral declivity seem to be the most useful diagnostic characters to differ both species, but only between mature exemplars (Faccoli, 2006). Concerning their distribution, *T. destruens* seems to be the predominant species in Spanish mainland, living in low an hot areas, whereas *T. piniperda* inhabit wet and cold areas of north-central Spain (Northern Plateau, the Pyrenees and perphaps in the Betic Mountains, at the South) (Gallego et al., 2004). However, there would be an overlapping of both species in the Atlantic Coast and the Bay of Biscay Coast, where they apparently coexist in sympatry. On the other hand, these authors consider *T. minor* as a less abundant species, with a fragmented distribution through high and wet areas. These potential distributions were suggested after applying predictive General Additive Models and Ecological Niche Factor Analysis models from 254 specimens of 81 different plots of Spanish mainland.

Although *P. sylvestris* is considered as the natural host of *T. piniperda*, its development in *P. radiata* (including maturation feeding in Monterrey pine shoots) is well reported. Host preference between both pine species has been tested in northern Spain. Amezaga (1996) observed that *T. piniperda* is able to exploit *P. radiata* as well as *P. sylvestris*. Even though brood production (number of progeny adults per gallery) was not significantly affected by tree species, the development was slowlier in *P. radiata*, and callows weighed less. In addition, after sampling two study areas at different altitudes (at 250 and 650 m) no sister generation was detected, so Amezaga (1996) hypothesized with the small chance of ocurring a second generation in northern Spain. She observed that attacks of *T. piniperda* began in March. Thus, according to her results *T. piniperda* might start its swarmming flight approximately in January over the entire altitudinal range in which pine stands are present in Northern Plateau.

Regardig to damage data, in 1989 massive outbreaks of *Tomicus piniperda* caused losses of 72 million € in *Pinus radiata* in Basque Country region (Northern Spain) (Amezaga, 1993). It has been estimated that a total of 200,000 ha have been affected by *T. piniperda* from 1990 to 1999 in Spain (Grégoire & Evans, 2004). The monitoring of *T. piniperda* populations is a common task carried out in different spanish provinces, but to our knowledge there has not been any control program with potential antiaggregant compounds until now. In contrast, researches aimed to *T. destruens* have been conducted in order to test the role of non host volatiles in its behaviour. Guerrero et al. (1997) showed by single-cell electrophysiological technique that *T. destruens* antennae possess specific olfactory cells capable of detecting benzyl alcohol. Furthermore, a ca. 700 mg/day release rate of this compound can significantlly reduce the attraction of *T. destruens* to host logs. Besides, the development of an effective lure for monitoring populations of *T. destruens* has been another objective. Gallego et al. (2008) tested different releases rates of ethanol and *a*-pinene, alone or in combination (with an upper threshold of 1800 and 900 mg/day respectively) in monospecific *P. halepensis* stands of southern Spain. The addition of *trans*-verbenol was also tested, but it did not affect the response. *a*-pinene alone did not showed a strong attraction effect, but a synergistic effect when adding it to ethanol. The most attractive blend appeared to when releasing 300 mg/day of *a*-pinene and 900 mg/day of ethanol.

4.3 Twig beetle *Pityophthorus pubescens* (Marsham) (Fig. 6)

Genus *Pityophthorus* has been typically not considered as a major pest species-complex, due to its life habits and development in branches of mainly dead and decaying trees (Bright, 1981; Wood, 1982).

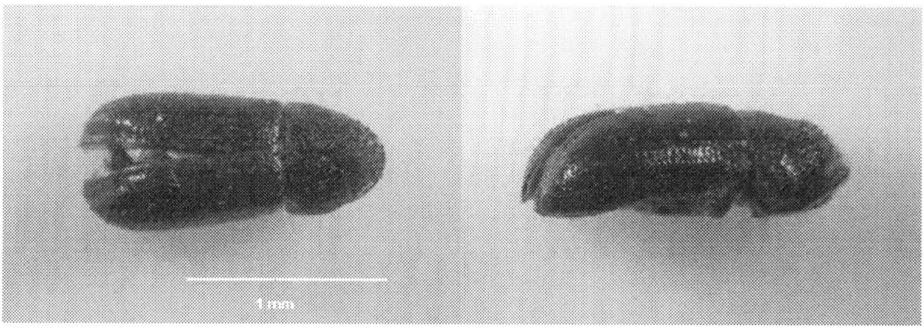

Fig. 6. *Pityophthorus pubescens*, dorsal and lateral view. Photographs taken from López et al., 2007a.

However, *Pityophthorus setosus* and *P. carmeli* has been associated with the causal agent of pitch canker *Fusarium circinatum* (Storer et al., 2004; Sakamoto et al., 2007). In Spanish mainland, it has been previously reported that *Pityophthorus pubescens* is also associated with *F. circinatum* in *P. radiata* stands of the Basque Country (northern Spain) (Romón et al., 2007b). Regarding the chemical ecology of *P. pubescens*, López et al. (2011) showed that both sexes emit (2*R*,5*S*)-2-(1-hydroxy-1-methylethyl)-5-methyltetrahydrofuran, also known as (*E*)-pityol, through different techniques of volatile collection (PORAPAK-Q and Solid Phase Microextraction/SPME). Positive enantiomer of this compound is also a component of the aggregation pheromone of other species of the genus, such as *P. pityographus* (Francke et al., 1987) and *P. carmeli, P. nitidulus* and *P. setosus* (Dallara et al., 2000), and the female-produced aggregation pheromone of the cone beetles *Conophthorus resinosae, C. coniperda* and *C. ponderosae* (Pierce *et al.* 1995; Birgersson *et al.* 1995; Miller *et al.*, 2000). However, in contrast with *P. pubescens* only one of the sexes of these species seems to emit (*E*)-(+)-pityol, as follows: males of *P. pityographus* and *P. carmeli* and females of *P. nitidulus* and *P. setosus*. Electroantennographic assays has revealed that both males and females of *P. pubescens* are able to detect (*E*)-(+)-pityol (López et al., 2011) (Fig. 7).

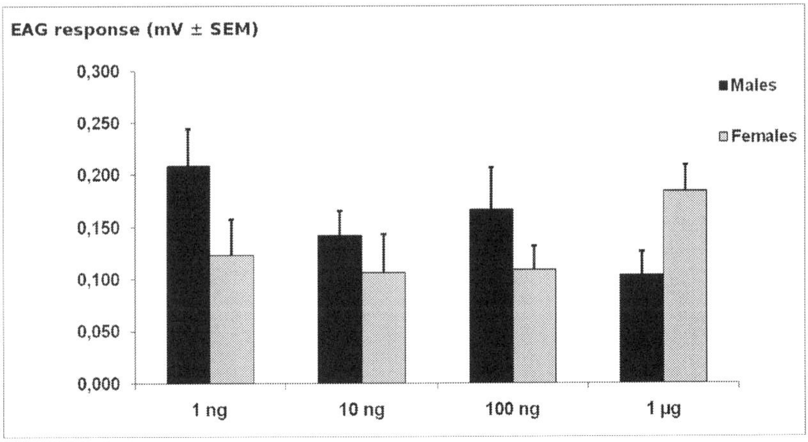

Fig. 7. Absolute EAG (mV± SE) responses of *P. pubescens* males (dark grey) and females (light grey) to serial dilutions containing 1 ng, 10 ng, 100 ng and 1 µg of (*E*)-(+)-pityol. Means followed by different letters were significantly different (Two-way ANOVA followed by Tukey multiple range test ($P \leq 0.05$), n = 8).

Moreover males were more attracted to (*E*)-(+)-pityol and (*E*)-(±)-pityol in olfactometric bioassays when testing three different doses (from 1 to 100 ng in decadeic steps) (López et al., 2011) (Fig. 8). In addition, sex-ratio appears to be male-biased in field trapping performed in different *Pinus* spp. stands of the Basque Country with multiple funnel traps baited with (*E*)-(+)-pityol and racemic pityol (López et al., unpublished data). Thus, the use of (*E*)-(+)-pityol or its cheaper racemate form might be an useful tool for monitoring *P. pubescens* populations and even to trap out male beetles in *P. radiata* stands.

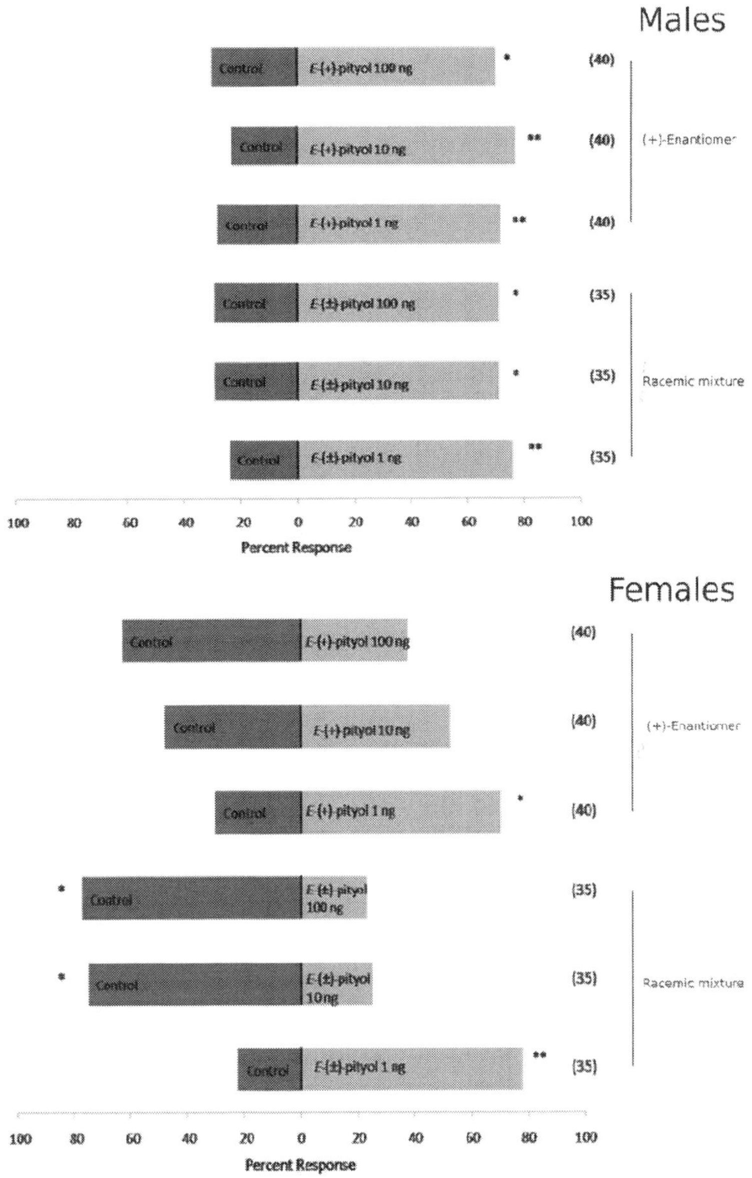

Fig. 8. Response of *P. pubescens* males (upper panel) and females (lower panel) at different doses of (*E*)-(+) and racemic (*E*)-pityol in Y-tube olfactometer trials. One and two asterisks indicate significant differences at $P<0.05$ and $P<0.01$, respectively (Chi-square test, with a significance level of $a = 0.05$). Number in parentheses indicates number of beetles responding.

On the other hand, verbenone shows promise as a disruptant of the aggregation of *P. pubescens*. Field studies has been undertaken in *P. radiata* stands testing four different release rates (0.01, 0.20, 1.80 and 3.10 mg/day) (at 22-24ºC) of this compound (Romón et al., 2007b). This work revealed a significant negative dose-dependent relationship in captured insects when comparing with control traps baited with a racemic pityol-releasing device at 0.14 mg/day (Fig. 9).

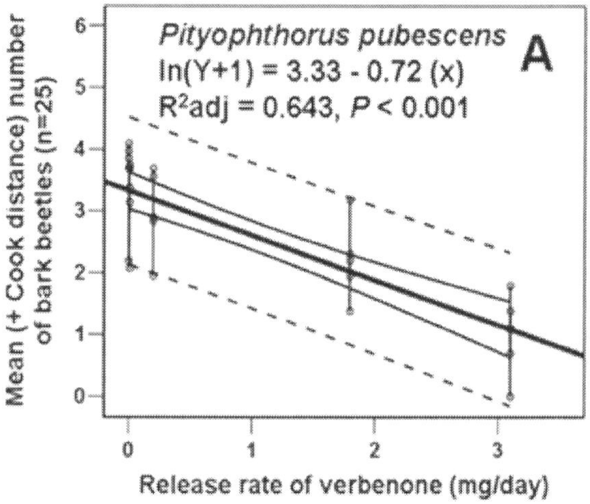

Fig. 9. Effect of four different release rates of verbenone on the attraction of *Pityophthorus pubescens*. Slope of regression line is significantly different from zero (t-test, $P < 0.001$). Confidence limits (95%) (thin solid lines) are associated with the regression line (thick solid line). Dashed lines represent confidence limits (95%) for catches in control trap.

4.4 *Hylurgops palliatus* (Gyllenhal) and *Hylaster ater* (Paykull)

Even though these two species (Fig 10 & 11) are considered as secondary, due to their colonization of dying or decaying trees, it is remarkable to mention their association with different species of pathogenic fungi (see subsection 2, table 3).

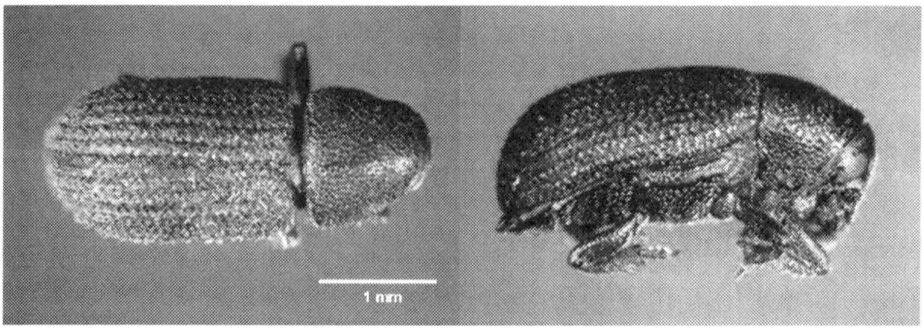

Fig. 10. *Hylurgops palliatus*, dorsal and lateral view. Photographs taken from López et al., 2007a.

Fig. 11. *Hylastes ater*, dorsal and lateral view. Photographs taken from López et al., 2007a.

5. Conclusions

At the light of what has been exposed, it is apparent that further extensive studies are needed to determine many gaps that still remain unclear. Excluding taxonomic and faunistic composition, which have been widely studied, little is known about the chemical ecology (including insect-host and insect-non host detailed interactions) of the species perceived as pest, although substantial progresses have been made. A deeper study of populations dynamics would be neccesary, in order to characterize better the brood production, flight periods and number of generations per year. This information would aid to a major understanding of which control methods should be applied. In addition, the development of useful IPM strategies, especially in the field of semiochemicals which might act as effective anti-aggregants, represents an important research line with many questions to be responded. Moreover, proper forestry management should be also recommended to be combined with, due to its relative influence on favoring the generation of populations outbreaks.

6. References

Alonso Zarazaga, M.A. (2002). Lista preliminar de los Coleoptera Curculionoidea del área ibero-balear, con descripción de *Melicius* gen. nov. y nuevas citas. *Boletín de la Sociedad Entomológica Aragonesa*, Vol.31, (October 2002), pp. 9-33, ISSN 1134-6094

Alonso Zarazaga, M.A. & Goldarazena, A. (2005). Presencia en el País Vasco de *Rhyephenes humeralis* (Coleoptera, Curculionidae), plaga de *Pinus radiata* procedente de Chile. *Boletín de la Sociedad Entomológica Aragonesa*, Vol.36, (April 2005), pp. 143-152, ISSN 1134-6094

Amezaga, I. (1993). The ecology and pest status of *Tomicus piniperda* L. (Coleoptera: Scolytidae) on pines. Ph.D. dissertation. University of London. United Kingdom.

Amezaga, I. (1996). Monterrey Pine (*Pinus radiata* D.Don) suitability for the pine shoot beetle (*Tomicus piniperda* L.) (Coleoptera: Scolytidae). Vol.86, pp. 73-79, ISSN 0378-1127

Ayres, M.P., Lombardero, M.J. (2000). Assessing the consequences of global change for forest disturbance from herbivores and pathogens. *The Science of Total Environment*, Vol.262, No.3, (November 2000), pp. 263-286, ISSN 0048- 9697

Bakke, A. (1983). Host tree and bark beetles interactions during a mass outbreak of *Ips typographus* in Norway. *Zeitschrift Fuer Angewandte Entomologie,* Vol.96, pp. 118-125, ISSN 0044-2240

Berryman, A.A., Dennis, B., Raffa, K.F. & Stenseth, N.C. (1985). Evolution of optimal group attack with particular reference to bark beetles (Coleoptera: Scolytidae). *Ecology,* Vol.66, No.3 (June 1985), pp.898-903, ISSN 0012-9658

Birgersson, G., Schlyter, F., Löfqvist, J. & Bergström, G. (1984). Quantitative variation of pheromone components in the spruce bark beetle *Ips typographus* from different attack phases. *Journal of Chemical Ecology,* Vol.10, No.7, (July 1984), pp. 1029-1055, ISSN 0098-0331

Birgersson, G. & Leufvén, A. (1988). The influence of host tree response to *Ips typographus* and fungal attack on production of semiochemicals. *Insect Biochemistry,* Vol. 18, pp. 761-770, ISSN 0020-1790

Birgersson, G., Byers, J.A., Bergström, G. & Löfqvist, J. (1990). Production of pheromone components, chalcogran and methyl (E,Z)-2,4-decadienoate, in the spruce engraver *Pityogenes chalcographus. Journal of Insect Physiology,* Vol. 36, No.6, pp. 391-395, ISSN 0022-1910

Birgersson, G., DeBarr, G.L., de Groot, P., Dalusky, M.J., Pierce Jr, H.D., Borden, J.H., Meyer H., Francke, W., Espelie, K.E. & Berisford, C.W. (1995) Pheromones in white pine cone beetle, *Conophthorus coniperda* (Schwarz) (Coleoptera: Scolytidae). *Journal of Chemical Ecology,* Vol.21, No.2, (February 1995), pp. 143-167, ISSN 0098-0331

Borden, J.H. (1982). Aggregation pheromones. In: *Bark Beetles in North American Conifers,* B. Mitton and K.B. Sturgeon (Eds.), 74-139, University of Texas Press, ISBN 0292707444, Austin, Texas

Bright, D.E. 1981. Taxonomic monograph of the genus *Pityophthorus* Eichhoff in North and Central America (Coleoptera: Scolytidae). *Memoirs of the Entomological Society of Canada,* Vol.118, pp. 1-378, ISSN 1920-3047

Bueno, A., Díez, J.J. & Fernández, M.M. (2010) Ophiostomatoid fungi transported by *Ips sexdentatus* (Coleoptera: Scolytidae) in *Pinus pinaster* in NW Spain. *Silva Fennica,* Vol.44, No.3, pp. 387-397, ISSN 0037-5330

Byers, J.A. (1983). Influence of sex, maturity and host substances on pheromones in the guts of the bark beetles, *Dendroctonus brevicomis* and *Ips paraconfusus. Journal of Insect Physiology,* Vol.29, No.1, (January 1983), pp. 5-13, ISSN 0022-1910

Byers, J.A. (1992). Attraction of bark beetles, *Tomicus piniperda, Hylurgops palliatus,* and *Trypodendron domesticus* and other insects to short-chain alcohols and monoterpenes. *Journal of Chemical Ecology,* Vol.18, No.12, (December 1992), pp. 2385-2402, ISSN 0098-0331

Byers, J.A. (1995). Host tree chemistry affecting colonization in bark beetles. In: *Chemical ecology of insects 2,* R.T. Cardé and W.J. Bell (Eds.), 154-213, Chapman and Hall, ISBN 978-0412039614, New York, United States.

Byers, J.A. (2004). Chemical ecology of bark beetles in a complex olfactory landscape In: *Bark and wood boring insects in living trees in Europe, a synthesis,* F. Lieutier, K.R. Day, A. Battisti, J.C. Grégoire and H.F. Evans (Eds.), 89-134, Kluwer Academic Press, ISBN 1-4020-2240-9, Dordrech, The Netherlands.

Byers, J.A., Wood, D.L., Browne, L.E., Fish, R.H., Piatek, B. & Hendry, L.B. (1979). Relationship between a host plant compound, myrcene and pheromone production in the bark beetle, *Ips paraconfusus*. *Journal of Insect Physiology*, Vol.25, pp. 477-482, ISSN 0022-1910

Byers, J.A., Wood, D.L., Craig, J. & Hendry, L.B. (1984). Attractive and inhibitory pheromones produced in the bark beetle, *Dendroctonus brevicomis*, during host colonization: Regulation of inter- and intraspecific competition. *Journal of Chemical Ecology*, Vol.10, No.6, (June 1984), pp. 861-877, ISSN 0098-0331

Byers, J.A., Zhang, Q.H. & Schlyter, F. (1998). Volatiles from non-host birch trees inhibit pheromone response in spruce bark beetles. *Naturwissenschaften*, Vol.85, No.11, (December 1998), pp. 557-561, ISSN 0028-1042

Carter, D.R., O'Laughlin, J. & McKinney, C. (1991). Southern pine beetle impacts and control policy in the 1982-1986 Texas epidemic. *Southern Journal of Applied Forestry*, Vol.15, No.3, (August 1991), pp. 145-153, ISSN 0148-4419

Correll, J.C., Gordon, T.R., McCain, A.H., Fox, J.W., Koehler, C.S., Wood D.L. & Schultz, M.E. (1991). Pitch Canker Disease in California: pathogenicity, distribution, and canker development on Monterrey pine (*Pinus radiata*). *Plant Disease*, Vol.75, No.7, (July 1991), pp. 676-682, ISSN 0191-2917

Czokajlo, D. (1998). Semiochemicals for the larger pine shoot beetles (*Tomicus piniperda* L.) and its clerid predators. Ph.D. Dissertation, State University of New York, Syracuse, NY. United States.

Dallara, P.L., Seybold, S.J., Meyer, H., Tolasch, T., Francke, W. & Wood, D.L. (2000). Semiochemicals from three species of *Pityophthorus* (Coleoptera: Scolytidae): Identification and field response. *The Canadian Entomologist*, Vol.132, No.6, (November 2000), pp. 889-906, ISSN 0008-347X

De Groot, P., DeBarr, G.L. & Birgersson, G. (1998). Field bioassays of synthetic pheromones and host monoterpenes for *Conophthorus coniperda* (Coleoptera: Scolytidae). *Environmental Entomology* Vol.27, No.2, (April 1998), pp. 382-387, ISSN 0046-225X

De Groot, P. & DeBarr, G.L. (2000). Response of cone and twig beetles (Coleoptera: Scolytidae) and a predator (Coleoptera: Cleridae) to pityol, conophthorin, and verbenone. *The Canadian Entomologist*, Vol.132, No.6, (November 2000), pp. 843-851, ISSN 0008-347X

Dwinell, L.D., Barrows-Broaddus, J.B. & Kuhlman, E.G. (1985). Pitch canker: a disease complex of southern pines. *Plant Disease*, Vol.69, No.3, (March 1985), pp. 270-276, ISSN 0191-2917

Etxebeste, I. & Pajares, J.A. (2011). Verbenone protects pine trees from colonization by the six-toothed pine bark beetle, *Ips sexdentatus* Boern. (Col.: Scolytinae). *Journal of Applied Entomology*, Vol.135, No.4, (May 2011), pp. 258-268, ISSN 0931-2048

Faccoli, M. (2006). Morphological separation of *Tomicus piniperda* and *T. destruens* (Coleoptera: Curculionidae: Scolytinae): new and old characters. *European Journal of Entomology*, Vol.103, No.2, pp. 433–442, ISSN 1210-5759

Fernández, M.M.F. (1996). Estructura biológica y ecológica del barrenillo de pino *Tomicus minor* (Coleoptera, Scolytidae). Ph.D. dissertation. University of León, Spain.

Fernández, M.M.F., García A.E. & Lieutier, F. (2004). Effects of various densities of *Ophiostoma ips* inoculations on *Pinus sylvestris* in north-western Spain. *Forest Pathology,* Vol.34, No.4, pp. 213–223, (August 2004), ISSN 1437-4781

Fox, J.W., Wood, D.L., Koehler, C.S. & O'Keefe, S.T. (1991). Engraver beetles (Scolytidae: *Ips* species) as vectors of the pitch canker fungus, *Fusarium subglutinans*. *The Canadian Entomologist,* Vol.123, No.6, (November, 1991), pp. 1355-1367, ISSN 0008-347X

Francke, W., Hindorf, G. & Reith, W. (1978). Methyl-1,6-dioxaspiro[4.5]decanes as odors of *Paravespula vulgaris* (L.). *Angewandte Chemie International Edition,* Vol.17, No.11, (November 1978), pp. 862, ISSN 0044-8249

Francke, W., Hindorf, G. & Reith, W. (1979). Alkyl-1,6-dioxaspiro[4.5]decanes- a new class of pheromones. *Naturwissenschaften,* Vol. 66, No.12, (December 1979), pp. 618-619, ISSN 0028-1042

Francke, W., Pan, M.L., König, W.A., Mori, K., Puapoomchareon, P., Heuer, H. & Vité, J.P. (1987). Identification of 'pityol' and 'grandisol' as pheromone components of the bark beetle, *Pityophthorus pityographus*. *Naturwissenschaften,* Vol.74, pp. 343-345, ISSN 0028-1042

Gallego, D. & Galián, J. (2001). The internal transcribed spacers (ITS1 and ITS2) of the rDNA diferenciate the bark beetle forest pests *Tomicus destruens* and *T. piniperda*. *Insect Molecular Biology,* Vol.10, No.5, (October 2001), pp. 415-420, ISSN 0962-1075

Gallego, D., Cánovas, F, Esteve, M.A. & Galián, J. (2004). Descriptive biogeography of *Tomicus* (Coleoptera: Scolytidae) species in Spain. *Journal of Biogeography,* Vol.31, No.12, (December 2004), pp. 2011-2024, ISSN 0305-0270

Gallego, D., Galián, J., Díez, J.J. & Pajares, J.A. (2008). Kairomonal responses of *Tomicus destruens* (Col., Scolytidae) to host volatiles α-pinene and ethanol. *Journal of Applied Entomology,* Vol.132, No.8, (September 2008), pages 654–662, ISSN 0931-2048

Gil L.A. & Pajares J.A. (1986). *Los Escolítidos de las Coníferas de la Península Ibérica*. Instituto Nacional de Investigaciones Agrarias, Ministerio de Agricultura, Pesca y Alimentación. ISBN 84-7498-250-2, Madrid, Spain.

González, R. (1990). Estudio bio-ecológico de *Phloeotribus scarabaeoides* (Bernard, 1788) en la provincia de Granada. Ph.D. dissertation. University of Granada, Spain.

Grégoire, J.C. & Evans H.F. (2004). Damage and control of BAWBILT organisms an overview. In: *Bark and Wood Boring Insects in Living Trees in Europe, A Synthesis,* F. Lieutier, K.R. Day, A. Battisti, J.C. Grégoire and H.F. Evans (Eds.), 19-37. Kluwer Academic Press, ISBN 1-4020-2240-9, Dordrech, The Netherlands.

Guerrero, A., Feixas, J., Pajares, J.A., Wadhams, L.J., Pickett,. J.A & Woodcock, C.M. (1997). Semiochemically Induced Inhibition of Behaviour of *Tomicus Destruens*. *Naturwissenschaften,* Vol.84, No.4, (April, 1997), pp. 155-157, ISSN 0028-1042

Hendry, L.B., Piatek, B., Browne, L.E., Wood, D.L., Byers, J.A., Fish, R.H. and Hicks, R.A. (1980). In vivo conversion of a labelled host plant chemical to pheromones of the bark beetle *Ips paraconfusus*. *Nature,* Vol.284, pp. 485, 0028-0836

Hoover, K., Wood, D.L., Fox, J.W. & Bros W.E. (1995). Quantitative and seasonal association of the pitch canker fungus, *Fusarium subglutinans* f.sp. *pini* with *Conophthorus radiatae* (Coleoptera: Scolytidae) and *Ernobius punctulatus*

(Coleoptera: Anobiidae) which infest *Pinus radiata*. *The Canadian Entomologist*, Vol.127, No.1, (January 1995), pp. 79-91, ISSN 0008-347X

Huber, D.P.W., Gries, R., Borden, J.H. & Pierce Jr., H.D. (1999). Two pheromones of coniferophagous bark beetles found in the bark of nonhost angiosperms. *Journal of Chemical Ecology*, Vol.25, No.4, (April 1999), pp. 805–816, ISSN 0098-0331

Huber, D.P.W., Borden, J.H., Jean-Willians, N.L. & R. Gries. (2000). Differential bioactivity of conophthorin on four species of North American bark beetles (Coleoptera: Scolytidae). *The Canadian Entomologist*, Vol.132, No.5, (September 2000), pp. 649-653, ISSN 0008-347X

Huber, D.P.W., Borden, J.H. & Stastny, M. (2001). Response of the pine engraver, *Ips pini* (Say) (Coleoptera: Scolytidae), to conophthorin and other angiosperm bark volatiles in the avoidance of non-host. *Agricultural and Forest Entomology*, Vol.3, No.3, (August 2001), pp. 225-232, ISSN 1461-9563

Hughes, P.R. (1974). Myrcene: a precursor of pheromones in *Ips* beetles. *Journal of Insect Physiology*, Vol.20, pp. 1271-1275. ISSN 0022-1910

Inventaire Forestier Nationale (2009) Estimations pour l'ensemble de la zone évaluée. In: *Tempête Klaus du 24 janvier 2009* , 01.08.2011. Available from http://www.ifn.fr/spip/spip.php?article618

Jactel, H., Van Halder, I., Menassieu, P., Zhang, Q.H. & Schlyter, F. (2001). Non-host volatiles disrupt the response of the stenographer bark beetle, *Ips sexdentatus* (Coleoptera: Scolytidae), to pheromone-baited traps and maritime pine logs. *Integrated Pest Management Reviews* Vol.6, No.3-4, (September 2001), pp. 197-207, ISSN 1353-5226

Kangas, E., Oksanen, H. & Perttunen, V. (1970). Responses of *Blastophagus piniperda* L. (Col., Scolytidae) to trans-verbenol, cis-verbenol, and verbenone, known to be population pheromones of some American bark beetles. *Annals of Entomologica Fennica*, Vol.36, pp. 75-83, ISSN 0785-8760

Kerdelhué, C., Roux-Morabito, G., Forinchon, J., Chambon, J.M., Robert, A. & Lieutier F. (2002). Population genetic structure of *Tomicus piniperda* L. (Coleoptera: Curculionidae: Scolytinae) on different pine species and validation of *T. destruens* (Woll.). *Molecular Ecology*, Vol.11, No.3, (March 2002), pp. 483-495, ISSN 0962-1083

Kirisits, T. (2004). Fungal associates of european bark beetles with specian emphasis to ophiostomatoid fungi. In: *Bark and Wood Boring Insects in Living Trees in Europe, A Synthesis*, F. Lieutier, K.R. Day, A. Battisti, J.C. Grégoire and H.F. Evans (Eds.), 181-235. Kluwer Academic Press, ISBN 1-4020-2240-9, Dordrech, The Netherlands.

Kirkendall, L.R. & Faccoli, M. (2010). Bark beetles and pinhole borers (Curculionidae, Scolytinae, Platypodinae) alien to Europe. In: *Sixty years of discovering scolytine and platypodine diversity: A tribute to Stephen L. Wood*. A.I. Cognato and M. Knížek (Eds), pp. 227-251. *ZooKeys*, Vol.56, ISSN 13132989

Kohnle, U., Densborn, S., Kölsch, P., Meyer, H. & Francke W. (1992). E-7-methyl-1,6-dioxaspiro[4.5]decane in the chemical communication of European Scolytidae and Nitidulidae. *Journal of Applied Entomology*, Vol.114, pp. 187-192, ISSN 0931-2048

Landeras, P., García, Y., Fernández, L. & Braña, M. (2005). Outbreak of pitch canker caused by *Fusarium circinatum* on *Pinus* spp. in Northern Spain. *Plant Disease,* Vol.89 No.9, pp. 1015, ISSN 0191-2917

Lanne, B.S., Schlyter, F., Byers, J.A., Löfqvist, J., Leufvén, A., Bergström, G., Van Der Pers, J.N.C., Unelius, R., Baeckström, P. & Norin. T. (1987). Differences in attraction to semiochemicals present in sympatric pine shoot beetles, *Tomicus minor* and *T. piniperda. Journal of Chemical Ecology,* Vol.13, No.5, (May 1987), pp. 1045-1067, ISSN 0098-0331

Lieutier, F., Yart, A., Garcia, J., Ham, M.C., Morelet, M. & Levieux, J. (1989). Champignons phytopathogènes associés à deux coléoptères scolytidae du pin sylvestre (*Pinus sylvestris* L.) et étude préliminaire de leur agressivité envers l'hôte. *Annals of Forest Science,* Vol.46, No.3, pp. 201-216, ISSN 1286-4560

Lindgren, B.S. (1994) Research needs for anti-aggregation pheromones of bark beetles. In: *Research, Development, and Commercialization of Semiochemicals in Insect Pest Management in Canada. Canada Pest Management Alternatives Office and Pest Management Alternatives Program* pp. 62-67. Ottawa, Canada.

Lindgren, B.S., Borden, J.H., Cushon, G.H., Chong, L.J. & Higgins C.J. (1989). Reduction of mountain pine beetle (Coleoptera: Scolytidae) attacks by verbenone in lodgepole pine stands in British Columbia. *Canadian Journal of Forest Research,* Vol.19, No.1 (January 1989), pp. 65-68, ISSN 1208-6037

Lindgren, B.S., Nordlander G. & Birgersson, G. (1996). Feeding deterrency and acute toxicity of verbenone to the pine weevil, *Hylobius abietis. Journal of Applied Entomology,* Vol.120, No.5, (December 1996), pp. 397-403. ISSN 0931-2048

Lindgren, B.S. & Miller, D.R. (2002). Effect of verbenone on five species of bark beetles (Coleoptera: Scolytidae) in lodgepole pine forests. *Environmental Entomology,* Vol.31, No.5, (October 2002), pp. 759-765, ISSN 0046-225X

Lombardero, M.J. (1994). Estudio de los Scolytidae (O. Coleoptera) de Galicia. Ph.D. dissertation. University of Santiago de Compostela, Spain.

Lombardero, M.J. & Novoa, F. (1994). Datos faunísticos sobre escolítidos ibéricos (Coleoptera: Scolytidae). *Boletín de la Sociedad Entomológica Aragonesa,* Vol.18, No.1-2, pp. 181-186, ISSN 1134-6094

López, S., Romón, P., Iturrondobeitia, J.C. & Goldarazena A. (2007a). *Los escolítidos de las coníferas del País Vasco. Guía práctica para su identificación y control.* Servicio Central de Publicaciones del País Vasco. ISBN 978-84-457-2650-1. Vitoria, Spain.

López S., Iturrondobeitia J.C. and Arturo Goldarazena A. (2007b). Primera cita de la Península Ibérica de *Gnathotrichus materiarius* (Fitch, 1858) y *Xylosandrus germanus* (Blandford, 1894) (Coleoptera: Scolytinae). *Boletín de la Sociedad Entomológica Aragonesa,* Vol.40, (October 2007), pp. 527-532, ISSN 1134-6094

López, S., Quero, C., Iturrondobeitia, J.C., Guerrero, A. & Goldarazena A. (2011). Evidence for (*E*)-Pityol as an aggregation pheromone of *Pityophthorus pubescens* (Coleoptera: Curculionidae: Scolytinae). *The Canadian Entomologist* Vol.143, No.5, pp. 447-454, ISSN 0008-347X

Michel, M. (2004). El pino radiata (*Pinus radiata* D.Don) en la historia de la Comunidad Autónoma de Euskadi. Análisis de un proceso de forestalismo intensivo. Ph.D. dissertation. University of Madrid, Spain

Miller, D.R, Pierce Jr, H.D., de Groot, P., Jean-Williams, N., Bennett, R., & Borden, J.H. (2000). Sex pheromone of *Conophthorus ponderosae* (Coleoptera: Scolytidae) in a coastal stand of western white pine (Pinaceae). *The Canadian Entomologist,* Vol.132, No.2, (March 2000), pp. 243-245, ISSN 0008-347X

Muramoto, M., Minamihashi,. H., Mitsuishi, K. & Kobayashi, T. (1989). Occurrence of resiniferous damage on *Pinus luchuensis* Mayr - Symptom and analysis of the damage. In: *Proceedings, Transnational 99th annual meeting of Japan Forestry Society*

Nowak, J. (2005). Southern pine beetle prevention and restoration program. USDA Forest Service Technical Report, 02.08.2011, Available from http://www.fs.fed.us/foresthealth/publications/spb_success_story.pdf.

Pajares JA, 1987. Contribución al conocimiento de los escolítidos vectores de la grafiosis en la Península Ibérica. Ph.D. dissertation. University of Madrid, Spain.

Perttunen, V. (1957). Reactions of two bark beetle species, *Hylurgops palliatus* Gyll. and *Hylastes ater* Payk. (Col., Scolytidae) to the terpene alpha-pinene. *Annals of Entomologica Fennica,* Vol.23, pp. 101-110, ISSN 0785-8760

Pierce, H.D., Conn, J.E., Oehlschlager, A.C. & Borden, J.H. (1987). Monoterpene metabolism in female mountain pine beetles, *Dendroctonus ponderosae* Hopkins attacking *ponderosa* pine. *Journal of Chemical Ecology* Vol.13, No.6, (June 1987), pp. 1455-1480, ISSN 0098-0331

Pierce, H.D., de Groot, P., Borden, J.H., Ramaswamy, S., & Oehlschlager, A.C. (1995). Pheromones in red pine cone beetle, *Conophthorus resinosae* Hopkins, and its synonym, *C. banksianae* McPherson (Coleoptera: Scolytidae). *Journal of Chemical Ecology,* Vol.21, No.2, (February 1995), pp. 169-185, ISSN 0098-0331

Plaza, E. (1983). Los representantes españoles de las tribus *Crypturgini* y *Pityophthorini* (Col. Scolytidae). *Eos,* Vol.59, pp. 223-241, ISSN 0013-9440

Plaza, E. & Gil. L. (1982). Los *Ipini* (Col. Scolytidae) de la Península Ibérica. *Eos,* Vol.58, pp. 237-269, ISSN 0013-9440

Poland, T.M., de Groot, P., Burke, S., Wakarchuk, D., Haack, R.A., Nott, R. & Scarr, T. (2003). Development of an improved attractive lure for the pine shoot beetle, *Tomicus piniperda* (Coleoptera: Scolytidae). *Agricultural and Forest Entomology,*Vol.5, No.4 (November 2003), pp. 293-300, ISSN 1461-9563

Raffa, K.F. (2001). Mixed messages across multiple trophic levels: the ecology of bark beetle chemical communication systems. *Chemoecology,* Vol.11, No2, (April 2001), pp. 49-65, ISSN 0937-7409

Raffa, K.F. & Smalley, E.B. (1988). Response of red and jack pines to inoculations with microbial associates of the pine engraver, *Ips pini* (Coleoptera: Scolytidae). *Canadian Journal of Forest Research,* Vol.18, No.5, (May 1988), pp. 581-586, ISSN 1208-6037

Rappaport, N.G., Stein, J.D., del Rio Mora, A.A., DeBarr, G., de Groot P. & Mori S. (2000). Responses of *Conophthorus* (Coleoptera: Scolytidae) to behavioral chemicals in field trials: A transcontinental perspective. *The Canadian Entomologist,* Vol.132, No.6, (November 2000), pp. 925-937, ISSN 0008-347X

Renwick, J.A.A. & Vité, J.P. (1968). Isolation of the population aggregating pheromone of the southern pine beetle. *Contributions of the Boyce Thompson Institute,* Vol.24, pp. 65-68, ISSN 0006-8543

Riba, J.M. (1994). Bioecología de los Scolytidae (Coleoptera) que nidifican en los abetales del Valle de Arán (Pirineos Orientales). Ph.D. dissertation. University of Barcelona.

Rodriquez, R.G. (1989). Pitch canker on *Pinus douglasiana*, pines indigenous to San Andres Milpillas, Municipal of Huajicori, Nay. Forest. In: *Proceedings, Parasitology Symposium, City of Juarez, Chihuahua*, Mexico, 4–6 October 1989

Romón, P., Zhou, X.D., Iturrondobeitia J.C., Wingfield, M.J. & Goldarazena, A., (2007a). *Ophiostoma* species (Ascomycetes: Ophiostomatales) associated with bark beetles (Coleoptera: Scolytinae) colonizing *Pinus radiata* in Northern Spain. *Canadian Journal of Microbiology*, Vol.53, No.6, (2007a), pp. 756-767, ISSN 0008-4166

Romón, P., Iturrondobeitia, J.C., Gibson, K., Lindgren, B.S. & Goldarazena A. (2007b). Quantitative association of bark beetles with pitch canker fungus and effects of verbenone on their semiochemical comunication in Monterey pine forests in Northern Spain. *Environmental Entomology*, Vol.36, No.4, (2007b), pp. 743-750, ISSN 0046-225X

Rudinsky, J.A., Morgan, M.E., Libbey, L.M. & Putnam, T.B. (1974). Antiaggregative-rivalry pheromone of the mountain pine beetle, and a new arrestant of the southern pine beetle. *Environmental Entomology*, Vol. 3, No.1, (February 19874), pp. 90-98, ISSN 0046-225X

Sakamoto, J.M., Gordon, T.R., Storer, A.J. & Wood, D.L. 2007. The role of *Pityophthorus* spp. as vectors of pitch canker affecting Monterey pines (*Pinus radiata*). *The Canadian Entomologist*, Vol.139, No.6, (November 2007), pp. 864-871, ISSN 0008-347X

Salom, S.M., Carlson, J.A., Ang, B.A., Grossman, D.M. & Day, E.R. (1994). Laboratory evaluation of biologically based compounds as antifeedants for the Pales weevil, *Hylobius pales* (Herbst) (Coleoptera: Curculionidae). *Journal of Entomological Science*, Vol.29, pp. 407-419. ISSN 0749 -8004

Sánchez, G., González E., Vela, A., Ayuso, S., (2008). *Control de la población de Ips sexdentatus en el área afectada por el incendio del Rodenal.* Consejería de Medio Ambiente y Desarrollo Rural de Castilla la Mancha. ISBN 978-84-7788-521-4. Castilla la Mancha, Spain.

Schroeder L.M., 1992. Olfactory recognition of nonhosts aspen and birch by conifer bark beetles *Tomicus piniperda* and *Hylurgops palliatus*. *Journal of Chemical Ecology*, Vol.18, No.9, (September 1992), pp. 1583-1593, ISSN 0098-0331

Scriber, J.M. (1984). Host-plant suitability. In: *Chemical Ecology of Insects*, W.J. Bell and R.T. Cardé (Eds.), 159-202. Chapman and Hall, ISBN 978-087-8930-708, London, United Kingdom.

Seifert, K.A. (1993). Sapstain of commercial lumber by species of *Ophiostoma* and *Ceratocystis*. In: *Ceratocystis and Ophiostoma: Taxonomy, Ecology, and Pathogenicity.* M.J. Wingfield, K.A. Seifert and J.F. Webber (Eds.), 141-151, APS Press, ISBN 978-0-89054-156-2, St. Paul, United States.

Serez, M. (1987). Use of the aggregation pheromone preparation 'Ipslure' against the Mediterranean pine bark-beetle *Ips* (*Orthotomicus*) *erosus* (Woll.) (Col., Scolytidae). *Anzeiger für Schädlingskunde, Pflanzenschutz und Umweltschutz*, Vol.60, pp. 94-95, ISSN 0340-7330

Song, L.W., Ren, B.Z., Sun, S.H., Zhang, X.J., Zhang, K.P. & Gao, C.Q. (2005). Field trapping test on semiochemicals of pine shoot beetle *Tomicus piniperda* L. *Journal of Northeast Forestry University*, Vol.33, No,1, (January 2005), pp. 38-40, ISSN 1000-5382

Storer, A.J., Wood, D.L. & Gordon, T.R. (2004). Twig beetles, *Pityophthorus* spp. (Coleoptera : Scolytidae), as vectors of the pitch canker pathogen in California. *The Canadian Entomologist*, Vol.136, No.5, (September 2004), pp. 685-693, ISSN 0008-347X

Viljoen, A., Wingfield, M.J. & Marasas, W.F.O. (1994). First report of *Fusarium subglutinans* f.sp. *pini* on pine seedlings in South Africa. *Plant Disease*,Vol. 78, No.4, (April 1994), pp. 309-312, ISSN 0191-2917

Villarreal, M., Rubio, V., De Troya, M.T. and Arenall, F. (2005). A new *Ophiostoma* species isolated from *Pinus pinaster* in the Iberian Peninsula. *Mycotaxon*, Vol.92, No., pp. 259-268, ISSN 0093-4666

Vité, J.P., Bakke, A. & Renwick, J.A.A. (1972). Pheromones in *Ips* (Coleoptera: Scolytidae): occurrence and production. *The Canadian Entomologist*, Vol.104, No.12, (December 1972), pp. 1967-1975, ISSN 0008-347X

Vité, J.P., Volz, H.A., Paiva, M.R. & Bakke, A. (1986). Semiochemicals in host selection and colonization of pine trees by the pine shoot beetle, *Tomicus piniperda*. *Naturwissenschaften*, Vol.73, No.1 (January 1986), pp. 39-40, ISSN 0028-1042

Volz, H.A. (1988). Monoterpenes governing host selection in the bark beetles *Hylurgops palliatus* and *Tomicus piniperda*. *Entomologia Experimentalis et Applicata*, Vol. 47, No.1 (January 1988), pp. 31-35, ISSN 0013-8703

Westfall, J. (2006). Summary of forest health conditions in British Columbia. In: *Forest Health Conditions in 2006*, 01.08.2011. Available from http://www.for.gov.bc.ca/ftp/HFP/external/!publish/Aerial_Overview/2006/Aer_OV_final.pdf

Wood, D.L. (1982). The role of pheromones, kairomones, and allomones in the host selection and colonization behavior of bark beetles. *Annual Review of Entomology*, Vol.27, pp. 411-446, ISSN 0066-4170

Wood, D.L., Browne, L.E., Bedard, W.D., Tilden, P.E., Silverstein, R.M. & Rodin, J.O. (1968). Response of *Ips confusus* to synthetic sex pheromones in nature. *Science*, Vol.159, No.3860, pp. 1373-1374, ISSN 0036-8075

Wood, S.L. (1982). The bark and ambrosia beetles of North and Central America, a taxonomic monograph. *Great Basin Naturalist Memoirs*, Vol.6, pp. 1-1359, ISSN 0160-239X

Yamaoka, Y., Swanson, R.H. & Hiratsuka, Y. (1990). Inoculation of lodgepole pine with four blue-stain fungi associated with mountain pine beetle, monitored by a heat pulse velocity (HPV) instrument. *Canadian Journal of Forest Research*, Vol.20, No.1 (January 1990), pp. 31-36, ISSN 1208-6037

Zhang, Q.H., Liu, G.T., Schlyter, F., Birgersson, G., Anderson, P. & Valeur, P. (2001). Olfactory Responses of *Ips duplicatus* from Inner Mongolia, China to Nonhost Leaf and Bark Volatiles. *Journal of Chemical Ecology*, Vol.27, No.5, (May 2001), pp. 995-1009, ISSN 0098-0331

Zhang, Q.H. & Schlyter, F. (2003). Redundancy, synergism, and active inhibitory range of non-host volatiles reducing pheromone attraction in European spruce bark beetle *Ips typographus*. *Oikos*, Vol.101, No.2, (May 2003), pp. 299-310, ISSN 0030-1299

Zhang, Q.H. & Schlyter, F. (2004). Olfactory recognition and behavioural avoidance of angiosperm nonhost volatiles by conifer-inhabiting bark beetles. *Agricultural and Forest Entomology*, Vol.6, No.1, (January 2004), pp. 1-19, ISSN 1461-9563

15

Biological Studies and Pest Management of Phytophagous Mites in South America

Carlos Vásquez[1,*], José Morales-Sánchez[1], Fernando R. da Silva[2] and María Fernanda Sandoval[3]

[1]*Universidad Centroccidental Lisandro Alvarado, Decanato de Agronomía Departamento de Ciencias Biológicas. Barquisimeto, Estado Lara*

[2]*University of Amsterdam. Institute for Biodiversity and Ecosystem Dynamics (IBED) Research Group of Population Biology, Amsterdam*

[3]*Instituto Nacional de Salud Agrícola Integral (INSAI). Av. Principal Las Delicias. Edif. INIA – Maracay, Estado Aragua*

[1,3]*Venezuela*

[2]*The Netherlands*

1. Introduction

Mites are the most diverse representatives of an ancient lineage in phylum Arthropoda-subphylum Chelicerata- subclass Acari. Their body plan is strikingly different to that of other arthropods in not having a separate head, instead, an anterior region, the cephalothorax, combines the functions of sensing, feeding, and locomotion (Walter & Proctor 1999). Antennae, mandibles and maxillae are also absent; rather, a pair of often pincer-like mouthparts are present, the so-called chelicerae. Those members of the subclass Acari, which feed on plants, are known as phytophagous mites. Mites constitute one of the most heterogeneous cheliceran groups, since they are extremely diverse in their morphology, biology and ethnology, enabling them to colonize different environments. Their remarkable diversity in acarine morphology is reflected in the variety of ecological and behavioral patterns that mites have adopted (Krantz 2009). Thus, species can inhabit soil, litter (i.e. Cryptostigmata, Mesostigmata and Prostigmata), water (Hydrachnidia) or plants (Prostigmata or Mesostigmata). Phytophagy is widespread enough among the Trombidiform Acariformes so as to suggest that there was an early evolution commitment to plant feeding by several primitive predaceous and saprophagous trombidiform lineages (Krantz 2009). Some Prostigmatan mites, chiefly spider mites, false spider mites and eriophoid mites, use their specialized mouthparts to feed on the vascular tissues of higher plants and with their activity they can cause losses to field and protected crops (Evans 1992), becoming economically important pests.

This review summarizes more important phytophagous mites in tropical crops in South America, biological aspects, damage, and also main control strategies in tropical conditions.

*Corresponding Author

2. Economically important mite species

Species of agricultural importance can exhibit either phytophagous or predatory habits. Most important taxa including exclusively phytophagous mites are Eriophyoidea, Tetranychoidea (Tetranychidae, Tenuipalpidae). Tarsonemidae is also a family of mites which includes several pest species. All these taxa cover the most important crop pest species distributed worldwide, and several more geographically restricted species. The Eriophyoidea is a large superfamily of worldwide distribution. Over 3,000 species belonging to about 250 genera are known in the world. These worm-like or fusiform mites cause many forms of plant abnormalities such as galls, leaf blisters and rusts. Most eriophyid mite species are monophagous or are limited to plant species within a single genus. Some rust mites and gall mites are important pests on economic plants.

The Tetranychidae, also known as spider mites, is a large family including of some 1,200 species belonging to over 70 genera of worldwide distribution. Spider mites cause mechanical damage by sucking cell content from leaves. At first, it shows up as a stippling of light dots on the leaves; and sometimes leaves became bronze in color. As feeding continues, the leaves turn yellow and drop off. Often leaves, twigs, and fruit are covered with a large amount of webbing. In Tenuipalpidae, also known as false spider mites or flat mites, about 800 species have been described in over 25 genera. Only *Brevipalpus, Tenuipalpus* and *Raoiella* and few other genera become pests of economic plants, mainly on tropical fruit crops and ornamental plants. Virus transmission has only been documented in some Eriophyidae or Tenuipalpidae species.

Due to the economic losses caused by mite pests, management tactics need to be established to keep population levels under the economic threshold of infestation. This practice should be based on integrated pest management (IPM) including spraying chemical products, using biological control agents and/or resistant varieties. Currently, chemical control has to deal with serious control failures in mite populations, since the evolution of pesticide resistance in phytophagous mites is very common. Consequently, different chemical molecules are being currently developed to face this phenomenon, mainly in spider mites. However, the low level of immigration of susceptible individuals and the rapid reproductive rate associated to these mite groups have made it difficult to manage population in crops.

Biological control is an environmentally safe, cost-effective and energy efficient pest control, either on its own or as a component of integrated pest management. Although several mite species belonging to Bdellidae, Cheyletidae, Cunaxidae, Stigmaeidae and Tydeidae have shown predatory habits, Phytoseiidae mites have been more widely included in biological control programs, due to their capacity for surviving and reproducing on other arthropods. Additionally, some phytoseiid mites have shown to be resistant or less susceptible to chemical compounds commonly used to control pest mites in commercial crops, thus making them suitable for their use in integrated pest management programs.

3. Spider mites

This family consists of two subfamilies: Bryobinae and Tetranychinae. Spider mites occur on most of the major food crops and ornamental plants in almost all environments where mainly Tetranychinae species can potentially cause economic damage.

Spider mites are soft-bodied, medium-sized mites. They are often red, green, orange or yellow in color when alive. The gnathosoma has a capsule-like structure known as the stylophore, which is formed by the fusion of chelicerae. The movable digit of the chelicerae is very long, often whip-like and recurved proximally. A pair of stigmata is located near the base of the chelicerae, where the peritremes arise. The palps are five-segmented. The palp-tarsus and tibia often form a thumb-claw process. The tarsus often has an enlarged distal eupathidium (spinneret) in the Tetranychinae and this is used to spin webbing in many species. The size and shape of the spinneret is of taxonomic significance. The idiosoma is often covered with a striate cuticle. The pattern of the striation and the shape/density of lobes distributed on the striae are useful diagnostic characters.

There are three or four pairs of normal setae in two rows (*v*1-2, *sc*1-2) and two pairs of eyes on the dorsal propodosoma. On the opisthosomal dorsum, there are five rows of setae: *c, d, e, f* and *h*. The number, location, length and structure of dorsal setae are of taxonomic significance. Female genital pores are transverse and are bordered anteriorly by a genital flap and laterally by characteristic cuticular folds. The structures of the paired lateral claws and the medial empodium are of taxonomic importance. The empodium may be claw-like or pad-like with tenant hairs. Claws may bear dorsal or ventral hairs. The tarsi of legs I and II bear duplex setae (a long solenidion and a short normal tactile seta with their bases joined together. The number of duplex setae and their positions are of taxonomic significance.

Wedge-shaped males are smaller than ovoid females and have a tapering opisthosoma. Males have a protrudable aedeagus, the shape of which is very important in species identification.

Life history and biology: as other Prostigmata mites, the spider mite life cycle consists of egg, larva, protonymph, deutonymph and adult stages, except for some *Schizotetranychus* and *Eotetranychus* species, which may have one nymphal stage in males (Zhang 2003). Moulting takes place during the quiescent stages between each active stage. Development from egg to adult often takes one to two weeks or more, depending on mite species, temperature, host plants, humidity and other environmental factors. Males develop slightly faster than females and soon they search for and fight for quiescent deutonymph females. Unfertilized eggs produce only haploid males, while fertilized eggs produce diploid females.

3.1 Spider mite species of economic importance in South America

3.1.1 *Tetranychus urticae* Koch

The two-spotted spider mite (TSSM) is considered one of the most harmful tetranychid species in agriculture both in temperate or tropical countries (Cerna *et al.* 2009, González Zamora *et al.* 1993). Although a recent checklist includes some 920 host plant species in 70 genera (Bolland *et al.* 1998), only 150 of these host species have relevant economic value. The TSSM feeds on cell chloroplast on the underside of the leaves and symptoms become clearly visible on the upper side as characteristic whitish or yellowish punctures which, under high population levels, can join and become brownish o even cause leaf drop and plant death (Tomczyk & Krompczyńska 1985, Zhang 2003).

Due to its relevance and economic impact on tropical agriculture in South America, the TSSM has been the most extensively studied spider mite species. A number of papers related to biological aspects of the TSSM are available (Table 1).

Host Plants	Life cycle (days)	N° eggs/female/day	Longevity (days)	Reference
Sweet pepper	8.2 (27°C, 70% RH)	2.6	12.2	Gallardo *et al.* (2005)
Cotton	Non Bt 7.5 Bt 7.3 (25°C, 57.4% RH)		Non Bt: 16.7 Bt: 16.6	Esteves Filho *et al.* (2010)
Gerbera	21.6 (25°C, 70% RH)	3.75	8.83	Silva *et al.* (2009)

Table 1. Some biological parameters of the TSSM in various host plant and environmental conditions in South America.

Population management: For decades population management of tetranychid mites has been based mostly on chemical control in South America. However, biological control is being increasingly used under greenhouse conditions, mainly in Argentina, Brazil, and Colombia. In Colombia reductions of TSSM population on rose trees have been observed after periodical release of *Neoseiulus* sp. (Forero *et al.* 2008). Even though a lower number (19.4%) of phytophagous mites was found in plots chemically treated, a lower percentage (damage index 1 and 3, 8 and 13% less, respectively) of leaf damage was observed when *Amblyseius* sp. was released. On the other hand, the consumption rate (functional response) after a 24 h. period was 6.66 eggs, 18.06 larvae, and 19.15 nymphs under laboratory conditions and 4.56 eggs, 12.65 larvae, and 15.71 nymphs under greenhouse conditions (Forero *et al.* 2008). Greco *et al.* (2005) found that *Neoseiulus californicus* (McGregor) is a promising agent for successful TSSM control through conservation techniques, on strawberry crops in La Plata, Argentina. Accordingly, the authors showed that initial relative densities had an important effect on system dynamics. Thus, when pest/predator ratio was 5/1 (at initial pest densities from 5 to 15 females/leaflet) the final number of active *T. urticae*/leaflet was significantly lower than the economic threshold level (ETL), while at 20 females/leaflet this number did not differ from the ETL. At 7.5/1 ratio, the final number of active *T. urticae*/leaflet, at initial pest densities from 5 to 15 females/leaflet, reached the ETL without surpassing it. At 10/1 and 15/1 ratios, pest densities exceeded the ETL only at 15 initial *T. urticae*/leaflet. Since *N. californicus* showed to be very effective in limiting pest densities, conservation of this predator promoting favorable pest/predator ratios may result in early control of *T. urticae*.

Furthermore, *Phytoseiulus macropilis* (Banks) demonstrated a positive functional response (increases prey consumption). However, all indices evaluated showed that *P. macropilis* was unavailable to control the TSSM efficiently when the population numbers were low, possibly to keep itself in an environment with low populations (Ferla *et al.* 2011). Biological control by this predator is more efficient when five or more prey items are present at a given time.

3.1.2 *Tetranychus cinnabarinus* (Boisduval)

This species is commonly known as the carmine spider mite and it is associated with more than 120 host plant species such as cotton, strawberry, tomato, eggplant, and also ornamental species and fruit trees (Biswas *et al.* 2004). In Chile, the carmine spider mite is commonly found on carnation, strawberry, melon and beans from Arica, Parinacota and dessert areas where chemical control is frequently used (Klein & Waterhouse 2000).

Tello *et al.* (2009) found that the life cycle (egg-adult) of *T. cinnabarinus* on carnation var. Celta lasted 12.8 days (29.4 °C, 42.3 RH and 14:10 h (L:D) photo phase) and female longevity and mean daily oviposition rate were 24.28 days and 3.92 eggs/female/day, respectively. Under these conditions life table parameters were as follows: the intrinsic rate of increase (r_m) 0.183; the finite rate of increase (λ) 1.201 individuals/female/day; the mean generation time (T) was 20.24 days; and the net rate of reproduction (R_0) was 40.809.

Control strategies: most efforts to control pest population densities under threshold levels have relied on chemical tactics; however, particular attention has been paid to other management strategies in last decades, including resistant cultivars. So far, kidney bean has shown moderate resistance to other arthropod pests such as thrips (Cardona *et al.* 2002) and it seems to be based on antixenosis and antibiosis mechanisms (Frei *et al.* 2003). In general, these defense mechanisms are considered plant responses to stressing conditions either abiotic (drought, salinity) or biotic (herbivore or pathogens attacks), which induce the development of physical barriers to prevent feeding or secondary metabolites affecting oviposition or survival (Tomczyk & Krompczyńska 1985, Gardner & Agrawal 2002). Vásquez *et al.* (2007) observed that mean oviposition of *T. cinnabarinus* females was significantly lower on 22-day-old-leaf disks, ranging from 0.98 to 1.17 eggs/female/day on ICA-Pijao and Coche beans cultivars, respectively (Table 2), while females reared on 55 day-old leaf disks oviposition increased about 58 or 95%

		Number of eggs/female	
		22 day-old	55 day-old
Cultivar	n		
Tacarigua	25	1.16 a ± 0.7342 (34)	1.66 b ± 0.8119 (72)
Coche	25	1.17 a ± 0.5905 (30)	1.86 b ± 0.7126 (86)
ICA-Pijao	25	0.98 a ± 0.3852 (15)	1.92 b ± 0.8907 (99)

Values in a file followed by same letter did not show significant differences (Tukey's Test; $P<0.05$) Number in parenthesis represent total egg number during a five day period. From Vásquez *et al.* (2007).

Table 2. Oviposition (mean ± S.D.) of *T. cinnabarinus* on 22 or 55 days-old leaf disks from different kidney bean cultivars.

In general *T. cinnabarinus* females showed a higher rate of survival on 22 day-old disks, ranging from 5.67 to 6.63 living individuals in Tacarigua and Coche leaves; while on 55 day-old leaves survival ranged from 5.42 to 1.80 on Coche and ICA-Pijao, respectively. A significant survival reduction was observed on mites reared on 55 day old leaf disks from ICA-Pijao cultivar (Fig. 1), suggesting that this cultivar could dissuade to mite from feeding, affecting thus mite survival.

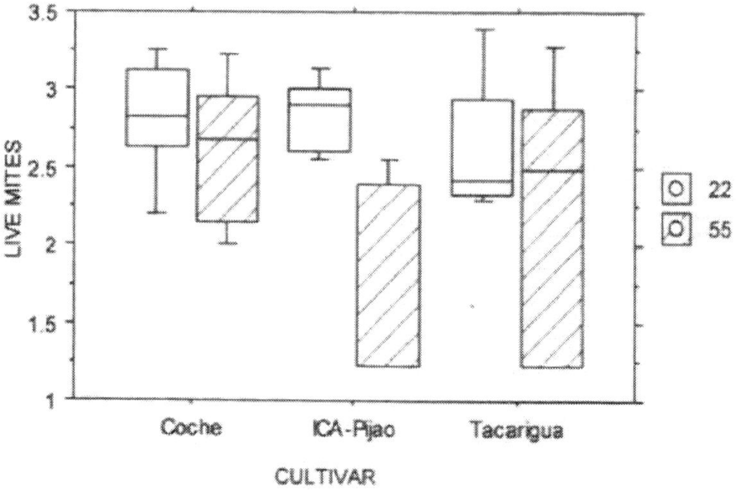

Fig. 1. Number of live mites on kidney bean cultivars on 22- or 55 day-old leaf disks. From Vásquez *et al.* (2007).

3.1.3 Red spider mites (*Tetranychus desertorum* Banks and *Tetranychus ludeni* Zacher)

The red spider mite, *T. desertorum* has been found on about 173 plant species in Argentina, Brazil, China, the United States of America, Mexico, Venezuela and some other countries (Bolland *et al.* 1998). This spider mite species might cause severe damage to bean plants, as *T. urticae* does (Moraes & Flechtmann 2008). In Venezuela, previous studies have demonstrated that bean cultivars are susceptible to *T. desertorum* attacks under irrigation conditions (Doreste 1984). Rivero & Vásquez (2009) showed that total developmental time of *T. desertorum* females on kidney bean 'Tacarigua' was 6.8 days, with partial duration of immature stages corresponding to 3.8, 1.4, 1.0 and 0.7 days for egg, larva, protonymph and deutonymph, respectively. Higher mean fecundity (6.93 eggs/female/day) was observed on day 4 and females lived during 10 days. The recorded life table parameters were as follows: net reproduction rate (Ro) = 41.10 individuals; generation time (T) = 11.15 days; intrinsic natural growth (rm) = 0.144 individuals/female/day, and finite natural increase rate (λ) = 1.155 individuals/female.

Furthermore, *T. ludeni* is a worldwide pest that attacks various plant species, such as *Phaseolus vulgaris* L., representing a severe threat in several countries, including Venezuela (Morros & Aponte 1994). According to these authors, the life cycle (egg-adult) of *T. ludeni* lasted 9.98 and 9.25 days for females and males, respectively (26.34± 3.92°C and 69.44± 19.44% HR). The life table also showed the following values: reproduction rate (Ro) 77.42; mean generation time (T) 19.63; intrinsic rate of natural increase (r) 0.2526 individuals/female/day; finite rate of natural increase (λ) 1.2874 individual/female/week. Morros & Aponte (1995a, b) evaluated the effect of two levels of mite infestation on vegetative and reproductive stages of the black bean *P. vulgaris* under greenhouse and field conditions. The authors observed greater reduction on number of leaves and internodes, leaf area, and dry weight of vegetative organs when mite infestation occurred in the early vegetative stage (Table 3).

	Greenhouse		Field	
	Development stage			
	Vegetative	Reproductive	Vegetative	Reproductive
Foliar area (cm^2)	263.65	289.84	30.76	36.13
Number of leaves	22.83	25.00	34.54	45.72
Number of internodes	12.77	14.50	13.13	14.63
Dry leaf weight (g)	2.80	3.35	3.28	6.10
Dry plant weight (g)	1.61	1.80	2.41	3.83
Dry pod weight (g)	1.82	1.88	0.36	0.56
Total dry weight (g)	7.26	8.27	6.06	10.50

From Morros & Aponte (1995a, b).

Table 3. Reductions of plant performance as effect of *T. ludeni* feeding on black beans in vegetative or reproductive stages under greenhouse and field conditions.

3.1.4 The avocado brown mite, *Oligonychus punicae* Hirst

The avocado brown mite (ABM) is considered an important tetranychid pest in southern California (USA), as high population densities can cause severe defoliation on several avocado cultivars (McMurtry & Johnson 1966; McMurtry 1985). This tetranychid mite predominantly feeds on the upper leaf surface, although feeding could extend to the lower leaf surface at high population levels (Tomczyk & Kropczynska 1985). Damage to host plants is shown by a bronze tone on the leaves, and is associated with the rates of oviposition and female production (McMurtry 1970).

In tropical America, the ABM has been reported in more than 20 plant species, such as *Mangifera indica* L., *Musa sapientum* L., *Punica granatum* L. and *Vitis vinifera* L. (Ochoa *et al.* 1994; Bolland *et al.* 1998). In Venezuela, it was previously recorded on *Musa* spp. from Sur del Lago, Zulia State (Freitez & Alvarado 1978; Quirós 1978). More recently, in Lara State, it has been observed as an occasional pest whose feeding on grape delays fruit ripening (Vásquez *et al.* 2008a). These authors provided information about life cycle, fecundity and longevity of the ABM on six grapevine cultivars (Tucupita, Villanueva, Red Globe, Sirah, Sauvignon and Chenin Blanc), under laboratory conditions at 27 °C, 80% RH, and L12:D12 photoperiod (Tables 4 and 5).

Periods of pre-oviposition, oviposition and post-oviposition in *O. punicae* females varied among grape cultivars. Shorter pre- and post-ovipositional periods were found on Chenin Blanc leaves (1.2 and 0.9 days, respectively), while on others cultivars these parameters ranged from 1.6-2.0 to 1.2-2.5 days, respectively (Table 5). Oviposition periods lasted from 6.7 days (on Villanueva) to 16.1 days (on Sauvignon). Average daily egg production was highest on Tucupita (2.8 eggs/female/day) and lowest on Sirah (0.9 eggs/female/day). Daily oviposition rate ranged from 2.0 to 6.1 eggs/female up to day 7 in females feeding on Tucupita leaves, while on Chenin Blanc, Red Globe, Sauvignon, Sirah and Villanueva leaves it varied from 1.1-6.1, 0.2-5.3, 0.3-5.0, 0.2-2.7 to 0.5-3.5 eggs/female, respectively (Fig. 2). Also, female longevity of *O. punicae* was affected by grape cultivar: females lived longest on Sauvignon (17.5 days), and shortest on Villanueva (8.1 days)

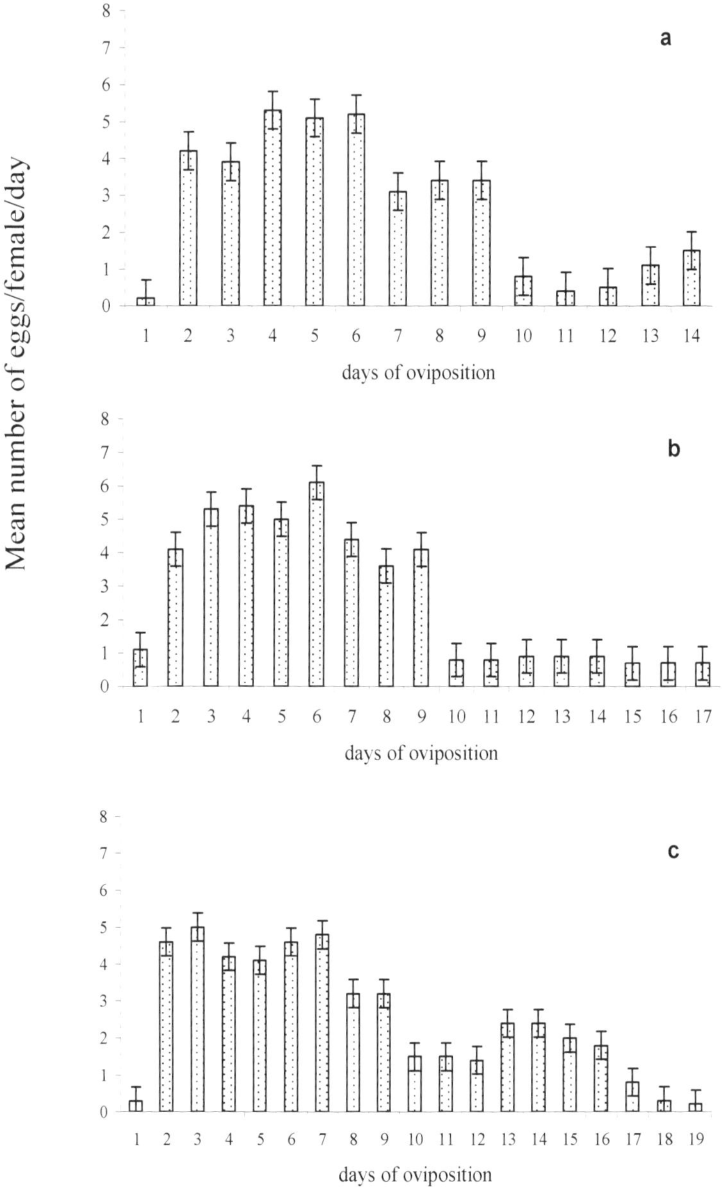
Mean number of eggs/female/day
a
b
c
days of oviposition
days of oviposition
days of oviposition

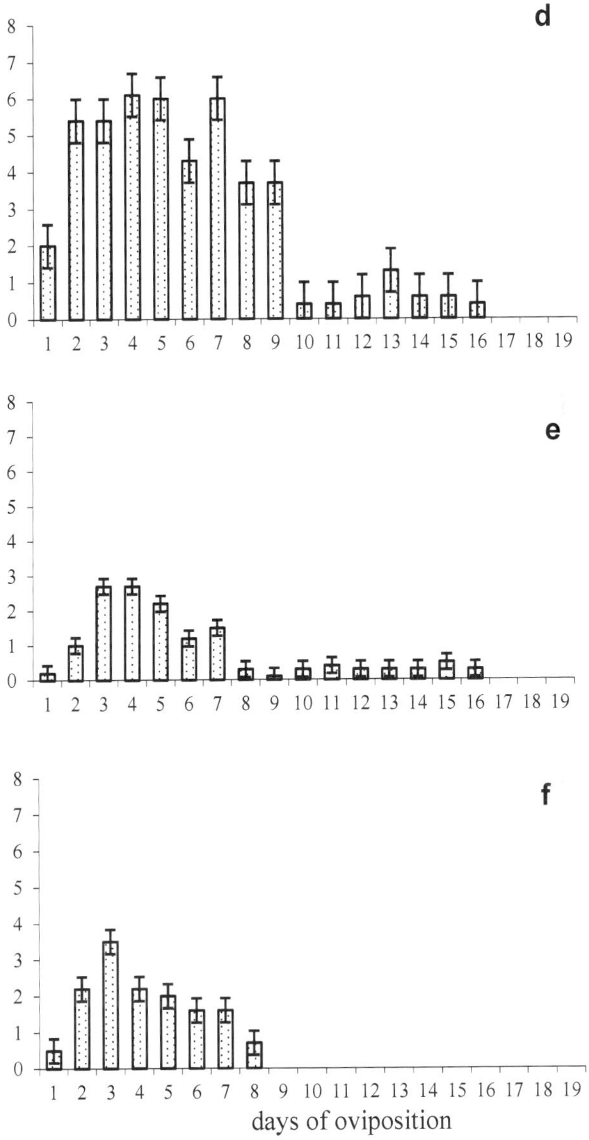

Fig. 2. Daily oviposition rate of *O. punicae* females feeding on grapevine leaves cultivars Red Globe (a), Chenin Blanc (b), Sauvignon (c), Tucupita (d), Sirah (e) and Villanueva (f).

	Mean duration (hours) ± SD					
Stages	Tucupita	Sauvignon	Villanueva	Red Globe	Chenin Blanc	Sirah
Egg	104.80± 5.9330a	106.50± 7.5498a	112.80± 5.215a	105.20± 8.438a	105.60± 6.066a	111.20± 3.347a
Larva	22.40± 1.3416ab	20.25± 4.1130b	22.40± 1.949ab	21.00± 6.042b	29.60± 4.775a	27.60± 4.506ab
Protochrysalis	11.80± 1.4832ab	15.75± 3.403a	10.80± 1.304b	13.80± 3.115ab	13.40± 1.517ab	11.60± 1.517ab
Protonymph	13.00± 2.8284b	14.75± 2.363ab	15.60± 6.229ab	19.00± 2.828ab	21.40± 4.930a	16.60± 5.413ab
Deutocrhysalis	11.80± 1.3038a	11.00± 0.8165a	11.80± 2.950a	12.20± 3.347a	13.50± 4.359a	11.60± 2.074a
Deutonymph	22.60± 2.7928a	17.00± 1.414a	23.60± 3.578a	21.80± 2.775a	20.00± 6.733a	24.40± 5.899a
Teliochrisalis	11.20± 1.9235ab	9.25± 3.403b	10.80± 1.924ab	15.00± 1.414a	11.25± 1.500ab	10.80± 2.683ab
Life cycle[a]	197.6± 4.5056b (8.16)	198.40± 8.5391b (8.27)	208.80± 4.5497ab (8.70)	209.20± 12.6770ab (8.72)	215.00± 13.7840ab (8.96)	217.20± 8.8148a (9.05)

Values in a row followed by the same letter are not significantly different according to the Tukey's multiple comparison differences (P=0.0036, df=5, F=4.78).
[a]Values in brackets, days to complete development. From Vásquez *et al.* (2008a).

Table 4. Developmental time of different life stages of *O. punicae* in various grapevine cultivars.

Cultivars	N	Longevity	Preoviposition[(1)]	Oviposition [(1)]	Postoviposition [(1)]	Daily fecundity[(2)]
Tucupita	10	12.60 ± 5.1897abc	1.20 ± 0.4216b	11.40 ± 3.2042bc	1.20 ± 0.4216ab	2.82 ± 2.4186a
Sauvignon	10	17.50 ± 4.8028a	1.80 ± 0.4216ab	16.10 ± 2.8460a	1.40 ± 0.8433ab	2.15 ± 1.4130bc
Red Globe	10	13.20 ± 3.3417abc	1.80 ± 0.4216ab	11.10 ± 2.6013bc	1.30 ± 1.0593ab	2.72 ± 1.9132ab
Chenin Blanc	10	14.40 ± 3.8930ab	1.20 ± 0.4216b	13.50 ± 3.3417ab	0.90 ± 1.1005b	2.16 ± 1.9949ab
Villanueva	10	8.10 ± 1.1972c	1.60 ± 0.6992ab	6.70 ± 1.1595d	1.40 ± 1.2649ab	1.79 ± 0.9433cd
Sirah	10	10.30 ± 3.4383bc	2.00 ± 0.6667a	7.90± 4.9318cd	2.50 ± 1.3540a	0.94 ± 0.9532d

[(1)]Transformed values by $y = \sqrt{x + 0.5}$ [(2)] Transformed values by $y = \sqrt{x + 1.5}$

Values in a column followed by different letter showed significant differences. P<0.05 (preovip: F=4.21; ovip: F=12.8; and postovip: F=2.36; df=5); P<0.001 (longevity: F= 7.05; df=5). From Vásquez *et al.* (2008a).

Table 5. Adult longevity, preoviposition, oviposition and postoviposition times (days) and fecundity (eggs/female/day) of *O. punicae* on several grapevine cultivars.

An effect of host plant on mite reproduction has been established for several Tetranychid species (e.g., de Ponti 1977; Ribeiro *et al.* 1988; Hilker & Meiners 2002; Praslička & Huszár 2004). Previous studies have demonstrated that grape leaves and fruits synthesize phenolic compounds in response to fungal attacks or abiotic factors (Morrissey & Osbourn 1999). Furthermore, Harborne (1994) hypothesized that low molecular weight phenol compounds could act synergistically with tannins to provide plant resistance. Thus, reproductive parameters of *O. punicae* seemed to be associated with flavonoid content of grape cultivars; the higher the flavonoids content, the lower the mites' fecundity (Fig. 3). These findings could be considered an ecological approach for sustainable pest management programs in grapevine in tropical areas.

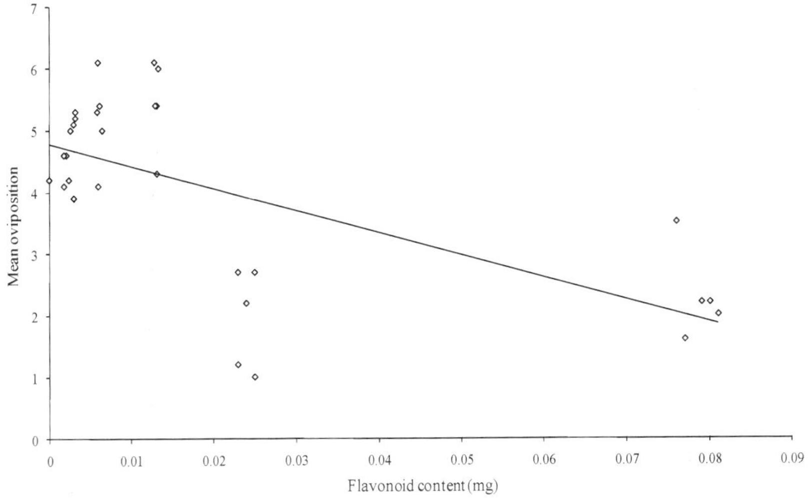

Fig. 3. Lineal regression between flavonoid content and mean oviposition of *O. punicae* on grapevine cultivars (y = 4.78 - 36.42x). From Vásquez *et al.* (2008a).

4. Tenuipalpidae mites

The false spider mites include about 800 described species belonging to over 25 genera, mostly found in tropical and subtropical areas (Jeppson *et al.* 1975, Baker & Tuttle 1987). This family consists of three subfamilies: Tegopalpinae, Brevipalpinae and Tenuipalpinae, the two latter ones including most of the described species: *Brevipalpus* and *Tenuipalpus* are the two largest genera and also the most economically important pest on citrus (Kitajima *et al.* 1972), coffee (Chagas *et al.* 2000), passion-fruit (Kitajima *et al.* 1997) and ornamentals (Smith-Meyer 1979). Since 2004, *Raoiella indica* Hirst is also becoming more important in the Caribbean islands and northern countries in South America (Vásquez *et al.* (2008b).

The false spider mites are also known as flat mites because most species are dorsoventrally flattened. They are slow-moving and are usually found on the lower surface of the leaves near the midrib or veins. Some species feed on the bark while others live in flower heads, under leaf sheaths or in galls. Only a small number of species belonging to a few genera have become pests of economic plants and they are most commonly found on tropical fruit crops and ornamental plants.

Life history and biology: Thelytoky is commonly observed in *Brevipalpus* mites, since female offspring consist in females and rarely males are found (Childers *et al.* 2001), being males and females haploid (Pijnacker *et al.* 1980). Life cycle of *Brevipalpus* is shown in figure 4, consisting in four active stages: larva, protonymph, deutonymph and adult; an inactive stage being observed between each active one. According to Goyal *et al.* (1985), developmental rate depends on temperature, relative humidity and host plant species.

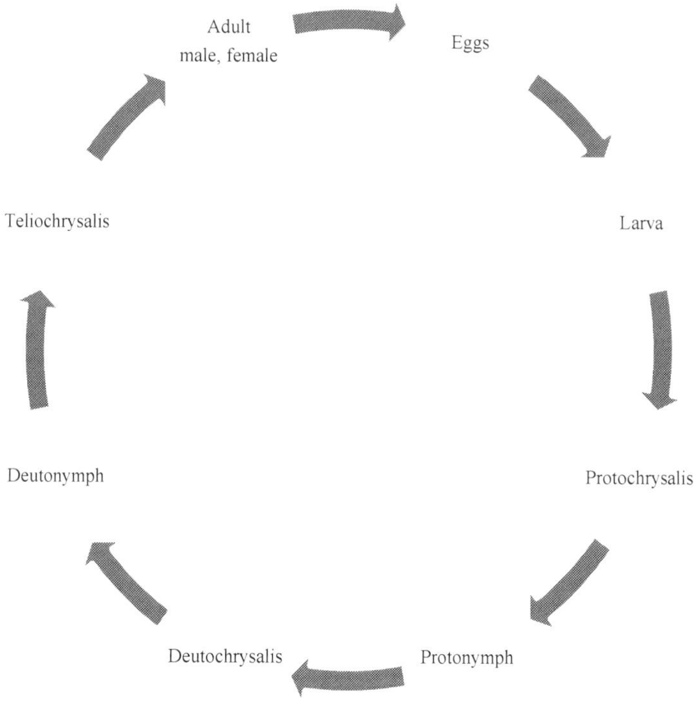

Fig. 4. Life cycle of *Brevipalpus* mites. From Childers *et al.* (2001).

4.1 Tenuipalpidae mite species of economic importance in South America

4.1.1 *Brevipalpus and Tenuipalpus* species

Brevipalpus californicus (Banks), *B. obovatus* (Donnadieu) and *B. phoenicis* (Geijskes) constitute the major economically important species in this genus. These mites usually feed on lower leaf surface and aggregate along the mid-vein or major lateral veins and they inject toxic saliva into fruits, leaves, stems, and bud tissues of their host plant, while *Tenuipalpus pacificus* Baker and

B. phoenicis have been observed feeding on the upper leaf surface of orchids (Childers *et al.* 2003). Ochoa *et al.* (1994) reported 177 host plant species in Central America, including 114 host plants for *B. phoenicis*, 29 for *B. californicus* and 34 for *B. obovatus*.

Probably the major threat related to these three *Brevipalpus* species is that concerning their ability to vectoring a virus borne disease called leprosis (Kitajima *et al.* 1996). Virus particles of citrus leprosis are short, bacilliform, 120–130 nm long (occasionally up to 300 nm), and 50–55 nm wide and occur mostly in parenchyma cells of the lesion area in affected orange leaves, fruits, or stems (Rodrigues *et al.* 2003). Leprosis is a serious disease on citrus from Brazil, Argentina, Paraguay and Venezuela, and probably Colombia and Uruguay (Childers *et al.* 2001). More recently, leprosis has been recorded in Panama (Dominguez *et al.* 2001) and it still represents a long term problem and causes severe damage on trees, yield reduction, and even death of trees if mite population control measurements are not taken into account (Rodrigues 2000).

4.1.2 The red palm mite, *Raoiella indica* Hirst

The red palm mite (RPM) was firstly reported in the Caribbean region in 2004 (Flechtmann & Etienne 2004). It is currently widely distributed in most of the Caribbean islands (Kane *et al.* 2005, Rodrigues *et al.* 2007). Subsequent mainland colonization was verified in Venezuela (Vásquez et al. 2008b) and Florida (Peña *et al.* 2008); and more recently, the pest has been reported to occur in Mexico (NAPPO 2009), Brazil (Návia *et al.* 2011) and Colombia (Carrillo *et al.* 2011). The RPM may cause severe damage to Arecaceae, especially coconut (*Cocos nucifera* L.), but also on Musaceae and other plant families (Flechtmann & Etienne 2004, Flechtmann & Etienne 2005, Etienne & Flechtmann 2006).

In coconut nurseries, mite feeding can provoke death of younger plants; meanwhile damage in adult plants is evidenced by a strong yellowish color in mature leaves (Peña *et al.* 2006, Welbourn 2005). After the red palm mite detection in the Americas, 70–75% fruit production decreasing has been estimated in Trinidad and Tobago and Venezuela (Návia 2008). Although the RPM primarily infests palms and bananas, some wild heliconia, gingers and other ornamental plant species growing under heavily infested coconut trees revealed small and sporadic colonies of the red palm mite (Roda 2008). However, according to Hoy *et al.* (2006), it is not clear whether these are valid host plants or whether the enormous mite populations on coconuts and other palms have temporarily moved on to substory plants under the palms.

Currently, the governments and researchers from those countries where the RPM has been recorded are making efforts to implement an alternative to pest management different from that of chemical control, since application of chemical products is not only hazardous to environment and public health, but also expensive. Thus, research is devoted to test the effectiveness of the predatory mite, *Amblyseius largoensis* Muma, the most frequent phytoseiid mite species found in association to the RPM in the Caribbean islands and Venezuela. Preliminary research has shown that this predator seems to be a promising agent of biological control (Carrillo *et al.* 2010, Rodríguez *et al.* 2010). However, when being considered in a population management program, other ecological strategies should also be included, such as using plant resistance from different genotypes which have been obtained in northern Venezuela by Dr. Antonio Ruiz (Instituto Nacional de Investigaciones Agrícolas [INIA-Irapa]), organic or inorganic fertilization programs in order to reestablish soil fertility and lastly, irrigation to avoid abiotic stress in old and young coconut plantations.

5. Eriophyid mites

The Eriophyoidea is a large superfamily of worldwide distribution. About 3,000 species belonging to over 250 genera are known in the world and a number of new species have been recently described from natural vegetation in South America (Table 6). The superfamily consists of three families: Phytoptidae, Diptilomiopidae and Eriophyidae. About three-quarters of the described species in Eriophyoidea belong to the Eriophyidae.

Species	Host plant	Country	Reference
Abacarus nectandrae *Aceria megalops*	*Nectandra membranacea* (Lauraceae) *Guapira opposita* (Nyctaginaceae)	Brazil	Flechtmann & Moraes (2002b)
Acalitus santibanezi	*Ipomoea murucoides* (Convolvulaceae)	Mexico	García-Valencia & Hoffmann (1997)
Aceria anisodorsum	*Caesalpinia peltophoroides* (Leguminosae)	Brazil	Flechtmann & Santana (2007)
Aceria coussapoae *Shevtchenkella desmodivagus* *Cosella ceratopudenda*	*Coussapoa microcarpa* (Cecropiaceae) *Desmodium barbatum* (Leguminosae) *Piptadenia gonoacantha* (Leguminosae).	Brazil	Flechtmann & Moraes (2002a)
Aceria inusitata	*Caesalpinia echinata* (Leguminosae)	Brazil	Britto *et al.* (2008)
Amrineus cocofolius	*Cocos nucifera* (Arecaceae)	Brazil	Flechtmann (1994)
Dichopelmus ibapitanga	*Eugenia uniflora* (Myrtaceae)	Brazil	Reis *et al.* (2010)
Epitrimerus torus *Epitrimerus angustisternalis*	*Acalypha reptans* (Euphorbiaceae) *Bougainvillea spectabilis* (Nyctaginaceae)	Brazil	Flechtmann (2010)
Juxtacolopodacus phalakros *Procalacarus perporosus* *Scolotosus centrolobii* *Scolotosus hartfordi* *Metaculus tanythrix*	*Mollinedia clavigera* (Monimiaceae) *Randia armata* (Rubiaceae) *Centrolobium robustum* (Leguminosae) *Centrolobium tomentosum* (Leguminosae) *Dicksonia sellowiana* (Dicksoniaceae)	Brazil	Flechtmann & De Queiroz (2010)
Tetra tarabanensis	*Bulnesia arborea* (Zygophyllaceae)	Venezuela	Flechtmann & Vásquez (2007)

Table 6. Some eriophyoid mite species recently described from South America.

Eriophyoid mites are tiny worm-like or fusiform mites and they form galls or live freely on various host-plants (Royalty & Perring 1996; Westphal & Manson 1996). The wounding and injecting of specific salivary secretions into host-cells result in a specific response of the affected leaf, stem, or bud tissues; such as gall differentiation, hypersensitive reaction, or non-distortive feeding effects and in some cases complex symptoms, considered as syndromes (Petanović & Kielkiewicz 2010). Most species are monophagous and many species are limited to plant species within a single genus, with few exceptions (Zhang 2003). Most species cause little harm to their host plants, however, some rust mites and gall mites are important pests on economic plants.

Life cycle: Eriophyid mites passes through the egg, larva, nymph and adult stages. Both females and males complete their life cycle in about a week at an average temperature of around 25°C. The mating process is indirect, since male deposits spermatophores on host plants, and then the genital flap in the female presses the spermatophore into the body and crushes it. Females lay up to three eggs per day for up to a month, with a total of up to 87 eggs per female.

5.1 Eriophyid mite species of economic importance in South America

5.1.1 The coconut mite, *Aceria guerreronis* Keifer

The coconut mite is an invasive mite pest that has been disseminated and established rapidly in main coconut (*C. nucifera*) production areas (Návia *et al.* 2009) causing copra yield reductions, premature dropping of fruit (Moore & Howard 1996). *A. guerreronis* is mostly found in the meristematic region of fruit, under the perianth (Fernando *et al.* 2003). Development of colonies usually starts at earlier stages of fruit formation, resulting in discolored areas that frequently become larger and necrotic, longitudinal cracks and eventually producing an exudate (Haq *et al.* 2002). In addition to fruit damage, this mite attacks the growing tip of plantlets, turning it dark brown and often causing death of attacked plants in Brazil (Aquino *et al.*1968).

Despite the great economic importance of the coconut mite, information on basic aspects is scarce. Acording to Moraes & Zacarias (2002), the use of biological control agents based on the location of the *A. guerreronis* are crucial tools to manage this mite pest.

5.1.2 Tomato russet mite, *Aculops lycopersici* (Masse)

The tomato russet mite (TRM) is cosmopolitan in distribution and widespread in almost all areas where solanaceous crops are grown (Jeppson *et al.* 1975). This eriophyoid is one of the most common key pests of the commercially grown tomato, *Lycopersicon esculentum* Mill on a worldwide scale. In addition, the TRM host range includes tomatillo, potato, eggplant, poha (cape gooseberry), wild blackcurrant, popolo, wild gooseberry, blackberry, tobacco, bell pepper, cherry pepper, tolguacha, eggplant, Jerusalem cherry, hairy nightshade, black nightshade, horse nettle, morning glory, Jimson weed, Chinese thorn apple, petunia, nightshade, small flowered nightshade, amethyst, field bindweed, and Brinjal (Perring 1996). According to Duso *et al.* (2010), it is important to investigate: (a) TRM bioecological data useful to improve control strategies; (b) the different specialized intimate interactions that TRM establishes with different plants and/or different areas of the same plant; (c) the strength of the biochemical and physiological mechanisms/steps determining the intensity of closeness/dependence with the host plant.

6. Tarsonemid mites

The Tarsonemidae includes about 545 species of 45 genera widely distributed. This family consists of three subfamilies: Pseudotarsonemoidinae, Acarapinae and Tarsoneminae; the latter including most of the described species in the two large genera: *Tarsonemus* (over 270 species) and *Steneotarsonemus* (over 70 species). Tarsonemid mites exhibit various feeding habits, some species feed on fungus, algae, plants, and some of them can prey on parasite insects (Moraes & Flechtmann 2008). Some plant-feeding tarsonemid mites are pests of agricultural crops, most of them belonging to a few genera in the Tarsoneminae, except for the *Polyphagotarsonemus*. Since Tarsonemid mites feed on surface cells, more significant damage is observed in young tissues of the host plant. Symptoms are characterized by leaf discoloration with a silver aspect. Expanding leaves became shriveled or curled and eventually shed and plants severely attacked stop growing. Occasionally, plant tissue ontogeny is altered due to toxins injected.

Life history and biology: Tarsonemid mites are haplodiploid, being males produced by arrhenotoky and females by sexual reproduction; however, thelytoky has also been observed (Norton *et al.* 1993). Its life cycle consists of egg, larva, "pupa" and adult (male and female) stages, but "pupa" is an inactive stage in which the nymph stage takes place.

Polyphagotarsonemus latus (Banks) is undoubtedly the most important pest of many crops and ornamentals in field or greenhouses worldwide. This species disperses by wind, human transport of infested products and also through insects living on plants. *P. latus* females have been observed as phoront on *Bemisia tabaci* (Gennadius) on beans (*P. vulgaris*) in Colombia and on cucumber (*Cucumis sativus* L.) var. poinsett-76 and sesame (*Sesamun indicum* L.) var. INIA-1 in Venezuela (Bautista *et al.* 2005).

Life-history traits depend on temperature, host plants and even on varieties. On pepper, the developmental period (egg - adult) is, as an average, 4.1 days at 25°C for males and females, respectively. Adult female and male longevity is 11 and 15 days, respectively. The female/male sex ratio is 2.8 in the laboratory, and 2.3 on seedlings in a greenhouse. The intrinsic rate of increase is 0.359, the finite rate of increase is 1.43 individuals/female/day, the mean generation time 10.34 days and the net reproductive rate 41.0.

Tarsonemid mites of the genus *Steneotarsonemus* are phytophagous and specialized on monocotyledon plants (Lindquist 1986; Almaguel *et al.* 2000). In America, *Steneotarsonemus spinki* Smiley is considered the most destructive pest mite in rice ever (Rodríguez *et al.* 2009). In 1997, the rice mite was first reported in Cuba (Ramos & Rodríguez 1998), causing about 30 - 90% yield (Almaguel *et al.* 2000). Soon, *S. spinki* spread to the Caribbean (Haiti, Dominican Republic and Puerto Rico), Central America (Panama, Costa Rica, Nicaragua, Honduras and Guatemala), South America (Colombia and Venezuela) and North America (Mexico and USA) (Rodríguez *et al.* 2009, Sandoval *et al.* 2009). Although the rice mite has been detected in rice fields in Venezuela, important mite infestation focuses have not been observed to affect rice production in this country.

Pest control: Most information about control strategies of this species has been compiled in Cuba. In this country, nine predator species have been found, 4 Phytoseiid, 3 Ascid and 2 Laelapid species (Table 7).

Mite family	Species
Phytoseiidae	*Neoseiulus baraki*
	Neoseiulus paraibensis
	Proprioseiopsis asetus
	Neoseiulus paspalivorus
Ascidae	*Aceodromus asternalis*
	Asca sp.
	Proctolaelaps sp.
Laelapidae	*Aceodromus asternalis*
	Hypoaspis sp.

From Rodríguez *et al.* (2009).

Table 7. Predatory mites associated to *S. spinki* in rice fields in Cuba.

Another important tarsonemid species, *Steneotarsonemus furcatus* De Leon has been found on coconut, damaging fruit in Central America and in Brazil (Ochoa *et al.* 1994, Návia *et al.* 2005) and various gramineous species in Cuba (La Torre *et al.* 2005), and more recently, *Steneotarsonemus concavuscutum* Lofego & Gondim was examined on coconut in northeastern Brazil (Lofego & Gondim 2006).

7. Conclusions

In South America, there is a significant volume of literature dealing with taxonomy, biology and crop damage caused by phytophagous mites on agricultural crops. Research has provided a great deal of valuable information about their impact on several important crops in Brazil, Colombia, Mexico and Venezuela. At present, a multi-institutional and multi-disciplinary project on the biological control of the coconut mite in Tropical America is being developed. As a preliminary step, some explorations have been conducted in target countries to assess the abundance of the coconut mite, the diversity and prevalence of the natural enemies associated with this pest, and to evaluate the potential of these natural enemies as biological control agents for their use in the IPM programs. In addition, much work has been carried out to determine other host plants of the coconut mite and its natural enemies. Furthermore, some Government and Research Institutions from Brazil, Trinidad and Tobago and Venezuela are making efforts to find some ecological strategies to control the Red Palm Mite *Raoiella indica* Hirst (Acari: Tenuipalpidae), and thus minimize its impact on commercial coconut farms and other important crops in tropical areas.

8. References

Almaguel L, Hernández, J, Torre P de la, Santos A, Cabrera RI, García A, Rivero LE, Báez L, Cáceres I, Ginarte A. (2000) Evaluación del comportamiento del ácaro

Steneotarsonemus spinki (Acari: Tarsonemidae) en los estudios de regionalización desarrollados en Cuba. Fitosanidad, 4, 15 - 19.

Aquino MLN, Flechtmann CHW, Arruda GP (1968) Contribuição ao estudo do ácaro causador da necrose do olho do coqueiro em Pernambuco e seu controle. Recife, IPA, 17 pp. (Boletim Técnico, 34).

Baker AE, Tuttle M (1987) The false spider mites of Mexico (Tenuipalpidae: Acari) U.S. Department of Agriculture, Technical Bulletin No, 1706, 237 pp.

Bautista L, E Arnal, O Aponte (2005) Relación forética de *Polyphagotarsonemus latus* (Banks) (Acari: Tarsonemidae) y adultos de *Bemisia tabaci* (Gennadius) (Hemiptera: Aleyrodidae). Entomotropica, 20(1): 79 - 80.

Biswas GC, Islam W, Haque MM, Saha RK, Hoque KMF, Islam MS, Haque ME (2004) Some biological aspects of carmine mite, *Tetranychus cinnabarinus* Boisd. (Acari: Tetranychidae) infesting egg-plant from Rajshahi. Journal of Biological Sciences, 4 (5): 588 - 591.

Bolland HR, Gutierrez J, Flechtmann CHW (1998) World catalogue of the spider mite family (Acari: Tetranychidae). Leiden, Koninklijke Brill NV, 408 pp.

Britto EPJ, Gondim Jr. MGC, Návia D (2008) A new deuterogynous eriophyid mite (Acari: Eriophyidae) with dimorphic males from *Caesalpinia echinata* (Caesalpiniaceae) from Brazil: description and biological observations. International Journal of Acarology, 34 (3): 307 - 316.

Cardona C, Frei A, Bueno JM, Díaz J, Gu H, Dorn S (2002) Resistance to *Thrips palmi* (Thysanoptera: Thripidae) in beans. Journal of Economic Entomology, 95: 1066 - 1073.

Carrillo D, Návia D, Ferragut F, Peña JE (2011) First report of *Raoiella indica* (Acari: Tenuipalpidae) in Colombia. Florida Entomologist, 94(2): 370 - 371.

Carrillo D, Peña JE, Hoy MA, Frank JH (2010) Development and reproduction of *Amblyseius largoensis* (Acari: Phytoseiidae) feeding on pollen, *Raoiella indica* (Acari: Tenuipalpidae), and other microarthropods inhabiting coconuts in Florida, USA. Experimental and Applied Acarology, 52(2): 119 - 129.

Cerna E, Landeros J, Ochoa Fuentes YM, Luna Ruiz J, Vázquez Martínez O, Ventura López O (2009) Tolerancia del ácaro *Tetranychus urticae Koch* a cuatro acaricidas de diferente grupo toxicológico. Investigación y Ciencia, 44 (4-10): 4-10.

Chagas CM, Rossetti V, Colariccio A, Lovisolo O, Kitajima EW, Childers CC (2000) *Brevipalpus* mites (Acari: Tenuipalpidae) as vectors of plant viruses. *In:* Int. Congress Acarology (10, 2000, Melbourne). Halliday, RB; Walter, DE; Proctor, HC; Norton, RA; Colloff, MJ. (Eds) Proceedings. CSIRO Pub.

Childers CC, French JV, Rodrigues JCV (2003) *Brevipalpus californicus, B. obovatus, B. phoenicis,* and *B. lewisi* (Acari: Tenuipalpidae): a review of their biology, feeding injury and economic importance. Experimental and Applied Acarology, 30 (1-3): 5 - 28.

Childers CC, Kitajima EW, Welbourn WC, Rivera C, Ochoa R (2001) *Brevipalpus* como vectores de la leprosis de los cítricos. Manejo Integrado de Plagas, 60: 61 - 65.

de Ponti O (1977) Resistance in *Cucumis sativus* L. to *Tetranychus urticae* Koch. II. Designing a reliable laboratory test for resistance based on aspects of the host-parasite relationship. Euphytica, 26: 641 - 654.

Domínguez FS, Bernal A, Childers CC, Kitajima EW (2001) First report of the citrus leprosis virus in Panama. Plant Disease (Disease Notes), 85(2): 228.

Doreste E (1984) Acarología. San José, Instituto Interamericano de Cooperación para la Agricultura, 391pp.

Duso C, Castagnoli M, Simoni S, Angeli G (2010) The impact of eriophyoids on crops: recent issues on *Aculus schlechtendali*, *Calepitrimerus vitis* and *Aculops lycopersici*. Experimental and Applied Acarology, 51 (1-3): 151 - 168.

Esteves Filho, AB, Oliveira JV de, Torres, JB, Gondim Jr, MGC (2010) Biologia comparada e comportamento de *Tetranychus urticae* Koch (Acari: Tetranychidae) e *Phytoseiulus macropilis* (Banks) (Acari: Phytoseiidae) em algodoeiro bollgardTM e Isolinha não-Transgênica. Neotropical Entomology, 39 (3): 338 - 344.

Etienne J, Flechtmann CHW (2006) First record of *Raoiella indica* (Hirst, 1924) (Acari: Tenuipalpidae) in Guadalupe and Saint Martin, West Indies. International Journal of Acarology, 32: 331-332.

Evans GO (1992) Principles of acarology. CAB International. Wallindford, UK. 563 pp.

Ferla NJ, Marchetti M, Johann L, Haetinger C (2011) Functional response of *Phytoseiulus macropilis* under different *Tetranychus urticae* (Acari: Phytoseiidae, Tetranychidae) population density in laboratory. Zoologia (Curitiba), 28 (1): 17 - 22.

Fernando LCP, Aratchige NS, Peris TSG (2003) Distribution patterns of coconut mite, *Aceria guerreronis*, and its predator *Neoseiulus aff. paspalivorus* in coconut palms. Experimental and Applied Acarology, 31: 71 - 78.

Flechtmann CHW (1994) *Amrineus cocofolius* n.g., n.sp. (Acari: Eriophyidae) from Brazil. International Journal of Acarology, 20(1): 57 - 59.

Flechtmann CHW (2010) Two new *Epitrimerus* Nalepa (Eriophyidae, Prostigmata) species from Brazil and further remarks on the immature stages of a few eriophyoid species. Systematic & Applied Acarology, 15: 71 - 80.

Flechtmann CHW, de Queiroz DL (2010) New taxa in the Eriophyidae (Acari, Prostigmata) from forest trees in southern Brazil. Zootaxa, 2337: 18 - 30.

Flechtmann CHW, Etienne J (2004) The red palm mite, *Raoiella indica* Hirst, a threat to palms in the Americas (Acari: Prostigmata: Tenuipalpidae). Systematic and Applied Acarology, 9:109-110.

Flechtmann CHW, Moraes de, GJ (2002a) New Brazilian eriophyid mites (Acari: Eriophyidae). Zootaxa, 75: 1 - 12.

Flechtmann CHW, Moraes de, GJ (2002b) Three new species of eriophyid mites (Acari: Eriophyidae) from the State of São Paulo, Brazil. Zootaxa, 23: 1 - 8.

Flechtmann CHW, Santana DLQ (2007) A new deuterogynous eriophyid mite (Acari: Eriophyidae) from a semideciduous tree in Southern Brazil. International Journal of Acarology, 33(2): 129 - 132.

Flechtmann CHW, Vásquez C (2007) A new species and new records of plant mites of *Bulnesia arborea* (Zygophyllaceae) from Venezuela. Boletín del Centro de Investigaciones Biológicas, 40 (1): 57 - 67.

Flechtmann, CHW, Etienne J (2005) Un nouvel acarien ravageur des palmiers: En Martinique, premier signalement de *Raoiella indica* pour les Caraïbes. Phytoma, 548: 10-11.

Forero, G, Rodríguez M, Cantor F, Rodríguez D, Cure JR (2008) Criterios para el manejo de *Tetranychus urticae* Koch (Acari: Tetranychidae) con el ácaro depredador *Amblyseius*

(*Neoseiulus*) sp. (Acari: Phytoseiidae) en cultivos de rosas. Agronomía Colombiana, 26 (1): 78 - 86.

Frei A, Gu H, Bueno J, Cardona C, Dorn S (2003) Antixenosis and antibiosis of common beans to *Thrips palmi* Karny (Thysanoptera: Thripidae). Journal of Economic Entomology, 96: 1577 - 1584.

Freitez F, Alvarado G (1978) Contribución al conocimiento de ácaros en Musaceae de Venezuela (Acarina). Revista de la Facultad de Agronomía, 26:197–207.

Gallardo A, Vásquez C, Morales J, Gallardo S (2005) Biología y enemigos naturales de *Tetranychus urticae* en pimentón. Manejo Integrado de Plagas y Agroecología, 74: 34 - 40.

García-Valencia AS, Hoffmann A (1997) Especie nueva de ácaro eriófido en México (Prostigmata: Eriophyidae). Anuales del Instituto de Biología de la Universidad Nacional Autónoma de México, serie Zoológica, 6 (2): 253 - 260.

Gardner SN, Agrawal AA (2002) Induced plant defense and the evolution of counter-defense in herbivores. Evolutionary Ecology Research, 4: 1131 - 1151.

González Zamora JE, García-Marí F, Ribes A, SaquésJ, Masiello L, Orenga S (1993) Métodos de muestreo binomial y secuencial para *Tetranychus urticae* Koch (Acari: Tetranychidae) y *Amblyseius californicus* (McGregor) (Acari: Phytoseiidae) en fresón. Boletín de Sanidad Vegetal y Plagas, 19: 559-586.

Goyal M, Sadana GL, Sharma NK (1985) Influence of temperature on the development of *Brevipalpus obovatus* (Acarina:Tenuipalpidae). Entomon, 10: 125 - 129.

Greco NM, Sanchez NE, Liljesthrom GG (2005) *Neoseiulus californicus* (Acari: Phytoseiidae) as a potential control agent of *Tetranychus urticae* (Acari: Tetranychidae): effect of pest/predator ratio on pest abundance on strawberry. Experimental & Applied Acarology, 37:57 - 66.

Haq MA, Sumangala K, Ramani N (2002) Coconut mite invasion, injury and distribution, pp. 41 - 49. *In*: LCP Fernando, GJ Moraes & IR Wickramananda (Eds.), Proceedings of the International Workshop on Coconut Mite (*Aceria guerreronis*). Sri Lanka, Coconut Research Institute, 117pp.

Harborne J (1994) Do natural plant phenols play a role in ecology? Acta Horticulturae, 381:36 - 43.

Hilker M, Meiners T (2002) Induction of plant responses to oviposition and feeding by herbivorous arthropods: a comparison. Entomologia Experimentalis et Applicata, 104:181 - 192.

Hoy M, Peña J, Nguyen R (2006) Featured Creatures. University of Florida. Florida Department of Agriculture and Consumer Services, Division of Plant Industry. Available in: http://entomology.ifas.ufl.edu/creatures/orn/palms/red_palm_mite.htm#dist. July 19, 2010.

Jeppson H, Keifer H, Baker E (1975) Mites injurious to economic plants. Univ. of California. Press. Riverside. 614 pp.

Kane, E., R. Ochoa, G. Mathurin & E. Erbe. 2005. *Raoiella indica* Hirst (Acari: Tenuipalpidae): An island-hopping mite pest in the Caribbean. The ESA Annual Meeting and exhibition. Ft. Lauderdale, Florida, USA. Available in: http://www.sel.barc.usda.gov/acari/PDF/Raoiella%20indica-Kane%20et%20al.pdf.

Kitajima EW, Muller GW, Costa AS, Yuki VA (1972) Short, rodlike particles associated with citrus leprosis. Virology, 50:254-258.

Kitajima EW, Rezende JAM, Freitas JC (1996) Two types of particles associated with lesions induced by *Brevipalpus* mites in different plant hosts. VII. Encontro Nacional de Virologia. S. Lourenço, MG. 275 (abstract).

Kitajima EW, Rezende JAM, Rodrigues JCV, Chiavegato LG, Piza Jr CT, Morozini W (1997) Green spot of passion fruit, a possible viral disease associated with infestation by the mite *Brevipalpus phoenicis*. Fitopatologia Brasileira, 22: 555 - 559.

Klein C, Waterhouse DF (2000) The distribution and importance of arthropods associated with agriculture and forestry in Chile. ACIAR Monograph 68, Canberra, Australia. 234 p.

Krantz GW (2009) Habits and habitats. pp. 64 - 82. *In*: A manual of acarology. GW Krantz & DE Walter (Eds.) Texas Tech University Press. Lubbock, Texas, USA.

La Torre, P de; Almaguel L, Botta E, Cáceres I (2005) Plantas hospedantes de *Steneotarsonemus furcatus* de leon (Acari: Tarsonemidae) en Cuba. Neotropical Entomology, 34 (3): 517 - 519.

Lindquist EE (1986) The world genera of Tarsonemidae (Acari: Heterostigmata): a morphological, phylogenetic, and systematic revision, with a reclassification of family-group taxa in the Heterostigamata. Memoirs of the Entomological Society of Canada, 136, 1 - 517.

Lofego AC, Gondim Jr MGC (2006) A new species of *Steneotarsonemus* (Acari: Tarsonemidae) from Brazil. Systematic & Applied Acarology, 11: 195 - 203.

McMurtry JA (1970) Some factors of foliage condition limiting populations growth of *Oligonychus punicae* (Acarina: Tetranychidae). Annals of Entomological Society of America, 63:406-412.

McMurtry JA (1985) Avocado. pp. 327 - 338 In: Helle W, Sabelis M (Eds) Spider mites: their biology, natural enemies and control, vol. 1B. Elsevier Science Publishers B. V, Amsterdam, The Netherlands,

McMurtry JA, Johnson HG (1966) An ecological study of the spider mite *Oligonychus punicae* (Hirst) and its natural enemies. Hilgardia 37(11):363-402

Moore, D, Howard FW (1996) Coconuts. pp. 561 - 570. In: Lindquist EE, Sabelis MW & Bruin J (Eds.) Eriophyoid mites: their biology, natural enemies and control. Amsterdam, Elsevier.

Moraes GJ de, Flechtmann CHW (2008) Manual de acarologia: acarologia básica e ácaros de plantas cultivadas no Brasil. Ribeirão Preto, Holos, 288 pp.

Moraes GJ de, Zacarias MS (2002) Use of predatory mites for the control of eriophyid mites. pp. 78-88. *In*: Fernando LCP, Moraes GJ de, Wickramananda IR (Eds.), Proceedings of the International Workshop on Coconut Mite (*Aceria guerreronis*), 6-8 January 2000, Coconut Research Institute, Lunuvila, Sri Lanka.

Morrissey J, Osbourn AE (1999) Fungal resistance to plant antibiotics as a mechanism of pathogenesis. Microbiology and Molecular Biology Reviews, 63:708 - 724.

Morros ME, Aponte O (1994) Biología y tabla de vida de *Tetranychus ludeni* Zacher en caraota *Phaseolus vulgaris* L. Agronomía Tropical, 44(4): 667 - 677.

Morros ME, Aponte O (1995a) Efecto de dos niveles de infestación de *Tetranychus ludeni* Zacher sobre las fases de desarrollo de la caraota I: Nivel de campo. Agronomía Tropical, 54(2): 189 - 194.

Morros ME, Aponte O (1995b) Efecto de dos niveles de infestación de *Tetranychus ludeni* Zacher sobre las fases de desarrollo de la caraota II: Nivel de invernadero. Agronomía Tropical, 54(2): 195 - 202.

NAPPO (2009) Detecciones del ácaro rojo de la palma (*Raoiella indica*) en Cancún e Isla Mujeres, Quintana Roo, México. Notificación oficial de Plaga. Published on Nov 20, 2009.

Návia D (2008) Riesgo del "ácaro rojo de la palma", *Raoiella indica* Hirst, para Brasil. En: Acta del VI Seminario Científico Internacional de Sanidad Vegetal. La Habana, Cuba. 22-26 sept.

Návia D, Marsaro Jr AL, Silva FR da, Gondim Jr MGC, Moraes GJ de (2011) First report of the red palm mite, *Raoiella indica* Hirst (Acari: Tenuipalpidae), in Brazil. Neotropical Entomology, 40(3): 409 - 411.

Návia D, Moraes GJ de, Lofego AC, Flechtmann CHW (2005) Acarofauna associada a frutos de coqueiro (*Cocos nucifera* L.) de algumas localidades das Américas. Neotropical Entomology, 34(2), 349 - 354.

Návia D, Moraes, GJ, Querino, RB (2009) Geographic pattern of morphological variation of the coconut mite, *Aceria guerreronis* Keifer (Acari: Eriophyidae), using multivariate morphometry. Brazilian Journal of Biology, 69 (3): 773 - 783.

Norton RA, JB Kethley, DE Johnston & BM OConnor (1993) Phylogenetic perspectives on genetic systems and reproductive modes of mites. pp. 8-99. In: DL Wrensch & MA Ebbert (Eds.) Evolution and diversity of sex ratio in insects and mites. Chapman & Hall Publications, New York.

Ochoa R, Aguilar H, Vargas C (1994) Phytophagous mites of Central America: an illustrated guide. CATIE, Turrialba, Costa Rica, 234 pp.

Peña JE, Mannion C, Howard FW, Hoy MA (2008) Red palm mite, *Raoiella indica* Hirst (Acari: Prostigmata: Tenuipapidae). *In*: Encyclopedia of Entomology. Capinera J, Editor. Kluwer Academic Publishers, Dordrecht, The Netherlands.

Peña JE, Mannion CM, Howard FW, Hoy MA (2006) *Raoiella indica* (Prostigmata: Tenuipalpidae): The red palm mite: a potential invasive pest of palms and bananas and other tropical crops of Florida. University of Florida IFAS Extension, ENY-837. 2006 (July 12, 2007). Available in: http://edis.ifas.ufl.edu/BODY_IN681.

Perring TM (1996) Vegetables. pp. 593-606. *In*: Lindquist EE, Sabelis MW, Bruin J (Eds.) Eriophyoid Mites: their biology, natural enemies and control. Elsevier, Amsterdam, The Netherlands.

Petanović R, Kielkiewicz M (2010) Plant-eriophyoid mite interactions: specific and unspecific morphological alterations. Part II. Experimental and Applied Acarology, 51: 81 - 91.

Pijnacker LP, Ferwerda MA, Bolland HR, Helle W (1980) Haploid female parthenogenesis in the false spider mite *Brevipalpus obovatus* (Acari: Tenuipalpidae). Genetica, 51: 211 - 214.

Praslička J, Huszár J (2004) Influence of temperature and host plants on the development and fecundity of the spider mite *Tetranychus urticae* (Acarina: Tetranychidae). Plant Protection Science, 40(4):141-144

Quirós M (1978) Presencia del 'ácaro marrón del aguacate', *Oligonychus punicae* (Hirst) en hojas de plátano del Distrito Baralt, estado Zulia. Revista de la Facultad de Agronomía (LUZ), 5(1):422 - 424.

Ramos M, Rodríguez H (1998) *Steneotarsonemus spinki* Smiley (Acari: Tarsonemidae): Nuevo Informe para Cuba. Revista de Protección Vegetal, 13 (1): 25-30.

Reis AC, Gondim Jr. MGC, Návia D, Flechtmann CHW (2010) Eriophyoid mites (Acari: Prostigmata: Eriophyoidea) from fruit trees in Northeastern Brazil - a new genus, three new species and a redescription. Zootaxa, 2694: 43–56.

Ribeiro LG, Villacorta A, Foerster LA (1988) Life cycle of *Panonychus ulmi* (Koch, 1836) (Acari: Tetranychidae) in apple trees, cultivar Gala and Golden Delicious. Acta Horticulturae, 232:228

Rivero E, Vásquez C (2009) Biologia e tabela de vida de *Tetranychus desertorum* (Acari: Tetranychidae) sobre folhas de feijão (*Phaseolus vulgaris*). Zoologia (Curitiba), 26 (1): 38 - 42.

Roda A, Dowling A, Welbourn C, Peña JE, Rodrigues JCV, Hoy MA, Ochoa R, Duncan RA, De Chi W (2008) Red palm mite situation in the Caribbean and Florida. Proceedings of the Caribbean Food Crops Society, 44(1): 80-87.

Rodrigues JCV (2000) Relações patógeno-vetor-planta no sistema leprose dos citros. Thesis Ph.D. Piracicaba, Brasil,Centro de Energia Nuclear na Agricultura, da Universidade de São Paulo.

Rodrigues JCV, Kitajima EW, Childers CC, Chagas CM (2003) Citrus leprosis virus vectored by *Brevipalpus phoenicis* (Acari: Tenuipalpidae) on citrus in Brazil. Experimental and Applied Acarology, 30: 161–179.

Rodrigues JCV, Ochoa R, Kane EC (2007) First report of *Raoiella indica* Hirst (Acari: Tenuipalpidae) and its damage to coconut palms in Puerto Rico and Culebra Island. International Journal of Acarology, 33(1): 3–5.

Rodríguez H, Miranda I, Louis J, Hernández J (2009) Comportamiento poblacional de *Steneotarsonemus spinki* Smiley (Acari: Tarsonemidae) en el cultivo del arroz (*Oryza sativa* L.). Temas de Ciencia y Tecnología, 13 (39): 55 - 66.

Rodríguez H, Montoya A, Flores-Galano G (2010) Conducta alimentaria de *Amblyseius largoensis* (Muma) sobre *Raoiella indica* Hirst. Revista de Protección Vegetal, 25(1): 26 - 30.

Royalty RN, Perring TM (1996) Nature of damage and its assessment. pp. 493-512. In: Lindquist EE, Sabelis MW, Bruin J (Eds) Eriophyoid mites—their biology, natural enemies and control. Elsevier, Amsterdam.

Sandoval MF, Almaguel L, Fréitez F, Vásquez C (2009) Situación actual del ácaro del arroz, *Steneotarsonemus spinki* Smiley (Acari: Tarsonemidae) en Venezuela. Entomotropica, 24(3): 135 - 139.

Silva EA, Reis, PR, Carvalho, TMB, Altoé, BF (2009) *Tetranychus urticae* (Acari: Tetranychidae) on *Gerbera jamesonii* Bolus and Hook (Asteraceae). Brazilian Journal of Biology, 69(4): 1121 - 1125.

Smith-Meyer MKP. 1979. The Tenuipalpidae (Acari) of Africa with keys to the world fauna. Republic of South Africa. Dept.Agric.Tech.Serv. Entomology Memoir 135 pp.

Tello V, Vargas R, Araya J (2009) Parámetros de vida de *Tetranychus cinnabarinus* (Acari: Tetranychidae) sobre hojas de clavel, *Dianthus caryophyllus*. Revista Colombiana de Entomología, 35(1): 47 - 51.

Tomczyk A; D Krompczyńska (1985) Effects on the host plant. pp. 317 - 329. *In*: Spider mites: their biology, natural enemies and control, vol 1A. W. Helle and M. W. Sabelis. (Eds.) Elsevier, Amsterdam. 405 pp.

Vásquez C, Colmenárez M, Valera N, Díaz L (2007) Antibiosis of kidney bean cultivars to the carmine spider mite, *Tetranychus cinnabarinus* (Boisduval) (Acari: Tetranychidae). Integrated Control of Plant-Feeding Mites. IOBC wprs Bulletin, 30 (5): 133 - 138.

Vásquez C, Morales J, Aponte O, Sanabria ME, García G (2008a) Biological studies of *Oligonychus punicae* (Acari: Tetranychidae) on grapevine cultivars. Experimental and Applied Acarology, 45: 59 - 69.

Vásquez C, Quirós de G M, Aponte O, Sandoval MF (2008b) First report of *Raoiella indica* Hirst (Acari: Tenuipalpidae) in South America. Neotropical Entomology, 37 (6): 739 - 740.

Walter D, H Proctor (1999) Mites: ecology, evolution and behaviour. CABI Publishing. University of New South Wales Press, Sydney, New South Wales. 322 pp.

Welbourn C. (2005) Red palm mite *Raoiella indica* Hirst (Acari: Tenuipalpidae). Pest Alert, 2005. (on line). (Accessed on August 17th, 2011) Available in: htpp://www.daocs.state.fl.us/pi/enpp/ento/r.indica.html.

Westphal E, Manson DCM (1996) Feeding effects on host plants: gall formation and other distortion. pp. 231-241. In: Lindquist EE, Sabelis MW, Bruin J (Eds) Eriophyoid mites—their biology, natural enemies and control. Elsevier, Amsterdam.

Zhang ZQ (2003) Mites of greenhouses: identification, biology and control. CABI Publishing. Wallingford UK. 244 pp.

16

Research on One Kind of Essential Oil Against Drugstore Beetle *Stegobium paniceum* (L.)

Can Li
Department of Biology and Engineering of Environment, Guiyang University, Guiyang People's Republic of China

1. Introduction

Insect pest management in stored products is facing a crisis due to several serious drawbacks of using insecticides, such as the development of resistance in the treated pest, toxic residues, and the increasing cost of application (Tapondjou et al 2002). Hence it is necessary to develop alternative pest control techniques for protecting stored commodities (Gunasekaran and Rajendran 2005). Toward this end, intensified efforts have lead to an increasing number of research studies to find safe, effective and viable alternatives (Tapondjou et al. 2002). Carbon dioxide gas (CO_2) and essential oils from plants have received considerable attention for the control of stored products insects, because of their relative safety to the non-target organisms (AliNiazee 1972; Bekele et al 1996; Juliana and Su 1983; Sudesh et al 1996; Bouda et al 2001).

Chinese medicinal materials (CMM) are widely available in China. While most of these products are stored prior to use for health protection or disease treatment, great losses often occur during storage due to infestation by insect pests. Drugstore beetle *Stegobium paniceum* (L.) is the most widely encountered insect causing serious damages to stored products. It is one of the dominant species found on stored CMM in Hubei, Guizhou and Shandong, causing huge economic losses (Can et al 2004; Guilin and Wangxi 1996; Zhaohui and Fangqiang 2001). A series of measures have been taken to control infestation (Nielsen 2001; Hashem 2000, Toh 1998; Platt et al 1998). However, concerns over health and safety associated with traditional synthetic insecticides have prompted the development of plant-based insecticides.

The present study explores the efficacy of essential oil from *Z. bungeanum* Maxim against larvae and adults of *S. paniceum*. The contact toxicity and fumigant toxicity were investigated, and the subsequent development of treated insects was recorded.

2. Materials and methods

2.1 Extraction of essential oil

The fruits of *Z. bungeanum* were collected from the mountains in the west of China. After cleaning and pounding into powder, the materials were soaked in clear water (1:5, w:v) at

60℃ for 24h. After a 14 hour hydro-distillation, the crude oil was collected, dehydrated with anhydrous sodium sulfate and stored at 4℃ until use.

2.2 Collecting and rearing insects

The insects were obtained from a local CMM storehouse. They were mass reared on *Euphorbia* kansui Liou in a growth chamber under controlled conditions (29±1°C; 75±5% r.h.; light for 14 h).

2.3 Fumigant toxicity of crude oil to drugstore beetle

The oil was introduced onto a filter paper (7cm diameter, surface 38.5cm^2) placed in the center of a 1L glass jar. To prevent insects from contacting the oil, the insects and cultural medium were separated by a piece of gauze placed around the filter.

Larvae (25 to 30 days-old, n=20 in each of the three replicates) were exposed to concentrations of 12, 24, 48, 96 and 192 ul crude oil in the glass jars. Clear water was used as control. The jars were covered with fine gauze to prevent the insects from fleeing. The insects were incubated in the growth chamber at 29±1°C and 75±5% r.h.. During the exposure, the test insects were fed on *E. kansui* Liou. After exposure of 6 to 144 h, the insect mortalities were determined respectively. Mortality was corrected by Abbotts (1925) formula when the mortality of control insects is above 10%.

Similarly, adult insects (2 to 3 days-old, 20 per replicate) were exposed to concentrations of 12, 24, 48, 96 and 196 ul in 1L glass jars respectively. The glass jars were covered by gauze to prevent the insects from flying away and they were kept in the same conditions as the larvae. Insect mortalities were determined after exposure to oil for 3 to 72 h.

2.4 Contact toxicity of oil to drugstore beetle

The crude oil was diluted with clear water to 200, 400, 667, 1000, and 2000 ppm respectively for this test. One *ul* of each of these solutions was introduced onto the back of one insect. The glass jars were placed into growth chamber at 29±1°C and 75±5% r.h.. Three replicates of each dose and three controls treated with water were tested (n=20 per replicate). The toxicity to insects was assessed after treating for 12 to 144 h.

2.5 Observation of subsequent development

After LT_{50} (The median mortality time) treatment, the living insect samples were transferred to similar sized jars and were incubated in the growth chamber, and untreated insects were transferred as a control group. They were under regular observation to assess their subsequent development in another day. The development time of the larvae and the number of eggs laid by each adult were recorded.

3. Results

3.1 Fumigant toxicity of essential oil to drugstore beetle

The percentages of mortality for insects exposed to different doses of oil are shown in Figures 1 and 2. In general, the higher mortality was achieved when the insects were exposed to a higher dose of oil or a longer exposure period. The dose of 192 ul/L of oil achieved complete mortality of larvae after the pests were exposed for 96 h and the dose of 96 ul/L achieved the same result after exposure of 120 h. The dose of 48 ul/l needed 144 h or longer to achieve complete mortality. The lowest dosage of 12ul/l induced little mortality (<10%) when the pests were exposed for 6h or shorter.

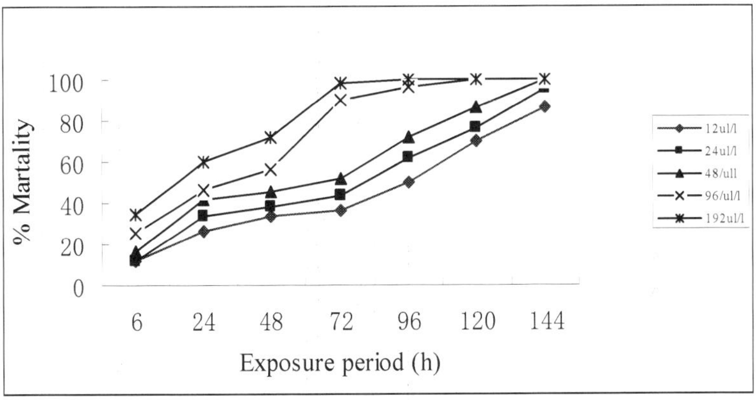

Fig. 1. Fumigant toxicity of essential oil to the larvae of drugstore beetle *S. paniceum*. Mortality was NOT corrected by Abbotts (1925) formula in all calculations because the mortality of control insect is <5%.

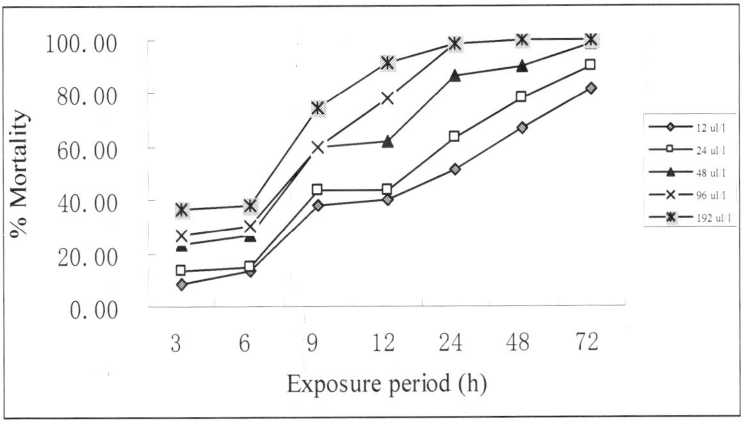

Fig. 2. Fumigant toxicity of essential oil to the adults of drugstore beetle *S. paniceum*.

Figure 2 illustrates the fumigant toxicity of essential oil on the adults. The trend is similar to that shown in Figure 1. With the increase of concentration and exposure time, the percentage of mortality improved. It reached 100% at the concentration of 96 or 192 ul/L of oil with the exposure time of 48 h. After 6 h treatment, the mortality rate was significantly higher than control group at all doses.

Significant differences of susceptibility were noted between adults and larvae under the same conditions. In fumigant test for 12h the calculated regression line equation was Y=3.24+0.66X for larva and Y=3.05+1.40X for adult, respectively (Probit analysis). Similar analysis of 24h treatment generated an equation Y=3.61+0.70X for larvae and Y=2.80+1.99X for adult. Comparison of LD_{50} values for the two life stages showed that the adult was more susceptible (LD_{50}=24.57 ul/l for 12h; LD_{50}=12.64 ul/l for 24h) than larva (LD_{50}=485.84 ul/l for 12h; LD_{50}=98.92 ul/l for 24h) (Table 1).

Life stages	Fumigant toxicity 24h		Fumigant toxicity 12h	
	LD_{50}(ul/l)	Regression model	LD_{50}(ul/l)	Regression model
Larva	98.92	y=3.61+0.70x,r=0.996	485.84	y=3.24+0.66x,r=0.988
Adult	12.64	y=2.80+1.99x, r=0.968	24.57	y=3.05+1.40x,r=0.976

Table 1. LD_{50} and regression model calculated for mortality within exposure period of12 and 24 h to fumigant toxicity.

3.2 Contact toxicity of essential oil to drugstore beetle

Figure 3 represents contact toxicity of the larvae of different concentrations of essential oil solution. The dose of 2000 ppm of solution achieved complete mortality of larvae after the pests were exposed for 120 h. The dose of 1000 ppm or 667 ppm of the solution also achieved complete mortality after 144 h exposure (Fig 3). The percentage of mortality of adult reached 100% at the concentration of 2000 ppm and exposure of 144 h. It also achieved complete mortality at 1000 ppm of the solution for 144h (Fig. 4).

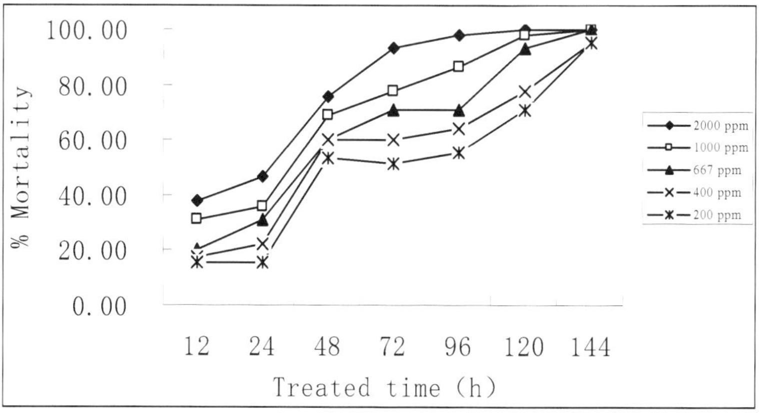

Fig. 3. Contact toxicity of essential oil to the larvae of drugstore beetle *S. paniceum*.

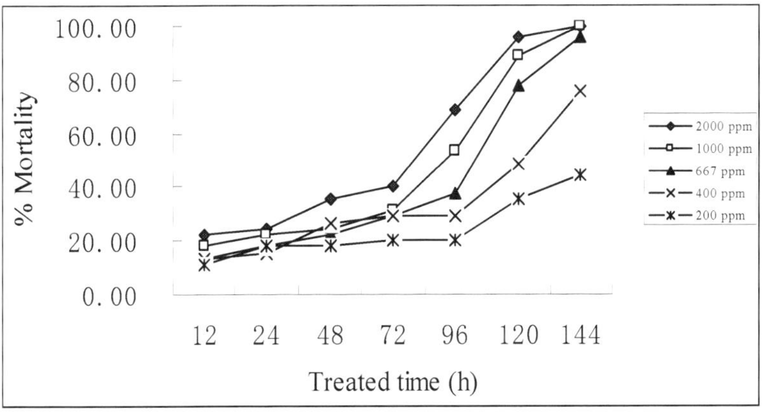

Fig. 4. Contact toxicity of essential oil to the adults of drugstore beetle *S. paniceum*.

The mortality of larvae was significantly higher than that of adults (Fig 3 and Fig 4) when treated for the same period and concentration. According to Probit analysis, the calculated regression line equation of 24 h for larva was $Y=1.82+0.91X$ and for adult was $Y=2.35+0.60X$. Similar trend was found with the 48 h data. Comparison of LD_{50} values for life stages showed that the larva was more susceptible (LD_{50}=3175.91 ppm for 24h; LD_{50}=292.13ppm for 48h) than adult (LD_{50}=23610.74 ppm for 24h; LD_{50}=6784.18 ppm for 48 h) (Table 2).

Life stages	Contact toxicity 24h		Contact toxicity 48h	
	LD_{50}(ppm)	Regression model	LD_{50}(ppm)	Regression model
Larva	3175.91	y=1.82+0.91x r=0.948	292.13	y=2.92+0.85x r=9.981
Adult	23610.74	y=2.35+0.60x r=0.967	6784.18	y=2.09+0.76x r=0.986

Table 2. LD_{50} and regression model calculated for mortality within treated time of 24 and 48 h for contact toxicity.

3.3 The subsequent development of test insects

Subsequent observation on the adult pests revealed that the essential oil also prevented adults from laying eggs normally. Significant differences were observed on the number of eggs and development time of larvae between the treated and untreated pests ($p<0.05$). The development time of the treated larvae was 29.77±6.27 (d) (n=13), and untreated was 16.77±2.17 (d) (n=13) (mean ±SE). The mean number of eggs laid by treated adult was 14.21±3.28 (n=27), comparing with 63.50±23.94 (n=27) (mean ±SE) by the untreated insects.

4. Discussion

Contact and fumigant insecticidal actions of plant essential oils have been well demonstrated against pests in stored products (Murray B. Isman 2000). Several investigations have demonstrated contact, fumigant and antifeedant effects of a range of essential oil constituents against the red flour beetle *Tribolium castaneum* and the maize weevil *Sitophilus zeamais* (Ho et al 1994, 1995, 1997; Huang and Ho 1998). Essential oils from many plants have been developed as pest control agents. Perhaps the most attractive benefit of using essential oils (or their constituents) as protectants is their low mammalian toxicity (Murray B. Isman 2000).

Insecticidal activity of *Z. bungeanum* Maxim volatile oil was reported before . The results showed that *Z. bungeanum* Maxim volatile oil had high repellent and insecticidal activity against *Sitophilus* zeamais and *Tribolium* castaneum. The volatile constituents of the essential oil were identified by their retention index and mass spectrum in comparison with those of standard synthetic compounds (Zhi'an et al 2001). The major chemical constituents are C_9, C_{10} alcohols and alkenes. The most abundant constituent was 3-cyclohexen-1-ol, 4-methyl-1-(1-methylethyl) (33.578%), followed by Eucalyptol (15.656%) and Benzene,1-methoxy-4-(1-propenyl) (8.33%). Among those chemical components, beta-Phellandrence (23.2%), ailha-Pinene (1.89%) (Jun et al 2001) could be the key factors to control the pests.

The infestation of stored Chinese medicinal materials by drug store beetle is a serious problem. The use of insecticides is avoided because of toxicity. In this study, we investigated the effects of the essential oil from *Z. bungeanum* on the control of drug store beetle and found that it not only prolonged the development time of larvae but can also prevented adults from laying eggs successfully. Effective contact and fumigant insecticidal actions of such oil were demonstrated, and the essential oil may help control expansion of drugstore beetle populations. Adult drugstore beetles were more susceptible than larvae to fumigant action of the oil. Adults showed higher mortalities at the same concentration than larvae, likely cause by suffocation. However, larvae were more susceptible than adults to contact treatment. These results are likely because the respiration of adults is higher than that of larvae and the adults' hard elytra may have prevented the infiltration of oil. The distinct responses to oil at these two life stages may be attributed to the morphological and behavioral differences (Tapondjou 2002). It is important to note that the susceptibility varies during the life cycle when using these oils as pest control regime.

Essential oil from *Z. bungeanum*, as a plant product, is unlikely to affect the nutritional or medicinal composition of stored CMM. Its use can be beneficial for human health and environment safety (Murray B. Isma 2000). The effectiveness and safety of the essential oil supported its use as a natural pest control agent.

5. Acknowledgements

This research was sponsored in part by Guizhou provincial Modernization of Chinese Material Medical project (No.QKHZYZ[2011]5049),Organization Department of CPC Guizhou Committee [TZJF-2009-02] and Guiyang city science and technology bureau [No. 2010(1-4)].

6. References

AliNiazee, M.T. (1972). Susceptibility of the confused and red flour beetles to anoxia produced by helium and nitrogen at various temperatures. *Journal of Economic Entomology*, Vol.65,No1, pp. 60-64, ISSN 0022-0493

Bekele, J. A.; Obeng-Ofori, D.; Hassanali, A. (1996). Evaluation of Ocimum suave (Wieed) as a source of repellents toxicants and protectants in storage against three stored product insect pests. *International Journal of Pest Management* ,Vol.42, pp.139-142, ISSN 0967-0874

Bouda, H.; Tapondjou, L.A.; Fontem, D.A. & Gumedzoe, M.Y.D.(2001), Effect of essential oils from leaves of Ageratum conyzoides, Lantana sitophilus, Chromolaena odorata on the mortality of Sitophilus zeamais (Coleoptera, Curculionidae), *Journal of Stored Products Research*, Vol.37, No. 2, (April 2001) , pp. 103-109, ISSN 0022-474X

Can, L.; Zizhong, L.; & Youlian, Y.(2004). Analysis of the structure of insect community on the stored Chinese Medicinal Materials in Guiyang. *Journal of mountain agriculture and biology*, Vol.23,No.1 (February 2004),pp. 41-45 ISSN 1008-0457

Guilin, L. & Wangxi, D.(1996). The community quantitative characters of insects on the stored TCM materials. *Journal of huazhong agriculture university* Vol.15,No.4, (December 1996), pp.338-345. ISSN 1000-2421

Gunasekaran, N. & Rajendran, S. (2005). Toxicity of carbon dioxide to drugstore beetle Stegobium paniceum and cigarette beetle Lasioderma serricorne. *Journal of Stored Products Research*, Vol.41,No.3,pp.283-294. (September 2005),ISSN: 0022-474X

Hashem, M.Y. (2000). Suggested procedures for applying carbon dioxide (CO_2) to control stored medicinal plant products from insect pests. *zeitschrift fur pflanzenkrankheiten und pflanzenschutz-journal of plant diseases and protection*,Vol. 107,No.2,(June 2000),pp. 212-217. ISSN 0340-8159

Ho, S.H.; Cheng, L.P.L.; Sim, K.Y. & Tan, H.T.W.(1994). Potential of cloves (*Syzygium aromaticum* [L.] Merr. and Perry) as a grain protectant against *Tribolium castaneum* (Herbst) and *Sitophilus zeamais* Motsch. *Postharvest Biology and Technology*,Vol.4, No.1-2.(April 1994), pp.179-183,ISSN 0925-5214

Ho, S.H.; Ma,Y. & Huang, Y. (1997). Anethole, a potential insecticide from Illicium verum Hook F., against two stored product insects. *International Pest Control*, Vol.39, No.2, (June 1997), pp.50-51, ISSN 0020-8256

Ho, S.H.; Ma, Y.; Goh, P.M. & Sim, K.Y. (1995). Star anise, Illicium verum Hook F. as a potential grain protectant against *Tribolium castaneum* (Herbst) and *Sitophilus zeamais* Motsch. *Postharvest Biology and Technology*, Vol.6, No.3-4 (October,1995), pp 341-347. ISSN 0925-5214

Huang, M.T.; Ferraro, T. & Ho, C.T.(1994). Cancer chemoprevention by phytochemicals in fruits and vegetables. Food Phytochemicals for Cancer Prevention. *IACS Symposium Series*,Vol.546, 2-16(December 1993) ISBN13: 9780841227682eISBN: 9780841214101

Huang, Y.; Hee, S.K.; & Ho, S.H. (1998). Antifeedant and growth inhibitory effects of a-pinene on the stored-product insects, Tribolium castaneum (Herbst) and Sitophilus zeamais Motsch. *International Pest Control* .Vol.40, No.1, (Appil 1998),pp. 18-20, ISSN 0020-8256

Huang, Y. & Ho, S.H. (1998). Toxicity and antifeedant activities of cinnamaldehyde against the grain storage insects, *Tribolium castaneum* (Herbst) and *Sitophilus zeamais* Motsch. Journal of Stored Prodducts Research. Vol.34, No.1,(January 1998)pp.11-17 ISSN 0022-474X

Juliana, G. & Su, H.C.F. (1983). Laboratory studies on several plant materials as insect repellents for protection of cereal grains. *Journal of Economic Entomology*. Vol.76, pp. 154-157.ISSN 0022-0493

Jun,Q.;Tong, C. & Qing, L. (2001). Simultaneous Distillation and Solvent Extraction and GC/MS Analysis of Volatile Oils of *Zanthoxylum Bungeagum* Maxim. *Journal of Guizhou University of technology* (Natural Science Edition) Vol.30,No.6,(November 2001) pp.1-6 ISSN：1009-0193

Murray,B. Isman (2000). Plant essential oils for pest and disease management. *Crop Protection*, Vol.19, No.8-10, (September 2000) pp.603-608. **ISSN:** 0261-2194

Nielsen, P.S. (2001).The effect of carbon dioxide under pressure against eggs of Ephestia kuehniella Zeller and the adults of S. paniceum (L.) and Oryzaephilus surinamensis (L.). *Anzeiger fur Schadlingskunde* Vol.74, No.3,(June 2002),pp.85-88. ISSN 1436-5693

Platt, R.R.; Cuperus, G.W.; Payton, M.E. Bonjour, E.L. & Pinkston, K.N. (1998). Integrated pest management perceptions and practices and insect populations in grocery stores in south-central United States. *Journal of stored products research,* Vol.34, No.1 (January 1998), pp.1-10 ISSN 0022-474X

Sudesh, J.; Kapoor, A.C. & Ram, S. (1996). Evaluation of Some Plant Products Against Trogoderma granarium Everts in Sorghum and Their Effects on Nutritional Composition and Organoleptic Characteristics. *Journal of stored products research,* Vol.32,No.4(October 1996,),pp.345-392 ISSN 0022-474X

Tapondjou, L.A.; Adler, C.; Bouda, H. & Fontem, D.A. (2002). Efficacy of powder and essential oil from Chenopodium ambrosioides leaves as post-harvest grain protectants against six-stored product beetle. *Journal of stored products research,* Vol.38,No.4.(November 2002), pp.395-402 ISSN 0022-474X

Toh, C. A. (1998). Studies on pest insects of crude drugs. *Korean Journal of Pharmacognosy,* Vol.29,pp.365-373 ISSN, 0253-3073

Zhaohui, L. & Fangqiang, Z. (2001). Research on numerical characteristic of the structure of insect community on the stored Chinese Medicinal Materials in Shangdong province. *Stored Food stuff* Vol.30, No.3, (June 2001),pp.12-16. ISSN 1000-6958

Zhi'an G.; Jingchuan. Z. & Zhihai, X. (2001) Study on Chemical Constituents of the Essential Oil from *Zanthoxylum bungeanum* Maxim by Gas Chromatography-Mass Spectrometry. *Chinese journal of chromatography,* Vol.19, No.6,(November 2001) pp.567-568. ISSN 1000-8713

17

Integrated Pest Management of Eucalypt Psyllids (Insecta, Hemiptera, Psylloidea)

Dalva Luiz de Queiroz[1], Daniel Burckhardt[2] and Jonathan Majer[3]
[1]*Embrapa Florestas, Colombo-PR*
[2]*Naturhistorisches Museum, Basel*
[3]*Department of Environment and Agriculture Biology, Curtin University of Technology, Perth*
[1]*Brazil*
[2]*Switzerland*
[3]*Australia*

1. Introduction

Eucalypts, which are native to Australia, grow rapidly and have multiple uses. Because of these properties, they are planted on all continents except for Antarctica. In Brazil, many eucalypt species find suitable weather and environmental conditions to develop and today they are commercially grown on a large scale. In 2010, plantations covered a surface of over 4.75 million hectares (ABRAF, 2011). In Australia, eucalypts host an abundant fauna of phytophagous insects, among which the jumping plant-lice (Hemiptera: Psylloidea) are particularly species rich.

Psyllids are tiny sap-sucking insects resembling minute cicadas and they generally develop on woody dicotyledons (Hodkinson, 1974; Burckhardt, 1994; Hollis, 2004). Most species have very restricted host plant ranges. Larvae can be free-living or develop in open or closed galls, whereas others build waxy coverings, called lerps, under which they develop (Hodkinson, 1984, 2009; Hollis, 2004; Burckhardt, 2005). The subfamily Spondyliaspidinae (Aphalaridae *sensu* Burckhardt & Ouvrard, 2011) is almost exclusively restricted to Australia and to host plants of the Myrtaceae, in particular *Eucalyptus* species (Burckhardt, 1991; Hollis, 2004). Unsurprisingly, several Australian spondyliaspidine species have been introduced into other continents where they have become pests (Burckhardt, 1994; Burckhardt et al., 1999; Hollis, 2004).

In Brazil, the first spondyliaspidine found was *Ctenarytaina spatulata*, when it was observed in 1994 in a *Eucalyptus grandis* plantation in Paraná State (Iede et al., 1997). Another three species infesting eucalypts were introduced into Brazil within a decade: *Blastopsylla occidentalis* in 1997, *Ctenarytaina eucalypti* in 1998, and *Glycapis brimblecombei* in 2003 (Maschio et al., 1997; Burckhardt et al., 1999; Wilcken et al., 2003).

In this chapter, we present information on eucalypt psyllids that are pests outside Australia with particular emphasis on the situation in Brazil. This information constitutes the necessary base to control these pests. Some of the control options are also discussed.

2. Jumping plant-lice

Psyllids, or jumping plant-lice, are small phloem-feeding insects, measuring 1–10 mm body length. Together with the white flies, aphids and scale insects they constitute the suborder Sternorrhyncha within the Hemiptera. Currently, 3850 species have been described world-wide (Li, 2011), which is probably less than half of the existing number of species. Bekker-Migdisova (1973) suggested that psyllids are a very old group already represented in the Permian by the extinct family Protopsyllidiidae. According to Klimaszewski (1964), the major diversification of Psylloidea occurred between the Middle Jurassic and the Middle Cretaceous, which would coincide with the major diversification of angiosperms. More recent studies, however, imply that origin and diversification of Psylloidea may be much younger. Grimaldi (2003) showed that Protopsyllidiidae may constitute the sister-group of all Sternorrhyncha. Host plant and biogeographic evidence presented by White and Hodkinson (1985) further suggests that the modern psyllids evolved in Gondwana from an ancestor associated with the plant order Sapindales. The Mesozoic Liadopsyllidae form a potential sister-group of modern Psylloidea and the latter are represented in the fossil record only from the Eocene and later (Ouvrard et al., 2010).

As with other hemipterans, psyllids have piercing-sucking mouth-parts. When feeding, the mandibular and maxillary stylets are inserted into the host tissue, saliva is injected and then the liquid food is absorbed. Before feeding, the insects probe more or less extensively. The probing also involves injection of saliva, which is particularly relevant in species which transmit bacterial or viral pathogens. As with other Sternorrhyncha and many Fulgoromorpha, psyllids are specialised phloem-feeders and display several adaptations for coping with this unbalanced, nitrogen-poor diet. The anterior and posterior portions of the mid-gut form a loop which permits water to pass directly from the fore- to the hind-gut. The excess water with dissolved sugars is excreted as honeydew. The often copious production of waxy secretions is a result of the hydrocarbon-rich phloem sap. Psyllids possess bacterial endosymbionts which are situated in the mycetome, a specialised organ in the abdomen. Thao et al. (2000, 2001) showed that psyllids and their primary prokaryotic endosymbionts co-speciated.

Unlike in the related aphids, the life cycle of psyllids usually consists of the egg stage, five larval instars and sexually reproducing adults. The ratio of males/females is near to 1:1 (Hodkinson, 2009). Most psyllid species complete their larval development on one or a few closely related plant species. Adult psyllids are always winged and are easily dispersed by wind. They also have large metacoxae, fused to the metathorax, containing strong muscles enabling them to jump, hence their names 'psylla' from Greek 'flea', or 'jumping plant-louse'. Most psyllids are well camouflaged on the substrate they live and remain generally unnoticed. As a result of these properties, they are predisposed to be accidentally transported by humans together with their host plants or accidentally dispersed by wind over large distances. Psyllids are attracted by yellow colours, a fact which is exploited in monitoring pests with yellow water or sticky traps.

Free living psyllid larvae, depending on the species, sit in the leaf or flower buds or along the veins of young leaves. Some species are covered in copious waxy secretions, which provide defence against soiling with honeydew, desiccation, and also protection against predators and parasitoids. An extreme form of this is found in some species of Spondyliaspidinae, Pachypsyllinae or Macrocorsinae which construct lerps, i.e. shields consisting of wax and sugar which often display a very characteristic shape, structure and chemical composition (Moore, 1961; Hollis, 2004). Many psyllids develop in galls (Hodkinson, 1984; Burckhardt, 2005), which are generally induced by the first instar larva. The insertion of the stylets triggers increased cell growth around the larva, which may cover it completely. In *Apsylla cistellata*, the feeding of the first instar larva on leaves induces cell growth in the leaf buds farther away. Later, the second instar migrates into the already formed gall (Raman et al., 2009). Galls can be open, called pit galls, or closed and resemble small nuts, discs or globes (Hodkinson, 1984; Yang & Mitter, 1994). Yang et al. (2000) described a case of inquilinism. Newly hatched first instar larvae of *Pachypsylla cohabitans*, which are unable to induce galls, feed next to larvae of several other gall-inducing *Pachypsylla* species and become incorporated into the growing gall.

Depending on the climatic conditions, psyllid species of temperate regions tend to be univoltine or bivoltine, while those of the tropics are often polyvoltine, with several overlapping generations per year (Burckhardt, 1994; Hollis, 2004; Santana et al., 2010). Hodkinson (2009) analysed the life history characteristics of 342 psyllid species from all over the world and concluded that environmental temperatures and water availability acting on the psyllids, directly or via the host plant, are the major determinants of psyllid life cycles. The phenology of psyllids is, therefore, well synchronised with that of their hosts. Psyllid populations are generally controlled by a whole range of predators and parasitoids (Hodkinson, 1974; Hollis, 2004). Birds and small mammals are known to occasionally eat psyllids. In agricultural systems, anthocorids, syrphids, coccinellids and chrysopids are sometimes capable of controlling psyllid populations, but parasitoids such as encyrtids and eulophids are usually more efficient. Psyllids are also affected by entomopathogenic fungi, with which they may be controlled (Dal Pogetto et al., 2011).

Psyllid host plants are mostly perennial dicotyledonous angiosperms. There are only a few psyllids associated with monocots, such as the holarctic genus *Livia*. Even less species develop on gymnosperms, most notably some triozids in New Zealand and two species of *Ehrendorferiana* from southern Chile (Hodkinson, 2009). Restricted host ranges characterise not only psyllid species but, to a certain extent, also higher taxonomic ranks. Related psyllid species tend to breed on related plant species. Members of the tropical family Homotomidae are all associated with *Ficus* species and other Moraceae (Hollis & Broofield, 1989) and those of its putative sister-group, the Carsidaridae, are associated with members of the Malvales (Hollis, 1987). This pattern suggests that there may be co-speciation between angiosperms and psyllids. Detailed phylogenetic studies, however, show that the observed species richness in psyllids is better explained by geographic vicariance than by co-speciation with the host-plants (Burckhardt & Basset, 2000; Percy, 2003; Burckhardt & Ouvrard, 2007).

Within modern Psylloidea, three probably monophyletic lineages exist (Burckhardt & Ouvrard, 2011), namely: 1. Phacopteronidae; 2. Aphalaridae + (Carsidaridae + Homotomidae); and 3. Liviidae + (Calophyidae + (Psyllidae + Triozidae). Host associations with the plant family Myrtaceae occur in the following taxa: Aphalaridae (Spondyliaspidinae), Liviidae (Diaphorininae) and Triozidae. This suggests that psyllids colonised Myrtaceae at least three times independently. The Spondyliaspidinae comprises 24 genera (Burckhardt, 1991; Hollis, 2004), with species which breed almost exclusively on Myrtaceae and which are restricted to the Australian region. The monotypic Oriental genus *Eurhinocola* is an exception, as it is not represented in the Australian fauna. The genus *Boreioglycaspis* has species in Australia and southeast Asia and is associated with Myrtaceae and the related Lythraceae. *Ctenarytaina* has an even wider distribution, including species native to New Zealand, southeast Asia, China, the Indian subcontinent, and possibly even tropical Africa. It is also wider in its host associations and has some hosts outside the Myrtales.

3. Eucalyptus

Members of the eucalypt group (Myrtaceae) of trees and mallees (shrub forms) represent an old lineage that can be related back to the Late Cretaceous, indicating their Gondwanan origin. There are well in excess of 700 species, with most being confined to Australia, although 15 species naturally occur outside of Australia in southeast Asia, as distant as Sulawesi and the Philippines, and related genera occur in New Caledonia. Australian representatives of the group were originally regarded as two genera, *Eucalyptus* and *Angophora* (Burbidge, 1960). More recently, seven genera within the group have been recognised (Ladiges, 1997), with the main genus being split into *Eucalyptus* L'Hér. *sensu stricto* (+600 species) and the bloodwoods, *Corymbia* Hill & Johnson (+100 species), which are of equal taxonomic rank to *Angophora*. Within the genus *Eucalyptus*, a number of subgenera are recognised (Brooker 2000), namely *Acerosae, Alveolata, Cruciformes, Cuboidea, Eucalyptus, Eudesmia, Idiogenes, Minutifructus, Primitiva,* and *Symphyomyrtus*. The other genera within the eucalypt group which, with one exception, are monotypic, are *Arillastrum* (New Caledonia), *Allosyncarpia* (northern Australia), *Eucalyptopsis* (New Guinea, Moluccan Archipelago, Woodlark Island) and *Stockwellia* (northern Australia).

Considering the antiquity of the 'eucalypts', it is not surprising that invertebrates have co-evolved to feed on or utilize other resources from these trees. Working in the canopy alone, Majer et al. (2000) sampled 641 and 726 invertebrate species from *Eucalyptus moluccana* and *E. crebra* in NSW, eastern Australia and 448 and 444 from *E. marginata* and *Corymbia calophylla* in WA, western Australia. In all instances, the percentage of herbivores in the count ranged from 20.5–25%. The richness of herbivores on 'eucalypts' is accompanied by high degrees of specificity of certain groups to 'eucalypt' species or subgenera. This has been observed in gall-forming eriococcid scale insects (Cook, 2001; Cook & Gullen, 2004), where particular species have been found to be specific to certain species of *Eucalyptus* or *Corymbia*. Another example is the Australian psyllids (Aphalaridae: Spondyliaspidinae), which are renowned for their association with 'eucalypts', with 67% of species utilising 'eucalypts' as host plants (Majer et al., 1997). Working on the 'eucalypt'-feeding genus *Glycaspis*, Moore (1961, 1970) was able to divide the genus into subgenera on the basis of morphology and host plant association. This is but one example of the high degree of host-specificity of Australian psyllids on 'eucalypts'.

Hosts	*C. spatulata*	*C. eucalypti*	*B. occidentalis*	*G. brimblecombei*
E. benthamii		x		
E. bicostata		x		
E. blakelyi			x	
E. brassiana			x	
E. bridgesiana				x
E. camaldulensis		x		x
E. camphora				x
E. dealbata				x
E. diversicolor				x
E. dunnii		x		
E. forrestiana			x	
E. globulus		x	x	x
E. gomphocephala			x	
E. grandis	x		x	
E. maidenii		x		
E. mannifera				x
E. mannifera maculosa				x
E. microneura			x	
E. microtheca			x	
E. nicholii			x	
E. nicholsii			x	
E. nitens		x		x
E. oleosa			x	
E. pellita	x			
E. pulverulenta		x		
E. robusta	x			
E. rudis			x	
E. saligna			x	
E. sideroxylon				x
E. spathulata			x	
E. tereticornis			x	x
E. urophylla	x		x	

Table 1. Lists the four Australian psyllids that have become established in Brazilian eucalypt plantations, showing their main hosts in Brazil (in bold) and also the other species that they can exploit (sources: Burckhardt et al., 1999; Brennan et al., 2001; Hollis, 2004; Meza & Baldini, 2001). All host species are members of the *Symphyomyrtus* subgenus.

4. Psyllid pests and eucalypts

Due to the close association with their hosts, some psyllids are of economic relevance. While most of these are minor pests (Burckhardt, 1994), a few species are responsible for huge economic losses, such as the species transmitting bacterial phytopathogens (e.g. *Diaphorina citri*, some *Cacopsylla* spp., *Bactericera cockerelli*) (Hodkinson, 2009) or the eucalypt psyllids (Santana & Burckhardt, 2007).

Over 350 species of Psylloidea are reported from Australia (Hollis, 2004), of which 79% are associated with Myrtaceae and 71% with eucalypts. In Australia most populations of eucalypt psyllids are in balance (Collet, 2001a). However, when introduced into other continents, their populations may increase and become a serious problem (Paine & Millar, 2002). This has been the case with *Ctenarytaina eucalypti* (Dahlsten et al., 1998a; Pinzón et al., 2002) and *Glycaspis brimblecombei* (Bouvet et al., 2005). Currently, seven Australian species are known from other continents (*Blastopsylla occidentalis, Cardiaspina fiscella, Cryptoneossa triangula, Ctenarytaina eucalypti, C. spatulata, Eucalyptolyma maideni*, and *Glycaspis brimblecombei*). An eighth species probably also originates from Australia but has not yet been found there. It is *Ctenarytaina peregrina*, which was described from Eire (Hodkinson, 2007) and has later been found in Germany (K. Schrameyer, pers. comm.).

4.1 *Ctenarytaina spatulata* Taylor, 1997 (rose gum psyllid)

Ctenarytaina spatulata originates from southeast Australia and has been introduced into New Zealand, the USA (California), Uruguay, Brazil, Portugal and Spain (Burckhardt et al., 1999; Santana et al., 1999; Hollis, 2004; Valente et al., 2004). It was observed for the first time in Brazil in Arapoti - Paraná State in 1994, in an *E. grandis* plantation (Burckhardt et al., 1999). It is now commonly found in the states of São Paulo, Paraná, Santa Catarina and Rio Grande do Sul.

The adults of *C. spatulata* (Fig. 1A) are yellowish or orange with dark spots or dark brown stripes. The forewings are transparent and yellowish with slightly darker veins. They remain most of the time on the leaves and new apical shoots, where they feed and reproduce. The females lay their eggs (Fig. 1B) on the newly growing leaf axils (Santana & Zanol, 2006).

All five larval instars have a dorso-ventrally flattened body. They live in colonies and feed on young plants or shoots. They secrete large amounts of honeydew which is secreted as small wax-covered globules together with a large amount of flocculent waxy secretions along the sides of the abdomen, which spread out all over the colony (Santana & Zanol, 2005). Honeydew and waxy secretions harm the development of the young plants, particularly in the first two years of planting (Collet, 2001a). In the first larval instar, the body is entirely yellow, except for the small red eyes. In the final instar larva, the length of the body is 1.40–1.43 mm. The body is brown to yellowish with brown patches (Fig. 1C). The caudal plate bears five lanceolate setae on each side of the anus. The wing buds are well-developed; the forewing buds lack a humeral lobe, and their fore margin lies posterior to the posterior eye margin (Taylor, 1997; Santana & Zanol, 2005).

C. spatulata is associated with several *Eucalyptus* spp., in particular with *E. grandis, E. saligna, E. robusta, E. pellita, E. resinifera* and *E. urophylla*, but it has also been observed in lower numbers on *E. deanei, E. saligna, E. tereticornis, E. microcorys, E. viminalis, E. camaldulensis, E. alba* and *E. nitens* (Santana, 2003). Among the species of *Corymbia*, only eggs of *C. spatulata* have been observed on adult leaves of *C. citriodora*. Neither eggs nor larvae of *C. spatulata* have been observed on native Myrtaceae in Brazil (Santana, 2003). Valente et al. (2004) found *C. spatulata* mostly on *E. globulus*, the main eucalypt species planted on the Iberian Peninsula. This eucalypt is also a major host of *C. spatulata* in California (Brennan et al., 2001). These authors also observed large populations of the psyllid on *E. nitens, E. dalrympleana* and *E. maidenii*. Other host species mentioned in the literature are *E. leucoxylon, E. mannifera maculosa, E. pauciflora, E. longifolia, E. rodwayi, E. ovata, E. nitida* (Taylor, 1997), *E. amplifolia, E. dunnii, E. robusta, E. rostrata* and *E. tereticornis* (Burckhardt et al., 1999).

In Brazil, this species has many generations per year, leading to a higher number of individuals during the cold and dry months (Santana & Burckhardt, 2007).

The damage of this pest was observed on *E. grandis* in Paraná state and hybrids of *E. grandis x E. urophylla* in São Paulo, with symptoms including sooty mould on the leaves and tips, dieback, loss of apical dominance, super sprouting and decrease of growth (Santana et al., 2005). *C. spatulata* completes its life cycle from egg to adult at temperatures around 20 °C in approximately 45 days (Santana & Zanol, 2006). In São Paulo and Paraná all life stages of the species can be observed during the entire year (Santana, 2003). The first damage by *C. spatulata* to *E. grandis* is caused by oviposition (Santana et al., 2005). A small black spot appears where the egg is inserted, which may increase and lead to withering of the terminal buds. The larvae secrete large amounts of honeydew, which accumulates on leaves and buds, allowing the growth of sooty mould and phytopathogenic fungi. The successive piercing of the substrate and the extraction of plant sap causes deformations and curling of the leaf, thus reducing the leaf surface. High populations weaken the plant further by the extraction of plant sap, causing the death of the terminal buds, the loss of apical dominance and super sprouting of the lateral buds (similar to formation of witches' brooms by excessive growth of lateral buds) (Cadahia, 1980; Zondag, 1982; Meza & Baldini, 2001; Santana et al., 2005).

In addition, the sucking of *C. spatulata* can reduce the increment of stem diameter and of the internodes, resulting in more branches on the stem and rendering the wood of the stem more fragile (Santana et al., 2005). The damage caused by *C. spatulata* in Brazil has been estimated in a green-house (Santana et al., 1999). By studying the nutritional stress, the authors observed that there is an interaction between the insects and Mg deficiency. Both factors together may stop growth and production of biomass, in addition to affecting root development. The research done in the greenhouse and the field shows that *C. spatulata* appeared in all samples, with 100% occurrence of larvae and eggs. Its presence was observed during all months of the year, in every stage, which is typical for a polyvoltine life cycle. It also has a population peak in the colder months with low rainfall (Santana et al., 2005).

Water stress is one of the environmental factors that may improve the development of the psyllid population (White, 1969) due to higher nitrogen concentration in the plants. Santana et al. (2003a) simulated water stress in *E. grandis* plants, with and without the presence of *C. spatulata*, and noticed that the insects may cause a 20% loss of height growth in *E. grandis*. The combination of water stress and presence of the insects may cause considerable damage.

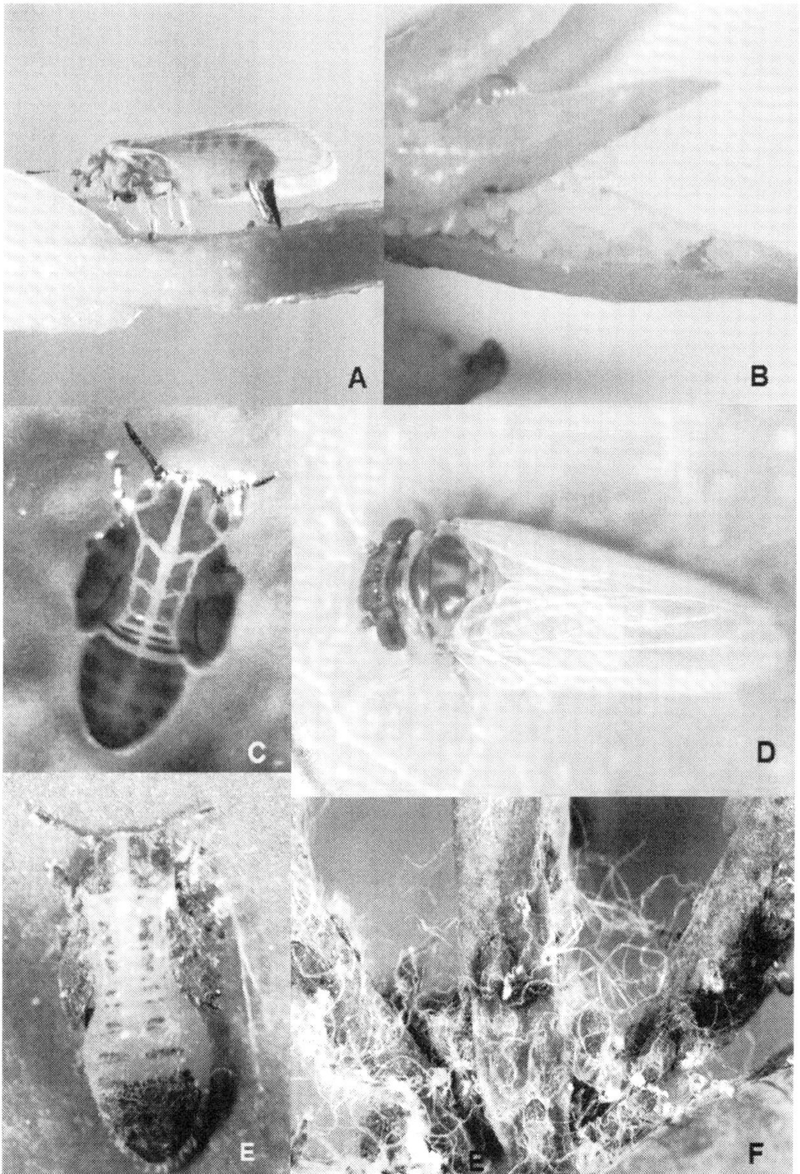

Fig. 1. *Ctenarytaina spatulata* (A - adult, B - eggs, C - larva); *C. eucalypti* (D - adult, E - larva, F - colony of larvae).

4.2 *Ctenarytaina eucalypti* (Maskell, 1890) (blue gum psyllid)

This is a species from southeast Australia that has been introduced into many countries throughout the world. The first mention of *C. eucalypti* from Brazil is by Burckhardt et al. (1999), who reported it from Colombo - Paraná on seedlings of *E. dunnii*. Subsequently, it has also been detected in São Paulo, Santa Catarina and Rio Grande do Sul. Its geographical distribution, apart from its origin in Australia, includes Bolivia, Brazil, Chile, Colombia, Eire, France, Germany, Italy, New Zealand, Papua New Guinea, Portugal, South Africa, Spain, Sri Lanka, Switzerland, UK, Uruguay, and the USA (Burckhardt, 1998; Hodkinson, 1999; Durán & Urrutia, 2001; Burckhardt & Mühlethaler, 2003).

The adults of *C. eucalypti* (Fig. 1D) measure from 1.5–2.0 mm and have dirty whitish forewings with contrasting brown veins and a clear membrane which are normally folded over the body (Burckhardt et al., 1999). The body colour is usually dark brown to black (Burckhardt et al., 1999), with darker transverse stripes on the abdomen, both dorsally and ventrally. The antennae are yellow with dark apices of the individual segments. The compound eyes are dark brown and prominent. The legs are dark yellow. The female proctiger is longer and more slender in comparison with *C. spatulata*, with relatively acute apico-lateral peg setae and strongly curved valvula ventralis (Burckhardt et al., 1999). The female lays 20 to 100 eggs, normally in groups, on the leaf buds and in the axils of young leaves in nurseries or on young trees. Eggs are also laid in small fissures between the bud and the leaf pedicel. Several females may contribute to one egg mass. In summer, the incubation lasts about one week, whereas it is longer in the cold periods (Cadahia, 1980).

The first instar larvae are light yellow with red eyes and thick legs. There are thick setae along the margin of the caudal plate (Zondag 1982). The last instar larvae are yellow with dark patches. The eyes, antennae and wing buds are reddish-brown (Fig. 1E).

C. eucalypti has been observed in Brazil on many *Eucalyptus* spp., such as *E. globulus*, *E. maidenii*, *E. bicostata*, *E. dunnii* and *E. nitens*. Of particular importance are *E. dunnii* and *E. benthamii*, which are planted in temperate region on the south of Brazil where a high psyllid attack has been observed (Santana et al., 1999).

Hodkinson (1999) described the damage of *C. eucalypti*, including direct effects of psyllid feeding such as severe shoot dieback, leaf curl and leaf discoloration. A further factor damaging the plants is sooty mould growing on the honeydew, which is secreted in large amounts.

In Brazil, the development is continuous throughout the year, with several overlapping generations. All developmental stages have been observed to occur together at any particular time in the same population. All five larval instars excrete large amounts of honeydew and white waxy secretions (Fig. 1F).

4.3 *Blastopsylla occidentalis* Taylor, 1985 (eucalypt shoot psyllid)

Blastopsylla occidentalis originates from South and West Australia. Its geographical distribution is Argentina, Brazil, Cameroon, Chile, Hong Kong, Kenya, Mexico, New Zealand, Paraguay, South Africa and USA (Hodkinson, 1991; Burckhardt et al., 1999; Burckhardt & Elgueta, 2000; Hollis, 2004; Bouvet et al., 2005; anonymous, 2007; Tamesse et al., 2011). According to the original description by Li (2011), *Blasopsylla barbara* from China may be a synonym of *B. occidentalis*.

The adults are small insects, measuring 1.5-2.0 mm. They have a yellow head and thorax with dark pattern, the head is as large as the thorax and strongly inclined relative to the longitudinal body axis, with short antennae (Fig. 2A). The forewings have brown veins and a grey membrane. The male terminalia are yellow, while those of the females are dark brown with a yellow base of the subgenital plate. The males are usually yellow, the females are darker coloured. The last instar larvae are yellow with a dark brown antennal tip; lacking humeral lobe on forewing buds or specialized setae; with 9-segmented antennae (Burckhardt & Elgueta, 2000; Meza & Baldini, 2001; Durán & Urrutia, 2001). *Blastopsylla* differs from other spondyliaspidines by the very long posterior lobes on the basal segment of the male proctiger and often in the presence of a single spur on the metabasitarsus (Burckhardt & Elgueta 2000; Hollis, 2004). *B. occidentalis* can be easily separated from *Ctenarytaina* species by the lack of an outer apical comb of bristles on the mesotibia, and the apically curved vein Rs in the forewings (Taylor, 1990; Burckhardt et al., 1999; Burckhardt & Elgueta, 2000). Within *Blastopsylla*, *B. occidentalis* is similar to *B. moorei* from which it differs in the shape of the apical portion of the aedeagus, which is spherical rather than elongate and weakly curved as in *B. moorei*, and the fewer black setae on the apical portion on the inner surface of the paramere and the 4-6 dark sclerotised setae which form a line along hind margin, rather than a group of 5-7 as in *B. moorei* (Taylor, 1985). The females of *B. occidentalis* lay the eggs on the shoots, the leaf axils, on small branches and young leaves.

The larvae secrete large amounts of small wax covered globules containing honeydew and copious white flocculence (Fig. 2B), which enhances growth of sooty mould (Taylor, 1985; Meza & Baldini, 2001). The wax often sticks to the end of the larval abdomen.

B. occidentalis was observed for the first time in Brazil in Goiás in 1997 on hybrids of *Eucalyptus urophylla* and *E. grandis* (Burckhardt et al., 1999). Other hosts are *E. microtheca, E. rudis, E. gomphocephala, E. camaldulensis, E. microneura, E. nicholsii, E. spathulata, E. forrestiana, E. oleosa, E. rudis, E. tereticornis, E. saligna, E. globulus* and *E. nicholii* (Meza & Baldini 2001; Hollis, 2004).

Contrary to the two *Ctenarytaina* spp., *B. occidentalis* occurs in central Brazil, where the dry season is longer than in southern Brazil, with over four months without rain.

4.4 *Glycaspis brimblecombei* Moore, 1964 (red gum lerp psyllid)

Glycaspis brimblecombei (Fig. 2C), the red gum lerp psyllid, has an Australian origin and was introduced into the USA in June 1998 (Brennan et al., 1999). In 2000 it was observed in Mexico, in Chile in 2002 (Dahlsten, 2003), in Brazil in 2003, in Argentina in 2004 (Bouvet et al., 2005), in Ecuador in 2007 (Onore & Gara, 2007), in 2007 in Venezuela (Rosales et al., 2008), in Peru in 2008 (Burckhardt et al., 2008), in 2008 on the Iberian Peninsula (Hurtado & Reina, 2008; Valente & Hodkinson, 2009; Prieto-Lillo et al. in 2009) and in Italy in 2010 (Laudonia & Garonna, 2010; Peris-Felipo et al., 2011). In Brazil, *G. brimblecombei* was first detected in São Paulo and now it is wide-spread in the whole country in areas where its host is planted.

G. brimblecombei develops on various *Eucalyptus* spp., especially *E. camaldulensis* and *E. tereticornis* but also on *E. blakelyi, E. brassiana, E. bridgesiana, E. camphora, E. dealbata, E. diversicolor, E. globulus, E. mannifera, E. mannifera maculosa, E. nitens* and *E. sideroxylon* (Brennan et al., 2001; Hollis, 2004).

The eggs are laid in groups, usually in rows or circles (Fig. 2D). The larvae of *G. brimblecombei* (Fig. 2E) differ from those of the other three species discussed above, as they are not free-living. They live under a shield of wax, called lerp (Fig. 2F). Adults can easily be differentiated from *Blastopsylla* and *Ctenarytaina* spp. by the long genal processes, which are much longer than the vertex along the mid-line (Olivares et al., 2004; Burckhardt et al., 2008).

The damage of this species is similar to that made by the other species discussed above, but *G. brimblecombei* is a more aggressive exploiter of resources. There are reports of outbreaks in many South American countries were the species caused the death of the trees, resulting in serious production losses.

4.5 *Cardiaspina fiscella* Taylor, 1962 (brown lace lerp or brown basket lerp psyllid)

Cardiaspina ficella is native to Australia, where it occurs naturally in the Australian Capital Territory, New South Wales and Victoria (Hollis, 2004). It uses as hosts *Eucalyptus botryoides, E. saligna, E. robusta, E. grandis* (Collet, 2001a), *E. camaldulensis, E. blakelyi* (Collet, 2001b), *E. globulus* and *E. tereticornis*. According to Withers (2001), 57 insect species originating from Australia have been introduced in New Zealand, among them, *C. fiscella*. This species was detected in New Zealand in 1996 and it quickly diffused to most of the North Island, with severe infestations, successive defoliation, dieback and general decline of plantings of *Eucalyptus botryoides* and *E. saligna*. According to Berry (2006), *C. fiscella* is the only introduced species of Australia that has become a significant pest on *E. botryoides, E. grandis* and *E. saligna* in New Zealand.

There are many species of *Cardiaspina* using eucalyptus as a host, among which *C. fiscella* is one which most often reaches high populations (Campbell, 1992). Adults of this Australian psyllid species are 3.4 mm in length, from head to wing tip. They have a short, more robust body than the other species mentioned previously. Antennae are almost equal to width to the head. Legs are stout with two very small claws. The general colour is light brown; the head straw coloured; the thorax bears several medium brown patches; the abdomen is brownish black, with a yellow or red caudal margin on each segment. The forewings are transparent with uniformly light brown veins (Fig. 3A).

The larvae (Fig. 3B) build very elaborate lerps, which are shell-shaped (Fig. 3C), consisting of a tangled lattice, usually on the dorsal surface of the leaves (Hollis, 2004). Dimensions are approximately 4.1 mm from hinge to apex, and approximately 5.4 mm across. General colour is light brown, darker near the hinge with a darker band where the ribs begin to fan out. It is moderately convex. Larvae can move freely underneath the lerp and can sometimes be seen moving on the undersides of leaves outside of the lerp (Appleton, 2009). The larvae of *C. fiscella* have glands that secrete wax filaments to produce the lerp. Larvae of all *Cardiaspina* spp. form fan-shaped basket lerps, with unique characteristics for each species. Often, necrosis on foliage of mature leaves occurs beneath the lerp (Fig. 3D, E), as can be found on many eucalypt species, especially of *E. camaldulensis* and *E. blakelyi* (Collet, 2001a).

Fig. 2. *Blastopsylla occidentalis* (A - adult, B – colony of larva); *Glycaspis brimblecombei* (C - adult, D - eggs, E - larvae, F - lerps on leaf).

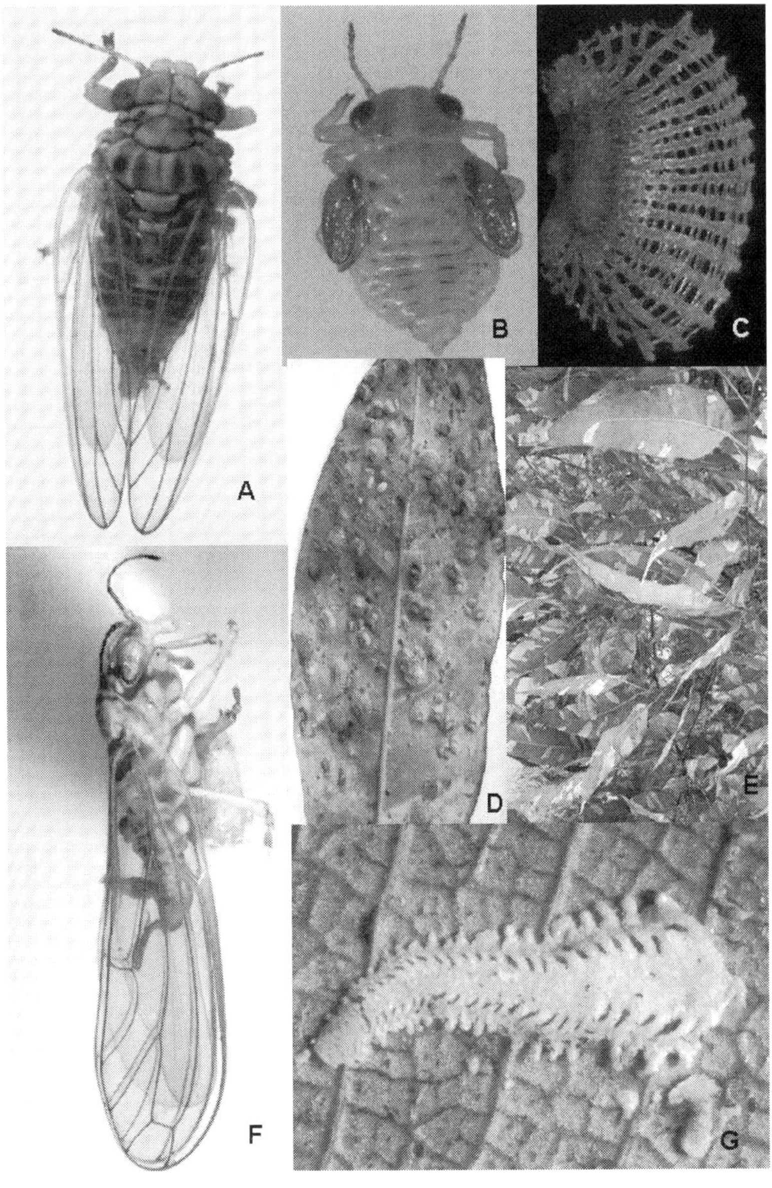

Fig. 3. *Cardiaspina fiscella* (A - adult, B- larva, C - lerp, D, E - damage on leaves); *Eucalyptolyma maideni* (F - adult, G - lerp on leaf).

4.6 *Cryptoneossa triangula* Taylor, 1990 (lemon gum psyllid)

Cryptoneossa triangula was described from *Corymbia citriodora* and *C. maculata* trees collected in the Australian Capital Territory and New South Wales (Taylor, 1990). This species is morphologically similar to *C. minuta and C. leptospermi* and derives its name from the triangular shape of the parameres. Adults have a general ochraceous to brown colour and the vertex has a narrow black margin in some specimens. The forewings are relatively longer and narrower than those of *C. minuta* and the marginal cell cu_1 is not as flat as in *C. leptospermi*. Males of *C. triangula* are 1.8 to 2.2 mm length. The head bears spheroid genal processes and anterior margins of vertex expanded on each side of median ocellus; anteriorbital lobes small, conical. The basitarsus has two black spurs. Females are 2.3 to 2.6 mm long with rounded anterorbital lobes (Taylor, 1990).

In August 2000, this psyllid was discovered on California's lemon-scented gum (*C. citriodora*) and spotted gum (*C. maculata*) trees at Disney Resort and surrounding areas of Anaheim, California. The lemon gum psyllid (*C. triangula*) is a free-living psyllid causing leaf damage and inducing leaf drop, which may stress trees and make them susceptible to fatal attack by other insects. The psyllids also produce sticky honeydew, which soils the ground, cars and sidewalks (Dahlsten, 2001; Daane & Paine, 2005).

4.7 *Eucalyptolyma maideni* Froggatt, 1901 (spotted gum lerp psyllid)

Eucalyptolyma maideni was described from *Eucalyptus* spp. and later redescribed by Taylor (1987) from *C. citriodora, C. gummifera* and *C. maculata,* with records in New South Wales, Australian Capital Territory and South Australia. The adults have general bright green colour tinged with yellow (Fig. 3F). The lerp is white, corniculate, with laciniate margins (Taylor, 1987). According to Morgan (1984), this psyllid is widespread in eucalypt forests from Queensland and South Australia to Tasmania. It is trivoltine and prefers fully mature leaves upon which to feed, oviposit and develop.

This psyllid was introduced in USA in 2000 and collected from *C. citriodora* in Los Angeles, California. Since then, it was also discovered near Anaheim (at Disney Resort) and heavy infestations have been reported from many locations within Los Angeles, Orange County and San Diego. Although this insect does not cause the death of the plants, it is considered to be an ornamental pest of lemon scented gum and spotted gum in California. Larvae build a flattened and elongated triangular lerp (Fig. 3G) and produce a copious amount of honeydew which soils the ground beneath trees, similar to that mentioned for *C. triangula,* where a blackish sooty mould grows (Garrison, 2001).

5. Integrated Pest Management (IPM)

The integrated management of the psyllids should start with monitoring, which should be continuous, with more attention being paid in peak periods. The monitoring of the psyllids can be done by installing sticky traps or by manual sampling. These should be run continuously and at regular intervals. Yellow sticky traps are the best way to monitor psyllids infesting eucalypts. Adult psyllids and psyllid parasites are attracted by the yellow colour and become stuck to the surface. The traps should be inspected once a week and the number of adult psyllids and their parasites should be counted and recorded (Paine et al., 2007).

Adult psyllids can also be monitored by shaking or beating plants over a collecting sheet to knock them onto the collecting surface, where they can be easily seen and counted. The sample should be taken about once a week during the season of new plant growth, when adult psyllids are expected (Dreistadt & Dahlsten, 2007).

Through such monitoring, it is possible to determine the moment of population peak, the occurrence of natural enemies and other factors that affect the insect population. In an Integrated Pest Management Programme (IPM), monitoring the pest's development is one of the main components, which uses different techniques to quantify the population and predict outbreaks of the pest. Besides, it can be also used to determine the geographic distribution of the pest, to detect a risk to the area before any damage actually occurs and to determine the effectiveness of the control treatment.

5.1 Biological control

The use of parasitism is a major component of psyllid IPM, with several biological control programme having been successfully carried out for different species of introduced psyllid. A biological control programme of *C. eucalypti* was supported by the University of California IPM project. The parasitoid *Psyllaephagus pilosus* (Hymenoptera: Encyrtidae) was collected in Australia, reared at UC Berkeley and released in California. The parasitoids quickly established in the release sites, and sampling in 1994 revealed parasitism rates of 50–100%. By 1995, the parasitoids had become broadly distributed throughout many parts of the State. An economic analysis of the benefit to the cut-foliage industry alone indicated that the biological control programme generated a benefit-cost ratio ranging from a minimum of 9:1 to a maximum of 24:1, based solely on elimination of pesticide treatments (Dahlsten et al., 1998a). A similar IPM programme was adopted to manage the populations of the lerp psyllid *G. brimblecombei* in California (Paine et al., 2007).

In Brazil, in the end of 2000, beginning of 2001, a sudden population decrease was observed, because of the hymenopterous parasitoids which attacked *C. eucalypti*. The parasitoid was identified as *Psyllaephagus* sp. (Hymenoptera: Encyrtidae), the same genus which has been introduced into many European and American countries for the biological control of *C. eucalypti* (Santana et al., 2002). The parasitoids of *C. eucalypti* (*P. pilosus*) and *G. brimblecombei* (*P. bliteus*) were introduced accidentally together with their hosts in Brazil (Santana et al., 2002; Berti Filho et al., 2003). These parasitoids adapted and dispersed over the whole Brazilian territory. Now they can be found in practically all the areas where the pest is present. The natural control of these psyllids has been successfully facilitated using these specific parasitoids (Santana & Burckhardt, 2007).

The decrease of the *G. brimblecombei* population observed in São Paulo, Brazil in the end of spring, beginning of summer in 2003, was attributed to the rainfall and certainly because of the presence of parasitoids and predators (Fig. 4). The parasitoid *P. bliteus* (Fig. 4 A, B) was detected in the same year of introduction of *G. brimblecombei* in São Paulo State, Paraná and Minas Gerais. With the detection of this parasitoid, it is expected that the population of the psyllid will also stabilize (Santana et al., 2003b). Now, the management of the pest and its natural enemies is established.

Fig. 4. Psyllid IPM. A, B - Parasitoid *P. bliteus*; C-F – predators, G - clone plantation: *G. brimblecombei* resistant (left) susceptible (right); H – detail of infestation on susceptible clone.

Withers (2000) reported the occurrence of *Psyllaephagus gemitus* in *C. fiscella*. According to this author, the health of eucalypt trees in the North Island of New Zealand can be expected to improve as populations of *C. fiscella* continue to be killed by *P. gemitus*, a nymphal parasitoid. This same parasitoid, *P. gemitus*, was tested in 1997 as a possible biological control agent for *C. fiscella* by entomologists funded by the New Zealand Farm Forestry Association. A parasitic microwasp associated with brown lace lerp (*C. fiscella*) has been recorded from New Zealand. This microwasp, *Coccidoctonus psyllae*, is a hyperparasitoid and its presence will probably contribute to an increase in brown lace lerp populations (Berry, 2006).

The eucalypt psyllids are also attacked by many predators, including the ladybird beetles (*Cycloneda sanguinea* , *Olla v-nigrum* (Fig. 4C), *Hippodamia convergens, Eriopsis connexa* and *Harmonia axyridis* (Fig. 4D), the green lacewings (*Chrysoperla* spp.) (Fig. 4E), syrphid flies, pirate bugs (*Anthocoris* sp.) (Fig. 4F) and spiders (Santana, 2003; Santana et al., 2004). Although predators do not provide complete biological control, they can reduce psyllid abundance. Whenever possible, management efforts should be selected that have less adverse effects on these beneficial species (Paine et al., 2007). Santana (2003) observed the following potential natural enemies for *C. spatulata*: *Coccinella ocelligera, Curinus coeruleus, Cycloneda pulchella, Cycloneda sanguinea, Eriopsis connexa, Harmonia axyridis, Hyppodamia convergens, Hyperaspis* sp., *Scymnus* (*Pullus*) sp. and *Olla v-nigrum* (Coleoptera: Coccinellidae), *Chrysoperla externa* (Neuroptera: Chrysopidae); *Allograpta exotica, Pseudodorus clavatus, Syrphus phaeostigma* (Diptera: Syrphidae), as well as spiders and the fungus *Verticilium lecanii*. The same predators observed for *C. spatulata* also attack *B. occidentalis*, including Syrphidae, Dolichopodidae (both Diptera), Chrysopidae (Neuroptera) and Coccinellidae (Coleoptera) (Santana & Burckhardt; 2007).

Azevedo and Figo (1979) list the following natural enemies of *C. eucalypti* in Portugal: *Syrphocotonus abdominator* (Hymenoptera: Ichneumonidae); *Haematopota ocelligera* (Diptera: Tabanidae); *Sphaerophoria scripta, Meliscaeva cinctellus, Pipizella* sp., *Eumerus* sp. (Diptera: Syrphidae); and *Bradysia* sp. (Diptera: Sciaridae). The predator complex feeding on *C. eucalypti* includes hoverfly larvae, ladybirds, lacewings, anthocorids and spiders (Hodkinson, 1999). According to Zondag (1982), *C. eucalypti* is frequently attacked by a small black wasp in New Zealand. In Tasmania, a small ladybird (*Cleobora mellyi*), used to control a beetle, has been observed feeding on larvae and eggs of psyllids, showing a high potential to control *C. eucalypti*.

5.2 Plant resistance

The search for tolerant or resistant plants has been an economically viable alternative to contain populations of these insects in large plantations. Tolerance varies among eucalypt species and location of plants. Prevention of planting susceptible species is the best way to avoid damage (Dreistadt et al., 2007). According to White (1970), who studied aspects of the life history of *Cardiaspina densitexta*, the physiological and physical characteristics of plants influence the selection of hosts.

Brennan and Weinbaum (2001a) studied the performance of psyllid adults on leaves of *E. globulus* and noted that the epicuticular wax of juvenile leaves plays an important role in resistance to *C. spatulata* and *G. brimblecombei*; these species tend to avoid more waxy leaves. Brennan and Weinbaum (2001b) suggested that the tarsi of *C. eucalypti* are more adapted for adhering to the epicuticular wax-coated surfaces than those of the others two psyllid species. Continuing these studies, Brennan and Weinbaum (2001c) concluded that the epicuticular wax on juvenile *E. globulus* leaves reduces stylet probing by *C. spatulata* and *G. brimblecombei* and this psyllid avoids oil glands on the leaves.

In breeding programmes for improving physical and chemical properties of plants, all aspects related to resistance should be analysed. In Brazil, where most eucalypt plantations are clonal, the forestry companies already have some materials selected in the search for resistance (Fig. 4G, H).

In order to verify the preference of *C. spatulata* for laying eggs and feeding, Santana et al. (2010) carried out a green-house trial on 19 *Eucalyptus* species, one *Eucalyptus* hybrid (Cambiju), three *Corymbia* species and four native Myrtaceae species (*Hexaclamys edulis, Plinia edulis, Plinia trunciflora* and *Psidium* sp.). As a result of this trial, they found that the largest populations of *C. spatulata* were observed on *E. robusta* and *E. pellita*. *E. grandis* and *E. resinifera*, however, presented the largest number of plants with symptoms of damage. *E. cinerea, E. cloeziana, E. dunnii, E. benthamii, E. nitens, E. viminalis, E. pilularis, E. camaldulensis* and *E. dunnii* did not suffer infestations of *C. spatulata*. Among the *Corymbia* species, eggs of *C. spatulata* were only observed on *C. citriodora* with adult leaves. None of the native Myrtaceae had eggs or larvae of *C. spatulata*.

Camargo et al. (2009) evaluated the resistance of different clones of *E. camaldulensis* to attack by *G. brimblecombei* and observed that the largest mean for eggs and larvae were observed on commercial clones 7, 58, 62, 10 and 6. The same ones did not differ statistically, all being considered as highly susceptible to the attack of the red gum lerp psyllid. The clones GG100, 36, 2, I042 and I224 did not differ statistically to each other and they were classified as resistant to psyllid attacks. Clone 19 presented an intermediate average among the two groups, being classified as susceptible. Clone 58 is known as one of the most productive in the Cerrado region of Minas Gerais but, because of its high susceptibility to red gum lerp psyllids, companies are being discouraged from planting it in areas of greatest occurrence of this pest.

5.3 Cultural control

According to White (1986), the decline of some eucalypt forests in Australia is primarily caused by changes in rainfall patterns, which induce physiological stress in plants. In physiologically stressed plants, the amount of nitrogen available as a food source for psyllids is higher. The eucalypts are damaged more by the psyllids as the increase in nitrogen content increases the chance of survival of the psyllids and their populations grow rapidly (White, 1969). The requirements of insects for mineral salts is not well defined but it is known that they are very important for the ionic balance and cell membrane permeability of insects, acting as activators of enzymes (Panizzi & Parra, 1991). Thus, different types of fertilizers can affect insect populations positively or negatively. Some nutrients are very important in plant-insect relationships and have been thoroughly researched, such as nitrogen (White, 1969), magnesium (Santana et al., 1999), silicon (Camargo et al., 2011) and others.

To minimize plant stress, certain cultural practices have been recommended as a measure to strengthen the plant and provide higher resistance to psyllids. As excess nitrogen in the leaves leads to an increase in insect populations, a balanced fertilisation and irrigation in the dry season is recommended to avoid the concentration of this nutrient in the leaves (Garrison, 2001).

Although silicon is not considered an essential element for plants, it is absorbed and is involved in the formation of structures of defences such as trichomes and spines. It also contributes to greater leaf toughness by forming polymers (crystals) that are immobilised in the leaf tissue (Camargo et al., 2011). Thus, the application of industrial ashes rich in silicates has been recommended for commercial planting of eucalypts in regions poor in this element.

5.4 Chemical control

Adult psyllids should be monitored before damage becomes evident and the numbers of adults present should be recorded on a weekly basis. During subsequent seasons control action should be taken, if necessary, when populations or damage approach the levels that were previously identified to be intolerable. The foliar damage is primarily caused by larvae, but sprays are generally aimed at killing eggs or newly hatched larvae before the damage occurs, which is why the adults should be monitored. Therefore, a decision to spray should be based on the numbers of adults infesting the plants several weeks before larval damage becomes intolerable (Dreistadt et al., 2007).

The overlapping generations of the psyllids makes chemical control even harder because it requires successive spraying of insecticides. This increases production costs and requires additional work, which demmands the definition of a management strategy to be more critical. Usually the control is made through IPM, mainly biological control using predators or parasitoids. In most case the chemical control is not recommended because it is expensive, less efficient and may cause environmental damage (Santana & Burckhardt, 2007).

There are, however, instances when chemical control is necessary. When *C. eucalypti* was introduced into the USA, it caused up to 30% of production loss in commercial plantations of *E. pulverulenta* in California (Dahlsten et al., 1998b), making it necessary to apply chemical control. Nevertheless, considering the fast dispersal and establishment of this psyllid, eradication methods are not effective and chemical control is expensive and efficient for a short time. Its ready adaptation to tropical climatic conditions, its fast dispersal and the large areas planted with eucalypts suggest that this psyllid should be controlled by a programme of Integrated Pest Management, based on the management of the pest and its interaction with the ambient environment and other organisms.

6. Modelling Psylloidea dispersion

Invasive species can be introduced accidentally by humans or by natural dispersal. To provide the right conditions for the occurrence of a particular species, it is necessary to understand the factors that circumscribe their niche, such as abiotic conditions that define the physiological limits to the persistence of species, biological factors that influence the survival of populations (which may be negative in the case of competition, predation and parasitism, or positive, in the case of mutualism), and dispersal ability, which reflects what sites are accessible to individuals of a species (Soberón & Peterson, 2005).

Far from home and their environment and free of predatory and competitive processes, invasive alien species have favorable conditions for expansion and occupation, especially if its space or ecosystem has been altered by successive processes of human interventions. Possible areas of risk of introduction, spread and future distribution of invasive species are often estimated by bioclimatic modelling, also known as ecological niche modelling.

The process of modelling the potential geographic distribution of biological species can be summarized in the following steps: 1) a set of points of occurrence (georeferenced) is combined with a set of environmental variables, creating a niche group of points. Each point in this niche is formed by the values assumed by the environmental variables at each point of occurrence; (2) a modelling algorithm is used to create a niche model from the set of points niche; and (3) the model of niche created by the algorithm is applied on a certain geographic region, taking into account the same environmental variables used to create this niche model (Rodrigues, et al. 2010). Work is currently underway to model the potential distribution of introduced psyllids in Brazil so that plantations can be planned and managed in order to minimize the threat from these pests.

7. Conclusions

Jumping plant-lice are a major threat to large scale commercial eucalypt plantations. Of the approximately 350 described Australian species, over 250 are associated with eucalypts (Hollis, 2004). Of these, only seven species have been introduced into other continents and four have become economically important pests. However, there is large potential for other Australian spondyliaspidine species to be accidentally or intentionally exported into other continents. Improving the existing taxonomic base would help in the rapid recognition of new spodyliaspidine introductions. Incidentally, two of the four major psyllid pests on eucalypts have been described, in part, on the basis of non-Australian material (*Blastopsylla occidentalis* and *Ctenarytaina spatulata*). *Ctenarytaina peregrina,* which is almost certainly of Australian origin, is currently known only from Europe. A detailed taxonomic revision of *Ctenarytaina* is particularly important if we are to understand the threat from members of this genus.

Of the eucalypt species observed as hosts of *C. fiscella, E. grandis* and its hybrids are particularly important for Brazilian national forestry. Besides these, other species such as *E. camaldulensis* and *E. urophylla* are also susceptible and they are planted over large areas, or are the base for the production of clones. These species are already attacked by the red gum lerp psyllid (*G. brimblecombei*) and the introduction of one more pest can make the use of some genetic materials of great productivity unviable. For all these reasons, the acquisition of good taxonomic knowledge, and the amassing of control tools for incorporation into IPM programme are essential if Brazilian forestry is to be protected from this important group of insects.

8. References

ABRAF (2011). Anuário estatístico da ABRAF 2011 ano base 2010/ ABRAF. Brasília,130 pp.

Anonymous (2007). *Blastopsylla occidentalis*: another new *Eucalyptus* pest in South Africa. *Plant Protection News*, vol. 72, p. 2

Appleton, C. (2009). NZ FRI http://www.nzffa.org.nz/images/design/Pests/Cardiaspina-fiscella/Cardiaspina-fiscellaFHNews64.html, Access in November/30, 2009

Azevedo, F. & Figo, M.L. (1979). *Ctenarytaina eucalypti* Mask. (Homoptera, Psyllidae). *Boletino Servicio de Plagas Forestales*, vol. 5, pp. 41-46

Bekker-Migdiscova, E.E. (1973). Systematics of the Psyllomorpha and the position of the group within the order Homoptera. In: Narchuk, E.P. (ed.), Doklady na Dvadsat Chetvertom Ezheghodnom Chtenii Pamyati N.A. Kholodkovskogo, 1-2 aprelya 1971, p. 90-117. (British Lending Library Translation RTS 8526)

Berry, J. (2006). Brown lace lerp hyperparasitoid found in New Zealand. *Biosecurity*, No. 68, pp. 18-19

Berti Filho, E.; Costa, V.A.& Zuparko, R.L. & LaSalle, J. (2003). Ocorrência de *Psyllaephagus bliteus/quadricyclus* Riek (Hymenoptera: Encyrtidae) no Brasil. Revista de Agricultura, vol. 78, pp. 304-304

Bouvet, J.P.R.; Harrand, L. & Burckhardt, D. (2005). Primera cita de *Blastopsylla occidentalis* y *Glycaspis brimblecombei* (Hemiptera: Psyllidae) para la República Argentina. *Revista da la Sociedad Entomológica Argentina*, vol. 64, pp. 99-102

Brennan, E.B.; Gill, R.J.; Hrusa, G.F. & Weinbaum, S.A. (1999). First record of *Glycaspis brimblecombei* (Moore) (Homoptera: Psyllidae) in North America: Initial observations and predator associations of a potentially serious new pest of *Eucalyptus* in California. *Pan-Pacific Entomol*, vol. 75, pp. 55-57

Brennan, E.B.; Hrusa, G.F.; Weinbaum, S.A. & Levison, Jr. W. (2001). Resistance of *Eucalyptus* species to *Glycaspis brimblecombei* (Homoptera: Psyllidae) in the San Francisco by area. *Pan-Pacific Entomologist*, vol. 77, pp. 249-253

Brennan, E.B. & Weinbaum, S.A. (2001a). Performance of adult psyllids in no-choice experiments on juvenile and adult leaves of *Eucalyptus globulus*. *Entomologia Experimentalis et Applicata*, vol. 100, pp. 179–185

Brennan, E.B. & Weinbaum, S.A. (2001b). Effect of epicuticular wax on adhesion of psyllids to glaucous juvenile and glossy adult leaves of *Eucalyptus globulus* Labilliardière. *Australian Journal of Entomology*, vol. 40, pp. 270–277

Brennan, E.B. & Weinbaum, S.A. (2001c). Stylet penetration and survival of three psyllid species on adult leaves and 'waxy' and 'de-waxed' juvenile leaves of *Eucalyptus globulus*. *Entomologia Experimentalis et Applicata*, vol. 100, pp. 355–363

Brooker, M.I.H. (2000). A new classification of the genus *Eucalyptus* L'Hér. (Myrtaceae). *Australian Systematic Botany*, vol. 13, pp. 79-148

Burbidge, N. (1960). The phytogeography of Australia. *Australian Journal of Botany*, vol. 8, 75-209

Burckhardt, D. (1991). *Boreioglycaspis* and Spondyliaspidine classification (Homoptera: Psylloidea). *Raffles Bulletin of Zoology*, vol. 39, pp. 15-52

Burckhardt, D. (1994). Psylloid pests of temperate and subtropical crop and ornamental plants (Hemiptera, Psylloidea): a review. *Trends in Agricultural Sciences, Entomology*, vol. 2, pp. 173-186

Burckhardt, D. (1998). *Ctenarytaina eucalypti* (Maskell) (Hemiptera, Psylloidea) neu für Mitteleuropa mit Bemerkungen zur Blattflohfauna von *Eucalyptus*. *Mitteilungen der Entomologischen Gesellschaft Basel*, vol. 48, pp. 59-67

Burckhardt, D. (2005). Biology, ecology and evolution of gall-inducing psyllids (Hemiptera: Psylloidea), 143-157. In Raman, A.; Schaefer, C.A. & Withers, T.M. (eds) Biology, ecology and evolution of gall-inducing arthropods. Science Publishers, Inc., Enfield (NH), USA

Burckhardt, D. & Basset, Y. (2000). The jumping plant-lice (Hemiptera, Psylloidea) associated with *Schinus* (Anacardiaceae): systematics, biogeography and host plant relationships. *Journal of Natural History*, vol. 34, pp. 57-155

Burckhardt, D. & Elgueta, M. (2000). *Blastopsylla occidentalis* Taylor (Hemiptera: Psyllidae), a new introduced eucalypt pest in Chile. *Revista Chilena de Entomología*, vol. 26, pp. 57-61

Burckhardt, D.; Lozada, P.W. & Diaz, B.W. (2008). First record of the red gum lerp psyllid *Glycaspis brimblecombei* (Hemiptera: Psylloidea) from Peru. *Mitteilungen der Schweizerischen Entomologischen Gesellschaft*, vol. 81, pp. 83-85

Burckhardt, D. & Mühlethaler, R. (2003). Exotische Elemente der Schweizer Blattflohfauna (Hemiptera, Psyllidae) mit einer Liste weiterer potentieller Arten. *Mitteilungen der Entomologischen Gesellschaft Basel*, vol. 53, pp. 98-110

Burckhardt, D. & Ouvrard, D. (2007). The taxonomy, biogeography and host plant relationships of jumping plant-lice (Hemiptera: Psyllidae) associated with creosote bushes (*Larrea* spp., Zygophyllaceae). *Systematic Entomology*, vol. 32, pp. 136-155

Burckhardt, D. & Ouvrard, D. (2011). A revised classification of the jumping plant-lice (Hemiptera: Psylloidea). *Zootaxa*, in preparation

Burckhardt, D.; Santana, D.L.Q.; Terra, A.L.; Andrade, F.M.; Penteado, S.R.C.; Iede, E.T. & Morey, C.S. (1999). Psyllid pests (Hemiptera, Psylloidea): in South American eucalypt plantations. *Mitteilungen der Schweizerischen Entomologischen Gesellschaft*, vol. 72, pp. 1-10

Cadahia, D. (1980). Proximidad de dos nuevos enemigos de los *Eucalyptus* en España. *Boletino Servicio de Plagas Forestales*, vol. 6, pp. 165-192

Camargo, J.M.M.; Moraes, J.C.; Zanol, K.R.M. & Queiroz, D. L. (2011). Interação silício e insetos-praga: Defesa mecânica ou química? *Revista de Agricultura (Piracicaba)*, vol. 85, pp. 10-12

Camargo, J.M.M.; Santana, D.L.Q.; Dedecek, R.; Zanol, K.R.M. & Melido, R.C.N. (2009). Avaliação da resistência de clones de *Eucalyptus camaldulensis* Dehn. Ao psilideo de concha *Glycaspis brimblecombei* Moore (Hemiptera: Psyllidae). In: 1° Congresso Brasileiro sobre florestas energéticas, 2009, Belo Horizonte

Campbell, K.G. (1992). The biology and population ecology of two species of *Cardiaspina* (Hemiptera: Psyllidae) in plague numbers on *Eucalyptus grandis* in New South Wales. *Proceedings of the Linnean Society of New South Wales*, vol. 113, pp.135-150

Collet, N. (2001a). Biology and control of psyllids, and the possible causes for defoliation of *Eucalyptus camaldulensis* Dehnh. (river red gum) in south-eastern Australia - a review. Australian Forestry, vol. 64, pp. 88-95

Collett, N. (2001b). Insect pests of young eucalypt plantations. Forest Science Centre, Heidelberg. Note Number: AG0799. December, 2001

Cook, L.G. (2001). Extensive chromosomal variation associated with taxon divergence and host specificity in the gall-inducing scale insect *Apiomorpha munita* (Schrader) (Hemiptera: Sternorrhyncha: Coccoidea: Eriococcidae). *Biological Journal of the Linnean Society*, vol. 72, 265–278

Cook, L.G. & Gullen, P.J. (2004). The gall-inducing habit has evolved multiple times among the eriococcid scale insects (Sternorrhyncha: Coccoidea: Eriococcidae). *Biological Journal of the Linnean Society*, vol. 83, 441-452

Daane, K. & Paine, T.D. (2005). Biological Control of psyllids on lemon-scented and spotted gum in California. Latest news - updated graphs of pysllid populations through 2004. University of California. Accessed in: http://nature.berkeley.edu/biocon/dahlsten/lemon_gum/

Dahlsten, D.L. (2001, modified 2002). Berkeley University. Biological control of psyllids on lemon-scented and spotted gum in California by). http://www.cnr.berkeley.edu/biocon/dahlsten/lemon_gum/
Los Angeles County Agricultural Commissioner's Office New Agricultural Pest for Southern California - Spotted Gum Lerp Psyllid, *Eucalyptolyma maideni*. http://acwm.co.la.ca.us/pdf/SpoGum.pdf

Dahlsten, D.L. (2003). Biological control of the red gum lerp psyllid, aspects of eucalyptus species in California. Center of biological control, University of California. 5 pp. http://www.cnr.berkeley.edu/biocon/dahlsten/rglp/ Accessed in 02/Aug/2005.

Dahlsten, D.L.; Hansen, E.P.; Zuparko, R.L. & Norgaard, R.B. (1998a). Biological control of the blue gum psyllid proves economically beneficial. *California Agriculture*, vol. 52, pp. 35-40

Dahlsten, D.L.; Rowney, D.L.; Copper, W.A.; Tassan, R.L.; Chaney, W.E.; Robb, K.L.; Tjosvold, S.; Bianchi, M. & Lane, P. (1998b). Parasitoid wasp controls blue gum psyllid. *California Agriculture*, vol. 52, pp. 31-34

Dal Pogetto, M.H.F.A.; Wilcken, C.F.; Christovam, R.S.; Prado, E.P. & Gimenes, M.J. (2011). Effect of formulated entomopathogenic fungi on red gum lerp psyllid *Glycaspis brimblecombei*. *Research Journal of Forestry*, vol. 5, no. 2, pp. 99-106

Dreistadt, S.H. & Dahlsten, D.L. (2007). Pest Notes: Psyllids. UC ANR Publication 7423. Accessed April 10, 2007, in: http://www.ipm.ucdavis.edu/PMG/PESTNOTES/pn7423.html

Durán, P.M. & Urrutia, A.B. (2001). Dos nuevos psilidos en Chile: *Ctenarytaina eucalypti* y *Blastopsylla occidentalis*. Santiago: CONAF, GEDEFF.

Garrison, R.W. (2001). New agricultural pest for Southern California. Spotted Gum Lerp Psyllid, *Eucalyptolyma maideni*. San Diego County Agricultural Commissioner's office. http://www.sdcounty.ca.gov/reusable_components/images/awm/Docs/ipd_spottedpsyllid.pdf Accessed in: August/10/2011

Grimaldi, D.A. (2003). First amber fossils of the extinct family Protopsyllidiidae, and their phylogenetic significance among Hemiptera. *Insect Systematics & Evolution*, vol. 34, pp. 329-344

Hodkinson, I.D. (1974). The biology of the Psylloidea (Homoptera): a review. *Bulletin of Entomological Research*, vol. 64, pp. 325-338

Hodkinson, I.D. (1984). The biology and ecology of the gall-forming Psylloidea (Homoptera), 59-77. In: Ananthakrishnan (ed.), Biology of gall insects. Oxford and IBH Publishing, New Dehli, India.

Hodkinson, I.D. (1991). First record of the Australian psyllid *Blastopsylla occidentalis* Taylor (Homoptera; Psylloidea) on *Eucalyptus* (Myrtaceae) in Mexico. *Pan-Pacific Entomologist*, vol. 67, p. 72

Hodkinson, I.D. (1999). Biocontrol of eucalyptus psyllid *Ctenarytaina eucalypti* by the Australian parasitoid *Psyllaephagus pilosus*: a review of current programmes and their success. *Biocontrol News Information*, vol. 20, pp. 129-134

Hodkinson, I.D. (2007). A new introduced species of *Ctenarytaina* (Hemiptera: Psylloidea) damaging cultivated *Eucalyptus parvula* (=*parvifolia*) in Europe. Deutsche Entomologische Zeitschrift, vol. 54, pp. 27–33

Hodkinson, I.D. (2009). Life cycle variation and adaptation in jumping plant lice (Insecta: Hemiptera: Psylloidea): a global synthesis. *Journal of Natural History*, vol. 43, No. 1-2, pp. 65-179

Hollis, D. (1987). A review of the Malvales-feeding psyllid family Carsidaridae (Homoptera). *Bulletin of the British Museum (Natural History) (Entomology)*, vol. 56, No. 2, pp. 87–127

Hollis, D. (2004) Australian Psylloidea: jumping plantlice and lerp insects. CSIRO, Camberra, Australia

Hollis, D. & Broomfield, P. S. (1989). *Ficus*-feeding psyllids (Homoptera), with special reference to the Homotomidae. *Bulletin of the British Museum (Natural History) (Entomology)*, vol. 58, pp. 131–183

Hurtado A. & Reina I. (2008). Primera cita para Europa de *Glycaspis brimblecombei* Moore (Hemiptera: Psyllidae), una nueva plaga del eucalipto. *Boletín Sociedad Entomológica Aragonesa*, vol. 43, pp. 447-449

Iede, E.T.; Leite, M.S.P.; Penteado, S.R.C. & Maia, F. (1997). *Ctenarytaina* sp. (Homoptera: Psilidae) associada a plantios de *Eucalyptus* sp. em Arapoti, PR. In: CONGRESSO BRASILEIRO DE ENTOMOLOGIA, 16, Salvador. Resumos. p. 253

Klimaszewski, S.M. (1964). Studies on systematic of the suborder Psyllodea. Annales Zoologici, vol. 22, pp. 81-138

Laudonia, S. & Garonna, A.P. (2010). The red gum lerp psyllid, *Glycaspis brimblecombei* (Hem.: Psyllidae) a new exotic pest of *Eucalyptus camaldulensis* in Italy. *Bulletin of Insectology*, vol. 63, no. 2, pp. 233-236

Li, F.S. (2011). Psyllomorpha of China. 2 volumes Science Press, Beijing, China.

Ladiges, P.Y. (1997). Phylogenetic history and classification of eucalypts. In: (eds. J. E. Williams & J. H. Z. Woinarski) *Eucalyptus Biology*. Pp. 16-29. Cambridge University Press, Cambridge

Majer, J.D.; Recher, H.F. & Ganesh, S. (2000). Diversity patterns of eucalypt canopy arthropods in Eastern and Western Australia. *Ecological Entomology*, vol. 25, pp. 295-306

Majer, J.D.; Recher H.F.; Wellington, A.B.; Woinarski, J.C.Z. & Yen, A.L. (1997). Invertebrates of eucalypt formations. In: (eds. J.E. Williams & J.H.Z. Woinarski) *Eucalyptus Biology*. Pp. 278-302. Cambridge University Press, Cambridge.

Maschio, L.M.A.; Andrade, F.M.; Leite, M.S.P.; Bellote, A.F.J; Ferreira, C.A.; Iede, E.T.; Nardelli, A.M.B.; Auer, C.G.; Grigolleti, Jr.A. & Wiechetek, M. (1997). Seca dos ponteiros do eucalipto em Arapoti-PR./ Dryout of eucalipt buds in Arapoti, PR. In: *Conferência IUFRO sobre silvicultura de eucalipto*, pp. 353-359

Meza, P.A. & Baldini A.R. (2001). El psilido de los eucaliptos *Ctenarytaina eucalypti* Maskell (1890) (Hemiptera, Psyllidae). Chile, CONAF, vol. 21, no. 39

Moore, K.M. (1961). Observations on some Australian forest insects. 7. The significance of the *Glycaspis* species (Hemiptera: Homoptera: Psyllidae) association with their

Eucalyptus species hosts: erection of a new subgenus and descriptions of thirty eight new species of *Glycaspis*. *Proceedings of the Linnaean Society of New South Wales*, vol. 86, pp. 128-167

Moore, K.M. (1970). Observations on some Australian forest insects. 24. Results from a study of the genus *Glycaspis* (Homoptera: Psyllidae). *Australian Zoologist*, vol. 15, pp. 343-376

Morgan, F.D. (1984). Psylloidea of South Australia. D.J. Woolman, Government Printer, South Australia, 136 pp.

Olivares, T.S.; Burckhardt, D.H. & Cerda L.A. (2004). *Glycaspis brimblecombei* Moore, "Psyllido de los eucaliptos rojos" (Hemiptera: Psyllidae: Spondyliaspidinae): caracteres taxonómicos. *Revista Chilena de Entomología* vol. 30, no. 1, pp. 5-10

Onore, G. & Gara, R.L. (2007). First record of *Glycaspis brimblecombei* (Hemiptera: Psyllidae) in Ecuador, biological notes and associated fauna. *4th European Hemiptera Congress Ivrea, Turin, Italy*, extended abstracts, pp. 41-42

Ouvrard, D.; Burckhardt, D.; Azar, D. & Grimaldi, D. (2010). Non-jumping plant-lice in Cretaceous amber (Hemiptera: Sternorrhyncha: Psylloidea). *Systematic Entomology*, vol. 35, pp. 172-180

Paine, T.D. & Millar, J.G. (2002). Insect pests of eucalyptus in California: Implications of managing invasive species. *Bulletin of Entomological Research*, vol. 92, pp. 147-51

Paine, T.D.; Dreistadt, S.H.; Garrison, R.W. & Gill, R.J. (2007). *Eucalyptus* redgum lerp psyllid. http://www.ipm.ucdavis.edu/PMG/PESTNOTES/pn7460.html. Accessed April 10, 2007

Panizzi, A.R. & Parra, J.R.P. (1991). Ecologia nutricional de insetos e suas implicações no manejo de pragas. *São Paulo: Manole*, 259 pp.

Percy, D.M. (2003). Radiation, diversity, and host-plant interactions among island and continental legume-feeding psyllids. *Evolution*, vol. 57, pp. 2540-2556

Peris-Felipo F.J.; Mancusi G.; Turrisi, G.F. & Jiménez-Peydró, R. (2011). New corological and biological data of the Red Gum Lerp Psyllid, *Glycaspis brimblecombei* Moore, 1964 in Italy (Hemiptera, Psyllidae). *Biodiversity Journal*, vol. 2, pp. 13-17

Pinzón, F.O.P.; Guzmán, C.M. & Navas, N.F. (2002). Contribución al conocimiento de la biología, enemigos naturales y daños del pulgón del eucalipto *Ctenarytaina eucalypti* (Homoptera: Psyllidae). *Revista Colombiana de Entomología*, vol. 28, pp. 123-128

Prieto-Lillo, E.; Rueda, J.; Hernández, R. & Selfa, J. (2009). Primer registro del psílido rojo del eucalipto, *Glycaspis brimblecombei* (Homoptera: Psyllidae), en la Comunidad Valenciana. *Boletín Sanidad Vegetal Plagas*, vol. 35, pp. 277-281

Raman, A.; Burckhardt, D. & Harris, K.H. (2009). Biology and adaptive radiation in the gall-inducing Cecidomyiidae and Calophyidae (Hemiptera) on *Mangifera indica* (Anacardiaceae) in the Indian subcontinent. Tropical Zoology, vol. 22, no. 1, pp. 27-56

Rodrigues, F.A.; Rodrigues, E.S. da; Corrêa, P.L.P.; Rocha, R.L. de A. & Saraiva, A.M. (2010). Modelagem da biodiversidade utilizando redes neurais artificiais. www.inf.pucminas.br/sbc2010/anais/pdf/wcama/st02_04.pdf Accessed 1 February 2011

Rosales, C.J.; Lobosque, O.; Carvalho, P.; Bermúdez, L. & Acosta, C. (2008). *Glycaspis brimblecombei* Moore (Hemiptera: Psyllidae). "Red Gum Lerp". Nueva plaga forestal en Venezuela. *Entomotropica*, vol. 23, pp. 103-104

Santana, D.L.Q. (2003). *Ctenarytaina spatulata* Taylor, 1997 (Hemiptera: Psyllidae): morfologia, biologia, dinâmica, resistência e danos em *Eucalyptus grandis* Hill. ex Maiden, Doctorate Thesis.

Santana, D.L.Q., Andrade, F.M.; Bellote, A.F.J. & Grigoletti, Jr.A. (1999). Associação de *Ctenarytaina spatulata* e de teores de Magnésio foliar com a seca dos ponteiros de *Eucayptus grandis*. *Boletim de Pesquisa Florestal, Colombo, PR*, vol. 39, pp. 41-49

Santana, D.L.Q., Bellote, A.F.J. & Dedecek, R.A. (2003a). *Ctenarytaina spatulata*, Taylor: Água no solo, Nutrientes minerais e suas interações com a seca dos ponteiros de eucalipto. *Boletim de Pesquisa Florestal, Colombo, PR*, vol. 46, pp. 57-67

Santana, D.L.Q. & Burckhardt, D. (2007). Introduced *Eucalyptus* psyllids in Brazil. *Journal of Forest Research*, vol. 12, pp. 337-344

Santana, D.L.Q., Carvalho, R.C.Z.; Favaro, R.M. & Almeida, L.M. (2004). *Glycaspis brimblecombei* (Hemiptera: Psyllidae) e seus inimigos naturais no Paraná. In: XX Congresso Brasileiro de Entomologia p. 450

Santana, D.L.Q., Menezes, A.O. & Bizzi, R.M. (2002). Ocorrência de *Psyllaephagus* sp. (Hymenoptera: Encyrtidae) parasitando *Ctenarytaina eucalypti* (Maskell) (Homoptera Psyllidae) no Brasil. In: 19° Congresso Brasileiro de Entomologia, Manaus, Resumos, p. 149

Santana, D.L.Q., Menezes, A.O.; Silva, H.D.; Bellote, A.F.J. & Favaro, R.M. (2003b). O psilídeo-de-concha (*Glycaspis brimblecombei*) em eucalipto. *Embrapa Florestas, Comunicado Técnico*, 3 pp.

Santana, D.L.Q. & Zanol, K.M.R. (2006). Biologia de *Ctenarytaina spatulata* Taylor, 1997 (Hemiptera, Psyllidae) em *Eucalyptus grandis* Hill ex Maiden. *Acta Biologica Paranaense*, vol. 35, pp. 47-62

Santana, D.L.Q., Zanol, K.M.R.; Botosso, P.P.C. & Mattos, P.P. (2005). Danos causadas por *Ctenarytaina spatulata* Taylor, 1977 (Hemiptera: Psyllidae) em *Eucalyptus grandis* Hill. Ex Maiden. *Boletim de Pesquisa Florestal, Colombo, PR*, vol. 50, pp. 11-24

Santana, D.L.Q. & Zanol, K.M.R. (2005). Morfologia externa das ninfas e adultos de *Ctenarytaina spatulata* Taylor (Hemiptera, Psyllidae). *Revista Brasileira de Entomol ogia*, vol. 40, pp. 340-346

Santana, D.L.; Zanol, K.M.R.; Oliveira, E.B.; Anjos, N. & Majer, J. (2010). Feeding and oviposition preferences of *Ctenarytaina spatulata* Taylor (Hemiptera: Psyllidae) on *Eucalyptus* spp and other Myrtaceae growing in Brazil. *Revista Brasileira de Entomologia*, vol. 54, pp. 149-153,

Soberon, J. M., Peterson A. T. (2005). Interpretation of models of fundamental ecological niches and species' distributional areas. *Biodiversity Informatics*, vol. 2, pp. 1-10

Tamesse, J.L; Laurentine, S.; Wenceslas, Y. & Joly, D.V. (2011). First record of *Blastopsylla occidentalis* Taylor, 1985 (Hemiptera: Psyllidae), a *Eucalyptus* psyllid in Cameroon, Central Africa. Entomological Research, vol. 40, no. 4, pp. 211–216

Taylor, K.L. (1985). Australian psyllids: A new genus of Ctenarytainini (Homoptera: Psylloidea) on *Eucalyptus*, with nine new species. *Journal of the Australian entomological Society*, vol. 24, pp. 17-30

Taylor, K.L. (1987). *Ctenarytaina longifolia* sp. n. (Homoptera: Psylloidea) from *Lophostemon confertus* (R. Brown) in Australia and California. *Australian Journal of Entomology,* vol., 26, no. 3, pp. 229–233

Taylor, K.L. (1990). The tribe Ctenarytainini (Hemiptera: Psylloidea): A key to known Australian genera, with new species and two new genera. *Invertebrate Taxonomy,* vol. 4, pp. 95-121

Taylor, K.L. (1997). A new Australian species of *Ctenarytaina* Ferris and Klyver (Hemiptera: Psyllidae: Spondyliaspidinae) established in three other countries. *Journal of the Australian entomological Society,* vol. 36, pp. 113-115

Thao, L.M.; Moran, N.A.; Abbot, P.; Brennan, E.B.; Burckhardt, D.H. & Bauman, P. (2000). Cospeciation of psyllids and their primary prokaryotic endosymbionts. *Applied and environmental microbiology,* July 2000, pp. 2898-2905.

Thao, L.M.; Clark M.A.; Burckhardt, D.H., Moran, N.A. & Bauman, P. (2001). Phylogenetic analysis of vertically transmitted psyllid endosymbionts (*Candidatus* Carsonella ruddii) based on atpAGD and rpoC: Comparisons with 16S-23S rDNA-derived phylogeny. *Current microbiology,* vol. 42, pp. 419-421.

Valente, C. & Hodkinson I.D. (2009). First record of the Red Gum Lerp Psyllid, *Glycaspis brimblecombei* Moore (Hem.: Psyllidae), in Europe. *Journal of Applied Entomology,* vol. 133, pp. 315-317

Valente, C.; Manta, A. & Vaz, A. (2004). First record of the Australian psyllid *Ctenarytaina spatulata* Taylor (Homoptera: Psyllidae) in Europe. *Journal of Applied Entomology,* vol. 128, pp. 369–370

White, I.M. & Hodkinson, I.D. (1985). Nymphal taxonomy and systematics of the Psylloidea (Homoptera). *Bulletin of the British Museum (Natural History) (Entomology),* vol. 50, pp. 153–301

White, T. C. R. (1968). Uptake of water by eggs of *Cardiaspina densitexta* (Homoptera: Psyllidae) from leaf of host plant. *Journal of Insect Physiology,* vol. 14, pp. 1669-1683

White, T.C.R. (1969). An index to measure weather-induced stress on trees associated with outbreaks of psyllids in Australia. *Ecology,* vol. 50, pp. 905-909

White, T. C. R. (1970). Some aspects of the life history, host selection, dispersal, and oviposition of adult *Cardiaspina densitexta* (Homoptera: Psyllidae). *Australian Journal of Zoology,* vol. 18, pp. 105-17

Wilcken, C.F.; Couto, E.B.; Orlato, C.; Ferreira-Filho, P.J. & Firmino, D.C. (2003). Ocorrência do psilídeo-de-concha (*Glycaspis brimblecombei*) (Hemiptera: Psyllidae) em florestas de eucalipto do Brasil. *Circular Técnica IPEF,* No. 201, pp. 1–11

Withers, T. 2000. Parasitoid on *Cardiaspina fiscella.* Access in 01/August/2011, in: http://www.nzffa.org.nz/farm-forestry-model/the-essentials/forest-health-pests-and-diseases/Pests/Cardiaspina-fiscella/Parasitoid-Cardiaspina-FHnews95. From Forest Health News No. 95, April 2000

Withers, T.M. (2001). Colonization of eucalypts in New Zealand by Australian insects. *Austral Ecology,* vol. 26, No. 5, pp. 467–476

Yang, M.M.; Mitter, C. & Miller, D.R. (2001). First incidence of inquilinism in gall-forming psyllids, with a description of the new inquiline species (Insecta, Hemiptera, Psylloidea, Psyllidae, Spondyliaspindinae). *Zoologica Scripta,* vol. 30, pp. 97-113

Yang, M.M. & Mitter, C. (1994). Biosystematics of hackberry psyllids (*Pachypsylla*) and the evolution of gall and lerp formation in psyllids (Homoptera: Psylloidea): a preliminary report. *In*: Price PW, Mattson WJ, Baranchikov YN, eds. The Ecology and Evolution of Gall-forming Insects. *U.S.D.A. Forest Service (North Central Forest Experiment Station) General Technical Report*, NC-174, St. Paul, Minnesota, pp. 172-185

Zondag, R. (1982) *Ctenarytaina eucalypti* (Maskell), (Hemiptera, Psyllidae). Blue-gum psyllid. *New Zealand Forest Service, Forest and Timber Insects in New Zealand*, No. 53, 4 pp.

18

Biological and Ecological Studies on Land Snails and Their Control

Ahmed Sallam[1] and Nabil El-Wakeil[2,*]
[1]*Plant Protection Dept. Faculty of Agriculture, Sohag University, Sohag,*
[2]*Pests & Plant Protection Dept. National Research Centre, Dokki, Cairo*
Egypt

1. Introduction

Mollusca are the second largest phylum of the animal kingdom, forming a major part of the world fauna. The Gastropoda is the only class of molluscs which have successfully invaded land. They are one of the most diverse groups of animals, both in shape and habit. Among gastropods, land snails (subclass: Pulmonata) are one of the most numerous with almost 35,000 described species of the world. The present study dealt also with the chemical analysis of the mucus of three common land snails, *Eobania vermiculata* (Müller), *Theba pisana* (Müller) and *Monacha obstructa* (Montagu), and identification of the chemical compositions by using GC-MS. Results revealed that several variations in composition were observed between all species. Oxime, methoxy-phenyl and cyclotrisiloxane, hexamethyl were major components found that in three species, the total areas detected were 86.23, 76.83 and 70.83, respectively. This different composition of mucus may be due to differences from one species to another; different mechanical properties (function) are influenced by external factors such as temperature, humidity, light intensity, soil conditions and food supply. On reviewing literature conducted on land snails of Egypt, as far as can be ascertained, most studies were focused on Lower Egypt but Upper Egypt was neglected. So, the present investigation was designed to fulfil this gap and to promote and enhance the studies of land snails, especially those which have economic importance.

The Phylum Mollusca is probably the third most important animal group after the arthropods and vertebrates (South, 1992) Snails and slugs belong to the class Gastropoda. Snails and slugs are molluscs, a group of invertebrate animals with soft unsegmented bodies. Slugs are often described as snails without a shell, while snail bodies are enclosed in calcareous shells (Barker, 2001; Ramzy, 2009). The terrestrial mollusca including snails and slugs are destructive agricultural pests causing economic damage to a wide variety of plants including horticulture, field crops, and forestry. In addition they are of importance in medical and veterinary practice, since they serve as intermediate hosts for certain

Three common land snails, *E. vermiculata, T. pisana* and *M. obstructa*, are important crop pests and cause considerable damage in agriculture and horticulture, especially in areas where they find the conditions necessary for rapid multiplication, as shown in Fig. (1).

*Corresponding Author

Damage caused by snails depends not only on their activity and population density, but also on their feeding habits, which differ from one species to another. Damage involving considerable financial loss is inflicted on cereal, vegetables, Egyptian clover as well as other agricultural and field crops. The land snails feed on leaves, roots, tubers and ornamental plants (Bishara et al., 1968; El-Okda, 1981). In addition, during movement snails cause an undesirable smell which prevents men and even animals from feeding on these contaminated plants (El-Okda, 1984; Kassab & Daoud, 1964). Crops contaminated by snail slime lose their marketability and hence their export potential in many countries (Baker & Hawke, 1990; Ittah & Zisman, 1992). Land snails cause also a heavy damage to seed of oil plants and leaves of ornamental plants, as well as, citrus, peach, plam and vegetable, i.e. cabbage, carrot and bean. (El-Deeb et al., 1999; El-Okda, 1979, 1981; Ismail et al., 2003; Lokma, 2007; Shahawy et al., 2008).

The snail movement is rather slow and sluggish for a short distance depending upon temperature, food and natural of soil. They were active during optimum temperature, Humidity and moistened soil. They aestivate during the hot summer and hibernate during the cold winter (Kassab & Daoud, 1964). In Egypt, the land snails dispersing in northern Governorate, i.e. Alexandria, Kafr El-Shikh, Behera, and Domiate, Egypt (El-Okda, 1980; Hashem et al., 1993; Kassab & Daoud, 1964) At the present time these snails distribute in Ismaellia, Sharkia, Monofia, Gharbia, Minia, and Assiut Governorates, Egypt (El-Deeb et al., 2004; El-Massry, 1997; Metwally et al., 2002; Ramzy, 2009; Shoieb, 2008).

Gastropods such as slugs and snails secrete a trail of mucus from their pedal gland while traveling across a surface (Denny, 1983). The unique mechanical properties of snail pedal mucus enable the animal's locomotion while also causing the mucus to function as an adhesive to the substrate. The mucus trail performs a number of other functions, including the provision of mechanisms for re-tracing a path and for finding a mate of the same species by following a trail (Al-Sanabani, 2008). An understand ding of the functionality of trail mucus, including its interactions with water vapour, can therefore lead to a means of controlling the reproduction of snails and thereby limiting their impact on the environment, especially vegetable crops. When freshly deposited by terrestrial snails, trails of pedal mucus are reported to be in the range of 10 - 20 mm thick (Denny, 1989). But since the mucus typically consists of between 90 and 99.7% water by weight (Denny 1983), the trails dry to leave a much thinner solid film. It is generally believed that the fundamental structure of mucus gels consists of giant protein- polysaccharide complexes. This complex is usually classified into the broad categories of mucopolysaccharides and glycoproteins (Davies & Hawkins, 1998; Denny, 1983); mucus secretions can function as effective adhesives due to their viscoelasticity (Abd El-wakeil, 2005; Daoud, 2004; Grenon & Walker, 1980).

It is clearly known that successful control methods of terrestrial mollusca depend greatly on the broad base of knowledge of biological and ecological aspects of mollusca particularly in integrated approaches (El-Deeb et al., 2003a; Gabr et al., 2006; Ramzy, 2009; Shoieb, 2008). In Egypt, there is currently no information on the biochemical structure of *E. vermiculata, T. pisana* and *M. obstructa* mucus. A part of this work focuses on identifying the chemical composition of the mucus of three species of mollusks, i.e. *E. vermiculata, T. pisana* and *M. obstructa* under Egyptian conditions, and to compare the compositions between them. This book chapter is an attempt to gain information on land snails as agricultural pest and their control.

(A) Shell of *Eobania vermiculata* (Müller, 1770).

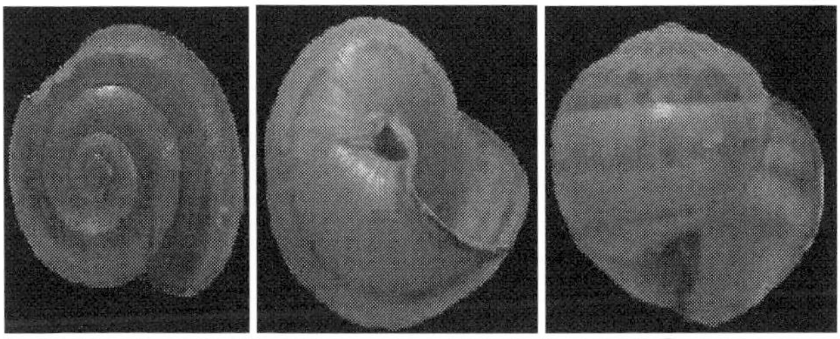

a b c

20 mm↔

(B) Shell of *Theba pisana* (Müller, 1774).

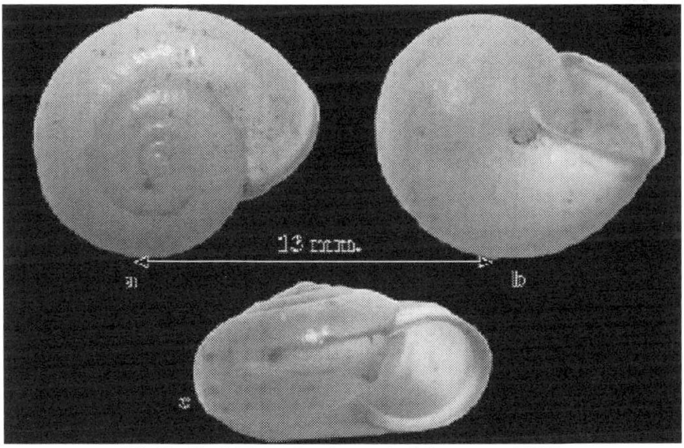

(C) Shell of *Monacha obstructa* (Montagu, 1803)

Fig. 1. Land snail types in Egyptian Agricultural fields (Cited from Ramzy, 2009).

2. Calssification of gastropods

The identification of terrestrial mollusca species could be classified according to the full description of Godan (1983) as follows:

Kingdom	:	Animalia
Sub Kingdom	:	Metazoa
Phylum	:	Mollusca
Class	:	Gastropoda
Sub class	:	Pulmonata
Order	:	Stylomatophora
Super Family	:	Helicoidae (Rafinesque, 1815)
Family	:	Helicidae (Rafinesque, 1815)
Genus	:	Cochlicella (Férussac, 1820)
Species	:	*Cochlicella acuta* (o.f. Müller, 1974)
	:	Cochlicella barbara (L., 1758)
	:	*Cochlicella ventricosa* (Draparnoud, 1801)
Genus	:	Helicella (Férussac, 1820)
Species	:	*Helicella obvia* (Hartman, 1840)
	:	Helicella bolenensis (Locard, 1884)
	:	*Helicella vestials* (Locard, 1882)
Sub family	:	Monacheae (Fitzinger, 1833)
Genus	:	Monacha (Fitzinger, 1833)
Species	:	*Monacha cantiana* (Montagu, 1803)
	:	*Monacha cartusiana* (o.f. Müller, 1774)
	:	*Monacha obstructa* (Pfeiffer, 1842)
Sub family	:	Helicinae (Rafinesque, 1815)
Genus	:	Theba (Riss, 1826)
Species	:	*Theba pisana* (o.f. Müller, 1774)
Genus	:	Cepaea (Held, 1837)
Species	:	*Cepaea hortensis* (o.f. Müller, 1774)
	:	*Cepaea silvatica* (Draparnoud, 1801)
	:	*Cepaea vindobonensis* (Férussac, 1821)
Genus	:	Eobania (Hesse, 1915)
Species	:	*Eobania Vermiculata* (o.f. Müller, 1774)
Genus	:	Helix (L., 1658)
Species	:	*Helix pomatia* (L., 1658)
	:	*Helix cantareus* (Risso, 1826)
	:	*Helix aperta* (Born, 1778)
	:	*Helix aspersa* (o.f. Müller, 1774)

Super family	:	Limacoidae (Rafinesque, 1815)
Family	:	Zonitidae (Mörch, 1864)
Sub family	:	Zonitinae (Mörch, 1864)
Genus	:	Oxychilus (Fitzinger, 1833)
Species	:	*Oxychilus alliarius* (Müller, 1822)
Family	:	Succinidae

Genus	:	Succinea
Species	:	Succinea ovalis
Family	:	Limacidae (Rafinesque, 1815)
Genus	:	Limax (L., 1758)
Species	:	*Limax maximus* (L., 1758)
	:	*Limax tenellus* (o.f. Müller, 1774)
	:	*Limax flavus* (L., 1758)
Genus	:	Deroceras (Rafinesque, 1820)
Species	:	*Deroceras reticulatum* (o.f. Müller, 1774)
	:	*Deroceras laeve* (o.f. Müller, 1774)
	:	*Deroceras caruanae* (Pollonera, 1891)
Family	:	Arionidae (Gray, 1840)
Sub family	:	Arioninae (Gray, 1840)
Genus	:	Arion (Férussac, 1819)
Species	:	*Arion ater* (L., 1758)
	:	*Arion rufus* (L., 1758)
	:	*Arion hortensis* (Ferussac, 1819)

3. Identification of snails

For identification of snails, the height and breadth of the shell as well as its shape and color are the main features (Fig. 2). The number of whards of the shell is established by observing the shell from above and counting downwards from the apex, which is clearly the beginning of the spire (Fig. 3). The dotted line indicates the extent of the first whorl. Compared with the shells of older snails, those of younger animals have a sharp edge, which is, moreover, neither thickened, folder back, nor enlarged. In those families in which these characteristics are not shown in the shells of older animals, the shell edge of younger snails is soft, flexible and without calcium deposits. The edge of the mouth (peristome) is the growing region of the shell and may be regarded as being composed of an outer lip (Godan, 1983).

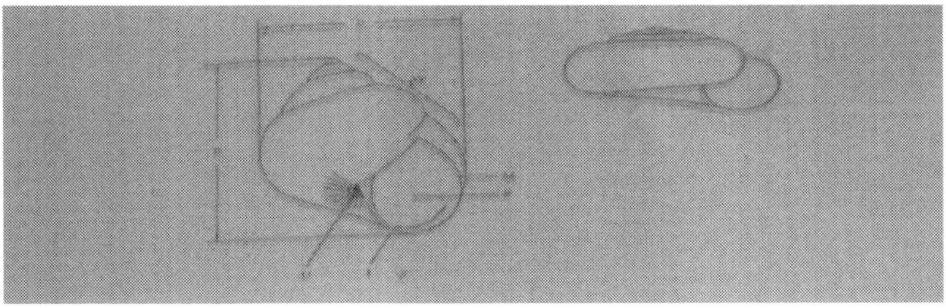

Fig. 2. Left shell concial to globular: right shell flat-conical. L. lip; *M. mouth* of shell (aperture), *P. peristome* (shell mouth edge); U. umbilicus; W. whorls; B. breadth; H. height (cited from Godan, 1983).

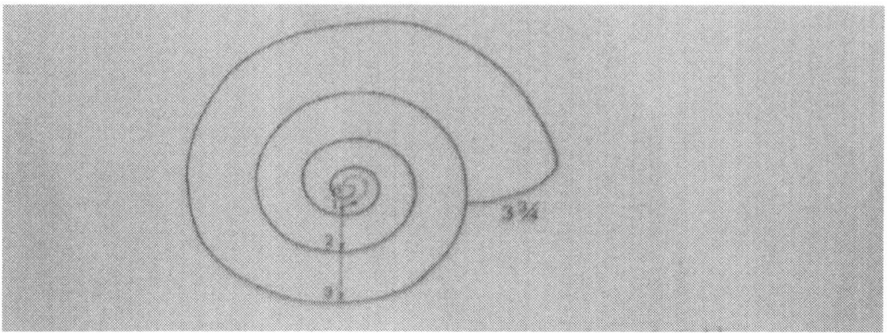

Fig. 3. Spire of the shell, Scheme, a apex; dotted line indicates extent of first whorl (cited from Godan, 1983).

4. Biology of terrestrial snails

Snails and slugs are simultaneously hermaphroditic, self incompatible. Five phases can be distinguished in the reproductive cycle of snails and slugs: Courtship, copulation, nest - building, egg - laying and embryonic development followed by egg hatching. Reproductive behaviour begins only when the humidity is high (80 - 85% for snail and 90 - 95% for slug).

Biological aspects of the land snails have been studied by many researches; for example, Kassab & Daoud (1964) found that the life cycle of *Helicella vestalis* was a relative simple. It laid the eggs in clutches, each contained from 25 to 30 eggs or more in the soft soil and were deposited in small cavities or holes in the soil. Eggs deposited at any time during spring season. The eggs were round and white in colour with calcareous or limy shell. Under normal conditions, the incubation period lasted from 12-15 days on the average. Soon after emerged from the eggs the young snails which were seen with their small and mucous shells began to move about in search for food. They were a little bit bigger than the head of a pin. Gradually, they added coils to the shell as they grew and rate of growth depended to some extent on the abundance of food and weather conditions. *H. lucorum* were sexually mature three years after hatching when the largest diameter of their shell was equal to greater than 25 mm as reported by Ramzy (2009) and Staikou et al. (1988).

Staikou & Lazaridou- Dimitriadou (1990) found that *M. cartusiana* (M.) reached maturity within one year at a size of 8 - 10 mm and could lay eggs immediately upon maturation. The reproductive period started in the beginning, middle or end of autumn depending on the weather condition which also effect on the growth of newly hatched individuals. While El-Massry (1997) revealed that *M. cartusiana* began to lay clutches from mid November to mid February. Number of clutches and clutch size were changed during breeding season.

Mohamed (1999) rearing snail *E. vermiculata* and slug *Limax flavus* (L.) at 20°C ± 1 and 80% R. H. for snail and 90% R. H. for slug and feeding on lettuce + cucumber and carrot. Investigated that some biological observation such as: Mating, egg laying and growth development.

Mating usually takes place at night, frequently on the soil surface. The snail *E. vermiculata* needs introductory behaviour (foreplay) with reciprocal lactile, oral contract and curving turns to reach on optimal position with respect to the genital opening of the partner (Fig. 4). This is followed by dart shooting, the pushing of a calcareous dart into the mating partner of body, which is assumed to facilitate meeting by increasing behavioural synchrony. Finally, the copulation is reciprocal; spermatophores are transferred after simultaneous intromission (Baur & Baur, 1999).

Fig. 4. Mating behaviour in snails (Cited from Godan, 1983).

Mating behaviour in slug *Limax flavous* is the two slugs twist around one another, like structure averted and intertwined to from a light spiral, and spermatophore are transferred. Courtship may last 3 - 4 hours in snail and slug (Fig. 4) (Godan, 1983).

Slug eggs are usually laid into soil holes and crevices into or on the soil, under stones and on decaying wood (Figs. 5, 6 and 7).

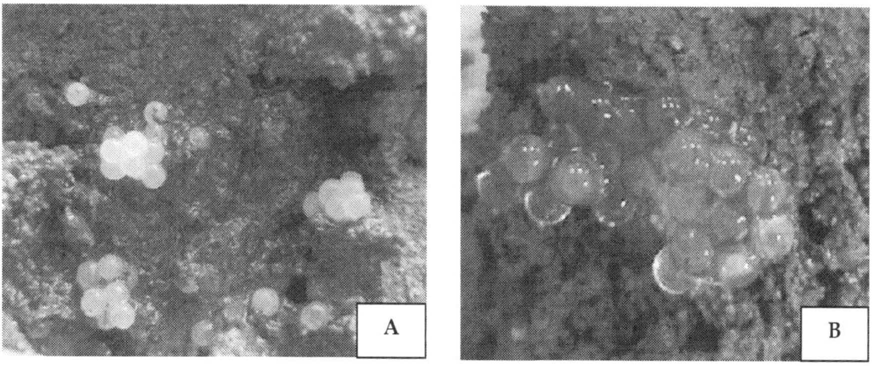

Fig. 5. The clutches of eggs of *Monacha obstructa* (A) and *Oxyloma elegans* (B) in soil (Cited from Ramzy, 2009).

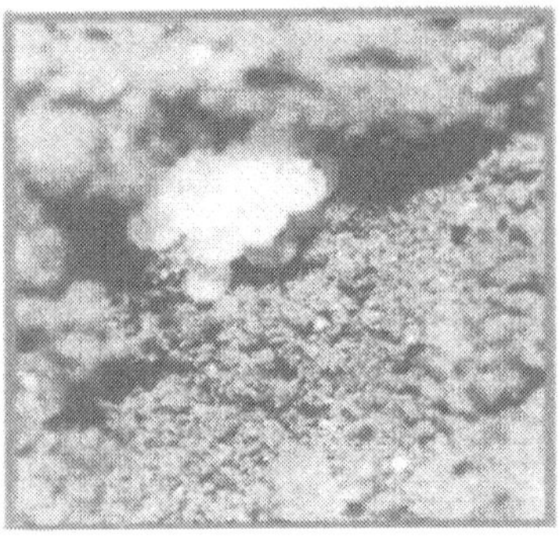

(1) Clutch of eggs of *E. vermiculata*

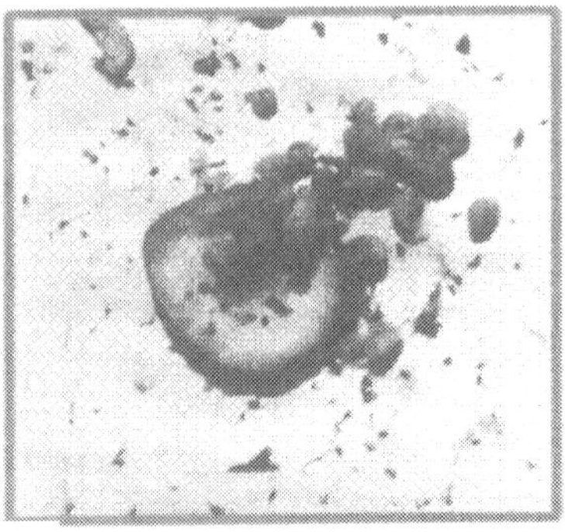

(2) Hatching

(3) Adult stage

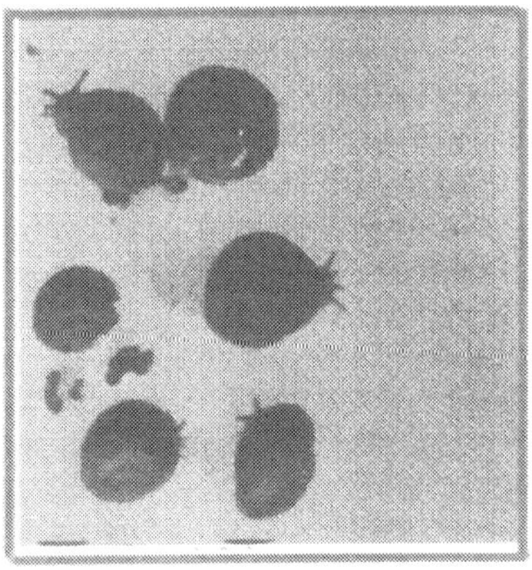

(4) The development of juveniles

Fig. 6. Life-cycle of *E. vermiculata* (cited from Mohamed, 1999).

Adult snail prepares to lay a clutch of eggs by excavating a deep hole in moist soil (Fig. 6). The animal pushes much effort in digging down with the anterior part to foot. Snails make a circular chamber about 5.7 - 6.1 cm. in deep. There is relationship between the body length and hole depth. This hole of egg - deposition is preparing after 10 hours of copulation.

During this time the snail is relatively conspicuous as the shell remains visible at the surface of nest. After that, egg deposition process takes place about 24 hours. Once egg - lying has finished the snail with draws, its foot and covers over the entrance of the hole with soil.

Snail eggs are whitish in colour and spherical shaped (2.9 - 3.0 mm diameter), while slug eggs are oval and strung along a thread (4.9 - 5.6 mm diameter). Snail and slug young does grow at a steady and relative fast. Growth continues at an almost constant rate even after animal maturation, by which time the slug has reached to weight 13 g. while the snail to 8.4 g. and the shell of snail is equal of 1/5 of a body weight. Animal snails and slugs arrive to adulthood after 321 to 364 days in average and their size and weight are more slowly increased (Figs. 6 and 7).

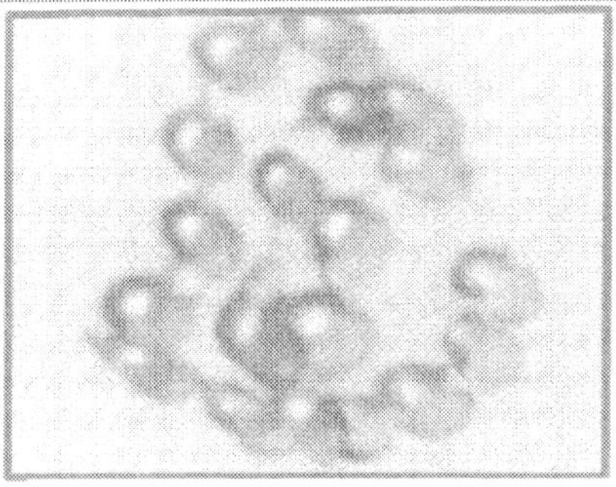

(1) Clutch of eggs of *L. flavus*

(2) Hatching

(3) The development of juveniles

(4) Adult stage

Fig. 7. Life-cycle of *L. flavus* (cited from Mohamed, 1999).

Generally, the highest number of eggs, hatchability percentage and longer life span were recorded in snail *E. vermiculata* and slug *L. flavus* in December followed by January, November, February and October, while these were decreased as temperature increased in March. The results are summarized in Table (1).

Type of land snail	Parameters			References
	No. of eggs/cultch	Hatchability	Life span(day)	
Eobnia vermiculata	80.92±5	91.43±3.2	1379±9.8	Mohamed, 1999
Limax flavus	62.23±2.7	91.882.1	677.87±10.4	Mohamed, 1999
Helicella vestalis	71.68	93.66	468.58	El-Massray, 1997
Helicella sp.	37.6	51.2	----------	Arafa, 1997
Eobania sp.	19	38.4	----------	Arafa, 1997
Monacha sp.	22.8	100	----------	Arafa, 1997
Monach cartus	22.55	99.1	----------	Ismail, 1997

Table 1. Some biological data of certain land snails.

5. Ecology of terrestrial snails

Terrestrial snails are mainly nocturnal, but following a rain may come out of hiding during the day. Temperature and moisture, rather than light, are the main factors to account for their nocturnal habits. Native snails may be found everywhere but prefer habitats offering shelter, adequate moisture, an abundant food supply and an available source of lime. Forested river valleys generally provide such habitats, and those with outcrops of limestone usually show the most abundant and varied mollusk faunas. Snails are very adaptable to times of drought and adverse climatic conditions. During these periods, the snails close the shell aperture with a mucus flap (epiphragm) which hardens and prevents desiccation. Snails can remain in this dormant state (aestivation) for years, breaking dormancy when climatic conditions are favorable again. Some ecological observation such as: Survey, population dynamic and movement, daily activity and dispersal of land snails, have been studied by many researchers (Bishara et al., 1968; Daoud 2004; El-Deeb et al., 1996, 2004; El-Okda, 1984; Metwally, et al., 2002; Ramzy, 2009).

5.1 Survey of land snails species

Bishara et al. (1986) found that *Euparypha pisana* (Müller), *Theba sp., E. vermiculata, Rumina decolata, Helicella sp.*, and *Cochlicella acuta* were common species in field and orchards of the northern Delta Nile in Egypt.

El-Okda (1984) stated that *Monacha* sp. and *Oxychilus* sp. were found in Ismaellia Governorate on the Egyptian clover (*Trifolium alexandrium*), mango orchards, citrus and ornamental nurseries, in addition to wheat fields. He added that beans, watermelon, maize and tomato were attacked by land snails. El-Deeb et al. (1996) and (2003b) surveyed different terrestrial snails on the field crops, vegetables, ornamental plants and in orchards at Kafr El-Shekh and Dakahlia Governorate, Egypt. The obtained results showed that in Kafr El- Shekh Governorate, *E. vermiculata, Succinia putris* and *Cepaea nemoralis* snails were common. In Demietta Governorate *M. cartusiana, E. vermiculata, Cepaea nemoralis, C. acuta, Oxychilus aliavus* and *Helicella* sp. were recorded on different host plants. *M. cartusiana, Succinia putris, E. vermiculata, C. acuta* and *C. nemoralis* were recorded on different host plants at Dakahlia Governorate.

Metwally et al. (2002) found that six species of terrestrial mollusca belonging to families Helicidae and limacidae were recorded on different crops at 23 locatities representing 10 districts at Monofia and Gharbia Governorate. These species were *M. cartusiana*, (the glassy clover snail), *E. vermiculata*, (the Brown garden snail), *C. acuta*, (the conical snail), *O. alliarus*, the slugs, *Limax flavus* and *Deroceras reticulatum*, (the gray garden slug). *M. cartusiana* snail have the upper hand on snail incidence compared to other species; similar results were obtained by El-Deeb et al. (2003b). In Upper Egypt, Ramzy (2009) identified nine land snail species in Assiut governorate, Egypt. All the species recorded belong to order: Pulmonata from eight families. These species were *Pupoides coenopictus, Vallonia pulchella, Oxyloma elegans, Vitrea pygmaea, E. vermiculata, T. pisana, M. obstructa, Helicodiscus singleyanus inermis, and Cecilioides acicula*. The results are summarized in Table (2).

Species	Governorate	Referenceq
1- *Monacha sp.*; and *Oxychilus sp.*	Ismaellia	El- Okda, 1984
2- *E. vermiculata; Succinia putris*; and *Cepaea nemoralis*	Kafer El-Shekh	El-Deeb et al., 1996
3- *M. cartusina.; E. vermiculata.; C. nemoralis,; C. acuta.; Oxchilus aliavus.*; and *Helicella sp*	Demietta	El-Deeb et al., 1996
4- *M. cartusina.; Succinia putris.; E. Vermiculata.; C. acuta.; and C.nemoralis*	Dakahlia	El-Deeb et al., 1996
5- *M. Cartusina.; E. Vermiculata.; C. Acuta.; O. Aliavus.; Limax flavus* and *Deoceras reticulatum*	Monofia & Gharbia	Metwally et al., 2002
6- *Pupoides coenopictus, Vallonia pulchella, Oxyloma elegans, Vitrea pygmaea, Eobania vermiculata, Theba pisana, Monacha obstructa, Helicodiscus singleyanus inermis,* and *Cecilioides acicula*	Assiut	Ramzy, 2009

Table 2. Survey of land snail species in Egypt Governorates

5.2 Population dynamics of land snails

The population of *Monacha obstructa* in Egyptian clover field began to increase gradually from the end March to the middle of April, whenever the suitable temperature and humidity (Kady, 1983).

The life cycle, population dynamics and secondary production of the land snail *M. cartusina* in northern Greece was studied by Staikou & Lazaridou- Dimitriabou (1990). Demographic analysis of populations of *M. cartusina* revealed that (a). From two to three cohorts existed in the field throughout the year. (b). The reproductive period started in the beginning middle or end of autumn depending upon weather conditions. (c). Growth of the newly hatched individuals was also influenced by weather conditions. Population fluctuation of the land snails, *T. pisana, H. vestalis* and *C. acuta* in citrus orchards were noticed by Hashem et al. (1992). The main activity season collapsed from February to November for *T. pisana* and *C. acuta*. They added that the land snail *H. vestalis* proved to be the most abundant species reaching the peak during March to June (Shoieb, 2008).

5.3 Movement, daily activity and dispersal of land snails

Snail activity differs from one species to another; it is influenced by external factors such as temperature, humidity, light intensity, soil conditions and food supply. The migratory behavior of land snails is greatly affected by the microclimatic conditions in their habitat. Activity increased as a result fall in temperature below 21°C and arise till 30°C. Moisture also influenced the activity of land snails. The mucus of land snails consists of 98% water. At high temperature activity is inhibited by water lack. Light influences the activity of land snails. They remain in their hides during the day and only after dusk; they emerge to go in search of food. A decrease in light intensity with a fall in temperature below 21°C and a rise in humidity through dew fall at dusk, resulted in the animal moving on to vegetation, while in the morning and during the day the snails returned to the upper soil or between the earth clods where it is cool and shady (Godan, 1983). The dispersal of the land snail *T. pisana* in South Australia was influenced by variation in habitat; snails moved on free way average where grassy vegetation was scattered. Snails moved out of a well grazed permanent pasture to adjacent weedy roadside vegetation and trees in early summer. They returned to the pasture in autumn. Average movement varied between 0.1 m. and 1.1 m. per day, some snails moved > 55 m in a month in spring - autumn and 75 m in three month in autumn - winter (Baker, 1988; El-Deeb et al., 2003b).

The effect of moon light, temperature and relative humidity on daily activity of the small sand snail, *H. vestalis* was studied by El-Massry (1997). It was found that the activity of *H. vestalis* investing navel orange trees was significant different from time to time during the day and from season to another during the year. Also, moon light showed a significant effect on the daily activity of *H. vestalis*. The lowest number of active snails was recorded in summer, while the highest one was recorded in spring (Lokma et al., 2007). On the other hand, the time from 10 to 2 o'clock represented the time of the lowest activity, while the highest activity was recorded between the mid night and the six o'clock in the morning (Ismail et al., 2003).

6. Chemical analysis of mucus land snails

6.1 Sample collection

Mature snails, i.e. *E. vermiculata, T. pisana and M. obstructa* were collected in the field from different locations in Egypt during the winter and spring of 2005. These animals were transported in white cloth bags to the laboratory. Healthy individuals were kept in round plastic boxes (13 cm in diameter) containing moistened soil and feeding on cabbage paper from the market for one year under laboratory conditions (25±5°C temperature and 70±5% R.H.). Mucus (5 ml) was collected from roughly 100 individuals by stimulating the surface of live snails by small plastic syringe (5 ml). The samples were stored at -20°C in a deep freezer until analysis according to Sallam et al. (2009).

6.2 Chemical analysis

Mucus from *E. vermiculata, T. pisana,* and *M. obstructa* were analysed by GC-MS which was performed with an agilent 6890 gas chromatograph equipped with a mass spectrometric detector (MSD) model agilen 5973. A fused silica capillary column (HP-5MS), 5% phenyl polysiloxane as non-polar stationary phase (30 m x 0.25 mm x i.d) and 0.25 μm film thickness was used. Operating conditions were as follows: injector port temperature, 2508C. Helium was used as a carrier gas at a flow rate of 1.0 ml/min pulsed splitless mode programmed at 8°C/min to 260°C, and held for 18 min. The total analysis time was 41 min. A 1 ml volume was injected splitless. The mass spectrometric detector (MSD) was operated in electron impact ionization mode with an ionizing energy of 70 eV, scanning from m/z 50 - 500. The ion source temperature was 230°C and the quadruple temperature was 150°C. The electron multiplier voltage (EM voltage) was maintained at 1100 V above autotune, and a solvent delay of 3 min was employed. The instrument was manually tuned using perfluorotributylamine (PFTBA). Identification was based on comparison with the MS computer library (NIST Software Package, Finnigan) and on the respective retention indices. The separated components were identified by matching data with those of the data published by Wiley7n.1.

6.3 Chemical components of mucus

Chemical constituents of mucus from *E. vermiculata, T. pisana* and *M. obstructa* GC-MS analysis of mucus from *E. vermiculata* detected the presence of the following compounds: iso-valeric acid, methyl, 3-methoxyamino-propanoate, oxime, methoxy-phenyl, pantanolic acid, 4-methyl, cyclotrisiloxane, hexamethyl, tetradecanal, furan, 2-isobutenyl-4vinyl, and di-n-octylphthalate (Table 3). Data showed that eight compounds were identified in this mucus after comparison with library data by Wiley7n.1. Both oxime, methoxy-phenyl, and iso-valeric acid were mainly characterized by a high concentration of total compounds (81.58 and 8.76%, respectively), while methyl, 3-methoxyaminopropanoate, and di-n-octylphthalate were characterized by low concentration of total compounds (0.63 and 0.56%), respectively.

Peak No.	Name of compounds	Chemical formula	Molecular weight (MW)	Area %	Retention time (RT)
1	Iso-valeric acid	$C_5H_{10}O_2$	102.07	8.76	3.45
2	Methyl, 3-methoxyamino-propanoate	$C_5H_{11}NO_3$	133.07	0.63	3.51
3	Oxime-, methoxy-phenyl	$C_8H_9NO_2$	151.06	81.58	4.01
4	Pantanolic acid, 4-methyl	$C_6H_{12}O_2$	116.08	1.26	4.36
5	Cyclotrisiloxane, hexamethyl	$C_6H_{18}O_3Si_3$	222.06	4.65	6.63
6	Tetradecanal	$C_{14}H_{28}O$	212.21	1.01	16.45
7	Furan, 2-isobutenyl-4-vinyl	$C_{10}H_{16}O$	152.12	1.55	16.85
8	Di-*n*-octylphthalate	$C_{24}H_{38}O_4$	390.28	0.56	26.64
9	Total			100.00	

Table 3. Chemical composition of mucus from *E. vermiculata* (cited from Sallam et al., 2009).

Data in Table 4 show the chemical constituents of mucus from *T. pisana*. Ten compounds were identified in this mucus after comparing with library data from Wiley7n.1. Both oxime, methoxy-phenyl, and cyclotrisiloxane, hexamethyl were mainly characterized by a high concentration of total compounds (56.28 and 20.55%, respectively). In contrast, 1,2,4-trichloroacetophenone and pyridine, 1-acetyl-5-(3.4.-dihydro-2H-pyrrol-5-yl) 1,2,3.4-tetrahydro were characterized by a low concentration of total compounds (1.36 and 1.10%), respectively.

Peak No.	Name of compounds	Chemical formula	Molecular weight (MW)	Area %	Retention time (RT)
1	Butanoic acid, 2-methyl	$C_5H_{10}O_2$	102.07	2.89	3.62
2	Oxime-, methoxy-phenyl	$C_8H_9NO_2$	151.06	56.28	4.11
3	Cyclotrisiloxane, hexamethyl	$C_6H_{18}O_3Si_3$	222.06	20.55	6.69
4	1-Butyl-2,4,6-trimethyl benzene	$C_{13}H_2O$	176.16	5.81	7.60
5	Thymyl methyl ether	$C_{11}H_{16}O$	164.12	2.98	8.02
6	Pyridine,1-acetyl-5-(3,4-dihydro-2H-pyrrol-5-yl) 1.2.3.4-tetrahydro	$C_{11}H_{16}N_2O$	192.13	1.10	8.41
7	Cyclotetrasiloxaneoctamethyl	$C_8H_{24}O_4Si_4$	296.07	2.41	9.38
8	Benzene, 1,4-Bis(trimethylsilyl)	$C_{12}H_{22}Si_2$	222.13	3.87	10.22
9	Thiosulfuric acid, *S*-(2-aminoethyl) ester	$C_2H_7NO_3S_2$	156.99	2.74	20.18
10	1,2,4-trichloroacetophenone	$C_8H_5Cl_{30}$	221.94	1.36	26.45
11	Total			100.00	

Table 4. Chemical composition of mucus from *T. pisana* (cited from Sallam et al., 2009).

Also, data in Table 5 show that the chemical constituents of mucus from *M. obstructa*. Seven compounds were identified in this mucus, bothoxime, methoxy-phenyl, and diethyl phthalate were mainly characterized by a high concentration of total compounds (57.95 and 14.20%, respectively), while pentadecane and hexatriacontane were characterized by a low concentration (3.55 and 2.22%), respectively.

Peak No.	Name of compounds	Chemical formula	Molecular weight (MW)	Area %	Retention time (RT)
1	Oxime-, methoxy-phenyl	$C_8H_9NO_2$	151.06	57.95	3.97
2	Cyclotrisiloxane, hexamethyl	$C_6H_{18}O_3Si_3$	222.06	12.88	6.62
3	Pentadecane	$C_{15}H_{32}$	212.25	3.55	13.19
4	Diethyl phthalate	$C_{12}H_{14}O_4$	222.09	14.20	15.04
5	10-Methylnonadecane	C_2OH_{42}	282.33	4.51	16.62
6	Hexatriacontane	$C_{36}H_{74}$	506.58	2.22	18.04
7	1,2-Benzene dicarboxylic acid, dibutyl ester	$C_{16}H_{22}O_4$	278.15	4.70	20.23
8	Total			100.00	

Table 5. Chemical composition of mucus from *M. obstructa* (cited from Sallam et al., 2009).

It is obvious that the different composition of mucus from different species of land snails, oxime, methoxy-phenyl, and cyclotrisiloxane, hexamethyl were major components found in all species. This difference in composition of mucus may be due to differences from one species to another, different mechanical properties (function) and are influenced by external factors such as temperature, humidity, light intensity, soil conditions and food supply. These data agreement with Meikle et al. (1988) found substantial differences between the mucus of six coral species. It should not be surprising that different forms of mucus have different compositions and different mechanical properties. There is wide variation in the gross composition of mucus secretions (Davies & Hawkins, 1998) and in their function (Denny, 1989). As Davies & Hawkins (1998) point out, relatively little is known about the structure of invertebrate mucus secretions. In the study by Smith & Morin (2002), the composition of the adhesive form of march periwinkle mucus was compared to the trail mucus used during locomotion. They found that the trail mucus consists primarily of large, carbohydrate-rich molecules with some relatively small proteins. In contrast, the adhesive mucus has 2.7 times as much protein with no significant difference in carbohydrate concentration. This change in composition corresponds to an order of magnitude increase in tenacity with little clear change in overall concentration.

Previous research on marine mucus secretion found that roughly 50% of the dry weight was inorganic residue (Connor, 1986; Davies et al., 1990). This value can also be estimated from the percentage of inorganic salts in seawater (3.56%) (Schmidt-Nielsen 1990); for typical marine mucus containing 96 - 98% water, we would predict that inorganic material would make up 46 - 64% of the dry weight (Sallam et al., 2009).

7. Damage and feeding behaviour of land snails

Land mollusca pests are serious problem, every year; damage involving considerable financial losses is inflicted on cereal, potatoes, vegetables, lettuce, carrots, cabbage, maize, clover as well as other agricultural and horticultural crops. They eat leaves, root and tuber of nearly all vegetables, field crops, ornamental plants as well as fruits in field, garden and green house. Land snails cause heavy damage, especially to seeds and seedlings of cereals and seeds of oil plants. Damage was manifested in chewing soft vegetative growth, flowers and fruits, beside eating seeds, roots and tubers after sowing or during repining. On the other hand, land molluses left viscous liquids upon the plant on which they had been fed giving bright trace films. Moreover, unpleasant garlic odour was smelt on Egyptian clover which infested with *M. abstracta*, making farm animals refuse it. (El-Okda, 1980; Imevore & Ajayi, 1993; Ismail et al., 2003; Ramzy, 2009).

El-Okda (1980) mentioned that land mollusca attacked raw Succulent vegetables and proffered soggy parts. These pests attacked seeds, seedlings, roots and tuber crops. The more succulent raw leaf vegetables, fruits and buds were extra ordinarily attacked in addition to flower damage when land mollusca become abundant. Also, these land mollusca leave unpleasant slimy tracks on the injured parts.

Imevbore & Ajayi (1993) reported that when mature snails fed once daily at 2 % of body weight on diet containing 20 different feeds categorized as leaves, fruits, and household waste. Feed intake data indicated that the African giant snail a definite preference for fresh fruits and low preference for household waste.

The food preference and consumption of certain vegetable plants and field crops leaves for three land snails: *M. Cantiana, Succina Putris* and *T. pisana* was studied by El- Deeb et al. (2001). Results showed that Egyptian clover was the most preferred crop for *M. contiana* followed by lettuce, cucumber, carrot, cabbage and squash, while carrot fruits were the most preferable for *S. putris* followed by Egyptian clover, lettuce, cabbage, cucumber and squash. On the other hand, the lettuce was the most preferable for *T. pisana* followed by cabbage, carrot, cucumber, clover and squash, see Fig (8). In the same time, the results indicated that the bran was the most preferred bait for the three snails species followed by crushed wheat, crushed bread, crushed rice and crushed maize. The relative susceptibility of five fruits species (apple, orange, pear, plum trees and banana plants) to the four land snail species infestation was studied under laboratory conditions by identification of fruit leaf cells in their excrement. Results indicated that *T. pisana, H. vestalis* and *C. acuta* highly prefer pear, while *E. vermiculata* prefers banana. Orange ranks second in their favorability. On the contrary, plum, and apple are not preferred for the snail species, beside banana for *C. acuta*

Moreover, pear and orange are mostly attacked by *T. pisana* and banana by *E. vermiculata* (Shahawy et al., 2008; Tadros et al., 2001).

Fig. 8. Feeding damage by terrestrial snails on different plants (cited from Ramzy 2009).

8. Control of terrestrial mollusca

Molluscan pests can be destroyed in several different ways: by chemical means (Molluscicides), through the use of biotechnical measures, by mechanical methods and by interference with the environment. These different methods can, and sometimes have to be used simultaneously.

The control of pulmonates by whatever method is not simple, and each of the above-mentioned methods has its associated problems. A prerequisite from species to species is even in the same area.

8.1 Legislative control

Different ways of dispersal of economic pest mollusca may occur by man, plants, seeds, soil, ships, trains, airplane, cars, cargo, baggage, postal packages and food stuffs (Herzberg & Herzberg, 1962). For example, most of gastropds entering the USA with baggage come from Europe, Asia, Africa, Hawaii and the Philippines (Godan, 1983). Most Countries have now established quarantine regulations allowing important of plant and baggage under certain strictly specified conditions. Land mollusca among the economic pest species are listed in the rules and regulations of plant pest control in USA, the plant quarantine treatments contains dosages for fumigation of containers and ships to control gastropods.

Several investigators used the fumigants to combat the quarantinable snails. Richardson & Roth (1963) used ethyleue oxide 10 %, carbondioxide 90 % mixture (carboxide) instead of methyl bromide for fumigation against the very resistant aestivated *C. barbara* (L.) and *T. pisana* on military cargo from Mediterranean areas, and it showed good efficiency, penetration, and stability as a suitable substitute for methyl bromide.

Ittah & Zisman (1992) evaluated the volatile allylalcohol derivatives for control of *T. pisana* snails on cut roses for export. They found that allyl alcohol esters (propionate, formate, acetate) to be very effective in the fumigation of snails infesting flowers of roses. Concentration of allyl acetate of 0.022 mM at 25°C for 3 hrs was found to cause 100 % snail mortality without any fear of phytotoxicity towards roses.

8.2 Mechanical control

1. Hand collection with subsequent squashing of the slugs and snails is the oldest mechanical methods and was the only control measure used up to the middle ages (Carman, 1965; Mahrous et al., 2002; Shah, 1992). Common salt was then employed this causes an over reproduction of mucus by snails. Later, traps and protective barriers were used. The flower pots tiles, cabbage leaves, flat stones towels and also drain pipes, these hand to be searched for snails in early morning such dark shelters are especially effective as traps when they are baited with raw fresh potato slices of fresh lettuce leaves. It is necessary, however, to collect the snails daily and to kill them with a strong solution of common salt or in boiling water.
2. Burning -over- burning vegetation on which aestivating snails attach will reduce throwers or commercial weed- burners is effective in reducing snail populations along fence rows and in areas where other measures may not be practical. (Joubert & walters, 1951)
3. Protective barriers of dehydrating substance such as cattle salt, caustic soda, kainite or completely dry quick lime can act as barrier. Nakhla (1995) studied a mechanical method to protect orchard trees from land snails by using a band of metal sheets around the tree trunk. The rings tested were in the shape of cornea and were made of copper sheet, aluminum sheet, wire screen gauze (14 mesh) and a fiber cord, (Rape). The results revealed that wire screen rings gave the highest percentage of production against the different snails, i.e. *E. vermiculota, C. acuta, H. vestalis* and *T. pisana* followed by the copper sheet rings. Aluminum sheet rings gave intermediate protection while the rape rings gave poor protection

8.3 Agricultural control

Plough of the soil before sowing seeds of wheat, in early spring, is the most effective means of ecological control, that protect wheat from damage caused by slugs. This way results in exposure of the slugs and their eggs to predators and the weather (Wouters, 1970). Slugs need moist soil for egg-laying and cool shelter for protection during the day opening up the garden to more light and air movement will reduce the amount of favorable habitat (Mahrous et al., 2002; Shetlar, 1995). There are many methods which are effective in the agricultural practices:

1. Plowing-In open fields, adjacent to outside storage and on base housing areas, plowing the soil twice a year has been found to reduce both snails and slugs populations. Cultivating the soil in late autumn destroys many of the immature and adult snails, as well as eggs that have been deposited in the soil.
2. Disking and culti-packing. This is helpful in reducing land snail populations in areas where plowing may be not practical because of thin top soil or where erosion may be a serious problem. The mechanical action of the disc and culti-packr will eliminate many adult snails, while stirring the soil will destroy many eggs.
3. Improvement of soil structure, results in exposure of the sensitive egg and juvenile slugs to direct sunlight and they die of desiccation.
4. Clearing of the edges of fields and irrigation ditches after the harvest can reduce the level of infestation, but does not eliminate it altogether.

8.4 Biological control

In the following section the possibilities are discussed of using such environmentally harmless methods of control of molluscs, whether by diseases, parasites, predators or biotechnical procedures (Baker, 1989; Godan, 1983).

A. Fungi

Fungi attack mainly eggs of gastropods, for examples those, of *Deroceras reticulatum* by *verticullium chlamydosorium*. This fungus is not suitable for use in biological control, despite a high infection rat of the eggs (Mead, 1961). While fungal infections are important when rearing molluscs in the laboratory, since the may destroy the whole stock, in the field, fungi seldom cause total elimination or even a reduction in gastropod number.

B. Bacteria

Bacteria of the genera Escherichia, Alcaligenes and Bacillius, in France an aerobacter infection occurred on a *Helix fram* (Mead, 1961).

The combination between "victoback12" as (*Bacillus thuringiensis*) and the parasitic nematoda *Rhabditis* sp. can play an effective role in controlling *E. vermiculuta* as well as the other terrestrial snail and slugs in Egypt (Azzam & Belal, 2002)

C. Viruses

Very little is known about the effect of viruses on molluscs. This is a most neglected area of research.

D. Nematodes

The nematodes are the third large group of worms which parasitize molluscs; they are mostly of interest to veterinary medicine, but in a few cases also affect man. A comprehensive review of snails considered them as intermediate hosts of nematodes (Azzam, 1999; Azzam & Hegab, 2000; Wilson et al., 1996).

The slug nematodae *Phasmarhabditis hermaphrodita* (Schiender) has been successfully used control slugs in field experiments (Wilson et al., 1996). The snail parasitic nematode *Rhabditis* sp. was recorded for the first time in Egypt and showed high infectivity different snails, slugs and insects in laboratory (Azzam, 1998). Production of this nematode from different pests was investigated and it was found that *L. flavus* slug and *E. vermiculata* snail was the most adequate host producing high numbers of this nematode, as shown in Table (6) (Azzam, 1999).

Azzam & Hegab (2000) exposed the terrestrial snail, *E. vermiculata* to different concentrations (10-300 I.S/ Snail) of the snail parasitic nematode, *Rhabditis* sp. under laboratory conditions. They are found that the period needed for snail death decreased by increasing concentration.

E. Parasites and predators

E.1. Parasites

The most important of these parasitic include protozoa, flat worms, lung worms, carabid beetles, the glowworm larvae of lampyrid bettles as well as the larvae of Sciomyzidae (Diptera), as shown in tables (6 and 7). Protozoa associated with molluscs both parasitically and symbiotically (Baker, 1989; Godan, 1983). The sprrozoan *klossia* sp. has been found in the kidney in various slugs and snails, and an amoeba is found as on endo-parasite in *Arion rufus*

Tetrhymena rostrata (Kahl) was isolated from the soil in which the slugs live, showing that contamination occurs through the soil. Slug is probably infected by contact with cysts in the soil during the winter. It is also possible that the parasites are transferred by way of the eggs; the some eggs laid by *Deroceras reticulatum* were found to be infested with ciliates.

Field slugs infested with *T. rostrata* show the following symptoms: mantle shield swollen, posterior end of the body some what elongated and laterally compressed; after death the body shortened, and a tumor-like swelling of the mantle shield.

E. 2. Predators

Attempts have understandably been made to use predators for control of pest gastropods, especially for the biological control of land snails. For example *Byfo melanosticus, Rantigran tigrian* used in control *Lavicaulis alte* (Muthumani et al., 1992). Insects from order diptera, family sciomyzidae including 84 species as predators of Molluscs (Knutson, 1966; Neff, 1964; Verbeke, 1964). Beetles from order coleoptera, family carabidae, species, i.e. *Thermophilum hexaasticum* (Gerstaeker), *Carabus violaceus* (L.), and *Scophinorus striatopunctatus* (Choudoir) were used as biological control agents for *Achatina fulica.*

Fouad et al. (2000) illustrated the predation potential of some vertebrates, i.e. rodents, (*Rattus rattus rattus, Rattus norvegicus, Arvicanthes niloticus* and *Acomys cahirinus*) and birds (*Ardola ibis ibis* and *Bubulcus ibis*) for the different stages of three land snail species, *M. abstructa, T. pisana* and *E. vermiculata* under laboratory conditions.

It is concluded that the predation potential for the vertebrate animals markedly differed according to predator species and size of the prey or tested snail species as the rodent *R. norvegicus* exhibited a comparatively high potential for all stages of the tested gastropod species followed by *R. rattus* when compared with the *A. niloticus*, while *A. cahirinus* showed the lowest effect against all tested gastropod species. On the other hand, *A. ibis ibis* occupied the first order between bird species when compared with *B. ibis* birds. In the same time, the obtained data revealed that all tested predators exhibited a high predacious effect against all land snails particularly their juvenile stages as animals characterized by soft shell were more vulnerable for predator attacking, while the adults of snails were enable to protect themselves either by disappearing inside the hard shell or by releasing extensive mucus, as *M. obstructa* more acceptance to both vertebrate species which have small size than *T. pisana* and *E. vermiculata.*

Species	Host	
	Snail	Slug
Plathelmiuthes		
Trematoda		
Brachycoelium obesum (Nicolli)		*Deroceras agreslo* (L.)
Brachylaema helices (Pomaliae)	*Cepaea hortensis* ((Müller)	
Brachyiaema nicolli (Witenberg)		*Deroceras agreslo* (L.)
Brachyiaema obesum (Nicolli)		*Deroceras agreslo* (L.)
Dicrocoelium dendriticum (Rudolphi)	*Helicella obvia* (Hartmann)	*Arion fasciatus* (Nilsson)
(*D. lanceolatum* (Rudolphi)	*Helicella itala* (L.)	*Arion subfuscus* (Draparneud)
(Lancet liver fluke)(Grazing stock)	Monacha cartisicne (Müller)	*Deroceras reticulatum* (Müller)
	Cochlicelle acuta (Müller)	*Limax tenellus* (Müller)
	Cochlicopa lubrica (Müller)	
Leucochloridlum paradoxum	*Succinea putris* (L.)	
Monocerus Sp.		*Arion* spp.
Cestods		
Anomotaenia arionis (Siebold)		*Arion ater* (L.)
Aporina delafondi Railliet		*Arion rufus* (L.)
Choanotaenia crassiscolex (Linstow)	*Oxychilus cellarius* (Müller)	*Deroceras* spp., *Arion* spp.
Cyslicercoidos dukae (Holland)	*Succinea Putris* (L.)	
Davainea proglottina (Davaine)	*Helicella itala* (L.)	*Arion hortensis* (Ferussac)
	Helicella obvia (Hartmann)	*Arion intermedius* (Normand)
	Succinnea putris (L.)	*Deroceras caruanae* (Pollonera)
	Succineidae	*Deroceras leave, D. reticulatum* (Müller)
		Lehmannia marginata (Müller)
		Limax flaves (L.), *Limax maximus* (L.)
Hemenolepis multiformis (Creplin)		*Arion* spp.
Raillietina bonini (Megnin)	*Cepaea hortensis* (Müller)	*Arion* sp., *D. reticulatum* (L.)
		Marginata
Teania bothrioplitis (Filippi)	*Monacha cartusiane* (Müller)	
Nemathelminthes		
Nematoda		
Aelurostrongytus abstrusus (Railliet) (Lungworm of catts)	*Helix* spp.	*Arion circumscri* *Deroceras agreste., D. reticulatum*
Alloionema appendiculatum (Schneider)		*Arionater* (L.) *A. tufus* (L.) *D. agreste*
		D. agreste
Angiostoma helices (Conte & Bonnet)	*Helix asperea* (Müller)	
Angiostoma limacis (Dujardin) *Leptodera angiostoma* (Oujardin)		*Arion* spp., *Limax* spp.
Angiostrongylus cntonensis (Chen)	*Achatina fulica*	*Deroceras leave* (Müller), *Limax* spp.
Crenosoma vulpis (Dujardin) (Cat, Dog, Fox)	*Cepaea* spp., *Helix* spp., *Succinea* spp.	*Arion* spp., *D. reticulatum*
Cystocaulus ocreatus (Davtian)	*Cochilicella acuta* -Helicidae	*D. reticulatum* (Müller)
Hexamermis albicans (Siebold)	*Succinea patrus* (L.)	*D. agreste., D. reticulatum*
Mermis nigrescens (Dujardin)	*S. patrus* (L.)	*D. agreste., D. reticulatum*
Mullerius capillaries (Müller) (Lungworm of Sheep & Cattle)	*Cepaea vindobonensis, C. acuta* *H. obvia, H. Pomatia, M. cartusiane, T. pisana*	*A. ater, A.hortensis, D. laeve* *D. reticulatum, Limax* sp.
Protostrongylux nufescens (Longworm of Sheep)	*Cepaea* sp., *Helicella* sp.,*Helix* sp., *Monacha* spp.	

Table 6. Terrestrial gastropods intermediate hosts of worm parasites.

Molluscan predators have two ways of seizing their prey. If the prey is a snail, it is reached through the mouth of the shell, the predator penetrating deeper and devouring its prey as it does so: Ganaxis feed on terrestrial snails in this way. *Edentuline ovoidea* (Brugier) predatory snails used for control of *Achatina fulica* (Bowdied). Other predators on terrestrial gastropods, *Edentulina affiris,* and *Haplotrema minimum* (Ancey) are used to control the *H.*

aspersa (Müller). The latter species reduced the population and succeeds in keeping is number down. (Godan, 1983; Zeidan, 2001).

Species	Host	
	Snail	Slug
Amoeba		
Acanehamoebe sp		*Arion fasciatus* (Nilsson)
Amoeba sp		*Arion rufus* (L.)
Ryplobia hellcis (Trypanoplasma)	*Cepaea hortensis* (Müller), *C. nemorallis* (L.) *Helix pomatia* (L.) (in Receptaculum semnis)	
Trichomonas liracis	*Cernuella Virgata* (Dacosta)	*Limax* sp.
Sporozoa		
Isopora incerta (Schneider)		*Limex cinereoniger* (Wolf)
Isopora rera Aime (Schneider)		*Limax* sp.
Klossia loosi (Nabin)	*Helix, Cepaea and Succinea* sp.	*Arion* spp., *Limax* sp.
Pfeifferirclla impudica (Leger & Hollande)		*Lehmannia marginata* (Müller)
Plistiphora husseyi (Michaud)	*Achating zobra* (Sganzin)	
Trichodina echauinae	*Achatina zebra* (Sganzin)	
Ciliata		
Colpoda aspersa (Kahl)		Deroceras agreste (L.)
Colopoda steini (Maupas)		*Deroceras agreste* (L.) *Deroceras reticulatum* (Müller) *Lehmannia marainata* (Müller)
Concophthirus steenstrupi (Stein)	*Helix* spp.	*Arion ater* (L.) *Deroceras agreste* (L.)
Semitricholina sphaeronuclea	Zonitidae	------------------------------
Tetrahymena limacis (Warren)	*Trichia lubomirskii* (Slosarski)	*Arion hortensis* (Ferussac), *Deroceras reticulatum* (Müller), *D. leave* (Müller), *Lehmannia marginata* (Müller), *Limax flavusl, L. maximus* (L.) *Milax gagates* (Dr.)
Tetrahymena rostrata	*Zonitoides nitidus* (Müller)	*Arion intermedius* (Normand) *Deroceras reticulatum* (Müller)

Table 7. Parasitic protozoan in terrestrial gastropods (cited from Zeidan, 2001).

Predation by the carnivorous snail, *oxychilis* sp. on the newly hatched and youngsters of *Monacha* sp. snails was observed by El-Okda (1984). Another predatory decollates snails (*Rumina decolata*). These snails are used very successfully in commercial citrus groves in California and provide excellent control to the brown garden snails (Fisher & Orth, 1985) and to slugs (Allikas, 1997).

8.5 Chemical control

Chemical control of exotic snails typically employs metaldehyde, methiocarb (Mesurol), salt, or combinations of these chemicals with other molluscicides in a myriad of bait formulations or foliar sprays.

8.5.1 Methods of application

Spraying, Dusting and use of pellet bait. Today the use of bait is the most common method of gastropod control in both agriculture and horticulture, where as spraying and dusting are less frequently used. El-Massry (1997) studied the effectiveness of certain pesticides namely, i.e. methomyl (Lannate); paraquate (Garamoxone); oxyfluorfen (Goal); Glyphosate (Lansar) and pendimethalin (Stomp) against adult stage of three species of land snails (*H. vestalis, M. contiana* and *E. vermiculata*) under laboratory conditions using three methods for testing, i.e. direct spray, dipping and poisonous bait technique. He found that toxicity of any tested compound was varied according to the method of application. Mothomyl was the most effective toxicant against adults of the three tested species (Gabr et al., 2006).

8.5.2 Chemical components as snailcides

A. Metaldehyde treatments

They are applied during dry climatic conditions are usually more successful than the degree of control achieved during damp, high humidity conditions, at which time snails are likely to be more active (Dax1, 1970; Moens, 1970). The principal toxic effect of metaldehyde is through stimulation of the mucous gland which cause excessive sliming, leading to death by dehydration. Metaldehyde is broken down into acetaldehyde by sunlight, so where possible the pellet should be put in shady places, particularly under the leaves of the affected plants (Henderson, 1970; Henderson & Triebskorn, 2002).

B. The pesticidal properties of methiocarb

They are similar to the toxic action of other carbamates which prevent effective nerve transmission by inhibiting the enzyme acetyl cholinesterase. The methyl carbamate most widely used as amolluscicide is methiocarb. This compound is more poisonous than metaldehyde pellets and less active as a contact killer, acting more as stomach poison when ingestion the baits for controlling slugs *D. reticulatum* (Getzin & Cole, 1964). The effectiveness of methiocarb is compromised less by low temperatures and high humidity than that of metaldehyde which is amagor advantage (Mallet & Bougran, 1971).

C. Differences in the symptoms of poisoning by metaldehyde and methiocarb

The symptoms shown by gastropods poisoned with metaldehyde differ markedly from those following carbamate poisoning (Godan, 1965). The metaldehyde can affect molluscs either by contact, with absorption through the skin, or through the gut when eaten. The main effect is that of an irritant, causing the molluscs to produce masses of mucus, leading to dehydration and sometimes death. Loss of mucus also means that the animals can no longer move around, so that death and dying animals are found close to the baiting site. Molluscs that have been poisoned by methiocarb can however, move around for a while, but then swell up with fluid and become immobile dying shortly afterwards. In dry conditions this swelling can be reduced, and some animals may recover, although generally recovery rates are lower than with metaldehyde (Abd El-Wakeil, 2005).

D. In addition to these molluscicides, sodium chloride (common table salt)

It is an effective dehydrating agent. It may be applied as ainch barrier application on the perimeter of known/ suspected snail infested area. During periods of rain or high relative humidity, salt are aestivating.

E. Further chemical compounds

They are being tested for their possible molluscicidal effects, even substance with other uses, such as fertilizer, fungicides, insecticides and herbicides (Van der Gulk & Springett, 1980). Fox (1964) used herbicides among the substance used for plant protection in agriculture and horticultural plants. The results indicated that herbicides were effective against land snails. Mode of action was in contact over long period of time with snail or as direct feeding on treated plants.

El-Massry et al. (1998) tested fertilizers, urea, ferrous sulfate and calcium super phosphate against many species of land snails; *H. vestalis, M. Contiana* and *E. vermiculata* in the laboratory. He found that urea the highest toxic effect against the three species of land snails, respectively, followed by ferrous sulfate, while calcium super phosphate showed the last toxic.

The evaluation of a molluscicide is based on testes of its toxicity, persistence of effectiveness, attractiveness to gastropods and also the chances of recovery of affected slugs or snails studies by many researchers (i.e. Arafa, 1997; El-Okda et al., 1989; El-Sebae et al., 1982; Ghamry et al., 1994; Mahrous et al., 2002; Radwan & El-Zemity, 2001).

El-Sebae et al. (1982) tested locally formulated bran baits containing aldicarb, methomyl or Dupont- 1642 against land snails, *H. vestalis, E. vermiculata* and *T. pisana*. Different wheat and rice brains are containing 0.5% aldicarb or methomyl showed high attractant action and toxicity for land snails, represented by their high mortality percentages.

El-Okda et al. (1989) evaluated the efficacy of the formulated local 0.5 %, aldicarb, oxamyl, methiocarb, Lannat and metaldhyde in controlling the land molluscs ; *H. aspersa, Eobania sp., Theba sp., Rumina* sp. and *oxychilus* sp. The results indicated that, aldicarb, oxamyl and Lannat gave the highest toxicity against the most snails and slugs species, while methiocarb and metaldehyde were less toxic.

Ghamry et al. (1994) evaluated fourteen insecticides against tow land snails; *M. contiana and E. vermiculata*. Results from bait testes revealed that, methomyl, dithiocarb, carbaryl, chlorpyrifos and dimethoate were effective for killing snails after 12 days under laboratory conditions. On the other hand, the same trend was observed with those insecticides under field conditions.

Arafa (1997) tested sucmate- granules, Mesural and Nuvacron against the land snail *E. vermiculata* under field conditions. The results revealed that sucmate- granules gave the highest mortality (100%) when used as poison baits while Novacron gave 46% mortality when it used as direct spray after two weeks. Mesural gave 48.5 % mortality percentage when it used as poison baits.

Radwan & El-Zemity (2001) synthesized a new series of 1,2,4- triazol derivatives and screened for their molluscicidal activity against two type of terrestrial snail, *H. aspersa* and *T. pisana*, by two methods of application, either as contact or as bran baits. Several of the tested compounds exhibited good molluscicidal activity, and *T. pisana* was more sensitive than *H. aspersa*. Substitution at the 0 and or P- positions of the phenyl ring with chlorine or bromine gave higher molluscicidal activity than unsubstituted compound, with O, P- dichloro substitution being optimum. In addition compounds containing two triazole moieties showed higher molluscicidal activity, particulary as stomach poisons, that the contact toxic effect of the corresponding compound with one triazole ring. In general, carbamate derivatives were more active than their corresponding 1,2,4- triazol derivatives.

Mahrous et al. (2002) tested seven pesticides to evaluate their molluscicidal activity as poisonous baits against *M. Cartusiana* in Sharkia Governorate, Egypt. The obtained results that the molluscicidal efficiency of the tested pesticides after 15 day- treatment could be arranged as follows: fenamiphos> sethoxydim> oxamyl> monocrotophos> butachlor> biofly and seed grad. On the other hand, carrier or attractive materials usually used in poisonous baits showed insignificant effect on the molluscicidal activity of fenamiphos in controlling *M. cartusiana*.

8.6 Botanical pesticides against molluscan pests

Plant products known to possess molluscicidal activity against the snails of agricultural importance are presented in Table (8) along with details of their forgeneric names, plant parts tested to possess biological activity and their formulations (Prakash & Rao, 1997). Azardirachtin, and active component isolated from neem kernel extract, was also reported to show mulluscicidal activity against *Lymnea luteda* (Ramesh, 1983). The molluscicidal activity of saponins isolated from saponaria roots was investigated (Kadey et al., 1982). Also, Kady et al. (1986) attributed the molluscicidal action of the wild herb, *Peganum harmala* (L.) (seeds) to its alkaloidal constituents which affect the respiration and/ or the nervous sytem of the snails. Kishor & Sati (1990) reported "spirostanol glycoside" from an ornamental plant, *yacca aloifolia* to be 100 % toxic at 10 ppm when tested against the snails like *Biumphalaria glabrata*. El- Hwashy et al. (1996) showed that the ethanolic leave extracts of Cauliflower, Oshar and pergularia were most effective against *E. vermiculata* snails when tested as residue film technique with mortality percentage of 88.8, 88.8 and 77.7 respectively. Molluscicidal activity and repellent properties of thirteen monoterpenoidal compounds were studied against the snail, *H. aspersa*. Camphor, thymol, (R)-carvone and carvacrol proved to be potent molluscicides of the compounds tested only citronellol, geraniol, (±) methanol and thymol were highly effective as repellents (El-Zemity et al., 2001). Repellency effect of 28 plant extracts obtained from different parts of 13 indigenous plants, i.e. Damsisa, Halfa barr, Colocynth, khella, Harmala, Datura, Santonica, Sucalyptus (L.), Eucalyptus (S), Enab eddhib, Calotropis, Alocasia, Halouk and Geranium, was investigated against *M. cantiana* land snails, using one and two choice feeding methods (Abd El- All et al., 2002). All plant materials either extracted with hexane and/ or ethanol showed a considerable snail repellency effect, when their crude extracts were tested using one choice feeding methods. In contrast results of two choice feeding test method Indicated that all the tested plant extracts showed snail repellant effect except Damsisa, Enabeddip, Calotropis and Geranium hexane crude extracts, which failed to achieve 50 % or more repellency level.

Generic name	Part of plant and its formulations	Biological activity	Reference
Azardirachta indica A. Juss. (Meliaceae)	Kernel extract of neem and neem cake extract	Toxicity to *Lymnaea matalensis* and *L.auricularea.*	Reynaud, 1986; Sasmal, 1991
Balantisa egyptica (L.) Delite (Simaroubaceae)	Nut and leaf powders	Toxicity to the snails	Belen, 1982
Citrus mitis (Linn.) Macf. (Rutaceae)	Dried fruit powder and its aqueous extract	Toxicity to the snails	Belen, 1982
Coryzadenia balsamifera Griff. (Hernandiaceae)	Fruit and leaf aqueous extracts	Toxicity to the snails	Belen, 1982
Croton figluim (L.) (Euphorbiaceae)	Leaf extract in water	Toxicity to the snails	Belen, 1982
Entade gigas(L.) Fawc and Rendle (Mimosaceae)	Leaf seed and bark extracts in water	Toxicity to the snails	Belen, 1982
Jatropha curcas (L.) (Euphorbiaceae)	Fruit and seed aqueous extract	Toxicity to the snails	Belen, 1982
Menispermum coculus (L.) (Menispermaceae)	Leaf, branch, fruit and seed extracts in water	Toxicity to the snails	Belen, 1982
Prago pabularia Hybrid (Cactaceae)	Aqueous leaf extract	Toxicity to the snails	Belen, 1982
Thevetia nerifolia Juss.ex Steud (Apocynaceae)	Alcoholic fruit extract	Toxicity to the garden snail	Johri et al., 1993

Table 8. Molluscicidal properties of plant products (Cited from Parakash & Rao, 1997).

8.7 Integrated mollusca control

Integrated pest management (IPM) is an economic necessity, and is vital for our modern agriculture, this approach requires a good understanding of all biological and ecological aspects of the mollusca in question (Snails or slugs). These methods include using all control procedures to suppress molusca populations to non-damaging levels as follows:

1. Plough of the soil before sowing seeds, results in exposure of the sensitive egg and Juvenile slugs and snails to direct sunlight and they die of desiccation.
2. Hand collection of the snails and slugs daily and to kill them with a strong solution of common salt or in boiling water (Mahrous et al., 2002)

3. Burning over is a quick method of clearing land of pests before sowing crops.
4. Beer - baited traps have been used to trap and down slugs and snails, and scrape off the accumulated snails and slugs daily and destroy them by crushing (Olhendorf, 1996)
5. Protective barriers of dehydrating substance, will keep snails and slugs out of planting beds.
6. The use of poisonous bait is the most common method of gastropod control.

9. References

Abd El-A11, S.M., Zedan, H.A., El-Deeb, H.I., & Soliman, A.M. (2002). Repellent and toxic effect of various indigeneous plant extracts and pesticides against *Monacha cantiana* land snails. *Journal of Agricultural Science Mansoura University* 27: 4167-4181.

Abd El-Wakeil, K. F. (2005): Ecotoxicolgical studies on terrestrial isopods (Crustacea) in Assiut, Egypt. *Ph. D. Thesis, Assiut University Egypt,* 271pages.

Allikas, G. (1997). Orchid Online. *AOS magazine orchids*, Austuralia, pp. 32-38.

Al-Sanabani, A. S. M. (2008): Biological studies on terrestrial slugs (Gastropoda) in Assiut, Egypt with special reference of their ecology. *Ph. D. Thesis, Fac. of Sci., Thamar Univ., Yemen* 174 pages.

Arafa, A.A.I. (1997). Studies on some land molluscs at Sharkia Governorate. *M.Sc. Thesis Faculty Agricultural Al-Azhar University* 137pages.

Azzam K.M. (1998). First record of the snail parasitic nematode *Rhabditis* sp. isolated from Egyptian terrestrial snails and its compability to infect other pests. *Egyptian Journal of Biological Pest control* 8: 27-29.

Azzam, K.M. (1999). Production of the snail parasitic nematode *Rhabiditis* sp. from different pests. *Proceedings of 1st regional symposium for Applied Biological Control in Mediterranean Countries Cairo*, Egypt, 25-29.

Azzam, K.M. & Belal, M.H. (2002). Molluscicidal effect of a bacterial exotoxin, victoback12 as *Bacillus thuringiensis*, and the snail parasitic nematode, *Rhabditis* sp. as biocontrol agents against *Eobania vermiculuta* (Müller) snail. *Proceedings of 1st Conference of the central Agricultural Pesticide Laboratory*,pp. 499-501.

Azzam, K. M. & Hegab, A.M. (2000). Potential of the snail parasitic nematode, *Rhabditis* sp. in controlling the snail *Eobania vermiculata* (Müller) and its effect on Albino rat, *Rattus norvegicus*. *Journal of Agricultural Science Mansoura Unversity* 25: 1825-1829.

Baker, G.H. (1988). Dispersal of *Theba pisana* (Mollusca: Helicidea). *Journal of Applied Ecology* 25: 889-900.

Baker, G.H. (1989). Damage, population dynamics, movement and control of pest helicid snails in Southern Australia. *BCPC Mono. No. 41* slugs and snails in world Agric.

Barker, G. M. (2001). The biology of terrestrial Molluscs. *CABI publishing, New Yourk, USA*: 558pp.

Baker, G.H. & Hawke, B.G. (1990). Life history and population dynamics of *Theba pisana* (mollusca: Helicidae), in a cereal pasture rotation. *Journal of Applied Ecology* 27: 16 - 29.

Baur, B. & Baur A. (1991). Effect of courtship and repeated copulation on egg production in the simultaneously hermaphroditic land snail *Arianta abustorum* (L.). *Invertebrate Reproduction and Development* 21: 201-206.

Belen, E.M (1982). Plant extracts possessing rodenticidal and molluscicidal activities. *Monitor* 10: 6-7

Bishara, S.I., Hassan, M.S. & Kalliny, A.S. (1968). Studies on some land snails injurious to agriculture in U.A.R. Revue de zoologie et de botanique africaines Lxx VII (3-4): 239-252.

Carman, G.E. (1965). Electrical trapping device for land snails. *Journal of Economic Entomology* 58: 786-787.

Connor VM. 1986. The use of mucous trails by intertidal limpets to enhance food resources. Biol Bull 171: 548 - 564.

Daoud, M. I. A. (2004). Ecological and morphological studies on some terrestrial snails in Dakahlia governorate. *M. Sc. Thesis Agriculture Faculty Al-Azhar University* 177pages.

Davies, M.S. & Hawkins S.J. (1998). Mucus from marine mollusks. Advances in Marine Biology 34: 101 - 112.

Davies MS, Jones HD, Hawkins SJ. 1990. Seasonal variation in the composition of pedal mucus from *Patella vulgate* L. *Journal* of Experimental Marine Biology and Ecology 144: 101 - 112.

Daxl, R. (1970). Der Einflub von Temperature and relativer Luftfeuchte die Molluskizide Wirkung die Metaldehyde. Isolan and Ioxynil auf *Limax flavus* und dessen Eier. *Zeitschrift Angewandte Entomologie* 67: 57-87.

Denny, M.W. (1983). Molecular biomechanics of mollucan mucous secretions. *In*: Wilbur, K., Simkiss, K., Hochka, P.W., (eds.) Mollusca, Vol. 1, New York: Academic Press. pp 341 - 465.

Denny, M.W. (1989). Invertebrate mucous secretions: functional alternatives to vertebrate paradigms. *In*: Chantler, E., Ratcliffe, N.A. (eds.) Mucus and Related Topics. Symposia of the Society for Experimental Biology 43: 337 - 366.

El-Deeb, H. I., Abdel-Halim, A., Koutb, I., Khidr, F. K. & Edress, N. M. (2004): Studying some ecological aspects associated with the prevalent land snails at Kafr El-Sheikh governorate. *Journal of Agricultural Science Mansoura University* 29: 2847-2853.

El-Deeb, H. I., Ewies, E. A., Kandil, M. A., Gabr, W. M. & Mobarak, S. A. (2003a). Toxicity and biochemical studies of methomyl and diazinon on different ages of the land gastropod species *Monacha obstructa*. *Journal of Agricultural Science Mansoura University* 28: 7011-7023.

El-Deeb, H. I., Zidan, Z. H., Fouad, M. M. & Asran, F. D. (2003b). Survey Of Terrestrial Snails And Their Malacophagous Insects At Three Governorates In Egypt. *Egyptian Journal of Applied Science* 18: 355-361.

El-Deeb, H.I., Foda, M.E., Zedan, H. & Edress, N.M. (2001). Bait preference and food consumption of some land snails in Egyptian new reclaimed lands. *Journal of Environment Science* 3: 90-104.

El-Deeb, H.I., Wilson, M. & Eshara, E.H. (1999). Ecological studies on certain land snails infest some Economic crops at Beheira Governorate, Egypt. *Proceedings of 2nd Intentional Conference Pest Control*, Mansoura, Egypt, 19-28.

El-Deeb, H.I; El-Ghamry, E.M; El-Hwashy, N. & Essa, N. (1996). Relative abundance of some land snails in certain governorate of Egypt. *Journal of Agricultural Science Mansoura University* 21(8): 2977-2983.

El-Hawashy, N.M., Zedan, H.A. & Abd-All, S.M. (1996). Toxicity effect of certain indigenous plant extracts on *Eobania vermiculata* land snails in Egypt. *Journal of Agricultural Science Mansoura University* 21: 4133-4138.

El-Massry, S.A.A. (1997). Studies on the control of some land snails infesting certain fruit trees. *Ph.D. Thesis, Agriculture Faculty Zagazig University* 123-131 pp.

El-Massry, S.A.A., Ghamry, El.M., Hegab, A.M. & Hassan, A.I. (1998). Effect of certain fertilizers against some snails infesting navel orange trees. *Egyptian Journal of Applied Science* 13: 654-662.

El-Okda, M.K. (1979). Land snails of economic importance at Alexandria region with some notes on the morphological features classification, economic damage & population on ornamental plants. *Agricultural Research Review* 57: 125-131.

El-Okda, M.K. (1980). Land snails of economic importance on vegetable at Alex&ria & neighboring region. *Agricultural Research Review* 58: 79-85.

El-Okda, M.M.K. (1981). Locomation activity and infestation abundance of certain terrestrial mollusca in fruit orchard, Alexandria province, ARE. *Proceedings of 4th Arab Pesticides Conference*, Tanta University, Egypt 2: 279 - 287.

El-Okda, M.K. (1984). Land mollusca infestation & chemical control in El-Ismaelia Governorate. *Agricultural Research Review*, Egypt 62: 87-92.

El-Okda, M.M.K., Emara, M.M., & Selim, A.M. (1989). The response of the harmful & useful terrestrial mollusca towards several toxicant. Efficacy of six toxicants under laboratory conditions. *Alexandria Science Exchage* 10: 375 -385.

El-Sebae, A.H., El-Okda, M.M.K. & Marei, A.S. (1982). Local formation of three carbamated oxims as terrestrial molluscicides. *Agricultural Research Review* 60: 85-91.

El-Zemity, S.R., Mohamed, Sh. A., Radwan, M.A. & Sherby, Sh.M. (2001). Molluscicidal efficacy & repellency of some naturally occurring monoterpenoids against the land snail, *Helix aspersa*, Muller (Mollusca: Pulmonata). *Annals Agricultural Science Ain Shams University* Cairo, 46: 339-349.

Fisher, T.W. & Orth, R.E. (1985). Biological control of snails: observations of the snails, *Rumina decollate* (L.), 1758 (Subulinidae) with particular reference to its effectiveness in the biological control of *Helix aspersa* (Muller), 1974 in Californai occasional paper/ *Department Entomology, California- Riverside University* (1): 111 p., USA.

Fouad, M.M., Ibrahim, I.K., Khidr, F.K. & El-Deeb, H.I. (2002). The predational potential of some vertebrates for three land snail species. *Proceedings of 2nd International Conference, Plant Protection Research Institute*, Cairo, Egypt, 85-87.

Fox, C.J. (1964). The effect of five herbicides on the numbers of certain invertebrate in grass land soil control. *Journal of plants Science* 44: 405-409.

Gabr, W. M., Youssef, A. S. & Khidr, F. K. (2006): Molluscicidal effect of certain compound against two land snail species, *Monacha obstructa* and *Eobania vermiculata* under laboratory and field conditions. *Egyptian Journal Agricultural Research* 4: 43-50.

Getzin, L.W. & Cole, S.G. (1964). Evaluation of potential molluscicides for slug control. *Washington Agricultural Experiment Washington State University Bulltein* 658: 1-9.

Ghamry, E.M., Amer, k.Y. & Wilson, B.M. (1994). Screening test of some insecticides against two land snails *Monacha cantiana* & *Eobania vermiculata* in Sharkia Governorate. *Zagazig Journal of Agricultural Research* 21: 1539-1545.

Godan, D. (1983). Pest slugs & snails -Biology & control. *Springer Verlag Berlin Heidelberg*, 73-74: 287-293.

Godan, D. (1965). Untersuchungen über die Molluskizide wirkung der Carbamate. I. Teil: Ihre Toxizitat auf Nacktschnecken. *Zeitschrift Pflanzenkrank- und Schutz*, 72: 398-410.

Grenon, J.F., Walker, G. (1980). Biomechanical and rheological properties of the pedal mucus of the limpet, *Patella vulgata* L. *Comparative Biochemistry and Physiology* B 66: 451 - 458.

Hashem, A. G., Nakhla, J.M., Tadros, A.W., & Korashy, M.A. (1993). Monitoring land snails on sweet orange trees in Behera governorate. *Egyptian Journal of Agricultural Research* 20: 699-707.

Hashem, A.G., Nakhla, J.M. & Tadros, A.W. (1992). The seasonal fluctuation in the land snails population on citrus trees in the northern reclaimed lands. *Al-Azhar Journal of Agricultural Research* Cairo, Egypt 16: 325-337.

Henderson, I.F. (1970). The Fumigant effect of metaldehyde on slugs. *Annals Applied of Biology* 56: 507-510.

Henderson, H. & Triebskorn, R. (2002). Chemical control of Terrestrial gastropods. CAB International, Molluscs as crop pests Barker, G.M. (ed.), 31 pages.

Herzberg, F. & Herzberg, A. (1962). Observations on reproduction in *Helix aspersa*. *American Midland Naturalist Journal*, 68: 297-306.

Imevbore, E.A. & Ajayi S.S. (1993). Food preference of the African giant snail, Archachatina *marginata* in captivity. African J. Ecol., 31 (3): 265-267. C.F. Review Agricultural Entomology, 1994, 82: 891pp.

Ismail, S.A.A. (1997). Ecology, biology & control of certain terrestrial snails infesting some vegetables & field crops in Sharkia Governorate. *Ph.D. Thesis, Agriculture Faculty, Zagazig University* 107-116 pp.

Ismail, Sh. A. A., El-Massry, S. A. A., Khattab, M. M. & Hassan, A. Sh. (2003). Daily activity and damage caused by *Eobania vermiculata* Müller (Gastropoda) in citrus orchards. *Egypttain Journal of Agricultural Research* 18: 1-6.

Ittah, Y. & Zisman, U. (1992). Evaluation of volatile allyl alcohol derivatives for control of snails on cut roses for export. *Pest Science* 35: 183-186.

Johri, R., Katiyar, A.k. & Johri, P.K. (1983). Indigenous plant as molluscicide. *Journal of Applied Zoological Research* 4: 48-50.

Joubert, C.J. & Walters, S.S. (1951). Control of snails & slugs. *Farming in South Africa*, 26: 379-380.

Kady, M.M., El-Sebae, A.H., Abd El-Ghany, A.A. & Zedan, H.A.(1986). Alkaloids with molluscicidal activity from the wild herb *peganum harmala* (L.). *Journal of Agricultural Science Mansoura University* 11: 1246-1251.

Kady, M.M., Ghanim, A.A., El-Adl,A.A. & Nassar, O.A. (1983). Seasonal abundance of the terrestrial snail *Monacha obstructa* & its chemical control. *Journal of Agricultural Science Mansoura University* 8: 377-383.

Kady, M.M., Hssan, R.A. & Said, A.A. (1982). Chemical & toxicological studies on saponins isolated from saponaria roots. *Journal of Agricultural Science Mansoura University* 7: 370pp.

Kassab, A. & Daoud, H. (1964). Notes on the biology & control of land snail of economic importance in the U.A.R. *Jouranl of Agricultural Research Review*, Cairo, 42: 66-98.

Kishor, N. & Sati, O.P. (1990). A new molluscicidal spirostanol glycocide of *Yacca aloifolia*. *Journal of Natural Products* 53: 1557-1559.

Knutson, L.V. (1966). Biology & immature stages of malacophagous flies: *Antichaeta analis, A. brevipennis, A. obliviosa* (Sciomyzidae) *Transactions of the American Entomological Society* 92: 67-191.

Lokma, M. H. E. (2007). Studies on some terrestrial gastropods inguious to field crops at Sharkia governorate. *M. Sc. Thesis Faculty of Agriculture Zagazig University*, Egypt 147pp.

Mahrous, M.E., Ibrahim, M.H. & Abd-El-Aal, E.M. (2002). Control of certain land snails under field conditions in Sharkia Governorate. *Egypttain Journal of Agricultural Research* 29: 1041-1045.

Mallet., C. & Bougaran, H. (1971). Action molluscicide de divers carbamates. *In: XXième Symposium International de Phytopharmacie et de Phytiatrie*, 12: 207-215.

Mead, A.R. (1961). The Giant African snail: A problem in economic malacology 1-257. *Chicago University press, Chicago, Illinosi, USA*. C.F.Godan, 1983.

Meikle P., Richards G.N. & Yellowless, D. (1988). Structural investigations on the mucus from six species of coral. Marine Biology 99:187–193.

Metwally, A.M., Zedan,H.A., El-Saeid, A.B. & El-Akra, T.M.M. (2002). Ecological studies on certain land snails in Monofia & Gharbia Governorate. *Proceedings of 2nd International Conference, Plant Protection Research Institute*, Cairo, Egypt: 65-79.

Moens, R. (1970). Techniques chimiques de destruction de gites a limnees dans les prairies. Colloque information Scient., L" Assainissement des Prairies.par les Triatements molluscicides" *Gembloux centre de Recherches agronomiques de l'Etat*, 232-55, Gembloux.

Mohamed, M.F.M. (1999). Ecological & biological studies on land snails & slugs in Egypt. *Ph.D. Thesis, Faculty Agriculture. Cairo, University* 166-170 pp.

Muthumani, K., Jeyakumar, E., Mani, A. & Mathavan, S. (1992). Biological control of veronicellid slug, *Lavicaullius alte* (stylomatophorn :veronicollidae) using frog as a predator. *Environment and Ecology* 10: 649-652.

Nakhla, J.M. (1995). A mechanical method for controlling land snails on pear trees in Egypt. *Journal of Agricultural Research* 73: 357-363.

Neff, S.E. (1964). Snail-Killing Sciomyzid flies: Application in biological control. *Verhandlungen des Internationalen Verein Limnologie* 15: 933-939.

Prakash, A. & Rao, J. (1997). Botanical pesticides in agriculture. *CRC press, Inc. Lewis publishers*, 345- 354 pp.

Radwan, M.A. & El-Zemity, S.R. (2001). Synthesis & molluscicidal structure- activity relationships of some novel 1,2,4-triazole N-methyl carbamates. *Pest Management Science* 57: 707-712.

Ramesh, B. G. (1983). Neem extract destroys snails. *Science Reporter* (India) Dec., (Issue):23

Ramzy, R.R. (2009). Biological and ecological studies on land snails at Assiut, Egypt. *M. Sc. Thesis, Faculty of Science, Assiut University*, Egypt, 164 pages.

Reynaud, P.A. (1986). Control of the azolla pest, *Lymnaea natalensis* with molluscicides of plant origin. *International Rice News* 11: 27-28.

Richardson, H.H. & Roth, H. (1963). Ethylene oxide Fumigants to eliminate Quarantinable snails Cochlicelle or Theba in Cargo. *Journal of Economic Entomology*, 56: 836-839.

Sallam A.A.A., El-Massry S.A. & Nasr I.N. (2009). Chemical analysis of mucus from certain land snails under Egyptian conditions. *Archives of Phytopathology and Plant Protection* 42: 874–881.

Sasmal, S. (1991). Bionomics & control of pests of *Azolla pinnata* R.Br. *Ph.D.Thesis Utkal University Bhubaneswar, India* p168. C.F. prakash & Rao, 1997.

Schmidt-Nielsen K. (1990). Animal physiology: adaptation and environment. *Cambridge University Press* p 607.

Shah, S. (1992). Management of the African snail. *Indian, farming*, 41 (11): 21 pp, C.F. Rev. Agric. Entomol., 81 (7): 744 pp.

Shahawy, W. A., Hendawy, A. S., Abada, A. E. & Kassem, A. A. (2008). Land snails infesting rice plants and their accompanied parasitoids and predators at Kafr El-Sheikh governorate, Egypt. *Egyptian Journal of Agricultural Research* 86: 971-980.

Shetlar, D.J. (1995). Slugs & their management, Ohio State Univ. Ext. Factsheet HYG-2010-95.

Shoieb, M.A. (2008). Occurrence and distribution of terrestrial mollusks in Suez canal Governorates and south of Sinai, *Egyptian Journal of Agricultural Research* 86: 989-994.

Smith, A.M. & Morin, M.C. (2002). Biochemical differences between trail mucus and adhesive mucus from marsh periwinkle snails. *Biological Bulletin* 203: 338 - 346.

South, A. (1992). Terrestrial slugs. Biology, & control. Dept. of Biological Sciences, City of London polytechnic Chapman & Hall London New York, *Tokoy of Melbourne, Marras*, C.F. Zedan, *et al.* (2002)

Staikou, A. & Lazaridou- Dimitriadou, M. (1990). Aspects of the life cycle, population dynamics, growth & secondary production of the snail *Monacha cartusiana* (Muller) (Gastropoda: pulmonata) in Greece. *Malacologia* 31: 353-362.

Staikou, A., Lazaridou-Dimitriadou, M. & Farmakis, N. (1988). Aspect of the life cycle, population dynamics, growth & secondary production of the edible snail *Helix Lucorum*, (L.) (Gastropoda, pulmonta) in Greeca. *Journal of molluscan studies* 54: 139-155.

Tadros, A.W., Nakhla, J.M. & Rizk, M.A. (2001). The relative susceptibility of fruit tree species to land snail infestation as indicated by cells in their excrement. *Egyptian Journal of Agricultural Research* 79: 57-68.

Van der Gulik, J. & Springette, J.A. (1980). The effect of commonly used biocides on slugs. *Proceedings of 3rd N.Z. weed & pest control Conference*, pp. 225-229.

Verbeke, J. (1964). Contribution a etude des dipteres malacophages. II. Donnecs nouvelles sur la taxonomic et la repartition geographique des sciomyzidae palearctiques. *Bulletin de l'Institut Royal des Sciences Naturelles de Belgique Entomologie* 36: 1-15. C.F. Godan, 1983

Wilson, M.J., Hughes, L.A., Hamacher, G.M., Barahona, L.D. & Glen, D.M. (1996). Effect of soil incorporation on the efficacy of the Rhabditia nematode, *Phasmarhabditis hermaphrodita*, as a biological control agent for slugs. *Annals of Applied Biology* 128: 117-126.

Wouters, L. (1970). Schneckenbekampfungi in Oost-Flevoland. *Pflanzenschutz Nachrichten*, Bayer, 23, 173-177. C.F. Godan, 1983.

Zeidan, H.A. (2001). Biological control of land snails. *Proccedings of 3rd scientific conference on role of natural enemies in the biological control of agricultural pests*. Faculty of Agriculture, Mansoura University, Egypt pp. 46-60.

19

Bacillus thuringiensis – Based Biopesticides Against Agricultural Pests in Latin America

Ricardo Antonio Polanczyk, Sergio Antonio De Bortoli
and Caroline Placidi De Bortoli
Universidade Estadual Paulista (UNESP)
Brazil

1. Introduction

Twenty years after its discovery, the first attempts to use the entomopathogenic bacterium *Bacillus thuringiensis* (*Bt*) to control pests were made in Europe. Due to the successes achieved in some laboratory trials, commercial production of this pathogen began in France, where commercial formulations entered the market in 1938 as the product Sporeine. In the U.S., the use of this pathogen increased after 1950, mainly for the control of lepidopteran pests, resulting in the production of a formulation called Thuricide (Beegle & Yamamoto, 1992; Peferoen & Lambert, 1992; Lord, 2005).

Bt-based biopesticides have stimulated pathology and microbial control studies, and *Bt* is now one of the main pathogens used in agricultural pest control (Lord, 2005; Brar et al., 2006; Bravo et al., 2011). More than 200 *Bt*-based biopesticides account for 53% of the worldwide market for biopesticides, generating annual revenues of 120–140 million dollars (CAB, 2010). The Americas are responsible for almost 50% of this consumption, particularly the US and Canada, with Latin America accounting for only 8–10% of the total consumption (Guerra et al., 2001).

It is noteworthy that the public's concern of pesticide residues in food and the effect of pesticides on the environment has encouraged the use of microbial products in the production of vegetables and fruits of high commercial value in particular. The production of Dipel (*Bt* kurstaki-based biopesticide) began in 1970. This product proved to be 20–200-fold more potent than other *Bt*-based biopesticides (Beegle & Yamamoto, 1992). Dipel is currently used to control more than 167 lepidopteran pests (Glare & O'Callagham, 2000).

The insecticidal action of *Bt* is mainly due to crystalline inclusions (crystals or "Cry") that contain proteins synthesized during the second phase of the growth cycle, coinciding with the formation of spores (Bravo et al., 2001; Copping & Menn, 2000). These crystals become protoxins when dissolved in alkaline medium in the digestive system of insects. In the presence of digestive enzymes, protoxines are converted into 4 or more toxic polypeptides (δ endotoxins). The hydrolyzed Cry toxins cross the peritrophic membrane and bind to specific receptors present in the apical microvilli membrane of the midgut columnar cells, forming pores that increase membrane permeability, thereby affecting ion gradients and osmotic balance in the apical membrane. The increase in water absorption leads to the lysis and eventual rupture and disintegration of the midgut cells, which represses feeding and results in death due to starvation. The insect may also die due to an environment favorable for spore germination after cell lysis in the midgut. In this new environment, because of the mixing of the hemolymph and midgut contents, the pH becomes acidic, spores multiply, and eventually septicemia and insect death occur (Hofte & Whiteley, 1989; Knowles, 1994; Copping & Menn, 2000).

Although the detection of parasporal inclusions (Cry) in *Bt* cells and speculation about their importance in pathogenicity date back to the 1950s, their formulations were based on the number of spores, with inconsistent results, until the 1970s. However, since then, new formulations of *Bt* have considered the presence of δ-endotoxins. This fact, together with the potential of rapid growth and sporulation in relatively cheap substrates, increased the potential of *Bt* to control insects and enhanced market success.

The purpose of this chapter is to report the use of *Bt* in biological control programs in several countries of Latin America as well as to discuss the ongoing research to emphasize the potential efficacy of these pathogens in pest management. Other issues such as characterization, toxins, isolation, production, formulation effects on natural enemies, plants expressing *Bt* genes, and the evolution of resistance are discussed using previous publications (Brar et al., 2006; Estruch et al., 1997; Glare & O'Callagham, 2000; Charles et al., 2000; Valicenti & Zanasi, 2005).

2. Argentina

Despite the restricted use of *Bt* to control pests in this country, a number of projects have been developed by research institutes (Botto, 1996). *Bt*-based biopesticides were first used in the early 1950s. The target insect was *Colias lesbia pyrrothea* (alfalfa caterpillar/clouded yellow), and the use of biopesticides increased crop yields by approximately 72%. However, efforts aimed at increasing the use of this pathogen were not resumed until 1984 (Sosa-Gómez & Moscardi, 1991).

The Centro de Investigaciones Biologicas, belonging to the Universidad Nacional de Mar del Plata, performs the isolation, selection, morphological and biochemical characterization, and genetic toxicologic investigation of new *Bt* strains. Studies conducted at this center are aimed at developing and characterizing the efficacy of *Bt*-based agents against *Anticarsia gemmatalis, Spodoptera frugiperda* (Lepidoptera), *Anthonomus grandis, Diabrotica speciosa,* and *Tenebrio molitor* (Coleoptera) (Beron et al., 1999; Beron & Salerno, 2000).

Currently, imported products based on *Bt* are used in Argentina to control *Rachiplusia nu* and *A. gemmatalis* (Bac-Tur, Dipel, and Gale BT-PM). However, their use is not extensive since the market is dominated by chemical pesticides, particularly pyrethroids, and research of these products is still hampered by the severe economic crisis faced in recent years. In view of this, in the short term, it is unlikely that the use of entomopathogens, including *Bt,* will increase. The lack of policies at the national level and the isolated or discontinuous nature of the developed projects are factors that restrict the advancement of biological control in this country (Botto, 1996).

3. Brazil

Early studies in Brazil regarding the use of *Bt*-based products were conducted by Figueiredo et al. (1960) and Pigatti et al. (1960). These authors highlighted the high potential of this pathogen to control many pests such as *Ascia monuste orseis*, *Sylepta silicalis*, *Dirphia sabina*, *Azochis gripusalis*, *Alabama argillacea*, *Mocis repanda*, *Xanthopastis timais*, *Musca domestica*, and *Erinnyis ello*. The first results in relation to the use of *Bt*-based products were considered promising and spurred further research on *Bt*. The first project that aimed to control an important agricultural pest in Brazil, the fall armyworm (*S. frugiperda*), with *Bt* was started in 1993 by Embrapa Milho and Sorgo (CNPMS in Sete Lagoas - MG).

Until the early 1990s, only 3 commercial products were available in the Brazilian market, all based on *Bt kurstaki* (Dipel, Thuricide, and Bactospeine) (Habib & Andrade, 1991). The introduction of *Bt*-based products has encountered problems related to use strategies, marketing, as well as negative opinions advertised by the manufacturers and retailers of chemical insecticides, which diffused the idea that the insects should be rapidly controlled (Alves et al., 1998). Despite these issues, the number of *Bt*-based products on the market has recently increased.

This increase occurred mainly due to changes in the marketing strategies of the companies that sell these products, which emphasized some advantages, such as the maintenance of populations of parasitoids, predators, and pollinators (Alves et al., 1998). However, despite the environmental and social advantages of microbial products, the area treated with *Bt*-based biopesticides is approximately 150,000 hectares (Souza, 2001). This low use is due to a number of factors such as competition with chemicals (higher cost), specificity (reduced action spectrum), and low persistence of most commercially available formulations in the field (inactivation by ultraviolet radiation). New products based on *Bt* should be available in the Brazilian market in the coming years, increasing the availability of biopesticides. The cost of treating 1 hectare ranges from US$7.5 to US$15.0, depending of the crop (Alves et al., 1998).

The incidence of *Ecdytolopha aurantiana* (citrus borer) parasitism has risen sharply in citrus orchards in Brazil since the late 1990s (Gallo et al., 2002). This is mainly due to the incidence of variegated chlorosis, the insect vector, and its chemical control with pyrethroid insecticides through the fog application of non-selective insecticides, which in most cases have high shock power or contact, causing high mortality among natural enemies. The citrus borer limits production in many citrus growing regions of São Paulo State, causing losses of up to 60%, which translates to losses of up to $50 million per year; moreover, citrus borer infestation creates export barriers for some countries. *Bt*-based biopesticides effectively control this pest (Alves et al., 2001b) when the first application is made with more than 6 females caught per trap; the second application occurs 20–30 days later. The bacterium has been used in over 50,000 hectares of citrus orchards in Brazil, mainly in São Paulo State. Among the most important restrictions in the use of this method is the difficulty in synchronizing the short period of larvae exposure (just 4 h on the surface before it penetrates the fruit) and product application and its low persistence in the field due to the action of ultraviolet radiation.

The persistence of the product in the field can be increased by adding vegetable or mineral oil (Gallo et al., 2002) or using formulations containing microencapsulated factors that protect against ultraviolet rays, as reported by Dunkle & Shasha (1989). These authors reported that spores and crystals microencapsulated with Congo red or folic acid exhibited persistent viability of at least 50% of its toxic activity for 12 days. To reduce the impact of this pest on citrus, the Group of Advisers on Citrus recommends, among others measures, the use of products based on *Bt* plus a half dose of insecticide (pyrethroid).

Lepidopteran larvae that cause major defoliation in soybean include *A. gemmatalis* (velvetbean caterpillar), which consumes approximately 90 cm^2 of leaves over the course of its development (Gallo et al., 2002). Despite the high efficiency of *Bt* against this pest (Habib & Amaral, 1985; Bobrovski et al., 2002), its use is restricted mainly due to the widespread use of *Baculovirus anticarsia* (nuclear polyhedrosis virus). This biopesticide has a lower cost, ranging from US$1.3 to US$2.0 per hectare (Alves et al., 1998) and high efficiency, controlling over 80% of the pest population (Moscardi, 1998). Its fast growth can be attributed to the simple strategies used, including the amenability of its multiplication in the field by producers themselves, and the private sector's interest in producing and marketing the product (Moscardi, 1998).

The "Associação Riograndense de Empreendimentos de Assistência Técnica e Extensão Rural" (EMATER-RS), the rural extension agency of the Rio Grande do Sul State, one of the main producers of soybean, led a program aimed at the biological control of *A. gemmatalis* that sprayed approximately 13,000 hectares with Dipel in 2001 and 2002. However, the total area treated with this biopesticide is possibly higher because the data refer only to cities where EMATER-RS works, and do not represent the entire state.

In Brazil, a major pest of pastures is *Mocis latipes* (striped grass looper). This pest occurs cyclically (of some years); when in large numbers, it can significantly reduce the amount of forage available by completely consuming plant leaves. *Bt*-based products are used against small caterpillars in initial outbreaks at concentrations of 0.6–1 kg/ha. An important aspect is that these products are selective, being harmless to humans, natural enemies, and other animals (Gallo et al., 2002), thus indicating its usefulness in pastures and forage crops.

An economic viability study of *Thyrinteina arnobia*, the most important pest of *Eucalyptus*, was performed using *Bt* in approximately 4,000 hectares of *Eucalyptus grandis* and *Eucalyptus urophylla* (White, 1995). The cost of control was US$12.50/ha (equivalent to 1% of the timber produced), and this remained uniform even after pest infestation. A return on investment occurred in the first year after the control was instituted, because the measures adopted led to an income of US$34.20/ha/year. By contrast, a lack of control measures can result in losses of US$460.70–878.91/ha, or 37–70 times the cost of control, depending of the infestation level. Although studies comparing *Bt*-based products with chemicals have not been conducted for this pest, the economic feasibility of using *Bt* to control *T. arnobia* is obvious.

The pickleworm moths *Diaphania nitidalis* and *Diaphania hyalinata* are pests of cultivated cucurbits, and their larvae feed on all plant parts, preferring the fruit . The new leaves and branches become dry after being attacked. In fruits, the larvae open galleries and destroy the pulp, causing decay. *D. nitidalis* attacks fruits of any age, whereas *D. hyalinata* attacks the leaves and fruit skin. When using Dipel to control *D. nitidalis*, it is essential to determine the age of larvae because Dipel must be applied to early-stage larvae, i.e., before the larvae enter the fruit, as it is ineffective once larvae enter the fruit.

Other pests controlled with *Bt*-based produced are the imperial moth (*Eacles imperialis magnifica*) and cassava hornworm (*E. ello*), which can completely defoliate coffee and cassava plants, respectively. For the tomato pinkworm (*Tuta absoluta*), applications should be made weekly, in the evening or at night, simultaneously with the use of the egg parasitoid *Trichogramma pretiosum*. This pest can attack all parts of the tomato plant at any stage of development, consuming the leaves, branches, and mainly the apical buds and destroying new shoots, resulting in depreciation of the commercial value of the fruit (Gallo et al., 2002).

The fall armyworm in Brazil can reduce corn yield by 20%, and under dry conditions, it cuts off the plants close the ground, causing serious crop damage (Gallo et al., 2002). Although Thuricide is cited as a product registered for *S. frugiperda* control (Compêndio de Defensivos Agrícolas, 2010), its use is restricted due to low efficacy in most regions. Preliminary results from laboratory experiments conducted in the Laboratory of Pathology and Microbial Control of Insects, Piracicaba, SP, Brazil (ESALQ-USP) indicate that populations of fall armyworm from different geographic regions have different levels of susceptibility to the same strain of *Bt* (Polanczyk et al., 2005). These initial results confirm the need for comprehensive monitoring to verify variation among populations, which would increase the effectiveness of this strategy.

Bt-based biopesticides can be used in combination with other substances such as molasses and corn bran (Gravena et al., 1980a) to reduce *Diatraea saccharalis* (sugarcane borer) infestation in sugarcane crops by more than 50%. Parasitoids such as *Trichogramma pretiosum* enhance the effectiveness of *Bt kurstaki* in the control of *T. absoluta* (Marques, 1993). Similarly, pesticides, including insecticides (Assunção et al., 1980; Gravena et al., 1980b; Gravena et al., 1983; Zanuncio et al., 1992), have the same effect on biopesticides in controlling certain pests.

Published research by EMBRAPA indicates the importance of this *Bt* in agricultural pest control (Batista et al., 2005; Praça et al., 2004). In addition to EMBRAPA reports, several other federal and state institutions have performed research related to *Bt*. Among them, the Laboratory of Microbiology and Genetic of the UNISINOS University investigated the selection of *Bt* strains against coleopteran and lepidopteran pests (Fiuza et al., 2002, Pinto & Fiuza, 2002; Schünemann et al., 2005). ESALQ-USP studied the production, formulation, and selection of *Bt* strains for insect pests such as *T. absoluta*, *S. frugiperda*, and *Sitophilus oryzae* (Marques & Alves, 1996; Alves et al., 1997; Alves et al., 1999; Alves et al., 2001; Giustolin et al., 2001; Polanczyk et al., 2005). The sublethal effects of this pathogen on target insects and its interaction with other biological control agents have been studied (Polanczyk & Ahmed, 2005a, b). In the Laboratory of Entomology and Plant Pathology, projects are underway to select *Bt* strains with efficacy against lepidopteran pests and investigate their effects on natural enemies (Grecco et al., 2006; Pratissoli et al., 2006).

However, according to IBGE (2006), the 19 major crops of economic importance (cotton, peanuts, rice, oats, potatoes, cocoa, coffee, sugarcane, onions, barley, beans, oranges, castor oil plant, cassava, corn, soybeans, sorghum, wheat, and triticale) occupy an area of 146.6 million hectares. However, only 150,000 hectares are treated with *Bt*-based biopesticides for pest control, although their potential use may be extended up to 6 million hectares.

4. Colombia

Despite the prevalence of insect parasitoids in biological control programs in Colombia, some *Bt*-based products, albeit on a small scale, are used in pest control (Botto, 1996). However, a few decades ago, Revelo (1965) emphasized the potential of this pathogen in the control of *S. frugiperda, Agrotis ipsilon,* and *D. saccharalis,* the 3 major pests of maize in Columbia. In 1990, US$5.8 million was spent to control *S. frugiperda* in cotton crops, and US$4.2 million was spent to control the pest in corn and sorghum. In area, these amounts represent 430,000 and 440,000 treated hectares, respectively (Bosa & Cotes, 1997).

Important pests of many vegetable crops such as *Leptophobia aripa, Ascia monuste,* and *Plutella xylostella* have been well controlled by *Bt*-based biopesticides; thus, the damage caused by these insects is economically insignificant (Ruiz, 1998). However, Ruiz also states that in most cases, control measures involving the use of biological control must be accompanied by other tactics to maintain the pest population level below the economic injury level.

In 1989, the Laverlam Company began research of biopesticides to protect the environment and integrate tactics of integrated pest management in Colombia. These efforts resulted in the production of Turilav (*Bt kurstaki*), which is used to control *Heliothis* spp., *A. argillacea, Agrotis* spp., and *Spodoptera* spp., particularly in cotton crops.

In Valle del Cauca, the use of *Bt* to control *T. absoluta* is part of an integrated pest management program that has successfully reduced the costs of control by more than 54% to approximately US$650 per hectare (Belotti et al., 1990; Garcia, 1992). In the same region, *Bt*-based biopesticides are used against *Caligo ilioneus,* an important sugarcane pest that reduces the weight of the plant by 26–56% and its sugar content by 8–18%. In cotton crops, *Bt* is applied to control occasional outbreaks of *A. argillaceae* when defoliation exceeds 30%.

Sánchez et al. (1999) verified the high efficacy of *Bt kurstaki* and *Bt aizawai* to control *D. hyalinata* and *D. nitidalis.* These pests are the most destructive parasites of melon crops, in which they defoliate the crop and feed on the branches and fruits, causing annual losses of up to 25% of the yield. The use of *Bt* control has reduced the amount of insecticides used, which were sprayed up to 24 times in just 1 season in some cases.

Bosa & Cotes (1997) reported the high efficacy of 2 *Bt* strains (*Bt*-127 and HD-137) against *S. frugiperda.* Arango et al. (2001) used 1,100 native *Bt* isolates to study their selectivity to control the fall armyworm. Among the tested material, 32 isolates exhibited activity against this pest. The most powerful strains expressed *cry1Aa, cry1Ab, cry1Ac, cry1B, cry1C,* and *cry1D.* Serotyping revealed the isolates to be *Bt kurstaki, Bt thuringiensis, Bt canadensis,* and *Bt indiana.*

Maduell et al. (2002) isolated *Bt* from 13 species of the *Piper* genus in Colombian forests, detecting *Bt* in 74% of the samples. Regarding the presence of *cry* genes, *cry1* was amplified from 70% of the material (usually toxic to lepidopterans), and 60% exhibited some toxicity (12–100% mortality) to *S. frugiperda* larvae. In similar research, Yaro et al. (1999) isolated and characterized this bacterium in different parts of the country.

Recent studies by several research institutes revealed that 3 native isolates (IBUN6.4, IBUN3.3, and IBUN2.6) exhibit promising efficacy against *Heliothis virescens,* and 2 others (IBUN4.0 and IBUN3.8) were selected for *S. frugiperda.* For *H. virescens,* Cry 1F is the most effective toxin, and Cry1B and Cry1E are effective against *S. frugiperda.*

δ-Endotoxin expression is directly associated with sigma factors responsible for different sporulation stages. Studies with mutagenic agents found a direct relationship between the mutation of σ^E and σ^K factors with changes in *cry1Aa*, *cry1Ba*, *cry2Aa*, and *cyt1Aa* expression and changes in the specificity of insecticidal toxins. Research undertaken by the IWC sought to develop asporogenic *Bt* strains via chemical agent-induced mutagenesis. A strain was obtained that is currently being characterized in bioassays against *S. frugiperda* to determine the effect of the mutations on the integrity of genes, proteins, and toxin efficiency. The great interest in asporogenic *Bt* strains is justified even though the generation of these strains is difficult. However, because they are mutants produced in the laboratory, these strains should be utilized only after careful assessment of their environmental impact.

Corporación Nacional para Investigaciones Biologicas, Corporación Colombiana de Investigación Agropecuária, and Microbiological Research Center da Universidad de los Andes e Biotechnology Institute da Universidad Nacional de Colombia have conducted *Bt* studies with the following objectives: pathogen isolation, δ-endotoxin toxic activity determination against lepidopterans, toxic protein characterization, and DNA segment amplification of *cry* genes.

5. Cuba

Since 1990, due to the drastic reduction in the assistance provided by the former Soviet Union, Cuba has instituted necessary reforms to meet the basic needs of its citizens. The country made several transformations in many areas, including agriculture, leading to the development of more sustainable technologies in integrated pest management programs (Rosset & Moore, 1999). In the 1960s, the importation of some products based on *Bt* to control *H. virescens* and *M. latipes* stimulated the search for native pathogen strains (Pérez & Vasquéz, 2001). The use of chemical insecticides, which were mainly provided by the former Soviet Union, prevailed in Cuban agriculture, with methyl parathion being widely used in pest control before 1990.

The Ministry of Agriculture of Cuba has accelerated and significantly expanded plans to increase the production biological control agents to replace imported chemical insecticides. Since 1990, there has been an 89% reduction in the import of chemical insecticides and fertilizers. At the end of 1991, approximately 56% of insecticides used in Cuban agriculture were organic, representing a savings of US$15.6 million per year. In 1994, with 222 labs, the Centros de Producción de Entomófagos y Entomopatógenos (CREE) began operations, producing insects, nematodes, and pathogens in 15 provinces of Cuba. Currently, CREE operates 280 labs (53 in cultivated sugarcane areas and 227 in fruit areas and other crops), and *Bt* is primarily used (Pérez & Vasquéz, 2001).

Due to the prevalence of these bacteria-based products in Cuba, researchers seek to discover new native strains with high potential for pest control. Several institutions such as Instituto de Investigaciones de Sanidad Vegetal, Instituto de Investigaciones Fundamentales de la Agricultura Tropical, Centro National de Sanidad Agropecuária, Instituto de Investigaciones de Cítricos y Frutales, Instituto Cubano de Investigaciones Derivados de la Caña de Azúcar, and Instituto de Ecología y Systemática have contributed to the development of biopesticides in several areas.

The first biopesticides produced in Cuba were based on *Bt*, and they have been used to control *H. virescens*, *M. latipes*, *P. xylostella*, *S. frugiperda*, *E. ello*, and *D. hyalinata* (Table 3).

Among the registered products, those belonging to the line Thurisav are used to control *M. latipes*, *P. xylostella*, *Heliothis* spp., *Trichoplusia ni*, *Spodoptera* spp., and mites.

M. latipes is one of the most important agricultural pests in Cuba, causing losses of approximately 86,000 tons per year (equivalent to 7,800 hectares). Pérez et al. (1991) tested several products based on different subspecies of *Bt* (*Bt thuringiensis*, *Bt kurstaki*, and *Bt dendrolimus*) against this pest in pastures, and *Bt dendrolimus* was the most effective, with mortality rates of 86–90%.

According to Blanco (2006), *Brassica oleracea capitata* is one of the most important vegetable crops in Cuban agriculture, and among the factors limiting its production is *P. xylostella* (diamondback moth), which reduces its yields by 75–95%. To avoid these losses, many governmental and nongovernmental organizations have worked together to develop control strategies focusing the context of integrated pest management (IPM). The tactics employed should include *Bt*-based biopesticides to manage this pest and others such as *Ascia monuste eubotea* and *T. ni* to significantly reduce the control costs. Due to the success of these products in Cuba, the area treated with *Bt* to control *P. xylostella* increased from 17,400 hectares in 1988–89 to 53,000 hectares in 1993–94. Initial entomopathogen studies to control pest mites (*Phyllocoptruta oleivora*, *Tetranychus tumidus*, *Polyphagotarsonemus latus*) began in 1980. For products based on *Bt*, the mortality varies from 70% to 100% depending on the species, preventing the unnecessary use of chemical insecticides (Pérez, 1996). Between 1994 and 1996, almost 1,000 hectares were sprayed with *Bt* (LBT-13) to control *P. latus* (broad mite) in potato crops, with an efficacy exceeding 85% at a dose of 3–5 L/ha. The same strain caused between 83% and 90% mortality in adults and 100% in nymphs of *P. oleivora* (citrus rust mite) in citrus crops 72 h after application (20 L of fermented solution/ha). Moreover, it is selective to natural enemies such as predatory *Cycloneda sanguinea* in all its stages of development (Rojas, 2006).

The LBT-24 strain is one of the most effective biological agents for *S. frugiperda* control, and the mortality observed in the field ranged from 70 to 90% (Montesbravo, 2006), resulting in an increase of 15% in crop yield. The use of *Bt* formulations to control *S. frugiperda* in Cuba is very important because this pest can reduce corn yield by up to 40%, and important results have been obtained using this tactic together with others, such as the nuclear polyhedrosis virus and parasitoids (*Telenomus* sp. and *Chelonus insularis*).

The efficacy of LBT and LBT-3-21 *Bt kurstaki* isolates has exceeded 84% in potato, tomato, and tobacco crops. In Cuba, more than 20 species of insects are potato pests, and some of them such as *P. latus*, *Spodoptera* spp. and *T. ni* are controlled by *Bt*. In potato and tomato, the use of *Bt* favors the preservation of the natural enemies of *Liriomyza trifolii*. The utilization of this pathogen allows the action of *Opius* spp. and *Heteroschema* sp., which induce greater than 70% mortality (Morales, 2006).

The lepidopteran *H. virescens* is the most important pest of cotton in Cuba, and the efficacy of *Bt*-based products is similar to that of chemicals (14% for chemicals vs. 15.13% for biopesticides) (Jiménez, 1996).

Forestry is an important component of the Cuban economy, and 18 species of harmful insects greatly limit production. The most important pest of pine forests is *Rhyacionia frustrana*, the larvae of which invade the sprouts and buds of pine. *Bt*-based insecticides are widely used to control this pest with satisfactory results (Vásquez et al., 1999).

6. México

The use of entomopathogenic agents to control pests in Mexico began in the 1950s. In 1968-69, tests demonstrated the effectiveness of products formulated with *Bt* (Thuricide 90T and Thuricide 90TS) in controlling *P. xylostella* (Carrillo, 1971). However, the use of these insecticides increased sharply after 1990 due to the great interest of the public and private institutions in the use of pathogens in pest control.

In 1999, the use of *Bt*-based insecticides in Mexico increased from 15% to 20% and became the most commonly used and accepted biopesticide in that country. It was employed in 100,000 hectares of corn, 174,000 acres of cotton, and 200,000 hectares of vegetables and other crops. In the Bajio region of Guanajuato, 100 tons of *Bt*-based biopesticides were applied in 2001. In the Aguascalientes State, products made from *Bt kurstaki* and *Bt aizawai* are used against pests of vegetables, spinach, and potatoes, treating approximately 1,000 hectares (Vallés, 1998). In the entire country, it was estimated that 4–10% of the insecticides used contain *Bt* bacterium as the active ingredient. The leading *Bt*-producing companies are multinational, and the cost of control is US$19/ha, which is competitive with the chemicals in the market (Guerra et al., 2001).

Among the pests reported in Table 5, some of them are very important in several crops. For example, *S. frugiperda* can cause losses of up to 58% in plants and reduce corn yield by 1,148 kg/ha at a plant density of 45,000/ha (Castro-Franco et al., 1995). This species, together with *Helicoverpa zea, Diatraea grandiosella,* and *D. saccharalis,* may be responsible for losses up to 30% in Mexican corn crops.

In addition to the aforementioned pests, Rodríguez et al. (1991) reported the effectiveness of *Bt* (DL_{100} 75 mg/kg diet) against *Galleria mellonella,* an apiculture pest that causes losses of honey, pollen, and wax production of up to 10%. In addition, Rodriguez & Trumble (1993) reported that this pathogen was useful in controlling tomato pests in the IPM context without harmful effect on natural enemies such as *Trichogramma* spp.

Edwards et al. (1999) obtained different CL50 estimates of the same isolate for several S. frugiperda populations collected in 5 regions of Mexico. The authors emphasized that variation may be due to geographical isolation, which results in reproductive isolation and physiologically different populations showing differential susceptibility to Bt.

The Centro de Investigación y de Estudios Avanzados, the International Maize and Wheat Improvement Center, and the Institute of Biotechnology, National Autonomous University of Mexico (UNAM) work together to locate new *Bt* toxins to control corn pests. Other research obtained promising results against *D. grandiosella, D. saccharalis, S. frugiperda, H. zea,* and *T. ni* (Bohorova et al. 1996; Bohorova et al. 1997; Del Rincón-Castro et al., 2006; Magdalena et al., 2001).

Researchers at the University of Guanajuato and Centro de Investigación de Estudios Avanzados del IPN investigated the characterization and selection of *Bt* against *Manduca sexta* and *T. ni* (Rosas et al., 1994; Corona et al., 1998; Rosales-Reyes et al., 2003).

At the Institute of Biotechnology UNAM, research has being conducted on the interaction between receptors and Cry1Ab toxins (lepidoptera-specific) to identify the regions responsible for toxicity. At the same institute, research for new insecticidal proteins that can potentially control insect pests has been performed using *Bt* (Bravo et al., 1998, Guerra et al., 2007). This project constructed a database composed of 500 *Bt* strains isolated from Mexican soil. Some of these were effective in controlling pests such as *Epilachna varivestis, Tapinoma melanocephalum, Phyllophaga* spp., *Anomala* spp., and *Bemisia tabaci.*

Another approach taken by researchers is to study the interaction of the toxins Cry1C and Cry1D with the intestinal membranes of insects such as *S. frugiperda, Rhopalosiphum maidis,* and *D. grandiosella.* More recently, work has been conducted to analyze pore formation in the insect midgut apical membrane and the subsequent effects on *Manduca sexta* gut cells.

7. Concluding remarks

The entomopathogenic bacterium *B. thuringiensis* is used against a wide range of pests in several crops in Latin America. Its widespread use against *P. xylostella* is noteworthy, even in countries not covered in this paper due to a lack of accurate information, as well as the use of *Bt*-based products produced in Cuba by small labs funded by the federal government.

Regional production is restricted to Cuba, and more recently, to Mexico, whereas imported products are used in other countries, which directly increases the cost of pest control and indirectly increases the final cost of production. The ability to formulate these products on a commercial scale should be considered by the governments of Latin American countries and/or private industries. The decrease in cost to the farmer would boost the use of these microbial products and increase the producers' profit, as the "products of biological origin," or those grown without pesticides, have better acceptance and price on the market in several countries.

The use of *Bt*-based products in Latin American countries depends more on political and economic aspects than technological approaches. Considering the basic economic differences in Brazil and Cuba, it would be logical to assume that the use of microbial products in Brazil, including those formulated with *Bt*, would be much greater than that in Cuba. However, the Cubans, under U.S. blockade, produced technology and developed strategies to use entomopathogens that exemplify how it is possible to make microbial control one of the pillars of sustainable agriculture. Brazil created a program that represents the most successful microbial control in the world—the control of *A. gemmatalis* with nuclear polyhedrosis virus. Brazil also undertook programs using bioinsecticides based on *Bt* frequent to control dipteran vectors. However, in Brazil, the availability of large numbers of pesticide formulations, supported by an intensive propaganda financed by multinational companies, together with the lack of a structure to implement advancements of alternative methods of control in production, has limited the use of entomopathogens. The solution to this impasse involves greater investment in technical assistance and research, principally in the production and formulation of general entomopathogens, aiming to convince the farmer that the biological method may in many cases be an important tactic of control that can minimize or even replace some pesticides to control certain pests.

Apparently, the use of plants expressing *Bt* genes has a negative impact on the use of products formulated with this bacterium, but this is negligible and unlikely to affect the national economy significantly. These crops, in addition to performing a vital role in economic development, have great potential to increase the use of bioinsecticides to control pests.

Other aspects such as knowledge of the biology and/or behavior of the target insects are very important, as they are directly related to their field efficiency. This is true for *E. aurantiana* and *T. absoluta*, 2 very important pests of citrus and tomato crops, respectively. Larvae of *E. aurantiana*, after emerging, spend little time on the fruit surface, and the larvae of *T. absoluta* remain inside the leaf for the most part of the larval stage. In both cases, the immature insects are exposed to *Bt* quickly. Therefore, although the efficacy in relation to some pests depends on the formulation, pathogenicity, and virulence of the pathogen, other aspects should be considered and studied to better enable the control of pests.

8. References

Alves, S.B.; Moino Jr., A.; Almeida, J.E.M. 1998. Desenvolvimento, potencial de uso e comercialização de produtos microbianos. In: *Controle microbiano de insetos*, S.B. Alves (ed.), Piracicaba: FEALQ, p.1143-1163.

Alves, L.F.A.; Alves, S.B.; Pereira, R.M.; Capalbo, D.M.F. 1997. Production of *Bacillus thuringiensis* var. *kurstaki* Berliner grow in alternative media. *Biocontrol Science and Technology*, 7: 377-383.

Alves, L.F.A.; Alves, S.B.; Augusto, N.T. 1999. Seleção de inertes e adjuvantes visando a elaboração de formulação granulada de *Bacillus sphaericus* Neide. *Arquivos do Instituto Biológico*, 66(1): 77-81.

Alves, L.F.A.; Alves, S.B.; Lopes, R.B.; Augusto, N.T. 2001a. Estabilidade de uma formulação de *Bacillus sphaericus* armazenada sob diferentes temperaturas. *Scientia Agricola*, 58(1): 21-26.

Alves, S.B.; Medeiros, M.B.; Tamai, M.A.; Lopes, R.B. 2001b. Trofobiose e microrganismos na proteção de plantas. *Biotecnologia Ciência e Desenvolvimento*, 21: 16-21.

Altieri, M.A. 1999. *Agroecologia: bases científicas para uma agricultura sustenable*. Montevidéu: Nordan-Comunidad. 325p.

Arango, J.A.; Romero, M.; Orduz, S. 2001. Diversity of *Bacillus thuringiensis* strains from Colombia with insecticidal activity against *Spodoptera frugiperda* (Lepidoptera: Noctuidae). *Journal of Applied Microbiology*, 91:1-10.

Assunção, M.S.; Campos, A.R.; Gravena, S. 1980. Efeito de Amitraz, Amitraz + *Bacillus thuringiensis* e Amitraz + inseticidas no controle de *Heliothis* spp. em algodão. *Anais da Sociedade Entomológica do Brasil*, 9(2): 249-251.

Batista, A.C.; Melatti, V.M.; Demo, C.; Martins, E.S.; Praça, L.B.; Gomes, A.C.M.M.; Falcão, R.; Brod, C.S.; Monnerat, R.G. 2005. *Prospecção de estirpes de Bacillus thuringiensis efetivas para o controle de Anticarsia gemmatalis*. Brasília: EMBRAPA, 19p. (Boletim de Pesquisa e Desenvolvimento, 82).

Bravo, A.; Likitvivatanavong, S.; Gill, S. S.; Soberón, M. 2011. *Bacillus thuringiensis*: a story of a successful bioinseticide. *Insect Biochemistry and Molecular Biology*, 41: 1-9.

Beegle, C.C.; Yamamoto, T. 1992. Invitation paper (C.P. Alexander Fund): history of *Bacillus thuringiensis* Berliner reseach and development. *Canadian Entomologist*, 124: 587-616.

Belloti, A.C.; Cardona, C.; Lapointe, S.L. 1990. Trends in pesticide use in Colombia and Brazil. *Journal of Agricultural Entomology*, 7: 191-201.

Béron, C.; Arcas, J.; Salerno, G.L. 1999. Uso de *Bacillus thuringiensis* en el control de insectos plaga de cultivos de interés agronômico. In: *IV Simpósio Nacional de Biotecnologia Vegetal*, Resumos, Argentina, Buenos Aires, p.77.

Béron, C.; Arcas, J.; Salerno, G.L. 2000. Characterization of a *Bacillus thuringiensis* strain toxic against lepidopterans and coleopterans. In: *XXI International Congress of Entomology*, Abstratcs, Brasil, Rio de Janeiro, p.515.

Blanco, E. 2006. Control biologico de la polilla de la col *Plutella xylostella* (L.). INISAV, Cuba. 13 p. (http://www.aguascalientes.gob.mx/agro/produce/PLUT-BIO.htm). Acessado em 25/09/2006.

Bohorova, N.; Maciel, A.M.; Brito, R.M. 1996. Selection and characterization of mexican strains of *Bacillus thuringiensis* active against four major lepidopteran maize pests. *Entomophaga*, 41(2):153-165.

Bohorova, N.; Cabrera, M.; Abarca, C. et al. 1997. Susceptibility of four tropical lepidopteran maize pests to *Bacillus thuringiensis* Cry1-type insecticidal toxins. *Journal of Economic Entomology*, 90(2): 412-415.

Bobrowski, V.L.; Pasquali, G.; Bodanese-Zanettini, M.H.; Pinto, L.M.F.; Fiuza, L.M. 2002. Characterization of two *Bacillus thuringiensis* isolates from South Brazil and their toxicity against *Anticarsia gemmatalis* (Lepidoptera: Noctuidae). *Biological Control*, 25(2): 129-135.

Bosa, C. F.; Cotes, A. M. 1997. Evaluacion de la actividad insecticida de isolados nativas de *Bacillus thuringiensis* contra *Spodoptera frugiperda* (Smith). *Revista Colombiana de Entomologia*, 23(3-4): 107-112,

Botto, E.N. 1996. Controle biológico de plagas en la Argentina: informe de la situación actual. In: *El Control Biológico en America Latina*, M.C. Zapater (ed.), Argentina, Buenos Aires, p.1-8.

Branco, F.E. 1995. *Aspectos Econômicos do Controle de Thyrinteina arnobia (Stoll, 1782) (Lep.:Geometridae) com Bacillus thuringiensis (Berliner) em Povoamentos de Eucalyptus spp.* Dissertação: Mestrado. Universidade Federal do Paraná. 130p.

Brar, S.K.; Tyagi, V.R.D.; Valéro, J.R. 2006. Recent advances in downstream processes and formulations of *Bacillus thuringiensis* based biopectides. *Process Biochemestry*, 41: 323-342.

Bravo, A.; Sarabia, S.; Lopez, L. 1998. Characterization of *cry* genes in a mexican *Bacillus thuringiensis* strain collection. *Applied and Environmental Microbiology*, 64(12): 4965-4972.

CAB International. The 2010 Worldwide Biopesticides Market Summary. London: CPL Business Consultant, 2010. 40p.

Carbajal, J. 1995. Uso de *Bacillus thuringiensis* cepa LBT 13 en el control de *Phylocoptruta oleivora* en el cultivo de los cítricos. In: *III Encontro Nacional Científico-Técnico de Bioplaguicidas*, Anais, Cuba, Havana, p.18.

Carozzi, N.B.; Kramer, V.C.; Warren, G.W.; Evola, S.; Koziel, M.G. 1991. Prediction of insecticidal of *Bacillus thuringiensis* strains by polymerase chain reaction product profiles. *Applied and Environmental Microbiology*, 57: 3057-3061.

Carrillo, J.L. 1971. Pruebas de Thuricide (*Bacillus thuringiensis*) para combatir gusanos de la col en Chapingo, Mex. *Agricultura Técnica em México*, 3(2):58-60.

Castro-Franco, R.; García-Alvarado, J.S.; Galán-Wong, L.J.; Gallegos-Del Tejo, A. 1995. Efecto sinérgico de extractos de bagazo de *Agave lechuguilla* Torr. en la actividad insecticida de *Bacillus thuringiensis* contra *Spodoptera frugiperda* Smith. *Phyton*, 57(2): 113-119.

Cerón, J.; Ortiz; A.; Quintero; R.; Güereca; L.; Bravo, A. 1995. Specific PCR primers directed to identify cryI and cryIII genes within *Bacillus thuringiensis* strain collection. *Applied and Environmental Microbiology*, 61: 3826-3831.

Cerón, J.; Covarrubias L.; Quintero, R.; Ortiz, A.; Ortiz, M.; Aranda, E.; Lina, L.; Bravo, A. 1994. PCR analysis of the cryI insecticidal crystal family genes from *Bacillus thuringiensis*. *Applied and Environmental Microbiology*, 60: 353-356.

Charles, J.F.; Delécluse, A.; LeRoux, C.N. 2000. *Entomopathogenic bacteria: from laboratory to field application*. Dordrecht: Kluwer Academic Publishers, 524 p.

COMPÊNDIO DE DEFENSIVOS AGRÍCOLAS: Guia prático de produtos fitossanitáros para uso agrícola. 2005. 7 ed. São Paulo: Andrei. 1141p.

Copping, L. G.; Menn, J.J. 2000. Review biopesticides: a review of their action, applications and efficacy. *Pest Management Science*, 56: 651-676.

Corona, J.E.B.; Meza, J.E.L.; Ibarra, J.E. 1998. Caracterización de una cepa mexicana de *Bacillus thuringiensis* ssp. *kenyae*: un análises de sua baja toxicidad hacia lepidópteros. *Vedalia*, 5(1): 3-12.

Cruz, P.F.N. 1977. Resultados preliminares sobre a eficácia do *Bacillus thuringiensis* Berliner no controle do "mandorová" da seringueira (*Erinnyis ello* L.) (Lepidoptera: Sphingidae), na Bahia. *Revista Theobroma*, 7: 93-98.

Del Rincón-Castro, M.C.; Méndez-Lozano, J.; Ibarra, J.E. 2006. Caracterización de isolados nativas de *Bacillus thuringiensis* con actividad inseticida hacia del gusano cogollero del maíz *Spodoptera frugiperda* (Lepidoptera: Noctuidae). *Folia Entomologica*, 45(2): 157-164.

Dunkle, R.L.; Shasha, B.S. 1989. Response of starch-encapsulated *Bacillus thuringiensis* containing ultraviolet screens to sunlight. *Environmental Entomology*, 18(6): 1035-1041.

Edwards, M.L.; Hernández, M.J.L.; Rubio, A.P. 1999. Biological differences between five populations of fall armyworm (Lepidoptera: Noctuidae) collected from corn in Mexico. *Florida Entomologist*, 82(2): 254, 262.

Estruch, J.J.; Carozzi, N.B.; Desai, N.; Duck, N.B.; Warren, G.W.; Koziel, M.G. 1997. Transgenic plants: an emerging approach to pest control *Nature Biotechnology*, 15: 137-141.

Fernández-Larrea, O. 1999. A review of *Bacillus thuringiensis* (*Bt*) production and use in Cuba. *Biocontrol News and Information*, 20(1): 47-48.

Ferrer, F. 2001. Biological control of agricultural insect pest in Venzuela; advances, achievements and future perspectives. *Biocontrol News and Information*, 22(3): 67-74.

Figueiredo, M.B.; Coutinho, J.M.; Orlando. A. 1960. Novas perspectivas para o controle biológico de algumas pragas com *Bacillus thuringiensis*. *Arquivos do Instituto Biológico*, 27: 77-88.

Fiuza, L.M.; Pinto, L.M.N.; Azambuja, A.O.; Steffens, C.; Menezes, V.G.; Vargas, J.O. 2002. Pathogenic effect of *Bacillus thuringiensis* isolate from south Brazil against *Oryzophagus oryzae* (Coleoptera, Curculionidae). In: *XXXV Annual Meeting of the SIP, VIII International Colloquim on Invertebrated Pathology and Microbial Control,VI*

International Conference on Bacillus thuringiensis, Program and Abstracts, Brasil, Foz do Iguaçu, p.60.

Gallo, D.; Nakano, O.; Silveira Neto, S.; Carvalho, R.P.L.; Baptista, G.C.; Berti Filho, E.; Parra, J.R.P.; Zucchi, R.A.; Alves, S.B.; Vendramim, J.D.; Marchini, L.C.; Lopes, J.R.S.; Omoto, C. 2002. Entomologia Agrícola, Piracicaba: FEALQ, 920p.

Garcia, F.R. 1992. Manejo integrado de plagas en cultivos del Valle del Cauca. *ICA Informa*, 26: 11-12.

Glare, T.R.; O'Callaghan, M. 2000. *Bacillus thuringiensis: biology, ecology and safety*. Chichester: John Wiley & Sons, 350 p.

Gravena, S.; Sanguino, J.R.; Bara, J.R. 1980a. Controle biológico da broca da cana *Diatraea saccharalis* (Fabricius, 1794) por predadores de ovos e *Bacillus thuringiensis* Berliner. *Anais da Sociedade Entomológica do Brasil*, 9(1): 87-95.

Gravena, S.; Campos, A.R.; Maia, O.S.; Paulaneto, G. T. 1980b. Eficiência de *Bacillus thuringiensis* Berliner e *Bacillus thuringiensis* + Methomyl no controle de lepidópteros no tomateiro. *Anais da Sociedade Entomológica do Brasil*, 9(2): 243-248.

Gravena, S.; Araújo, C.A.M.; Campos, A.R.; Villani, H.C.; Yotsumoto, T. 1983. Estratégias de manejo integrado de pragas do algodoeiro em Jaboticabal, SP, com *Bacillus thuringiensis* Berliner e artrópodos benéficos. *Anais da Sociedade Entomológica do Brasil*, 12(1): 243-248.

Grecco, E.D.; Polanczyk, R.A.; Pratissoli, D.; Andrade, G.S.; Malacarne, A.M. 2006. Potencial de *Bacillus thuringiensis* Berliner no controle de *Spodoptera eridania* (Lep.: Noctuidae). In: *XXI Congresso Brasileiro de Entomologia*, Resumos, Brasil, Recife, ID 186-1.

Guerra, P.T.; Wong, L.J.G.; Roldán, H.M. 2001. Bioinseticidas: su empleo, produción y comercialización en México. *Ciencia UANL*, 4(2): 143-152.

Guerra, P.T.; Iracheta, M.M.; Pereyra, R.A.; Galán, L.J.W.; Gómez, R.G.; Tamez, R.S.G.; Rodriguez, C.R. 2007. Caracterización de isolados mexicanas de *Bacillus thuringiensis* tóxicas para larvas de lepidópteros e coleópteros. *Ciencia UANL*, 3(4): 477-482.

Giustolin, T.A.; Vendramin, J.D.; Alves, S.B.; Vieira, S.A. 2001. Efeito associado de genótipo de tomateiro resistente a *Bacillus thuringiensis* var. *kurstaki* sobre o desenvolvimento de *Tuta absoluta* Meyrich (Lep., Gelechiidae). *Neotropical Entomology*, 30(3): 461 - 465.

Habib, M.E.M.; Amaral, M.E.C. 1985. Aerial application of *Bacillus thuringiensis* against the velvetbean caterpillar *Anticarsia gemmatalis* Huebner, in soybean fields. *Revista de Agricultura*, 60(2): 141-149.

Habib, M.E.M.; Andrade, C.F.S. de. 1991. Controle microbiano de insetos com o uso de bactérias. *Informe Agropecuário*, 15(167): 21-26.

Hofte, H.; Whiteley, H.R. 1989. Insecticidal crystal proteins of *Bacillus thuringiensis*. *Microbiological Reviews*, 53: 242-255.

Huerta, A.; Cogollor, G. 1995. Control de la polilla del brote del pino (*Rhyacionia buoliana* Den. *et* Schiff.) mediante isolados del bacterio *Bacillus thuringiensis* var. *kurstaki*. *Ciencias Florestales*, 10(1-2): 135-141.

IBGE. 2006. *Levantamento sistemático da produção agrícola*. http://www.ibge.gov.br/. Acessado em 27/09/2010.

Jiménez, J. 1996. Control de *Heliothis virescens* (Fabr.) com medios biológicos en el cultivo del tabaco. In: *Curso International de Sanidad Vegetal en Tabaco*, Cuba, INISAV, 14 p.

Jiménez, J.; Fernández, R.; Canzado, A.; Vásquez, M. 1997. Efectividad de biopreparados nacionales de *Bacillus thuringiensis* para la lucha contra *Heliothis virescens* en tabaco. In: *V Encuentro Nacional Técnico-Científico de Bioplaguicidas*, Resumos, Cuba, Havana, p.58.

Knowles, B.H. 1994. Mechanism of action of *Bacillus thuringiensis* insecticidal δ-endotoxins. *Advances in Insect Physiology*, 24: 275-308.

Lambert, B.; Peferoen, M. 1992. Insecticidal promise of *Bacillus thuringiensis. BioScience*, 42: 112-122.

Leréclus, D. 1988. Génétique et biologie moléculaire de *Bacillus thuringiensis. Bulletin de l'Institut Pasteur*, 86: 337-371.

Licor, L.; Almaguel, L.; López, A., 1995. Control de ácaros en toronja y nananja com *Bacillus thuringiensis* (LBT-13) en Ciego de Avila, in: *Encuentro Nacional Técnico-Científico de Bioplaguicidas*, 3, Resumos, Havana, p. 2-3.

Lord, J.C. 2005. From Metchnikoff to Monsanto and beyond: The path of microbial control. *Journal of Invertebrate Pathology*, 89:19-29.

Maduell, P.; Callejas, R.; Cabrera, K.R.; Armengol, G.; Orduz, S. 2002. Distribution and characterization of *Bacillus thruringiensis* on the phylloplene of species of *Piper* (Piperaceae) in three altitudinal levels. *Microbial Ecology*, 44: 144-153.

Iracheta, M.C.; Galán, L.W.;Ferré, J.M.;Pereyda, B.A. 2001. Selección de toxinas Cry contra *Trichoplusia ni. Ciência UANL*, 4(1): 55-62.

Marques, I.M.R. 1993. *Ação de Bacillus thuringiensis Berliner var. kurstaki sobre Scrobipalpuloides absoluta (Meyrick, 1917) (Lepidoptera: Gelechiidae) e sua interação com o parasitóide Trichogramma pretiosum Riley, 1879 (Hymenoptera: Trichogrammatidae)*. Piracicaba, SP. Tese Doutorado, ESALQ/USP, 75 p.

Marques, I.M.R.; Alves, S.B. 1996. Efeito de *Bacillus thuringiensis* Berl. var. *kurstaki* sobre *Scrobipalpuloides absoluta* Meyer (Lepidoptera: Gelechiidae). *Anais da Sociedade Entomológica do Brasil*, 25(1): 39-45.

Martínez, Z.; Tamara, P.; Lérida, A.; Brito, R.; Hernández, A.; Peñate, E.; Lluvides, J.; Marín, R. 1995. Utilización de la cepa LBT-13 y de *Phytoseiulus macropilis* para el control de *Tetranychus tumidus*. In: *III Encuentro Nacional Técnico-Científico de Bioplaguicidas*, Resumos, Cuba, Havana, p.7-8.

Monnerat R.; Martins, E.; Queiroz, P.; Ordúz, S.; Jaramillo, G.; Benintende, G.; Cozzi, J.; Dolores Real, M.; Martinez-Ramirez, A.; Rausell, C.; Cerón, J.; Ibarra, J.E.; Del Rincon-Castro, M. C.; Espinoza, A.M.; Meza-Basso, L.; Cabrera, L.; Sánchez, J.; Soberon, M.; Bravo, A. Genetic variability of Spodoptera frugiperda Smith (Lepidoptera: Noctuidae) populations from Latin America is associated with variations in susceptibility to Bacillus thuringiensis Cry toxins. Applied and Environmental Microbiology, no prelo.

Montero, J.F. 1991. Uso de *Bacillus thuringiensis* var. *kurstaki* para el control de plagas en cucúrbitas. In: *III Taller Centroamericano en Fitoprotección del Melón*, Memoria, Honduras, p.71-72.

Montesbravo, E.P. 2006. Control biologico de *Spodoptera frugiperda* Smith en maiz. INISAV, Cuba. 10 p. http://www.aguascalientes.gob.mx/agro/produce/SPODOPTE.htm. Acessado em 25/09/2006.

Morales, C.A.M. 2006. Control biologico dentro del MIP en solanaceas. INISAV, Cuba. 12p. http://www.aguascalientes.gob.mx/agro/produce/SOLAN-BI.htm. Acessado em 25/09/2006.

Moscardi. 1998. Utilização de vírus entomopatogênicos em campo. In: Controle Microbiano de Insetos, S.B. Alves (ed.), Piracicaba: FEALQ, p. 509-540.

Pérez A.R. 1996. *Elementos para el manejo integrado del ácaro rojo,* Tetranichus tumidus *Banks (Acari: Tetranychidae) en plátano y banano.* Tese de Doutorado, INISAV, La Habana, Cuba, 89p.

Pérez, N.; Vasquéz, L.L. 2001. Manejo ecológico de plagas. In: *Transformando el campo cubano. Avances de la agricultura sostenible,* F. Funes et al. (eds.,) Cuba, La Habana, p.191-226.

Pérez, T., Jimenez , J.; Pazos, R. 1991. Efectividad de *Bacillus thuringiensis* obtenido en medio liquido estatico contra *Mocis* spp. en areas de pastos. *Protección de Plantas,* 1(1): 15-20.

Pigatti, A.; Figueiredo, M.B.; Orlando, A. 1960. Experiências de laboratório sôbre a atividade de novos inseticidas contra o mandorová da mandioca. *Biológico,* 26: 47-51.

Pinto, L.M.N.; Fiuza, L.M. 2002. Toxicity of *Bacillus thuringiensis kurstaki* HD68 ICPs to the *Spodoptera frugiperda* (Lepidoptera, Noctuidae) larvae from south Brazil. In: *XXXV Annual Meeting of the SIP, VIII International Colloquim on Invertebrated Pathology and Microbial Control,VI. International Conference on Bacillus thuringiensis,* Program and Abstracts, Brasil, Foz do Iguaçu, p.90.

Polanczyk, R. A. 1999. *Isolamento e seleção de Bacillus thuringiensis Berliner para controle de Spodoptera frugiperda (Smith & Abbot, 1797) (Lepidoptera: Noctuidae).* Dissertação: Mestrado. Universidade Federal do Rio Grande do Sul. 64p.

Polanczyk, R.A.; Alves, S.B. 2005. Interação entre *Bacillus thuringiensis* e outros entomopatógenos no controle de *Spodoptera frugiperda. Manejo Integrado de Plagas Y Agroecologia,* 74: 24-33.

Polanczyk, R.A.; Alves, S.B. 2005. Interação entre *Bacillus thuringiensis* e outros entomopatógenos no controle de *Spodoptera frugiperda. Scientia Agricola,* 62(5): 464-468.

Polanczyk, R.A.; Alves, S.B.; Padulla, L.F. 2005. Screening of *Bacillus thuringiensis* against three brazilian populations of *Spodoptera frugiperda* (Lepidoptera: Noctuidae). *Biopesticides International,* 1(1,2): 114-124.

Pratissoli. D.; Polanczyk, R.A.; Vianna, U.R.; Andrade, G.S.; Oliveira, R.G.S. 2006. Desempenho de *Trichogramma pratissoli* Querino & Zucchi (Hymenoptera, Trichogrammatidae) em ovos de *Anagasta kuehniella* (Zeller) (Lepidoptera, Pyralidae) sob efeito de *Bacillus thuringiensis* Berliner. *Ciência Rural,* 36(2): 369-377.

Praça. L.B.; Batista, A.C.; Martins, E.S.; Siqueira, C.B.; Dias, D.G.S.; Gomes, A.C.; Falcão, R.; Monnerat, R.G. 2004. *Pesquisa Agropecuária Brasileira,* 39(1): 11-16.

Revelo, M.A. 1965. Efectos del *Bacillus thuringiensis* sobre algunas plagas Lepidópteras del maíz bajo condiciones tropicales. *Agricultura Tropical,* 27(1): 392-395.

Rodrigues, B.A.; Trumble, J.T. 1993. A successful IPM system for processed tomatoes in Sinaloa, Mexico using *Bacillus thuringiensis* as a key component. In: *XXVI Annual Meeting of the Society for Invertebrate Pathology,* Program and Abstracts, Asheville, p.67.

Rodríguez, M. C.; Trujillo, H. S. 1991. Control de la polilla mayor de la cera *Galleria melonella* por isolados de *Bacillus thuringiensis* el municipio de Tecomán, Colima, México. *Revista Latino Americana del Microbiologia*, 33: 203-207.

Rojas, L. A. 2006. Control biologico de acaros fitofagos em diferentes cultivos. INISAV, Cuba. 16 p. (http://www.aguascalientes.gob.mx/agro/produce/acarbio.htm). Acessado em 25/09/2006.

Rosales-Reyes, T.; Salcedo-Hernandez, R.; Ibarra, J.E.; Barobza-Corona, J.E. 2003. Identificación de los genes cry en isolados mexicanas de *Bacillus thuringiensis* con potencial inseticida. *Acta Universitaria*, 13:39-46.

Rosas, A.M.M.; Perez, V.M.J.; Henonin, L.A.; Ibarra, J. 1994. Aislamiento y caracterizacion de una cepa nueva de *Bacillus thuringiensis*, esencialmente inmovil y altamente toxica a *Manduca Sexta* (Lepidoptera: Sphingidae). *Vedalia*, 1(1): 3-12.

Rosset, P.; Moore, M. 1999. Food security and local production of biopesticides in Cuba. *ILEIA Newsletter*, 13(4): 18-23.

Ruiz, R.A.V. 1998. El control biológico de insectos plagas en hortalizas. In: *Nuevos aportes del control biológico en la agriocultura sostenible*, Lizárraga, T. A., Barreto, C.U., Hollands, J. (eds.), Bogotá: Editora Nacional, p.269-294.

Sáenz, C.E.; Salazar, J.D.; Rodriguez, A.; Alfaro, D.; Oviedo, R. 2002. Resultados de vinte anos da direção de investigação de extensão de cana-de-açúcar (DIECA) no manejo integrado de pragas em Costa Rosa. In: *VIII Congresso Nacioanl STAB*, Brasil, Pernambuco, p.58-66.

Sánchez, R.M.Y.; Vargas, B.H.L.; Sánchez, G.G. 1999. Evaluación de variedades de *Bacillus thuringiensis* para el control de *Diaphania* spp. (Lepidoptera: Pyralidae) en melón, *Revista Colombiana de Entomologia*, 25(1-2): 17-21.

Santos, G.P.; Zanuncio, T.V.; Neto, H.F.; Zanuncio, J.C. 1995. Suscetibilidade de *Eustema sericiea* (Lepidoptera: Notodontidae) ao *Bacillus thuringiensis* var. *kurstaki*. *Revista Ceres*, 42(242): 423-430.

Schnepf, E.; Crickmore, N.; Van Rie, J.; Lereclus, D.; Baum, J.; Feitelson, J.; Zeigler, D.R.; Dean, D.H. 1998. *Bacillus thuringiensis* and its pesticidal crystal proteins. *Microbiology and Molecular Biology Reviews*, 62: 775-806.

Schünemann, R.; Barbieri, C.; Fiuza, L.M. 2005. Toxicidade de delta-endotoxinas de *Bacillus thuringiensis* à *Anticarsia gemmatalis* (Lepidoptera, Noctuidae). In: *IX Simpósio de Controle Biológico*, Resumos, Brasil, Recife, p.80.

Souza, M.L. de. 2001. Utilização de microrganismos na agricultura. *Biotecnologia Ciência e Desenvolvimento*, 21: 28-31.

Sosa-Gómez, D.R.; Moscardi, F. 1991. Microbial control and insect pathology in Argentina. *Ciência e Cultura*, 43(5): 375-379.

Valicente, F.H.; Zanasi, R.F. 2005. *Uso de meio alternativos para produção de bioinseticida à base de* Bacillus thuringiensis. Sete Lagoas: EMBRAPA, 4p. (Circular Técnica, 60).

Vallés, S.S. 1998. Avances en el uso del control biologico en el estado de Aguascalientes. *Cuaderno de Trabajo*, 85: 1-22.

Vásquez, L.L.; Menéndez, J.M.; López, R. 1999. Manejo de insectos de importância forestal en Cuba. *Revista Manejo Integrado de Plagas*, 54: 4-14.

Zanuncio, J.C.; Guedes, R.N.C.; Cruz, A.P. da; Moreira, A.M. 1992. Eficiência de *Bacillus thuringiensis* e de Deltametrina, em aplicação aérea, para o controle de *Thyrinteina arnobia* Stoll, 1782 (Lepidoptera: Geometridae) em eucalipital no Pará. *Acta Amazonica,* 22(4): 485-492.

Yaro, E.; Cerón, J.; Uribe, D. 1999. Caracterización de aislamientos nativos de *Bacillus thuringiensis* provenientes de bosque natural y zonas de cultivos agrícolas Colombianos. *Revista Colombiana de Entomologia,* 25(3-4): 185-190.

20

Baculoviruses: Members of Integrated Pest Management Strategies

Vanina Andréa Rodriguez,
Mariano Nicolás Belaich and Pablo Daniel Ghiringhelli
LIGBCM-AVI (Laboratorio de Ingeniería Genética y Biología Celular y Molecular Area Virosis de Insectos)
Universidad Nacional de Quilmes/Departamento de Ciencia y Tecnología
Argentina

1. Introduction

Integrated Pest Management (IPM) can be defined as "an ecologically based pest control strategy that relies heavily on natural mortality factors and seeks out control tactics that disrupt these factors as little as possible" (Flint & Bosch, 1981). IPM began to be applied because the extensive use of chemical insecticides show different types of environmental damages, as development of resistant insects, the appearance of new pests, injury to bird and mammals populations and human health damage due to the release of toxic waste on environment and food contamination. The aims of IPM are to protect crops, with minimum cost and risk for humans, animals and ecosystems. The development and application of IPM requires the knowledge of how the ecosystem influences on pest insects and its natural control agents, and how to modify this environment to control particularly pest insects and avoid the related chemical agent's problems.

2. Integrated pest management (general overview)

IPM applies different tactics, like pest resistant plants, use of entomopathogens such as bacteria and viruses, and strategies that involves cultural, physical, mechanical, biological and chemical control. The use of these combined tactics reduces the chances of generating resistance and insect survival.

2.1 Pest resistant plants (transgenesis using plant genes)

Plants have a vast metabolic capability and produce many secondary chemicals which are toxic, anti-nutritional, or aversive to species might otherwise be potential predators (Norris & Kogan, 1980). Examples include the pyrethrins from chrysanthemums and alkaloids like nicotine from tobacco. Other compounds implicated in protection from insect attack include the terpenoids, steroids, flavonoids, phenolic, glucosinolates, cyanogenic glycosides, rotenoids, saponins and non protein amino acids (Gatehouse *et al.*, 1991). As secondary compounds are the products of multi-enzyme pathways which involve the interaction of many gene products, such defense system are in most cases too complex to be used in plant genetic engineering (Dawson *et al.*, 1989; Hallahan *et al.*, 1992). However, a few plant defense mechanisms are based on the product of a single gene, and the target site of many is the insect digestive system. Most of these types of single genes are suitable for gene transfer. Being of plant origin, they have the advantage that they are likely to have a high degree of compatibility with the metabolic system of the transgenic host plant.

Crops resistant to insect attack offer an alternative strategy of pest control upon chemical pesticides. Transgenic plant technology can be a useful tool in producing resistant crops, by introducing novel resistance genes into a plant species. Several different classes of plant proteins have been shown to be insecticidal towards a range of economically important insect pests. Genes encoding insecticidal proteins have been isolated from various plant species and transferred to crops by genetic engineering. Amongst these genes are those that encode:

1. Protease inhibitors (serine and cysteine). The damage of leaves of certain solanaceous plants, either by insect feeding or mechanical wounding, induced the synthesis of protease inhibitors (Green & Ryan, 1972; Shumway *et al.*, 1976; Walker-Simmons & Ryan, 1977; Brown *et al.*, 1985). The first gene of a plant successfully used to be transferred to another plant was a trypsin inhibitor (cowpea trypsin inhibitor, CpTI) (Pusztai *et al.*, 1992; Graham *et al.*, 1995; Xu *et al.*, 1996). As an example of a commercial deployment of a proteinase inhibitor transgene to date, could be mentioned the culture of genetically engineered cotton varieties in China. These varieties express two transgenes to improve cotton protection, Bt toxins against lepidopteran larvae and CpTI. In 2005, Bt/CpTI cotton was grown on over 0.5 million hectares (Gatehouse, 2011).
2. Inhibitors of α-amylase. This enzyme is able to catalyse the hydrolysis of α-1-4 glycosidic bonds, transforming polysaccharides into mono and disaccharides (Grossi-de-Sá & Chrispells, 1997; Franco *et al.*, 2002; Pelegrini *et al.*, 2006). Interference of the carbohydrate absorption could be a way to reduce the insect pest feeding (Yamada *et al.*, 2001). Leguminous seeds are known as rich sources of proteinaceous α-amylase inhibitors (α-AIs) (Payan, 2004). Expression of α-amylase inhibitors (*a*-AIs) from both scarlet runner bean (*Phaseolus coccineus*) and common bean (*Phaseolus vulgaris*) has been shown to be effective in transgenic plants, showing high protection against seed weevils in pea (Shade *et al.*, 1994; Schroeder *et al.*, 1995), azuki bean (Sarmah *et al.*, 2004), chickpea (Sarmah *et al.*, 2004; Ignacimuthu & Prakash., 2006), cowpea (Solleti *et al.*, 2008) and coffee (Barbosa *et al.*, 2010).
3. Lectins. This group of carbohydrate-binding proteins constitutes entomotoxic factors present in many plant species. During the last decade a lot of progress was made in the study of the properties of a few lectins that are expressed in response to phytophagous insects. Based on their activity towards pest insects, these proteins have a high potential for use in pest control strategies. For example, the use of plant lectins has been applied to control numerous pests: melon fruit fly larvae (Kaur *et al.*, 2009); *Aedes aegypti* larvae, which has developed tolerance to many other insecticides (Coelho *et al.*, 2009); and the cotton leafworm *Spodoptera littoralis,* an economically important caterpillar in agriculture and horticulture (Hamshou *et al.*, 2010).

2.2 Entomopathogen bacteria (*Bacillus thuringiensis*)

Bacillus thuringiensis is a Gram positive bacteria belonging to Eubacteria. It was isolated in the early twentieth century in Japan from dead larvae of the silkworm. This bacteria, produces spores and crystalline bodies composed of one or more proteins with insecticidal activity

(Schnepf *et al.*, 1998, Sedlak *et al.*, 2000). The crystalline toxins, named Cry δ-endotoxins, exist in a variety of forms: bipyramidal, spherical, rhomboidal, cuboidal and irregular, among others, and are active against a large number of insect groups as well as nematodes and protozoa. Today there are over 40 groups of Cry proteins (Crickmore *et al.*, 2009). The δ-endotoxins are synthesized as an inactive pro-Cry-toxin and when they are ingested by the larvae feeding on plant debris or soil, the inclusions are solubilized in the alkaline conditions of the digestive tract of the larvae and are converted by the action of the insect proteases in active peptides (Feitelson *et al.*, 1992; Schnepf *et al.*, 1998). Active toxin is recognized by a specific receptor; it binds to microvilli of intestinal cells (Gazit *et al.*, 1998; Gerber & Shai, 2000), and generates ion channels. The natural ion imbalance dissipates and the pH of medium diminishes causing osmotic cell lysis and larva ceases to feed (Schnepf *et al.*, 1998). Moreover, tissue destruction allows mixing the contents of gastrointestinal tract with hemolymph. Both phenomena favor the germination of bacterial spores, resulting in the death of the larva few days after ingestion crystals (Aranda *et al.*, 1996; Schnepf *et al.* 1998; Crickmore *et al.*, 2009).

There are two main approaches to use Bt as a pest control agent:

1. Preparations based on living or dead Bt containing spores and crystals are sprayed on crops, as if it were a conventional insecticide. This strategy are currently used in the United States, Europe, Argentina and Mexico as a biological control for insect and other invertebrate pests (mites, nematodes, flatworms and protozoa) that affect crops of corn, potato, tomato, sorghum, rice, coffee, beans, sugar cane, among others (Neppl, 2000). The application of Bt in insect control is not exempt from the emergence of resistance. Using of a combination of toxins reduce resistance to individual toxins, maintains a populational balance and prevents the prevalence of resistant variants (Georghiou & Wirth, 1997; Ives *et al.*, 2011; Yang *et al.*, 2011). Research and years of use have shown that the employment of Bt products is not hazardous to non-target arthropods, birds, fish, mammals, or environment (EPA, 2008).
2. The genes that encode different Cry proteins have been used to generate genetically modified (GM) plants. Transgenic plants are resistant to the attack by insect pests. The most widely GM crops commercialized so far include mainly maize, cotton and rice. The use of this modified crops has been approved in several countries, including United States, Brazil, Argentina, Canada, Australia, Spain, South Africa, among other (James, 2011). Considering that insecticidal crystal proteins can be released continuously into the soil in different forms during the growing period of Bt-plants (Zhou *et al.*, 2011), biosafety of the use of genetically modified plants is always questioned. However, data regarding the development and commercial use of transgenic Bt varieties have shown that the currently available Bt crops have no direct detrimental effects on non target organism due to their narrow spectrum of activity. In addition, the use of these modified crops, such as Bt maize and Bt cotton, results in significant reductions of insecticide application and has clear benefits on the environment and farmer health. Consequently, Bt crops can be a useful component of IPM systems to protect the crops from targeted pests (Yu *et al.*, 2011).

2.3 Entomopathogen viruses (*Baculoviruses*)

The insect viruses are intracellular parasites that can only reproduce inside a susceptible insect host. They are valuable natural control agents, providing a secure control, effective and sustainable in a variety of insect pests. The virus particles are present in the environment and usually can be found on the surface of plants or in the soil. Insects become infected by consuming plant material contaminated with viral particles on the surface or by contact with the soil. Baculoviruses are the most common type of insect viruses. It has been reported that infect over 600 species of insects worldwide. Most baculovirus infect caterpillars, the immature form of moths and butterflies. Naturally, these viruses are potent regulators of the population of many caterpillar pests, but the use as a tool for biological control in agriculture is limited by biological or technical reasons.

1. Advantages. Insect viruses are very safe to handle, since they are not infectious to organisms other than their natural hosts. Moreover, most insect viruses have a high specificity, so that the risk of affecting non-target beneficial insects is very low.
2. Disadvantages. Most insect viruses have a low speed of action on their insect host, during this time the plague is still eating and damaging. Insect death is also dose dependent, and very high doses are often necessary for adequate control. Usually, viruses are very effective against early larval stages; the late larval stages are less susceptible to virus infection. Virus particles exposed to sunlight or high temperatures are rapidly inactivated. In addition, some cultural practices can affect viral persistence, hiding the viral particles in the soil.

3. Baculovirus

3.1 Baculovirus biology

Baculoviridae is a viral family that infects insects. Their genomic material is composed by double strand, circular, super coiled DNA, with a size ranging from 80 to 180 kpb. These viruses are classified in four genera: *Alphabaculovirus* (nucleopolyhedroviruses -NPVs- that specifically infect insects of lepidopteran order), *Betabaculovirus* (granuloviruses -GVs- that specifically infect insects of lepidopteran order), *Gammabaculovirus* (NPVs that specifically infect insects of hymenoptera order) and *Deltabaculovirus* (NPVs that specifically infect insects of diptera order) (Jehle *et al.* 2006). The virus genome is packaged within a rod-shaped nucleocapsid which is further surrounded by a lipoprotein envelope to form the virus particle. This structure is then occluded at very late stages by a crystalline matrix (Occlusion Body or OB) largely comprising for a single occlusion protein (about 28 kDa), which serves to protect it in the environment. During the infection cycle two different phenotypes are generated: budded virus (BVs, responsible for the systemic cell to cell infection) and occluded-derived virus (ODVs, responsible for the host to host infection) (Figure 1).

The most common way of insect host primary infection is by ingestion during larval feeding of contaminated foliage (Figure 2). Following virus ingestion, the OBs are dissolved in the high pH conditions (pH 8.5 to 11) of the insect midgut, releasing the virus particles (ODVs) into the gut lumen (Granados & Lawler, 1981; Pritchett *et al.*, 1981; Pritchett *et al.*, 1984; Rohrmann, 2008). Released viruses bind to the columnar epithelial cells and enter the tips of the microvilli on the apical brush border of cells (Kuzio *et al.*, 1989; Faulkner *et al.*, 1997; Haas-Stapleton *et al.*, 2004). Following fusion between cell and virus membranes, the nucleocapsids are released into the cytoplasm and are transported to the nucleus, where viral DNA transcription and replication occurs. Into the cytoplasm, nucleocapsids are transported from the basal membrane to the haemocoel, acquiring the host derived membrane and virus encoded proteins (Washburn *et al.*, 2003). Secondary infection is achieved by BVs produced from the midgut cells. Thus, the viruses spread to other insect tissues including the fat body, endodermis, muscle sarcolemma and nerve ganglia (Harrap, 1970; Washburn *et al.*, 2003). Prior to death larvae become creamy in colour, cease feeding and show limited movement. In most baculovirus infections, the host tissues break down as a result of the expression of virus-encoded chitinase and cathepsin proteins (Ohkawa *et al.*, 1994; Hawtin *et al.*, 1995; Slack *et al.*, 1995). Then, OBs are released into the environment following the rupture of the insect cuticle; 10^9 OBs may be released from a single larva, and may remain viable in the environment for several years, until ingestion by another host larva resumes replication cycle (Evans & Harrap, 1982) (Figure 2).

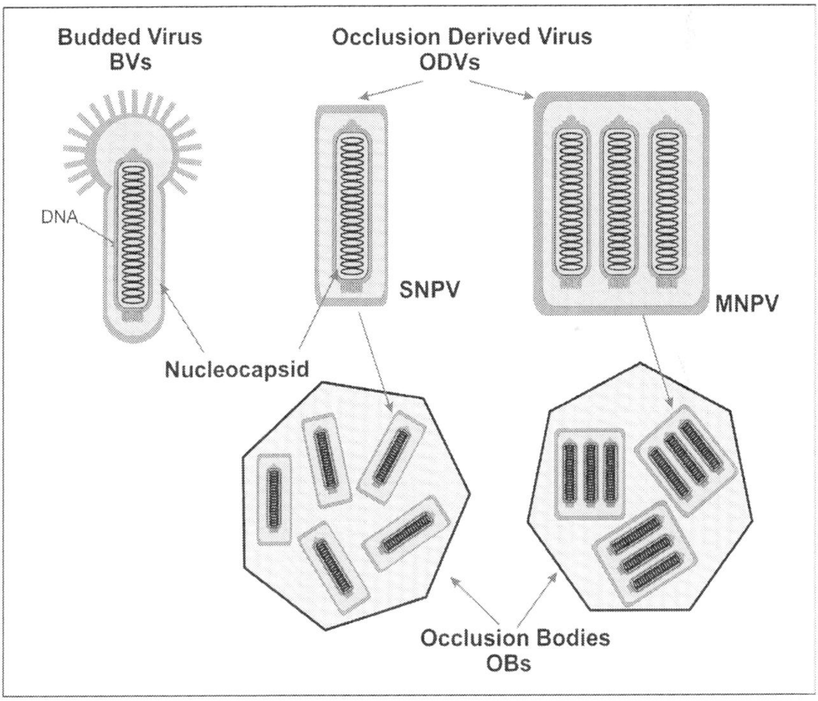

Fig. 1. **Alphabaculovirus phenotypes.** Occlusion bodies (OBs) containing occlusion derived virus (ODVs) responsibles for the primary infection in the midgut cells and Budded virus (BVs) that spread the infection to other larval tissues.

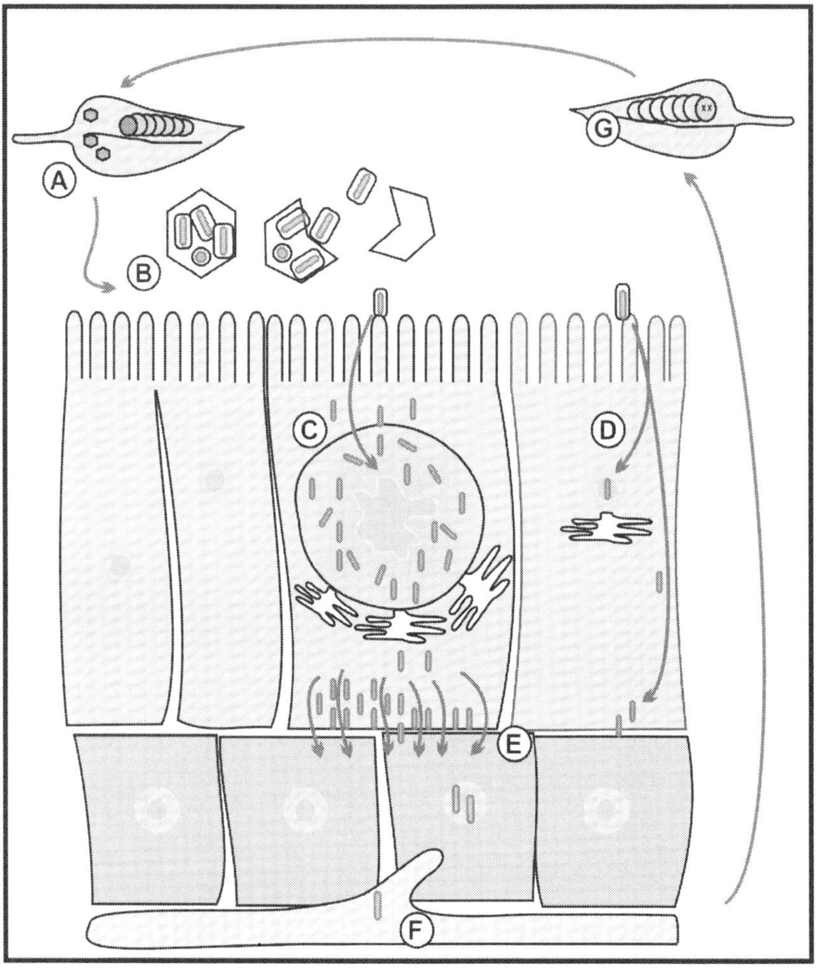

Fig. 2. **Infection cycle.** NPV biological cycle. **A**. Larva ingests a baculovirus-contaminated leaves. **B**. The virions are released and bind to midgut cell microvilli. **C**. Into columnar cells the nucleocapsids are transported to the nucleus, where the viral DNA is released and the BVs are generated. **D**. Nucleocapsids can pass through the cell to the haemocele. **E**. Viral particles enter into susceptible tissue cells and initiate a replication cycle to produce a new BV generation. **F**. Later, the OBs are generated into each infected cell. **G**. Cell lysis occurs, larval integument breaks, and the OBs are released to be eaten by other larva and thus restart the infection.

3.2 Baculovirus as control agents (Figure 3)

Wild type baculoviruses

A large number of Lepidoptera are pests of economically important crops such as soybean, sunflower and cotton, among others. Alpha- and Betabaculoviruses are very useful tools for control of lepidoteran pests. These viruses have a narrow host range, are safe for other insects and organisms in the environment (Szewczyk *et al*, 2006), and have strong pathogenicity and virulence. Other advantages are the stability in the environment for long periods and that can be applied using simple methods. Because of the great advantages of baculoviruses as biological control tools and their safety for the ecosystem, their use in IPM is accepted and is steadily increasing. Currently, different baculovirus are used in pest control management worldwide.

In North America, the use of baculoviruses as a pest control agent started as early as 1930 with the protection of pine trees with *Diprion hercyniae* NPV (Bird & Buek, 1961). Subsequently, this strategy was use to protect numerous commercial crops, including alfalfa, cabbage, corn, cotton, lettuce, soybean, tobacco and tomato (Granados & Federici, 1986). Another baculovirus that is currently used to control pest infection is *Helicoverpa zea* Simple Nucleopolyhedrovirus (HzSNPV). This virus infects several species belonging to the genera *Heliothis* and *Helicoverpa*, which include a wide range of crop pests (Chakrabarti *et al.*, 1999). HzSNPV provides a tool for control against the bollworm, soybean, sorghum, corn, tomatoes and beans. In the 90´s of the latest century the virus was registered under the name GemStar™ and marketed by Thermo Trilogy Company. This virus produced in United States is also marketed in Australia by Aventis Crop Science and it is now used to protect crops against *Helicoverpa armigera* (Mettenmeyer, 2002), which attack near 200 crops including cotton, soybeans, chickpeas, sunflower, snuff, pepper, and corn, among others. HaSNPV has been adopted for large-scale viral pesticide production in China and has been used extensively to protect crops of cotton (Zhang *et al.*, 1995).

In South America, probably the most successful program of using baculovirus in pest control is the program implemented in Brazil since 1980 to control the velvet bean caterpillar *Anticarsia gemmatalis* (pest of soybean) (Moscardi, 1989, 1999). In this program, the *Anticarsia gemmatalis* Multiple Nucleopolyhedrovirus (AgMNPV) is used as a biological insecticide against *Anticarsia gemmatalis* larvae, and by 2005 the treated area reached 2 million hectares (Szewczyk *et al.*, 2006). One advantage of using this virus is that it is highly virulent and only needs to be applied once, whereas chemical insecticides should be applied twice. In addition, the use of this virus is 20 to 30% less expensive than chemical insecticides. Recently, a pilot plant was constructed in Embrapa Soja, Londrina, in order to improve the laboratory process and train people in the production of virus. This laboratory will be able to inoculate 20,000 to 30,000 larvae per day (Szewczyk *et al.*, 2006).

Moreover, the *Cydia pomonella* Granulovirus (CpGV) has been used as pest control in crops of apples and pears. Its use has increased in Europe and North America since 2000 and is used in about 100,000 ha in those continents. Currently there are several commercial preparations that include Cyd-X, Virosoft CP4 (North America), Carpovirusine™ (France), Madex™ y Granupon™ (Switzerland), Granusal™ (Germany) and Virin-CyAP (Russia) (Rohrman, 2008). In Argentina, *Cydia pomonella* is a pest of pear, apple and walnut crops. The National Institute of Agricultural Technology (INTA; Argentina) in agreement with the

Natural Plant Protection company (France) developed a bio-insecticide called Carpovirus to control that pest, using a formulation based on CpGV.

For an additional example of baculovirus employed as tools of biological control and pest management, can be mentioned *Spodoptera exigua* Multiple Nucleopolyhedrovirus (SeMNPV). Formulations of this virus are being used to protect crops of sweet peppers in Spain against *Spodoptera exigua*. The company Biocolo SRL is responsible for the commercial production of this bio-insecticide with the capacity to treat 50,000 larvae per day. The introduction of this biological insecticide has helped to multiply more than 15 times the area of biologically protected crops grown, reaching 23,000 ha in 2008 (Caballero *et al.*, 2009).

Modified baculoviruses

Despite the advantages that baculovirus have, there are also some limitations for their use as pest control agents. Some of these limitations are the high costs of in vivo production and the low persistence in very sunny conditions (Ignoffo *et al*, 1977). Also, the virus must be used in early insect development stages, because in the late stages the insects are more resistant to infection (Washburn *et al.*, 2003). Another limitation is their slow speed to kill the pest (Ignoffo *et al*, 1992). To avoid this, recombinant baculovirus have been developed and offer attractive alternatives to broad-spectrum chemical control. These recombinant viruses can express specific toxins (Inceoglu *et al*, 2001), hormones (Elvira *et al*, 2010) or enzymes (Gramkow *et al*, 2010) and are much more efficient than the wild-type virus in speed to kill.

1. **Expression of insect-selective toxin.** The *aait* gene obtained from *Androctonus australis* scorpion is one of the most promising toxins to be expressed in baculovirus. A recombinant virus containing this gene showed to be 40% faster in killing larvae than the wild type and a reduction of host feeding by 60% (Cory *et al.*, 1994; Inceoglu *et al.*, 2001). The site of action of this neurotoxic polypeptide is one insect sodium channel. Lepidopterous larvae infected with an AaIT-expressing baculovirus reveal symptoms of paralysis identical to those induced by injection of the native toxin (Elasar *et al.*, 2001) and many of the physiological effects are very similar to those of pyrethroid insecticides which also act at the same target (Gordon *et al.*, 1992). Other useful insect-selective neurotoxins are SFI1 (obtained from a European spider *Segestria florentina*) and ButaIT (derived from the South Indian red scorpion *Mesobuthus tamulus* (Wudayagiri *et al.*, 2001). Some toxins could exert a cooperative effect when they are co-expressed, such as LqhIT1 and LqhIt2, obtained from *Leiurus quinquestriatus* scorpion (Regev *et al.*, 2003).
2. **Expression of insect hormone genes.** Disruption, over expression or inactivation of one or more insect hormones results in abnormal growth, feeding cessation and/or death. So, the insertion of genes that encode insect hormones were the first strategies used to generate genetically modified baculovirus. A recombinant virus of *Bombyx mori* MNPV (BmNPV) that encodes an active diuretic hormone (DH) showed to be 20% faster in killing larvae than wild type virus (Maeda *et al.*, 1989). Later, by the deletion of *egt* gene, which prevents the larval molt, the mutant virus resulted to be 30% faster in killing larvae and in a considerable reduction in food intake than wild type virus (O´Reilly & Miller, 1991). Also, this gene may be replaced by an exogenous gene and enhance the insecticidal activity (Arif, 1997; Popham *et al.*, 1997; Sun *et al.*, 2004).

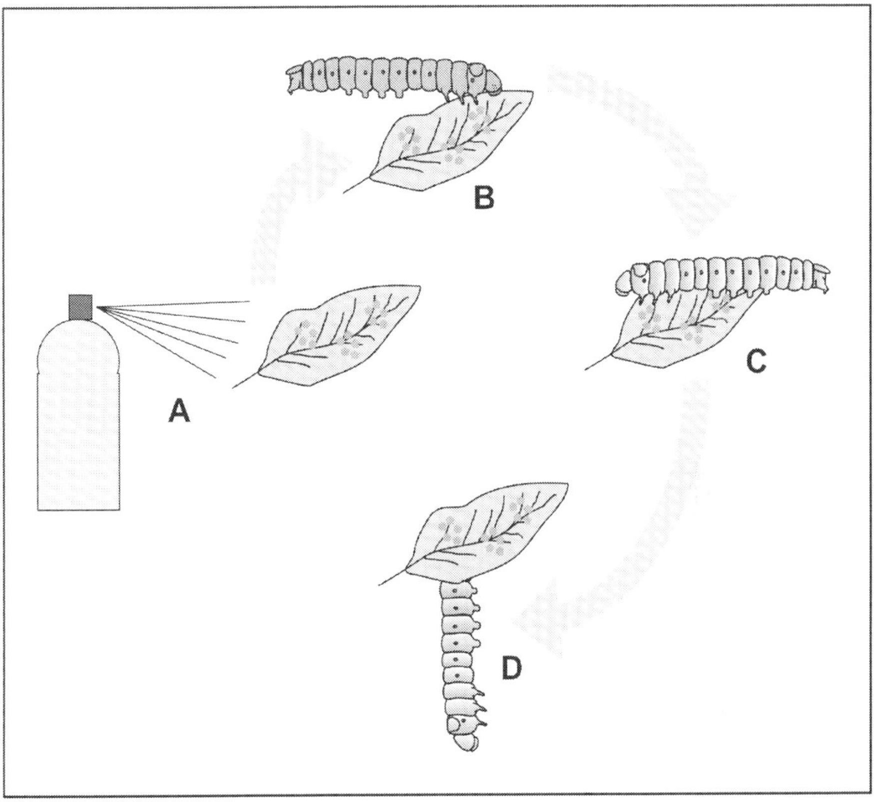

Fig. 3. **Baculoviral bioinsecticide. A**. Bioinsecticide spreading on leaves. **B**. Larvae feeding a contaminated leaf. **C**. Infected larvae. **D**. Dead larvae. Released virus will permit the resume of the infection cyle.

Baculovirus safety

Alpha- and Betabaculoviruses are not infectious for predatory or beneficial insects outside of the order Lepidoptera, or toward other non-targeted organisms (Black *et al.*, 1997; Szewczyk et al., 2006). Baculoviruses do not replicate in mammals, birds, fish, amphibians, reptiles and other vertebrates (Barsoum *et al.*, 1997). Moreover, different studies have concluded that the transgenes that encode toxins or other foreign proteins in recombinant baculovirus increase the speed to kill the specific target, but do not confer any selective ecological advantage compared with the wild-type baculovirus. (Cory *et al.*, 1997; Lee *et al.*, 2001). Another concern associated with GM organisms is the possibility of genetic recombination that results in the "jump" of the recombinant baculovirus foreign gene to another organism. If the pesticide based on a recombinant baculovirus is used long enough and in sufficiently high concentrations in the field, it is expected that genetic recombination could occur. In both the field and in laboratory conditions, this phenomenon is expected to be higher among highly homologous baculoviruses infectious to the same host (Hajos *et al.*, 2000). However, it is important to note that the recombinant virus with a phenotype more virulent than the wild type has disadvantages, since they produces fewer progeny and are rapidily out-competed in the ecosystem (Inceoglu *et al.*, 2001).

Future prospects

In order to improve the use of baculovirus in pest control management, various methodologies are being studied and developed. Among these we can mention:

1. Expression of fusion proteins to expand the host range. Because ButaIT toxin alone exhibits weak oral toxicity, and as an alternative to the use of recombinant baculovirus different strategies have been developed based on the generation of fusion proteins. One of the proteins used is a lectin, which functions as a carrier (GNA, *Galanthus nivalis* agglutinin) and the other protein is a toxin. A fusion protein comprising GNA and SFI1 has been shown to have insecticidal effects on both lepidopteran and homopteran plant pests (Fitches *et al.*, 2004; Down *et al.*, 2006) while a fusion protein combining GNA with ButaIT has also been shown to have insecticidal activity against lepidopteran larvae. The injection data reported show that fusion proteins containing SFI1 and ButaIT are insecticidal towards a range of insects including lepidopteran, dipteran, coleopteran and dictyopteran pests (Fitches *et al.*, 2010).
2. Antisense strategies. A new approach to the development of integrated pest control is the use of technologies based on iRNA (Gordon & Waterhouse, 2007; Price & Gatehouse, 2008). For efficient pest control methods using iRNA in the field, the major challenge is the development of easy and reliable methods for production and delivery of dsRNA. The dsRNA can be produced *in vivo* using a bacteria system or can be synthesized *in vitro*. Other way to produce dsRNA is using transgenic plants that express the desired antisense. The delivery of dsRNA can be done by spraying dsRNA on the crop plants or by feeding insects. In a recent study feeding heat-killed bacteria that produced dsRNA or *in vitro* synthesized dsRNA was used to silence five target genes in Colorado potato beetle, *Leptinotarsa decemlineata*. It was observed that the loss of function of these target genes caused larval mortality and significantly decreased insect growth (Zhu *et al.*, 2011). In a similar study, it was used a synthetic iRNA to inhibit a mitochondrial electron transport of *Plutella xylostela*, causing prominent insect mortality (Gong *et al.*, 2011). The use of different target genes could minimize the risk of resistance development. Offtarget effects and species specificity of dsRNA are two major potential issues that need to be addressed. However, the same way as was observed with *Caenorhabditis elegans* and other plant nematodes (Price & Gatehouse, 2008), in insects, described methods to feed dsRNA to a range of different insect species demonstrated that even highly conserved genes can be exploited to trigger species-specific iRNA without affecting non-target species (Whyard *et al.* 2009).
3. Combined control methods. The use of synergism between two different pathogens is a strategy that could be useful to conduct pest control strategies. To evaluate this possibility it have been conducted studies using *Spodoptera exigua* larvae and assessed the effect of using a baculovirus (*Spodoptera exigua* NPV) and a parasitoid (*Microplitis pallidipes*) to control this pest (Jiang *et al.*, 2011). The results obtained indicate that when *M. pallidipes*-parasitised *S. exigua* larvae were infected by SeMNPV, the virus did not affect the developmental period of *M. pallidipes* in the host, and most parasitoids

completed development, possibly because parasitism by *M. pallidipes* reduced larval sensitivity to the virus. The results of this study also indicate that *M. pallidipes* is an important vector of SeMNPV and contributes to natural epizootics of the virus. Female parasitoids that had developed or oviposited in virus-infected hosts, or that emerged from cocoons contaminated with virus, were able to transmit infective doses of virus to healthy host larvae (Jiang *et al.*, 2011).

4. Conclusions

Considering the amount of baculovirus species that have been isolated so far, its development as bio-pesticides has not been commensurate with all its potential. Most of the viruses found in commercial phase are produced by small or medium companies or by the users themselves, as is the case of USDA Forest Service (US Department of Agriculture), CIP (International Potato Center) or EMBRAPA (Empresa Brasilera de Pesquisa Agropecuaria). Among the factors limiting their commercialization it has been noted that they are too specific, they present a slow speed of action, they have a low persistence in the field, and are costly to produce infecting larva. Also, the approaches that involve *in vitro* production processes in insect cell cultures are still in development stage.

An entomopathogenic product will be considered as a viable alternative in pest control if it meets control with the same speed, ease of use, at the same cost than a chemical insecticide. This way, do not take into account their unique capabilities: the ability to replicate in their host and be dispersed in culture, the ability to synergistically act with natural enemies, and availability to be produced locally or regionally. Faced with this situation have been conducted many studies to increase the speed of action, to extend the host range and maintain their safety to non-target organisms. This involves the generation of genetically modified baculoviruses and the use of different viral formulations (Cherry & Williams, 2001).

Baculovirus survival in the environment can be affected by temperature, pH, moisture, the presence of additives, exposure to UV light, and by the action of some plant metabolites such as peroxidases, that generate free radicals (Hoover *et al.*, 1998; Zhou *et al.*, 2004).

Actually, there have been developed some protective agents against UV that have been included into the viral formulations. Some commercial bleach such as Phorwite AR, Blankophor, and Tinopal C1101 are very effective to protect *Lymantria dispar* Nucleopolyhedrovirus -LdNPV-, *Helicoverpa zea* Nucleopolyhedrovirus -HzNPV-, and *Spodoptera frugiperda* Multiple Nucleopolyhedrovirus -SfMNPV- (Shapiro *et al.*, 1994; Zou & Young, 1994, 1996; Mondragón *et al.*, 2007). Recent studies have shown that the addition of 1% (wt:vol) aqueous extracts of cocoa (*Theobroma cacao* L.) (Malvales: Malvaceae), coffee (*Coffea arabica* L.) (Gentianales: Rubiaceae), and green and black tea (*Camellia sinensis* L.) (Ericales: Theaceae) provided excellent UV radiation protection for the beet armyworm, *Spodoptera exigua* Multiple Nucleopolyhedrovirus under laboratory conditions (SeMNPV) (El-Salamouny *et al.*, 2009).

Regarding the safety of genetically modified baculovirus, a recombinant HzNPV carrying the *aait* gene was not pathogenic for bees, birds, fish, and other vertebrates (Szewczyk *et al.*, 2006). Natural enemies of the larvae, like parasitoids and predators, were not adversely affected by ingesting individuals infected with recombinant viruses (Li *et al.*, 1999; Smith *et al.*, 2000; Boughton *et al.*, 2003). So far, neither has been shown that the transgene be transferred from baculovirus to other organism or virus (Inceoglu *et al.*, 2001). On this basis, there is no evidence that a recombinant baculovirus is more dangerous than the corresponding wild type.

Finally, knowledge about the biology of baculoviruses suggests that bioinsecticides based on formulations containing these viruses have much lower risk to the environment than the classic chemical insecticides (Szewczyk *et al.*, 2006).

5. Acknowledgment

This work was supported by research funds from *Agencia Nacional de Promoción Científica y Técnica* (ANPCyT) and *Universidad Nacional de Quilmes*. PDG is member of the Research Career of CONICET (*Consejo Nacional de Ciencia y Tecnología*); MNB holds a postdoctoral fellowship of CONICET.

6. References

Aranda, E.; Sánchez, J.; Peferoen, M.; Güereca, L.; Bravo, A. (1996). "Interactions of the *Bacillus thuringiensis* crystal proteins with the midgut epithelial cells of *Spodoptera frugiperda* (Lepidoptera: Noctuidae)". J. Invertbr. Pathol. 68, pp. 203-212.

Barbosa, A.E.A.D.; Albuquerque1, E.V.S.; Silva1, M.C.M.; Souza1, D.S.L.; Oliveira-Neto1, O.B.; Valencia, A.; Rocha1, T.L.; Grossi-de-Sa, M.F. (2010). "α-amylases inhibitor-1 gene from *Phaseolus vulgaris* expressed in *Coffea arabica* plants inhibits α-amylases from the coffee berry borer pest". BMC Biotechnology. 10, pp. 44-52.

Barsoum, J.; Brown, R.; McKee, M.; Boyce, F.M. (1997). "Efficient transduction of mammalian cells by a recombinant baculovirus having the vesicular stomatitis virus G glycoprotein". Hum. Gene Ther., 8, pp. 2011-2018.

Bird, F.T.; Burk, J.M. (1961). "Artificial disseminated virus as a factor controlling the European spruce sawfly, *Diprion hercyniar* (Htg) in the absence of introduced parasites". Canad. Entomol. 96, pp. 228-238.

Black, B.C.; Brennan, L.A.; Dierks, P.M.; Gard, I.E. (1997). "Commercialization of baculoviral insecticides". In The baculovirus, ed by Miller LK, Plenum Press, New York, pp. 341-387.

Boughton, A.J., Obrycki, J.J.; Bonning, B.C. (2003). "Effects of a protease-expressing recombinant baculovirus on nontarget insect predators of *Heliothis virescens*". Biol. Control. 28, pp. 101-110.

Brown, W.E.; Takio, K.; Titani, K.; Ryan, C.A. (1985). "Wound-induced trypsin inhibitor in alfalfa leaves: identity as a member of the Bowman-Birk inhibitor family". Biochemistry 24, pp. 2195-2108.

Caballero, P.; Murillo, R.; Muñoz, D.; Williams, T. (2009). "El nucleopoliedrovirus de *Spodoptera exigua* (Lepidoptera: Noctuidae) como bioplaguicida: análisis de avances recientes en España". Revista Colombiana de Entomología 35, pp. 105-115.

Chakrabarty, S.; Monsour, C.; Teakle, R.; Reid, S. (1999). "Yield, biological activity, and field performance of a wild-type Helicoverpa nucleopolyhedrovirus produced in H. zea cell cultures". J. Invertebr. Pathol. 73, pp. 199-205.

Cherry, A.; Williams, T. (2001). "Control de insectos plaga mediante los baculovirus". En: Los baculovirus y sus aplicaciones como bioinsecticidas en el control biológico de plagas. P. Caballero, M. López-Ferber. & T. Williams Eds. Phytoma S.A., Valencia, España. pp. 389-450.

Coelho, J.S.; Santos, N.D.; Napoleão, T.H.; Gomes, F.S.; Ferreira, R.S.; Zingali, R.B.; Coelho, L.C.; Leite, S.P.; Navarro, D.M.; Paiva, P.M. (2009). "Effect of *Moringa oleifera* lectin on development and mortality of *Aedes aegypti* larvae". Chemosphere 77, pp. 934-938.

Cory, J.S.; Hirst, M.L.; Williams, T.; Halls, R.S.; Goulson, D.; Green, B.M.; Carty, T.M.; Possee, R.D.; Cayley, P.J.; Bishop, D.H.L. (1994). "Field trial of a genetically improved baculovirus insecticide". Nature (London). 360, pp. 138-140.

Crickmore, N.; Zeigler, D.; Feitelson, J.; Schnepf, E.; Van Rie, J.; Lereclus D. (2008). "*Bacillus thuringiensis* toxin database". Online: http://www.lifesci.sussex.ac.uk/home/Neil_Crickmore/Bt/index.html.

Dawson, G.W.; Hallahan, D.L., Mudd, A.; Patel, M.M.; Pickett, J.A.; Wadhams, L.J.; Wallsgrove, R.M. (1989). "Secondary plant metabolites as targets for genetic modification of crop plants for pest resistance". Pestic. Sci. 27. pp 191-201.

Down, R.E.; Fitches, E.C.; Wiles, D.P.; Corti, P.; Bell, H.A.; Gatehouse, J.A. et al.. (2006). "Insecticidal spider venom toxin fused to snowdrop lectin is toxicto the peach-potato aphid, *Myzus persicae* (Hemiptera: *Aphididae*) and the rice brown planthopper, *Nilaparvata lugens* (Hemiptera:Delphacidae)". Pest Manag. Sci. 62, pp. 77–85.

Elazar, M., Levi, R.; Zlotkin, E.J. (2001). "Targeting of an expressed neurotoxin by its recombinant baculovirus". Exp. Biol. 204, pp. 2637-2645.

El-Salamouny, S.; Ranwala, D.; Shapiro, M.; Shepard, B.M.; Farrar, R.R. Jr. (2009). "Tea, coffee, and cocoa as ultraviolet radiation protectants for the beet armyworm nucleopolyhedrovirus". J. Econ. Entomol. 102, pp. 1767-1773.

Elvira, S.; Williams, T.; Caballero, P. (2010). "Juvenile hormone analog technology: effects on larval cannibalism and the production of *Spodoptera exigua* (Lepidoptera: Noctuidae) nucleopolyhedrovirus". J. Econ. Entomol. 103, pp. 577-582.

EPA (Environmental Protection Agency) (2008). "Pesticide Registration eligibility decision (RED): *Bacillus thuringiensis*. Online: http://www.epa.gov/oppsrrd1/REDs/0247.pdf.

Evans, H.F.; Harrap, K.A. (1982). "Persistence of insect viruses". In Virus persistence. Vol 33, ed. BWJ Mahy, AC Minson & GK Darby. SGM Symposium Cambridge University Press, pp. 57-96.

Faulkner, P.; Kuzio, J.; Williams, G.V.; Wilson, J.A. (1997). "Analysis of p74, a PDV envelope protein of *Autographa californica* nucleopolyhedrovirus required for occlusion body infectivity in vivo". J. Gen. Virol. 78, pp. 3091-3100.

Feitelson, J.S.; Payne, J.; Kim, L. (1992). "*Bacillus thuringiensis*: insects and beyond". Nat. Biotech. 10, pp. 271-275.

Flint, M.L.; van den Bosch, R. (1981). "Introduction to Integrated Pest Management". Plenum Press. Nueva York, EEUU, pp. 240.

Fitches, E.C.; Bell, H.A.; Powell, M.E.; Back, E.; Sargiotti, C.; Weaver, R.J.; Gatehouse, J.A. (2010). "Insecticidal activity of scorpion toxin (ButaIT) and snowdrop lectin (GNA) containing fusion proteins towards pest species of different orders". Pest Manag. Sci. 66, pp. 74–83.

Fitches, E.; Edwards, M.G.; Mee, C.; Grishin, E.; Gatehouse, A.M.R.; Edwards, J.P.; Gatehouse, J.A. (2004). "Fusion proteins containing insect-specific toxins as pest control agents: snowdrop lectin delivers fused insecticidal spider venom toxin to insect haemolymph following oral ingestion". J. Insect Physiol. 50, pp. 61–71.

Franco, O.L.; Rigden, D.J.; Melo, F.R.; Grossi-de-Sá, M.F. (2002). "Plant α-amylase inhibitors and their interaction with insect α-amylases. Structure, function and potential crop protection". Eur. J. Biochem. 269, pp. 397-412.

Gatehouse, J.A.; Hilder, V.A.; Gatehouse, A.M.R. (1991). "Genetic engineering of plant for insect resistance". In Plant Genetic Engineering, Plant Biotechnology Series, Vol 1, ed. D. Grierson. Blackie & Sons Ltd, London/Chapman and Hall, New York, pp. 105-135.

Gatehouse, J.A. (2011). "Prospects for using proteinase inhibitors to protect transgenic plants against attack by herbivorous insects". Curr. Protein Pept. Sci. 12, pp. 409-416.

Gazit, E.; La Rocca, P.; Samson, M.S.; Shai, Y. (1998). "The structure and organization within the membrane of the helices composing the pore-forming domain of *Bacillus thuringiensis* δ-endotoxin are consistent with an "umbrella-like" structure of the pore". Proc. Natl. Acad. Sci. USA 95, pp. 12289-12294.

Georghiou, G.P.; Wirth, M.C. (1997). "Influence of exposure to single versus multiple toxins of *Bacillus thuringiensis* subsp. israelensis on development of resistance in the mosquito *Culex quinquefasciatus* (Diptera: Culicidae)". Appl. Environ. Microbiol. 63, pp. 1095-1101.

Gerber, D.; Shai, Y. (2000). "Insertion and organization within membranes of the δ-endotoxin pore-forming domain, helix 4-loop-helix 5, and inhibition of its activity by a mutant helix 4 peptide". J. Biol. Chem. 275, pp. 23602-23607.

Gong, L.; Yang, X.; Zhang, B.; Zhong, G.; Hu, M.Y. (2011). "Silencing of Rieske iron–sulfur protein using chemically synthesised siRNA as a potential biopesticide against *Plutella xylostella*". Pest Manag. Sci. DOI 10.1002/ps.2086

Gordon, D.; Moskowitz, H.; Eitan, M.; Warner, C.; Catterall, W.A.; Zlotkin, E. (1992). "Localization of receptor sites for insect-selective toxins on sodium channels by site-directed antibodies". Biochemistry 31, pp. 7622-7628.

Gordon, K.H.; Waterhouse, P.M. (2007). "RNAi for insect-proof plants". Nature Biotechnol. 25, pp. 1231-1232.

Graham, J.; McNicol, R.J.; Greig, K. (1995). "Towards genetic based insect resistance in strawberry using the cowpea trypsin inhibitor". Ann. Appl. Biol. 127, pp. 163-173.

Gramkow, A.W.; Perecmanis, S.; Sousa, R.L.; Noronha, E.F.; Felix, C.R.; Nagata, T.; Ribeiro, B.M. (2010). "Insecticidal activity of two proteases against *Spodoptera frugiperda* larvae infected with recombinant baculoviruses". Virol. J. 7, doi:10.1186/1743-422X-7-143.

Granados, R.R.; Federici, B.A. (1986). "Biology of baculoviruses". CRC, Boca Ratón, Florida.

Granados, R.R.; Lawler, K.A. (1981). "In vivo pathway of *Autographa californica* baculovirus invasión and infection". Virology 108, pp. 297-308.

Green, T.R. & Ryan, C.A. (1972). "Wound-induced proteinase inhibitor in plant leaves: a possible defence mechanism against insects". Science 175, pp. 776-777.

Grossi-de-Sá, M.F.; Chrispeels, M.J. (1997). "Molecular cloning of bruchid (*Zabrotes subfasciatus*) α-amylase cDNA and interaction of the expressed enzyme with bean α-amylase inhibitors". Insect Biochem. Mol. Biol. 27, pp. 271-281.

Haas-Stapleton, E.J.; Washburn, J.O.; Volkman, L.E. (2004). "P74 mediates specific binding of *Autographa californica* M nucleopolyhedrovirus occlusion-derived virus to primary cellular targets in the midgut epithelia of *Heliothis virescens* Larvae". J. Virol. 78, pp. 6786-6791.

Hajos, J.P.; Pijnenburg, J.; Usmany, M.; Zuidema, D.; Zavodszki, P.; Vlak, J.M. (2000). "High frequency recombination between homologous baculoviruses in cell culture". Arch. Virol. 145, pp. 159-164.

Hallahan, D.L., Pickett, J.A., Wadhams, L.J., Wallsgrove, R.M.; Woodcock, C.M. (1992). "Potential of secondary metabolites in genetic engineering of crops for resistance". In Plant genetic manipulation for crop protection-Biotechnology in agriculture No 7, ed. A.M.R. Gatehuse, V.A. Hilder and D. Boulter. CAB international, Wallingford, Oxon, UK, pp. 215-248.

Hamshou, M.; Van Damme, E.J.; Smagghe, G. (2010). "Entomotoxic effects of fungal lectin from *Rhizoctonia solani* towards *Spodoptera littoralis*". Fungal Biol. 114, pp. 34-40.

Harrap, K.A. (1970). "Cell infection by a nuclear polyhedrosis virus". Virology 42, pp. 311-318.

Hawtin, R.E.; Arnold, K.; Ayres, M.D.; Zanotto, P.M.; Howard, S.C.; Gooday, G.A.; Chappell, L.H.; Kitts, P.A.; King, L.A.; Posse, R.D. (1995). "Identification and preliminary characterization of a chitinase gene in the *Autographa californica* nuclear polyhedrosis virus genome". Virology 212, pp. 673-685.

Hoover, K.; Kishida, K.T.; DiGiorgio, L.A.; Workman, J.; Alaniz, S.A.; Hammon, B.D.; Duffey, S.S. (1998). "Inhibition of baculoviral disease by plant-mediated peroxidase activity and free radical generation". J. Chem. Ecol. 24, pp. 1949-2001.

Ignacimuthu, S.; Prakash, S. (2006). "Agrobacterium-mediated transformation of chickpea with alpha-amylase inhibitor gene for insect resistance". Journal of Biosciences 31, pp. 339–345.

Ignoffo, C.M. (1992). "Environmental factors affecting persistence of entomolpathogens". Florida Entomol. 75, pp. 516-525.

Inceoglu, A.B.; Kamita, S.G.; Hinton, A.C.; Huang, Q.; Severson, T.F.; Kang, K.; Hammock, B.D. (2001). "Recombinant baculoviruses for insect control". Pest Manag. Sci. 57, pp. 981-987.

Ives, A.R.; Glaum, P.R.; Ziebarth, N.L.; Andow, D.A. (2011). "The evolution of resistance to two-toxin pyramid transgenic crops". Ecol. Appl. 21, pp. 503-515.

James, C. (2011). "Global Status of Commercialized Biotech/GM Crops: 2010". Ithaca, NY, ISAAA www.isaaa.org.

Jehle, J.A.; Blissard, G.W.; Bonning, V.; Cory, J.S.; Herniou, E.A.; Rohrmann, G.F.; Theilmann, D.A.; Thiem, S.M.; Vlak, J.M. (2006). "On the classification and nomenclature of baculoviruses: A proposal for revision". Arch. Virol. 151, pp. 1257-1266.

Jiexian Jiang, Aiping Zeng, Xiangyun Ji, Nianfeng Wan and Xiaoqin Chen. (2011). "Combined effect of nucleopolyhedrovirus and *Microplitis pallidipes* for the control of the beet armyworm, *Spodoptera exigua*". Pest Manag. Sci. 67, pp. 705-713.

Kaur, M.; Singh, K.; Rup, P.J.; Kamboj, S.S.; Singh, J.J. (2009). "Anti-insect potential of lectins from Arisaema species towards *Bactrocera cucurbitae*". Environ. Biol. 30, pp. 1019-1023.

Kuzio, J.; Jaques, R.; Faulkner, P. (1989). "Identification of p74 a gene essential for virulence of baculovirus occlusion bodies". Virology 173, pp. 759-763.

Lee, Y.; Fuxa, J.R.; Inceoglu, A.B.; Alaniz, S.A.; Richter, A.R.; Reilly, L.M.; Hammock, B.D. (2001). "Competition between wild type and recombinant nucleopolyhedroviruses in a greenhouse microcosm". Biol. Control 20, pp. 84-93.

Maeda, S. (1989). "Increased insecticidal effect by a recombinant baculovirus carrying a synthetic diuretic hormone gene". Biochem. Biophys. Res. Commun. 165, pp. 1177-1183.

Mettenmeyer, A. (2002). "Viral insecticidas hold promise for bio-control". Farming Ahead 124, pp. 50-51.

Mondragón, G.; Pineda, S.; Martínez, A.; Martínez, A.M. (2007). "Optical brightener Tinopal C1101 as an ultraviolet protectant for a nucleopolyhedrovirus". Commun. Agric. Appl. Biol. 72, pp. 543-547.

Moscardi, F. (1989). "Use of viruses for pest control in Brazil: the case of the nuclear polyhedrosis virus of the soybean caterpillar, *Anticarsia gemmatalis*". Memorias do Instituto Oswakdo Cruz, Rio de Janeiro, 84, pp. 51-56.

Moscardi, F. (1999). "Assessment of the application of baculoviruses for control of Lepidoptera". Annu. Rev. Entomol. 44, pp. 257-289.

Neppl, C. (2000). "Management of resistance to *Bacillus thuringiensis* toxins". The Environmental Studies Program Education- University of Chicago (Chicago, U.S.A.).

Norris, D.M. & Kogan, M. (1980). "Biochemical and morphological bases of resistance". In Breeding Plants Resistant to Insects, ed. F.G. Maxwell and P.R. Jennings. John Wiley & Sons, London, pp. 23-60.

O´Reilly, D.R.; Miller, L.K. (1991). "A baculovirus blocks insecto molting by producing ecdysteroid UDP-Glucosyl Transferase". Science (Washington) 245, pp. 1110-1112.

Payan, F. (2004). "Structural basis for the inhibition of mammalian and insect α-amylases by plant protein inhibitors". Biochim. Biophys. Acta 1696, pp. 171-180.

Pelegrini, P.B.; Murad, A.M.; Grossi-de-Sa, M.F.; Mello, L.V.; Romeiro, L.A.; Noronha, E.F.; Caldas, R.A.; Franco, O.L. (2006). "Structure and enzyme properties of *Zabrotes subfasciatus* α-amylase". Arch. Insect Biochem. Physiol. 61, pp. 77-86

Popham, H.J.R.; Li, Y.; Miller, L.K. (1997). "Genetic improvement of *Helicoverpa zea* polyhedrosis virus as a biopesticide". Biol. Control 10, pp. 83-91.

Price, D.R.G.; Gatehouse, J.A. (2008). "RNAi-mediated crop protection against insects". Trends Biotechnol. 26, pp. 393-400.

Pritchett, D.W.; Young, S.Y.; Geren, C.R. (1981). "Proteolytic activity in the digestive fluid of larvae of *Trichoplusia ni*". Insect Biochem. 11, pp. 523-526.

Pritchett, D.W.; Young, S.Y.; Yearian, W.C. (1984). "Some factors involved in the dissolution of *Autographa californica* NPV by digestive fluids of *Trichoplusia ni* larvae". J. Invert. Pathol. 43, pp. 160-168.

Pusztai, A.; Grant, G.; Steward, J.C.; Bardocz, S.; Ewen, S.W.B.; Gatehouse, A.M.R. (1992). "Nutritional evaluation of the trypsin inhibitor from cowpea". British J. Nutr. 68, pp. 783-791.

Regev, A.; Rivkin, H.; Inceoglu, B.; Gershburg, E.; Hammock, B.D.; Gurevitz, M.; Chejanovsky, N. (2003). "Further enhancement of baculovirus insecticidal efficacy with scorpion toxins that interact cooperatively". FEBS Lett. 537, pp. 106-110.

Rohrmann, G. (2008). "Baculovirus Molecular Biology". Bethesda (MD): National Library of Medicine (US), NCBI.

Sarmah, B.K.; Moore, A.; Tate, W.; Molvig, L.; Morton, R.L.; Rees, D.P.; Chiaiese, P.; Chrispeels, M.J.; Tabe, L.M.; Higgins, T.J.V. (2004). "Transgenic chickpea seeds expressing high levels of a bean alpha-amylase inhibitor". Mol. Breed. 14, pp. 73-82.

Schnepf, E.; Crickmore, N.; Van Rie, J.; Lereclus, D.; Baum, J.; Feitelson, J.; Zeigler, D.R.; Dean, D.H. (1998). "*Bacillus thuringiensis* and its pesticidal crystal proteins". Microbiol. Mol. Biol. 62, pp. 775-806.

Schroeder, H.E.; Gollasch, S.; Moore, A.; Tabe, L.M.; Craig, S.; Hardie, D.C.; Chrispeels, M.J.; Spencer, D.; Higgins, T.J.V. (1995). "Bean alpha-Amylase lnhibitor Confers Resistance to the Pea Weevil (*Bruchus pisorum*) in Transgenic Peas (*Pisum sativum*)". Plant Physiol. 107, pp. 1233-1231. 1239.

Sedlak, M.; Walter, T; Aronson, A. (2000). "Regulation by overlapping promoters of the rate of synthesis and deposition into crystalline inclusions of *Bacillus thuringiensis* δ-endotoxins". J. Bacteriol. 182, pp. 734-741.

Shade, R.E.; Schroeder, H.E.; Pueyo, J.J.; Tabe, L.M.; Murdock, L.L.; Higgins, T.J.V.; Chrispeels, M.J. (1994). "Transgenic Pea-seeds expressing the alpha-amylase inhibitor of the common bean are resistant to bruchid beetles". Bio-Technology 12, pp. 793-796.

Shapiro, M.; Daugherty, E.M. (1994). "Enhacement in activity of homologous and heterologous viruses against gypsy moth (Lepidoptera, Lymantridae) by an optical brightener". J. Econ. Entomol. 87, pp. 361-365.

Shumway, L.K.; Yang, V.V.; Ryan, C.A. (1976). "Evidence for the presence of proteinase inhibitors I in vacuolar protein bodies of plant cell". Planta 129, pp. 161-165.

Slack, J.M.; Kuzio, J.; Faulkner, P. (1995). "Characterization of v-cath, a cathepsin, L-like proteinase expressed by the baculovirus *Autographa californica* multiple nuclear polyhedrosis virus". J. Gen. Virol. 76, pp. 1091-1098.

Smith, C.R.; Heinz, K.M.; Sansone, C.G.; Flexner, J.L. (2000). "Impact of recombinant baculovirus field applications on non-target heliothine parasitoid *Microplitis croce*ipes (Hymenoptera: Braconidae)". J. Econ. Entomol. 93, pp. 1109-1117.

Solleti, S.K.; Bakshi, S.; Purkayastha, J.; Panda, S.K.; Sahoo, L. (2008). "Transgenic cowpea (*Vigna unguiculata*) seeds expressing a bean alpha-amylase inhibitor 1 confer resistance to storage pests, bruchid beetles". Plant Cell Rep. 27, pp. 1841-1850.

Sun, X.; Sun, X.; van der Werf, W.; Vlak, J.M.; Hu, Z. (2004). "Field inactivation of wild type and genetically modified *Helicoverpa armigera* single nucleocapsid nucleopolyhedrovirus in cotton". Biocontrol Sci. Technol. 14, pp. 185-192.

Szewczyk, B., Hoyos-Carvajal, L.; Paluszek, M,; Skrzecz, I.; Lobo de Souza, M. (2006). "Baculoviruses-re-emerging biopesticides". Biotechnol. Adv. 24, pp. 143-160.

Walker-Simmons, M. & Ryan, C.A. (1977). "Immunological identification of proteinase inhibitor I and II in isolated tomato leaf vacuoles". Plant Physiol. 60, pp. 61-63.

Washburn, J.O.; Chan, E.Y.; Volkman, L.E.; Aumiller, J.J.; Jarvis, D.L. (2003). "Early synthesis of budded virus envelope fusion protein GP64 enhances *Autographa californica* multicapsid nucleopolyhedrovirus virulence in orally infected *Heliothis virescens*". J. Virol. 77, pp. 280-290.

Whyard, S.; Singh, A.D.; Wong, S. (2009). "Ingested double-stranded RNAs can act as species-specific insecticides". Insect Biochem. Mol. Biol. 39, pp. 824-832.

Wudayagiri, R.; Inceoglu, B.; Herrmann, R.; Derbel, M.; Choudary, P.V.; Hammock, B.D. (2001). "Isolation and characterization of a novel lepidopteran-selective toxin from the venom of South Indian red scorpion, *Mesobuthus tamulus*". BMC Biochem. 2, pp. 11-16.

Xu, D.P.; Xue, Q.Z.; McElroy, D.; Mawal, Y.; Hilder, V.A.; Wu, R. (1996). "Constitutive expression of a cowpea trypsin-inhibitor gene, CpTI, in transgenic rice plants confers resistance of two major rice insect pest". Mol. Breed. 2, pp. 167-173.

Yamada, T.; Hattori, K.; Ishimoto, M. (2001). "Purification and characterization of two α-amylase inhibitors from seeds of tepary bean (*Phaseolus acutifolius* A. Gray)". Phytochem. 58, pp. 59-66.

Yang, Z.; Chen, H.; Tang, W.; Hua, H.; Lin, Y. (2011). "Development and characterisation of transgenic rice expressing two *Bacillus thuringiensis* genes". Pest Manag. Sci. 67, pp. 414-422.

Yu, H.L.; Li, Y.H.; Wu, K.M. (2011). "Risk Assessment and Ecological Effects of Transgenic *Bacillus thuringiensis* Crops on Non-Target Organisms(F)". J. Integr. Plant Biol. 53, pp. 520-538.

Zhang, G.Y.; Sun, X.L.; Zhang, Z.X.; Zhang, Z.F.; Wan, F.F. (1995). "Production and effectiveness of the new formulation of Helicoverpa virus pesticide-emulsifiable suspension". Virologica Sinica 10, pp. 242-247.

Zhou, M.Z.; Sun, H.C.; Hu, Z.H.; Sun, X.L. (2004). "SOD enhances infectivity of *Helicoverpa armigera* single nucleocapsid nucleopolyhedrosis virus against *H. armigera* larvae". Virologica Sinica 18, pp. 506-507.

Zhou, X.Y.; Liu, N.; Zhao, M.; Li, H.; Zhou, L.; Tang, Z.W.; Cao, F.; Li, W. (2011). "Advances in effects of insecticidal crystal proteins released from transgenic Bt crops on soil ecology". Yi Chuan. 33, pp. 443-448.

Zhu, F.; Xu, J.; Palli, R.; Ferguson, J.; Palli, S.R. (2011). "Ingested RNA interference for managing the populations of the Colorado potato beetle, *Leptinotarsa decemlineata*". Pest Manag. Sci. 67, pp. 175-182.

Zou, Y.; Young, Y.S. (1994). "Enhancement of nuclear polyhedrosis virus activity in larval pest of Lepidoptera by a stilbene fluorescent brightener". J. Entomol. Sci. 29, pp. 130-133.

Zou, Y.; Young, Y.S. (1996). "Use of fluorescent brightener to improve *Pseudoplusia includens* (Lepidoptera: Noctuidae) nuclear polyhedrosis virus activity in the laboratory and field". J. Econ. Entomol. 89, pp. 92-96.

21

Recent Advances in Our Knowledge of Baculovirus Molecular Biology and Its Relevance for the Registration of Baculovirus-Based Products for Insect Pest Population Control

Renée Lapointe[1], David Thumbi[1] and Christopher J. Lucarotti[2]
[1]Sylvar Technologies Inc., New Brunswick,
[2]Natural Resources Canada, Canadian Forest Service-Atlantic Forestry Centre, New Brunswick
[1,2]Canada

1. Introduction

Public demand for safer, environmentally-benign alternatives to synthetic chemical pesticides and more stringent barriers put in place by regulatory agencies worldwide has led to increased interest in microbial pest control agents (MPCA) based on viruses, bacteria, fungi, protozoa and nematodes as the active ingredient. The MPCA market has recently experienced an increase of 47% between 2004/2007 with sales worth $396 million in 2007/2008 (CPL Consultants, 2010). Despite this increase, the microbial biopesticide market still only represents about 1% of the sales of chemical pesticides (CPL Consultants, 2010; Marrone, 2007). Factors impeding the establishment of strong MPCA markets are complex (Chandler et al., 2011; Marrone, 2007; Ravensberg, 2011) but include the burdensome costs associated with the registration of commercial products that are aimed at relatively small niche markets (Chandler et al., 2011; Ehlers, 2011). The main priority of regulatory agencies is to protect human health and safety and the environment from potential risks associated with the use of pest control products. A common feature of MPCA registration processes around the world is that they grew out of registration processes designed for chemical pesticides with adjustments allowing for the reduced risks of MCPAs (Chandler et al., 2011; Ehlers, 2011). Even though public attitudes to the use of biological control agents has been favourable (63%) a large proportion of the public (46%) has expressed concerns about the consumption of food treated with microbial pesticides (Cuddleford, 2006).

Baculoviruses only infect insects, are ubiquitous in the environment and are known to be important in the regulation of many insect populations. Baculoviruses are host specific, infecting only one or a few closely-related species, helping to make them good candidates for management of crop and forest insect pests with minimal off-target impacts (Cory, 2003; England et al., 2004; Hewson et al., 2011; Raymond et al., 2005). In fact, baculoviruses have been recognized as being amongst the safest of pesticides (Black et al., 1997; Gröner, 1986) and have been included on lists of "low risk" biocontrol agents by the European Union (Leuschner, 2010; Regulation of Biological Control Agents in Europe; http://www.rebeca-net.de). Since the start of their commercial use, baculoviruses have been tested extensively to assess their safety in order to meet registration requirements (reviewed in Burges et al., 1980a, 1980b; Gröner, 1986; Ignoffo, 1973). As of 2010, over 24 baculovirus species have been reported to be registered for use in insect pest management throughout the world (Kabaluk et al., 2010; Quinlan & Gill, 2006). Market share of baculoviruses is 6% of all microbial pesticides (Marrone, 2007; Quinlan & Gill, 2006) and millions of hectares have been treated with registered baculovirus products over the years (Kabaluk et al., 2010; Moscardi, 2011; Szewczyk et al., 2009). Despite many years of use and testing against nontarget organisms, no adverse effects have ever been attributed to baculoviruses (McWilliam, 2007; OECD, 2002). In this review of baculoviruses, we discuss how baculovirus evolution, host range determination and pathogenesis have contributed to their inherent safety for nontarget organisms including humans.

2. Classification and origins of baculoviruses

The virus family *Baculoviridae* is divided into four genera that are restricted to three insect orders: *Alphabaculovirus* (nucleopolyhedrovirus or NPV) in Lepidoptera, *Betabaculovirus* (granulovirus or GV) also in Lepidoptera, *Gammabaculovirus* (NPV) in Hymenoptera, and *Deltabaculovirus* (NPV) in Diptera (Jehle et al., 2006). The large, covalently-closed, double-stranded DNA genome is packaged into an enveloped, rod-shaped capsid. The virions of NPVs are enveloped either singly (SNPV) or in groups of multiple virions (MNPV) which are then occluded in a protein called polyhedrin to form the occlusion body (OB). Virions of GVs are enveloped and occluded singly in an ovicylindrical granule (also an OB) formed from the protein granulin. Baculoviruses are normally named for the initial host from which they were first isolated. The International Committee on Taxonomy of Viruses (ICTV) lists the designated members of the Baculoviridae (http://www.ictvdb.org/Ictv/index.htm).

To date, 58 baculovirus genomes have been sequenced; 41 alphabaculoviruses, 13 betabaculoviruses, three gammabaculoviruses and one deltabaculovirus. Baculovirus genome sizes range from the smallest gammabaculovirus, *Neodiprion lecontei* nucleopolyhedrovirus (NeleNPV), at 81,755 bp (Lauzon et al., 2004) to the largest betabaculovirus, *Xestia c-nigrum* granulovirus (XcGV) at 178,733 bp (Hayakawa et al., 1999). No matter the genus or genome size, all baculoviruses share 31 core genes in common (Miele et al., 2011). These are essential genes involved in oral infection (*pif-0 (p74), pif-1, pif-2, pif-3, pif-4/19kd/odv-e28, pif-5/ odv-e56*), cell cycle arrest (*odv-ec27, ac81*), replication (*dnapol, helicase, lef-1, lef-2*), late gene transcription (*lef-4, lef-5, lef-8, lef-9, p47*) and virus assembly, packaging and release (*38k/ac98, alk-exo, desmoplakin, gp41, odv-e18, odv-nc42, odv-ec43, p6.9, p33/ac92, p49, vlf-1, vp39, vp91, vp1054*) (Miele *et al.* 2011). Twenty of these core genes (*p47, lef-4, lef-5, lef-8, lef-9, vlf-1, pif-0, pif-1, pif-2, pif-3, pif-4, pif-5, vp91, vp39, 38k, ac68, ac81, p33, dnapol, helicase*) are also found in insect dsDNA viruses belonging to the genus *Nudivirus* (Wang et al., 2011) and a number (e.g., *lef-4, lef-5, lef-8, p47, 38k/ac98, vp91, pif-0, pif-1, pif-2, pif-3*) are also found in bracoviruses (polydnaviruses) associated with parasitic wasps belonging to the family Braconidae (Bézier et al., 2009). It has been suggested that bracoviruses arose from the insertion of a nudivirus ancestor into braconid wasps about 100 million years ago (mya) (Bézier et al., 2009). Nudiviruses and baculoviruses, however, are thought to have shared a common ancestral virus (Wang & Jehle, 2009). Deltabaculoviruses and gammabaculoviruses are thought to be more primitive than the alphabaculoviruses and betabaculoviruses because of their smaller genomes and tissue tropism which is limited to midgut epithelial cells (Lauzon et al., 2004) and, in the case of deltabaculoviruses, cells of the posterior midgut and gastic caeca (Moser et al., 2001). Gammabaculoviruses, however, are thought to be more closely (although still distantly) related to the alphabaculoviruses and betabaculoviruses than are the deltabaculoviruses (Herniou et al., 2004). The virions of *Culex nigripalpus* deltabaculovirus (CuniNPV) are occluded in a 90-kDa protein that bears no similarity to polyhedrin/granulin proteins or any other protein in available sequence databases (Perera et al., 2006).

The occlusion derived virions (ODVs) that emerge from OBs are the universal virion phenotype for all baculoviruses as they are responsible for the initial oral (*per os*) infection of host insect gut cells. In lepidopteran hosts, the initial, primary infection of midgut cells by ODVs is followed by secondary infection of tissues within the insect hemocoel that is effected by the budded virion (BV) phenotype. The genome content of ODVs and BVs is identical but differences in virion morphology, structural proteins, envelopes, antigenicity, and cellular site of maturation are the basis for their respective patterns of infectivity. In mosquito hosts, following the primary infection of the gastric caeca and posterior midgut by ODVs, deltabaculovirus BVs also spread the infection further from cell to cell but only to these same tissues (Moser et al., 2001). Sawfly gammabaculoviruses do not appear to have a BV phenotype (Duffy et al., 2006; Garcia-Maruniak et al., 2004; Lauzon et al., 2004) and OBs are only produced in the nuclei of midgut epithelial cells (Federici, 1997).

In Lepidoptera, NPVs have been reported from 28 families and GVs from 19 families (Martignoni & Iwai, 1981). In the Diptera, NPVs have been reported from the Calliphoridae, Chironomidae, Culicidae, Sciaridae, Tachinidae and Tipulidae (Martignoni & Iwai, 1981). Fewer families of sawflies (Argidae, Diprionidae, Pamphiliidae and Tenthredinidae) are reported to be infected by NPVs (Martignoni & Iwai, 1981). However, due to a general lack of viral isolates, sequence data and other information, most of the baculoviruses listed by Martignoni & Iwai (1981) are not yet considered as valid species by ICTV (http://www.ictvdb.org/Ictv/index.htm). For example, the NPV of the pamphiliidid sawfly, *Acantholyda erythrocephala,* has been reported to occur in the fat body (Jahn, 1967), something which is not characteristic of gammabaculoviruses.

Recent phylogenetic analyses have indicated that the Hymenoptera, not the Coleoptera, are basal to the holometabolous insects that also include the Lepidoptera and Diptera (Savard et al., 2006). When and how the four genera of baculoviruses came to infect their different insect hosts is not known but selection pressure and co-evolution with their respective hosts appears to have constrained each baculovirus genus to a single insect order (Herniou et al., 2004). Historically, the Hymenoptera have been subdivided into the more advanced Apocrita, including ants, bees and wasps, and the basal Symphyta that includes the sawflies. It now appears, however, that the evolution from the ancestral hymenopteran to the Euhymenoptera (Apocrita and Orussoidea) was monophylletic and that the different superfamilies of sawflies constitute separate branches off the lower end of this lineage (Farris & Schulmeister, 2011). In this light, gammabaculoviruses have only been confirmed in the Diprionidae and, considering the paraphyletic origins of the different groups of sawflies, it may be the case that gammabaculoviruses are restricted to the Diprionidae or closely related families within the superfamily, Tenthredinoidea.

3. Baculovirus pathogenesis and potential blocks to infection

As is the case for all viral pathogens, baculovirus replication is dependent upon the availability of permissive host cells. The accessiblity and susceptibility of host cells to viral invasion and replication is classified into three categories; permissive, semi-permissive and non-permissive. A permissive infection results in successful viral replication and subsequent production of infectious virions that can transmit the infection to other permissive cells and individuals. Semi-permissive infections result in limited viral progeny resulting from defects in some replication events, such as gene expression or viral DNA replication. In non-permissive infections, cells do not support viral replication and the process does not yield infectious progeny. Determining what factors influence the level of permissiveness of an insect cell to a particular baculovirus has proven to be challenging because baculovirus host range is affected not only by the interactions between the baculovirus and the host cell at the molecular level (reviewed in Miller & Lu, 1997; Thiem & Cheng, 2009) but also by aspects of insect behaviour and physiology (reviewed in Cory & Hoover, 2006).

As hosts for viruses, insects can present challenges because of their sporadic and/or episodic availability and their relatively short life spans. Long periodicity of population fluctuations, for example the spruce budworm (*Choristoneura fumiferana*) which can span over 30 years (Royama, 1992), indicates that baculoviruses must be able to persist in the environment for long periods while waiting for permissive hosts to become available. The OB and its surrounding polyhedral envelope (PE) (a protein/carbohydrate matrix) (Gross & Rohrmann, 1993; Gross et al., 1994; Russell & Rohrmann, 1990) help protect the ODVs from degradation by such environmental factors as dessication and ultraviolet radiation (UV) (reviewed in Slack & Arif, 2007). OBs and ODVs can be further protected from UV radiation by establishing natural reservoirs in sheltered environments such as those in and on plants and in soil (Raymond et al., 2005; Witt, 1984).

3.1 Midgut lumen and pH

Baculoviruses are predominantly diseases of the larval stages of insects. When a larval host consumes foliage or water that is contaminated with OBs, the alkaline pH (8-11) of the larval midgut (Fig. 1) dissolve the PE and OB matrix within minutes (Adams & McClintock, 1991) releasing ODVs into the midgut lumen. The gut environment, into which OBs enter, is a first deciding specificity factor as OBs will only dissolve in an alkaline environment. The dissolution of OBs is further facilitated by OB-associated alkaline proteases. While the PE lattice is sufficiently narrow to restrict access by large digestive enzymes of vertebrates, it does allow infiltration by anions from the alkaline midgut of insects. Midgut pH of Lepidoptera averages 10.5, while within the Diptera, only in the Culicidae does the gut pH reach 10 (Terra et al., 1996). Within the Hymenoptera, only sawflies are known to harbor baculoviruses and their midgut pH, although lower than those of lepidopteran and culicid mosquitoes, is between 6.7 and 8.7 (Heimpel, 1955), which is more alkaline than that of bees and wasps (Apidae) at pH 5.7 (Terra et al., 1996). When compared to the midgut of other coleopteran families, such as the Coccinelidae, Chrysomelidae and Cerambycidae (midgut pH 5.4 to 6.9), the midgut pH of Scarabeidae is 10.4 (Terra et al., 1996). Although not a baculovirus *per se*, the nudivirus of the scarab *Oryctes rhinoceros* (OrN) shares homologies with baculoviruses *per os* factors and other core genes (Wang et al., 2011). Although the nudiviruses have established more complex tissue tropisms and transmission routes than baculoviruses, their primary route of infection is also *per os* (Wang & Jehle, 2009).

In addition to protecting the ODVs against environmental factors, the stability of the crystalline structure of the OBs has been shown to assist in the dispersal of the virus by vertebrates. The acidic pH of the stomachs (from pH 1 to 7) of vertebrates (Fig. 1) helps to preserve the integrity of the OBs. Excreted OBs, recovered from the digestive tracts of non-host invertebrate and vertebrate animals (Lautenschlager et al., 1980; Vasconcelos et al., 1996) were found to remain infectious to their insect larval hosts, leading to the suggestion that the consumption of baculovirus-infected larvae by various non-target animals plays a role in the dissemination of OBs (Entwistle et al., 1977; Lautenschlager et al., 1980).

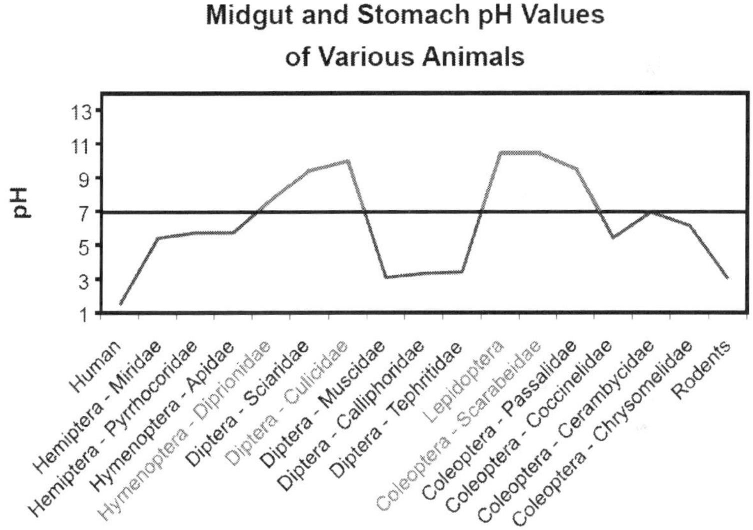

Fig. 1. Midgut and stomach pH values of various organisms. Red line shows alkaline values and blue line, acidic values. Insect orders and families that have been identified as baculovirus or nudivirus hosts are labeled in red. pH value of midguts of Hemiptera, Hymenoptera (Apidae), Diptera, Lepidoptera and Coleoptera (Terra et al., 1996), pH value of Hymenoptera (Diprionidae) (Heimpel, 1955) and pH values for stomach of human (Guyton & Hall, 2006) and rodents (McConnell et al., 2008).

3.2 Peritrophic membrane

The peritrophic membrane (PM) of insects is an acellular sleeve-like structure that lines and protects the gut. In insects, the PM consists of chitin sheets, proteoglycans and chitin-binding proteins such as peritrophins and intestinal mucins, that cross-link to form a thick, three-dimensional mesh (Barbehenn & Martin, 1995; Lehane, 1997; Peters, 1992; Tellam et al.,

1999). The chitin sheets provide the PM with strength and flexibility and the hydrating capacity of the proteoglycans is implicated in determining the permeability of the PM (Lehane, 1997). The pores, ranging from 21 to 36 nm (Barbehenn & Martin, 1995), allow bidirectional trafficking of small molecules such as digestive enzymes and the regulation of the flow of hydrolytic products (water and nutrients) between endoperitrophic and ectoperitrophic spaces (reviewed in Lehane, 1997). Insect intestinal mucins (IIM) consist of potentially-glycosylated, mucin-like domains and cysteine-rich chitin-binding domains (CBD) (Sarauer et al., 2003; Wang & Granados, 1997a), which are thought to have a similar lubricating and protective role as vertebrate mucins (Sarauer et al., 2003; Tellam et al., 1999; Wang & Granados, 1997a).

The small pore size of the PM limits the passage of larger materials such as pathogens, toxins and noxious phytochemicals (e.g. tannins) that are also ingested into the endoperitrophic space during feeding (Barbehenn, 2001). The size of baculovirus nucleocapsids, being between 40–70 nm in diameter and 250–400 nm in length (Boucias & Pendland, 1998), precludes them from crossing the PM. Disruptions in the links between PM chitin sheets and proteins can result in a decrease in the ability of the PM to act as a barrier to pathogens (Plymale et al., 2008; Wang & Granados, 1997b; 2000). ODVs passively cross the PM through physical breaches caused by mechanical abrasions, chemical means (Wang & Granados, 2000) or possibly by natural degradation of the PM. Diet has also been shown to affect the PM thickness. When fed cotton foliage rather than artificial diet, tobacco budworm larvae (*Heliothis virescens*) were shown to form a thicker PM that, by decreasing the number of primary infection foci on the midgut epithelium (Plymale et al., 2008), led to lower susceptibility to *Autographa californica* MNPV (AcMNPV).

An active mechanism of disruption of the PM has evolved in many GVs and a few lepidopteran group II NPVs whose genomes encode a metalloprotease called enhancin. Enhancins are occluded in the OB matrix either inside or on the surface of ODVs (Hashimoto et al., 1991; Lepore et al., 1996; Slavicek & Popham, 2005; Wang et al., 1994). In *Trichoplusia ni*, TnGV ODV enhancin results in disruption of the PM through the degradation of IIM structures (Derksen & Granados, 1988; Hashimoto et al., 1991; Wang & Granados, 1997b). LdMNPV enhancins are located on the ODV surface and were found to be essential for *per os* infectivity of *Lymantria dispar* larvae even when the PM was absent suggesting that enhancins may play a role in addition to PM breaching (Hoover et al., 2010).

Homologues of 11K protein are conserved in most baculoviruses and can be present in multiple copies (Dall et al., 2001). These proteins are characterized by the presence of C_6 chitin-binding motifs or peritrophin-A domains that are also present in chitin binding proteins such as mucins, peritrophins and chitinases of insect guts and basal laminae of trachea (Dall et al., 2001; Tellam et al., 1999). The AcMNPV 11K proteins were found to associate with BVs and OBs (Lapointe et al., 2004) but not with the ODVs themselves (Braunagel et al., 2003; Zhang et al., 2005) and were shown to enhance oral infection (Lapointe et al., 2004; Zang et al., 2005) but not hemocoelic infection. Although the function of Ac150 and Ac145 is not yet known (Lapointe et al., 2004; Zhang et al., 2005), they are important mediators of primary infection where they impact infectivity differentially in two different hosts of AcMNPV, *T. ni* and *H. virescens* (Lapointe et al., 2004).

3.3 Entry in midgut cells

The midgut tissue consists principally of midgut epithelial cells that are prone to apoptosis (Uwo et al., 2002) and slough off regularly (Engelhard & Volkman, 1995) which, when considering the obligate nature of a viral infection, makes them a sub-optimal environment for viral replication. The process of midgut epithelial cell infection is carried out by ODVs. Comparatively, BVs are 10,000 fold less efficient than ODVs at infecting midgut cells (Volkman & Summers, 1977; Volkman et al., 1976). On the other hand, AcMNPV ODVs have been shown not to infect or even penetrate insect tissue culture cells (Volkman & Summers, 1977; Volkman et al., 1976). In a cell culture assays, 2.3 x 10^5 ODVs were shown to be required to infect a single *T. ni* cell (TN-368) making them 1700 to 1900 times less infectious than BVs (Volkman et al., 1976). ODV infectivity to TN-368 cells was improved, however, in the presence of midgut juices from *Heliothis zea* and *Estigmene acrea* larvae (Elam et al., 1990), indicating that midgut lumen factors are important to the primary infection process.

Due to the complexity of *in vivo* studies and the lack of midgut cell lines that would simplify studies of ODV host cell entry, the exact mechanisms directing the primary infection processes are still relatively unclear. Once across the PM, ODVs bind to ODV-specific receptors and fuse to the brush border microvilli of the columnar epithelial cells (Haas-Stapleton et al., 2004; Horton & Burand, 1993). Nucleocapsids are then released into the cell cytoplasm to initiate primary infection of the midgut epithelium (reviewed in Slack & Arif, 2007). ODV-associated proteins of both NPVs and GVs are numerous (Braunagel et al., 2003; Wang et al., 2011) and include structural proteins responsible for encapsidation, unique ODV-envelope proteins involved in tissue tropism and proteins involved in viral gene expression and DNA replication (Braunagel et al., 2003). The ODVs unique association with proteins involved in viral gene expression and DNA replication is thought to optimize viral replication in midgut cells. Over 40 proteins were found to be associated with the ODVs of the alphabaculovirus AcMNPV (Braunagel et al., 2003), the betabaculovirus *Pieris rapae* GV (Wang et al., 2011) and the deltabaculovirus CuniNPV (Perera et al., 2007). Most of the proteins found in both AcMNPV ODVs and BVs are nucleocapsid proteins while most of the tissue-specific envelope proteins are different for each phenotype (Braunagel et al., 2003; Wang et al., 2010).

ODV-envelope proteins are highly complex and over 10 envelope proteins have been found to be associated with the ODVs of AcMNPV (reviewed in Slack & Arif, 2007; Braunagel & Summers, 2007). Only few of these ODV proteins, termed the *per os* infectivity factors (PIFs), have been shown to be essential for the AcMNPV *per os* infection process (Braunagel et al, 1996; Fang et al., 2009; Faulkner et al., 1997; Kikhno et al., 2002; Pijlman et al., 2003; Ohkawa et al., 2005; Sparks et al., 2011; Xiang et al., 2011) but more are expected to be identified by *in vivo* infectivity assays. Recently, three of these PIFs have been shown to form a very stable core complex (PIF-1, PIF-2 and PIF-3), held together by non-covalent bonds, that contributes to efficient entry and nucleocapsid delivery into midgut cells (Peng et al., 2010). The lepidopteran midgut is rich in trypsin and chymotrypsin (Johnston et al., 1995; Terra et al., 1996) and it has been suggested that the tight conformation of this PIF-complex might protect the active, internal PIF domains from the harsh chemical conditions of the midgut. In contrast, P74 (PIF-0) is only loosely associated with the core PIF-complex and actually requires the midgut environment to undergo functional activation by protease cleavage (Slack et al., 2008) by an ODV-associated host alkaline protease and by a host midgut trypsin

(Peng et al., 2011). Binding of ODVs to the tip of the midgut microvilli most likely occurs through a protein receptor binding mechanism (Horton & Burand, 1993; Yao et al., 2004) that relies upon PIF-0, PIF-1 and PIF-2 (Haas-Stapleton et al., 2004; Ohkawa et al., 2005). Supporting ODV adaptation to the midgut environment, binding efficiency has been shown to be optimal at alkaline pH (Horton & Burand, 1993).

Studies suggest that ODV binding proteins from different viruses interact with cell receptors that are specific to different virus/host systems (Haas-Stapleton et al., 2003, 2004, 2005, Horton & Burand, 1993; Ohkawa et al., 2005). Larvae of the fall armyworm *Spodoptera frugiperda* are highly resistant to oral infection by AcMNPV but are susceptible to infection by BVs when injected directly into the hemocoel (Haas-Stapleton et al., 2003). The resistance to *per os* infection is due to a lower level of ODV binding to the midgut cells (Haas-Stapleton et al., 2005) indicating that ODV interaction with specific midgut receptors is necessary for productive primary infection of midgut columnar epithelial cells. In addition, co-inoculation studies demonstrated that AcMNPV did not compete with SfMNPV for receptor binding indicating further that midgut receptor specificity is important for baculovirus host range determination.

As early as 30 min after BVs enter the cell, AcMNPV infection stimulates the formation of filamentous F-actin cables (Charlton & Volkman, 1993) and nucleocapsids are transferred to the nucleus by active actin polymerization (Lanier & Volkman, 1998; Ohkawa et al., 2010). The motile nucleocapsids encounter the nucleus and quickly enter through nuclear pores (Ohkawa et al., 2010) where the viral DNA genome is released. The nucleus is the site of replication for all baculoviruses and viral assembly for all NPVs is also carried out fully in the nucleus. In GVs, however, the occlusion step occurs following the disintegration of the nuclear envelope and the merging of the nucleoplasm and the cytoplasm (reviewed in Federici, 1997; Rohrmann, 2011; Winstanley and Crook, 1993). *Spodoptera frugiperda* and *T. ni* cells lines, Sf9 and High Five™, respectively, are permissive to AcMNPV but not to *Bombyx mori* NPV (BmNPV). Infection by BmNPV BVs was shown to be restricted, in part, by defective nuclear import of BVs where the virions entered the cell cytoplasm but were not able to enter nuclei of the cells (Katou et al., 2006). Mammalian cells lines are non-permissive to baculoviruses. AcMNPV ODVs were shown to bind to human carcinoma cells lines, A549 and HepG2, at the very low efficacy of 3×10^6 and 6×10^6 ODVs per cell, respectively (Mäkelä et al., 2008). Where binding occurred, the ODVs were inefficiently internalized into cytoplasmic vacuoles and were not released into the cytoplasm for intracellular trafficking and nuclear entry (Mäkelä et al., 2008). Thus, the non-permissive infection of HepG2 and A549 cells by AcMNPV ODVs is caused by ineffective binding to and internalization in the cell.

3.4 Disease progression in gammabaculoviruses

In hymenopteran sawfly NPVs, virions are singly enveloped and the host range of each virus is restricted to a single species (Federici, 1997). ODVs initiate midgut cell infection and once infected, the midgut cell nuclei become the site of viral gene expression and DNA replication in a manner that is consistent with that described for alphabaculoviruses (Duffy et al., 2007). In infected balsam fir sawfly larvae, *Neodiprion abietis* NPV (NeabNPV) viral DNA increased over 200% within the first 2 hours post infection (hpi) (Duffy et al., 2007). Progeny virions were occluded directly without the production of BVs (reviewed in

Federici, 1997; Slack & Arif, 2007; Rohrmann, 2011). Sequence data from gammabaculoviruses failed to identify homologues to the alphabaculovirus fusion proteins, GP64 or F-protein (Duffy et al., 2006; Lauzon et al., 2006), indicating the absence of the BV phenotype. Also lacking in gammabaculoviruses are viral fibroblast growth factor (vfgf) homologues (Jehle et al., 2006), that have been shown to accelerate the establishment of systemic infections in alphabaculoviruses (Detvisitsakun et al., 2007; Passarelli, 2011). (See section 3.7.) The NeabNPV replication cycle is rapid and efficient, with over 100 nucleocapsids being occluded in each OB (Duffy et al., 2007). Rather than using the BV phenotype to propagate the initial infection to other midgut cells, the released OBs have been suggested to serve as inoculum for other gut cells (Rohrmann, 2011). Midgut cell lysis in sawflies, rather than being a hindrance to gammabaculoviruses, is utilized as a mechanism for dispersal and only hymenopteran baculoviruses encode a trypsin-like protein (Duffy et al., 2006; Lauzon et al., 2006) that is thought to aid this process (Rohrmann, 2011). Much of the horizontal transmission occurs while infected larvae are still alive and feeding because OBs are excreted in a virus-laden diarrhea that is infectious to other larvae (Federici, 1997).

3.5 Disease progression in deltabaculoviruses

The replication of deltabaculoviruses is restricted to the cells of the posterior midgut and gastric caeca (Moser et al., 2001). Virions are singly enveloped prior to occlusion and the host range is restricted within suborders (Andreadis et al., 2003). While the primary infection is initiated by ODVs, the amplification of virus progeny within larvae occurs by cell-to-cell transmission that is carried out by BVs (Moser et al., 2001) and sequence data from CuniNPV identified an homologue to the F-protein of LdMNPV (Afonso et al., 2001). Restriction to the midgut is further corroborated by the lack of vfgf homologues in CuniNPV (Jehle et al., 2006). The production of OBs is rapid (14-48 hpi) and occluded progeny are released upon larval death (Becnel, 2007) for horizontal transmission to permissive hosts.

3.6 Disease progression in betabaculoviruses

Betabaculoviruses are only known to infect lepidopteran larvae. Primary infection is by OBs called granules and is initiated in midgut cells. The granules occlude a single virion and the host range of GVs is very specific (Cory, 2003). Some GV infections are restricted to the midgut (Type III), others cause systemic infections that progress either only to fat body tissues (Type I) or extend to most organs and tissues (Type II) (reviewed in Federici, 1997; Rohrmann, 2011). The time to host death caused by Type II GVs is similar to that of alphabaculoviruses (Lacey et al., 2002) and Type I GVs take the longest to kill the host larva (Federici, 1997). The transmission of Type II GVs occurs following the death and liquefaction of the infected larvae (Federici, 1997). The only identified Type III GV, *Harrisina brillians* GV (HabrGV) is transmitted horizontally by the release of OBs into the gut lumen of the infected larvae and out of the larva in the frass (Federici & Stern, 1990), similar to gammabaculoviruses. With the exception of the Type III midgut tropic HabrGV, the infection of the midgut epithelium is transient and does not produce OBs (Federici, 1997; Hess & Falcon, 1987). The cytopathology of GVs differs from that of other baculoviruses in that, although viral replication occurs and BV particles are produced in the nucleus of infected cells, OBs are formed in an area where the nucleoplasm and cytoplasm merge following the dissolution of the nuclear membrane (Hess & Falcon, 1987; Winstanley & Crook, 1993). Sequence data from GVs identified homologues of the F-protein and fgf orthologues (Liang et al., 2011; Miele et al., 2011) that are conserved in baculoviruses that cause systemic infections (Rohrmann, 2011).

3.7 Disease progression in alphabaculoviruses

The most studied baculovirus systems are those of lepidopteran alphabaculoviruses where ODVs enter midgut cells and BVs distribute the infection to other tissues in the hemocoel. In the well characterized system, AcMNPV-infected *T. ni* larvae, the infection progresses sequentially from midgut epithelial cells (4 - 12 hpi) to midgut-associated tracheoblasts and tracheal epithelial cells (12 - 24 hpi) (Engelhard et al., 1994). The infection then proceeds to hemocytes and tracheoblasts (36 hpi), later to fat body tissues (48 hpi) and finally to the majority of the remaining larval tissues by 70 hpi (e.g. cuticular epidermis, gonads, Malpighian tubules, midgut epithelia and salivary glands) (Engelhard et al., 1994). Although alphabaculovirus-infected midgut cells do not typically yield OBs (Granados & Lawler, 1981), they are crucial to the establishment of the systemic infection either by re-channeling ODV nucleocapsids into BV (without transiting through the nucleus) or by producing *de novo* BVs after a cycle of replication in infected midgut cell nuclei (reviewed in Rohrmann, 2011). While all other infected organs remain infected and produce viral particles, the infected midgut recovers by 48 hpi by sloughing off infected cells (Engelhard et al., 1995) and replacing them with new, healthy midgut cells (Keddie et al., 1989). Normally permissive to AcMNPV, *T. ni* larvae become resistant to AcMNPV infection as the larval development progresses. Active midgut cell sloughing is thought to play an important role in the developmental resistance that occurs in fourth-instar *T. ni* larvae (Engelhard & Volkman, 1995).

Once in the nucleus, the baculovirus genome utilizes the host transcriptional machinery to initiate the regulatory events that will result in the initial production of nucleocapsids that bud out of the cell by interacting with the cell plasma membrane and acquiring envelope fusion protein (EFP). Group I alphabaculoviruses (e.g. AcMNPV) encode for two main EFPs, GP64, and F-protein while group II alphabaculoviruses (e.g. LdMNPV) encodes only F-protein (Pearson et al., 2000). Deltabaculoviruses encode F-protein while gammabaculoviruses do not encode a discernable EFP (Miele et al., 2011). These EFPs are essential for BVs to exit infected cells and for cell-to-cell transmission (Monsma et al., 1996; Oomens & Blissard, 1999). An alternate mode for exiting the midgut cells rapidly has been described for AcMNPV (Washburn et al., 1999). In infected cells, some of the co-enveloped nucleocapsids enter the nucleus to initiate a round of replication while others, initially, remain in the cytoplasm. GP64 is produced early and modifies the basal membrane of the cell to mediate exit of BVs (Keddie et al., 1989). Even before *de novo* BVs are produced, ODV nucleocapsids in the cytoplasm bypass replication by reaching GP64-modified basal membranes and budding directly to initiate systemic infection of non-midgut tissues (Washburn et al., 2003a; Zhang et al., 2004). The fast shuttle of nucleocapsids through midgut cells is thought to accelerate the establishment of secondary infections before the midgut cells are sloughed off thereby potentially contributing to the wider host range of MNPVs (Washburn et al., 1999; Washburn et al., 2003b) and counteracting host defense mechanisms. This process of direct transformation of ODV to the BV phenotype has not been observed in SNPVs (Rohrmann, 2011).

ODVs must cross the PM to establish primary infections but BVs need to breach the basal lamina (BL) at the base of the midgut epithelium to initiate secondary infections. The BL is an extracellular layer of protein sheets that are secreted by epithelial cells lining the midgut trachea and other organs (Rohrbach & Timpl, 1993). The BL serves as structural support for regenerating epidermal cells that replace senescing cells that were sloughed off during development or physical assaults to the gut and as a separation between the sterile hemocoelic organs and the midgut, thus preventing the passage of natural microbiota and pathogens acquired during feeding. A model for AcMNPV breaching of the BL has recently been proposed (Means & Passarelli, 2010; Passarelli, 2011). The tracheal system is the insect respiratory system and the first cells to be infected by BVs are the tracheoblasts (Engerhard et al., 1994; Kirkpatrick et al., 1994; Washburn et al., 1995) which are highly motile, single-cell tracheal projections that respond to signaling from oxygen deficient cells and organs. One essential component to this response are the fibroblast growth factors (FGF) that, through a cascade of activation involving fibroblast growth factors receptors (FGFR), trigger tracheal cell motility. To branch to other cells and tissues, tracheoblasts are thought to degrade the BL by secreting enzymes. Baculoviruses are the only viruses known to encode FGF signaling molecules (Passarelli, 2011). Conserved only in alphabaculoviruses and betabaculoviruses, but absent in midgut-restricted gammabaculoviruses and deltabaculoviruses, viral FGF (vFGF) has been shown to accelerate BV exit from midgut cells and secondary infection by rerouting the host respiratory system to the midgut epithelium by mimicking host FGF. Although not essential for host infection *per se*, the difference in speed of establishment of systemic infection and speed of kill (Detvisitsakun et al., 2007; Katsuma et al., 2006) is thought to impact the capacity of any given virus to infect different hosts (Passarelli, 2011).

BV entry into insect cells is effected by GP64 via a clathrin-mediated, low-pH dependent, endocytic process (Long et al., 2006). Insect cells have different receptors for GP64 and F-protein and these two proteins appear to act separately (Hefferon et al., 1999; Westenberg et al., 2007; Wickham et al., 1999). GP64 tropism is so broad in fact, that BVs from AcMNPV and other baculovirus species such as BmNPV have been shown to be taken up by numerous non-lepidopteran cell lines including mammalian and dipteran cell lines (Carbonell et al., 1985; Shoji et al., 1997). Given the broad cellular tropism of GP64, receptors are thought to be common molecules present in invertebrates and vertebrates such as phospholipids (Tani et al., 2001). Therefore, GP64-mediated BV entry into cells is most likely not a restricting event. Receptor specificity for F-protein, however, is more restricted. Mammalian cells were shown not to possess F-protein receptors and could not to be transduced with a gp64-null AcMNPV pseudotyped with baculovirus F-protein (Westenberg et al., 2007). The lack of the *gp64* gene in LdMNPV, might contribute to its narrow host range (Barber et al., 1993; Glare et al., 1995).

Through their distribution in the hemolymph and systemic infection of a variety of tissues, BVs are largely responsible for the amplification of the virus within infected host larvae. Although the primary infection process is essential to the infectivity of baculoviruses in lepidopteran baculoviruses the systemic infection and ultimate death of larvae are dependent upon production of BVs. Horizontal transmission to neighboring larvae is dependent upon the release of OBs that occurs following larval death. Expression of viral proteases (cathepsin) and chitinase, late post infection, ensures that, in most alphabaculoviruses and in some betabaculoviruses, progeny OBs are released in the environment by lysing larval tissues and the exoskeleton of larvae following death (Hawtin et al., 1997; Hom et al., 2002). In addition, in silkmoth and gypsy moth, virally-produced proteins, tyrosine phosphatase (ptp) (Kamita et al., 2005) and ecdysteroid uridine 5´-diphosphate (UDP)-glucosyltransferase (egt) (Hoover et al., 2011), are responsible for behavioral changes that occur during the infection process where infected larvae leave their normal sheltered habitats and climb to exposed surfaces. This alteration in larval behaviour is thought to assist in virus distribution by facilitating predation by animals and by increasing exposure to elements.

4. Replication cycle

For all baculoviruses, once inside the nucleus, genome replication events follow a strictly-controlled cascade of temporal and sequential events (reviewed in Friesen, 2007; Rohrmann, 2011). Baculovirus genes are transcribed in three temporal phases (early, late and very late) where later steps of each phase are dependent on occurrence of earlier molecular events (Carstens et al., 1979; Guarino & Summers, 1986; Guarino & Summers, 1987).

4.1 Early phase

Upon viral DNA release into the cell nucleus, cellular transcription is harnessed for the expression of viral immediate early genes (Carstens et al., 1979; Guarino & Summers, 1986, 1987) and host-cell transcripts are decreased progressively from 12 - 18 hpi until complete shut down by 36 hpi (Nobiron et al., 2003). Baculoviruses depend on the cellular RNA polymerase II, to initiate the cycle of replication (Fuchs et al., 1983; Huh & Weaver, 1990) by recognizing and initiating transcription of viral early promoter sequences (Fuchs et al., 1983; Hoopes & Rohrmann, 1991; Huh & Weaver, 1990). Early viral transcripts can be detected as early as 0.5 hpi and through to 6 - 9 hpi in AcMNPV-infected susceptible cells (Chisholm & Henner, 1988; Guarino & Summers, 1986). The early promoter regions are conserved throughout baculoviruses and mimic those of the host RNA polymerase II with a consensus TATA element at about 30 bp upstream from the RNA start site (Pullen & Friesen, 1995) and a CAGT motif that acts as an initiator element (Blissard et al., 1992; Pullen & Friesen, 1995). Early genes encode mainly for polypeptides (IE1, IE0, IE2 and PE38) that have regulatory functions which are responsible for the transcriptional regulation of other viral genes, for the initiation of viral DNA replication (Lu & Carstens, 1991; Stewart et al., 2005; Todd et al., 1995), or to take control of the host cell for the purpose of viral multiplication (Possee & Rohrman, 1997). Early baculovirus gene expression is mostly regulated at the level of transcription and transfected viral DNA is infectious to permissive cells (Burand et al., 1980; Cartens et al., 1980) indicating that initial transcription of early genes does not require viral factors to be present at the start of the infection. The lack of requirements for viral proteins has been substantiated recently by proteomic analysis of AcMNPV BV particles that have been shown not to contain regulatory proteins (Wang et al., 2010). Larvae of *Anticarsia gemmatalis* (velvetbean caterpillar) are highly resistant to AcMNPV. Infection was shown to be blocked at early stages of replication where, even though ODVs successfully entered the midgut cells and were rechanelled to tracheal cells, the immediate early gene, *ie-1*, was not transcribed (Chikhalya et al., 2009). The inhibition of this essential transactivator resulted in the disruption of the gene expression cascade resulting in a failure to produce infectious viral particles.

4.2 Late phase – DNA replication

Baculovirus early gene products are required for viral DNA replication. In AcMNPV-infected cells, genome replication occurs between 6 hpi and to 18 hpi after which time it starts to decline (Erlandson et al., 1985; Erlandson & Carstens, 1983). Six AcMNPV genes were found to be essential for DNA replication in transient DNA replication assays (Kool et al., 1994; Lu & Miller, 1995a). The genes directly involved in DNA replication (reviewed in Rohrman, 2011) encode for a homologous-region (hr) binding protein and transcriptional activator (*ie-1*) (Leisy et al., 1995; Rodems & Friesen., 1995), a single-stranded DNA binding protein (*lef-3*) (Hang et al., 1995), DNA binding helicase (*p143*) (McDougal & Guarino, 2000), a putative primase (*lef-1*), a primase-associated protein (*lef-2*) (Mikhailov & Rohrmann, 2002) and a DNA polymerase (*dnapol*) (Hang & Guarino, 1999; McDougal & Guarino, 1999). Four of these genes, *dnapol, p143, lef-1,* and *lef-2* are core genes found in all baculoviruses (Okano et al., 2006). Other genes such as *ie-2, lef-7, pe38,* and *p35,* stimulate viral DNA replication in transient assays (Chen & Thiem, 1997; Lu & Miller, 1995a; Milks et al., 2003) and were found to be differentially required in cell lines from various lepidopteran origins as well as *in vivo* (Chen & Thiem, 1997; Lu & Miller 1995b; Milks et al., 2003; Prikhod'ko et al., 1999).

BmNPV and AcMNPV genomes are highly homologous (Gomi et al., 1999) but their host ranges are very different (Gröner, 1986). *Bombyx mori* larvae and cell lines such as BmN, are fully permissive to BmNPV but their infection by AcMNPV is non-permissive (Morris & Miller, 1993). Though delayed in BmN cells, AcMNPV temporal gene expression occurs as in the permissive *Spodoptera* cells until very late times post infection (Iwanaga et al., 2004; Morris & Miller, 1993). AcMNPV DNA replication also takes place in BmN cells but the infection is arrested before BV or OBs are produced (Morris & Miller, 1993). AcMNPV-infected BmN cells also showed marked cytopathic effects (Maeda et al., 1993) which led to a drop in gene expression. Though DNA replication seemingly occurred as in the permissive Sf-21 cells, the defect in AcMNPV-BmN cells was shown to be caused by differences in the *DNA helicase* gene (*p143*) (Maeda et al., 1993). A few amino acid changes in AcMNV P143 were sufficient to overcome the defect in *B. mori* cells and larvae (Argaud et al., 1998; Croizier et al., 1994). The cytotoxicity and block in AcMNPV infection of *B. mori* cells are suggested to stem from aberrant DNA replication (reviewed in Thiem & Cheng, 2009; Rohrmann, 2011).

4.3 Late phase – Late gene expression

The final step in the replication cycle of baculoviruses is the expression of late and very late genes that mainly code for structural proteins. AcMNPV genes encoding structural proteins of nucleocapsids and BVs are transcribed at their peak during the late phase (6 - 24 hpi), while occlusion-related genes are transcribed at very late times post-infection (18 - 76 hpi) (Thiem & Miller, 1990; Wu & Miller, 1989). The increase in late viral transcription parallels the decline in host and early viral transcription (Nobiron et al., 2003). Late promoter sequences are conserved in baculoviruses with the TAAG sequence being the essential component for the recognition of late and very late promoters by viral RNA polymerase factors with cis-acting sequences dictating the differential levels and temporal expression of late and very late genes (Ooi et al., 1989). In AcMNPV, 19 genes were found to be required for optimal transcription of late (*vp39* and *p6.9*) and very late promoters (*polh* and *p10*) but not early promoters (*etl* and *pcna*) (Li et al., 1993; Lu & Miller, 1995b; Passarelli & Miller, 1993a, 1993b, 1993c; Rapp et al., 1998; Todd et al., 1995). Being required for the transcription of late and very late genes, these 19 genes have been defined as late expression factors (lefs). Since the late transcriptional events are dependent upon early transcription and DNA replication, nine of these lefs have indirect effects on late transcription by being involved in early gene transcription and DNA replication (*ie-1, ie-2, lef-1, lef-2, lef-3, p143, dnapol, p35,* and *lef-7*) (Rapp et al., 1998). Only four of these lefs (*p47, lef-4, lef-8,* and *lef-9*) have been shown to form the viral RNA polymerase that is directly responsible for the *in vitro* transcription of baculovirus late and very late promoters (Guarino et al., 1998; Rapp et al., 1998). A unique feature of baculoviruses is the hyperexpression of the very late genes, *polyhedrin* and *p10.* Increases in transcription levels of RNA polymerase occurs through the binding of very-late expression factor (VLF-1) (McLachlin & Miller, 1994; Ooi et al., 1989; Yang & Miller, 1999) to very late promoters stimulating expression of the high levels of polyhedrin required for OB formation.

Gypsy moth Ld652Y cells are semi-permissive to AcMNPV and, although all temporal classes of virus genes are transcribed and viral DNA replication is detected, translation of both viral and host proteins is arrested by about 12 hpi (reviewed in Thiem & Cheng, 2009; Guzo et al., 1992; McClintock et al., 1986; Morris & Miller, 1993). An LdMNPV gene, named the host range factor (*hrf-1*) was found to rescue the translational arrest in AcMNPV-infected Ld652Y cells (Du & Thiem, 1997a; Thiem et al., 1996). AcMNPV does not encode for a *hrf-1* homologue. Gypsy moth *in vivo* resistance was also overcome by a hrf-1-bearing, recombinant AcMNPV (Chen et al., 1998).

5. Registration of baculoviruses

Agencies responsible for pesticide product regulations were initially put in place by governments to evaluate the efficacy and non-target safety of synthetic chemical pesticides. Their extension into environmental safety assessments came in response to concerns related to the increasing number of reports of environmental damage due to pesticide toxicity and the accumulation of chemical residues such as those from DDT (Hauschild et al., 2011). Gradually, mounting public pressure and the implementation and enforcement of stricter rules and regulations led to the ban and rejection of many chemical pesticides and the need for lower-impact pesticides such as those classified as biological control agents (BCA) and microbial pest control agents (MPCA). MPCAs are those products that have a microorganism (i.e., virus, bacterium, fungus, protozoan or nematode) as the active ingredient. When submitted for registration, MPCAs were initially evaluated using existing regulatory processes that were developed for broad-spectrum chemical pesticides. Since then, a trend has emerged that now favours the development of lower-risk products such as BCAs and MPCAs and specific regulations have been developed that are better suited to the requirements of MPCAs. To facilitate the registration process for MPCAs, many countries have established departmental branches that specialize in the registration of MPCAs and other low-risk product submissions (Hauschild et al., 2011; Kabaluk et al., 2010). Despite these efforts, the MPCA registration process in many countries still requires a number of toxicity tests that might not be necessary based on the biology of the microbe that acts as the active ingredient of a MPCA. Compared to chemical pesticides, MPCAs typically target small niche markets and unnecessary registration requirements can impose burdensome costs on the biopesticide industry (Chandler et al., 2011).

The most important function of any pesticide regulatory agency is to insure that unsafe pesticides are not registered for use. Although the registration procedures, costs and processing times differ between countries, data requirements are generally similar (Hauschild et al., 2011). Typically, numerous categories of data are required for the registration of MPCAs. These categories are designed so that applicants provide 1) product identity, physical, chemical and technical properties; 2) methods of analysis, manufacturing and quality control; 3) toxicological studies and exposure data geared towards human and veterinary health and safety; 4) product residue data; 5) product fate and behaviour in the environment; 6) environmental and non-target toxicity and 7) efficacy data (Hauschild et al., 2011).

The majority of the topics included in registration packages are essential and desirable. For baculoviruses in particular, it is essential that the active ingredient has been identified as belonging to the family, *Baculoviridae* and that the species of its primary host is known (e.g. *Neodiprion abietis* gammabaculovirus) (category 1), that the method of manufacturing, analysis and quality control is appropriate and reliable (category 2) and that it has been proven to be efficacious (category 7) for the purpose intended. The regulations as applied to some of the more recently registered baculovirus-based products may, however, demonstrate the redundancy of certain data requirements involving vertebrate toxicity (categories 3 and 6) and some aspects of environmental and non-target effects (categories 4 and 5). The Pest Management Regulatory Agency (PMRA) in Canada and the United States Environmental Protection Agency (EPA) accept that certain data requirements can be met using "waivers" that provide scientific arguments based on published, peer-reviewed, scientific literature and data while, in the European Union (EU), less formal procedures allow for similar science-based evidence to replace the actual, newly-generated data (Hauschild et al., 2011). A combination of fulfilling specific data requirements and the use of waivers were used in the successful registration of the baculovirus-based product, Abietiv™, for the suppression of balsam fir sawfly populations in Canada (Lucarotti et al., 2006, 2007).

5.1 Identity of baculovirus products

MPCAs are usually registered at the strain level where the active ingredient is derived from a single host, colony or spore. While this selection method is mostly feasible for MCPAs such as bacteria and fungi, it may not be feasible or desirable for baculovirus-based products. In nature, baculoviruses consist of mixtures of different genotypes of the same species (Cory et al., 2005) and it has been shown that this diversity is naturally favored in wild type virus populations (Clavijo et al., 2011) where these genomic variants are known to impact virulence in the target organism (López-Ferber, 2003; Simón et al., 2008). The different viral genotypes may compensate for variations that occur in the larval host and/or its environment (Berling et al., 2009; Hitchman et al., 2007; Hodgson et al., 2002). For this reason, the evaluation of baculoviruses should be carried out at the species level rather than at the level of a single isolated genotype (Hauschild et al., 2011).

5.2 Human toxicity and infectivity

The scientific literature on the health and environmental safety of baculoviruses is extensive and has been well reviewed (see reviews by Black et al., 1997; Burges et al., 1980a, 1980b; Gröner, 1986; Ignoffo, 1975; OECD, 2002) and more recently by the Food and Agriculture Organization of the United Nations (FAO) (McWilliam, 2007) and the European Food Safety Authority (EFSA) (EFSA, 2009; Leuschner et al., 2010). The host range of baculoviruses is restricted to terrestrial arthropods (Barber et al., 1993; Doyle et al., 1990; Cory, 2003; Cory & Hails, 1997; Miller & Lu, 1997; Thiem & Cheng, 2009). Baculovirus products that are commercially available for biological control of insect pests have been extensively tested to determine effect on humans and other non-target animals (Hauschild et al., 2011).

Data required for assessment of human infectivity and toxicity typically involve mammalian toxicological studies of the product in laboratory test mammals (e.g. mice, rats, rabbits) *in vivo* as well as in mammalian cell cultures. Baculovirus active ingredients and end products have been ingested and inhaled by, injected (intravenous, intraperitoneal, intramuscular) into, and applied to the skin and eyes of test animals without detrimental effects that could be attributed to the baculovirus tested (Table 1) (Ashour et al., 2007; Gröner, 1986; Ignoffo, 1975; Lightner et al., 1973). Many species of baculoviruses have been tested on numerous species of animals at doses that are many times those that could be acquired in the field. For example, for the registration of Abietiv™ (NeabNPV), typical toxicity data were presented where rats had been fed single dose of 1 x10^8 NeabNPV OBs (Health Canada PMRA, 2009; Lucarotti et al., 2006). All of the rats survived to the end of the observation period and showed no adverse clinical effects. At the application rate given on the Abietiv™ product label (1 x 10^9 OBs in 2.5 L of 20% aqueous molasses/ha), this would be the equivalent of a 70-kg man ingesting 16 L of the tank-mixed product. Thus, taking into account the label application rates and volumes at which the products are applied, the concentrations used for toxicity tests are well beyond what could be expected to be acquired in the field.

In vitro, mammalian and other vertebrate cells lines are non-permissive to baculoviruses (reviewed in Gröner, 1986; OECD, 2002). Although BV uptake has been observed, there has been no evidence that viral DNA replication, production of viral proteins or cytopathological effects have occurred. *In vivo*, the uptake of baculovirus OBs by various animals did not lead to the production of baculovirus-specific antibodies (reviewed in Gröner, 1986). Human carcinoma cell lines, HepG2 and A549, were recently challenged with AcMNPV ODVs that had been chemically extracted from OBs (Mäkelä et al., 2008). The non-permissive infection of HepG2 and A549 cells by AcMNPV ODVs was shown to be caused by the inefficient binding and internalization in the cell. The ODV-derived nucleocapsids did not reach the nucleus to release the viral genome. In addition to the lack of infection and replication in vertebrate cells, no evidence for baculovirus induced cytogenic, carcinogenic, mutogenic or teratogenic effects has ever been found (Gröner, 1986; Ignoffo, 1975; McWilliam, 2007; OECD, 2002).

Alphabaculovirus	Vertebrate Test Animals
Amscacta albistriga NPV	chicken
Autographa californica NPV	rat, guinea pig, rabbit, shrimp[a], fish[b]
Choristoneura fumiferana NPV	rat, rabbit, duck, quail, rainbow trout, white sucker
Erranis tilliara NPV	mouse
Galleria mellonella NPV	mouse
Heliothis zea NPV	mouse, rat, guinea pig, rabbit, dog, monkey, man, quail,chicken, sparrow, mallard, killifish, spotfish, rainbow trout, black bullhead, white sucker, sheepshead minnow
Lymantria dispar NPV	mouse, rat, guinea pig, rabbit, dog, blackcap chickadee,duck quail, sparrow, bluegill, brown trout
Malacosoma disstria NPV	guinea pig, rabbit, chicken
Mamestra brassicae NPV	mouse, guinea pig, pig, chicken
Orgyia pseudotsugata NPV	mouse, rat, rabbit, mule deer, duck pheasant, sparrow, chinook salmon, coho salmon, steelhead trout
Spodoptera exempta NPV	rat
Spodoptera exigua NPV	mouse, guinea pig
Spodoptera frugiperda NPV	mouse, guinea pig
Spodoptera littoralis NPV	Rat, fish[b]
Spodoptea litura NPV	chicken
Thymelicus lineola NPV	mouse, sheep, goldfish
Trichoplusia ni NPV	mouse, guinea pig, sparrow
Betabaculovirus	
Cydia pomonella GV[c]	mouse, rabbit
Estigmene acrea GV	mouse, guinea pig
Gammabaculovirus	
Neodiprion abietis NPV[d]	rat, mouse, rabbit
Neodiprion lecontei NPV	rat, hamster, rabbit, chicken, turkey, rainbow trout
Neodiprion sertifer NPV	rat, guinea pig, rabbit, duck, quail
Neodiprion swainei NPV	mouse, rat, guinea pig, rabbit, duck, quail

Table 1. List of vertebrate test animals exposed to baculoviruses from a variety of Lepidoptera and sawflies to which no adverse effects of exposure to the baculovirus could be attributed. (from Gröner, 1986; Ignoffo, 1975). a, b, c, d: Data obtained from Lightner et al. (1973), Ashour et al. (2007), and Health Canada PMRA Regulatory Notes REG2000-10 and RD2009-05, respectively.

Currently, all baculovirus-based MCPs are produced, *in vivo*, in permissive larval hosts. While baculovirus OBs are inert and non-allergenic, the larvae in which they are produced can produce dermatitis and contact urticaria where larval setae cause mechanical irritation or contain histamine or other irritating substances (Hossler, 2010a, 2010b). Although anaphylactic shock has not been reported to be caused by lepidopteran insects (Hossler, 2010a, 2010b), eye irritation studies on rabbit (Reardon et al., 2009) and limited skin eruptions have been reported from human exposure to gypsy moth larvae during heavy infestations (Tuthill et al., 1984). As a result, the Gypchek product label warns of potential eye irritation. While this is the case for some of the baculovirus products, the majority of products are not considered as sensitizers (Hauschild et al., 2011: Ignoffo, 1975).

5.3 Baculoviruses and biomedical applications

Additional evidence of the safety of baculoviruses to humans comes from their use in biomedical applications. The unique baculovirus properties, coupled with recent advances in molecular and cell biology, have broadened the scope of their application in basic and applied biomedical fields. To date, the prototype baculovirus, AcMNPV, is the most widely used baculovirus for the production of biologics for therapeutic purposes (Aucoin et al., 2010; Cox & Hollister, 2009; van Oers, 2006). This has been accomplished in part by use of baculovirus expression vector systems (BEVs) for heterologous recombinant protein production and gene transfer. The principle behind the use of BEVs is based on use of *polyhedrin* and *p10* promoters to drive the expression of foreign genes in cell culture or *in vivo* (van Oers, 2011; Summers, 2006). BEVs continue to evolve to new and robust systems. For instance, the latest vectors (flashBACultra/BacMagic3) can be semi-automated for high quality, yield, and stable recombinant protein (van Oers, 2011). Many advantages of using BEVs over other expressions systems have been reviewed in previous reports (Airenne et al., 2009, 2010; Hu, 2005). Also, there are available insect cell lines such as those derived from *S. frugiperda* (Sf9 cells) and *T. ni* (High Five™ cells), which have been extensively characterized for optimal, high quality recombinant protein production (Aucoin et al., 2010). Some of these cell lines have been adapted to grow as continous suspension cultures in serum-free media thus, allowing for high-throughput scale-up production in bioreactors (Elias et al., 2007; Feng et al., 2011). Furthermore, use of serum-free cell cultures have been recognized by regulatory agencies including the United States, Food and Drug Administration (FDA), and European Medicines Agency (EMA) as a standard approach for limiting potential adventitious agents in therapeuatic products (FDA, 2010). Many advantages of using insect cell substrate compared to embryonated eggs have been reported leading to simplified regulatory avenues for licensing baculovirus-based biologics (Cox & Hollister, 2009, Treanor et al., 2007). In addition to cell substrates, insect larvae such as *T. ni* (Chen et al., 2011), and *B. mori* (Kato et al., 2010) have been reported as potential biofactories for *in vivo* therapeutic production. For instance, *in vivo* production of antiviral agents including human interferon-γ against influenza virus H1N1 in *T. ni* larvae have been demonstrated (Chen et al., 2011; Gomez-Casado et al., 2011). Nevertheless, insect cell cultures and BEVs platforms continue to expand the applications of baculoviruses as novel tools for vaccine development, drug screening, and gene therapy (Airenne et al., 2010, 2011; Cox & Hollister, 2009, Kost et al., 2005, van Oers, 2011). The recent initiative of having a standard baculovirus reference material repository (BRM) will further boost their application and perhaps hasten the regulatory process for registering new baculovirus products (Kamen et al., 2011). This initiative was mainly proposed in order to have a proper standard that is acceptable to all researchers in academic institutions, regulatory agencies, and industries (Kamen, et al., 2011).

5.4 Baculoviruses and vaccine development

To date, there are different baculovirus-based vaccines for human and veterinary use (van Oers, 2011). Also, vaccines targeting highly pathogenic viruses that are transmitted by arthropod vectors (arboviruses) are being developed (Metz & Pijlman, 2011). Characteristics of baculovirus-based human vaccines that are currently approved or are in later phases of clinical trials are given in Table 2. The different strategies employed in the production of baculovirus-based vaccines include: (i) BEVs-based subunit vaccines; here, recombinant viral proteins or peptides are produced using BEVs in cell culture. Subunit vaccines can be efficiently produced in insect cells and have additional safety advantages over live attenuated vaccines (Madhan et al., 2010). A good example is the influenza vaccine (FluBlok), which is based on recombinant Hemagglutin (HA) proteins selected from three influenza virus strains as determined by the World Health Organization (WHO) and the Centre for Disease Control (CDC) (Airenne, 2009; Cox & Hollister, 2009). Subunit vaccines developed in BEVs have been approved based on the standards stipulated by various regulatory agencies especially clinical data on toxicology and efficacy assessment (Cox & Hollister, 2009; Cox & Hashimoto, 2011; FDA, 2009). (ii) BEVs-based virus like particles (VLPs); for example, a prophylactic, bivalent human papillomavirus vaccine for cervical cancer (Cervarix) consisting of C-terminally truncated HPV-16/18 L1 proteins is produced using BEVs in *T. ni* High- Five™ cells and assembled as VLP (Harper et al., 2006). VLPs mimic the real virus but are non-infectious due to lack of viral genome, and are safe for human use. Detailed safety data for human papillomavirus types 16 and 18 recombinant vaccine have been outlined by USA and Canada health regulatory agencies (FDA, 2009; Health Canada, 2010). (iii) Active cellular-based vaccine; a classical example and the first vaccine of this kind to be approved by FDA is Provenge® for prostate cancer. This vaccine is composed of fusion proteins consisting of a prostate cancer marker, prostatic acid phosphatase (PAP), linked to granulocyte macrophage colony stimulating factor (PAP-GM-CSF) and generated in insect cells via BEVs. The fusion protein is in turn loaded *ex vivo* in dendritic cells, the most potent antigen presenting cell (APC), leading to stimulation of cytotoxic T-cell immune response against patients cancer cells (Small et al., 2006; Vergati et al., 2010). The prostate cancer cells expressing these recombinant proteins are recognized by the patient's cell-mediated immune system. In addition to the aforementioned baculovirus strategies for vaccine development, there are other baculovirus technologies, such as baculovirus surface display technology, that are being considered for production of vaccines. Here, the desired foreign antigens are displayed on the surface of the baculovirus envelope or capsid (Mäkelä & Oker-Blom, 2006; Oker-Blom et al., 2003). More recently, a novel system based on the use of a defective baculovirus vector incapable of self assembly has been developed (Marek et al., 2010). In this approach, the baculovirus vector is engineered to produce biologics that are free from contaminating BVs and ODVs.

Vaccine	Producer	Disease	Status	Reference
Cervarix™	GlaxoSmithKlines, Rixensart, Belgium	Cervical cancer	Approved	Harper et al., 2004; 2006
Provenge	Dendreon Inc., Seattle, WA, USA	Prostate cancer	Approved	Kantoff et al., 2010 ; Small et al., 2006;
Chimigen	Virexx Medical Corp., Calgary, Canada	Hepatitis B and C	Clinical trial	
FluBlok	Protein Biosciences Corp., CT, USA	Influenza virus	Clinical trial	Cox & Holister, 2009, Cox & Hashimoto, 2011
Dyamid	Diamyd Medical AB, Stockholm, Sweden	Type-1 diabetes mellitus	Clinical trial	

Table 2. List of baculovirus-derived vaccines.

5.5 Baculovirus and mammalian gene delivery/ therapy platforms

Although baculoviruses replicate in the nucleus of specific insect hosts, mammalian cells have been shown to internalize baculoviruses, but no progeny virions are produced (Volkman & Goldsmith, 1983). Similarly, recombinant baculoviruses carrying a reporter gene under the control of human cytomegalovirus (CMV) and Rous sarcoma virus (RSV) promoters were shown to efficiently transduce mammalian cells and express foreign proteins (Boyce & Bucher 1996; Hofmann et al., 1995). These studies showed varying levels of reporter gene expression in mammalian cells of different origins. Although all cells were reported to internalize the same amount of virus, the block to expression or low expression observed in epithelial cells compared to human and rabbit hepatocytes was attributed to a specific receptor on the hepatocyte cell membranes and inhibition of endosomal maturation. Additional blocks have been linked to poor cytoplasmic transport or entry of nucleocapsid to the nucleus (reviewed in Airenne et al., 2009).

The basis of baculovirus gene delivery/ transfer in mammalian cells has been accomplished using BacMam vectors (Invitrogen Corporation, Carlesbad, CA). Unlike BEVs, which relies on baculovirus late promoters, the gene of interest in BacMam vectors is placed under the transcriptional control of mammalian active-promoters such as those of CMV, RSV, chicken beta-actin (CAG), among others (Madhan et al., 2010). Various cellular and viral promoters have been shown to affect transduction efficiency in different mammalian cells implying that promoter selection is critical to efficient use of baculovirus vectors (Kim et al., 2007; Shoji et al., 1997). Nonetheless, safety of baculovirus vectors in gene delivery is supported by extensive safety data that show lack of toxicity or pathogenicity in various mammalian species (reviewed in Airenne et al., 2009). Based on this, baculoviruses have been considered as ideal candidate for future gene therapy. Gene therapy is a novel approach for treating various forms of genetic diseases through the use of viral or non-viral shuttle vectors. This has been successfully demonstrated through *in vivo*, and *ex vivo* (human-derived tissues) studies as previously reviewed (Airenne et al., 2009, 2010; Hu, 2005). To date, viral-based vectors including those of DNA and RNA animal viruses are increasingly being tested as potential agents for gene therapy. Health Canada, like other regulatory bodies, recognizes various viral-based vectors for *in vivo* and *ex-vivo* gene therapy. Baculoviruses are included in these lists due to recent studies on their potential as tools for gene therapy and requirements for extensive preclinical studies. Baculovirus-based vectors remain promising candidates for gene delivery primarily due to the following attributes: (i) natural occurrence, (ii) host specificity (iii), well characterized genomes (Cohen et al., 2009), (iv) genetic stability due to lack of reversions or genome integration, (v) rapid and relatively low cost of production to high titers (~ 10^{12} pfu/ml) (Airenne, 2010), (vi) lack of cell substrates associated with animal serum, (vi) large transgene capacity (~100 kb), and allowance for multiple gene inserts, (viii) ability to transduce a myriad of dividing and non- dividing cells (Airenne et al., 2011; Kost & Condreay, 2002) and, (ix) lack of pre-existing immunity.

Although baculovirus gene therapy technology is relatively recent, there is a wealth of safety data in animal models (Airenne et al., 2009) and preclinical trials based on *ex-vivo* experiments (Georgopoulos et al., 2009). Their safety is augmented by early toxicity studies using intravenous, oral, intracerebral, and intramuscular inoculation of animal models, and feeding tests on voluntary humans (Gröner, 1986; Ignoffo, 1973, 1975). Similarly, techniques to assess the toxicity and transformation potential of baculovirus in mammalian cells have been developed (Hartig et al., 1989, Gonin & Gaillard, 2004). Here, quantitative PCR (qPCR), using SYBR Green or TaqMan probes is viewed as the standard tool for assessing the biodistribution of the transgene and expression of shuttle vectors (Gonin & Gaillard, 2004).

5.6 Environmental toxicity

The environmental and ecological impacts of baculovirus products are mirrored by characteristics of their pathogenesis and host range. In essence, every study that is required to assess their potential for environmental toxicity will be influenced by their limited host range and lack of infectivity to non-target animals. Data required for assessment of environmental toxicity typically involve environmental fate and environmental toxicological studies on birds, fish, plants, microorganisms, aquatic arthropods and non-target insects including beneficial insects. Baculoviruses are ubiquitous and persistent in aquatic, terrestrial and forest ecosystems (England et al., 2004; Hewsen et al., 2011; Podgwaite et al.,1979) yet, there has been no report of negative impact of baculoviruses on ecosystems other than the effect on the target host insect (Black et al., 1997; Cory, 2003; FAO, 2007; OECD, 2002). As a matter of fact, when applied in the context of pest control, the persistence and amplification of baculoviruses in the larval host population has been recognized as being an essential component of plant protection, in particular for forestry (Moreau et al. 2005; Moreau & Lucarotti 2007). Field application of baculoviruses into the environment, such as occurs when LdMNPV is applied for the control of gypsy moth, does not increase virus levels beyond those that would occur naturally (Reardon et al., 2009). Also, through water run off or direct deposit of contaminated material (insects, frass, etc.), aquatic systems are recipients of baculoviruses (Hewsen et al., 2011). None of the non-target arthropods, such as shrimps, *Daphnia* spp. and *Notonecta* spp., or fresh-water, estuarine and marine fishes that have been tested by exposure to several NPVs have shown evidence of infection, toxicity or mortality (Table 1) (Dejoux & Elouard, 1990; Lightner et al., 1973; Couch et al., 1984).

Non-target insect toxicity studies are complex and the results usually depend upon the natural host range of any given baculovirus. Fundamental studies that were aimed at determining putative host range factors generally found that a different barrier to infection occurred for every virus and non-host system examined (See sections 2 and 3). The determination of the host range for any given baculovirus has, therefore, proven to be difficult to predict. Most baculoviruses, however, are very specific to their host species or closely-related ones (reviewed in Miller & Lu, 1997; Thiem & Cheng, 2009) and cross-order infections do not occur. In addition, SNPV alphabaculoviruses, betabaculoviruses and gammabaculoviruses appear to be the most restricted in host range, while some of the MNPV alphabaculoviruses (e.g. AcMNPV and *Mamestra brassicae* NPV [MabrNPV]) can infect over to 30 species crossing over 10 families of Lepidoptera (reviewed in Miller & Lu, 1997; Thiem & Cheng, 2009). Even within the MNPVs, however, those infecting Lymantriidae hosts such as LdMNPV in the gypsy moth, *L. dispar,* appear to be truly specific to a single host (Barber et al., 1993; Cory, 2003; Cory & Myers, 2003; Glare et al., 1995). Most *in vivo* host range studies have been carried out in the laboratory and at extremely high baculovirus dosage rather than at a range of concentration that might allow for the determination of a range of lethal doses (e.g., LD_{50} - LD_{95}). This artificial system does not accurately reflect the field situation and additional caution must be given to older toxicology

Clavijo, G., Williams, T., Muñoz, D., Caballero, P., & López-Ferber, M. 2011. Mixed genotype transmission bodies and virions contribute to the maintenance of diversity in an insect virus. *Proceedings of the Royal Society B: Biological Sciences* 277: 943-951.

Cohen, D.P.A., Marek, M., Davies, B.G., Vlak, J.M., & van Oers, M.M. 2009. Encyclopedia of *Autographa californica* nucleopolyhedrovirus. *Virology Sinica* 24: 359-414.

Cory, J.S. 2003. Ecological impacts of virus insecticides. In: *Environmental Impacts of Microbial Insecticides: Need and Methods for Risk Assessment*, H.M.T., Hokkanen, & A.E., Hajek, (Eds.), pp. 73-91, Kluwer, ISBN 1-4020-0813-9, Dordrecht, The Netherlands.

Cory, J.S. & Hails, R.S. 1997. The ecology and biosafety of baculoviruses. *Current Opinion in Biotechnology* 8: 323-327.

Cory, J.S. & Myers, J.H. 2003. The ecology and evolution of insect baculoviruses. *Annual Review of Ecology, Evolution, and Systematics* 34: 239-272.

Cory, J.S., Green, B.M., Paul, R.K., & Hunter-Fujita, F. 2005. Genotypic and phenotypic diversity of a baculovirus population within an individual insect host. *Journal of Invertebrate Pathology* 89: 101–111.

Cory, J. S. & Hoover, K. 2006. Plant-mediated effects in insect–pathogen interactions. *Trends in Ecology and Evolution* 21: 278-286.

Couch, J.A., Martin, S.M., Thompkins, G., & Kinney, J. 1984. A simple system for the preliminary evaluation of infectivity and pathogenesis of insect virus in a nontarget estuarine shrimp. *Journal of Invertebrate Pathology* 43: 351-357.

Cox, M.M.J. & Hashimoto, J. 2011. A fast tract influenza virus vaccine produced in insect cells. *Journal of Invertebrate Pathology* 107: S31-S41.

Cox, M.M. & Hollister, J.S. 2009. FluBlok, a next generation influenza vaccine manufactured in insect cells. *Biologicals* 37: 182-189.

CPL Business Consultants. 2010. The 2010 Worldwide Biopesticides Market Summary, Volume 1, June 5, 2010 - Pub ID: BGEQ2703518. MarketResearch.com, Rockville, MD, USA.

Croizier, G., Croizier, L., Argaud, O., & Poudevigne, D. 1994. Extension of *Autographa-californica* nuclear polyhedrosisvirus host-range by interspecific replacement of a short DNA-sequence in the P143 helicase gene. *Proceedings of the National Academy of Sciences USA* 91: 48-52.

Cuddleford, V. 2006. Managing editor. Surveys gauge attitude of Canadian public. *Biocontrol Files* 5: 1-8.

Dall, D., Luque, T., & O'Reilly. D. 2001. Insect-virus relationships: sifting by informatics. *Bioessays* 23:184–193.

Dejoux, C. & Elouard, J.M. 1990. Potential impact of microbial insecticides on the freswater environment, with special reference to the WHO/UNDP/World Bank Onchocerciasis control program. In: *Safety of Microbial Insecticides*, M., Laird, L.A., Lacey, and E.W., Davidson, (Eds.), pp. 66-83, CRC Press, ISBN 978-0849347931, Boca Raton, FL, USA.

Derksen, A.C.G. & Granados,. R. 1988. Alteration of a lepidopteran peritrophic membrane by baculoviruses and enhancement of viral infectivity. *Virology* 167: 242–250.

Detvisitsakun, C., Cain, E.L., & Passarelli, A.L., 2007. The *Autographa californica* M nucleopolyhedrovirus fibroblast growth factor accelerates host mortality. *Virology* 365: 70–78.

Doyle, C.J., Hirst, M.L., Cory, J.S., & Entwistle, P.F. 1990. Risk assessment studies: detailed host range testing of wild-type cabbage moth, *Mamestra brassicae* (Lepidoptera: Noctuidae), nuclear polyhedrosis virus. *Applied and Environmental Microbiology* 56: 2704-2710.

Du, X.L. & Thiem, S.M. 1997. Characterization of host range factor 1 (hrf-1) expression in *Lymantria dispar* M nucleopolyhedrovirus- and recombinant *Autographa californica* M nucleopolyhedrovirus-infected IPLB- Ld652Y cells. *Virology* 227: 420-430.

Duffy, S.P., Young, A.M., Morin, B., Lucarotti, C.J., Koop, B.F., & Levin, D.B. 2006. Sequence analysis and organization of the *Neodiprion abietis* nucleopolyhedrovirus genome. *Journal of Virology* 80: 6952–63.

Duffy, S.P., Becker, E.M., Whittome, B.H., Lucarotti, C.J., & Levin, D.B. 2007. *In vivo* replication kinetics and transcription patterns of the nucleopolyhedrovirus (NeabNPV) of the balsam fir sawfly, *Neodiprion abietis*. *Journal of General Virology* 88: 1945-1951.

EFSA Panel on Biological Hazards (BIOHAZ). 2009. Scientific opinion on the maintenance of the list of QPS microorganisms intentionally added to food or feed (2009 update). *EFSA Journal*, 7: 1431. 92pp. www.efsa.europa.eu

Ehlers, R.-U. 2011. Regulation of biological control agents and the EU policy support action REBECA. In: *Regulation of Biological Control Agents*, R.-U. Ehlers (Ed.), pp. 3-23, Springer Dordrecht, ISBN 978-90-481-3663-6, Heidelberg, Germany.

Elam, P., Vail, P.V., & Schreiber, F. 1990. Infectivity of *Autographa californica* nuclear polyhedrosis virus extracted with digestive fluids of *Heliothis zea*, *Estigmene acrea*, and carbonate solutions. *Journal of Invertebrate Pathology* 55: 278-283.

Elias, C.B., Jardin, B., & Kamen, A. 2007. Recombinant protein production in large-scale agitated bioreactors using the baculovirus expression vector system. *Methods in Molecular Biology* 388: 225-46.

Engelhard, E.K., Kam-Morgan, L.N.W., Washburn, J.O., & Volkman, L.E. 1994. The insect tracheal system: a conduit for the systemic spread of *Autographa californica* M nuclear polyhedrosis virus. *Proceedings of the National Academy of Sciences USA* 91: 3224-3227.

Engelhard, E. K. & Volkman, L. E. 1995. Developmental resistance in fourth instar *Trichoplusia ni* orally inoculated with *Autographa californica* M nuclear polyhedrosis virus. *Virology* 209: 384–389.

England, L.S., Vincent, M.L., Trevors, J.T., & Holmes. S.B. 2004. Extraction, detection and persistence of extracellular DNA in forest litter microcosms. *Molecular and Cellular Probes* 18: 313–319.

Entwistle, P.F., Adams, P.H.W., & Evans, H.F. 1977. Epizootiology of a nuclear-polyhedrosis virus in European spruce sawfly, *Gilpinia hercyniae*: birds as dispersal agents of the virus during winter. *Journal of Invertebrate Pathology* 30: 15-19.

Erlandson, M.A. & Carstens, E.B. 1983. Mapping early transcription products of *Autographa californica* nuclear polyhedrosis virus. *Virology* 126: 398-402.

Erlandson, M.A., Gordon, J., & Carstens, E.B. 1985. Size and map locations of early transcription products on the *Autographa californica* nuclear polyhedrosis virus genome. *Virology* 142: 12-23.

Fang, M., Nie, Y., Harris, S., Erlandson, M.A., & Theilmann, D.A. 2009. *Autographa californica* multiple nucleopolyhedrovirus core gene ac96 encodes a *per os* infectivity factor (pif-4). *Journal of Virology* 83: 12569–12578.

Farris, S.M. & Schulmeister, S. 2011. Parasitoidism, not sociality, is associated with the evolution of elabrate mushroom bodies in the brains of hymenopteran insects. *Proceedings of the Royal Society B: Biological Sciences* 278: 940-951.

Faulkner, P., Kuzio, J., Williams, G. V., & Wilson, J. A. 1997. Analysis of p74, a PDV envelope protein of *Autographa californica* nucleopolyhedrovirus required for occlusion body infectivity *in vivo*. *Journal of General Virology* 78: 3091–3100.

FDA. 2009. *Review of MPL-Relevant Toxicology Information in Cervarix BLA (STN 125259)*, 1-41 pp., U.S. Department of Health and Human Services, FDA, Center for Biologics Evaluation and Research, Rockville, MD, USA.

FDA. 2010. *Guidelines for Industry. Characterization and qualification of cell substrate and other biological materials used in the production of viral vaccines for infectious disease indications.* 1-47 pp., U.S. Department of Health and Human Services, FDA, Center for Biologics Evaluation and Research, Rockville, MD, USA.

Federici, B.A. 1997. Baculovirus pathogenesis, In: *The Baculoviruses*, L.K. Miller, (Ed.), pp. 8-59, Plenum Press, ISBN 9780306456411, New York.

Federici, B.A. & Stern, V.M. 1990. Replication and occlusion of a granulosis virus in larval and adult midgut epithelium of the western grapeleaf skeletonizer, *Harrisina brillians*. *Journal of Invertebrate Patholology* 56: 401-414.

Feng, S.Z., Jiao, P.R., Qi, W.B., Fan, H.Y., & Liao, M. 2011. Development and strategies of cell-culture technology for influenza vaccine. *Applied Microbial Biotechnology* 89: 893-902.

Friesen, P.D. 2007. Insect Viruses, In: *Fields Virology 5th ed.*, D.M. Knipe and P.M. Howley (Eds.), pp. 707-736, Lippincott Williams & Wilkins, ISBN 9780781760607, Philadelphia, PA, USA.

Fuchs, L. Y., M. S. Woods, & R. F. Weaver. 1983. Viral transcription during *Autographa califomica* nuclear polyhedrosis virus infection: a novel RNA polymerase induced in infected *Spodoptera frugiperda* cells. *Journal of Virology* 48: 641-646.

Garcia-Maruniak, A., Maruniak, J.E., Zanotto, P.M., Doumbouya, A.E., Liu, J.C., Merritt, T.M., & Lanoie, J.S. 2004. Sequence analysis of the genome of the *Neodiprion sertifer* nucleopolyhedrovirus. *Journal of Virology* 78: 7036–7051.

Georgopoulos, L.J., Elgue, G., Sanchez, J., Dussupt, V., Magotti, P., Lambris, J.D., Totterman, T.H., Maitland, N.J., & Nilsson, B. 2009. Preclinical evaluation of innate immunity to baculovirus gene therapy vectors in whole human blood. *Molecular Immunology* 46: 2911-2917.

Glare, T., Newby, E., and Nelson, T. 1995. Safety testing of a nuclear polyhedrosis virus for use against gypsy moth, *Lymantria dispar*, in New Zealand. *Proceedings of the Forty-Eighth New Zealand Plant Protection Congress*, pp. 264-269, Hastings, New Zealand, August 8-10, 1995.

Gomez-Casado, E., Gomez-Sebastian, S., Núñez, M.C., Lasa-Covarrubias, R., Martínez-Pulgarín, S., & Escribano, J.M. 2011. Insect larvae biofactories as a platform for influenza vaccine production. *Protein Expression and Purification* 79: 35-43.

Gomi, S., Majima, K., & Maeda, S. 1999. Sequence analysis of the genome of *Bombyx mori* nucleopolyhedrovirus. *Journal of General Virology* 80: 1323–1337.

Gonin, P. & Gaillard C. 2004. Gene transfer vector biodistribution: pivotal safety studies in clinical gene therapy development. *Gene Therapy* 11: S98-108.

Granados, R.R. & Lawler, K.A. 1981. *In vivo* pathway of *Autographa californica* baculovirus invasion and infection. *Virology* 108: 297-308.

Gröner, A. 1986. Specificity and safety of baculoviruses. In: *The Biology of Baculoviruses Vol. II: Practical Application for Insect Control*, R.R. Granados & B.A. Federici (Eds.), pp.177-202, CRC Press, ISBN 0849359864, Boca Raton, FL, USA.

Gross, C. & Rohrmann, G.F. 1993. Analysis of the role of 50 promoter elements and 30 flanking sequences on the expression of a baculovirus polyhedron envelope protein gene. *Virology* 192: 273-281.

Gross, C.H., Russell, R.L.Q., & Rohrmann, G.F. 1994. *Orgyia pseudotsugata* baculovirus p10 and polyhedron envelope protein genes: analysis of their relative expression levels and role in polyhedron structure. *Journal of General Virology* 75: 1115-1123.

Guarino, L.A. & Summers, M.D. 1986. Functional mapping of a trans-activating gene required for expression of a baculovirus delayed-early gene. *Journal of Virology* 57: 563-571.

Guarino, L.A. & Summers, M.D. 1987. Nucleotide sequence and temporal expression of a baculovirus regulatory gene. *Journal of Virology* 61: 2091-2099.

Guarino, L.A., Xu, B., Jin, J., & Dong, W. 1998. A virus-encoded RNA polymerase purified from baculovirus-infected cells. *Journal of Virology* 72: 7985-7991.

Guyton, A.C., & Hall. J.E. 2006. *Textbook of Medical Physiology* (11 ed.). Elsevier Saunders ISBN 9780721602400, Philadelphia, PA, USA.

Guzo, D., Rathburn, H., Guthrie, K., & Dougherty E. 1992. Viral and host cellular transcription in *Autographa-californica* nuclear polyhedrosis virus-infected gypsy-moth cell-lines. *Journal of Virology* 66: 2966-2972.

Haas-Stapleton, E.J., Washburn, J.O., & Volkman, L.E. 2003. Pathogenesis of *Autographa californica* M nucleopolyhedrovirus in fifth instar *Spodoptera frugiperda*. *Journal of General Virology* 84: 2033-2040.

Haas-Stapleton, E.J., Washburn, J.O., & Volkman, L.E. 2004. P74 mediates specific binding of *Autographa californica* M nucleopolyhedrovirus occlusion-derived virus to primary cellular targets in the midgut epithelia of *Heliothis virescens* larvae. *Journal of Virology* 78: 6786-6791.

Haas-Stapleton, E.J., Washburn, J.O., & Volkman, L.E. 2005. *Spodoptera frugiperda* resistance to oral infection by *Autographa californica* multiple nucleopolyhedrovirus linked to aberrant occlusion-derived virus binding in the midgut. *Journal of General Virology* 86: 1349-1355.

Hang, X., Dong, W., & Guarino, L.A. 1995. The lef-3 gene of *Autographa californica* nuclear polyhedrosis virus encodes a single-stranded DNA-binding protein. *Journal of Virology* 69: 3924-3928.

Hang, X. & Guarino, L.A. 1999. Purification of *Autographa californica* nucleopolyhedrovirus DNA polymerase from infected insect cells. *Journal of General Virology* 80: 2519-2526.

Harper, D.M., Franco, E.L., Wheeler, C., Ferris, D.G., Jenkins, D., Schuind, A., Zahaf, T., Innis, B., Naud, P., De Carvalho, N.S., Roteli-Martins, C.M., Teixeira, J., Blatter, M.M., Korn, A.P., Quint, W., & Dubin, G. 2004. Efficacy of a bivalent L1 virus-like particle vaccine in prevention of infection with human papillomavirus types 16 and 18 in young women: a randomized controlled trial. *Lancet* 364: 1757-1765.

Harper, D.M., Franco, E.L., Wheeler, C.M., Moscicki, A.B., Romanowski, B., Roteli-Martins, C.M., Jenkins, D., Schuind, A., Costa, C.S.A., & Dubin, G. 2006. Sustained efficacy up to 4.5 years of a bivalent L1 virus-like particle vaccine against human papillomavirus type 16 and 18: Follow-up from a randomized control trial. *Lancet* 367: 1247-1255.

Hartig, P.C., Chapman, M.A., Hatch, G.G., & Kawanishi, C.Y. 1989. Insect virus assay for toxic effects and transformation potential in mammalian cells. *Applied and Environmental Microbiology* 55: 1916-1920.

Hashimoto, Y., Corsaro, B.G., & Granados, R.R. 1991. Location and nucleotide sequence of the gene encoding the viral enhancing factor of the *Trichoplusia ni* granulosis virus. *Journal of General Virology* 72: 2645–2651.

Hauschild, R., Speiser, B., & Tamm, L. 2011. Regulation according to EU Directive 91/414: Data requirements and procedure compared with regulation practice in other OECD countries. pp. 25-77. In: *Regulation of Biological Control Agents,* Ehlers, R.-U. (Editor). Springer Science+Business Media B.V. ISBN 978-90-481-3663-6, Dordrecht Heildelberg London New York.

Hawtin, R.E., Zarkowska, T., Arnold, K., Thomas, C.J., Gooday, G.W., King, L.A., Kuzio, J.A., & Possee, R.D. 1997. Liquefaction of *Autographa californica* nucleopolyhedrovirus-infected insects is dependent on the integrity of virus-encoded chitinase and cathepsin genes. *Journal of Virology* 238: 243–253.

Hayakawa, T., Ko, R., Okano, K., Seong, S., Goto, C., & Maeda, S. 1999. Sequence analysis of the *Xiestia c-nigrum* granulovirus genome. *Journal of Virology* 262: 277-297.

Health Canada. 2010. Summary Basis of Decision (SBD) CERVARIX®. Human papillomavirus Types 16 and 18 Recombinant AS04 adjuvanted vaccine GlaxoSmithKline Inc. Submission Control Number: 127987. Health Products and Food Branch (October 26, 2010), pp. 1-17.

Health Canada - Pest Management Regulatory Agency. 2009. *Neodiprion abietis* Nucleopolyhedrovirus Newfoundland Strain. Registration Decision RD2009-05. http:// www.hc-sc.gc.ca

Health Canada - Pest Management Regulatory Agency. 2000. Virosoft CP4 *Cydia pomonella* granulosis virus. Regulatory Note. REG2000-10. http:// www.hc-sc.gc.ca

Hefferon, K.L., Oomens, A.G., Monsma, S.A., Finnerty, C.M., & Blissard, G.W. 1999. Host cell receptor binding by baculovirus GP64 and kinetics of virion entry. *Journal of Virology* 258: 455–468.

Heimpel, A.M. 1955. The pH in the gut and blood of the larch sawfly, *Pristiphora erichsonii* (Htg.), and other insects with reference to the pathogenicity of *Bacillus cereus* Fr. and Fr.1. *Canadian Journal of Zoology* 33: 99-106.

Herniou, E.A., Olszewski, J.A., O'Reilly, D.R., & Cory, J.S. 2004. Ancient coevolution of baculoviruses and their insect hosts. *Journal of Virology* 78: 3244-3251.

Hess, R.T. & Falcon, L.A. 1987. Temporal events in the invasion of the codling moth, *Cydia pomonella* , by a granulosis virus: an electron microscope study. *Journal of Invertebrate Pathology* 50: 85-105.

Hewson, I., Brown, J.M., Gitlin, S.A., & Doud. D.F. 2011. Nucleopolyhedrovirus detection and distribution in terrestrial, freshwater, and marine habitats of Appledore Island, Gulf of Maine. *Microbial Ecology* 62: 48-57.

Hitchman, R.B., Hodgson, D.J., King, L.A., Hail, R.S., Cory, J.S., & Possee, R.D. 2007. Host mediated selection of pathogen genotypes as a mechanism for the maintenance of baculovirus diversity in the field. *Journal of Invertebrate Pathology* 94: 153–162.

Hochberg, M.E. 1991. Intra-host interactions between a braconid endoparasitoid, *Apanteles glomeratus*, and a baculovirus for larvae of *Pieris brassicae*. *Journal of Animal Ecology* 60: 51-63.

Hodgson, D.J., Vanbergen, A.J., Hartley, S.E., Hails, R.S., & Cory, J.S. 2002. Differential selection of baculovirus genotypes mediated by different species of host food plant. *Ecology Letters* 5: 512–518.

Hofmann, C., Sandig, V., Jennings, G., Rudolph, M., Schlag, P., & Strauss, M. 1995. Efficient gene transfer into human hepatocytes by baculovirus vectors. *Proceedings of the National Academy of Sciences USA* 92: 10099-10103.

Hom, L.G., Ohkawa, T., Trudeau, D., & Volkman, L.E. 2002. *Autographa californica* M nucleopolyhedrovirus ProV-CATH is activated during infected cell death. *Journal of Virology* 296: 212–218.

Hoopes, R. R., Jr. & G. F. Rohrmann. 1991. *In vitro* transcription of baculovirus immediate early genes: accurate mRNA initiation by nuclear extracts from both insect and human cells. *Proceedings of the National Academy of Sciences USA* 88: 4513-4517.

Hoover, K., Grove, M., Gardner, M., Hughes, D.P., McNeil, J. & Slavicek, J. 2011. A gene for an extended phenotype. *Science* 333: 1401.

Hoover, K., Humphries, M.A., Gendron, A.R., & Slavicek, J.M. 2010. Impact of viral *enhancin* genes on potency of *Lymantria dispar* multiple nucleopolyhedrovirus in *L. dispar* following disruption of the peritrophic matrix. *Journal of Invertebrate Pathology* 104: 150-152.

Horton, H. M. & J. P. Burand. 1993. Saturable attachment sites for polyhedron-derived baculovirus on insect cells and evidence for entry via direct membrane fusion. *Journal of Virology* 67: 1860–1868.

Hossler, E.W. 2010a Caterpillars and moths: Part I. Dermatologic manifestations of encounters with Lepidoptera. *Journal of the American Academy of Dermatology* 62: 1-10.

Hossler, EW. 2010b. Caterpillars and moths: Part II. Dermatologic manifestations of encounters with Lepidoptera. *Journal of the American Academy of Dermatology* 62: 13-28.

Hu, Y.C. 2005. Baculovirus as a highly efficient expression vector in insect and mammalian cells. *Acta Pharmacologica Sinica* 26: 405-16.

Huang, X.-P., Davis, T.R., Hughes, P., & Wood. A. 1997. Potential replication of recombinant baculoviruses in nontarget insect species: reporter gene products as indicators of infection. *Journal of Invertebrate Pathology* 69: 234-245.

Huh, N.E. & Weaver R.F. 1990. Identifying the RNA polymerases that synthesize specific transcripts of the *Autographa califomica* nuclear polyhedrosis virus. *Journal of General Virology* 71:195-201.

(QPS): a generic risk assessment approach for biological agents notified to the European Food Safety Authority (EFSA). *Trends in Food Science & Technology* 21: 425-435.

Li, Y., Passarelli, A.L., Miller, L.K. 1993. Identification, sequence, and transcriptional mapping of *lef-3*, a baculovirus gene involved in late and very late gene expression. *Journal of Virology* 67: 5260-8.

Liang, Z. Zhang, X., Yin, X., Sumei Cao, S., & Xu, F. 2011. Genomic sequencing and analysis of *Clostera anachoreta* granulovirus. *Archives of Virology* 156: 1185–1198.

Lightner, D.V., Procter, R.R., Sparks, A.A., Adams, J.R., & Heimpel, A.M. 1973. Testing of penaid shrimps for susceptibility to an insect nuclear polyhedrosis virus. *Environmental Entomology* 2: 611-614.

Long, G., Pan, X., Kormelink, R., & Vlak, J.M. 2006. Functional entry of baculovirus into insect and mammalian cells is dependent on clathrin-mediated endocytosis. *Journal of Virology* 80: 8830–8833.

Lòpez-Ferber, M., Simòn, O., Williams, T., & Caballero, P. 2003. Defective or effective? Mutualistic interactions between virus genotypes. *Proceedings of the Royal Society B: Biological Sciences* 270: 2249–2255.

Lu., A. & Carstens, E.B. 1991. Nucleotide sequence of a gene essential for viral DNA replication in the baculovirus *Autographa californica* nuclear polyhedrosis virus. *Virology* 181: 336-347.

Lu, A. & Carstens, E.B. 1992. Transcription analysis of the EcoRI-D region of the baculovirus *Autographa californica* nuclear polyhedrosis virus identifies an early 4-kilobase RNA encoding the essential *p143* gene. *Journal of Virology* 66: 655-663.

Lu, A. & Miller, L.K. 1995a. The roles of eighteen baculovirus late expression factor genes in transcription and DNA replication. *Journal of Virology* 69: 975–982.

Lu, A. & Miller, L.K. 1995b. Differential requirements for baculovirus late expression factor genes in two cell lines. *Journal of Virology* 69: 6265–6272.

Lucarotti, C.J., Kettela, E.G., & Mudryj, G. 2006. The registration of Abietiv™: A biological control product based on *Neodiprion abietis* nucleopolyhedrovirus for use against its natural host, the balsam fir sawfly. SERG International Report, 47 pp.

Lucarotti, C. J., Moreau, G., & Kettela, E. G. 2007. Abietiv™– a viral biopesticide for control of the balsam fir sawfly. In: *Biological Control: A Global Pespective*, C. Vincent, M. Goettel, G. Lazarovits (Eds.), pp. 353-361, CABI Publishing, ISBN-10: 184593265X, Wallingford, UK.

Madhan, S., Prabakaran, M., & Kwang, J. 2010. Baculovirus as vaccine vectors. *Current Gene Therapy* 1: 201-213.

Maeda, S., Kamita, S.G., & Kondo, A. 1993. Host-range expansion of *Autographa californica* nuclear polyhedrosis virus (NPV) following recombination of a 0.6- kilobasepair DNA fragment originating from *Bombyx mori* NPV. *Journal of Virology* 67: 6234-6238.

Mäkelä, A.R. & Oker-Blom, C. 2006. Baculovirus display: a multifunctional technology for gene delivery and eukaryotic library development. *Advances in Virus Research* 68: 91-112.

Mäkelä, A.R., Tuusa, J.E., Volkman, L.E., & Oker-Blom, C. 2008. Occlusion-derived baculovirus: interaction with human cells and evaluation of the envelope protein P74 as a surface display platform. *Journal of Biotechnology* 135: 145-156.

Marek, M., van Oers, M.M., Devaraj, F.F., & Vlak, J.M. 2010. Engineering of baculovirus vector for the manufacture of virion-free biopharmaceuticals. *Biotechnology and Bioengineering* 108: 1056-1067.

Marrone, P.G. 2007. Barriers to adoption of biological control agents and biological pesticides. *CAB Reviews: Perspectives in Agriculture, Veterinary Science, Nutrition and Natural Resources* 2007 2, No. 051. doi: 10.1079/PAVSNNR20072051 ISSN 1749-8848.

Martignoni, M.E., & Iwai, P.J. 1981. A catalogue of viral diseases of insects, mites and ticks, In: *Microbial Control of Pests and Plant Diseases*, H.D. Burgess, (Ed.), pp. 897-911, Academic Press, ISBN 9780121433604, London, UK.

McClintock, J. T., Dougherty, E. M., & Weiner, R. M. 1986. Semipermissive replication of a nuclear polyhedrosis virus of *Autographa californica* in a gypsy-moth cell-line. *Journal of Virology* 57: 197-204.

McConnell, E.L., Basit, A.W., & Murdan. S. 2008. Measurements of rat and mouse gastrointestinal pH, fluid and lymphoid tissue, and implications for *in-vivo* experiments. *Journal of Pharmacy and Pharmacology* 60: 63–70.

McDougal, V.V. & Guarino, L.A. 1999. *Autographa californica* nuclear polyhedrosis virus DNA polymerase: measurements of processivity and strand displacement. *Journal of Virology* 73: 4908–4918.

McDougal, V.V. & Guarino, L.A. 2000. The *Autographa californica* nuclear polyhedrosis virus p143 gene encodes a DNA helicase. *Journal of Virology* 74: 5273–5279.

McLachlin, J.R. & Miller, L.K. 1994. Identification and characterization of *vlf-1*, a baculovirus gene involved in very late gene expression. *Journal of Virology* 68: 7746–7756.

McWilliam, A. 2007. Environmental impact of baculoviruses. FAO. R7299_FTR_anx3. http://www.fao.org/docs/eims/upload/agrotech/2003/R7299_FTR_anx3.pdf.

Means, J.C. & Passarelli, A.L. 2010. Viral fibroblast growth factor, matrix metalloproteases, and caspases are associated with enhancing infection by baculoviruses. *Proceedings of the National Academy of Science USA* 107: 9825–9830.

Metz, S.W. & Pijlman, G.P. 2011. Arbovirus vaccines: opportunities for the baculovirus-insect cell expression system. *Journal of Invertebrate Pathology* 107: S16-S30.

Miele, S.A.B., Garavaglia, M.J., Belaich, M.N., & Ghiringhelli, P.D. 2011. Baculovirus: molecular insights on their diversity and conservation. *International Journal of Evolutionary Biology* 2011: article ID 379424, 15 pp. doi:10.4061/2011/379424.

Mikhailov, V.S. & Rohrmann, G.F. 2002. The baculovirus replication factor LEF-1 is a DNA primase. *Journal of Virology* 76: 2287-2297.

Milks, M.L., Washburn, J.O., Willis, L.G., Volkman, L.E., & Theilmann, D.A. 2003. Deletion of *pe38* attenuates AcMNPV genome replication, budded virus production, and virulence in *Heliothis virescens*. *Virology* 310: 224-34.

Miller, L.K. & Lu, A. 1997. The molecular basis of baculovirus host range. In: *The Baculoviruses*, L.K. Miller (Ed.), pp. 217-235, Plenum Press, ISBN 0-306-45641-9, New York, NY, USA.

Monsma, S.A., Oomens, A.G.P., & Blissard G.W. 1996. The gp64 envelope fusion protein is an essential baculovirus protein required for cell-to-cell transmission of infection. *Journal of Virology* 70: 4607–4616.

Moreau, G. & Lucarotti, C.J. 2007. A brief review of the past use of baculoviruses for the management of eruptive forest defoliators and recent developments on a sawfly virus in Canada. *The Forestry Chronicle* 83: 105–112.

Moreau, G., Lucarotti, C.J., Kettela, E.G., Thurston, G.S., Holmes, S., Weaver, C., Levin, D.B., & Morin, B. 2005. Aerial application of nucleopolyhedrovirus induces decline in increasing and peaking populations of *Neodiprion abietis*. *Biological Control* 33: 65–73.

Morris, T. D. & Miller, L, K. 1993. Characterization of productive and nonproductive AcMNPV infection in selected insect-cell lines. *Virology* 197: 339-348.

Moscardi, F., Lobo de Souza, M., Batista de Castro, M.Ee., Moscardi, L.M., & Szewczyk, B. 2011. Baculovirus pesticides - present state and future perspectives, In: *Microbes and Microbial Technology*, I. Ahmad, F. Ahmad & P. Pichtel, (Eds.), pp. 415-445, Springer, ISBN 978-1-4419-7930-8, New York, NY, USA.

Moser, B.A., Becnel, J.J., White, S.E., Afonso, C., Kutish, G., Shanker, S., & Almira, E. 2001. Morphological and molecular evidence that *Culex nigripalpus* baculovirus is an unusual member of the family Baculoviridae. *Journal of General Virology* 82: 283–297.

Nakai M. & Kunimi Y. 1997. Granulosis virus infection of the smaller tea tortrix (Lepidoptera: Tortricidae): effect on the development of the endoparasitoid, *Ascogaster reticulatus* (Hymenoptera: Braconidae). *Biological Control* 8: 74-80.

Nobiron, I., O'Reilly, D.R., & Olszewski, J.A. 2003. *Autographa californica* nucleopolyhedrovirus infection of *Spodoptera frugiperda* cells: a global analysis of host gene regulation during infection, using a differential display approach. *Journal of General Virology* 84: 3029-3039.

OECD (Organization for Economic Co-operation and Development). 2002. *Consensus Document on Information Used in the Assessment of Environmental Applications Involving Baculovirus*. OECD Environment, Health and Safety Publications, Series on Harmonization of Regulatory Oversight in Biotechnology Number 20. Paris, France. 79 pp. http://www.rebeca- net.de/downloads/report/deliverable %2012.pdf.

Ohkawa, T., Washburn, J.O., Sitapara, R., Sid, E., & Volkman, L.E. 2005. Specific binding of *Autographa californica* M nucleopolyhedrovirus occlusion-derived virus to midgut cells of *Heliothis virescens* larvae is mediated by products of pif genes *Ac119* and *Ac022* but not *Ac115*. *Journal of Virology* 79: 15258–15264.

Ohkawa, T., Volkman, L.E., & Welch. M.D. 2010. Actin-based motility drives baculovirus transit to the nucleus and cell surface. *Journal of Cell Biology* 190: 187–195.

Okano, K., Vanarsdall, A.L., Mikhailov, V.S., & Rohrmann, G.F. 2006. Conserved molecular systems of the Baculoviridae. *Journal of Virology* 344: 77–87.

Oker-Blom, C., Airenne, K.J., & Grabherr, R. 2003. Baculovirus display strategies: emerging tools for Eukaryotic libraries and gene delivery. *Briefings in Functional Genomics and Proteomics* 2: 224-253.

Ooi, B.G., Rankin, C., & Miller, L.K. 1989. Downstream sequences augment transcription from the essential initiation site of a baculovirus polyhedrin gene. *Journal of Molecular Biology* 210: 721–736.

Oomens, A.G. & Blissard, G.W. 1999. Requirement for GP64 to drive efficient budding of *Autographa californica* multicapsid nucleopolyhedrovirus. *Journal of Virology* 254: 297–314.

Passarelli, A.L. 2011. Barriers to success: how baculovirus establish efficient systemic infections. *Journal of Virology* 411: 383-392.

Passarelli, A.L. & Miller, L.K. 1993a. Identification and characterization of *lef-1*, a baculovirus gene involved in late and very late gene expression. *Journal of Virology* 67: 3481-3488.

Passarelli, A.L. & Miller, L.K. 1993b Three baculovirus genes involved in late and very late gene expression: *ie-1*, *ie-n*, and *lef-2*. *Journal of Virology* 67: 2149-2158.

Passarelli, A.L. & Miller, L.K. 1993c. Identification of genes encoding late expression factors located between 56.0 and 65.4 map units of the *Autographa californica* nuclear polyhedrosis virus genome. *Virology* 197: 704-714.

Pearson, M.N., Groten, C. & Rohrmann, G.F. 2000. Identification of the *Lymantria dispar* nucleopolyhedrovirus envelope fusion protein provides evidence for a phylogenetic division of the Baculoviridae. *Journal of Virology* 74: 6126- 6131.

Peng, K., van Lent, J.W., Vlak, J.M., Hu, Z., & van Oers, M.M. 2011. *In situ* cleavage of the baculovirus occlusion derived virus receptor binding protein P74 in the peroral infectivity complex. *Journal of Virology* 85: 10710-10718.

Peng, K., van Oers, M.M., Hu, Z., van Lent, J.W., Vlak, J.M. 2010. Baculovirus *per os* infectivity factors form a complex on the surface of occlusion-derived virus. Journal of *Virology* 84: 9497-504.

Perera, O.P., Valles, S.M., Green, T.B., White, S., Strong, C.A., & Becnel, J.J. 2006. Molecular analysis of an occlusion body protein from *Culex nigripalpus* nucleopolyhedrovirus (CuniNPV). *Journal of Invertebrate Pathology* 91: 35-42.

Perera, O., Green, T.B., Stevens, S.M., White, J.S., & Becnel, J.J. 2007. Proteins associated with *Culex nigripalpus* nucleopolyhedrovirus occluded virions. *Journal of Virology* 81: 4585–4590.

Peters, W. 1992. *Peritrophic Membranes*, Springer, ISBN 3540536353, New York, NY, USA.

Pijlman, G. P., Pruijssers, A. J., & Vlak, J. M. 2003. Identification of pif-2, a third conserved baculovirus gene required for per os infection of insects. *Journal of General Virology* 84: 2041–2049.

Plymale, R., Grove, M.J., Cox-Foster, D., Ostiguy, N., & Hoover. K. 2008. Plant-mediated alteration of the peritrophic matrix and baculovirus infection in lepidopteran larvae. *Journal of Insect Physiology* 54: 737–749.

Podgwaite, J., Shields, K., Zerillo, R., & Bruen, R. 1979. Environmental persistence of the nucleopolyhedrosisvirus of the gypsy moth. *Environmental Entomology* 8: 523-536.

Possee, R.D. & Rohrmann, G.F. 1997. Baculovirus genome organization and evolution. *The Baculoviruses,* L.K. Miller, (Ed.), pp. 109-140, ISBN 0-306-45641-9, Plenum Press, NewYork, NY, USA.

Prikhod'ko, E.A., Lu, A., Wilson, J.A., & Miller, L.K. 1999. *In vivo* and *in vitro* analysis of baculovirus ie-2 mutants. *Journal of Virology* 73: 2460–2468

Pullen, S.S., & Friesen, P.D. 1995. The CAGT motif functions as an initiator element during early transcription of the baculovirus transregulator ie-1. *Journal of Virology* 69: 3575–3583.

Quinlan, R, & Gill, A. 2006. *The World Market for Microbial Biopesticides. Overview Volume.* CPL Business Consultants, October 2006. 26 pp.

Rapp, J.C., Wilson, J.A., & Miller, L.K. 1998. Nineteen baculovirus open reading frames, including LEF-12, support late gene expression. *Journal of Virology* 72: 10197–10206.

Uwo, M.F., Ui-Tei, K., Park, P., & Takeda, M. 2002. Replacement of midgut epithelium in greater wax moth, *Galleria mellonella*, during larval-pupal moult. *Cell and Tissue Research* 308: 319-331.

van Oers, M.M. 2006. Vaccines for viral and parasitic diseases produced with baculovirus vectors. *Advances in Virus Research* 68: 193-253

van Oers, M.M. 2011. Opportunities and challenges for the baculovirus expression system. *Journal of Invertebrate Pathology* 107: 3-15.

Vasconcelos, S.D., Williams, T., Hails, R.S., & Cory, J.S. 1996. Prey selection and baculovirus dissemination by carabid predators of Lepidoptera. *Entomological Entomology* 21: 98-104.

Vergati, M., Intrivici, C., Huen, N.Y, Schlom, J., & Tsang, K.W. 2010. Strategies for cancer vaccine development. *Journal of Biomedicine and Biotechnology* 2010: 1-13 pp. doi:10.1155/2010/596432.

Volkman, L.E. & Summers, M.D. 1977. *Autographa californica* nuclear polyhedrosis virus: comparative infectivity of the occluded, alkali-liberated, and nonoccluded forms. *Journal of Invertebrate Pathology* 30: 102–103.

Volkman, L.E. & Goldsmith, P.A. 1983. *In vitro* survey of *Autographa californica* nuclear polyhedrosis virus interaction with nontarget vertebrate host cells. *Applied and Environmental Microbiology* 45: 1085-1093.

Wang, L., Salem, T.Z., Lynn, D.E., & Cheng, X.-W., 2008. Slow cell infection, inefficient primary infection and inability to replicate in the fat body determine the host range of *Thysanoplusia orichalcea* nucleopolyhedrovirus. *Journal of General Virology* 89: 1402–1410.

Wang, P. & Granados, R.R. 1997a. An intestinal mucin is the target substrate for a baculovirus enhancin. *Proceedings of the National Academy of Sciences USA* 94: 6977–6982.

Wang, P. & Granados, R.R. 1997b. An intestinal mucin is the target substrate for a baculovirus enhancin. *Proceedings of the National Academy of Sciences USA* 94: 6977–6982.

Wang, P. & Granados, R.R. 2000. Calcofluor disrupts the midgut defense system in insects. *Insect Biochemistry and Molecular Biology* 30: 135–143.

Wang, P., Hammer, D.A. & Granados, R.R. 1994. Interaction of *Trichoplusia ni* granulosis virus-encoded enhancin with the midgut epithelium and peritrophic membrane of four lepidopteran insects. *Journal of General Virology* 75: 1961–1967.

Wang P., Li G., & Granados R.R. 2004. Identification of two new peritrophic membrane proteins from larval *Trichoplusia ni*: structural characteristics and their functions in the protease rich insect gut. *Insect Biochemistry and Molecular Biology*. 34: 215-227.

Wang, R. R., Deng, F., Hou, D. H., Zhao, Y., Guo, L., Wang, H. L., & Hu, Z. H. 2010. Proteomics of the *Autographa californica* nucleopolyhedrovirus budded virions. *Journal of Virology* 84: 7233–7242.

Wang, X-F., Zhang,, B-Q. Xu, H-J., Cui, Y-J., Xu, Y-P., Zhang, M-J., Han, Y. S., Lee, Y.S., Bao, Y-Y., & Zhang, C-X. 2011. ODV-associated proteins of the *Pieris rapae* granulovirus. *Journal of Proteome Research* 10: 2817–2827.

Wang, Y. & Jehle, J.A. 2009. Nudiviruses and other large, double-stranded circular DNA viruses of invertebrates: new insights on an old topic. *Journal of Invertebrate Pathology* 101:187-193.

Wang, Y., Bininda-Emonds, O.R.P., Oers, M.M., Vlak, J.M., & Jehle, J.A. 2011. The genome of *Oryctes rhinoceros* nudivirus provides novel insight into the evolution of nuclear arthropod-specific large circular double-stranded DNA viruses. *Virus Genes* 42: 444-456.

Washburn, J.O., Kirkpatrick, B.A., & Volkman, L.E. 1995. Comparative pathogenesis of *Autographa californica* M nuclear polyhedrosis virus in larvae of *Trichoplusia ni* and *Heliothis virescens*. *Virology* 209: 561-568.

Washburn, J.O., Lyons, E.H., Haas-Stapleton, E.J., & Volkman, L.E. 1999. Multiple nucleocapsid packaging of *Autographa californica* nucleopolyhedrovirus accelerates the onset of systemic infection in *Trichoplusia ni*. *Journal of Virology* 73: 411-416.

Washburn, J.O., Chan, E.Y., Volkman, L.E., Aumiller, J.J., & Jarvis D.L. 2003a. Early synthesis of budded virus envelope fusion protein GP64 enhances *Autographa californica* multicapsid nucleopolyhedrovirus virulence in orally infected *Heliothis virescens*. *Journal of Virology* 77: 280-290.

Washburn, J.O., Trudeau, D., Wong, J.F., & Volkman, L.E. 2003b. Early pathogenesis of *Autographa californica* multiple nucleopolyhedrovirus and *Helicoverpa zea* single nucleopolyhedrovirus in *Heliothis virescens*: a comparison of the 'M' and 'S' strategies for establishing fatal infection. *Journal of General Virology* 84: 343-351.

Westenberg, M., Uijtdewilligen, P., & Vlak. J.M. 2007. Baculovirus envelope fusion proteins F and GP64 exploit distinct receptors to gain entry into cultured insect cells. *Journal of General Virology* 88: 3302-3306.

Wickham, T.J., Shuler, M.L., Hammer, D.A., Granados, R.R., & Wood, H.A. 1992. Equilibrium and kinetic analysis of *Autographa californica* nuclear polyhedrosis virus attachment to different insect cell lines. *Journal of General Virology* 73: 3185-3194.

Winstanley D. & Crook N.E. 1993. Replication of *Cydia pomonella* granulosis virus in cell cultures. *Journal General Virology* 74: 1599-1609.

Witt, D. J. 1984. Photoreactivation and ultraviolet-enhanced reactivation of ultraviolet irradiated nuclear polyhedrosis virus by insect cells. *Archives of Virology* 79: 95-107.

Wu, J.G. & Miller, L.K. 1989. Sequence, transcription and translation of a late gene of the *Autographa californica* nuclear polyhedrosis virus encoding a 34.8K polypeptide. *Journal of General Virology* 70: 2449-2459.

Xiang, X., Chen, L., Guo, A., Yu, S., Yang, R., & Wu, X. 2011. The *Bombyx mori* nucleopolyhedrovirus (BmNPV) ODV-E56 envelope protein is also a *per os* infectivity factor. *Virus Research* 155: 69-75.

Yang S. & Miller L.K. 1999. Activation of baculovirus very late promoters by interaction with very late factor. *Journal of Virology* 73: 3404-3409.

Yao, L., Zhou, W., Xu, H., Zheng, Y., & Qi, Y. 2004. The *Heliothis armigera* single nucleocapsid nucleopolyhedrovirus envelope protein P74 is required for infection of the host midgut. *Virus Research* 104: 111-121.

Zhang J.H., Washburn J.O., Jarvis D.L., & Volkman L.E. 2004. *Autographa californica* M nucleopolyhedrovirus early GP64 synthesis mitigates developmental resistance in orally infected noctuid hosts. *Journal of General Virology* 85: 833-842.

Zhang, J-H., Ohkawa, Washburn, T.J.O., & Volkman, L.E. 2005. Effects of Ac150 on virulence and pathogenesis of *Autographa californica* multiple nucleopolyhedrovirus in noctuid hosts. *Journal of General Virology* 86: 1619–1627.

Part 3

Integrated Pest Management – Future Challenges

Wheat Midges and Thrips Information System: Decision Making in Central Germany

Nabil El-Wakeil[1,2,*], Nawal Gaafar[1,2],
Mostafa El-Wakeil[3] and Christa Volkmar[2]
[1]Pests and Plant Protection Department, National Research Centre, Dokki, Cairo,
[2]Institute of Agric. and Nutritional Sciences, Martin-Luther-University Halle-Wittenberg,
[3]Kuwaiti Ministry of Interior – Information Systems Directorate, Dajij, Farwania,
[1]Egypt
[2]Germany
[3]Kuwait

1. Introduction

Wheat Midges and Thrips Expert System (WMTES) is constructive computer software, giving the users a recommendation based on pheromone and water traps catches as well as infestation levels. These results were collected from our field experiments which conducted in three locations in central Germany (Halle, Silstedt and Salzmünde) during three years 2007 to 2009. Computer programs can help in information recovery and decision support when dealing with pest problems. These decision support tools can provide farmers with easy, rapid access to accurate information that can help them to obtain the threshold to make adequate management decisions. Plans for future field testing and expert system implementation are also discussed. Using such as expert system for controlling wheat ear insects can be successfully applied to the solution of daily problems in plant protection programs for wheat producers. Finally, the obtained results would give a good guide for decision making which proved an efficient method of integrated plant protection for wheat ear insects as well as other insects in another crop.

An expert system is a computer program, which mimics behaviour of an expert in a particular area of knowledge. Expert systems (ES) have been developed and applied in many agriculture fields i.e. diagnose insects and diseases of various crops. Farmers across the world face problems like soil erosion, increasing cost of chemical pesticides, weather damage recovery, the need to spray, mixing and application, yield loses and pest resistance. On the other hand researchers in the field of agriculture are constantly working in Pakistan on new management strategies to promote farm success (Khan et al., 2008). Pest management is a highly challenging problem. Globally, annual losses from pests and diseases had increased year after year (Sharma, 2001).

The development of an agricultural expert system requires the combined efforts of specialists from many fields of agriculture, and must be developed with the cooperation of

*Correspondence Author

help is increasing rapidly. Diagnosis or diagnostic problem solving is the process of understanding what is wrong in a particular situation. Thus gathering of information and then interpreting the gathered information for determining what is wrong are of central importance in diagnostic problem solving (Lucus, 1997).

3.2 System structure

Figure (2) shows typical expert system structure we have created. Each of these blocks is explained below.

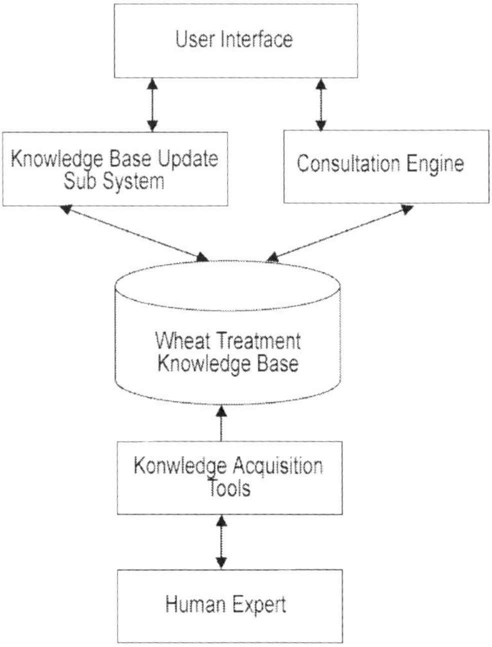

Fig. 2. Expert system structure of wheat ear insects (modified after Khan et al., 2008).

3.3 User interface

This is the interface the end user will use to interact with the system by providing parameters to it and having recommendations and consultation results out of it.

Knowledge base update sub system

As we know the utmost drawback of using expert system is that it has fixed knowledge base. If this base is not updated periodically, the results of consultation by time will be out of date. Thus, we developed this sub system to have the ability to update and enhance the knowledge base at any time easily and smoothly.

Consultation engine

Consultation Engine is the communication channel between the end user and the system; this is where user submits his consultation. Engine has 2 operation modes one is system wizard and the other is manual entry.

Wheat treatment knowledge base

This is the heart and the core of the system where it holds all the knowledge that we process to give the right decision to the user.

Knowledge acquisition tools

This is the ways we acquire knowledge from different sources and save it in the knowledge base.

Human expert

Everything in the end must return back to humans without the help of human expert we can not by any means have computerized expert system.

3.4 System user interface

Our system has 3 main modules:

1. System main data entry module
2. Knowledge base update module
3. Consultation engine module

3.4.1 System main data entry module: System main screen (fig. 3a)

Main Data Menu: (Fig. 3B)

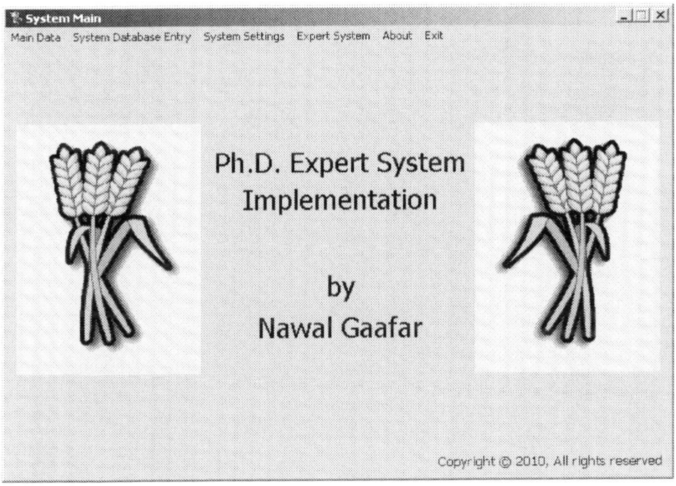

Fig. 3A. System Main Data Entry Module.

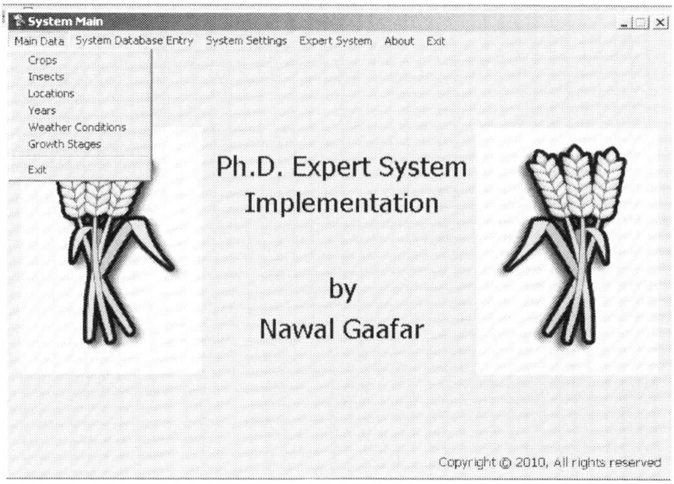

Fig. 3B. Main Data Menu.

Crops Window

Here we can add, update and delete any kind of crops that we are dealing with now or may be in need to deal with in the future (Fig. 4A).

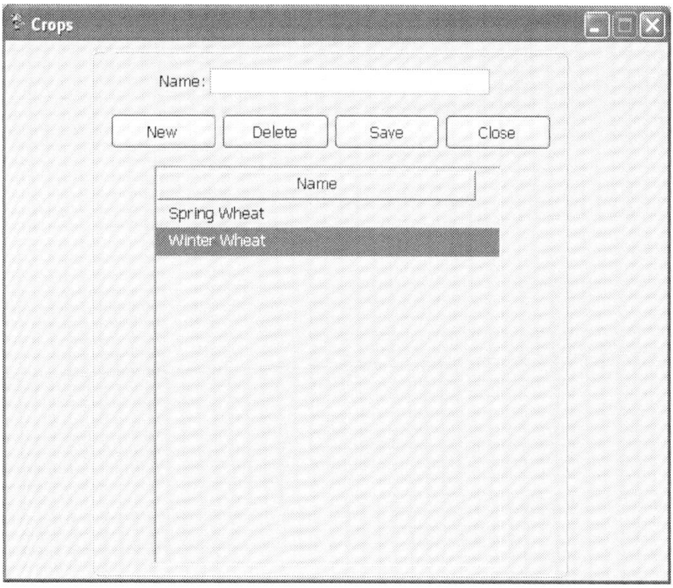

Fig. 4A. Crops window.

Insects Window

From this window we can add, update and delete any kinds of insects that we are dealing with now or may be in need to deal with in the future(Fig. 4B)

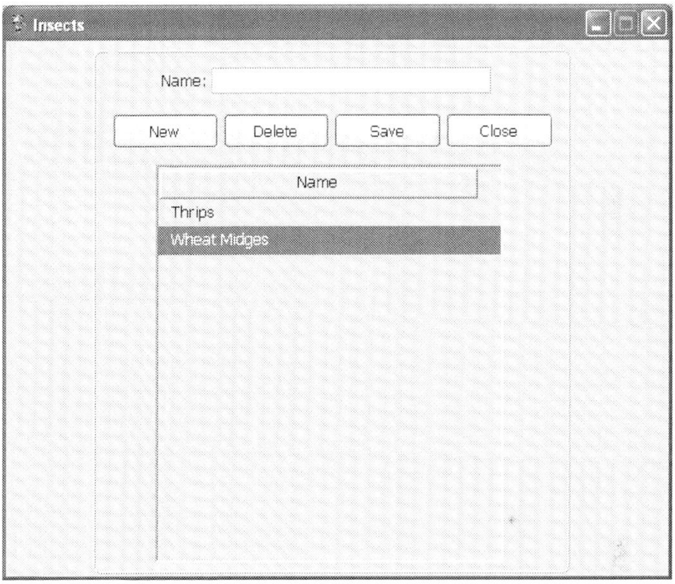

Fig. 4B. Insects window.

Locations Window

From this window we can add, update and delete any study locations that we are using now or may use in the future (Fig. 5A).

Weather Conditions Window

From this window we can add, update and delete any weather conditions that may be affected either on crop or insects (Fig. 5B).

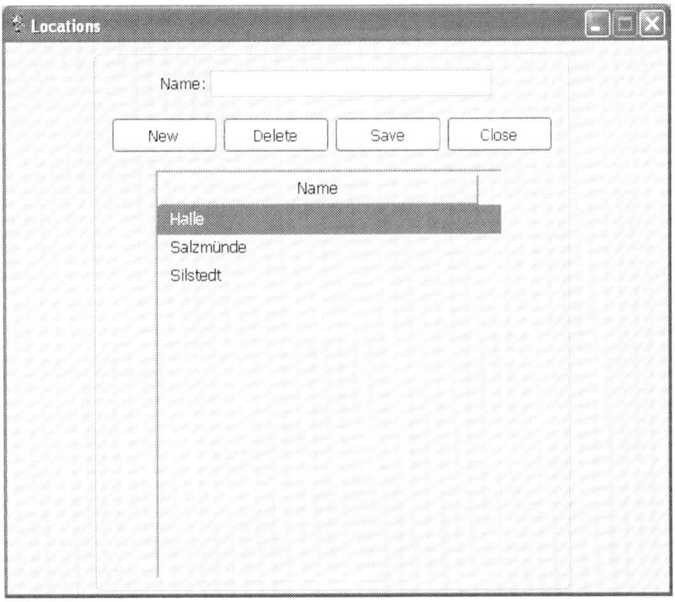

Fig. 5A. Locations window.

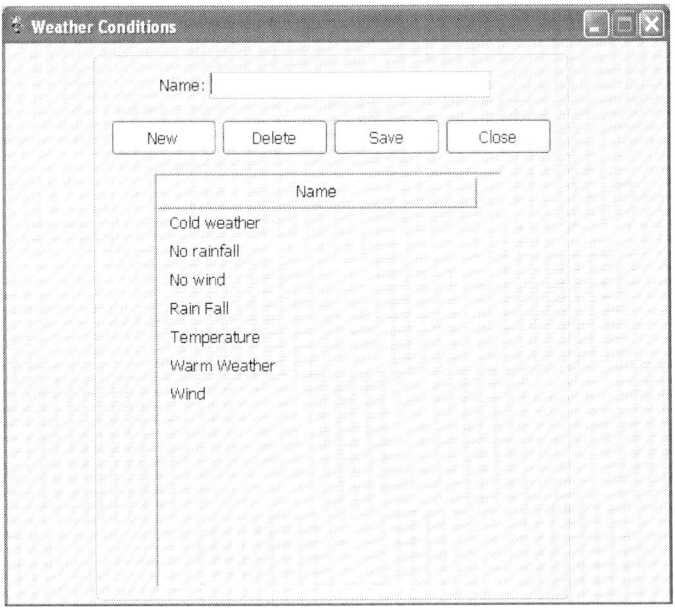

Fig. 5B. Weather conditions window.

Growth stages window

From this window we can add, update and delete any growth stages that we are interesting to study the population dynamic of insects (Fig. 6).

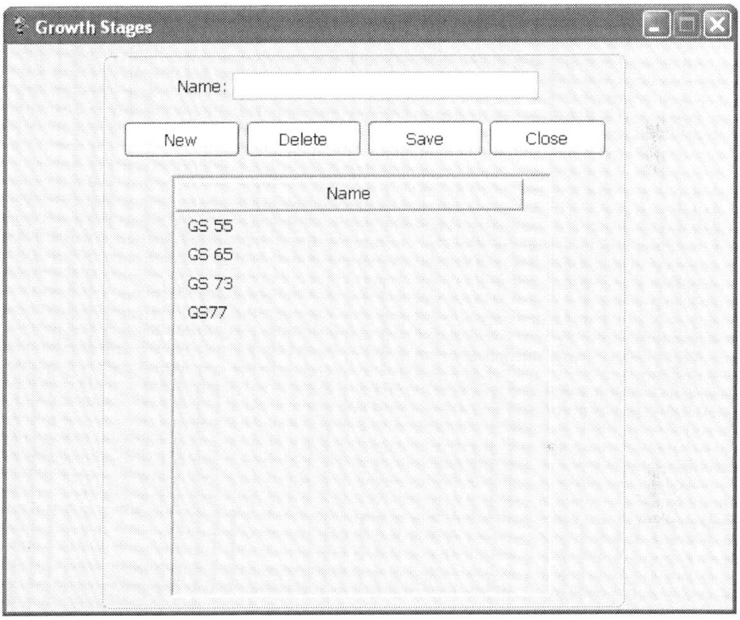

Fig. 6. Growth stages window.

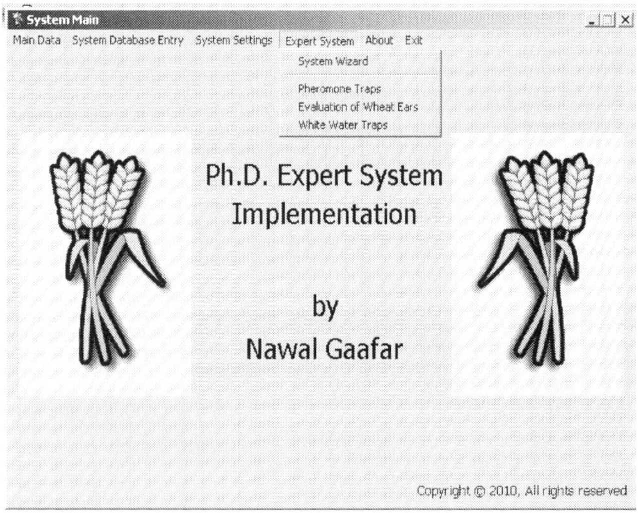

Fig. 9. System Wizard.

Summary window after gathering all the required information from the user (Fig. 10A)

Fig. 10A. All the required information.

This is an example for the consultation result out of the system (Fig. 10B).

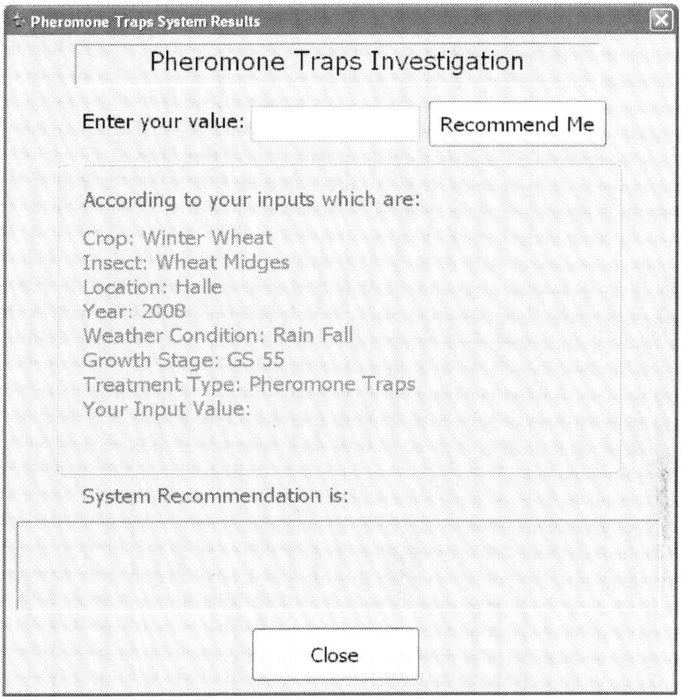

Fig. 10B. The consultation result out of the system.

Pheromone Traps: In this window system will use the defaults assigned in system settings in the consultation where user will only submit the OWBM value and click on recommend me and the system will process the value and give recommendation to user (Fig. 11 A&B)

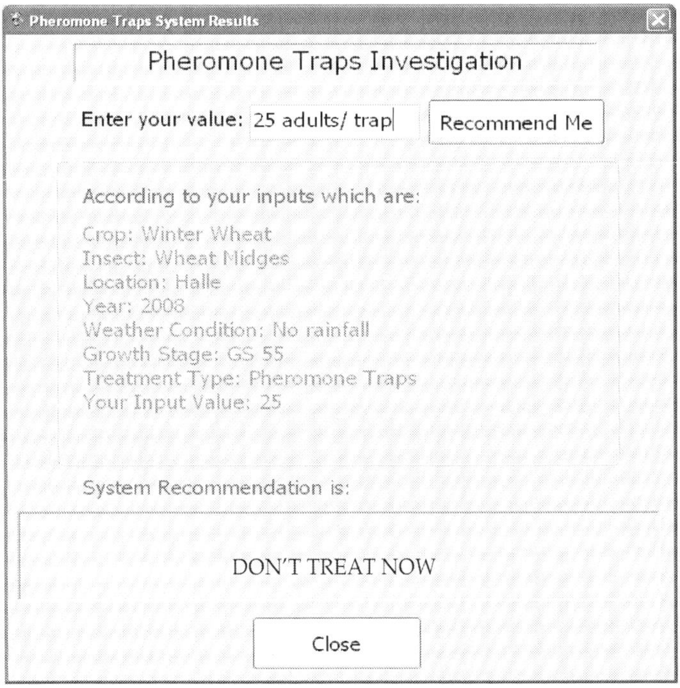

Fig. 11A. Recommendation without treatment.

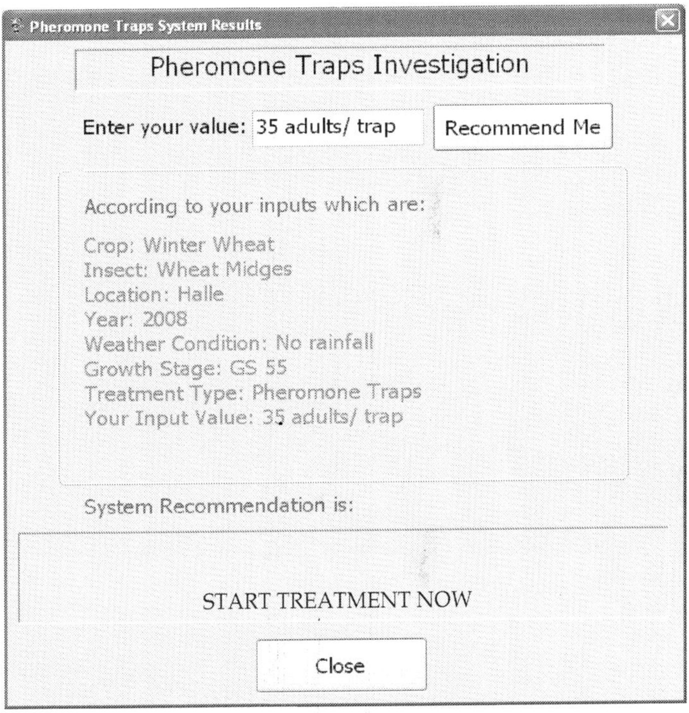

Fig. 11B. Recommendation with treatment.

Evaluation of wheat ears (wheat midges)

Here, this system will use the defaults assigned in system settings in the consultation where the user will only submit midge larvae value and click on recommend me and the system will process the value and give recommendation (Fig. 12 A&B).

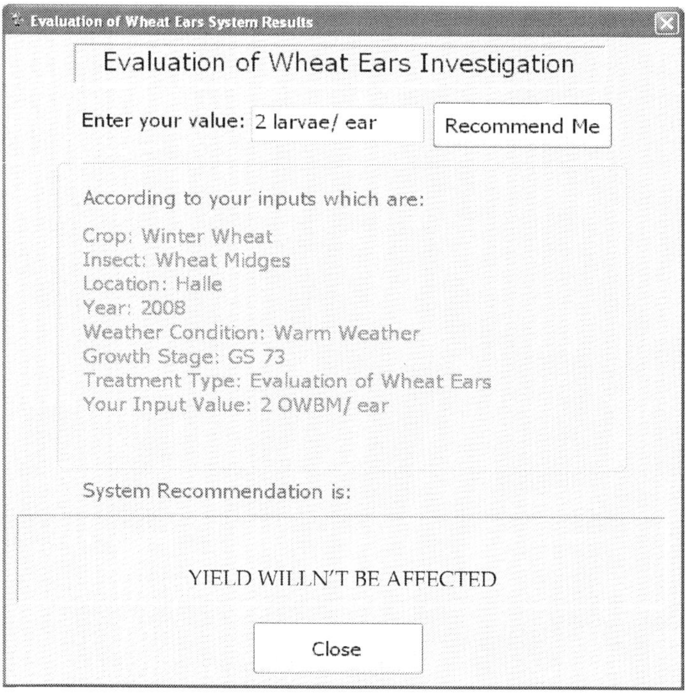

Fig. 12A. Expectation without yield losses.

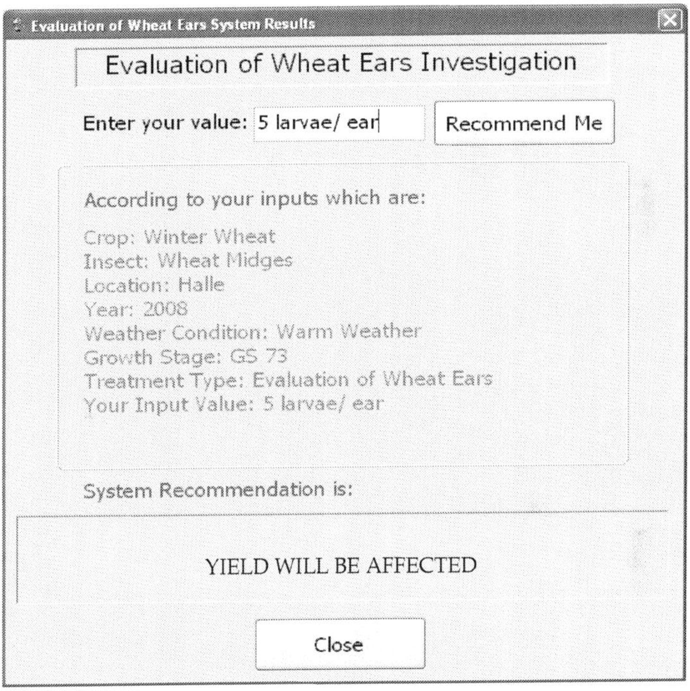

Fig. 12B. Expectation with yield losses.

Evaluation of wheat ears (thrips)

This system will use the defaults assigned in system settings in the consultation where the user will only submit thrips value and click on recommend me and the system will process the value and give recommendation to the user (Fig. 13 A&B).

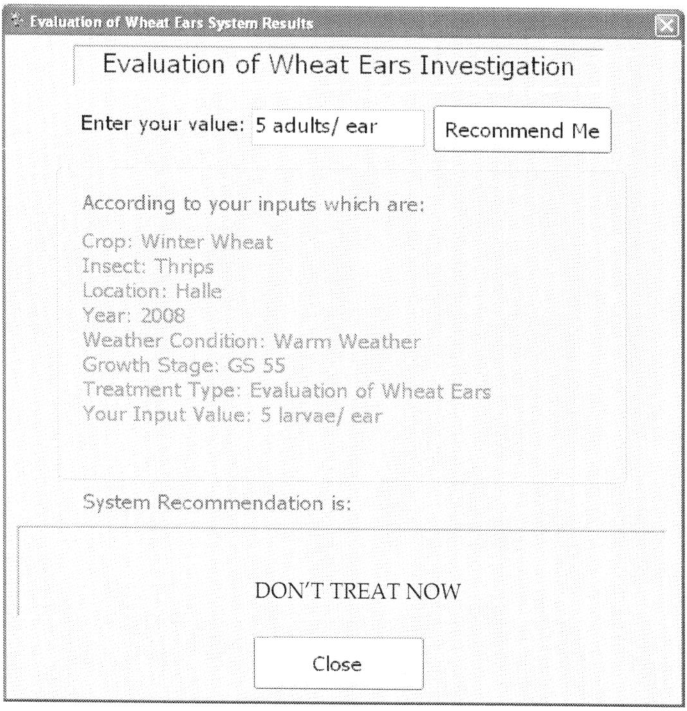

Fig. 13A. Recommendation without treatment.

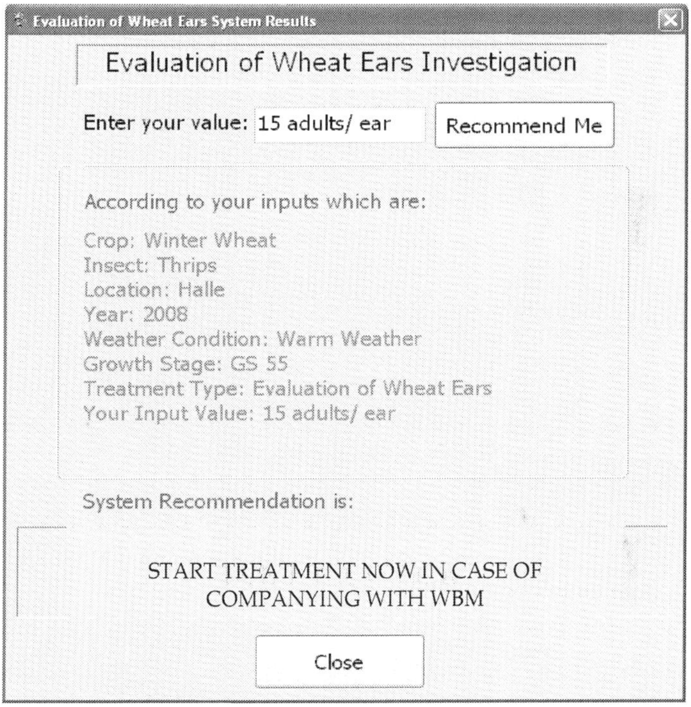

Fig. 13B. Recommendation with treatment.

White water traps

In this window, the system will use the defaults assigned in system settings in the consultation where the user will only submit the midge larvae value and click on recommend me and the system will process the value and give recommendation (Fig. 14 A&B).

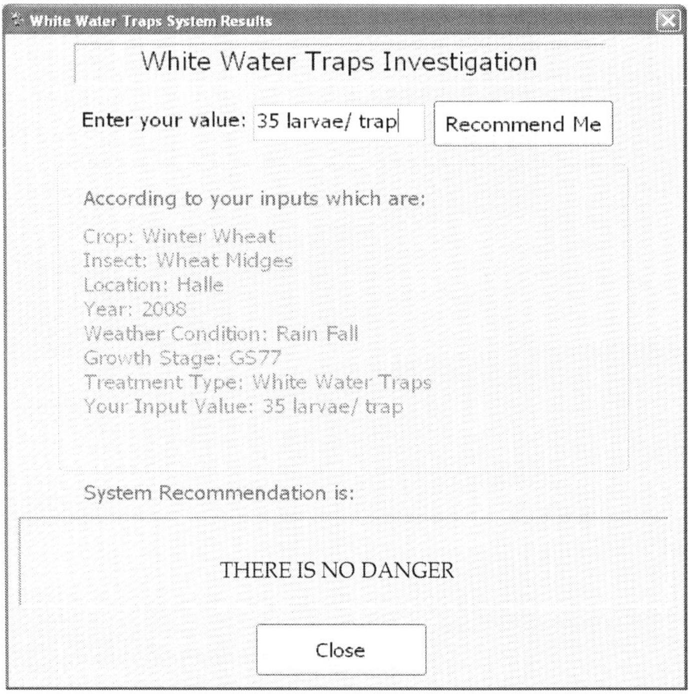

Fig. 14A. Recommendation without danger.

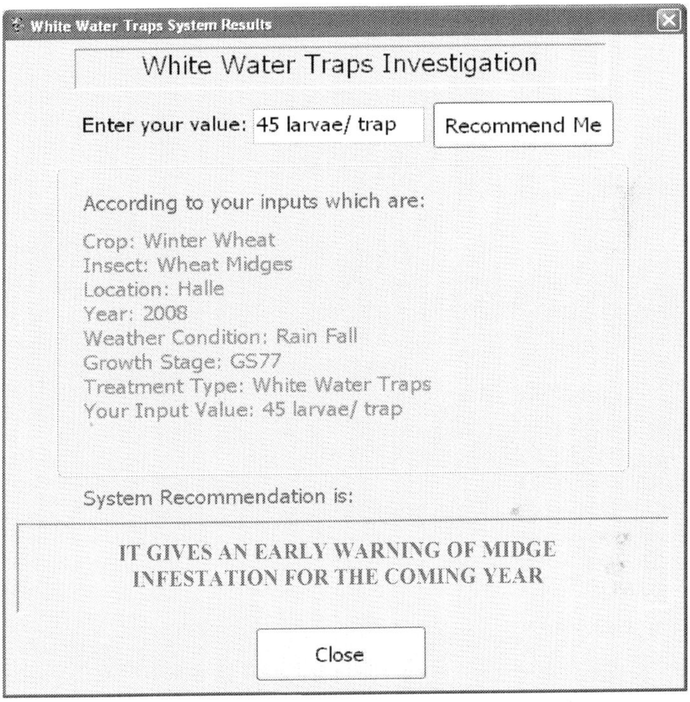

Fig. 14B. Recommendation with warning.

About

This is the about and copyright of the system (Fig. 15)

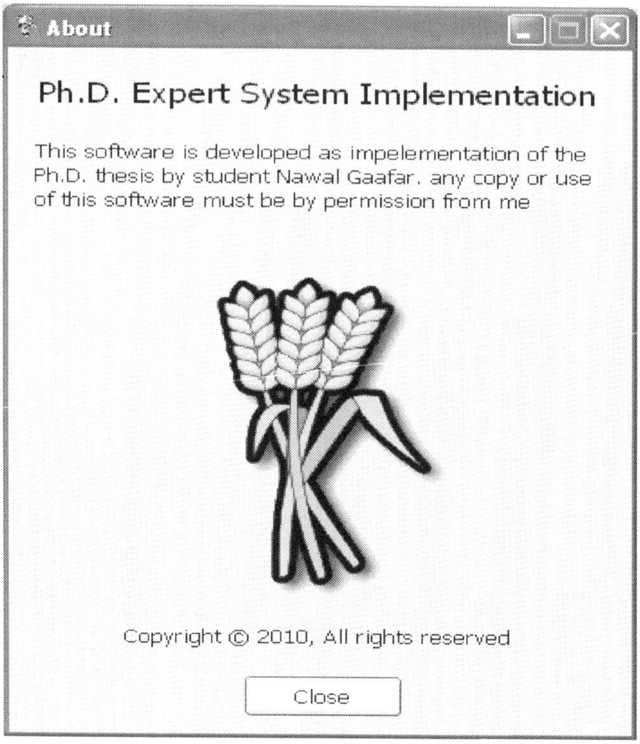

Fig. 15. The copyright of this system.

4. Testing and validation

WMTES field trials (based on crop samples) during 2007, 2008 and 2009 seasons, were done for the samples received from three different regions. As it is unwise to implement any ES from day one after completion of its development, parallel consultation from the IPM experts was found useful in improvement and validation of ES results. Also, real world ES need testing and validation in the real world environment i.e. field testing. Comparison of results produced from system as well as suggested by IPM experts has been used to improve the quality of inferences and consultation. Feedback forms have been used for preparation of validation case sheets.

5. Discussion

Pheromone traps monitor only male midge whereas it is female WBM that lay eggs from which the damaging stage of the pest emerges. Therefore, it is not possible to set a simple trap catch threshold above which economic damage occurs and below which it does not. A decision support model that can be used by farmers was developed using a stepwise decision tree involving yes/no answers to questions (Fig. 1). When growing a susceptible wheat variety pheromone traps need to be put out before ear emergence in fields where wheat was grown in previous years and provide a source of the pest. These traps should be monitored daily or at least twice weekly during the susceptible growth stage. When trap catches exceed threshold (30 midges/ day) crop inspections provide additional information to help decide whether to treat a field. It is advantageous that the pheromone traps are so sensitive and catch as many WBM as possible because they provide an early warning of midge flight during ear heading time and the suitable weather conditions, thus avoiding situations in which insecticide sprays are applied too late when they are needed. The threshold of midge larvae infestation is 3-4 larvae/ ear, where as in water traps, used to sample the migrated midge larvae is 30-40 larvae/ trap. Similar results were recorded by Ellis et al. (2009). The later should be also monitored carefully after the heavy rain, particularly at late milky stage; it gives an early caution of midge infestation for the coming year, especially intend cultivating wheat after wheat.

Midge infestation was higher in 2008 than in 2007 and 2009. The low levels of midge infestation hindered the verification of the decision flow tree as it was not possible to examine the impact of midge catches on infested ears. There can be some confidence in the proposed threshold of greater than 30 midges/ trap/ day to indicate a need to inspect crops for the pest as reported by Ellis et al. (2009).

The main objectives of WMTES have been met in allowing better data provision at all levels. However, the system can easily be changed and updated to meet new demands. For example, the current system automatically produces bulletins. This function could be developed further to produce automatically customized pest reports to include the interpretation of current pheromone or water traps data and to provide pertinent advice to growers and/or advisors. These could include comparisons between ear insect numbers in previous years and changes in the numbers between dates for specific regions, or a more forecast for weather conditions on the population dynamic of ear insect in any year. A knowledge base developed for a specific region in a problem can be tuned to make it more appropriate for other regions. Our results are consistent with Edrees et al. (2003) and Khan et al. (2008).

The current expert system has improved understanding of the WBM problem. Improving risk prediction: the decision flow chart proposes thresholds to help predict the need for insecticide treatment. The verification study suggested that this is a good basis for risk management. However, thresholds are based on data from a limited number of sites and years; further work is required to confirm the initial findings and improve the precision with which it is possible to predict the risk of pest attack. Risk of damage is also primarily dependent upon the coincidence among midge activity, the susceptible stage and weather conditions. Being able to predict the likely timing of the susceptible stage in relation to midge emergence would be a significant development, and help to limit unnecessary spray as stated by Freier et al. (1996).

The expert system will be used in training new experts. It will allow less experienced users to examine the reasoning process of an expert, to improve their understanding of how one takes control decision and to learn how to approach different situations to take the adequate decision. Agricultural extension services require more effective ways of handling, communicating, and using information. The program for the control of wheat ear insects is an example of one way that expert system technology can be successfully applied to daily problem solution in plant protection as recommended by Räder et al. (2007) and Suo & Shi (2008).

Much of the power and flexibility of expert systems are due to the fact that the knowledge base is separated from the inference mechanism (Rauscher, 1990; Waterman, 1985). The knowledge base can be modified without interfering with the operation of the system or the performance of other rules. In this way the program is now being enriched with rules for other insects, crops and sites, as well as by using of the user's feedback to improve the system. It is as important to develop the system as to maintain it. This task will be carried out for the extension services. WMTES is designed to enhance the quality and availability of knowledge required by decision makers in wheat insect management. It depends on a knowledge base that contains all the knowledge required to give useful, accurate and adequate consultations to wheat farmers.

6. References

Chakraborty, P. & Chakrabarti D.K. (2008). A brief survey of computerized expert systems for crop protection being used in India. *Progress in Natural Science* 18: 469-473.

Edrees, S.A., Rafea, A., Fathy, I. & Yahia, M. (2003). NEPER: a multiple strategy wheat expert system. *Computer & Electron in Agriculture* 40: 27-43.

ESICM (1994). Expert Systems for Improved Crop Management (FAO): A study of the needs assessment for expert systems in the agriculture sector in Egypt: *Technical Report No. TR-88*-024-33.

Ellis, S.A., Bruce, T.J.A., Smart, L.E., Martin, J.L., Snape, J. & Self, M. (2009). Integrated management strategies for varieties tolerant and susceptible to wheat midge. *HGCA Project Report number 451 May 2009*, 148 pages.

Freier, B., Triltsch, H. & Roßberg, D. (1996). GTLAUS – a model of wheat-cereal aphid-predator interaction and its use in complex agroecological studies. *Zeitschrift Pflanzenkrankh. Pflanzenschutz* 103: 543–554.

El-Wakeil, N., Gaafar, N. & Volkmar C. (2010). Susceptibility of spring wheat to infestation with wheat midges and thrips. *Journal of Plant Diseases and Protection* 117: 261–267.

Gaafar, N. (2010). Wheat midges and thrips information system: Monitoring and decision making in central Germany. PhD Dissertation Martin- Luther- University Halle-Wittenberg, 109 pages.

Gaafar, N. & Volkmar, C. (2010). Evaluation of wheat ear insects in large scale field in central Germany. *Agricultural Sciences* 1: 68-75.

Gaafar, N., Cöster, H. & Volkmar, C. (2009). Evaluation of ear infestation by Thrips and wheat blossom midges in winter wheat cultivars. In: Feldmann F., Alford D., Furk C., (eds.). *Proceedings 3rd International Symposium on Plant Protection & Plant Health*, pp. 349-359.

Gaafar, N., El-Wakeil, N. & Volkmar, C. (2011a). Assessment of wheat ear insects in winter wheat varieties in central Germany. *Journal of Pest Science* 84: 49-59.

Gaafar, N., Volkmar, C., Cöster, H. & Spilke J (2011b). Susceptibility of winter wheat cultivars to wheat ear insects in central Germany. *Gesunde Pflanzen* 62: 107-115.

Gonzalez-Andujar, J.L., Garcia De Ceca, J.L. & Fereres, A., (1993). Cereal aphids expert system" Identification and decision making. *Computer & Electron in Agriculture* 8: 293-300.

Gonzalez-Diaz, L., Martínez-Jimenez, P., Bastida, F. & Gonzalez-Andujar, J.L. (2009). Expert system for integrated plant protection in pepper (*Capsicum annuun* L.). *Expert System with Application* 36: 8975-8979.

Gosselke, U., Triltsch, H., Rossberg, D. & Freier, B. (2001). GETLAUS01 – the latest version of a model for simulating aphid population dynamics in dependence on antagonists in wheat. *Ecological Modelling* 145: 143–157.

Jörg, E., Racca, P., Preiß, U., Buttutini, A., et al. (2007). Control of Colorado potato beetle with the SIMLEP decision support system. *OEPP/EPPO* 37: 353-358.

Khan, F.S., Razzaq, S., Irfan, K., Maqbool, F., Farid, A., Illahi, I. &, Ul Amin, T. (2008). Dr. Wheat: a web-based expert system for diagnosis of diseases and pests in Pakistani wheat. World Congress on Engineering, London, July 2 - 4, 2008. *Proceedings of World Congress on Engineering* I: 549-554.

Kleinhenz, B. & Rossberg, D. (2000). Structure and development of decision-support systems and their use by the State Plant Protection Services in Germany. *OEPP/EPPO Bulletin* 30: 93-97.

Kleinhenz, B. & Zeuner, T. (2007). Introduction of GIS in decision support systems for plant protection. In Alord D., Feldman F., Hasler J., Tiedemann A. *Proceedings 2rd International Symposium on Plant Protection & Plant Health in Europe, Humboldt Universität, 2007, 82: 24-25.*

Knight, J. D., Tatchell, G. M., Norton, G. A. & Harrington, R. (1992). FLYPAST: an information management system for the Rothamsted Aphid Database to aid pest control research and advice. *Crop Protection* 11: 419-426.

Lucus, P. (1997). Symbolic diagnosis and its formalization, Knowle. *Engineering Review* 12: 109–146.

Mann, B.P., Wratten, S.D., Watt, A.D. (1986). A computer-based advisory system for cereal aphid control. *Computer Electron Agriculture* 1: 263-270.

Räder, T., Racca, P., Jörg, E. & Hau, B. (2007). PUCREC/PUCTRI - a decision support system for the control of leaf rust of winter wheat and winter rye. *OEPP/EPPO* 37: 378-382.

Rafea, A. & Shaalan, K. (1996). Using expert systems as a training tool in the agriculture sector in Egypt. *Expert Systems with Applications* 11: 343-349.

Rafea, A., El-Dessouki, A., Hassan, H. & Mohamed, S. (1993). Development and implementation of a knowledge acquisition methodology for crop management expert systems. *Computer & Electron in Agriculture* 8: 129-146.

Rauscher, H.M. (1990). Practical expert system development in Prolog. *Artificial Intelligence in Natural Resource Management* 4: 51-55.

Saini, H.S., Kamal, R. & Sharma, A.N. (2002). Web Based Fuzzy Expert System for Integrated Pest Management in Soybean. *International Journal Information Technology* 8: 54-74.

Sharma, M.C. (2001). Integrated pest management in developing countries: Special Reference to India. *IPM Mitr*, pp. 37-44.

Sivakami, S. & Karthikeyan, C. (2009). Evaluating the effectiveness of expert system for performing agricultural extension services. *Expert System with Application* 36: 9634-9636.

Suo, X. & Shi, N. (2008). Web-based expert system of wheat and corn growth management. *IFIP International Federation for Information Processing* 258: 111-119.

Tiedemann, A. & Kleinhenz, B. (2008). Prognose contra Praxis? *DLG- Mitteilungen* 3: 54-58.

Tottman, D.R. (1987). The decimal code for the growth stages of cereals, with illustrations. *Annual of Applied Biology* 110: 441-454.

Volkmar C., Werner C. & Matthes P. (2008). On the occurrence and crop damage of wheat blossom midges *Contarinia tritici* (Kby.) and *Sitodiplosis mosellana*. *Mitteilungen Deutschen Gesellschaft für Allgemeine und Angewandte Entomologie* 16: 305-308.

Volkmar C., Schröder A., Gaafar N., Cöster, H. & Spilke, J. (2009). Evaluierungsstudie zur Befallssituation von Thripsen in einem Winterweizensortiment. *Mitteilungen Deutschen Gesellschaft für Allgemeine und Angewandte Entomologie* 17: 227-230.

Waterman, D.A. (1985). A Guide to Expert Systems. *Addison-Wesley, Reading, MA*, 441 pp.

23

Role of GPI-Anchored Membrane Receptors in the Mode of Action of *Bacillus thuringiensis* Cry Toxins

Fernando Zúñiga-Navarrete, Alejandra Bravo,
Mario Soberón and Isabel Gómez
*Instituto de Biotecnología-UNAM. Depto.
Microbiología Molecular.Cuernavaca, Morelos
Mexico*

1. Introduction

Insect pests are the major cause of damage to commercially important agricultural crops. Chemical pesticides have long-term detrimental effects, leading to irreversible damage to the environment and elimination of natural predators. Also, several hundred insect species have developed resistance to one or more chemical insecticides. There is, therefore, a need for environmentally safe pest control to maintain sustainability of the environment. *Bacillus thuringiensis* (Bt) emerged as a valuable biological alternative in pest control, because of its advantages of specific toxicity against target insects, lack of polluting residues and safety to non-target organisms such as humans, other vertebrates and plants, and is completely biodegradable. Bt has been used as a biopesticide in agriculture, forestry and mosquito control and accounts for 95% of the 1% market share of biopesticides in the total pesticide market. However, insect resistance against Bt has been reported in many cases. Insects develop resistance to insecticides through mechanisms that reduce the binding of toxins to gut receptors (de Maagd et al., 2001).

Bacillus thuringiensis Cry toxins have been widely used in the control of insect pests either as spray products or expressed in transgenic crops. These proteins are pore-forming toxins with a complex mechanism of action that involves the sequential interaction with several toxin-receptors. Cry toxins are specific against susceptible larvae and although they are often highly effective, some insect pests are not affected by them or show low susceptibility. In addition, the development of resistance threatens their effectiveness, so strategies to cope with all these problems are necessary. In this chapter we will discuss and compare the different proteins that are involved in the mechanism of action of Cry toxins with special emphasis on GPI-receptors: Aminopeptidases and alkaline phosphatases. We will discuss how the mechanism of toxin-receptor interaction has an important role to design new strategies to improve insecticidal activity of Cry toxins. In addition we will discuss other insect gut proteins that have recently been shown to bind Cry toxins and that may be involved in Cry toxin action.

2. Cry toxins

Bacillus thuringiensis (Bt) is an endospore-forming bacterium characterized by the presence of a protein crystal within the cytoplasm of the sporulating cell. Individual Cry toxin has a defined spectrum of insecticidal activity, usually restricted to a few species in one particular order of Lepidoptera (butterflies and moths), Diptera (flies and mosquitoes), Coleoptera (beetles and weevils), Hymenoptera (wasps and bees), and also to nematodes, respectively (Rajamohan et al., 1998). A few toxins have an activity spectrum that spans two or three insect orders. For example, Cry1Ba is most notably active against the larvae of moths, flies, and beetles. The combination of toxins in a given strain, therefore, defines the activity spectrum of that strain. Cry proteins are defined as: a parasporal inclusion protein from Bt that exhibits toxic effects to a target organism, or any protein that has obvious sequence similarity to a known Cry protein (Schnepf et al., 1998).

To date, the tertiary structures of seven different Cry proteins, Cry1Aa, Cry2Aa, Cry3Aa, Cry3Bb, Cry4Aa, Cry4Ba and Cry8Ea have been determined by X-ray crystallography (Li et al., 1991; Grochulski et al., 1995; Galitsky et al., 2001; Morse et al., 2001; Boonserm et al., 2005; Boonserm et al., 2006; Guo et al., 2009). All these structures display a high degree of similarity with a three-domain organization (Figure 1), suggesting a similar mode of action of the Cry protein family even though they show very low amino acid sequence similarity. Cry toxins are classified by their primary amino acid sequence and more than 500 different *cry* gene sequences have been classified into 67 groups (Cry1–Cry67). They are globular molecules composed of three structural domains connected by single linkers. Domain I, a seven α-helix bundle, is implicated in membrane insertion, toxin oligomerization and pore formation. Domain II is a beta-prism of three anti-parallel β-sheets packed around a hydrophobic core with exposed loop regions that are involved in receptor recognition, and domain III, is a β-sandwich of two anti-parallel β-sheets. Both domain II and III are implicated in insect specificity by mediating specific interactions with different insect gut proteins (Bravo et al., 2007).

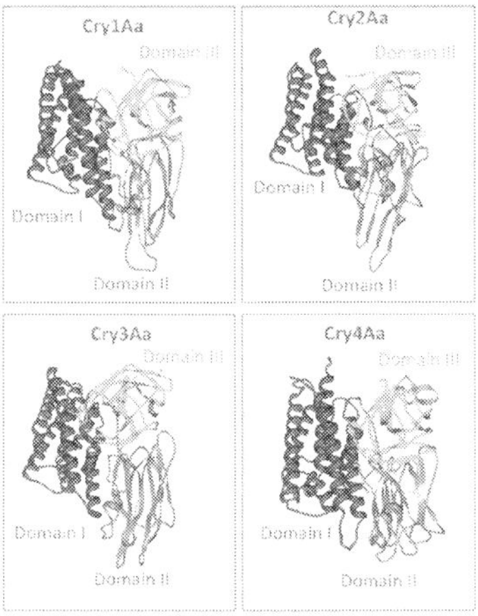

Fig. 1. Crystal structure of Cry1Aa (PDB code, 1CIY); Cry2Aa (PDB code, 1I5P); Cry3A (PDB code, 1DLC) and Cry4Aa (PDB code, 2C9K). Figures were generated using PyMol program.

2.1 Structure-function relationship

Domain I was immediately recognized as being equipped for pore formation, since shares structural similarities with other PFT like colicin Ia and N and diphtheria toxin, supporting the role of this domain in pore-formation. Isolated domain I fragments have been demonstrated to partition into model membranes to form pores (de Maagd et al., 2003); conversely, the domains II-III segment expressed without domain I was shown to bind to midgut membrane (Flores et al., 1997). Domain II was suspected to determine the specificity, because it represents the most divergent part of the toxin sequence, and exchanging domain II, or domains II and III, between closely related toxins has resulted in active hybrids showing altered specificity. Site-directed mutagenesis in its three hypervariable apical loops have identified residues involved in specific binding to membranes of several lepidopteran and coleopteran insects (Rajamohan et al., 1998; Schnepf et al., 1998). The specific binding consisted of reversible and irreversible steps, but only the irreversible binding is correlated with toxicity. More recent domain-exchange studies have found that toxicity of the hybrids to the insect host followed the movement of domain III, which would point to domain III as being responsible for the prerequisite step in toxicity, namely receptor binding. Site-directed mutagenesis in domain III located a small number of residues affecting membrane binding affinity and toxicity (de Maagd et al., 2001). However, direct observation of a toxin with a bound specificity determinant from the insect receptor is still needed to identify the receptor-binding site.

2.2 Molecular mode of action

The mode of action of Cry toxins has been characterized principally in lepidopteran insects and involves several steps and interactions with different receptors that depend on the oligomeric state of the toxin in a ping-pong binding mechanism. Cry1A toxins are produced as crystal inclusion bodies, which need to be ingested by the susceptible larvae to be toxic. These crystals are dissolved in the alkaline and reducing environment of the larval midgut, releasing soluble protoxins of 130 kDa. The inactive protoxins are then cleaved by midgut proteases yielding 60 kDa monomeric toxins (Soberon et al., 2009).

In figure 1, the activated monomeric toxins bind to highly abundant low affinity receptors, glycosylphosphatidylinositol (GPI)-anchored proteins, such as aminopeptidase N (APN) and alkaline phosphatase (ALP), localizing the toxin in the brush border microvilli. Specifically, loop 3 of domain II and β-16 of domain III are involved in this first interaction (Pacheco et al., 2009; Arenas et al., 2010). After this, the toxin binds to low abundant cadherin receptor, in a high affinity and complex interaction involving participation of loop 2, loop 3, and loop α-8 of domain II the toxin. Binding with cadherin facilitates additional protease cleavage of the N-terminal end of the toxin eliminating helix α-1 of domain I (Gomez et al., 2002). This cleavage induces assembly of an oligomeric form of the toxin. The conformational changes in toxin oligomers results in 100-fold increased binding affinity to APN and ALP receptors, through loop 2 (Bravo et al., 2004; Arenas et al., 2010). After the oligomers bind to these receptors they insert into membrane microdomains, creating pores in the apical membrane of midgut cells causing osmotic shock, bursting off the midgut cells and finally ending with the death of the insect (Soberon et al., 2009).

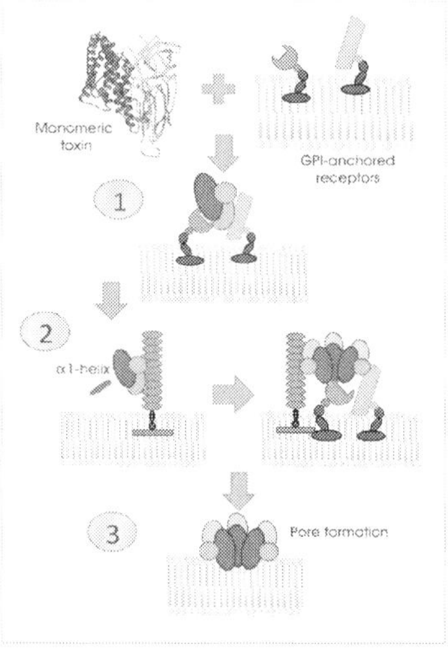

Fig. 2. Model of the mode of action of Cry toxins. 1) First interaction between monomeric toxin with low affinity-high abundant GPI-anchored receptors. 2) Second interaction occurs between monomeric toxin with cadherin receptor to induce oligomerization of the toxin. 3) The interaction of oligomeric toxin with GPI-anchored receptors result in pore formation.

2.3 Specifity=Binding receptor?

Although a given toxin has quite a narrow host range *in vivo, in vitro* many are non-specific and activated forms are even capable of forming pores in simple lipid bilayers. However, excessively high concentrations are required for lipid bilayer pore formation and the resulting pores have properties different than those formed in the presence of insect midgut membrane components. Many proteins have been identified in the midgut membrane that are capable of binding to Bt toxins but the ability to bind does not always correlate with susceptibility. For example, the pink bollworm (*Pectinophora gossypiella*) is susceptible to Cry1Ab and Cry1Ac toxins, both of which bind to a common site. A strain of pink bollworm (AZP-R) resistant to these two toxins could no longer bind Cry1Ab but, surprisingly, Cry1Ac binding was unaffected (Gonzalez-Cabrera et al., 2003). One explanation for such a result is that not all toxin binding is productive; that is, not all toxin binding results in pore-formation. Nevertheless, for several insect midgut proteins have been shown to have functional relevance to the toxic mechanism. The first is aminopeptidase N, whose expression in a *Drosophila* cell line resulted in an acquired susceptibility to toxin (Gill & Ellar 2002) and whose silencing by RNAi in a susceptible insect resulted in a reduced sensitivity to toxin (Rajagopal et al., 2002). Another membrane protein, cadherin, demonstrated a high toxin affinity, and mutations in cadherin genes result in a loss of sensitivity (Gahan et al., 2001; Morin et al., 2003).

3. Cry toxins receptors

As mentioned previously, Cry toxins are highly selective and kill only a limited number of insect species. This selectivity is mainly due to the interaction of Cry toxins with larval proteins located in the midgut epithelium cells. The crucial role of this receptor binding for toxicity is emphasized by the observation that insects selected for resistance to a Cry toxin often have no or reduced binding capacity for that toxin(Ferre & Van Rie 2002). A major research effort has taken place in the identification of insect proteins that bind Cry toxins and mediate toxicity. Among these, two major types of receptors have been identified: transmembrane proteins, such as cadherins, and proteins anchored to the membrane such as the GPI-anchored proteins that have been proposed to be involved in the action of Cry toxins (Gomez et al., 2007).

3.1 What kind of molecules are receptors for Cry toxins?

After it was demonstrated that specific high-affinity toxin binding sites are present in the midgut of susceptible insects, efforts to identify and clone this molecules has been intensified. Several putative Cry toxin receptors have since been reported, of which the best characterized are the aminopeptidase N (APN) receptors and the cadherin-like receptors identified in lepidopterans. In nematodes, glycolipids are believed to be an important class of Cry toxin receptors. Other putative receptors include alkaline phosphatases (ALPs), a 270-kDa glycoconjugate, and a 252-kDa protein (Pigott & Ellar 2007).

Protein	Insect order	Species
Cadherin	Lepidoptera	*Manduca sexta, Heliothis virescens, Ostrinia nubilalis, Helicoverpa armigera, Bombix mori, Pectinophora gossypiella*
	Diptera	*Anopheles gambiae, Aedes aegypti*
	Coleoptera	*Tenebrio molitor Diabrotica virgifera virgifera*
BTR-270	Lepidoptera	*Lymantria dispar*
Chlorophyllide-binding protein (252 protein)	Lepidoptera	*Bombix mori*
ADAM 10 metalloprotease	Coleoptera	*Leptinotarsa decemlineata*

Table 1. Receptors of Cry toxins (non-GPI anchored to the membrane) described in different insects.

3.1.1 Cadherins

Cadherin proteins represent a large family of glycoproteins that are responsible for inter-cellular contacts. These proteins are composed of an ectodomain formed by 11 to 12 cadherin repeats (CR), a transmembrane domain and an intracellular domain (Bel & Escriche 2006). In the case of CADR of the lepidopteran *Manduca sexta,* it was shown that this protein is located in the microvilli of midgut cells (Chen et al., 2005). Cry1A toxins binds to cadherin proteins of at least six lepidopteran species, *Manduca sexta*, *Bombyx mori*, *Heliothis virescens*, *Helicoverpa armigera*, *Pectinophora gossypiella* and *Ostrinia nubilalis* (Pigott & Ellar 2007).

The role of cadherin-like proteins as Cry1A toxin receptor was supported by the selection of a *Heliothis virescens* Cry1Ac-resistant line YHD2 that has a retrotransposon insertion mutation in the cadherin-like gene (Gahan et al., 2001). In addition, the characterization of CADR alleles in field-derived and in laboratory-selected Cry toxin-resistant strains of the cotton pest *Pectinophora gossypiella* (pink bollworm) revealed three mutated CADR alleles that were associated with Cry toxin resistance (Morin et al., 2003).

The interaction of Cry1A toxins with the CADR receptor is a complex process. Three regions in CADR proteins have been shown to interact with three domain II loop regions. Cry1Ab loop 2 interacts with CADR residues ^{865}NITIHITDTNN^{875} located in repeat 7, whereas loops α-8 and 2 interact with CADR residues ^{1331}IPLPASILTVTV^{1342} located in repeat 11. A third Cry1A binding region was located in CADR in the repeat 12. In the case of *H. virescens* cadherin, this binding epitope was narrowed to ^{1423}GVLTLNFQ^{1431} and demonstrated that it interacts with Cry1Ac domain II loop 3 (Gomez et al., 2007).

A cadherin protein from *T. molitor* was identified as a Cry3Aa binding protein, and it was shown to facilitate Cry3Aa oligomer formation. Moreover, silencing of the cadherin gene by feeding dsRNA showed that the silenced beetles were resistant to Cry3Aa indicating an active role of cadherin on Cry3Aa toxicity (Fabrick et al., 2009). A cadherin protein was also identified as a Cry3Aa receptor in *Diabrotica virgifera virgifera*. A fragment of this cadherin protein containing the membrane proximal cadherin repeats 8-10 bound Cry3Aa and Cry3Bb toxins with high affinity (Kd of 12 and 1.4 nM, respectively) and enhanced Cry3Aa and Cry3Bb toxicity to different coleopteran insects (Park et al., 2009).

As in lepidopteran insects, cadherin proteins have been identified in *Ae. aegypti* and *An. gambiae* showing binding to Cry11Aa and Cry4Ba respectively (Hua et al., 2008; Chen et al., 2009). In *Ae. aegypti*, cadherin also serves a receptor of Cry11Ba toxin that was isolated from the Bt var *jegathesan* strain but showed lower affinity to Cry4Ba protein (Gill et al., 2011). An *An. gambiae* cadherin fragment containing the Cry4Ba binding site enhanced the toxicity of Cry4Ba in both *An. gambiae* and *Ae. aegypti* larvae suggesting its active role as a receptor of Cry4Ba in these mosquitoes species. In the case of *Ae. aegypti* cadherin, it was shown that an anti-cadherin antibody competed binding of Cry11Aa to *Ae. aegypti* BBMV. In both *Ae. aegypti* and *An. gambiae*, cadherin is located in the microvilli of the caeca and in the microvilli of the posterior gut cells, that are the same sites where Cry11Aa and Cry4Ba bind (Hua et al., 2008; Chen et al., 2009; Park et al., 2009)

3.2 Other molecules

In lepidopteran insects, another proteins and molecules different from cadherin have been identified as a 270 kDa glycoconjugate and a 250 kDa protein named P252 (Pigott & Ellar 2007).

The 270 kDa glycoconjugate was identified as Cry1Ac binding protein in *L. dispar* (Valaitis et al., 2001). Recently *B. mori* P252 that binds Cry1Ac was identified as a choraphyllide-binding protein (Pandian et al., 2008). In addition, glycolipids were proposed to act as Cry toxin receptors in lepidopteran insects as was demonstrated for the nematode *Caenorhabditis elegans* (Griffitts et al., 2005). In the case of *L. decemlineata*, an ADAM 3 metalloprotease was identified as Cry3Aa receptor. Binding of Cry3Aa to ADAM-3 through domain II loop 1 enhanced Cry3Aa pore-formation activity suggesting that this binding interaction is important for Cry3Aa toxicity (Ochoa-Campuzano et al., 2007). The only GPI-anchored protein identified in coleopteran insects as a putative Cry receptor was an ALP from *A. grandis* that bound Cry1B toxin (Martins et al., 2010).

3.2.1 GPI-anchored receptors: APN, ALP and glucosidase

Glycosylphosphatidylinositol-anchored proteins (GPI proteins) are eukaryotic exoplasmic membrane proteins that play very diverse biological functions including hydrolytic enzyme activity, transmembrane signaling, intracellular sorting and cell adhesion interaction. Besides these biological functions, GPI proteins such as alkaline phosphatases (AP-GPI), 5-nucleotidase, dipeptidase, aminopeptidase P, have been used primarily as markers of plasma membranes during their purification procedure. Later, most of GPI proteins were found to be resistant to Triton X-100 solubilization at low temperature in kidney brush border membranes, whereas transmembrane enzymes were solubilized (Nosjean et al., 1997). Similar observations were done on brush border membranes from insects where APN and ALP are both GPI-anchored proteins; these proteins are proposed to be selectively included in lipid rafts that are conceived as spatially differentiated liquid-ordered microdomains in cell membranes. The APN and ALP in *M. sexta* and *H. virescens*, in contrast to the CADR receptors, were shown to be located in lipid rafts. The interaction of pore-forming toxins with lipid rafts could result in additional cellular events, including toxin internalization, signal transduction and cellular response (Zhuang et al., 2002; Bravo et al., 2004). In table 2 are presented the GPI-ancored proteins identified as Cry toxins receptors.

The first Cry1A toxin-binding protein that was described was an APN protein in the lepidopteran *M. sexta*. This protein was glycosilated and anchored to the membrane by a GPI anchor. Since then, other GPI-anchored APNs have been recognized as Cry toxin receptors in different lepidopteran species such as *H. virescens*, *Spodoptera litura*, *H. armigera*, *B. mori*, *Lymantria dispar*, *Plutella xylostella*, and in the dipteran *An. Quadrimaculatus*, and *A. aegypti*. Phylogenetic analyses suggest that in lepidopteran insects there are at least five different APN families and at least three of them have been shown to bind Cry1 toxins in different insect species (Gomez et al., 2007; Pigott & Ellar 2007).

The APN has been implicated in toxin insertion, since cleavage of APN by phosphatidyl-inositol specific phospholipase C treatment which cleaves out the GPI anchored proteins substantially decreased the levels of Cry1Ab incorporation into insoluble lipid raft membranes (Bravo et al., 2004) and drastically reduced the pore formation activity of the toxin assayed in BBMV from *Trichoplusia ni* (Lorence et al. 1997). In addition, the incorporation of APN into the lipid bilayer enhanced Cry1Aa pore formation activity (Schwartz et al., 1997). The sugar GalNAc in the APN receptor is an important epitope in the interaction with Cry1Ac toxin (Burton et al., 1999; de Maagd et al., 1999). In the case of the lepidopteran *Lymantria dispar*, it was proposed that the monomeric Cry1Ac toxin interacts with APN following a sequential binding model. In this model, APN is first recognized by domain III of Cry1Ac through the GalNAc moiety, followed by a protein-protein contact of the domain II loop region of Cry1Ac. The first contact is fast and reversible, and mutations close to a domain III cavity affect this initial binding, while mutations in domain II affect the rate constants of the second interaction step which is slower and tighter (Jenkins et al., 2000). Li et al. reported that the binding of GalNAc to monomeric Cry1Ac correlates with an increase of temperature factors in the pore-forming domain I. However, there was no indication of a clear conformational change in the monomeric-Cry1Ac toxin (Li et al., 2001). In contrast, the fluorescence spectroscopy studies of Cry1Ac in its oligomeric state showed that GalNAc induces a conformational change in domain III of the oligomeric structure of Cry1Ac in the vicinity of the sugar pocket. The interaction of Cry1A-oligomer with GalNAc enhanced membrane insertion of the soluble pre-pore oligomeric structure, supporting the model that interaction of the Cry1A pre-pore with GPI-anchored receptors facilitates membrane insertion and pore-formation. The APN-oligomer interaction may be especially critical when low toxin protein concentrations reach the midgut epithelium, conditions that may occur in vivo in the larvae gut where the Cry toxins are exposed to high concentration of proteases (Pardo-Lopez et al., 2006).

A resistant *S. exigua* population that is resistant to Cry1Ca toxin was shown to lack the RNA transcript of APN-1, suggesting that this APN is involved in Cry1C toxicity to this insect species (Herrero et al., 2005). Finally, in the case of *S. litura*, silencing an APN with dsRNA resulted in a lower susceptibility to Cry1Ca toxin, also indicating a role of APN in Cry1C toxicity in this insect species (Herrero et al., 2005).

Two APN isoforms (AaeAPN1 and AaeAPN2) were identified in *Ae. aegypti* by Cry11Aa pull down experiments. Protein fragments from both APN isoforms were produced in *E. coli* and shown to inhibit binding of Cry11Aa to BBMV, suggesting their active role in Cry11Aa binding to insect membranes (Chen et al., 2009). In the case of *An. gambiae* and *An. quadrimaculatus* larvae, two APN's were also identified as Cry11Ba binding proteins. Interestingly, Cry11Ba binds both *An. quadrimaculatus* and *An. gambiae* APN molecules with a very high binding affinity of 0.56 nM and 6.4 nM respectively (Abdullah et al., 2006; Zhang et al., 2008). These results suggest that APN may have a more important role in the toxicity of Cry11Ba in these two Anopheline species. In fact, it was recently shown that certain *A. gambiae* APN protein fragments enhanced Cry11Ba toxicity as has been shown for cadherin protein fragments (Zhang et al., 2008).

In regard to the APN binding epitopes involved in Cry toxin interaction, a region of 63 residues (I135-P198) involved in Cry1Aa binding was identified in *B. mori* APN1 (Nakanishi et al., 2002). The domain III of Cry toxins is involved in the APN–Cry interaction, as shown by the interchange of domain III between Cry1Ac and Cry1Ab toxins (de Maagd et al., 1999). Domain III residues of Cry1Aa ^{508}STRVN^{513} and ^{582}VFTLSAHV^{589} were shown to be involved in binding I135-P198 APN fragment (Atsumi et al., 2005).

ALPs are found in all animals and as is expected are mainly localized in microvilli of columnar cells and of insect midgut epithelium cells (Eguchi 1995). ALPs can be divided into two groups: soluble (s-ALP) and membrane-bound (m-ALP) (Eguchi et al., 1990; Itoh et al., 1991). In insects, both ALPs are found in larval midgut epithelium cells; however, they are expressed in different cell types. The s-ALP is found exclusively in the cavity of goblet cells and in the apical region of the midgut; whereas, m-ALP is localized in the brush border membrane of columnar cells and particularly restricted to the middle and posterior midgut. Moreover, s- and m-ALPs show distinct differences in enzymatic activity (such as optimal pH) and also the structure of sugar side chain, suggesting that they perform different functions *in vivo* (Eguchi et al., 1990).

GPI-anchored Cry toxin-binding receptors [a]				
ORDER	SPECIES	PROTEIN	BINDING TOXIN	IDENTIFICATION METHOD
Lepidoptera	*Manduca sexta*	Class 1 APN	Cry1Ac[11], Cry1Aa, Cry1Ab[14]	Affinity chromatography [11], Chromatography purification, ligand blot, SPR [14]
		Class 2 APN	Cry1Ab5 [5]	Affinity chromatography, ligand blot [5]
		ALP	Cry1Ac [15]	PLPC treatment, 2D-SDS-PAGE-Mass spectrometry [15]
	Bombix mori	Class 1 APN	Cry1Aa [23], Cry1Aa, Cry1Ab [16]	Ion exchange chromatography purification [23], Heterologous expression, toxin overlay assay [16]
		Class 3 APN, Class 4 APN	Cry1Aa, Cry1Ab [16]	Heterologous expression, toxin overlay assay [16]
	Helicoverpa armigera	Class 1 APN	Cry1Aa, Cry1Ab, Cry1Ac [18]	Heterologous expression, toxin overlay assay [18]
		Class 3 APN	Cry1Ac [18]	Heterologous expression, ligand blot [18]
		ALP	Cry1Ac [21]	Anion exchange chromatography [21]
	Heliothis virescens	Class 1 APN	Cry1Aa, Cry1Ab, Cry1Ac [13], Cry1Fa [3]	Affinity chromatography, SPR [13] Affinity chromatography, ligand blot [3]
		Class 3 APN	Cry1Ac [9]	Affinity chromatography, ligand blot [9]
		Class 4 APN	Cry1Fa [3]	Affinity chromatography, SPR [3]
		ALP	Cry1Ac [10]	PLPC treatment, western blot [10]
	Lymantria dispar	Class 1 APN	Cry1Ac [22]	Immunolocalization [22]
		Class 3 APN	Cry1Ac [8]	Chromatographic purification [8]
	Plutella xylostella	Class 1 APN, Class 2 APN, Class 3 APN, Class 5 APN	Cry1Aa, Cry1Ab [16]	Heterologous expression, toxin overlay assay [16]
	Spodoptera litura	Class 3 APN, Class 5 APN	Cry1C [2]	*In silico* identification, heterologous expression, toxin overlay assay [2]
	Epiphyas postvittana	Class 3 APN	Cry1Aa, Cry1Ba [19]	Ion exchange chromatography purification [19]
Diptera	*Aedes aegypti*	AaeAPN1, AaeAPN2	Cry11Aa [4]	Pull-down, heterologous expression, ligand blot [4]
		APN	Cry4Ba [4]	2D SDS-PAGE, ligand blot [4]
		APN2778, APN2783, APN5808	Cry4Ba [20]	*In silico* identification, RNA interference [20]
		ALP	Cry4Ba [4], Cry11Aa [7]	2D SDS-PAGE, ligand blot [4], PLPC treatment, ligand blot [7]
	Anopheles albimanus	α-amilase	Cry4Ba, Cry11Aa [8]	PLPC treatment, ligand blot [8]
	Anopheles quadrimaculatus	APN	Cry11Aa [1]	Anion-exchange chromatography, enzymatic activity, SPR [1]
	Anopheles gambiae	APN	Cry11Ba [24]	*In silico* identification, heterologous expression, PLPC treatment, anion- exchange chromatography [24]
Coleoptera	*Anthonomus grandis*	ALP	Cry8Ka5 [17]	2D-SDS-PAGE, ligand blot, mass spectrometry [17]

(a) Compiled from: (1) Abdulah et. al., 2006; (2) Agrawal et. al., 2002; (3) Banks et. al., 2001; (4) Bayyareddy et. al., 2009; (5) Chen et. al., 2009; (6) Denolf et. al., 1997; (7) Fernández et. al., 2006; (8) Fernández-Luna et. al., 2010; (9) Garner et. al., 1999; (10) Gill et. al., 1995; (11) Jurat-Fuentes & Adang 2004; (12) Knight et. al., 1994; (13) Luo et. al., 1997; (14) Masson et. al., 1995; (15) McNall & Adang 2003; (16) Nakanishi et. al., 2002; (17) Nakasu et. al., 2010; (18) Rajagopal et. al., 2003; (19) Simpson & Newcomb 2000; (20) Saengwiman et. al., 2011; (21) Santosh & Singh P, 2011; (22) Valaitis et. al., 1997; (23) Yaoi et. al., 1997; (24) Zhang et. al., 2008.

Table 2. Description of GPI-anchored receptors founded in insects that bind to Cry toxins.

A GPI-anchored ALP that binds Cry toxins has been described in the lepidopterans *M. sexta*, *H. virescens* and *H. armigera*; in the dipteran *Ae. aegypti*, and in the coleopteran *Anthonomus grandis* (McNall et al., 2003; Jurat-Fuentes et al., 2004; Martins et al., 2010; Ning et al., 2010). In the case of *M. sexta*, ALP protein that bind Cry1A toxins was shown to be located in the microvilli of epithelial cells that is the site of action of Cry1A toxins (Chen et al., 2005). Also, an *H. virescens* laboratory selected population, YHD2, contained a retrotransposon insertion in the cadherin gene (Gahan et al., 2001). However, the mutation in the cadherin gene only accounted for 40-80% of the resistance phenotype. Additional mutations were responsible for the rest of the resistant phenotype in YHD2. These additional mutations were shown to affect GPI-ALP production, indicating that ALP is likely a functional receptor of Cry1Ac toxin in *H. virescens* (Jurat-Fuentes & Adang 2004).

Blocking the interaction of Cry toxins with GPI-anchored receptors has been useful in some cases to show the role of these proteins in Cry insecticidal activity. In the case of *M. sexta*, a scFv-phage that bound Cry1Ab toxin through β16-β22 of domain III blocked binding of Cry1Ab with APN but not with Bt-R1 and inhibited the toxicity of Cry1Ab in bioassays (Gomez et al., 2006). Nevertheless, in *B. mori* detached midgut cells, an anti-APN antibody did not affect toxicity of Cry1Aa in contrast to an anticadherin antibody that inhibited toxicity, suggesting either that this APN may not be involved in toxicity or that other additional GPI-anchored proteins or lipids could substitute APN function (Hara et al., 2003; Ibiza-Palacios et al., 2008). In *Ae. aegypti*, a peptide-phage that bound the 65 kDa ALP competed binding of the Cry11Aa to BBMVof mosquito and inhibited Cry11Aa toxicity in bioassays, suggesting that GPI-anchored ALP is a functional receptor of Cry11Aa (Fernandez et al., 2006).

Recently, ligand blot assays indicated that a 70 kDa GPI-anchored protein present in midgut brush border membrane vesicles of *A. albimanus* interacts with Cry4Ba and Cry11Aa toxins. This protein was identified as an a-amylase by mass spectrometry and enzymatic activity assays (Fernandez-Luna et al., 2010).

The fact that similar Cry binding proteins are involved in the mechanism of action of Cry toxins in both lepidopteran and dipteran insects suggests the Cry toxins have a conserved mode of action. However, the precise role of the Cry toxins receptors identified in mosquitoes in the mode of action of Cry toxins still remains to be determined. As in lepidopteran insects cadherin binding might facilitate oligomer formation while binding of Cry oligomer to GPI-anchored ALP or APN receptors might be necessary to facilitate membrane insertion. Nevertheless, the high binding affinity of Anopheline APN's to Cry11Ba is substantially different from what has been reported in lepidopteran insects and further studies on the differential role of APN and cadherin in monomer/oligomer binding in mosquitoes are necessary to determine the precise role of toxin binding to these receptor molecules.

3.3 Other insect molecules that bind Cry toxins identified using proteomic approaches

Proteomic approaches based on two-dimensional (2D) gel electrophoresis and mass spectrometry have been used to discover novel Bt toxin binding proteins and elucidate changes in midgut proteins associated with Bt resistance. Using this approach ALP was identified as a Cry1Ac binding protein in brush border of *Manduca sexta* and *Heliothis virescens* (McNall & Adang 2003; Krishnamoorthy et al., 2007). This identification was validated in *H. virescens* when ALP was demonstrated as a functional receptor molecule and loss of the enzyme correlated with Bt resistance to Cry1Ac (Jurat-Fuentes & Adang 2004). Candas et al. used differential-in-gel electrophoretic (DIGE) analysis to compare Bt susceptible and resistant larvae of *Plodia interpunctella*. These authors detected increased levels of midgut enzymes associated with oxidative metabolism and altered migration of an F_1F_0-ATPase in resistant larvae when compared to susceptible proteins on 2D gels. Candas et al. also detected reduced levels of alkaline chymotrypsin in the resistant *P. interpunctella* larvae associated with reduced capacity for protoxin activation (Candas et al., 2003).

Additional Cry1Ac binding proteins in lepidopteran brush border preparations detected by 2DE ligand blots approach includes actin, aminopeptidase, vacuolar-ATPase subunit A and a desmocollin-like protein (McNall & Adang 2003; Krishnamoorthy et al., 2007). A proteomics-based approach using differential-in-gel electrophoretic (DIGE) analysis quantified altered levels of specific proteins in Bt susceptible and resistant larvae of *Plodia interpunctella* (Candas et al., 2003). Those authors detected changes in the levels of APN, V-ATPase and an F_1F_0-ATPase in resistant larvae.

The analysis of Cry4Ba binding proteins by mass spectrometry in *Ae. aegypti* BBMV, revealed two lipid rafts associated proteins, flotillin and prohibitin, as well as cytoplasmic actin, besides ALP and APN, thus suggesting that additional proteins as well as intracellular proteins may have an active role in the mode of action of Cry toxins in mosquitoes (Bayyareddy et al., 2009).

Protein	Insect	Reference
ABC transporter	*Heliothis virescens* (Lepidoptera)	Gahan *et. al.*, 2010
	Helicoverpa armigera (Lepidoptera)	Chen *et. al.*, 2010
V-ATPase	*Heliothis virescens* (Lepidoptera) (A subunit)	Krishnamoorthy *et. al.*, 2007
	Anthonomus grandis (Coleoptera) (A subunit)	Nakasu, 2010
	Helicoverpa armigera (Lepidoptera) (B subunit)	Chen *et. al.*, 2010
	Aedes aegypti (Diptera) (E subunit)	Bayyareddy *et. al.*, 2009
Heat Shock Proteins	*Helicoverpa armigera* (Lepidoptera)	Chen *et. al.*, 2010
	Anthonomus grandis (Coleoptera)	Nakasu, 2010
Actin	*Manduca sexta* (Lepidoptera)	McNall *et al.*, 2003
	Heliothis virescens (Lepidoptera)	Krishnamoorthy, *et al.*, 2007
	Helicoverpa armigera (Lepidoptera)	Chen *et. al.*, 2010
	Aedes aegypti (Diptera)	Bayyareddy *et al.*, 2009
ATP sintase	*Aedes aegypti* (Diptera)	Bayyareddy, *et al.*, 2009
Flotilin	*Aedes aegypti* (Diptera)	Bayyareddy *et al.*, 2009
Prohibitin	*Aedes aegypti* (Diptera)	Bayyareddy *et al.*, 2009

Table 3. Receptors of Cry toxins described in different insects using proteomic analysis.

3.4 How these receptors are involved in resistance to Cry toxins

Resistance to Cry toxins can be developed by mutations in the insect pests that affect any of the steps of the mode of action of Cry toxins. The most common mechanism of toxin resistance in insect pests until now is the reduction in toxin binding to midgut cells, that in different insect species include mutations in Cry toxin receptors (Gahan et al., 2001; Ferre & Van Rie 2002; Morin, Biggs et al. 2003; Zhang, Hua et al. 2008). In fact, the most frequently phenotype of insect resistance, denoted as "Mode 1 of Resistance", is characterized by the reduction of one Cry1A toxin binding, cross resistance of Cry1Aa, Cry1Ab and Cry1Ac and lack of resistance to Cry1C. In several lepidopteran insects, the mode 1 of resistance is linked to mutations in the cadherin gene (Gahan et al., 2001; Morin et al., 2003; Xu et al., 2005). In field conditions three lepidopteran insect pests have evolved resistance to formulated Bt products, *Plodia interpunctella, Plutella xylostella* and *T. ni* [McGaughey, 1985; Tabashnik, et al, 1994; (Janmaat & Myers, 2003). In recent years, at least four cases of resistance to Bt crops have been documented, *H. zea* to Bt-cotton expressing Cry1Ac in United States, *S. frugiperda* to Bt-corn expressing Cry1F in Puerto Rico, *Busseola fusca* to Bt-corn expressing Cry1Ab in South Africa and *P. gossypiella* to Bt-cotton expressing Cry1Ac in India (Gill et al., 2011).

Recently, a resistant allele of a *H. virescens* resistant population was identified as a mutation in a gene coding for an ABC transporter molecule. This mutation affected binding of Cry1A toxin to brush border membrane vesicles indicating that this ABC transporter molecule is a novel Cry1A toxin receptor probably involved in the later stages of oligomer membrane insertion (Gahan et al., 2010). Also it was reported the correlation between reduced ALP protein, activity, and mALP expression levels in strains of three species in the Noctuidae Family with diverse resistance phenotypes against Cry toxins (*Heliothis virescens, Helicoverpa armigera* and *Spodoptera frugiperda*) (Jurat-Fuentes et al., 2011). Finally, in a recent work it was showed the role of aminopeptidase the resistance to the Bt toxin Cry1Ac in the cabbage looper, *Trichoplusia ni,* evolved in greenhouses, is associated with differential alteration of two midgut aminopeptidases N, APN1 and APN6, conferred by a trans-regulatory mechanism (Tiewsiri & Wang, 2011).

4. Conclusion

The mode of action of Cry toxins is a multi-step process that involves the interaction with several receptor molecules leading to membrane insertion and cells lysis. The characterization of the mode of action of Cry toxins in susceptible organisms will be important to fully understand the mode of action of this family of proteins. In the case of GPI-anchored receptors, APN and ALP have been identified in different insect species as Cry toxin binding molecules and in several insects they have been shown to be important for toxin action. In the case of *M. sexta,* both APN and ALP have been shown to have a similar role since Cry1Ab toxin show similar binding affinities to both molecules depending on the oligomer state of the toxin. Thus it seems that APN and ALP have redundant roles in the action of Cry toxins. Nevertheless, it was shown that ALP could have a predominant role in toxicity since it was preferentially express in young instar larvae that are more sensitive to the toxin in contrast to APN that was expressed later in larval development. Nevertheless, it could still be possible that in certain insect species both APN and ALP act as a complex receptor or that APN or ALP could be differentially important in toxicity depending in the insect species.

5. Acknowledgments

This research was supported by CONACyT 83135 and PAPIIT/UNAM IN209011-3.

IG wants to special thankful to L'ORÉAL-UNESCO-AMC fellowship to support this work.

6. References

Abdullah, M. A., A. P. Valaitis, et al. (2006). "Identification of a *Bacillus thuringiensis* Cry11Ba toxin-binding aminopeptidase from the mosquito, *Anopheles quadrimaculatus*." *BMC Biochem* 7: 16.

Atsumi, S., E. Mizuno, et al. (2005). "Location of the *Bombyx mori* aminopeptidase N type 1 binding site on *Bacillus thuringiensis* Cry1Aa toxin." *Appl Environ Microbiol* 71(7): 3966-77.

Bayyareddy, K., T. M. Andacht, et al. (2009). "Proteomic identification of *Bacillus thuringiensis subsp. israelensis* toxin Cry4Ba binding proteins in midgut membranes from *Aedes (Stegomyia) aegypti Linnaeus* (Diptera, Culicidae) larvae." *Insect Biochem Mol Biol* 39(4): 279-86.

Bel, Y. and B. Escriche (2006). "Common genomic structure for the Lepidoptera cadherin-like genes." *Gene* 381: 71-80.

Boonserm, P., P. Davis, et al. (2005). "Crystal structure of the mosquito-larvicidal toxin Cry4Ba and its biological implications." *J Mol Biol* 348(2): 363-82.

Boonserm, P., M. Mo, et al. (2006). "Structure of the functional form of the mosquito larvicidal Cry4Aa toxin from *Bacillus thuringiensis* at a 2.8-angstrom resolution." *J Bacteriol* 188(9): 3391-401.

Bravo, A., S. S. Gill, et al. (2007). "Mode of action of *Bacillus thuringiensis* Cry and Cyt toxins and their potential for insect control." *Toxicon* 49(4): 423-35.

Bravo, A., I. Gomez, et al. (2004). "Oligomerization triggers binding of a *Bacillus thuringiensis* Cry1Ab pore-forming toxin to aminopeptidase N receptor leading to insertion into membrane microdomains." *Biochim Biophys Acta* 1667(1): 38-46.

Burton, S. L., D. J. Ellar, et al. (1999). "N-acetylgalactosamine on the putative insect receptor aminopeptidase N is recognised by a site on the domain III lectin-like fold of a *Bacillus thuringiensis* insecticidal toxin." *J Mol Biol* 287(5): 1011-22.

Candas, M., O. Loseva, et al. (2003). "Insect resistance to *Bacillus thuringiensis*: alterations in the indianmeal moth larval gut proteome." *Mol Cell Proteomics* 2(1): 19-28.

Chen, J., K. G. Aimanova, et al. (2009). "*Aedes aegypti* cadherin serves as a putative receptor of the Cry11Aa toxin from *Bacillus thuringiensis* subsp. israelensis." *Biochem J* 424(2): 191-200.

Chen, J., K. G. Aimanova, et al. (2009). "Identification and characterization of *Aedes aegypti* aminopeptidase N as a putative receptor of *Bacillus thuringiensis* Cry11A toxin." *Insect Biochem Mol Biol* 39(10): 688-96.

Chen, J., M. R. Brown, et al. (2005). "Comparison of the localization of *Bacillus thuringiensis* Cry1A delta-endotoxins and their binding proteins in larval midgut of tobacco hornworm, *Manduca sexta*." *Cell Tissue Res* 321(1): 123-9.

de Maagd, R. A., P. L. Bakker, et al. (1999). "Domain III of the *Bacillus thuringiensis* delta-endotoxin Cry1Ac is involved in binding to *Manduca sexta* brush border membranes and to its purified aminopeptidase N." *Mol Microbiol* 31(2): 463-71.

de Maagd, R. A., A. Bravo, et al. (2003). "Structure, diversity, and evolution of protein toxins from spore-forming entomopathogenic bacteria." *Annu Rev Genet* 37: 409-33.

de Maagd, R. A., A. Bravo, et al. (2001). "How *Bacillus thuringiensis* has evolved specific toxins to colonize the insect world." *Trends Genet* 17(4): 193-9.

Eguchi, M. (1995). "Alkaline phosphatase isozymes in insects and comparison with mammalian enzyme." *Comp Biochem Physiol B Biochem Mol Biol* 111(2): 151-62.

Eguchi, M., M. Azuma, et al. (1990). "Genetically defined membrane-bound and soluble alkaline phosphatases of the silkworm: their discrete localization and properties." *Prog Clin Biol Res* 344: 267-87.

Fabrick, J., C. Oppert, et al. (2009). "A novel *Tenebrio molitor* cadherin is a functional receptor for Bacillus thuringiensis Cry3Aa toxin." *J Biol Chem* 284(27): 18401-10.

Fernandez, L. E., K. G. Aimanova, et al. (2006). "A GPI-anchored alkaline phosphatase is a functional midgut receptor of Cry11Aa toxin in *Aedes aegypti* larvae." *Biochem J* 394(Pt 1): 77-84.

Ferre, J. and J. Van Rie (2002). "Biochemistry and genetics of insect resistance to *Bacillus thuringiensis*." *Annu Rev Entomol* 47: 501-33.

Flores, H., X. Soberon, et al. (1997). "Isolated domain II and III from the *Bacillus thuringiensis* Cry1Ab delta-endotoxin binds to lepidopteran midgut membranes." *FEBS Lett* 414(2): 313-8.

Gahan, L. J., F. Gould, et al. (2001). "Identification of a gene associated with Bt resistance in *Heliothis virescens*." *Science* 293(5531): 857-60.

Gahan, L. J., Y. Pauchet, et al. (2010). "An ABC Transporter Mutation Is Correlated with Insect Resistance to *Bacillus thuringiensis* Cry1Ac Toxin." *Plos Genetics* 6(12).

Galitsky, N., V. Cody, et al. (2001). "Structure of the insecticidal bacterial delta-endotoxin Cry3Bb1 of *Bacillus thuringiensis*." *Acta Crystallogr D Biol Crystallogr* 57(Pt 8): 1101-9.

Gill, M. and D. Ellar (2002). "Transgenic Drosophila reveals a functional in vivo receptor for the *Bacillus thuringiensis* toxin Cry1Ac1." *Insect Mol Biol* 11(6): 619-25.

Gill, S. S., S. Likitvivatanavong, et al. (2011). "Multiple Receptors as Targets of Cry Toxins in Mosquitoes." *Journal of Agricultural and Food Chemistry* 59(7): 2829-2838.

Gomez, I., I. Arenas, et al. (2006). "Specific epitopes of domains II and III of Bacillus thuringiensis Cry1Ab toxin involved in the sequential interaction with cadherin and aminopeptidase-N receptors in Manduca sexta." *J Biol Chem* 281(45): 34032-9.

Gomez, I., I. Arenas, et al. (2010). "Role of Alkaline Phosphatase from *Manduca sexta* in the Mechanism of Action of *Bacillus thuringiensis* Cry1Ab Toxin." *Journal of Biological Chemistry* 285(17): 12497-12503.

Gomez, I., L. Pardo-Lopez, et al. (2007). "Role of receptor interaction in the mode of action of insecticidal Cry and Cyt toxins produced by *Bacillus thuringiensis*." *Peptides* 28(1): 169-73.

Gomez, I., J. Sanchez, et al. (2002). "Cadherin-like receptor binding facilitates proteolytic cleavage of helix alpha-1 in domain I and oligomer pre-pore formation of *Bacillus thuringiensis* Cry1Ab toxin." *FEBS Lett* 513(2-3): 242-6.

Gonzalez-Cabrera, J., B. Escriche, et al. (2003). "Binding of *Bacillus thuringiensis* toxins in resistant and susceptible strains of pink bollworm (*Pectinophora gossypiella*)." *Insect Biochem Mol Biol* 33(9): 929-35.

Griffitts, J. S., S. M. Haslam, et al. (2005). "Glycolipids as receptors for *Bacillus thuringiensis* crystal toxin." *Science* 307(5711): 922-5.

Grochulski, P., L. Masson, et al. (1995). "*Bacillus thuringiensis* CryIA(a) insecticidal toxin: crystal structure and channel formation." *J Mol Biol* 254(3): 447-64.

Guo, S., S. Ye, et al. (2009). "Crystal structure of *Bacillus thuringiensis* Cry8Ea1: An insecticidal toxin toxic to underground pests, the larvae of *Holotrichia parallela*." *J Struct Biol* 168(2): 259-66.

Hara, H., S. Atsumi, et al. (2003). "A cadherin-like protein functions as a receptor for *Bacillus thuringiensis* Cry1Aa and Cry1Ac toxins on midgut epithelial cells of *Bombyx mori* larvae." *FEBS Lett* 538(1-3): 29-34.

Herrero, S., T. Gechev, et al. (2005). "Bacillus thuringiensis Cry1Ca-resistant *Spodoptera exigua* lacks expression of one of four Aminopeptidase N genes." *BMC Genomics* 6: 96.

Hua, G., R. Zhang, et al. (2008). "*Anopheles gambiae* cadherin AgCad1 binds the Cry4Ba toxin of *Bacillus thuringiensis* israelensis and a fragment of AgCad1 synergizes toxicity." *Biochemistry* 47(18): 5101-10.

Ibiza-Palacios, M. S., J. Ferre, et al. (2008). "Selective inhibition of binding of *Bacillus thuringiensis* Cry1Ab toxin to cadherin-like and aminopeptidase proteins in brush-border membranes and dissociated epithelial cells from *Bombyx mori*." *Biochem J* 409(1): 215-21.

Itoh, M., S. Takeda, et al. (1991). "Cloning and sequence analysis of membrane-bound alkaline phosphatase cDNA of the silkworm, *Bombyx mori*." *Biochim Biophys Acta* 1129(1): 135-8.

Jenkins, J. L., M. K. Lee, et al. (2000). "Bivalent sequential binding model of a *Bacillus thuringiensis* toxin to gypsy moth aminopeptidase N receptor." *J Biol Chem* 275(19): 14423-31.

Jurat-Fuentes, J. L. and M. J. Adang (2004). "Characterization of a Cry1Ac-receptor alkaline phosphatase in susceptible and resistant *Heliothis virescens* larvae." *Eur J Biochem* 271(15): 3127-35.

Krishnamoorthy, M., J. L. Jurat-Fuentes, et al. (2007). "Identification of novel Cry1Ac binding proteins in midgut membranes from *Heliothis virescens* using proteomic analyses." *Insect Biochem Mol Biol* 37(3): 189-201.

Li, J., D. J. Derbyshire, et al. (2001). "Structural implications for the transformation of the *Bacillus thuringiensis* delta-endotoxins from water-soluble to membrane-inserted forms." *Biochem Soc Trans* 29(Pt 4): 571-7.

Lorence, A., A. Darszon, et al. (1997). "Aminopeptidase dependent pore formation of *Bacillus thuringiensis* Cry1Ac toxin on *Trichoplusia ni membranes*." *FEBS Lett* 414(2): 303-7.

McNall, R. J. and M. J. Adang (2003). "Identification of novel *Bacillus thuringiensis* Cry1Ac binding proteins in *Manduca sexta* midgut through proteomic analysis." *Insect Biochem Mol Biol* 33(10): 999-1010.

Miranda-Rios, J., M. T. Fernandez-Luna, et al. (2010). "An alpha-amylase is a novel receptor for *Bacillus thuringiensis* ssp israelensis Cry4Ba and Cry11Aa toxins in the malaria vector mosquito *Anopheles albimanus* (Diptera: Culicidae)." *Environmental Microbiology* 12(3): 746-757.

Morin, S., R. W. Biggs, et al. (2003). "Three cadherin alleles associated with resistance to *Bacillus thuringiensis* in pink bollworm." *Proc Natl Acad Sci U S A* 100(9): 5004-9.

Morse, R. J., T. Yamamoto, et al. (2001). "Structure of Cry2Aa suggests an unexpected receptor binding epitope." *Structure* 9(5): 409-17.

Nakanishi, K., K. Yaoi, et al. (2002). "Aminopeptidase N isoforms from the midgut of *Bombyx mori* and *Plutella xylostella* -- their classification and the factors that determine their binding specificity to *Bacillus thuringiensis* Cry1A toxin." *FEBS Lett* 519(1-3): 215-20.

Nosjean, O., A. Briolay, et al. (1997). "Mammalian GPI proteins: sorting, membrane residence and functions." *Biochim Biophys Acta* 1331(2): 153-86.

Ochoa-Campuzano, C., M. D. Real, et al. (2007). "An ADAM metalloprotease is a Cry3Aa *Bacillus thuringiensis* toxin receptor." *Biochem Biophys Res Commun.* 362(2):437-42.

Pacheco, S., I. Gomez, et al. (2009). "Domain II loop 3 of *Bacillus thuringiensis* Cry1Ab toxin is involved in a "ping pong" binding mechanism with *Manduca sexta* aminopeptidase-N and cadherin receptors." *J Biol Chem* 284(47): 32750-7.

Pandian, G. N., T. Ishikawa, et al. (2008). "*Bombyx mori* midgut membrane protein P252, which binds to *Bacillus thuringiensis* Cry1A, is a chlorophyllide-binding protein, and the resulting complex has antimicrobial activity." *Appl Environ Microbiol* 74(5): 1324-31.

Pardo-Lopez, L., I. Gomez, et al. (2006). "Structural changes of the Cry1Ac oligomeric pre-pore from *Bacillus thuringiensis* induced by N-acetylgalactosamine facilitates toxin membrane insertion." *Biochemistry* 45(34): 10329-36.

Park, Y., M. A. Abdullah, et al. (2009). "Enhancement of *Bacillus thuringiensis* Cry3Aa and Cry3Bb toxicities to coleopteran larvae by a toxin-binding fragment of an insect cadherin." *Appl Environ Microbiol* 75(10): 3086-92.

Pigott, C. R. and D. J. Ellar (2007). "Role of receptors in *Bacillus thuringiensis* crystal toxin activity." *Microbiol Mol Biol Rev* 71(2): 255-81.

Rajagopal, R., S. Sivakumar, et al. (2002). "Silencing of midgut aminopeptidase N of *Spodoptera litura* by double-stranded RNA establishes its role as *Bacillus thuringiensis* toxin receptor." *J Biol Chem* 277(49): 46849-51.

Rajamohan, F., M. K. Lee, et al. (1998). "*Bacillus thuringiensis* insecticidal proteins: molecular mode of action." *Prog Nucleic Acid Res Mol Biol* 60: 1-27.

Schnepf, E., N. Crickmore, et al. (1998). "*Bacillus thuringiensis* and its pesticidal crystal proteins." *Microbiol Mol Biol Rev* 62(3): 775-806.

Schwartz, J. L., Y. J. Lu, et al. (1997). "Ion channels formed in planar lipid bilayers by *Bacillus thuringiensis* toxins in the presence of *Manduca sexta* midgut receptors." *FEBS Lett* 412(2): 270-6.

Soberon, M., S. S. Gill, et al. (2009). "Signaling versus punching hole: How do *Bacillus thuringiensis* toxins kill insect midgut cells?" *Cell Mol Life Sci* 66(8): 1337-49.

Valaitis, A. P., J. L. Jenkins, et al. (2001). "Isolation and partial characterization of gypsy moth BTR-270, an anionic brush border membrane glycoconjugate that binds *Bacillus thuringiensis* Cry1A toxins with high affinity." *Arch Insect Biochem Physiol* 46(4): 186-200.

Xu, X., L. Yu, et al. (2005). "Disruption of a cadherin gene associated with resistance to Cry1Ac {delta}-endotoxin of *Bacillus thuringiensis* in *Helicoverpa armigera*." *Appl Environ Microbiol* 71(2): 948-54.

Zhang, R., G. Hua, et al. (2008). "A 106-kDa aminopeptidase is a putative receptor for *Bacillus thuringiensis* Cry11Ba toxin in the mosquito *Anopheles gambiae*." *Biochemistry* 47(43): 11263-72.

Zhuang, M., D. I. Oltean, et al. (2002). "*Heliothis virescens* and *Manduca sexta lipid* rafts are involved in Cry1A toxin binding to the midgut epithelium and subsequent pore formation." *J Biol Chem* 277(16): 13863-72.

24

Evaluating Surface Seals in Soil Columns to Mitigate Methyl Isothiocyanate Volatilization

Shad D. Nelson[1,2], Catherine R. Simpson[2], Husein A. Ajwa[3] and Clinton F. Williams[4]

[1]*Texas A&M University-Kingsville, Kingsville*

[2]*Texas A&M University -Kingsville Citrus Center, Weslaco*

[3]*University of California-Davis, Salinas*

[4]*US Department of Agriculture-Agricultural Research Service, Maricopa*

USA

1. Introduction

The banning of methyl bromide (MeBr) as a pre-plant soil fumigant due to its implication as an ozone depleting substance, has led to increased interest in finding alternative soil fumigants to replace MeBr (United States Environmental Protection Agency [USEPA], 2009). One of the promising alternatives for certain crops is methyl isothiocyanate (MITC). Several MITC generating compounds, such as metam sodium®, metam potassium®, and dazomet® are being used to control a wide variety of fungal pathogens, weeds, and nematodes in soils. The physiochemical characteristics of MITC are significantly different than that of MeBr, such as that its effectiveness in regards to dissipation and movement in the soil is altered by multiple factors, such as soil type, texture, and soil moisture content. The largest challenge to soil fumigation is the prevention of fumigant loss to the atmosphere and especially to the nearby communities and homes adjacent to farm land. Rapid off-gassing or non-target release of the fumigant to the atmosphere can lead to poor pesticide performance and ineffective pest control. To combat this problem that is common to all soil fumigants currently on the market, various methods have been employed to reduce chemical off-gassing. A few of these methods are tarping the soil surface immediately following chemical application with high density polyethylene plastic, incorporation of organic matter to the soil surface to absorb the fumigant, or altering chemical formulations. Another method of reducing fumigant loss can be applying a surface water application as a means of sealing the soil surface to prevent chemical volatilization. On-farm field scale studies have been performed to evaluate all of these methods to better evaluate the potential for reducing fumigant loss to the atmosphere. However, field-scale studies are expensive to perform, and experimental error is challenging to control and replicate due to diurnal temperature fluctuations, varying soil physical properties, and air current differences. Thus, the volatilization loss in one study will not represent the typical fumigant loss from site to site. A more controlled laboratory environment is needed to more adequately predict fumigant loss under specific conditions. Laboratory-scale columns can be used to study soil fumigant release from soils under a wide array of conditions and under controlled circumstances. The aim of this study was to evaluate the amount of water applied to the surface of a specific soil type to reduce MITC volatilization in soil columns. Furthermore, evaluating the impact of various soil physical properties have on MITC loss is important, such as varied soil type, soil bulk density, organic matter additions and various MITC generating compound formulas have on MITC loss and mitigation. In short, the results of these studies will summarize the effectiveness of the use of soil columns to adequately assess MITC loss at the laboratory scale as a tool to predict chemical fate prior to the expense of large-scale on-farm studies.

1.1 History of fumigants

Soil fumigants are commonly used in high-value horticultural crop production to control soil originating pests such as plant-parasitic nematodes, soil-borne pathogens, insects and weeds. The intrinsic volatility of a fumigant is essential for a chemical to disperse laterally and vertically throughout the soil profile in order to control soil-borne diseases. Fumigants are typically applied via shank/chisel injection directly into the soil. After being applied, the fumigants quickly change into a gaseous phase whereby it is dispersed within the soil and results in pest control. Many compounds are classified as soil fumigants, with various rates of efficacy, with MeBr considered the most effective broad-spectrum pest control fumigant due to its high efficacy level. MeBr was one of the most widely used soil fumigants until, under the Montreal Protocol; it was officially phased out in 2005 as an ozone depleting compound (USEPA, 2009). MeBr is still used in developing countries, but must be phased out by 2015 (United States Government Printing Office [USGPO], 2005).

In effort to meet the challenge to find a suitable replacement for MeBr that has similar efficacy capabilities for crop protection a concentrated effort of research and funding has occurred. Although these studies on alternative fumigants to MeBr have been occurring for approximately two decades, there is still no fumigant replacement as effective in almost all soil types like MeBr. Currently there are still several instances where MeBr can be used; such as critical use exemptions (CUE), quarantine and pre-shipment (QPS), and emergency exemption (EE). However, these uses are highly restricted and subjected to strict regulation. In recent years, there has been a movement to find alternatives to MeBr that are as effective but less harmful to the atmosphere and environment. While this has proved a formidable challenge to scientists, there are several fumigants used in agriculture today that are effective under specific soil and cropping conditions. Table 1 shows the five most used soil fumigants in the United States.

1.2 Methyl bromide

MeBr has been the most effective soil fumigant for most soil borne pathogens and pests since it was introduced as a pesticide in 1932. Due to its harmful effects on the atmosphere as an ozone depletor, MeBr production has been phased out in most developed and developing countries in accordance with the Montreal Protocol (USEPA, 2009). MeBr can still be used under critical use and emergency exemptions but its use is strictly regulated by state and governmental agencies.

MeBr is a volatile gas at room temperature and 1 atm pressure and can be produced commercially or by plants and algae (National Pesticide Information Center [NPIC], 2000).

MeBr is a odorless, gaseous chemical above 4°C that is highly toxic to humans and vertebrate animals that can result in death under acute exposure. Thus, commercial formulations of MeBr include a certain percent of chloropicrin (tear-gas) added to act as a warning agent to indicate presence of MeBr to prevent overexposure. MeBr is applied under pressure as a liquid using shank injection into the soil, usually in conjunction with covering the soil with plastic tarps to suppress and prevent volatilization loss of the gas to the atmosphere (Papiernik et al., 2001; Wang et al., 1997). The gas then diffuses through soil pores and cracks and allows for control of soil borne pests and pathogens.

Rank	Fumigant	Formulations	Application	Amount Used per Year
1	Metam sodium/ Metam potassium	Liquid, soluble concentrate	Shank injection, chemigation	51-55 million lbs/ 1-2 million lbs (2002)
2	Methyl bromide	Pressurized gas	Shank injection, hot gas	14.76 million lbs (2007)*
3	Chloropicrin	Liquid, pressurized gas, pressurized liquid, emulsifiable concentrate	Shank injection, drip irrigation	10 million lbs (2007)
4	1,3- Dichloropropene	Liquid	Soil injection, deep drip irrigation	40,420 lbs (1998 estimate)
5	Dazomet	Granule, pellet, liquid, water soluble solids	Spreader	15,000 lbs (2003)

*Critical use exemption and emergency exemption usage
(USEPA, 2009).

Table 1. Top five most commonly used soil fumigants in the United States.

1.3 Methyl bromide alternatives

While no fumigant has proven as effective as MeBr for the control of soil-borne pests and pathogens, the reasons why the four most widely used fumigant alternatives are currently in use today are discussed below.

1.3.1 Metam sodium and metam potassium

Metam sodium (MS) is among the most widely used soil fumigant available for use (USEPA, 2008b; Sullivan et al., 2004). MS and metam potassium (MK) are broad-spectrum soil fumigants and are also used in sewers, drains, and ponds to control weeds and roots (USEPA, 2008b). MS is a sodium salt formulation of methyldithiocarbamate which breaks down into the active ingredient methyl isothiocyanate (MITC) when injected into the soil. MK is a potassium salt of N-methyldithiocarbamate and breaks down into MITC similarly

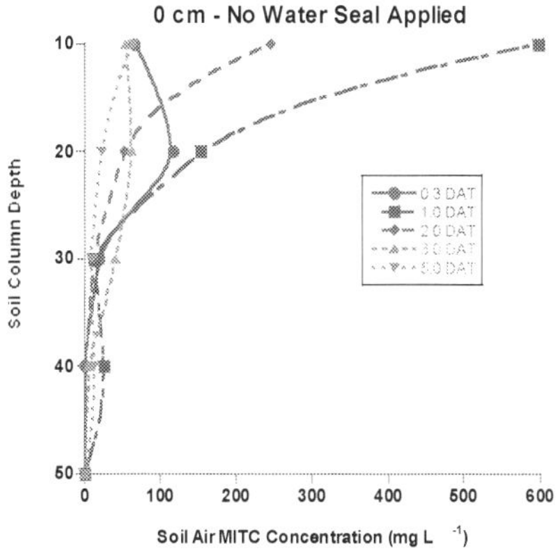

Fig. 4a.

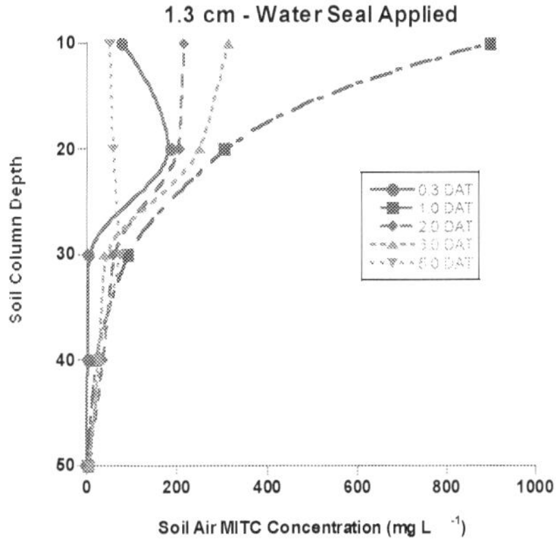

Fig. 4b.

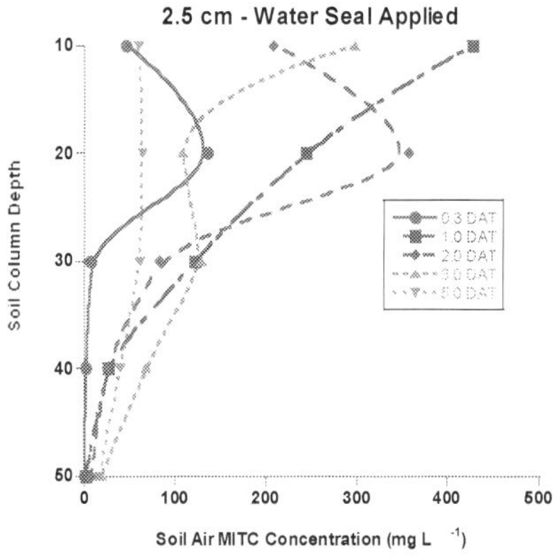

Fig. 4c.

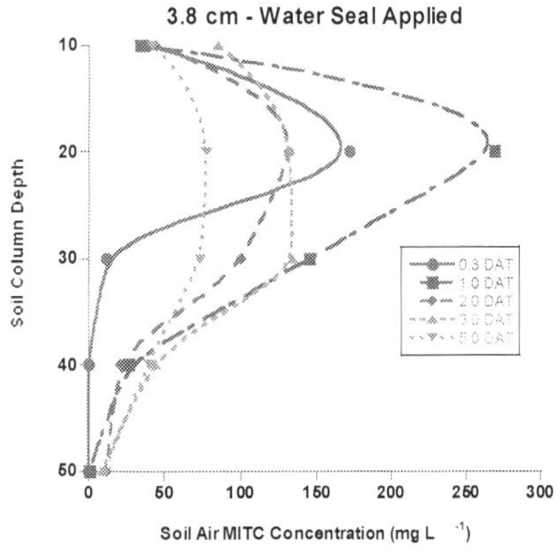

Fig. 4d.

Fig. 4. MITC distribution in soil airspace throughout the soil profile of columns over time (DAT=days after treatment, or after injection of metam sodium at 10-cm column depth); data shown represent the mean of three replications for each water seal treatment [0-cm (4a), 1.3-cm (4b), 2.5-cm (4c), and 3.8-cm (4d) water seal application depth].

The real impact of the water seal treatment at suppressing MITC volatilization was observed in the 2.5-cm water seal treatment (Fig. 4c). This is especially apparent when looking at the level of MITC over time at the 10 cm soil depth. The concentration of MITC at the 10 cm soil depth was lower 1.0 DAT for the 2.5-cm than the 0-cm and 1.3-cm water seal treatments (Fig. 4a-c), suggesting a restriction in the volatilization loss of MITC through the soil surface. Furthermore, the bulk amount of MITC resided at the 20 cm soil depth for a longer period of time after chemical application when compared to the lower water seal treatments, allowing the MITC to distribute vertically throughout the soil column with MITC concentrations observed at the 50 cm by 5.0 DAT (Fig. 4c).

Application of a 3.8-cm water seal resulted in the longest retention of MITC within the soil profile, along with the greatest suppression of MITC from the soil surface as evident by low MITC soil-air phase levels at the 10 cm soil depth up to 5.0 DAT (Fig. 4d). The extra water applied to the soil surface in the 3.8-cm treatment moved the MS further down the soil profile resulting in high MITC concentrations at both the 20 and 30 cm soil depths 1.0 to 5.0 DAT. The higher water amount within the soil profile was confirmed at the end of the study as soil moisture levels were higher at the 25 cm soil depth of the 3.8-cm than the 2.5-cm water seal treatments (data not shown).

3.2 Soil surface flux of MITC

The highest amount of MITC volatilized through the soil surface was observed from soil columns with no (0-cm) water seal applied after MS application (Fig. 5). The greatest MITC flux was observed within the initial 36 h after chemical application and decreased over time thereafter. A similar trend was observed for the 1.3-cm water seal treatment, but the amount of MITC evolved was substantially less than that from the 0-cm treatment. The lowest amount of MITC flux observed occurred from both the 2.5-cm and 3.8-cm water seal treatments, with the 2.5-cm treatment releasing slightly more MITC by 120 h after MS application.

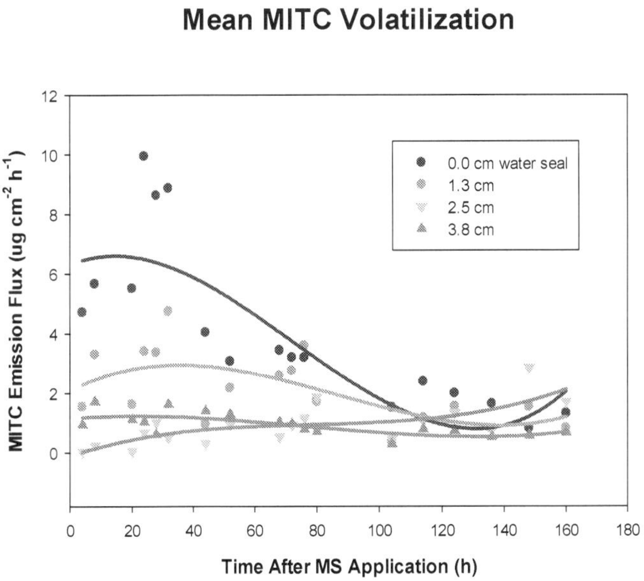

Fig. 5. The mount of MITC volatilized and captured on charcoal filters over time. Data represents mean of three replicates per water seal treatment.

In order to determine to total amount of MITC volatilized from the surface of the soil columns, cumulative MITC levels were calculated and plotted (Fig. 6). In this respect it is easily apparent that the highest MITC emissions occurred from soil columns without a water seal treatment. But more importantly, for soil columns that received additional surface water irrigation, total MITC volatilization decreased with increasing water seal depth (Fig. 6a). The total mean MITC volatilization loss from the 0-, 1.3-, 2.5- and 3.8-cm water seal treatments was respectively 24, 14, 9 and 6% of the total initial MITC applied (Fig. 6b). The highest variability in MITC loss was observed in the low to no water seal treatments, suggesting that neither of these treatments would be acceptable for suppressing MITC fumigant loss from soils. Whereas a low amount of variability (small error bars) was observed for the higher water seal treatments, with no statistical difference in total MITC loss (Fig. 6b).

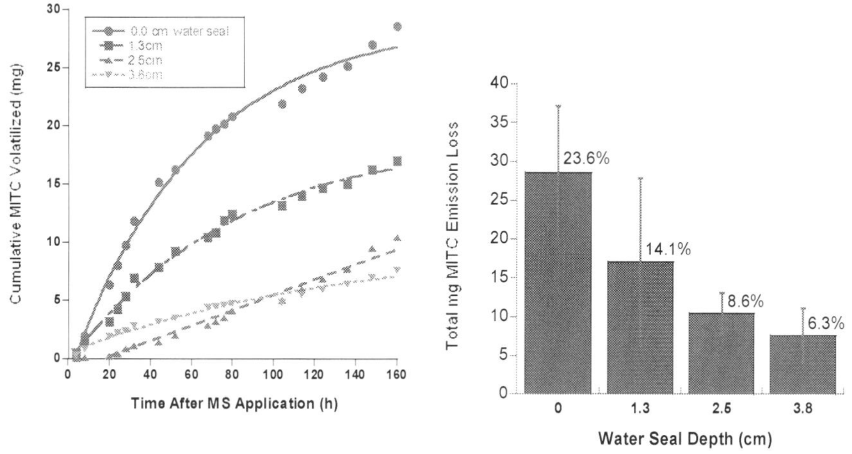

Fig. 6a. Mean MITC emitted from soil columns. Fig. 6b. MITC release ± std error mean.

Fig. 6. Total cumulative MITC evolved from soil columns as captured on charcoal filters.

4. Conclusion

These findings illustrate how effective bench-scale soil column studies are at assessing the volatilization potential of MS after varying surface water seal treatments. Keeping in mind that this study represents specific and restricted conditions, it does provide a good estimate of the proper water seal depth needed for a sandy clay loam soil type. Although a 3.8-cm water seal led to the least amount of fumigant loss, it is recommended that a 2.5 -cm water seal be applied in the field for similar soil types. This is suggested due to the fact that applying large amounts of water can significantly alter chemical behavior by further diluting the MS to a level below the critical threshold for MITC to be effective for pest control. Furthermore, in areas where water tables are high, adding too much water via supplemental overhead irrigation may lead to groundwater contamination and result in other environmental concerns. The 2.5-cm water seal application suppressed MITC volatilization to level statistically equivalent to that of the 3.8-cm treatments and therefore, it is a good practice to reduce fumigant emissions to the atmosphere while minimizing excessive chemical movement beyond the crop rooting depth. On-farm field investigations will be needed to back up these laboratory scale findings to provide confirmation that the suppressive loss of MITC is ultimately achievable.

5. Acknowledgement

The authors would like to thank the financial support of the US Department of Agriculture-CSREES Methyl Bromide Transition Program, grant award # 2002-51102-1922, and USDA-CSREES Hispanic Serving Institution program, grant award # 2004-38422-14608, and the dedicated work by student research aides like Mr. LeeRoy Rock (shown in Fig. 3c, 3d with permission) for the long hours required for data collection.

6. References

Ajwa, H. A.; Trout, T.; Mueller, J.; Wilhelm, S.; Nelson, S. D.; Soppe, R. & Shatley, D. (2002). Application of alternative fumigants through drip irrigation systems, *Phytopathology*, Vol.92, pp. 1349-1355.

Chase, C.A.; Gilreath, J.P.; Locascio, S.J.; Sinclair, T. R.; & Shilling, D. G. (1998). Light effects on rhizome morphogenesis in nutsedges (Cyperus spp.): implications for control by soil solarization, *Weed Science*, Vol. 46, No. 5, pp. 575-580.

Dungan, R. S.; Gan, J., & Yates S. R. (2001). Effect of temperature, organic amendment rate and moisture content on the degradation of 1,3dichloropropene in soil, *Pest Management Science*, Vol.57, pp. 1107-1113.

El Hadiri N.; Ammati, M.; Chgoura, M. & Mounir, K. (2003). Behavior of 1,3-dichloropropene and methyl isothiocyanate in undisturbed soil columns. *Chemosphere*, Vol. 52, pp. 893-899.

Gamliel, A.; Grinstein, A. & Katan, J. (1997). Improved technologies to reduce emission of methyl bromide from fumigated soil, *Phytoparasitica*, Vol. 25, pp. 21S-30S.

Gan, J.; Yates, S. R.; Crowley, D. & Becker, J. O. (1998a). Acceleration of 1,3dichloropropene degradation by organic amendments and potential application for emissions reduction, *Journal of Environmental Quality*, Vol.27, pp. 408-414.

Gan, J.; Yates, S. R. Wang, D. & Ernst, F. F. 1998b. Effect of application methods on 1,3-dichloropropene volatilization from soil under controlled conditions, *Journal of Enviromental. Quality*, Vol.27, pp. 432-438.

Gan, J.; Becker, J. O.; Ernst, F. F.; Hutchinson, C.; Knuteson, J. A. & Yates, S. R. (2000a). Surface application of ammonium thiosulfate fertilizer to reduce volatilitzation of 1,3-dichloropropene from soil, *Pest Management Science*, Vol.56, pp. 264-270.

Gan, J.; Hutchinson, C.; Ernst, F. F.; Becker, J. O. & Yates, S. R. (2000b). Column system for concurrent assessment of emission potential and pest control of soil fumigants, *Journal of Environmental Quality*, Vol.29, pp. 657-661.

Gan, J.; Yates, S. R.; Knuteson, J. A. & Becker, J. O. (2000c). Transformation of 1,3-dichloropropene in soil by thiosulfate fertilizers, *Journal of Envionmental Quality*, Vol.29, pp. 1476-1481.

National Pesticide Information Center. (2000). Methyl bromide, *National Pesticide Telecommunications Network Technical Fact Sheet*. 26.06.2011. Available from http://npic.orst.edu/factsheets/MBtech.pdf

Nelson, S.D.; Allen, L.H. Jr.; Gan, J.; Riegel, C.; Dickson, D.W. & Locascio, S. J. (2000) Can virtually impermeable films reduce the amount of fumigant required for pest-pathogen management in high value crops? *Proceedings of Soil and Crop Science Society of Florida*, Vol. 59 pp. 85-89.

Noling, J. W. & Becker, J. O. (1994). The challenge of research and extension to define and implement alternatives to methyl bromide, *Journal of Nematology Supplemental*, Vol.26, pp. 573-586.

Papiernik, S. K.; Yates, S. R. & Gan, J. (2001). An approach for estimating the permeablility of agricultural films, *Environmental Science and Technology*, Vol.35, pp. 1240-1246.

Schneider, R. C.; Green, R. E.; Wolt, J. D.; Loh, R. K. H.; Schmitt, D. P. & Sipes, B. S. (1995). 1,3-dichloropropene distribution in soil when applied by drip irrigation of injection in pineapple culture. *Pesticide Science*, Vol.43, pp. 97-105.

Shaw, D.V. & Larson, K. D. (1999). A meta-analysis of strawberry yield response to preplant soil fumigation with combinations of methyl bromide-chloropicrin systems. *HortScience*, Vol 34, No. 5, pp. 839-845.

Simpson, C.R.; Nelson, S.D.; Stratmann, J.E. & Ajwa, H.A. (2010). Surface water seal application to minimize volatilization loss of methyl isothiocyanate from soil columns, *Pest Management Science*, Vol. 66, pp. 686-692.

Sullivan, D. A.; Holdsworth, M. T. & Hlinka, D. J. (2004). Control of off-gassing rates of methyl isothiocyanate from the application of metam-sodium by chemigation and shank injection, *Atmospheric Environment*, Vol.38, pp. 2457-2470.

United States Government Printing Office. (2005). Final rule on continued production of methyl bromide for export to developing countries, In: *Federal Register*, Vol. 70, No. 249, pp. 1-7.

U.S. Environmental Protection Agency. (1998). RED fact sheet: 1,3-dichloropropene, In: *US Environmental Protection Agency Office of Pesticide Programs*. 26.06.2011. Available from http://www.epa.gov/pesticides/reregistration/REDs/factsheets/0328fact.pdf

U.S. Environmental Protection Agency. (2008a). RED fact sheet: chloropicrin, In: *US Environmental Protection Agency Office of Pesticide Programs*. 26.06.2011. Available from http://www.epa.gov/pesticides/reregistration/REDs/factsheets/chloropicrin-fs.pdf

U.S. Environmental Protection Agency. (2008b). RED fact sheet: Methyldithiocarbamate salts-metam sodium/potassium and MITC, In: *US Environmental Protection Agency Office of Pesticide Programs*. 26.06.2011. Available from http://www.epa.gov/oppsrrd1/REDs/factsheets/metamsodium-fs.pdf

U.S. Environmental Protection Agency. (2009). The phaseout of methyl bromide: the phaseout schedule for methyl bromide, In: *US Environmental Protection Agency, Ozone Depletion Rules and Regulations section*. 25.06.2011. Available from http://www.epa.gov/ozone/mbr

Wang, D.; Yates, S. R.; Ernst, F. F.; Gan, J. & Jury, W. A. (1997). Reducing methyl bromide emission with a high barrier plastic film and reduced dosage, *Environmental Science and Technology* 31:3686-3691.

Wang, Q.; Gan, J.; Papiernik, S. K. & Yates, S. R. (2001). Isomeric effects on thiosulfate transformation and detoxification of 1,3-dichloropropene, *Journal of Environmental Toxicology and Chemistry*, Vol.20, pp. 960-964.

25

Transgenesis, Paratransgenesis and Transmission Blocking Vaccines to Prevent Insect-Borne Diseases

Marcelo Ramalho-Ortigão and Iliano Vieira Coutinho-Abreu
Kansas State University, Department of Entomology USA

1. Introduction

Insect-borne diseases are responsible for severely affecting human life around the world, causing significant morbidity and mortality. Malaria alone is responsible for 1-2 million deaths annually, and approximately 300 million are at risk of becoming infected. Insect-borne diseases also are responsible for an estimated 50% of all neglected tropical diseases (NTDs), which affect over 1 billion people - one sixth of the world population - and include such diseases as leishmaniasis, filariasis, Chagas' disease, African trypanosomiases, onchocerciasis, schistosomiasis, trachoma and others. Such NTDs cause severe morbidity and are frequently referred to as "poverty causing diseases". The lack of effective vaccines or drugs for many insect-borne diseases makes control mainly dependent on insecticides. However, the appearance of insecticide resistance requires the development of new strategies to reduce pathogen transmission in the field [1]. Among the research themes with potential to generate new tools to control vector borne diseases, major efforts have been carried out to establish transgenesis, paratransgenesis, and transmission-blocking vaccines (TBVs) as new weapons to reduce vector competence.

While vector competence encompasses the intrinsic genetic factors that define the ability of a vector to transmit a pathogen (and it is a component of vectorial capacity), vectorial capacity is a measurement of the efficiency of vector-borne disease transmission (i.e., total number of infective bites delivery to a single host in one day), and influenced by vector density and longevity [2]. Regarding vector competence, several molecular techniques, such as quantitative trait loci (QTL) mapping, and gene knock-down, can be used to identify the intrinsic genetic factors (i.e., molecules expressed by the vector) involved with the ability of vectors to transmit pathogens. Molecules involved in vector competence can be directly targeted by antibodies (as in the case of TBVs), or overexpressed in transgenic insects or paratransgenic symbionts in order to reduce pathogen development and transmission.

In the late 1990's, the establishment of stable lines of genetically modified mosquitoes opened new avenues for studying molecules with potential to reduce vector competence [3]. Transgenic mosquitoes expressing dsRNAs (i.e., to induce RNAi pathways) targeting RNAs associated with mosquito immune-related proteins [4], or overexpressing microbial peptides [3, 5], or expressing a truncated transcription factor to generate a dominant-negative phenotype [6] were generated in order to investigate the role of these molecules in vector competence. Understanding such mechanisms is considered a pre-requisite for the development of molecular strategies to control vector-borne diseases.

For a heterologous protein (exogenous protein introduced into a disease vector) to be used to reduce vector competence, a gene drive mechanism is required to spread the gene encoding the protein throughout the targeted insect vector population. A gene drive system is spread within a population by increasing its frequency to ratios greater than those expected by traditional Mendelian rules. Thus, the combination of a given transgene (expressing a heterologous protein) with a gene drive system also can increase the frequency of the transgene in a population. Gene drive systems currently known include transposons, homing endonuclease, engineered under-dominance, meiotic drive, endosymbionts, and Medea element [7, 8]. Yet, only transposons are currently available to be used in the genetic transformation of insect disease vectors [8], and Wolbachia endosymbionts are thought to be a feasible way to spread paratransgenic symbionts in natural populations of these insect vectors [9].

The use of transposons to generate stably transformed insect germ lines, i.e., with exogenous DNA inserted into the genome and capable of being transferred into following generations (as depicted in Figure 1A), is well established [10] for a couple of insect vector species. Different species of mosquitoes, representing *Aedes* [11-14], *Anopheles* [15-24], and *Culex* [25] genera, have been genetically altered or transformed (Table 1), and, in some cases, the transformed mosquitoes expressed proteins targeting pathogen development [3, 20, 21, 26, 27]. Here, the common goal is to transform insect vectors with gene(s) whose protein(s) impair(s) pathogen development. As indicated above, genes that reduce pathogen development are to be associated with a gene drive system that increases the frequency of the transgenic vector when they are released into their natural habitats.

Paratransgenesis is an alternate approach to reduce vector competence via the genetic manipulating of symbionts commonly found in insect disease vectors (Figure 1B). The main characteristics of paratransgenesis are the simplicity with which symbionts are transformed (through viral or bacterial genetic transformation), the feasibility of the transformed symbiont to be spread across a population (maternally or via coprophagy), and the reduced fitness cost associated with the transformation of symbionts [28]. Symbionts currently targeted for paratransgenesis include bacteria that inhabit triatomine hindguts [28-30] and tsetse fly tissues [31], and densoviruses infecting *An. gambiae* and *Ae. aegypti* mosquitoes [32, 33]. To date, insect vector symbionts have been genetically modified to express antimicrobial peptides [28], single chain antibodies [29, 30, 34], and dsRNAs [35-38]. In all three approaches the expressed molecules proved harmful to the pathogens transmitted by each vector.

Transmission-blocking vaccines (TBVs) are intended to prevent the transmission of pathogens from infected to uninfected hosts (Figure 1C) by a disease vector. Such vaccines do not protect an individual from infection but rather can reduce transmission. TBVs target molecule(s) that are expressed on the surface of parasites during their developmental phase

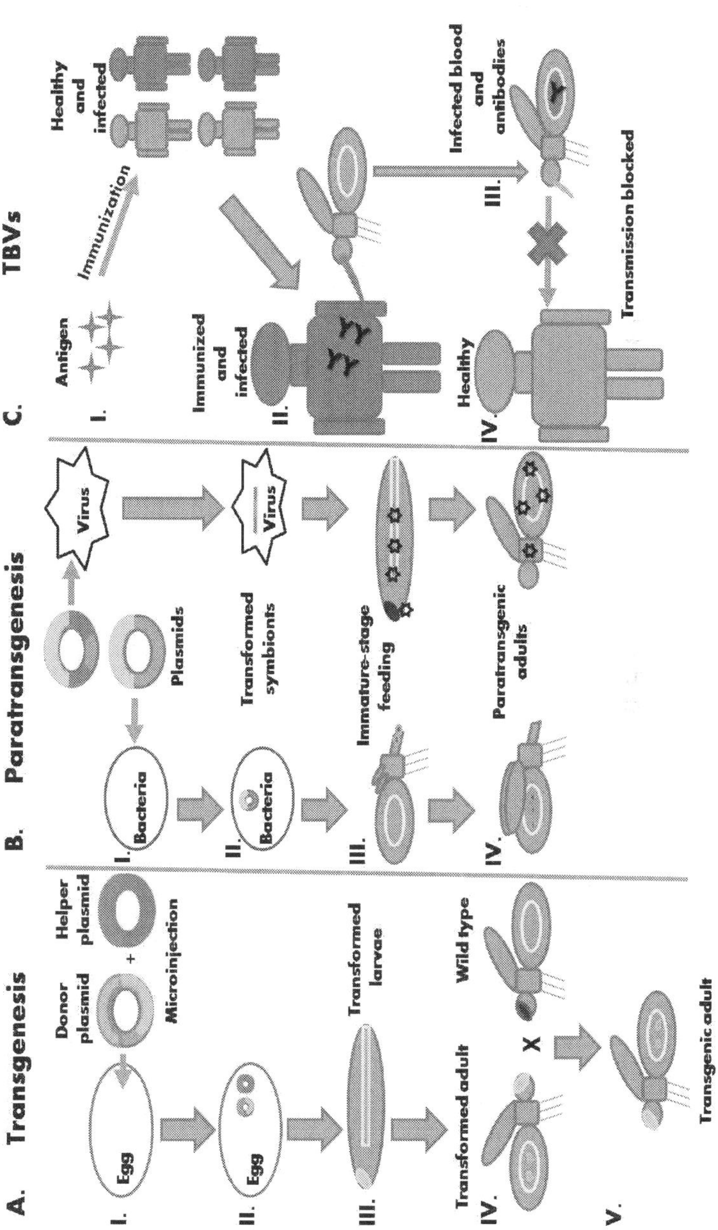

Fig. 1. Transgenesis, paratransgenesis, and transmission blocking vaccines (TBVs) (modified from (Coutinho-Abreu & Ramalho-Ortigao 2010; Coutinho-Abreu et al. 2010)). (**A**)

Transgenesis; the general technique for production of transgenic insects by germ line transformation is shown. (I) Insect eggs are microinjected with a donor plasmid [expressing a transgene (orange) and a reporter gene (green)], and with a helper plasmid, [expressing a transposase (purple)]. (II) After inoculation into the eggs, plasmids are taken up by some (or all) of the germ line cells. (III) transgenic and non-transgenic larvae can be separated by the expression of the reporter gene (green eye phenotype), controlled by an eye specific promoter. (IV) Transgenic insects are crossed with wild type to confirm that the transposon-carrying transgene was inserted into the chromosome. (V) Transgenic adult insect expressing the transgene (e.g., orange proteins in the insect midgut) is shown. (B) Paratransgenesis; the general technique used to obtain insect transformation via their symbionts is shown. Two insect orders are represented: Hemiptera/reduviids (left panel) and Diptera/mosquitoes (right panel). (I) Bacteria or viruses symbionts can be genetically modified to express a gene blocking parasite development in vectors' tissues. (II) Symbiotic bacteria are transformed with plasmids (blue) expressing a gene (green) to inhibit parasite development in insect gut. Alternatively, viral genome (red) is inserted into a plasmid (blue) and manipulated to express a transgene (green). Viral particles can be generated by expression of such a plasmid in insect cells. (III) The transformed symbionts are acquired by insect hosts through larvae or nymph feeding, or through thoracic injection. (IV) Once insects acquire the transformed symbionts, these microorganisms can express proteins to inhibit pathogen development. (C) Transmission-blocking vaccines (TBVs). TBV is a strategy to prevent transmission of a pathogen by the bite of an infected vector. Frequently, TBVs rely on generating antibodies against vector molecules that are involved in pathogen development. (I) Healthy (blue) and infected (red) individuals are immunized with a TBV antigen; (II) Insect-vectors take an infected blood meal containing TBV antigen-specific antibodies; (III) Specific antibodies produced against the antigen inhibit pathogens development within the insect vector, (IV) preventing transmission to uninfected host(s).

within the insect vector (Table 2) [39-48], or that are expressed on the surface of vector tissues with which pathogens may be required to interact during their development within the vector (Table 3) [48-52]. The potential application of proteins (antigens) expressed on the surface of the malaria parasites *Plasmodium falciparum* and *Plasmodium vivax* as TBV has been tested [39-48]. Two of these proteins (Pfs25 and Pvs25) were deemed safe following phase-1 human trials [53, 54]. Specific antibodies against molecules expressed in midgut tissues of *An. gambiae* and *Phlebotomus papatasi* are also capable of reducing parasite loads in these vectors, pointing to their potential as TBVs candidates [48-52].

In the following pages, details of each of these three approaches are provided. Technical aspects of the strategy utilized as well as results from studies *in vitro* and *in vivo*, or depending on the case animal and human tests, as well as semi-field or field release of modified insects are also indicated. Due to the massive amount of information that has been generated in recent years regarding some of the topics (e.g., TBVs), we intend to limit our analyses to some specific points we find critical for the readers information.

2. Transgenesis in insect disease vectors

Generally, the goal for vector transgenesis is the interruption of pathogen transmission through introduction of exogenous DNA fragment (i.e., gene) into the genome of a disease vector, followed by expression of the gene to inhibit pathogen development within the vector. Various mosquito species, vectors of different parasites and viruses, have been transformed (Table 1). Some of the transformed mosquitoes were shown capable of blocking pathogen development via tissue-specific expression of molecules that impair pathogen attachment to the midgut (Ito et al. 2002), or activate biochemical pathways detrimental to pathogen survival (Franz et al. 2006). However, vector transgenesis is a complex approach, highlighted by the fact that insect germ line transformation technique is only successfully performed by a handful of laboratories. Various issues related with transgenic vectors, including stability of the transgene in the genome, and fitness of the transformed insect in the field, need to be fully resolved prior to the successful application of transgenic insects fat body and hemocoel against insect-borne diseases.

A few aspects of development of transgenic vectors, how interference or blockage of parasite is achieved, and what may lie ahead for vector transgenesis are discussed below.

2.1 Germ line transformation

Stable genetic transformation of insects is accomplished by inserting an exogenous gene into the insect genome via the inoculation of plasmids containing the transgene (donor plasmids) into insect eggs (Figure 1). Donor plasmids are constructed to carry an engineered transposon element lacking the gene that encodes the transposase, the enzyme that mediates transposon activity by a cut and paste mechanism. *Hermes*, *Mariner* (*Mos1*), *minus*, and *piggyBac*, for the most part, have been the transposable elements of choice. Transposases are then supplied *in trans* (expressed in a separate plasmid) by the co-inoculation of a transposase-encoding helper plasmid along with the donor plasmid. Transposase expression is usually driven by a heat shock protein promoter that is activated upon raising the temperature, as it is frequently performed with injected mosquito eggs (Wimmer 2003).

Transformed offspring can be identified by a specific phenotype alteration, mediated by the expression of a reporter protein encoded by the donor vector, such as eye color (Wimmer 2003). However, this strategy is restricted to insect species displaying polymorphic eye phenotypes, such as *Ae. aegypti*. Alternatively, phenotypic markers such as firefly luciferase, or green (EGFP), red (dsRED) and cyan fluorescent (CFP) proteins can be used as transformation markers (Moreira et al. 2000; Kokoza et al. 2001; Nolan et al. 2002; Perera et al. 2002; Nirmala et al. 2006).

2.2 Tissue-specific transgene expression

Within the insect vector, a pathogen may interact with specific tissues, such as the midgut (as is the case for *Trypanosoma cruzi* and *Leishmania sp.*), or midgut, hemocoel and salivary glands (as is the case for *Plasmodium sp*). Frequently, molecules that can block pathogen development within its vector can be expressed in a tissue-specific manner to increase effectiveness. Tissue-specific expression of a transgene is accomplished by the use of a tissue-specific promoter (Table 1). Most promoters used in vector transgenesis drive protein expression specifically within the midgut, the hemocoel, or the salivary glands, as these are sites where pathogens are commonly found within an infected vector (Kokoza et al. 2000; Moreira et al. 2000; Abraham et al. 2005; Lombardo et al. 2005; Yoshida & Watanabe 2006; Chen et al. 2007b; Rodrigues et al. 2008).

In mosquitoes, promoters for carboxypeptidase and peritrophin have been widely used to drive midgut-specific expression of several transgenes. In *Ae. aegypti,* a vitellogenin promoter driving expression of transgenes in the fat body was used to express innate immune defense-related genes (Kokoza et al. 2000; Kokoza et al. 2001; Shin et al. 2003a; Bian et al. 2005), as well as dsRNA targeting *REL1* transcripts (Bian et al. 2005). The same promoter was used to express CFP in *An. stephensi* (Nirmala et al. 2006). The robustness of the vitellogenin promoter was confirmed by its capacity to function following multiple gonotrophic cycles in transgenic *An. stephensi* (Chen et al. 2007b).

Salivary gland-specific promoters also have been used in germ-line transformation of mosquitoes. *D7* and *apyrase* promoters from *An. gambiae* and *antiplatelet* from *An. stephensi* were used to transform *An. stephensi* (Lombardo et al. 2005; Yoshida & Watanabe 2006). *Ae. aegypti* also was transformed and successfully expressed luciferase within salivary glands using *maltase-like 1* and *apyrase* promoters (Coates et al. 1999).

2.3 Transgenes targeting *Plasmodium* development

In spite of the many mosquito species successfully transformed, only a handful has been transformed with molecules that impair pathogen development (Ito et al. 2002; Moreira et al. 2002; Kim et al. 2004; Abraham et al. 2005; Franz et al. 2006; Rodrigues et al. 2006). A list of genetically modified mosquitoes obtained to date, including the transposon and reporter genes used, tissue-expression specificity and target pathogens, among others is seen on Table 1.

In one example of a transgene targeting *Plasmodium,* expression of phospholipase-2 (PLA2) in *An. stephensi* (Moreira et al. 2002) led to an 87% reduction of *P. berghei* oocyst intensity compared to non-PLA2-expressing controls. When a peritrophin promoter was used to drive the expression of PLA2 in *An. stephensi,* inhibition of *P. berghei* oocyst intensity ranged from 74% to 94% (Abraham et al. 2005). Nevertheless, expression of PLA2 in the mosquito midgut did not exert a direct effect on the parasite, but rather led to structural damage of the midgut epithelium (Moreira et al. 2002; Abraham et al. 2005).

Synthetic peptides that block *Plasmodium* development in mosquitoes also have been identified and tested. SM1, identified using a phage display library (Ghosh et al. 2001), blocked *P. berghei* invasion of *An. stephensi* midgut and salivary glands. In *An. stephensi* transformed with *piggyBac* expressing a four tandem repeat of SM1 under a carboxypeptidase promoter, *P. berghei* intensity was inhibited by 81.6%. Interestingly, these transgenic mosquitoes even with sporozoites in their salivary glands were unable to transmit *P. berghei* to mice (Ito et al. 2002).

Another molecule tested was the C-type lectin (CELIII) from sea cucumber. When expressed in *An. stephensi,* CELIII is was shown to be cytotoxic to *P. berghei* ookinetes, reducing prevalence and intensity by 84% and 90%, respectively (Yoshida et al. 2007). CELIII expression was driven by a midgut-specific *An. gambiae*-derived carboxypeptidase promoter.

Besides PLA2, SM1, and CELIII, mosquitoes also were transformed with antimicrobial peptides (Kokoza et al. 2000; Kim et al. 2004; Bian et al. 2005). *An. gambiae* transformed with a *cecropin A* (driven by a carboxypeptidase promoter from *Ae. aegypti*) inhibited *P. berghei* intensity by 61% on average when compared with non-transformed controls (Kim et al. 2004). Recently, *An. stephensi* were transformed with the *piggyBac* transposon expressing single-chain antibodies (scFvs) targeting *P. falciparum* proteins. Some of these scFv genes have been attached to a *Cecropin A* gene so as to improve the effectiveness of the antibody against *P. falciparum*. In fact, the scFv-Cecropin A construct (m2A10) targeting the *P. falciparum* circunsporozoite protein reduced *P. falciparum* intensity by 97%, prevalence by 86%, and sporozoites load by 84% (Isaacs et al. 2011).

2.4 Transgenes inducing gene silencing and targeting viral transmission

Stable transformation of mosquitoes with two inverted repeats of the same gene to induce assembly of double-strand RNAs (dsRNAs) and activation of the RNAi pathway has also been obtained (Brown et al. 2003; Bian et al. 2005). This strategy takes advantage of the RNAi mechanism to block expression of insect molecules associated with vectorial competence (Bian et al. 2005), or it can directly target viral replication within insect tissues (Franz et al. 2006). In spite of the fact that both approaches are technically feasible, only the latter has led to substantial reduction in the development of any human pathogen (Franz et al. 2006).

Gene silencing via dsRNA was first demonstrated with *An. stephensi* expressing sense and anti-sense RNAs targeting EGFP (Brown et al. 2003). EGFP dsRNA-expressing mosquitoes were crossed with a transgenic mosquito line that expresses EGFP. The double transgenic offspring displayed lower level of EGFP expression than the parental line expressing it, indicating the effect of the RNAi machinery reducing the expression of the EGFP transgene (Brown et al. 2003). *Ae. aegypti* expressing dsRNA targeting *REL1*, a gene involved in innate immune response, also inhibited expression of *REL1* via RNAi (Bian et al. 2005).

As indicated, the activation of the RNAi pathway is intended to affect the replication of infecting RNA viruses transmitted by mosquitoes. *Ae. aegypti* expressing DEN2 sense and antisense RNAs reduced viral load by fivefold, confirming the effectiveness of the RNAi in controlling virus replication in disease vectors (Franz et al. 2006).

2.5 Future of vector transgenesis

Despite advances in the development of stable lines of genetically modified disease vectors (Moreira et al. 2000; Ito et al. 2002; Lobo et al. 2002; Perera et al. 2002), many challenges exist to the application of transgenesis to control vector-borne diseases outside the laboratory. Beyond issues dealing with social and environmental impact(s) that are inherent to the potential use of genetically modified organisms (not the scope of this chapter), of significance also is the fact that, for example, none of the most import human malaria vectors (i.e. *An. gambiae s.l.* and *An. funestus*) has been successfully transformed and displayed reduced vector competence. Moreover, only one strain of transgenic mosquitoes blocked the development of *P. falciparum*, the principal human malaria parasite (Yoshida et al. 2007). The only exceptions of transgenic insect lines robustly impairing development of a human pathogen in its human vector are the transgenic strain of *Ae. aegypti* capable of inhibiting DEN virus development (Franz et al. 2006; Mathur et al. 2010).

Insect species	Transposon	Promoter	Reporter gene	Transgene	Targeted pathogen	Reference
Ae. aegypti	*Hermes*	*D. melanogaster Cinnabar*	*D. melanogaster Cinnabar*	*Cinnabar*	-	(Jasinskiene et al. 1998)
	Mos1	*D. melanogaster Cinnabar*	*D. melanogaster Cinnabar*	*Cinnabar*	-	(Coates et al. 1998)
	Hermes	*Ae. aegypti maltase-like 1 / apyrase*	Firefly *luciferase*	Firefly *luciferase*	-	(Coates et al. 1999)
	Hermes	*D. melanogaster Actin 5C*	*EGFP*	*EGFP*	-	(Pinkerton et al. 2000)
	Hermes	*Ae. aegypti vitellogenin*	*D. melanogaster Cinnabar*	*Defensin A*	*Micrococcus luteus*	(Kokoza et al. 2000)
	Hermes	*Ae. aegypti carboxypeptidase*	Firefly *luciferase*	Firefly *luciferase*	-	(Moreira et al. 2000)
	Mos1	*An. gambiae carboxypeptidase*	Firefly *luciferase*	Firefly *luciferase*	-	(Moreira et al. 2000)
	piggyBac	*Ae. aegypti vitellogenin*	*EGFP*	*Defensin A*	*Enterobacter cloacae/ P. gallinaceum*	(Kokoza et al. 2001; Shin et al. 2003b)
	Minos	*D. melanogaster Actin 5C*	*dsRED*	*dsRED*	-	(Nolan et al. 2002)
	piggyBac	*D. melanogaster Actin 5C*	*dsRED*	*dsRED*	-	(Nolan et al. 2002)
	piggyBac	*D. melanogaster Cinnabar*	*D. melanogaster Cinnabar*	*Cinnabar*	-	(Lobo et al. 2002)
	Mos1	*D. melanogaster Cinnabar/hsp70*	*D. melanogaster Cinnabar*/dsRED	*Cinnabar/dsRED*	-	(Wilson et al. 2003)
	piggyBac	*D. pseudoobscura hsp82*	*EGFP*	*Mos1 transposase*	-	(Wilson et al. 2003)
	piggyBac	*Ae. aegypti vitellogenin*	*EGFP*	*ΔREL*[a]	-	(Shin et al. 2003a)
	piggyBac	*Ae. aegypti vitellogenin*	*EGFP*	*Cecropin A*	*Enterobacter cloacae*	(Shin et al. 2003a)
	piggyBac	*Ae. aegypti vitellogenin*	*EGFP*	*ΔREL1-A*[b]	-	(Bian et al. 2005)
	piggyBac	*Ae. aegypti vitellogenin*	*EGFP*	dsRNA against *REL1* transcripts	-	(Bian et al. 2005)

Insect species	Transposon	Promoter	Reporter gene	Transgene	Targeted pathogen	Reference
Ae. aegypti	*Mos1*	*An. gambiae carboxypeptidase*	*EGFP*	dsRNA against *prM* of DENV-2	Dengue-2 virus	(Franz et al. 2006; Franz et al. 2009)
	Mos1	*Ae. aegypti nos* *Ae. aegypti D7*	*dsRED*	*Mos1 transposase*	-	(Adelman et al. 2007)
	Mos1		*EGFP*	*Gal4*	-	
	Mos1	*Ae. aegypti* *30K*	*EGFP*	*Mnp*	Dengue-2 virus	(Mathur et al. 2010)
Ae. fluviatilis	*piggyBac*	*An. gambiae peritrophin*	*EGFP*	*mPLA2*	*P. gallinaceum*	(Rodrigues et al. 2008)
An. albimanus	*piggyBac*	*D. melanogaster polyubiquitin*	*EGFP*	*EGFP*	-	(Perera et al. 2002)
An. gambiae	*piggyBac*	*Baculovirus hr5-ie1*	*EGFP*	*EGFP*	-	(Grossman et al. 2001)
	piggyBac	*Ae. aegypti carboxypeptidase*	*EGFP*	*Cecropin A*	*P. berghei*	(Kim et al. 2004)
An. stephensi	*minos*	*D. melanogaster actin 5C*	*EGFP*	*EGFP*	-	(Catteruccia et al. 2000)
	piggyBac	*D. melanogaster actin 5C*	*dsRED*	*dsRED*	-	(Nolan et al. 2002)
	piggyBac	*An. gambiae carboxypeptidase*	*EGFP*	*PLA2*	*P. berghei*	(Moreira et al. 2002)
	piggyBac	*An. gambiae carboxypeptidase*	*EGFP*	*[SM1]*$_4$	*P. berghei*	(Ito et al. 2002)
	minos	*D. melanogaster actin 5C*	*dsRED*	dsRNA against *EGFP*	-	(Brown et al. 2003)
	minos	*An. gambiae apyrase An. gambiae D7r4*	*EGFP*	*LacZ*	-	(Lombardo et al. 2005)
	piggyBac	*An. gambiae peritrophin*	*EGFP*	*PLA2*	*P. berghei*	(Abraham et al. 2005)

Insect species	Transposon	Promoter	Reporter gene	Transgene	Targeted pathogen	Reference
An. stephensi	*piggyBac*	*D. melanogaster β2-tubulin*	*dsRED*	*EGFP*	-	(Catteruccia et al. 2005)
	piggyBac	*An. stephensi vitellogenin*	*EGFP*	*CFP*	-	(Nirmala et al. 2006)
	minos	*An. gambiae carboxypeptidase*	*EGFP*	CELIII	*P. berghei P. falciparum*	(Yoshida et al. 2007)
	minos	*An. stephensi antiplatelet*	*EGFP*	*dsRED*	-	(Yoshida & Watanabe 2006)
	piggyBac	*An. gambiae vitellogenin*	*dsRED*	*EGFP*	-	(Chen et al. 2007b)
	piggyBac *minos*	*D. melanogaster hsp70* *D. melanogaster actin 5C*	*dsRED* *EGFP*	*minos transposase* *EGFP*	- -	(Scali et al. 2007)
	piggyBac + *attP* docking site	3XP3	*ECFP*	-	-	(Amenya et al. 2010)
Cx. quinquefasciatus	*Hermes*	*D. melanogaster actin 5C*	*EGFP*	*EGFP*	-	(Allen et al. 2001)

([a]) *Drosophila* Relish-related gene lacking the transactivator domain.
([b]) REL1-A lacking a C-terminal domain.

Table 1. Germ line transformed mosquitoes (Modified from (Coutinho-Abreu et al. 2010)).

3. Paratransgenesis to reduce vector competence

Paratransgenesis usually refers to the use of genetically modified symbiotic organisms expressing molecules that can block pathogen development or transmission by vectors. Bacteria symbionts of blood sucking bugs (Durvasula et al. 1997; Durvasula et al. 1999; Durvasula et al. 2008), tsetse flies (Cheng & Aksoy 1999), and mosquitoes (Favia et al. 2007; Jin et al. 2009), and symbiotic viruses of *An. gambiae* (Ren et al. 2008) and *Ae. aegypti* (Carlson et al. 1995; Ward et al. 2001; Carlson et al. 2006), have been used (Figure 2). Recently, genetically modify entomopathogenic fungi strains have induced high levels of P. falciparum mortality in An. gambiae (Fang et al. 2011). Current data indicate that symbionts expressing molecules targeting pathogen development have the potential to reduce transmission in endemic regions, and appear unrelated to any fitness load (Durvasula et al. 1997; Cheng & Aksoy 1999). As with transgenesis, spread of transformed symbionts also would benefit from the availability of a gene drive system to replace non-transformed symbionts present in natural vector populations.

3.1 Transformation of reduviids, tsetse, and mosquitoes with bacterial symbionts

Paratransgenesis in disease vectors was demonstrated through the expression of cecropin A by *Rhodococcus rhodnni* within the midgut of the kissing bug (reduviid) *Rhodnius prolixus* (Durvasula et al. 1997). A 99% reduction in the intensity of *Trypanosoma cruzi* infection in the hindgut of *R. prolixus* was observed without interfering with insect fitness. Additionally, transformed symbionts were shown to be horizontally transmitted to *R. prolixus* carrying non-transformed symbionts via reduviid coprophagic habits (Durvasula et al. 1997). Subsequently, functionally active antibody fragments also were successfully expressed in the guts of *R. prolixus* (Durvasula et al. 1999) and *Triatoma infestans* (Durvasula et al. 2008) utilizing symbionts. Transformed symbionts were stably maintained within the gut of the insects without need for antibiotic selection (Durvasula et al. 1997; Durvasula et al. 1999; Durvasula et al. 2008).

Paratransgenesis seems to be a promising strategy to reduce African trypanosomes transmission by tsetse flies. Genetically transformed *Sodalis,* a symbiont of tsetse flies commonly found in the midgut and hemolymph of *Glossina m. morsitans, Glossina p. palpalis, Glossina austeni,* and *Glossina brevipalpis,* and the salivary glands of *Gl. p. palpalis,* is transmitted vertically via the female milk glands (Cheng & Aksoy 1999; Weiss et al. 2006; Aksoy et al. 2008). In addition, when *Sodalis* originally isolated from of *Gl. m. morsitans* and *Gl. fuscipes* was transformed with GFP, the *recSodalis* obtained colonized septic non-native tsetse host species at a density similar to a native colonization and without reducing host fitness (Weiss et al. 2006).

Symbiotic bacteria also have been isolated from *An. stephensi* (Favia et al. 2007; Riehle et al. 2007). One such symbiont, *Asaia sp.,* was successfully transformed with plasmids expressing GFP (Damiani et al. 2008) or with dsRED gene cassette inserted into bacterium genome (Damiani et al. 2008). *Asaia* was found in mosquito tissues, such as midgut and salivary gland, which are sites for pathogen development, as well as in male and female reproductive tracts, supporting bacteria transovarial and venereal transmission (Riehle et al. 2007; Damiani et al. 2008). Additionally, larval stages can acquire such bacteria strain from the environment (Riehle et al. 2007).

3.2 Transformation of viral symbionts

Symbiotic densovirus also can be genetically manipulated to express molecules to reduce vector competence. Densoviruses are linear single-stranded DNA viruses with the genome packaged in a non-enveloped particle. These viruses are suitable vectors for expression of foreign genes in mosquitoes because they are highly specific, environmentally stable, kill mosquito larvae in a dose-dependent manner, decrease lifespan of surviving adults, and are transmitted vertically (Carlson et al. 1995; Carlson et al. 2006). In *Ae. aegypti*, densoviruses can spread to fat body, muscles, and nerves (Ward et al. 2001) following infection through the anal papillae. Densoviruses infecting *Ae. aegypti* (AeDNV) and *An. gambiae* (AgDNV) were isolated and modified to express GFP (Ward et al. 2001; Ren et al. 2008). The green phenotype obtained by the expression of GFP in recombinant AgDNV-infecting *An. gambiae* was observed in 20% of F2 and F3 generations, suggesting that transformed densoviruses may be used to express molecules targeting pathogen development in mosquitoes (Ren et al. 2008).

3.3 Transformation of entomopathogenic fungi

Several studies demonstrated that entomopathogenic fungi are capable of reducing mosquito life span as well as vector competence (Carlson et al. 2006; Thomas & Read 2007; Cook et al. 2008; Read et al. 2009). Blanford et al. (2005) observed a high mortality rate (55-80%) in mosquitoes 7-14 days following infection with the fungus *Beauveria bassiana*, which was suggestive of the ability of entomopathogenic fungi to drastically reduce pathogen transmission in endemic areas. However, fungus-mediated killing is a slower process compared to chemical approaches, and critics have suggested that the use of entomopathogenis fungi alone is incapable of controlling mosquitoes in malaria-endemic areas.

More recently, genetically modified entomopathogenic fungus *Metarhizium anisopliae* expressing molecules that affect the development of *P. falciparum* in *An. gambiae* were generated (Fang et al. 2011). Specifically, *M. anisopliae* expressing the SM1 peptide ($[SM]_8$), the anti-microbial peptide scorpin, and a single chain antibody targeting a plasmodium surface protein were shown to reduce sporozoite load by 90% without affecting mosquito fitness. Moreover, co-expression of a $[SM]_8$-scorpin fusion protein along with scorpin led to nearly elimination of sporozoite infection in the salivary glands of the infected mosquitoes (Fang et al, 2011). The results using the transgenic *Metarhizium* suggest that this could be a powerful approach to control malaria transmission.

3.4 Future of vector paratransgenesis

Although a low number of bacteria symbionts have been transformed to date, a potential advantage of this approach over transgenesis is lack of fitness load (Durvasula et al. 1997; Weiss et al. 2006). Also, the alternate use of genetically modified symbiotic viruses (instead of bacterial symbionts) may provide additional tools against pathogen development. Densoviruses efficiently expresses heterologous proteins in *An. gambiae* and *Ae. aegypti* and are transmitted vertically (Carlson et al. 2006; Ren et al. 2008). Viral symbionts can be engineered to express single chain antibodies (scFv), blocking pathogen development (de Lara Capurro et al. 2000). Recombinant Sindbis expressing transcripts from an infecting virus genome can reduce viral load (of the infecting virus) in mosquitoes (Powers et al. 1996; Travanty et al. 2004). The results from Sindbis expressing transcripts from LaCrosse (LAC), dengue (DEN), or yellow fever (YF) viruses indicated a substantial interference with the replication of these viruses in *Aedes triseriatus* (LAC) and in *Ae. aegypti* (DEN and YF) (Olson et al. 1996; Powers et al. 1996; Higgs et al. 1998; Adelman et al. 2001). Such viral replication inhibition is accomplished by the mosquito RNAi machinery (Cirimotich et al. 2009).

Despite early successes with transformation of insect vector symbionts, it is not known if transformed symbionts can replace non-transformed in natural insect populations, and potentially affect pathogen development and transmission in natural habitats. Symbionts seem to have no fitness load on insect hosts and are capable of being transmitted vertically (via trans-ovarian transmission) or laterally (due to feeding habits). Thus, a strong gene drive system can potentiate the effectiveness of paratransgenesis. *Wolbachia* endosymbionts have been proposed as such gene drive system (Aksoy et al. 2008).

Wolbachia are intracellular, maternally inherited bacteria that manipulate reproduction of insects via cytoplasmatic incompatibility (CI) (Sinkins & Gould 2006). Due to the effects of CI, a *Wolbachia*-uninfected female will not breed with infected males successfully, reducing the frequency of uninfected individuals and increasing the frequency of *Wolbachia*-infected insects in a population (Sinkins & Gould 2006). Thus, other maternally inherited transformed symbionts would be spread within an insect population in association with *Wolbachia* (Aksoy et al. 2008), increasing the frequency of the transformed symbiont. This mechanism has been observed in *Ae. aegypti*, *Aedes albopictus*, and *Culex quinquefasciatus* (Sinkins & Gould 2006), representing a potential manner to spread transformed symbionts, such as densovirus, in natural populations of mosquitoes.

A life-shortening strain of *Wolbachia* (*w*MelPop) identified in *D. melanogaster* was recently introduced into *Ae. aegypti* (McMeniman et al. 2009) and *An. gambiae* (Jin et al. 2009). Beyond promoting the spread of the transformed symbiont across the mosquito population (i.e., acting as a gene drive mechanism), this strain of *Wolbachia* was also thought to reduce the time frame available (i.e., mosquito life span) for pathogen development within the mosquito (known as the extrinsic incubation period or EIP) (McMeniman et al. 2009). Thus, the application of *w*MelPop to eliminate disease vectors may lead to a reduction of the pathogen developmental time (or EIP) within the vector (Read & Thomas 2009). Counter to this argument is the possibility that, *Wolbachia*, as well as densovirus and entomopathogenic fungi potentially target older mosquitoes over younger ones, and are considered evolution-proof mosquitocidal biocontrols agents (Read et al. 2009). For *Anopheles*, due to loss of fecundity per gonotrophic cycle in natural conditions (from 20% to 40% per gonotrophic cycle, (Killeen et al. 2000)), a selective pressure on pathogen developmental time already exists, especially in the case of *Plasmodium*-infected mosquitoes (Read et al. 2009). Thus, such selective pressure from the addition of *Wolbachia* is likely not strong enough to shorten the parasite life cycle within the vector.

wMelPop infection reduces the feeding ability of old mosquitoes [16] and activates the mosquito immune system (particularly antimicrobial peptides), leading to reduction of filarial worms [17], dengue virus, and plasmodium parasite load (Moreira et al. 2009), including *P. falciparum* (Hughes et al. 2011).

As wMelPop and entomopathogenic fungi are capable of reducing vector competence (linear parameter) and vector survivorship (exponential parameter), these two effects combined may significantly reduce vectorial capacity and human malaria burden in endemic areas.

4. Transmission-blocking vaccines (TBVs)

Transmission-blocking vaccines (or TBVs) aim at interfering and/or blocking pathogen development within the vector, halting transmission to non-infected vertebrate host (depicted in Figure 1C). TBVs usually rely on immunization of vertebrate hosts (either infected or uninfected) with molecules derived from the pathogen or the vector in order to reduce pathogen transmission from infected to uninfected hosts. Such molecules (i.e., antigens) may be inoculated into the vertebrate host as purified proteins inducing the host immune system to produce specific antibodies (Singh & O'Hagan 1999). Alternatively, antibodies can be raised by inoculating the host with recombinant DNA plasmids containing the gene encoding such molecules (Lobo et al. 1999; Coban et al. 2004; Kongkasuriyachai et al. 2004; LeBlanc et al. 2008). The expression and secretion of the specific protein into host tissues induce the immune system to produce antibodies against such proteins (Abdulhaqq & Weiner 2008). To boost the immune response of the vertebrate, antigens are usually inoculated in conjunction with adjuvants. The mechanisms by which adjuvants improve the immune response are still poorly understood (Singh & O'Hagan 1999; Aguilar & Rodriguez 2007). The specific antibodies produced against pathogen and/or vector antigens will interfere with the development of the pathogen within the vector following a blood meal on a vaccinated and infected individual. Other insect-based vaccines, such as sialome-based vaccines (Valenzuela 2004; Oliveira et al. 2009) and insecticidal vaccines (Foy et al. 2003) that are, in some cases, dependent on cell-mediated immune-response in order to prevent vertebrate host infection and reduce insect lifespan, respectively, are not discussed in details this chapter. These can be found in (Willadsen 2004; Billingsley et al. 2006; Titus et al. 2006; Billingsley et al. 2008; Dinglasan & Jacobs-Lorena 2008; Oliveira et al. 2009). Insecticidal vaccines, due to their potential to reduce vectorial capacity exponentially (Billingsley et al. 2006), are briefly discussed.

For a molecule to be an effective TBV candidate it has to induce high antibody titers in order to block completely pathogen development within the insect (Kubler-Kielb et al. 2007). Additionally, the antigen/adjuvant combination has to be safe enough to the vertebrate host so as to avoid side effects after immunization (Saul et al. 2007; Wu et al. 2008). Ideally, a TBV candidate antigen will display low levels of polymorphisms (in field isolates) so that a unique antigen may be used to produce a TBV capable of recognizing all the field variants of that specific antigen (Kocken et al. 1995; Drakeley et al. 1996; Duffy & Kaslow 1997; Sattabongkot et al. 2003). Alternatively, an effective TBV may need to combine different antigens because the combined action of the antibodies against such antigens may produce a more efficient transmission-blocking result (Duffy & Kaslow 1997; Gozar et al. 1998; Kongkasuriyachai et al. 2004).

4.1 Parasite antigen-based TBVs

Most of the studies on TBVs to date were conducted using antibodies targeting antigens expressed on the surface of sexual stage of malaria parasites (Figure 1C; Table 2). *P. falciparum* proteins Pfs25, Pfs28, Pfs48/45, and Pfs230, and their orthologs in *Plasmodium vivax*, have been tested in transmission-blocking assays (Quakyi et al. 1987; Kaslow et al. 1988; Duffy & Kaslow 1997; Hisaeda et al. 2000; Sattabongkot et al. 2003; Malkin et al. 2005; Outchkourov et al. 2008).

4.1.1 *P. falciparum*-derived TBV candidate – Pfs25

Pfs25 is a 25kDa protein expressed on the surface of zygote and ookinete stages of *P. falciparum* and consists of four tandem epidermal growth factor (EGF) domains (Kaslow et al. 1988). The TBV potential of Pfs25 was demonstrated using the Vaccinia virus as delivery systems of this antigen to mammalian hosts (Kaslow et al. 1991), or using recombinant Pfs25 expressed in yeast (Barr et al. 1991; Kaslow et al. 1991).

4.1.2 *P. falciparum*-derived TBV candidate – Pfs28

Pfs28 is a 28kDa *P. falciparum* conserved protein expressed on the surface of retorts, a transitional stage between zygote and ookinete. This antigen also was tested in transmission blocking activity assays. Antibodies produced by the injection of yeast-expressed Pfs28 (yPfs28), in the presence of alum, significantly reduced the infectivity of *An. freeborni* mosquitoes with *P. falciparum*. Lower infectivity was exhibited when vaccination was carried out with yPfs28 and yPfs25 antigens injected together (Duffy & Kaslow 1997).

Transmission blocking activity against *P. falciparum* was further improved when Pfs25 and Pfs28 were expressed as a unique chimeric protein in yeast, the 25-28c recombinant protein. Vaccination with the 25-28c recombinant protein led to complete arrest of oocyst development earlier, using a lower dose and for a greater amount of time, than vaccination with either Pfs25 or Pfs28 alone or a combination of both (Gozar et al. 1998).

4.1.3 *P. falciparum*-derived TBV candidate – Pfs48/45

Another TBV candidate to control spread of *P. falciparum* is Pfs48/45. The *Pfs48/45* gene encodes a unique protein that migrates as a double band under non-reducing conditions (Milek et al. 2000). This protein is expressed on *P. falciparum* gametocyte and gamete surfaces and has a central role in male gamete fertility (van Dijk et al. 2001). Immunization of mice with this recombinant protein led to production of antibody titers that were capable of reducing *P. falciparum* oocyst intensity in *An. stephensi* by at least 88% in 11 out of 12 assays (Outchkourov et al. 2008).

Furthermore, with regards to the application of Pfs48/45 as a potential TBV against malaria, the variability of Pfs48/45 from culture and field isolates from many countries was analyzed (Kocken et al. 1995; Drakeley et al. 1996). The results obtained indicated low levels of polymorphism in the overall gene among either *in vitro* cultures or field isolates (Kocken et al. 1995; Drakeley et al. 1996).

4.1.4 *P. falciparum*-derived TBV candidate – Pfs230

Another *P. falciparum* protein tested in TBV assays was Pfs230, a 230kDa protein expressed on the surface of gametocytes. Although antibodies against Pfs230 blocked the development of *P. falciparum* in the midguts of *An. freeborni*, the transmission blocking activity of anti-Pfs230 monoclonal antibodies was completely lost when complement was inactivated. Thus, the blocking activity of anti-Pfs-230 antibodies was detected only when complement proteins were present (Quakyi et al. 1987).

4.1.5 *P. vivax*-derived TBV candidates – Pvs25 and Pvs28

P. vivax sexual stage surface proteins, orthologs of *P. falciparum* TBV candidates, also have been isolated and tested in transmission blocking experiments. Pvs25, a Pfs25 ortholog, is expressed on the surfaces of the insect-stages, zygotes and mature ookinetes, whereas Pvs28, a Pfs28 ortholog, is mainly expressed on retorts and mature ookinetes (Hisaeda et al. 2000).

Transmission blocking experiments using antibodies against either Pvs25 or Pvs28 were tested (Hisaeda et al. 2000). Four species of mosquitoes were artificially fed on a mixture of *P. vivax*-infected chimpanzee blood in the presence of antibodies (raised in mice co-injected with alum). *P. vivax* ookinete development was completely blocked by the anti-serum against Pvs25 (Hisaeda et al. 2000). Vaccination against Pvs25 and Pvs28 also presented efficient transmission blocking activity against *P. vivax* isolated from human patients, despite polymorphism in these proteins (Sattabongkot et al. 2003).

Transmission blocking activity of Pvs25 has been evaluated in phase 1 human trials. The results from the study revealed significant interference in *P. vivax* development within mosquito midgut caused by the human anti-Pvs25 sera. Additionally, long lasting antibody titers were elevated, and no reactogenicity (side effects) was observed (Malkin et al. 2005). Nevertheless, higher antibody titers are necessary for successful control of *P. falciparum* transmission by mosquitoes in endemic areas (Malkin et al. 2005). Unfortunately, a second phase 1 trial, using Pvs25 as a potential TBV using Montanide ISA 51 as an adjuvant was halted due to induced local and systemic reactions in the vacinees (Wu et al. 2008).

4.2 Other pathogen molecule-based TBV candidates

In regard to proteins expressed on the surface of parasites (other than *Plasmodium*) transmitted to humans by insect vectors (Table 2), only a limited number has been tested as potential TBVs (Tonui et al. 2001a; Saraiva et al. 2006).

In *Leishmania major*, the two most abundant surface antigens, LPG and gp63, were tested as transmission blocking vaccines. *Phlebotomus dubosqci* sand flies were partially fed on mice immunized with purified native LPG, recombinant gp63 (rgp63) expressed in bacteria, crude *L. major* lysate (WPA), or a cocktail of LPG and rgp63. The sand flies were subsequently fed on *L. major*-infected mice. The results indicated that serum against WPA and the two protein-cocktail exhibited greater *L. major* blocking activity than sera against either LPG or gp63 (Tonui et al. 2001a). However, blocking of *L. major* development was due to damage of the midgut epithelial layer, probably caused by immune-active substances present in the blood of the pre-vaccinated mice (Tonui et al. 2001b).

Interestingly, a commercially available treatment for canine visceral leishmaniasis (Leishmune®) was recently shown to function as a TBV in sand flies (Saraiva et al. 2006). Leishmune® (FML-vaccine) is a protective vaccine made of *L. donovani* fucose-manose ligand and the adjuvant saponin, which was successfully tested in a phase III vaccine trial (da Silva et al. 2000). Although the surface molecule (FML) was isolated from *L. donovani*, Leishmune® exhibited transmission blocking activity in the New World sand fly *Lutzomyia longipalpis* when infected with *Leishmania infantum chagasi* (Saraiva et al. 2006). Antibodies produced in dogs following Leishmune® injection reduced *Lu. longipalpis* infectivity by 79.3% and parasite load by 74.3% even after 12 months of immunization (Saraiva et al. 2006).

4.3 Vector-based TBVs

Proteins expressed within insect vector tissues and that may interact with pathogens also have been tested as TBV candidates (Table 3). Vector-based TBV candidates include (structural) proteins that are expressed by the insect midgut (Lal et al. 2001), midgut enzymes that play a role in blood digestion (Lavazec et al. 2007), and parasite receptors expressed by the epithelial cells lining the midgut (Kamhawi et al. 2004; Dinglasan et al. 2007)

In mosquitoes, polyclonal antibodies against *An. gambiae* midgut proteins nearly completely reduced the intensity of *P. falciparum* oocysts (98%) and sporozoites (96%) within *An. stephensi* tissues. Also, *An. gambiae*-derived anti-midgut monoclonal antibodies_inhibited development of *P. falciparum* and *P. vivax* in different *Anopheles* species (Lal et al. 2001). Additionally, these antibodies also can be used to reduce insect vector densities (vector-blocking vaccines) because they reduce vector survivorship and fecundity (Lal et al. 2001). Antibodies against carboxypeptidase cpbAg1 from *An. gambiae* reduced *P. falciparum* infectivity by more than 92% seven days after an infectious artificial blood feeding (Lavazec et al. 2007). In addition to the effect on the number of oocysts per infected mosquito, anti-cpbAg1 strongly reduced mosquito progeny (Lavazec et al. 2007). Antibodies to a midgut aminopeptidase (AgAPN1), which is one of the *P. falciparum* receptors in the *An. gambiae* midgut, were used to reduce *P. falciparum* oocyst intensity in *An. gambiae* and *An. stephensi* by 73% and 67%, respectively (Dinglasan et al. 2007).

Another molecule expressed on the surface of midgut cells that may serve as receptor for parasite attachment has also been assessed as a TBV candidate (Kamhawi et al. 2004). The galectin-like PpGalec characterized from the midgut of the sand fly *P. papatasi* is a receptor for *L. major* lipophosphoglycan (LPG). *P. papatasi* artificially fed on blood mixed sera from PpGalec-immunized mice displayed a reduction of 86% of *L. major* midgut infection. Moreover, no infectious metacyclic forms were detected from the flies fed on anti-PpGalec sera (Kamhawi et al. 2004).

4.4 Future of TBVs

In addition to identifying TBV candidates that are effective and may span different insect vector species, studies on TBV development must include antigenic variability present in field isolates (Kocken et al. 1995; Drakeley et al. 1996; Duffy & Kaslow 1997; Sattabongkot et al. 2003), immunogenicity of such antigens (Kubler-Kielb et al. 2007), reactogenicity caused by adjuvants (Saul et al. 2007; Wu et al. 2008), non-specific responses (Quakyi et al. 1987; Tonui et al. 2001a), and improper folding of antigens (Kaslow et al. 1994; Milek et al. 1998a; Milek et al. 1998b; Milek et al. 2000). Natural antigenic boosting is another important issue that must be dealt with (Arevalo-Herrera et al. 2005).

Antigens expressed on the surface of insect-stage parasites have been postulated as TBV candidates because they seem not to be under the selective pressure mediated by the vertebrate immune system.

Another interesting aspect of TBVs is the possibility of natural boosting of the immune response of animals infected with a pathogen (i.e., pre-immunized) (Milek et al. 1998a; Arevalo-Herrera et al. 2005). Hence, candidate TBV proteins expressed on the surface of both insect-stage and blood-stage pathogens may induce activation of the immune response in infected hosts vaccinated with the same antigens (Arevalo-Herrera et al. 2005). However, this approach may not be suitable to every TBV, such as Pvs25 which displays low expression in blood-stage *P. vivax* (Arevalo-Herrera et al. 2005), and has yet to be demonstrated for the *Plasmodium* TBV-antigen candidates that are expressed during gametocytogenesis, for example, Pfs230 (Quakyi et al. 1987) and Pfs48/45 (Milek et al. 1998a). Proper folding of the TBV candidate protein following expression via recombinant techniques also may affect the efficacy of the vaccinating antigen. Thus, the system of choice for recombinant expression can significantly affect the outcome of the TBV candidate.

In regards to vector-based TBV candidate molecules, the number of TBV antigens available is limited and must to be increased to target other vector species. In addition to assessing a TBV candidate molecule that prevents pathogen development within insect vector tissues, an effect on the vector survivorship is also one of the main objectives.

Reduction of vector survival is thought to interfere exponentially with vectorial capacity (Black IV & Moore 2004; Billingsley et al. 2006; Billingsley et al. 2008), as the time available for pathogen development within the vector is significantly shortened. Despite several studies showing that insect feeding on blood of animals immunized with insect tissue homogenates exhibit reduced survivorship, most of these studies suffered from high experimental variability (Billingsley et al. 2006). However, one study has shown that immunization with a unique insect molecule (mucin) can induce an immune response capable of killing insect vectors via a cell-mediated response (Foy et al. 2003). Thus, an ideal TBV antigen should reduce parasite development, reducing vector competence (a linear parameter in the vectorial capacity equation), as well as vector survivorship (the exponential parameter). These two effects associated can lead to thorough reduction of vectorial capacity and disease burden in endemic areas.

TBV could also be able to reduce survivorship of different species of insect vectors, via immunization with conserved antigens, as proposed by (Canales et al. 2009), providing protection to pathogens transmitted by different vectors. However, significant cross-species effects have yet to be demonstrated.

TBV antigen (antigen origin)	Antigen production	Antigen name	Vaccinated animal (adjuvant)	Sera dilution (concentration)	Targeted pathogen	Insect vector	Mean oocyst number	Infectivity %	Reference
Pfs25 (*P. falciparum*)	Virus (Vaccinia virus)	rPfs25^	Mouse (Ribi)	1:2 (200μg/ml) 1:4 (100μg/ml) 1:8 (50μg/ml) 1:16 (25μg/ml)	*P. falciparum*	*An. freeborni*	0.1 1 1.4 3.5	1% 11% 16% 39%	(Kaslow et al. 1991)
	Yeast (*S. cerevisiae*)	Pfs25-B	Mouse (FCA!) (MTP*) Monkey@ (MTP*)	 1:2 1:2 1:2	*P. falciparum*	*An. freeborni*	 0 0 0-2.6#	 <1% 0% 0-10%	(Barr et al. 1991)
	Yeast (*S. cerevisiae*)	Pfs25-B	Mouse (Alum) Monkey@ (Alum)	Neat Neat	*P. falciparum*	*An. freeborni*	0 0-3.8& 0-12.8$ 19.9-28.2$$	0% - - -	(Kaslow et al. 1994)
	DNA vaccination	VR1020/25	Mouse	1:5 1:10	*P. falciparum*	*An. stephensi*	0.17-0.26 0.19-0.39	3.4-3.8% 2.6-4.3%	(Lobo et al. 1999)
	DNA vaccination/ Yeast (*S. cerevisiae*)	VR1020/25 and Pfs25	Monkey< (Montanide ISA 720)	1:2 1:4 1:8	*P. falciparum*	*An. stephensi* *An. gambiae*?	0.9-3.1 2.4-6.4 3.2-10.4	5-17% 14-38% 19-62%	(Coban et al. 2004)
	Yeast (*P. pastoris*)	Pfs25	Mouse (cholera toxin)	1:2 1:8 1:32	*P. falciparum*	*An. dirus*	0 0 10	0% 0% 21.3%	(Arakawa et al. 2005)
	Yeast (*P. pastoris*)	Pfs25	Human (Montanide ISA 51)	-	*P. falciparum*	*An. stephensi*	-	<10%	(Wu et al. 2008)
	DNA vaccination (*in vivo* electroporation)	Pfs25	Mouse	1:2> 1:4> 1:8> 1:16>	*P. falciparum*	*An. gambiae*	1.0 3.4 9.5 51.9	2.5% 9% 24% 100%	(LeBlanc et al. 2008)

TBV antigen (antigen origin)	Antigen production	Antigen name	Vaccinated animal (adjuvant)	Sera dilution (concentration)	Targeted pathogen	Insect vector	Mean oocyst number	Infectivity %	Reference
Pfs28 (*P. falciparum*)	Yeast (*S. cerevisiae*)	yPfs28	-		*P. falciparum*	*An. freeborni*			(Duffy & Kaslow 1997)
			(FCA+Ribi)	Neat!!			0-0.33&	0-8%	
			(Alum)				0.21&	3%	
		yPfs25+yPfs28+	-						
			(FCA+Ribi)	1:40			0.047-0.16	7.6-9%	
Pfs25-Pfs28 (*P. falciparum*)	Yeast (*S. cerevisiae*)	25-28C	Mouse (Alum)	Neat!!	*P. falciparum*	*An. freeborni*	0	0%	(Gozar et al. 1998)
Pfs45/48 (*P. falciparum*)	Bacteria (refolded *in vitro*)	Pfs45/48-10C	Mouse (FCA!)	1:2 (10μg/ml)	*P. falciparum*	*An. stephensi*	0.45	0.06%	(Outchkourov et al. 2007)
	Bacteria (w/chaperonins)	Pfs45/48-10C	Mouse (FCA!)	1:2 (10μg/ml)	*P. falciparum*	*An. stephensi*	0-5.1	0-12%	(Outchkourov et al. 2008)
Pfs230 (*P. falciparum*)	Purification	Pfs230^	Rabbit (FCA!)	- (100μg/ml)	*P. falciparum*	*An. freeborni*	0.2-4	0.3-5.8%	(Quakyi et al. 1987)
Pvs25 (*P. vivax*)	Yeast (*S. cerevisiae*)	Pvs25	Mouse (Alum)	1:2	*P. vivax*	*An. stephensi*	0	0%	(Hisaeda et al. 2000)
	Yeast (-)	Pvs25	Mouse	1:2	*P. vivax*	*An. dirus*	0.18	-	(Sattabongkot et al. 2003)
			(Alum)	1:8			1.26	-	
			Rabbit	1:2			4.25	-	
			(Alum)	1:8			4.06	-	
	DNA vaccination	DV25	Mouse		*P. vivax*				(Kongkasuriyachai et al. 2004)
				1:8		*An. freeborni*	3.17	86%	
				1:10			0.4	87%	
		DV28		1:10		*An. freeborni*	0.4	87%	
				1:10		*An. gambiae*	1.5	86%	
		DV25+DV28		1:8		*An. freeborni*	0.8	93%	
				1:10			0.04	99%	
				1:10		*An. gambiae*	0.8	93%	
	Yeast (*S. cerevisiae*)	Pvs25	Monkey[1] (Montanide ISA 720)	1:4	*P. vivax*	*An. albimanus*	0.0-0.04	-	(Arevalo-Herrera et al. 2005)
	Yeast (*S. cerevisiae*)	Pvs25H	Human	Neat!!	*P. vivax*	*An dirus*	-	-	(Malkin et al. 2005)
			(Alhydrogel®)	1:1			-	-	

TBV antigen (antigen origin)	Antigen production	Antigen name	Vaccinated animal	Sera dilution (concentration)	Targeted pathogen	Insect vector	Mean oocyst number	Infectivity %	Reference
Pvs25 (*P. vivax*)	Yeast (*S. cerevisiae*)	Pvs25	Human (Montanide ISA 51)	-	-	-	-	-	(Wu et al. 2008)
	Yeast (*S. cerevisiae*)	Pvs25	Monkey< (Montanide ISA 720)		*P. vivax*	*An. freeborni*			(Collins et al. 2006)
				1:2			0++	0	
				1:8			1.41	1.6	
				1:32			1.63	32.8	
			(Alum)	1:2			1.71	2.9	
				1:8			1.3	10.6	
				1:32			2.05	34.2	
Pvs28 (*P. vivax*)	Yeast (*S. cerevisiae*)	Pvs28	Mouse (Alum)	1:2	*P. vivax*	*An. freeborni*	0.91	0.7%	(Hisaeda et al. 2000)
	Yeast (*S. cerevisiae*)	Pvs28	Mouse		*P. vivax*	*An. dirus*			(Sattabongkot et al. 2003)
			(Alum)	1:2			0.11	-	
				1:8			1.31	-	
			Rabbit	1:2			10.73	-	
			(Alum)	1:8			4.79	-	
WPA (*L. major*)	Purified	Whole cell lysate	Mouse	-	*L. major*	*P. duboscqi*	-	25%	(Tonui et al. 2001a)
rgp63 (*L. major*)	Bacteria	gp63	Mouse	-	*L. major*	*P. duboscqi*	-	40%	(Tonui et al. 2001a)
LPG (*L. major*)	Purified	LPG	Mouse	-	*L. major*	*P. duboscqi*	-	43.3%	(Tonui et al. 2001a)
LPG+rgp63 (*L. major*)	Purified + Bacteria	LPG+rgp63	Mouse	-	*L. major*	*P. duboscqi*	-	37.5%	(Tonui et al. 2001a)
Leishmune® (*L. donovani*)	Purification	FML	Dog (Saponin)	1:1	*L. major*	*L. longipalpis*	-	30.6%	(Saraiva et al. 2006)

(^) Monoclonal antibodies were used in transmission blocking assays; (!) Freund's complete adjuvant; (*) Muramyl tripeptide; (@) *Aotus trivirgatus;* (#) Oocysts present in midguts of mosquitoes fed on sera from monkeys immunized 22 weeks before challenge. No oocysts were present in mosquitoes that fed on sera from animals immunized 12 weeks before challenge; (-) Undetermined; (&) Seven days after 3rd immunization; ($) Sixty one days after 3rd immunization; ($$) Eighty nine days after 3rd immunization; (<) *Macaca mulatta;* (?) Similar results using *An. stephensi;* (>)Immunization with 20µg of plasmid; (+) Sera diluted 1:40; (++) TBV assayed 204 days after immunization; (1) *Aotus lemurinus griseimembr;* (") Sera were not previously diluted prior to mixing with equal or greater amount of blood for insect feeding.

Table 2. **Transmission blocking vaccines based on pathogens molecules** (Modified from (Coutinho-Abreu & Ramalho-Ortigao 2010)).

Milek, R. L., W. F. Roeffen, et al. (1998b). "Immunological properties of recombinant proteins of the transmission blocking vaccine candidate, Pfs48/45, of the human malaria parasite Plasmodium falciparum produced in Escherichia coli." *Parasite Immunol* 20(8): 377-85.

Milek, R. L., H. G. Stunnenberg, et al. (2000). "Assembly and expression of a synthetic gene encoding the antigen Pfs48/45 of the human malaria parasite Plasmodium falciparum in yeast." *Vaccine* 18(14): 1402-11.

Moreira, L. A., M. J. Edwards, et al. (2000). "Robust gut-specific gene expression in transgenic Aedes aegypti mosquitoes." *Proc Natl Acad Sci U S A* 97(20): 10895-8.

Moreira, L. A., J. Ito, et al. (2002). "Bee venom phospholipase inhibits malaria parasite development in transgenic mosquitoes." *J Biol Chem* 277(43): 40839-43.

Moreira, L. A., I. Iturbe-Ormaetxe, et al. (2009). "A Wolbachia symbiont in Aedes aegypti limits infection with dengue, Chikungunya, and Plasmodium." *Cell* 139(7): 1268-78.

Nirmala, X., O. Marinotti, et al. (2006). "Functional characterization of the promoter of the vitellogenin gene, AsVg1, of the malaria vector, Anopheles stephensi." *Insect Biochem Mol Biol* 36(9): 694-700.

Nolan, T., T. M. Bower, et al. (2002). "piggyBac-mediated germline transformation of the malaria mosquito Anopheles stephensi using the red fluorescent protein dsRED as a selectable marker." *J Biol Chem* 277(11): 8759-62.

O'Brochta, D. A., N. Sethuraman, et al. (2003). "Gene vector and transposable element behavior in mosquitoes." *J Exp Biol* 206(Pt 21): 3823-34.

Oliveira, F., R. C. Jochim, et al. (2009). "Sand flies, Leishmania, and transcriptome-borne solutions." *Parasitol Int* 58(1): 1-5.

Olson, K. E., S. Higgs, et al. (1996). "Genetically engineered resistance to dengue-2 virus transmission in mosquitoes." *Science* 272(5263): 884-6.

Outchkourov, N., A. Vermunt, et al. (2007). "Epitope analysis of the malaria surface antigen pfs48/45 identifies a subdomain that elicits transmission blocking antibodies." *J Biol Chem* 282(23): 17148-56.

Outchkourov, N. S., W. Roeffen, et al. (2008). "Correctly folded Pfs48/45 protein of Plasmodium falciparum elicits malaria transmission-blocking immunity in mice." *Proc Natl Acad Sci U S A* 105(11): 4301-5.

Palatnik-de-Sousa, C. B., F. Barbosa Ade, et al. (2008). "FML vaccine against canine visceral leishmaniasis: from second-generation to synthetic vaccine." *Expert Rev Vaccines* 7(6): 833-51.

Perera, O. P., I. R. Harrell, et al. (2002). "Germ-line transformation of the South American malaria vector, Anopheles albimanus, with a piggyBac/EGFP transposon vector is routine and highly efficient." *Insect Mol Biol* 11(4): 291-7.

Pinkerton, A. C., K. Michel, et al. (2000). "Green fluorescent protein as a genetic marker in transgenic Aedes aegypti." *Insect Mol Biol* 9(1): 1-10.

Powers, A. M., K. I. Kamrud, et al. (1996). "Molecularly engineered resistance to California serogroup virus replication in mosquito cells and mosquitoes." *Proc Natl Acad Sci U S A* 93(9): 4187-91.

Quakyi, I. A., R. Carter, et al. (1987). "The 230-kDa gamete surface protein of Plasmodium falciparum is also a target for transmission-blocking antibodies." *J Immunol* 139(12): 4213-7.

Read, A. F., P. A. Lynch, et al. (2009). "How to make evolution-proof insecticides for malaria control." *PLoS Biol* 7(4): e1000058.

Read, A. F. and M. B. Thomas (2009). "Microbiology. Mosquitoes cut short." *Science* 323(5910): 51-2.

Ren, X., E. Hoiczyk, et al. (2008). "Viral paratransgenesis in the malaria vector Anopheles gambiae." *PLoS Pathog* 4(8): e1000135.

Ribeiro, J. M. and M. G. Kidwell (1994). "Transposable elements as population drive mechanisms: specification of critical parameter values." *J Med Entomol* 31(1): 10-6.

Riehle, M. A., C. K. Moreira, et al. (2007). "Using bacteria to express and display anti-Plasmodium molecules in the mosquito midgut." *Int J Parasitol* 37(6): 595-603.

Robinson, A. S., G. Franz, et al. (1999). "Genetic sexing strains in the medfly, *Ceratitis capitata:* Development, mass rearing and field application." *Trends in Entomology* 2: 81-104.

Rodrigues, F. G., S. B. Oliveira, et al. (2006). "Germline transformation of Aedes fluviatilis (Diptera:Culicidae) with the piggyBac transposable element." *Mem Inst Oswaldo Cruz* 101(7): 755-7.

Rodrigues, F. G., M. N. Santos, et al. (2008). "Expression of a mutated phospholipase A2 in transgenic Aedes fluviatilis mosquitoes impacts Plasmodium gallinaceum development." *Insect Mol Biol* 17(2): 175-83.

Saraiva, E. M., A. de Figueiredo Barbosa, et al. (2006). "The FML-vaccine (Leishmune) against canine visceral leishmaniasis: a transmission blocking vaccine." *Vaccine* 24(13): 2423-31.

Sattabongkot, J., T. Tsuboi, et al. (2003). "Blocking of transmission to mosquitoes by antibody to Plasmodium vivax malaria vaccine candidates Pvs25 and Pvs28 despite antigenic polymorphism in field isolates." *Am J Trop Med Hyg* 69(5): 536-41.

Saul, A., M. Hensmann, et al. (2007). "Immunogenicity in rhesus of the Plasmodium vivax mosquito stage antigen Pvs25H with Alhydrogel and Montanide ISA 720." *Parasite Immunol* 29(10): 525-33.

Scali, C., T. Nolan, et al. (2007). "Post-integration behavior of a Minos transposon in the malaria mosquito Anopheles stephensi." *Mol Genet Genomics* 278(5): 575-84.

Shin, S. W., V. Kokoza, et al. (2003a). "Relish-mediated immune deficiency in the transgenic mosquito Aedes aegypti." *Proc Natl Acad Sci U S A* 100(5): 2616-21.

Shin, S. W., V. A. Kokoza, et al. (2003b). "Transgenesis and reverse genetics of mosquito innate immunity." *J Exp Biol* 206(Pt 21): 3835-43.

Singh, M. and D. O'Hagan (1999). "Advances in vaccine adjuvants." *Nat Biotechnol* 17(11): 1075-81.

Sinkins, S. P. and F. Gould (2006). "Gene drive systems for insect disease vectors." *Nat Rev Genet* 7(6): 427-35.

Thomas, D. D., C. A. Donnelly, et al. (2000). "Insect population control using a dominant, repressible, lethal genetic system." *Science* 287(5462): 2474-6.

Thomas, M. B. and A. F. Read (2007). "Can fungal biopesticides control malaria?" *Nat Rev Microbiol* 5(5): 377-83.

Titus, R. G., J. V. Bishop, et al. (2006). "The immunomodulatory factors of arthropod saliva and the potential for these factors to serve as vaccine targets to prevent pathogen transmission." *Parasite Immunol* 28(4): 131-41.

Tonui, W. K., P. A. Mbati, et al. (2001a). "Transmission blocking vaccine studies in leishmaniasis: I. Lipophosphoglycan is a promising transmission blocking vaccine molecule against cutaneous leishmaniasis." *East Afr Med J* 78(2): 84-9.

Tonui, W. K., P. A. Mbati, et al. (2001b). "Transmission blocking vaccine studies in leishmaniasis: II. Effect of immunisation using Leishmania major derived 63 kilodalton glycoprotein, lipophosphoglycan and whole parasite antigens on the course of L. major infection in BALB/c mice." *East Afr Med J* 78(2): 90-2.

Travanty, E. A., Z. N. Adelman, et al. (2004). "Using RNA interference to develop dengue virus resistance in genetically modified Aedes aegypti." *Insect Biochem Mol Biol* 34(7): 607-13.

Valenzuela, J. G. (2004). Blood-feeding arthropod salivary glands and saliva. *Biology of Disease Vectors*. W. C. Marquardt, Black IV, W. C., Freier, J. E., Hagedorn, H. H., Hemingway, J., Higgs, S., James, A. A., Kondratieff, B., Moore, C. G. Burlington, Elsevier Academic Press: 785.

van Dijk, M. R., C. J. Janse, et al. (2001). "A central role for P48/45 in malaria parasite male gamete fertility." *Cell* 104(1): 153-64.

Vreysen, M. J., K. M. Saleh, et al. (2000). "Glossina austeni (Diptera: Glossinidae) eradicated on the island of Unguja, Zanzibar, using the sterile insect technique." *J Econ Entomol* 93(1): 123-35.

Ward, T. W., M. S. Jenkins, et al. (2001). "Aedes aegypti transducing densovirus pathogenesis and expression in Aedes aegypti and Anopheles gambiae larvae." *Insect Mol Biol* 10(5): 397-405.

Weiss, B. L., R. Mouchotte, et al. (2006). "Interspecific transfer of bacterial endosymbionts between tsetse fly species: infection establishment and effect on host fitness." *Appl Environ Microbiol* 72(11): 7013-21.

Willadsen, P. (2004). "Anti-tick vaccines." *Parasitology* 129 Suppl: S367-87.

Wilson, R., J. Orsetti, et al. (2003). "Post-integration behavior of a Mos1 mariner gene vector in Aedes aegypti." *Insect Biochem Mol Biol* 33(9): 853-63.

Wimmer, E. A. (2003). "Innovations: applications of insect transgenesis." *Nat Rev Genet* 4(3): 225-32.

Wu, Y., R. D. Ellis, et al. (2008). "Phase 1 trial of malaria transmission blocking vaccine candidates Pfs25 and Pvs25 formulated with montanide ISA 51." *PLoS ONE* 3(7): e2636.

Yoshida, S., Y. Shimada, et al. (2007). "Hemolytic C-type lectin CEL-III from sea cucumber expressed in transgenic mosquitoes impairs malaria parasite development." *PLoS Pathog* 3(12): e192.

Yoshida, S. and H. Watanabe (2006). "Robust salivary gland-specific transgene expression in Anopheles stephensi mosquito." *Insect Mol Biol* 15(4): 403-10.

26

Essential Oils of Umbelliferae (Apiaceae) Family Taxa as Emerging Potent Agents for Mosquito Control

Epameinondas Evergetis[1], Antonios Michaelakis[2] and Serkos A. Haroutounian[1,*]
[1]*Chemistry Laboratory, Agricultural University of Athens, Athens*
[2]*Department of Entomology, Benaki Phytopathological Institute, Athens*
Greece

1. Introduction

Warm-humid areas around the globe constitute the cradle of humanity, providing their inhabitants the most favorable environments for living and agricultural production. In this "Garden of Eden", which spreads within the globe's temperate and tropical zones, is also thriving an annoying but dangerous daemon, the mosquito. This little devil constitutes the main vector of malaria and human encephalitis, both infectious diseases that account as major threats of public health (Becker et al., 2003). Recently, these threats have been spread to a broader geographical area, as a consequence of their vectors (*Aedes* sp., *Anopheles* sp. and *Culex* sp.) introduction into metropolitan areas of northern hemisphere, such as Chicago (Tedesco et al., 2010), New York (Peterson et al., 2006) and Paris (Delaunay et al., 2009). Since mosquito breeding habitats in both urban and rural areas are man-made (Imbahale et al., 2010), there are several restrictions limiting the efforts towards the development of an integrated vector management system. Todate, the history of evolutions of malaria vector interventions is directly connected with the mosquito control tools development, concerning either environmental modifications/manipulations or their chemical and/or biological control (Kilama, 2009).

In respect the chemical control, a significant milestone was the DichloroDiphenyl-Trichloroethane (DDT) synthesis by Zeidler in 1874. The DDT success was followed by the fast introduction of numerous chlorinated hydrocarbons, which were used in massive amounts for the control of mosquito-borne diseases (Ray, 2010). Despite their efficiency, the use of organochlorines had severe environmental impacts which were publicly (and dramatically) addressed by Carson (1962) in Silent Spring, initiating the development of insecticide resistant mosquito populations. These undesirable characteristics, in combination with concerns on public health risks, derived from the organochlorine residues detected in humans and animals, led to their ban in early 70's. Thus, they were replaced by less persistent chemicals, such as organophosphates, pyrethroids and avermectin derivates,

*Corresponding Author

substitutes that also display the major disadvantage of resistance development (Alves et al., 2010; Daaboub et al., 2008; Lima et al., 2011).

Recent research trend on mosquito chemical control mainly focuses on currently used compounds, aiming to enhance their potency and circumvent the problems connected with their application. In this respect, the so far developed pyrazole derivatives are quite efficient exhibiting however adverse environmental effects (Stevens et al., 2011), while the corresponding pyrroles display the desirable efficiency (Raghavendra et al., 2011) but adequate research on their environmental side effects is still underway. Amides, such as methazolamide and acetazolamide, were also evaluated as potent mosquito larvicides but were found to display significant bioaccumulation properties (Del Pilar Corena et al., 2006) which discourage their broad use. Finally, various novel pyrimido-quinolione molecules have been developed and assessed as highly toxic for other organisms (Rajanarendar et al., 2010). Ray (2010) recognized that the insecticide treated nets, in connection with the long lasting insecticidal nets, have resurrected the chemical control of malaria's mosquito vector. This may be rationalized considering that their targeted application resolves the problems connected with the environmental impacts of chemical control agents since limit their expansion, availability and environmental penetration.

Despite the numerous efforts and progress achieved, the efficacy of insecticidal nets in malaria prevention still constitutes a hot issue, since depends strongly upon a plethora of additional factors (Killeen & Smith, 2007). In particular, despite efforts (Pennetier et al., 2010) to overcome the recognized for longtime resistance development issues of insecticidal nets, todate these problems have not been resolved (Yadouleton et al., 2011). An additional drawback derives by the combined impact of herbicide application that promotes the cross-resistance to mosquito populations (Boyer et al., 2006; Riaz et al., 2009).

The corresponding biological control has dictated the development of novel-alternative mosquito control tools, including the sterile males technique (Patersson et al., 1968), the genetically modified mosquitoes (Gu et al., 2011; Lavialle-Defaix et al., 2011), the entomopathogenic fungi (Van Breukelen et al., 2011; Kanzok & Jacobs-Lorena, 2006) and bacteria. Among the tools developed the bacterial pathogens application is considered as the most prominent intervention, displaying species selective insecticidal ability (Hayes et al., 2011) which is considered as an efficient means for mosquito control without harmful impacts for the environment (Caquet et al., 2011). Major thresholds limiting the wider application of this technique are related with the induced pathogen introduction among the natural mosquito populations (Hancock et al., 2011) and the threats connected with bioaccumulation and resistance development (Tilquin et al., 2008). In general, the biological control tools are still under development, presenting todate a low degree of maturity for large-scale interventions.

Temephos was considered as one of the most potent-safe insecticides. Its recent exclusion from Annex I of the Directive 98/8/EC resulted in the discontinuation of its application in mosquito control programs by the European, emerging the development and use of new-safer insecticides. Thus, relative research directed towards the discovery-development of novel molecules, capable to control the mosquito populations without exhibiting the disadvantages of synthetic pesticides. In this respect, the plant originated natural compounds constitute a large deposit of such molecules, inherently allowing the retrieval of various commercially successful molecules like pyrethrins. Todate, the search for novel, potent and safer pesticides from this deposit has already provided several candidates, either as pure compounds and/or their extracts. Specifically, various organic acids such as lactic and orthophosporic acids (Chakraborty et al., 2010), alkaloids (Talontsi et al., 2011) and plant proteins (Chowdhury et al., 2008) have been identified as efficient mosquito control agents. Furthermore, several plants were used as the maternal material to produce bio-products which were applied against mosquitoes with hopeful results (Shaalan et al., 2005; Sukumar et al., 1991). On the other hand, the plant derived Essential Oils (EOs) constitute a special category of natural products that exhibit the major advantage -for the mosquito control endeavor- of exhibiting an insect oriented mode of action with low penetrability to the ecosystems that does not affect larger animals. In addition, the natural diversity of their constituents addresses effectively the problem of resistance development (Isman, 2000).

2. Literature review

2.1 Umbelliferae (Apiaceae) family: A source of potent natural agrochemicals

Many EOs originated from diverse plant families have been considered and studied as potential sources of natural agrochemicals. In this respect, previous research results on Umbeliferae (Apiaceae) family plant materials revealed the significant acaricidal activities of butylidenepthalides isolated from *Angelica acutiloba* Kitagawa var. *sugiyame* Hikino (Kwon & Ahn, 2002) and the similar activity of the EO of *Foeniculum vulgare,* attributed to the presence of *p*-anisaldehyde and (+,−)-fenchone in the EO (Lee, 2004). These EOs were practically inactive in fumigant toxicity tests against *Lycoriella mali* though they are known to contain the active monoterpenes *α*-pinene and *β*-pinene (Choi et al., 2006), which are common constituents of many Umbelliferae EOs. Methanolic extracts of *Angelica dahurica, Cnidium officinale,* and *Foeniculum vulgare* were also tested against the Coleoptera *Lasioderma sericorne, Sitophilus oryzae* and *Callosobruchus chinensis* exhibiting a moderate activity only the second extract (Kim et al., 2003a; Kim et al., 2003b). Other EOs of this family screened as inactive against coleoptera were originated from the species *Anethum graveolens* L., *Apium graveolens* Houtt., *Coriandrum sativum* L., *Cuminum cyminum* L. and *Petroselinum sativum* L. (Regnault-Roger & Hamraoui, 1994; Papachristos & Stamopoulos, 2002). On the contrary, the EOs of *Pimpinella anisum* L. and *Cuminum cyminum* L. displayed excellent ovicidal and insecticidal activities against the *Tribolium confusum* du Val and the *Ephestia kuehniella* Zeller (Tunc et al., 2000). In addition, the aqueous extract of *Pimpinella anisum* exhibited good repellent effect against the adults of sweet potato whitefly *Bemisia tabaci* (Ateyyat et al., 2007).

These rather controversial results are not connected with the impressive activities that Umbelliferae EOs were found to exhibit against the Diptera, with the EO of *Ammi visnaga* displaying -among 19 EOs- the most potent ovicidal activity against *Mayetiola destructor* (Lamiri et al., 2001). In addition, tests against *Droshophila melanogaster* of furanocoumarins and pthalides isolated from *Angelica acutiloba* Kitagawa var. *sugiyame* Hikino revealed the hypothesis that the insecticidal properties of the plant extracts are connected with the acetylcholinesterase inhibition (Miyazawa et al., 2004). Finally, alkylpthalides originated from *Cnidium officinale* Makino were tested as extremely effective against *Droshophila melanogaster* (Tsukamoto et al., 2005).

2.2 Umbelliferae (Apiaceae) family: A strong focal point for mosquito control

Table 1 summarizes the test results against various mosquito species reported for all extracts and EOs derived from plants belonging to the Umbeliferae family. Same table also contains the test results of fourteen EOs, which appear herein for the first time. Results indicate that the organic phase of the *Cryoptaenia canadensis* extract is the most active against fourth instars of *Culex pipiens*, leading to the isolation -from the extract- of the acetylated very toxic (LC_{50} values lower than 10 mgl^{-1}) molecules of falcarinol and falcarindiol (Eckenbach et al., 1999). The larvicidal properties of hexane soluble fraction of *Apium graveolens* seeds -a plant with pleasant aroma- and three isolated compounds (sedanolide, senkyunolide-N, senkyunolide-J) against *Aedes aegypti* mosquitoes highlighted sedanolide as very active (100% mortality at 50 mgl^{-1}, Momin & Nair, 2001). As a consequence, a gel containing 5% of the *Apium graveolens* hexane extract was developed, providing full protection to volunteers from mosquito bites for two hours (Tuetun et al., 2009), while the ethanolic formulations from the same plant also provided protection against *Aedes aegypti*. Another formulation containing the aforementioned hexane extract and 5% vanillin showed strong repellent activities against different mosquito species (Tuetun et al., 2005, see also **Table 1** for details). The crude seed extract had no adverse effects on human volunteers skins when tested for several anti-mosquito properties (Choochote et al., 2004). This plant's EO exhibits potent larvicidal activity against two laboratory-reared mosquito species, the malaria vector *Anopheles dirus* and the vector of dengue *Aedes aegypti* (Pitasawat et al., 2007).

Species	Part used	Mosquito species	Bioactivity	Reference
Ammi visnaga	Seeds	*Culex quinquefasciatus*	Larvicidal	Pavela, 2008
Anethum graveolens	(not mentioned)	*Anopheles stephensi, Aedes aegypti, Culex quinquefasciatus*	Larvicidal	Amer and Mehlhorn, 2006
Anethum graveolens	Leaves, twigs	*Aedes aegypti*	Larvicidal, effects on growth and development	Promsiri et al., 2006
Angelica archangelica	Fruits	*Culex quinquefasciatus*	Larvicidal	Pavela, 2009
Angelica sylvestris	Aerial parts	*Culex pipiens*	Larvicidal	(present study)
Apium graveolens	Seeds	*Aedes aegypti*	Larvicidal	Momin & Nair, 2001
Apium graveolens	Seeds	*Aedes aegypti*	Larvicidal, adulticidal, repellent	Choochote, 2004
Apium graveolens	Seeds	*Aedes aegypti, Aedes gardnerii, Aedes lineatopennis, Anopheles barbirostris, Armigeres subalbatus, Culex tritaeniorhynchus, Culex gelicus, Culex vishnui* group, *Mansonia uniformis*	Repellent	Tuetun et al., 2005

Species	Part used	Mosquito species	Bioactivity	Reference
Apium graveolens	Seeds	*Aedes aegypti, Anopheles ditrus*	Larvicidal	Pitasawat et al., 2007
Apium graveolens	Seeds	*Aedes, Anopheles, Armigeres, Culex, Mansonia*	Repellent	Tuetun et al., 2009
Athamanta densa	Aerial parts	*Culex pipiens*	Larvicidal	(present study)
Bupleurum fruticosum	Aerial parts	*Culex pipiens*	Larvicidal	Evergetis et al., 2009
Carum carvi	Fruits	*Aedes aegypti, Culex quinquefasciatus*	Larvicidal	Lee, 2006
Carum carvi	Seeds	*Aedes aegypti, Anopheles ditrus*	Larvicidal	Pitasawat et al., 2007
Carum ptroselinum	(not mentioned)	*Culex pipiens*	Larvicidal	Khater and Shalaby, 2008
Chaerophyllum heldreichii	Aerial parts	*Culex pipiens*	Larvicidal	(present study)
Conium divaricatum	Aerial parts	*Culex pipiens*	Larvicidal	(present study)
Conopodium capillifolium	Aerial parts	*Culex pipiens*	Larvicidal	Evergetis et al., 2009
Coriander sativum	Seeds	*Ochlerotatus caspius*	Larvicidal	Knio et al., 2008
Cryptotaenia canadensis	Fresh foliage, root, fruits	*Culex pipiens*	Larvicidal	Eckenbach et al., 1999
Daucus carota	Roots	*Aedes aegypti, Culex quinquefasciatus*	Larvicidal	Lee, 2006
Eleoselinum asclepium	Aerial parts	*Culex pipiens*	Larvicidal	Evergetis et al., 2009
Ferula assa-foetida	Stems	*Culex quinquefasciatus*	Larvicidal	Pavela, 2009
Ferula galbaniflua	(not mentioned)	*Anopheles stephensi, Aedes aegypti, Culex quinquefasciatus*	Larvicidal	Amer and Mehlhorn, 2006
Ferula lancerottensis	Stems	*Culex quinquefasciatus*	Larvicidal	Pavela, 2008
Ferulago nodosa	Aerial parts	*Culex pipiens*	Larvicidal	(present study)
Foeniculum vulgare	Fruits	*Aedes aegypti*	Repellent	Kim et al., 2002
Foeniculum vulgare	(not mentioned)	*Aedes aegypti*	Larvicidal	Orozco & Lentz, 2005
Foeniculum vulgare	Flowers	*Culex pipiens*	Larvicidal, repellent	Trabousli et al., 2005
Foeniculum vulgare	Fruits	*Aedes aegypti, Anopheles ditrus*	Larvicidal	Pitasawat et al., 2007
Foeniculum vulgare	Stems, inflorencences, leaves	*Culex pipiens*	Larvicidal	Manolakou et al., 2009

Species	Part used	Mosquito species	Bioactivity	Reference
Foeniculum vulgare	Leaves	*Aedes albopictus*	Larvicidal	Conti et al., 2010
Heracleum sphondylium	Aerial parts	*Culex pipiens*	Larvicidal	Evergetis et al., 2009
Imperatoria ostruthium	Roots	*Culex quinquefasciatus*	Larvicidal	Pavela, 2009
Laserpitium pseudomeum	Aerial parts	*Culex pipiens*	Larvicidal	(present study)
Oenanthe pimpinelloides	Aerial parts	*Culex pipiens*	Larvicidal	Evergetis et al., 2009
Petroselimum crispum	Seeds	*Ochlerotatus caspius*	Larvicidal	Knio et al., 2008
Peucedanum neumayeri	Aerial parts	*Culex pipiens*	Larvicidal	(present study)
Peucedanum officinale	Aerial parts	*Culex pipiens*	Larvicidal	(present study)
Pimpinella anisum	Seeds	*Anopheles stephensi, Aedes aegypti, Culex quinquefasciatus*	Larvicidal, adulticidal, ovicidal, oviposition-deterrent, repellent	Prajapati et al., 2005
Pimpinella anisum	Seeds	*Ochlerotatus caspius*	Larvicidal	Knio et al., 2008
Pimpinella peregrina	Aerial parts	*Culex pipiens*	Larvicidal	(present study)
Pimpinella rigidula	Aerial parts	*Culex pipiens*	Larvicidal	(present study)
Pimpinella tragium ssp tragium	Aerial parts	*Culex pipiens*	Larvicidal	(present study)
Scaligeria cretica	Aerial parts	*Culex pipiens*	Larvicidal	(present study)
Seseli montanum	Aerial parts	*Culex pipiens*	Larvicidal	Evergetis et al.,2009
Seseli pallasii	Stems	*Culex quinquefasciatus*	Larvicidal	Pavela, 2009
Seseli parnassicum	Aerial parts	*Culex pipiens*	Larvicidal	(present study)
Seseli tortuosum	Stems	*Culex quinquefasciatus*	Larvicidal	Pavela, 2008
Smyrnium rotundifolium	Aerial parts	*Culex pipiens*	Larvicidal	(present study)
Thamnosciadium junceum	Aerial parts	*Culex pipiens*	Larvicidal	(present study)
Trachyspermum ammi	Seeds	*Anopheles stephensi*	Larvicidal, oviposition-deterrent, vapor toxicity, repellent	Pandey et al., 2009

Table 1. Reported phytochemicals derived from plants belonging to Apiaceae family against various mosquito species.

Another EO found to possess potent larvicidal, oviposition-deterrent, vapor toxicity and repellent activities against *Aedes aegypti* was isolated from ajowan (*Tachyspermum ammi*, Pandey et al. 2009). *Anethum graveolens* extract exhibited larval toxicity with LC_{50} values from 27 to 20 mgl^{-1} (for 24 and 48 hours exposures respectively), while on growth survival and prolongation tests of the various instar larvae of *Aedes aegypti*, the second instar larvae was determined as the more susceptible. The lowest concentration of crude extracts of *Anethum graveolens* used (caused more than 50% larval mortality) was not toxic to guppy fish (*Poecilia reticulata*) at concentrations of 12.5 mgl^{-1} (Promsiri et al., 2006).

Among all EOs tested for mosquito control, the most potent was derived from *Foeniculum vulgare*, which caused the highest mortality against *Aedes albopictus* (Conti et al., 2010) and moderate against *Anopheles dirus* and *Aedes aegypti* (Pitasawat et al., 2007). Main component of this EO is methyl chavicol (more than 43%), while its methanolic extract (*trans*-anethole chemotype) was moderately active against *Aedes aegypti*, the yellow fever mosquito (Orozco & Lentz, 2005). The hexane fraction from its fruit-derived parts showed 99% repellency against *Aedes aegypti*, while the other fractions (chloroform, ethyl acetate and water: 37, 37 and 17% respectively) were practically inactive (Kim et al., 2002). Repellency and toxicity were also studied against *Culex pipiens* (Traboulsi et al., 2005), indicating that the EO of *Foeniculum vulgare* was the most effective, while the repellency assays revealed protection time for almost one hour when applied at concentration of 3%.

Pimpinella anisum L. EO proved to possess equally potent larvicidal and ovicidal activities against *Anopheles stephensi*, *Aedes aegypti*, *Culex quinquefasciatus* and only larvicidal against *Ochlerotatus caspius* (Prajapati et al., 2005; Knio et al., 2008). Similar larvicidal activity results were also observed when the EOs of *Coriander sativum* and *Petroselinum crispum* were tested against *Ochlerotatus caspius* (Knio et al., 2008). The larvicidal tests of EOs of genus *Carum* were performed for *Carum carvi* against *Anopheles dirus* and *Aedes aegypti* and for *Carum petroselinum* against *Culex pipiens* (Lee, 2006; Pitasawat et al., 2007; Khater & Shalaby, 2008). The results were directly similar to those of the EO of *Daucus carota* (against *Anopheles dirus* and *Aedes aegypti*) proving their inability to cause 100% mortality at the lowest concentration (Lee, 2006).

Among the methanolic extracts of 118 Euroasiatic plants, tested for their larvicidal effects against *Culex quinquefasciatus*, the species *Ammi visnaga* and *Seseli pallasii* were determined as two of the most toxic materials tested, with LC_{50} values lower than 10 mgl^{-1} (Pavela, 2008, 2009). On the other hand, the extracts of *Angelica archangelica* and *Imperatoria ostruthium* exhibited LC_{50} values lower than 70 mgl^{-1}, while *Seseli tortuosum* and *Ferula lancerottensis* displayed moderate larvicidal activity (LC_{50} values around 430 mgl^{-1}). The only inactive Apiaceae plant tested was *Ferula assa-foetida* (LC_{50} value higher of 1000 mgl^{-1}), with the EO of *Ferula galbaniflua* exhibiting the weakest activity against *Culex quinquefasciatus* and *Anopheles stephensi* (mortality level less than 14% of dead larvae after 48 hours, Amer & Mehlhorn, 2006a). The same authors also reported that *Anopheles stephensi* was the most resistant to dill (*Anethum graveolens*), while the *Culex quinquefasciatus* the more sensitive. Dill was also evaluated for persistency to larvicidal effects under different conditions for 1 month after the preparation of its solutions. In all cases (open, closed, in light or in dark) the EO was active only when was used immediately after preparation (Amer & Mehlhorn, 2006b).

Finally, an interesting result was obtained during the study of several EOs using coupled gas chromatography-electroantennographic detection (GC-EAD), on the hypothesis that compounds can be detected by the antennae of the yellow fever mosquito, *Aedes aegypti*. Thus, cumin aldehyde and cumin alcohol the *Cuminum cynimum* EO components were identified as such molecules. It must be noted that for both components, their EO (cumin oil) was also EAD-active (Campbell et al., 2011)

2.3 Greek Umbelliferae (Apiaceae) plants extract activities against Culex pipiens mosquitoes

The larvicidal activity of the EO obtained from the stem of Greek *Foeniculum vulgare* was determined against *Culex pipiens* larvae, while methyl chavicol was determined as its main component (more than 32%). Although the LC_{50} value of methyl chavicol was more than 80 mgl^{-1}, the respective EO was determined as 2.1-fold more toxic (Manolakou et al., 2009). *Culex pipiens* larvae were also used to test the mosquito control properties of EO from various naturally growing plants throughout Greece, belonging to the following six different Apiaceae family taxa: *Heracleum sphondylium, Seseli montanum, Conopodium capillifolium, Bupleurum fruticosum, Oenanthe pimpinelloides, Eleoselinum asclepium.* All EOs tested displayed good larvicidal activities with LC_{50} values ranging from 40.26-96.96 mgl^{-1} (Evergetis et al., 2009)

As a continuation of our ongoing efforts to exploit the use of natural products for the development of environmentally friendly means for the mosquito population control, our interest was stimulated on the investigation of Umbeliferae (Apiaceae) plants EOs. In this context, we report herein the chemical composition and larvicidal activity results for 14 EOs originated from different taxon obtained during Greek Umbelliferae biodiversity studies (**Table 1**).

3. Materials and methods

3.1 Plant material

Fourteen different taxa of the Umbeliferae (Apiaceae) family, Apioideae subfamily belonging to seven tribes and twelve different genera have been collected during the present study. Representatives of the Apieae Tribe are *Pimpinella peregrina* L., and 5 Greek endemics, namely *Athamanta densa* Boiss. & Orph., *Pimpinella tragium* ssp *tragium* Vill., *Pimpinella rigidula* (Boiss. & Orph.) H. Wolf, *Seseli parnassicum* Boiss. & Heldr. and *Thamnosciadium junceum* (Sibth. & Sm.) Hartvig.; of Smyrnieae tribe *Scaligeria cretica* (Miller) Boiss. and *Smyrnium rotundifolium* Miller; of Angeliceae tribe *Angelica sylvestris* L.; of Scandiceae tribe the Greek endemic *Chaerophyllum heldreichii* Orph. Ex Boiss.; of Peucedaneae tribe *Ferulago nodosa* (L.) Boiss., *Peucedanum neumayeri* (Vis.) Reichenb, *Peucedanum officinale* L., and of Laserpitieae tribe the Greek endemic *Laserpitium pseudomeum* Orph., Heldr. & Sart. Ex Boiss (Pimenov & Leonov, 1993; Tutin et al., 1968).

Full collection details are provided in **Table 2.** A voucher specimen of each plant is deposited in the herbarium of the Agricultural University of Athens, Athens, Greece.

3.2 Essential oils isolation

The freshly collected plant materials (steams, leaves and flowers) were washed thoroughly, chopped off finely and subjected to steam distillation in a Clevenger-type apparatus, using the Microwave Accelerated Reaction System (MARS 5) at 1400 W for 40 min with 3 L of H_2O in order to obtain their EOs. The resulting oils were dried over anhydrous sodium sulphate and stored at 4 °C. The EO yield of each plant is included in **Table 3**.

Species	**Abbreviation**	**Vegetative Stage**	**Date**	**Location**
Angelica sylvestris L.	AS	Flowering	05.09.2004	Mt. Parnon, Peloponnisos, forest streams
Athamanta densa Boiss. & Orph. *	AD	Flowering	15.06.2005	Mt. Parnassos, Sterea Hellas, vertical cliffs
Chaerophyllum heldreichii Orph. Ex Boiss. *	CH	Flowering	25.07.2004	Mt. Parnon, Peloponnisos, forest clearings
Ferulago nodosa (L.) Boiss.	FN	Flowering	02.05.2005	Antikyra, Sterea Hellas, olive groves
Laserpitium pseudomeum Orph., Heldr. & Sart. Ex Boiss.*	LP	Flowering	15.07.2004	Mt. Oiti, Sterea Hellas, rocky slopes
Peucedanum neumayeri (Vis.) Reichenb	PN	Flowering	28.08.2004	Mt. Smolikas, Hepiros, forest clearings
Peucedanum officinale L.	PO	Flowering	15.07.2004	Mt. Oiti, Sterea Hellas, rocky slopes
Pimpinella tragium ssp tragium Vill. *	PT	Flowering	15.07.2004	Mt. Oiti, Sterea Hellas, rocky slopes
Pimpinella peregrina L.	PP	Flowering	14.05.2005	Iraklio, Is. Crete, olive groves
Pimpinella rigidula (Boiss. & Orph.) H. Wolf *	PR	Flowering	17.08.2004	Molai, Peloponnisos, roadside
Scaligeria cretica (Miller) Boiss.	SC	Flowering	22.05.2005	Vouliagmeni, Sterea Hellas, seaside
Seseli parnassicum Boiss. & Heldr. *	SP	Flowering	15.07.2004	Mt. Oiti, Sterea Hellas, forest clearings
Smyrnium rotundifolium Miller	SR	Flowering	02.05.2005	Distomo, Sterea Hellas, roadside
Thamnosciadium junceum (Sibth. & Sm.) Hartvig *	TJ	Flowering	25.07.2004	Mt. Parnnassos, Sterea Hellas, alpic ravine

*=Greek Endemic.

Table 2. Collection data.

Species	Part distilled	Weight of aerial parts (g)	Volume of oil (mL)
Angelica sylvestris L.	Aerial	920	0,5
Athamanta densa Boiss. & Orph. *	Aerial	450	0,5
Chaerophyllum heldreichii Orph. Ex Boiss. *	Aerial	530	0,7
Ferulago nodosa (L.) Boiss.	Aerial	400	0,7
Laserpitium pseudomeum Orph., Heldr. & Sart. Ex Boiss. *	Aerial	270	0,9
Peucedanum neumayeri (Vis.) Reichenb	Aerial	600	0,5
Peucedanum officinale L.	Aerial	*180*	*0,8*
Pimpinella tragium ssp tragium Vill. *	Aerial	650	1,5
Pimpinella peregrina L.	Aerial	500	0,4
Pimpinella rigidula (Boiss. & Orph.) H. Wolf *	Aerial	235	0,7
Scaligeria cretica (Miller) Boiss.	Aerial	200	0,5
Seseli parnassicum Boiss. & Heldr. *	Aerial	*200*	*0,5*
Smyrnium rotundifolium Miller	Aerial	530	0,9
Thamnosciadium junceum (Sibth. & Sm.) Hartvig *	Aerial	600	2,0

*=Greek Endemic.

Table 3. Essential oils yields.

3.3 Gas Chromatography-Mass Spectrometry (GC-MS) analyses

Gas Chromatography (GC). All GC analyses were carried out on a Agilent Technologies 7890A gas chromatograph, fitted with a HP 5MS 30m x 0.25mm x 0.25μm film thickness capillary column and FID. The column temperature was programmed from 60 to 280 °C at a initial rate of 3 °C/min. The injector and detector temperatures were programmed at 230 and 300 °C, respectively. Helium was used as the carrier gas at a flow rate 1 ml/min.

Gas Chromatography-Mass Spectrometry (GC-MS). The GCMS analyses were performed on the same instrument using the Agilent 5957C, VL MS Detector with Triple-Axis Detector system operating in EI mode (equipped with a HP 5MS 30m x 0.25mm x 0.25μm film thickness capillary column), using He (1 ml/min) as the carrier gas. The initial temperature of the column was 60 °C. The column was heated gradually to 280 °C with a 3 °C/min rate. The identification of the compounds was based on comparison of their retention indices (RI) (Van den Dool & Kratz, 1963), obtained using various n-alkanes (C9-C24). Also, their EI-mass spectra were compared with the NIST/NBS and Wiley library spectra and the literature (Adams, 1995; Massada, 1976). Additionally, the identity of the indicated phytochemicals was confirmed by comparison with available authentic samples.

3.4 Mosquito rearing

A colony of the species *Culex pipiens* biotype *molestus* is maintained for more than 25 years in the laboratory of Entomology of the Benaki Phytopathological Institute, Kifissia, Greece. Adult mosquitoes are kept in wooden framed cages (33x33x33 cm) with a 32x32 mesh at 25±2 °C, 80±2% relative humidity and photoperiod of 14:10 (L:D) h. Cotton wicks saturated with 10% sucrose solution are used as food source. Females lay eggs in round, plastic containers (10 cm diameter x 5 cm depth) filled with 150 ml of tap water. Egg rafts are removed daily and placed in cylindrical enamel pans (with diameter of 35 cm and 10 cm deep), in order to hatch. Larvae are reared under the same conditions of temperature and light and are fed daily with baby fish food (TetraMin, Baby Fish Food) at a concentration of 0.25 gl^{-1} of water until pupation. Pupae are then collected and introduced into the adult rearing cages.

3.5 Larvicidal bioassays

Stock solutions of EOs tested were prepared in ethanol and maintained in a freezer as 1% mgl^{-1}solutions. They were dissolved in double distilled water to produce solutions of the tested materials in concentrations ranging from 5 to 150 mgl^{-1}. Prior to biological determinations the toxicity of each EO was evaluated (data not shown).

The larval mortality bioassays were carried out according to the test method for larval susceptibility, proposed by the World Health Organization (WHO, 1981). Twenty 3rd to 4th instar larvae of the species *Culex pipiens* biotype *molestus* were collected from the colony, placed in a glass beaker with 250 ml of aqueous suspension of the tested material at various concentrations and an emulsifier was added in the final test solution (less than 0.05%). Four replicates were made per each concentration and a control treatment with tap water and emulsifier was also included. Beakers with larvae were placed at 25±2 °C, 80±2% relative humidity and photoperiod of 14:10 h (L:D).

3.6 Data analysis

Larvicidal effect was recorded 48 h after treatment. Data obtained from each dose-larvicidal bioassay (total mortality, mgl^{-1} concentration in water) were subjected to probit analysis in which probit-transformed mortality was regressed against log_{10}-transformed dose; LC_{50}, LC_{90} values, and slopes were calculated (SPSS 11.0).

4. Results and discussion

4.1 Phytochemical analysis

Fourteen distinct Umbeliferae taxa (twelve genera) are studied herein, one of which is endemic to Greece (*Thamnosciadium* Hartvig). It must be noted that there are no literature reports and studies on the EOs and their chemical compostitions for the material obtained from the plants *Athamanta densa* Boiss. & Orph. (AD), *Chaerophyllum heldreichii* Orph. Ex Boiss. (CH), *Laserpitium pseudomeum* Orph., Heldr. & Sart. Ex Boiss. (LP), *Peucedanum neumayeri* (Vis.) Reichenb (PN), *Pimpinella tragium* ssp tragium Vill. (PT), *Pimpinella rigidula* (Boiss. & Orph.) H. Wolf (PR), *Scaligeria cretica* (Miller) Boiss. (SC), *Seseli parnassicum* Boiss. & Heldr. (SP) and *Smyrnium rotundifolium* Miller (SR). In addition, the discussion section on the related *taxa* EOs compositions includes ten (out of twelve) genera studied herein, since there are also no previous reports on the composition of EOs obtained from *Conium* L. and *Thamnosciadium* Hartvig genera.

In total seventy phytochemicals, representing 76.64 to 99.83 % of the respective EOs samples have been identified as their constituents using combined GC and GC/MS analyses and in certain occasions verified by NMR studies. The detailed qualitative and quantitative analytical data of the main constituents of steam volatiles (and their respective retention indices) are presented in **Table 4**.

Components	RI	PN	AS	TJ	SP	PO	CH	LP	PT	SC	PP	PR	FN	SR	AD	Identification
trans-2-hexanal	803												0.96		0.27	a, b
α-pinene	939	21.27	24.65	2.80		2.14	1.67	49.58	1.21	8.76			30.85	0.93	0.46	a, b, c
camphene	954	2.99	3.32			1.72							4.36			a, b, c
sabinene	975	2.76				1.15	71.76	24.73	4.30	13.74			1.96	0.93		a, b, c
β-pinene	979	2.66	1.33					8.51					1.79		8.86	a, b, c
myrcene	991	3.93	4.75	1.52		0.81	1.51	1.82					6.68	11.25	0.92	a, b, c
α-phellandrene	1003	2.53	2.35	3.83									1.33			a, b, c
α-terpinene	1017		3.58										0.32			a, b, c
p-cymene	1025	4.71						0.74								a, b, c
o-cymene	1026			0.91												a, b, c
limonene	1029	4.71		40.75		2.78			1.42	1.43					0.66	a, b, c
β-phellandrene	1030	12.76	42.96		1.42		10.86	6.73					10.20			a, b, c
cis-ocimene	1037	4.78		18.59									2.76		1.66	a, b, c
trans-ocimene	1050			0.82									1.01		5.16	a, b, c
γ-terpinene	1060	32.25					2.54	1.42		1.40			0.41			a, b, c
cis-sabinene hydrate	1070							2.53								a, b, c
terpinolene	1089			12.97												a, b, c
linalool	1097														0.50	a, b, c
trans-limonene oxide	1137			0.46												a, b
geijerene	1143								10.23							a, b
α-terpineol	1189						3.35	2.43		0.78						a, b, c
pregeijerene	1287								5.13							a, b
1-bornyl acetate	1289		3.84			81.13							0.52			a, b
2,3,4-trimethyl benzaldehyde	1359			2.16		4.68										a, b
2,3,6-trimethyl benzaldehyde	1371			0.62												a, b
isoledene	1376										0.69					a, b
α-copaene	1377									0.67			0.42		0.26	a, b
β-cubebene	1388													0.86		a, b
β-elemene	1391				10.85					0.50			0.51	2.09	0.33	a, b
aristolene	1407										19.92					a, b
calarene	1411										3.40				0.40	a, b
β-caryophyllene	1419		1.73		2.76				0.92	3.07			2.10			a, b, c
α-bergamontene	1435										62.15					a, b, c
γ-elemene	1437				3.22									0.72		a, b, c
β-humulene	1439				3.30					0.40						a, b, c
β-farnesene	1457									29.27		2.47			1.17	a, b, c
$C_{14}H_{30}O$ (m/z: 189, 147, 105, 91, 204)	1483														19.80	b
α-amorphene	1484											0.32				a, b
$C_{14}H_{28}O$ (m/z: 119, 91, 105, 145, 131)	1485								20.39							b
germacrene D	1487	2.55	4.42	0.89	13.02		2.53	0.71	0.87	28.37	0.85		6.42		1.21	a, b, c
β-selinene	1490				5.14						3.78	23.80			3.95	a, b, c
α-selinene	1498					0.95						3.53		5.28		a, b, c
α-zingiberene	1499											7.75				a, b
bicyclogermacrene	1500				6.25					3.51	1.21		4.04			a, b
α-farnesene	1506		1.84													a, b, c
β-bisabolene	1506				2.85						1.18	4.16			12.72	a, b, c
myristicin	1519											6.72			4.32	a, b, c
β-sesquiphellandrene	1523				30.39						1.04	1.80		0.92		a, b, c
δ-cadinene	1524	1.30	3.02							0.38						a, b, c
germacrene B	1561				10.64				19.28					2.13	0.98	a, b, c
spathulenol	1578				1.52					0.46						a, b, c
caryophyllene oxide	1583				1.02					0.61						a, b, c
β-elemenone	1601								1.72							a, b
isofuranogermacrene	1648													1.28		a, b
furanodiene	1649													11.81		a, b
α-bisabolol	1686		2.04													a, b
germacrone	1694								23.33					5.62		a, b
trans-isomyristicin	1721			10.14								7.74				a, b
trans-pseudoiso-eugenyl 2-methylbutyrate	1774											7.74				a, b
trans-epoxypseudoisoeugenyl 2-methylbutyrate	1783											26.72				a, b
furanoeremophil-1-one	1880													6.42	0.91	a, b
1β-acetoxyfuranoeudesm-4(15)-ene	1889													8.87		a, b
1β-acetoxyfuranoeudesm-3 ene	1911													20.72		a, b
$C_{12}H_{25}O_2N$ (m/z: 91, 55, 115, 129, 77)	1923														2.94	b
$C_{12}H_{25}O_2N$ (m/z: 91, 115, 55, 129, 77)	1943														8.58	b
$C_{13}H_{27}O_2N$ (m/z: 91, 115, 55, 129, 159)	2030														12.17	b
n-heneicosane	2100														0.59	a, b, c
tricosane	2300														0.32	a, b, c
pentacosane	2500														0.25	a, b, c
Total		99.20	99.83	96.46	92.38	95.36	94.22	99.20	88.80	93.35	94.22	92.75	76.64	79.83	89.39	

[a]Comparison of mass spectra with MS libraries and retention times
[b]Comparison of experimental RI with reported RI
[c]Comparison with authentic compounds
RI: Retention indices calculated against C_8 to C_{24} n-alkanes on the HP 5MS column.

Table 4. Chemical constituents of the essential oils tested.

The determined chemical composition of the EO from the aerial part of *Angelica sylvestris* L. (AS) is consistent with the literature reports for EOs obtained from its seeds (Bernard, 2001) and roots (Bernard & Clair, 1997), with α-pinene and β-phellandrene being the major components. Same compounds were reported as the prevailing phytochemicals in the EOs of *A. archangelica* L. *sensu lato* (Bernard, 2001; Nykanen et al., 1991; Bernard & Clair, 1997; Chalcat & Garry, 1997; Nivinskiene et al., 2005), while the EO of *A. glauca* is reported to contain β-phellandrene as major component and only small portions of α-pinene (Aghinotri et al., 2004; Kaul et al., 1996). Other *Angelica* L. taxa, such as *A. sinensis* (Dung et al., 1996; Kim et al., 2006), *A. gigas* (Kim et al., 2006), *A. acutiloba* (Kim et al., 2006), *A. heterocarpa* (Bernard, 2001; Bernard & Clair, 1997) and *A. tenuissima* (Ka et al., 2005) display a completely different, both qualitative and quantitative, EO composition profile.

In addition to *a*-pinene, which is the main constituent as previously reported by Demetzos et al. (2000), the studied EO of *Ferulago nodosa* (L) Boiss. (FN) was found to contain thirteen new components for the *taxon's* EO. More specifically, the molecules of *trans*-2-hexenal, myrcene, α-phellandrene, α-terpinene, β-phellandrene, *cis*-ocimene, *trans*-ocimene, γ-terpinene, bornyl acetate, β-elemene, β-caryophylene, germacrene D and bicyclogermacrene were also determined as constituents of this EO. With the exception of *trans*-2-hexenal all the abovementioned compounds have been assayed in the EOs of the following *Ferulago* W.D.J. Koch taxa; *F. asparagifolia* (Baser et al., 2001), *F. phialocarpa* (Masoudi et al., 2004b), *F. macrocolea* (Rustaiyan et al., 2005), *F. galbaniflua* (Rustaiyan et al., 2002a), and *F. thirkeana* (Baser et al., 2002).

The EO of *Peucedanum officinale* L. (PO) is dominated by bornyl acetate, which was previously found only in *P. scoparium* (Masoudi et al., 2004a). It is also characterized by the presence of 2,3,4-trimethyl benzaldehyde, which has not been previously reported as constitutent of *Peucedanum* L. EOs. In addition, the EO tested was found to contain five molecules, namely *a*-pinene, sabinene, myrcene, limonene and β-selinene, never reported in a EO of *P. officinale* (Jaimand et al., 2006). These five phytochemichals are abundant in the general profile of *Peucedanum* L. EOs, as reported for *P. scoparium* (Masoudi et al., 2004a), *P. zenkeri* (Menut et al., 1995), *P.vertcillare* (Fraternale et al., 2000), *P. petiolare* (Rustaiyan et al., 2001) and *P. cervariifolium* (Bazgir et al., 2005).

The EOs of *Pimpinella* L. have been thoroughly studied, mainly because the application of their several taxa as culinary herbs and/or spices. Though the EOs of fourteen (14) *taxa* were studied, only one (Tabanca et al., 2005) refers to PP (*Pimpinella peregrina* L.) and none to PT (*Pimpinella tragium* ssp tragium VIII) and PR (*Pimpinella rigidulla* Boiss. & Orph. H. Wolf). The main constituent of EO of PO is *a*-bergamontenene, reported so far only for *P. anagodendron* (Velasco-Negueruela et al. 2005) and *P. anisum* (Santos et al., 1998). Two additional components determined herein, β-bisabolene and β-sesquiphellandrene, have not been reported in previous studies for PP but are well documented for *P. anagodendron* (Velasco-Negueruela et al. 2005), *P. junoniae* (Velasco-Negueruela et al. 2003), *P. anisum* (Santos et al., 1998), *P. anisetum* (Baser et al., 1999; Tepe et al., 2006) and *P. tragioides* (Askari & Sefidcon, 2007). New entries, for this genera EO components list, are isoledene, aristolene, calarene and β-selinene which were also assayed in the EO of PP. On the contrary, the EO of PR is characterized by the complete absence of monoterpenes, advocating previous record of β-selinene and introducing *a*-amorphene, *a*-selinene and *trans*-isomyristicin as components of the *Pimpinella* L. EOs. Finally, the EO of PT has only two differences as compared to the genus EO components, an unidentified component and *β*-elemenone. In general, its composition is in accordance with the phytochemical profiles reported for the EOs of *P. aromatica* (Baser et al., 1996), *P. serbica* (Ivanic et al., 1983), *P. flabellifolia* (Tepe et al., 2006), *P. aurea* (Tabanca et al., 2005; Assadian et al., 2005), *P. acuminata* (Melkani et al., 2006), *P. barbata* (Fakhari & Sonboli, 2006), *P. rupicola* (Velasco-Negueruela et al. 2005), *P. corymbosa* and *P. puberula* (Tabanca et al., 2005).

Major components of the EO of *Scaligeia cretica* (Miller) Boiss (SC) are *α*-pinene, *β*-farnesene and germacrene D, which have also been detected in previous studies on the EOs of *Scaligeria* DC. In this respect, the EO of *S. lazica* contains *β*-farnesene as major and *α*-pinene, germacrene D as minor components (Baser et al., 1993). On the contrary, the EO of *S. tripartite* contains *β*-farnesene and germacrene D as minor compounds, while *α*-pinene is absent (Tabanca et al., 2007). Compounds assayed herein and never reported before in the EOs of *Scaligeria* DC are *α*-terpineol, *β*-elemene and *β*-humulene.

The EOs of *Laserpitium pseudomeum* Orph. Heldr. & Sart Ex Poiss. (LP) contains *α*-pinene, *β*-pinene, sabinene and *β*-phellandrene as major components, all well known constitutents of the EOs of *Laserpitium* L. Previous literature reports indicated that the EOs of *L. latifolium* contains *α*-pinene and *β*-pinene as major components (Borg-Karlson et al., 1994), the *L. petrophilum* *α*-pinene and sabinene (Baser et al., 1997), while the molecule of *β*-phellandrene is present in traces in both EOs. On the contrary, the phytochemical profile of the EO of *L. siler* is completely different containing mainly limonene and perillaldehyde (Chizzola et al., 1999)

The EO composition of *Smyrnium* L. has also been scarcely investigated, since only three *taxa's* EOs, namely *S. perfoliatum* (Molleken et al., 1998a; Tirillini et al., 1996; Tirillini & Tosi, 1992), *S. cordifolium* (Amiri et al., 2006) and *S. olusatrum* (Molleken et al., 1998b), have been studied todate. The studied of EO of *Smyrnium rotundifolium* Miller (SR) contains 7 major components, with the molecule of *α*-selinene reported for the first time as EO component of *Smyrnium* L.. Other compounds present in large quantitites are furanodiene (reported as major constitutent in *S. olusatrum*), myrcene, furanoeremophil-1-one, 1*β*-acetoxyfuranoeudesm-4(15)-ene, 1*β*-acetoxyfurano eudesm-3-ene (detected in *S. olusatrum* and *S. perfoliatum*, Molleken et al., 1998) and germacrone (present in *S. cordifolium*).

The phytochemical profile of *Chaerophyllum* L. EOs was studied previously for *C. macropodum* (Baser et al., 2006), *C. crinitum* (Baser et al., 2006; Nematollahi et al., 2005), *C. macrospermum* (Sefidcon & Abdoli, 2005; Rustaiyan et al., 2002b, Mamedova, 1994), *C. bulbosum sensu lato* (Mamedova & Akhmedova, 1991; Kokkalou & Stefanou, 1989), *C. aksekiense* (Baser et al., 2000b), *C. coloratum* (Vajs et al., 1995), *C. azoricum* (Pedro et al., 1999) and *C. prescotii* (Letchamo et al., 2005). The more significant differentiation among the literature results and the assayed herein EO of *Chaerophyllum heldreichi* Orph. Ex Boiss (CH) comprises the identification for first time of *α*-terpineol as main component of EO of *Chaerophyllum* L..

The EO of *Seseli parnassicum* Boiss. & Heldr. (SP) was found to contain three new compound entries, *β*-humulene, *β*-selinene and *β*-sesquiphellandrene, as compared with the EO of the Seseli L. *taxa* (also including the synonymous *Lomatopodium* Fisch. et C.A. Mey *taxa*). The remaining components are in accordance with the EO content of same *taxa* plants, such as *S. montanum* (Evergetis et al., 2009), *S. campestre* and *S. peucedanoides* (Baser et al. 2000a;

Bulatovic et al. 2006) and in *S. buchtormence*. These compounds were also present in the EOs of *S. resinosum* and *S. tortuosum*, obtained from the fruits and not the herbal part of the plants (Dogan et al. 2006). The *L. khorassanicum* and *L. staurophyllum* EOs were assayed to contain mostly aliphatic terpenes, while the corresponding cyclic terpenes were present in smaller amounts compared to EOs of Seseli L. (Sedghat et al. 2003; Sefidkon et al. 1997).

Finally, the investigated EO of *Athamanta densa* Boiss. & Orph., contains as major constituents myristicin and various unidentified alkaloids, which account for almost 24 % of its weight. The literature reports of EOs of *Athamanta* L. indicate that they mainly contain either myristicin, such as the EOs of *A. sicula* (Camarda & Di Stefano, 2003), *A. turbith sensu lato* (Tomic et al., 2009), *A. macedonica* (Verykokidou et al., 1995) and *A. haynaldi* (Zivanovic et al., 1994), or apiole as in *A. sicula* (Camarda & Di Stefano, 2008).

4.2 Larvicidal assays

The investigated EOs were evaluated —for the first time— in respect to their larvicidal activities against 3rd- 4th instar larvae of *Culex pipiens*. The relative results expressed as the respective LC_{50} and LC_{90} values are included in **Table 5**. Among the EOs tested only two were rather inactive (AS and PP, displaying LC_{50} values above 150 mgl^{-1}), while the EOs of SC and SP were moderately active displaying LC_{50} values above 100 mgl^{-1} (111.99 and 122.54 mgl^{-1} respectively).

Essential Oils tested	**LC_{50} (95% CL)[a]**	**LC_{90} (95% CL)[a]**	**Slope (±SE)**
Athamanta densa	10.15 (9.49-10.73)	15.75 (14.52-17.76)	6.72±0.80
Pimpinella tragium ssp tragium	40.13 (32.43-45.95)	71.10 (61.51-91.00)	5.15±0.52[b]
Pimpinella rigidula	40.31 (34.75-43.64)	60.41 (55.66-70.57)	7.29±1.44[b]
Thamnosciadium junceum	44.17 (41.52-46.62)	64.42 (59.94-71.28)	7.82±0.86[b]
Peucedanum neumayeri	47.40 (40.25-54.15)	81.47 (68.63-113.57)	5.44±0.53[b]
Chaerophyllum heldreichii	53.61 (50.29-56.55)	75.96 (71.53-82.15)	8.46±0.87
Laserpitium pseudomeum	56.73 (53.50-59.60)	79.59 (75.18-85.71)	8.46±0.86
Ferulago nodosa	67.39 (64.17-70.41)	95.59 (89.90-103.94)	8.43±0.84
Smyrnium rotundifolium	80.32 (76.88-84.16)	105.30 (98.33-116.61)	10.89±1.29
Peucedanum officinale	86.46 (82.27-90.30)	125.05 (117.23-136.95)	7.99±0.84
Scaligeria cretica	111.99 (107.86-115.47)	133.83 (128.35-143.21)	6.58±0.73[b]
Seseli parnassicum	122.54 (115.54-141.06)	167.15 (143.83-268.76)	6.30±0.68
Angelica sylvestris	>150		
Pimpinella peregrina	>150		

[a] LC values are expressed in mgl^{-1} and they are considered significantly different when 95% CL fail to overlap.
[b] Since goodness-of-fit test is significant ($P<0.05$), a heterogeneity factor is used in the calculation of confidence limits (CL)

Table 5. LC_{50} and LC_{90} values for the tested essential oils against larvae of *Culex pipiens* biotype *molestus*.

The EO derived from the endemic in Greece plant *Athamanta densa* was determined as the most active since displayed the highest toxicity against mosquito larvae, with LC_{50} value 10.15 mgl^{-1}. The EO tested contains a series of compounds which were not found in the other EOs tested, such as bisabolene and the unidentified compounds $C_{14}H_{30}O$, $C_{12}H_{25}O_2N$ and $C_{13}H_{27}O_2N$, which have to study more thoroughly in order to determine their activities. The remaining EOs (PR, TJ, PT, PN, CH, LP, FN, SR and PO) displayed LC_{50} values ranging from 40.31 to 86.46 mgl^{-1}. No significant relationship between toxicity and phytochemical content was detected.

5. References

Adams, R. (1995). Identification of essential oil components by Gas Chromatography / Mass Spectroscopy. Carol Stream, Allured Publishing, Illinois, USA. ISBN 0931710421.

Aghinotri, V.K., Thappa R.K., Meena, B., Kapaphi, B.K. Saxena, R.K., Qazi, G.N. & Agarwal S.G. (2004). Essential oil composition of aerial parts of *Angelica glauca* growing wild in North-West Himalaya (India). *Phytochemistry*, 65, 2411-2413.

Alves, S.N., Serro, J.E. & Melo, A.L. (2010). Alterations in the fat body and midgut of *Culex quinquefasciatus* larvae following exposure to different insecticides. *Micron*, 41, 592-597.

Amer, A. & Mehlhorn, H. (2006a) Larvicidal effects of various essential oils against *Aedes*, *Anopheles*, and *Culex* larvae (Diptera: Culicidae). *Parasitology Research*, 99, 466-472.

Amer, A. & Mehlhorn, H. (2006b) Persistency of larvicidal effects of plant oil extracts under different storage conditions. *Parasitology Research*, 99, 473-477.

Amiri, H., Khavari-Nejad R.A., Masoud, S., Chalabian, F. & Rustaiyan, A. (2006). Composition and antimicrobial activity of the essential oil from stems, leaves, fruits and roots of *Smyrnium cordifolium* Boiss. from Iran. *Journal of Essential Oil Research*, 18, 574-577.

Askari, F. & Sefidkon, F. (2007). Essential oil composition of *Pimpinella tragioides* (Boiss.) Benth et Hook from Iran. *Journal of Essential Oil Research*, 19, 54-56.

Assadian, F., Masoudi, S., Nematollahi, F., Rustaiyan, A., Larijani, K. & Mazloomifar, H. (2005). Volatile constituents of *Xanthogalum purpurascens* Ave-Lall., *Eryngium caeruleum* M.B. and *Pimpinella aurea* DC. three Umbelliferae herbs growing in Iran. *Journal of Essential Oil Research*, 17, 243-245.

Ateyyat, M.A., Al-Mazra'awi, M., Abu-Rjai, T. & Shatnawi, M.A. (2009). Aqueous extracts of some medicinal plants are as toxic as Imidacloprid to the sweet potato whitefly, *Bemisia tabaci*. *Journal of Insect Science*, 9:15, 6pp.

Baser, K.H.C. Demirci, B., Demirci, F., Hashimoto, T., Asakawa, Y. & Noma, Y. (2002). Ferulagone: a new monoterpene ester from *Ferulago thirkeana* essential oil. *Planta Medica*, 68, 564-567.

Baser, K.H.C., Demirci, B. & Duman, H. (2001). Composition of the essential oil of *Ferulago asparagifolia* Boiss. from Turkey. *Journal of Essential Oil Research*, 13, 134-135.

Baser, K.H.C. & Duman, H. (1997). Composition of the essential oil of *Laserpitium petrophilum* Boiss. et Heldr. *Journal of Essential Oil Research*, 9, 707-708.

Baser, K.H.C., Ozek, G., Ozek, T. & Duran, A. (2006). Composition of the essential oil of *Chaerophyllum macropodum* Boiss. fruits obtained by microdistillation. *Journal of Essential Oil Research*, 18, 515-517.

Baser, K.H.C., Ozek, T., Kurkcuoglu, M.K. & Guner. A. (1993). The essential oil of *Scaligeria lazica* Boiss. *Journal of Essential Oil Research*, 5, 463-464.

Baser, K.H.C., Ozek, T., Duman, H. & Guner, A. (1996). Essential oil of *Pimpinella aromatica* Bieb. from Turkey. *Journal of Essential Oil Research*, 8, 463-464.

Baser, K.H.C., Ozek, T. & Tabanca, N. (1999). Essential Oil of *Pimpinella anisetum* Boiss. et Bal. *Journal of Essential Oil Research*, 11, 445-446.

Baser, K.H.C., Ozek, T., Kurkcuoglu, M. & Aytac, Z. (2000a) Essential oil of *Seseli campestre* Besser. *Journal of Essential Oil Research*, 12, 105-107.

Baser, K.H.C., Tabanka, N., Ozek, T., Demicri, B., Duran, A. & Duman, H. (2000b). Composition of the essential oil of *Chaerophyllum aksekiense* A. Duran et Duman, a recently described endemic from Turkey. *Flavour and Fragrance Journal*, 15, 43-44.

Bazgir, A., Shaabani, A. &, Sefidkon, F. (2005). Composition of the essential oil of *Peucedanum cervariifolium* C.A. Mey. from Iran. *Journal of Essential Oil Research*, 17, 380-381.

Becker, N., Petric, D., Zgomba, M., Boase, C., Dahl, C., Lane, J., Kaiser, A. Mosquitoes and their control, Kluwer Academic/Plenum Publishers: New York, 2003. ISBN 0306473607.

Bermard, C. (2001). Essential oils of three *Angelica* L. species growing in France. Part II: fruit oils. *Journal of Essential Oil Research*, 13, 260-263.

Bernard, C. & Clair, G. (1997). Essential oils of three *Angelica* L. species growing in France. I. Root oils. *Journal of Essential Oil Research*. 9, 289-294.

Borg-Karlson, A-K., Valterova, I. & Anders Nilson, L. (1994). Volatile compounds from flowers of six species in the family Apiaceae: Bouquets for different pollinators? *Phytochemistry*, 35, 111-119.

Boyer, S., Serandour, J., Lemperiere, G., Raveton, M. &, Ravanel, P. (2006). Do herbicide treatments reduce the sensitivity of mosquito larvae to insecticides? *Chemosphere*, 65, 721-724.

Bulatovic, V.M., Savikin-Fodulovic, K.P., Zdunic, G.M. & Popovic, M.P. (2006). Essential oil of *Seseli peucedanoides* (MB) Kos.-Pol. *Journal of Essential Oil Research*, 18, 286-287.

Camarda, L. & Di Stefano, V. (2003). Essential oil of leaves and fruits of *Athamanta sicula* L. (Apiaceae). *Journal of Essential Oil Research*, 15, 133-134.

Camarda, L., Di Stefano, V. & Pitonzo, R. (2008). Chemical composition of essential oils from *Athmanta sícula*. *Chemistry of Natural Compounds*, 44, 532-533.

Campbell, C., Gries, R. & Gries, G. (2011) Forty-two compounds in eleven essential oils elicit antennal responses from *Aedes aegypti*. *Entomologia Experimentalis et Applicata*, 138, 21-32.

Yadouleton, A., Martin, T., Padonou, G., Chandre, F., Asidi, A., Djogbenou, L., Dabiré, R., Aïkpon, R., Boko, M., Glitho, I. & Akogbeto, M. (2011). Cotton pest management practices and the selection of pyrethroid resistance in *Anopheles gambiae* population in Northern Benin. *Parasites & Vectors*, 4, 60, doi: 10.1186/1756-3305-4-60.

Zivanovic, P., Djokovic, D., Vajs, V., Slavkovska, V., Todorovic, B., & Milosavljevic, S. (1994). Essential oils flowers and fruits of *Athamanta haynaldii* Bort. Et Uchtr. (Apiaceae). *Pharmazie*, 49, 463-464.

27

Flourensia cernua DC: A Plant from Mexican Semiarid Regions with a Broad Spectrum of Action for Disease Control

Diana Jasso de Rodríguez[1], F. Daniel Hernández-Castillo[1], Susana Solís-Gaona[1], Raúl Rodríguez- García[1] and Rosa M. Rodríguez-Jasso[2]
[1]*Universidad Autónoma Agraria Antonio Narro (UAAAN), Buenavista, Saltillo, Coahuila*
[2]*Centre of Biological Engineering, University of Minho,Campus Gualtar*
[1]*México*
[2]*Portugal*

1. Introduction

Mexico has an extensive variety of plants, it is the world´s fourth richest country in this aspect. Some 25,000 species are registered, and it is thought that there are approximately 30, 000 not described. Particularly the regions of the north of Mexico, with their semiarid climate, have a great number and variety of wild plants grown under extreme climatic conditions. Wild species which have compounds with flavonoid structures, sesquiterpenoids, acetylenes, *p*-acetophenones, benzofurans, and benzopyrans grow in these regions. The polyphenolic compounds include tannins and flavonoids which have therapeutic uses due to their anti-inflammatory, antifungal, antibacterial, antioxidant, and healing properties.

Flourensia cernua is an endemic species which grows in semiarid zones of Mexico and contains polyphenolic, lactone, benzofuran, and benzopyran compounds which give it a potential use for disease control.

In this work, *F. cernua* is reviewed in terms of its geographical distribution in Mexico, traditional uses, bioactive compounds identified for controlling fungi, bacteria, and insects, as well as cytotoxic activity.

2. Common names

It is commonly known in different ways, as it is found in the United States of America as well as in Mexico. The names given in the United States of America are: tarbush, hojase, American-tarbush, black-brush, varnish-brush, and hojasen (Correl and Johnston, 1970; Vines, 1960). In Mexico, it is known as hojasen, tarbush, black-brush (Arredondo, 1981).

3. Geographical distribution

In Mexico, *F. cernua* is found in the Chihuahuan and Sonoran deserts, as well as in the states of Coahuila, Chihuahua, Durango, Hidalgo, Nuevo León, San Luis Potosí, Sonora, and Zacatecas (Valdés, 1988; Martínez, 1993) (Fig. 1). In the United States of America, it is found West of Texas and South of New Mexico and Arizona (Vines, 1960).

It is found in altitudes ranging from 1000 to 2000 masl. In studies carried out in Mexico, it is observed that the prevailing altitude for this species is 1900 masl and slopes from 1 to 6 per cent (Arredondo, 1981) although it is also found in altitudes from 300 to 400 Northeast of Coahuila.

4. Associated species

The main shrub species to which *Flourensia cernua* (Fig.2) is related to are: *Larrea tridentata, Yucca filifera, Atriplex canescens, Castela texana, Acacia fernesiana, Prosopis juliflora, Agave lechuguilla, Parthenium incanum, Fouquieria splendens*, and *Acacia constricta* (Comisión Técnico Consultiva para la Determinación de los Coeficientes de agostadero, 1979).

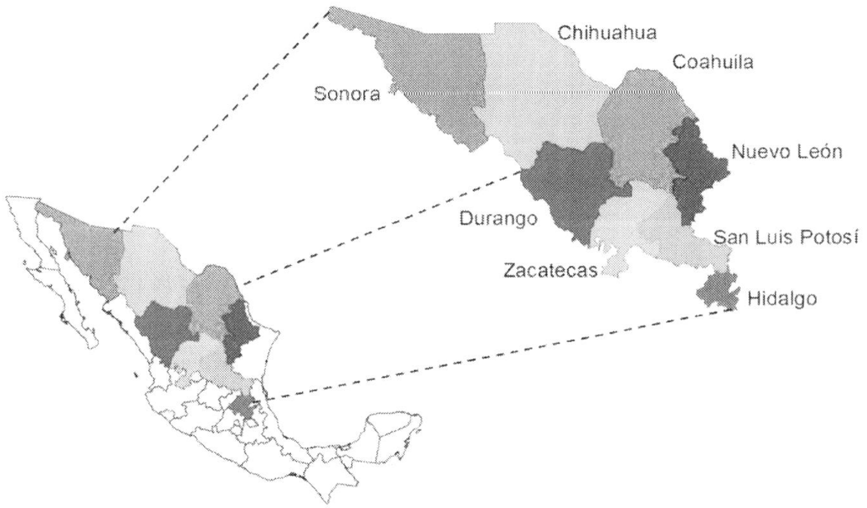

Fig. 1. Geographical distribution of *Flourensia cernua* DC in Mexico.

Fig. 2. *Flourensia cernua* plant at su in a wilderness site in Northeast Mexico.

5. Phytochemical analysis

The *Flourensia genus* is important due to the great amount of secondary metabolites it possesses; these are widely used for biological and ecological applications. Nine species of *Flourensia* have been reported, being *Flourensia cernua* the one with the highest number of chemicals (Aregullín and Rodríguez, 1983) with economical potential. The authors correlated the presence of benzofurans and benzopyrans with biological activity. The fact that these secondary metabolites are not present in other species, led to a correlation between the ecographical distribution and a possible chemical adaptation to the environment.

6. Properties and documented actions

Several medicinal properties have been reported for the tea obtained from the leaves or flowers for indigestion and gastrointestinal problems (Arredondo, 1981).

The green fruits are innocuous for cattle. However, dry fruits are toxic and when consumed at approximately 1% of the animal weight they cause death during the first 24 hours (Sperry et al., 1968).

7. Active chemicals

Fractionation of a CH_2Cl_2-MeOH (1:1) extract of the aerial parts of *Flourensia cernua* led to the isolation of three phytotoxic compounds, namely dehydroflourensic acid (Fig.3a), flourensadiol (Fig.3b), and methyl orsellinate (Fig.3c) and seven hitherto unknown γ-lactones were obtained (Fig.3d), these being tetracosane-4-olide, pentacosane-4-olide, hexacosane-4-olide, heptacosane-4-olide, octacosane-4-olide, nonacosane-4-olide and triacontane-4-olide. Besides, a previously known flavonoid, ermanin (Dominguez et al., 1973). Also there are benzopyrans (Fig.3e) and benzofurans (Fig.3f).

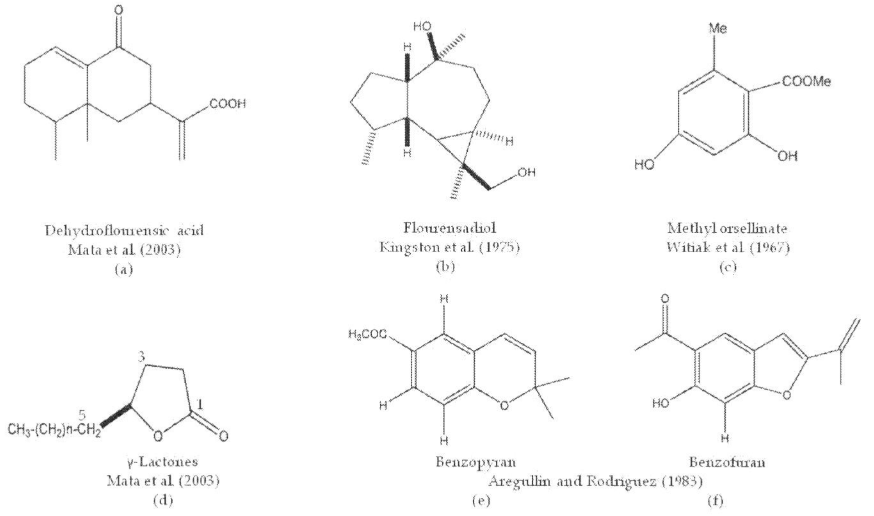

Fig. 3. Chemical of six active compounds from *Flourensia cernua*.

8. Plant fungicide, bactericide, and insecticide activity

8.1 Antifungal activity

Fungicide activity *in vitro* of leaves extracts at solution concentration of 1,000 mg L^{-1} on *Rhizoctonia solani, Pythium* sp. and *Fusarium oxysporum* was reported (Saeedi-Ghomi & Maldonado, 1982). The leaf fractions of hexane, diethyl ether and ethanol were active against *Colletotrichum fragariae* Brooks, *C. gloesporioides* Penz and Sacc. The essential oils from the hexane fraction were active at 1 μg doses, whereas the diethyl ether and ethanol fractions were active at 10 μg doses. The ethanol fraction was active against *C. accutatum* Simmons only at 400 μg. (Tellez et al., 2001).

Gamboa et al. (2003) used an extraction method by soxhlet to obtain methanolic extracts which were evaluated on soil pathogene *Rhizoctonia solani* and on phytopathogene algae *Phytophthora infestans*, 20,000 μl L^{-1} were the required dose for 86% pathogene inhibition.

Mata et al. (2003) reported that the fractionation of an extract of the aerial parts of *F. cernua* led to the isolation of three phytotoxic compounds namely: Flourensadiol, methyl orsellinate, and dehydroflourensic acid.

In a study carried out at our lab the inhibitory effect of ethanolic extracts was evaluated for three *Flourensia* species: *F. cernua, F. microphylla,* and *F. retinophylla* on three pathogens: *Alternia* sp., *Rhizoctonia solani* (Fig.4), and *Fusarium oxysporum,* which attack commercial cultivars (Jasso de Rodriguez et al., 2007). The variance analysis on the pathogen mycelial development showed highly significant differences ($p \leq 0.01$) on extract, dose, and extract interactions x dose. *F. microphylla* inhibited *Alternaria* sp 42.5% at a 10 μl L^{-1}, reaching 76.8% at a 100 μl L^{-1}. *F. cernua* and *F. retinophylla* showed a similar effect for high concentrations, however, inhibition was slightly lower than *F. microphylla* at low concentration. The highest inhibition level was 98.6%.

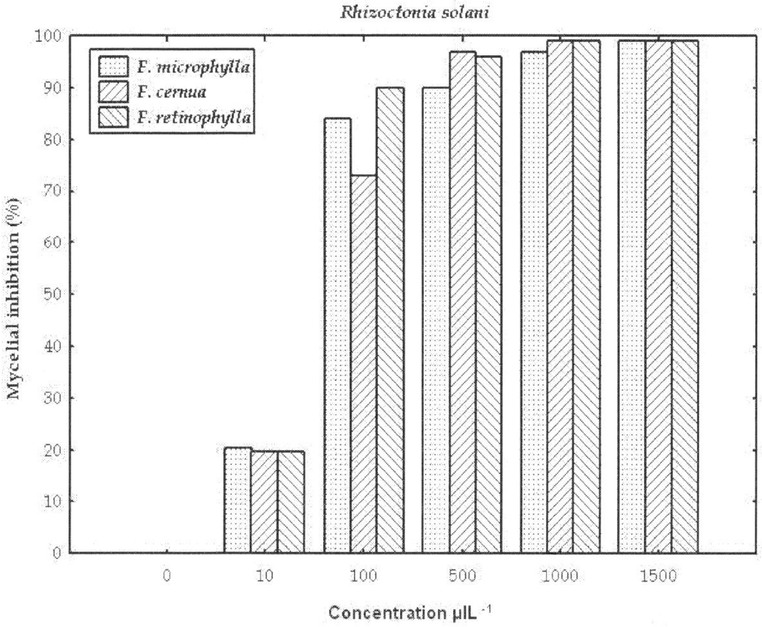

Fig. 4. Mycelial growth inhibition (percentage of *Rhrizoctonia solani*) as a function of *Flourensia* spp. extracts concentration.

In studies carried out by our research group, Guerrero-Rodriguez et al. (2007) proved the *F. cernua* effect on mycelial inhibition of *Alternaria alternata, Colletotrichum gloesporoides,* and *Penicillum digitatum,* where methanol:chloroform (1:1) solvents and sequential extractions with hexane, diethyl ether, and ethanol were used. *A. alternata* reported the highest mycelial inhibition, took place with hexane fractions (91.9%) and methanol: chloroform (88.4%) at 4000 mg L^{-1}. 2000 mg L^{-1} ethanol extract caused the lowest production of *C. gloesporoides* conidium. The four extracts reduced conidium production for *P. digitatum,* however they didn´t present statistical differences. In general, the ethanolic extract was the most efficient for inhibiting mycelial growth of *C. gloesporoides* and *P. digitatum,* (Fig. 5).

Studies carried out in the UAAAN by Castillo et al. (2010) on plant extract of *Larrea tridentata, Flourensia cernua, Agave lechuguilla, Opuntia* sp, and *Yucca* sp; obtained with alternative organic solvents (lanolin and cocoa butter) and water were tested against *Rhizoctonia solani* pathogens. The obtained results were as follows: The *L. tridentata* and *F. cernua* extracts by means of lanolin and cocoa butter at a 2000 and 1000 ppm total tannins inhibited the growth of *R. solani* 100%. Lanolin and cocoa butter solvents allowed a high recovery of polyphenolic molecules of strong antifungal activity against *R. solani* thus offering an alternative production of antimicrobial agents.

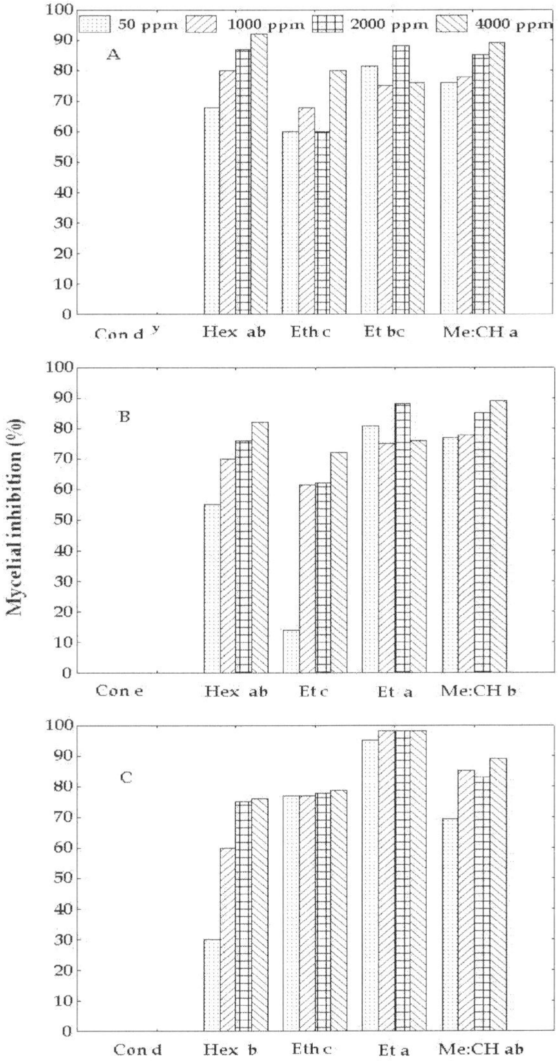

Fig. 5. Percentage of mycelial inhibition of four extracts from *Flourensia cernua* DC at four concentrations on three postharvest pathogens: A) *Alternaria alternata*, B) *Colletotrichum gloeosporioides*, and C) *Penicillium digitatum*. yP ≤0.01 (LSD). Con=Control; Hex=Hexane; Eth=Ether; Et=EtOH, and Me:CH=MeOH-$CHCl_3$.

8.2 Antibacterial activity

The mixtures of benzofurans and benzopyrans of *F. cernua*, were tested against Gram positive and Gram negative bacteria, fungi and *Saccharomyces* under two experimental conditions: One where the inoculated media was kept in darkness and the other one where the inoculated media was UV irradiated (280-400 nm) for 15 min previous to incubation in darkness. Bioactivity was greatly increased by the UV irradiation (Aregullin & Rodriguez, 1983; Towers et al., 1975).

Molina-Salinas et al. (2006), evaluated the crude extracts effect of methanol, acetone, and hexane of the aerial parts of *Artemisa ludoviciana* Nutt., *Chenopodium ambrosioides* L., *Murrubium vulgare* L., *Mentha spicata* L., and *Flourensia cernua* DC to inhibit the growth or death of *Mycobacterium tuberculosis* strains H37Rv and CIBIN:UMF:15:99. Results showed that from the evaluated plants, *F. cernua* was the only active plant which inhibited and killed *Mycobacterium tuberculosis* strains H37Rv and CIBIN:UMF:15:99. The hexane extract showed a minimal inhibitory concentration (MIC) 50 and 25 $\mu g\ L^{-1}$ against sensitive and resistant strains, respectively; acetone extract was active against and only CIBIN:UMF:15:99 (MIC=100 $\mu g\ L^{-1}$). It may be concluded that hexane extract of *F. cernua* leaves could be an important source of bactericidal compounds against multidrug-resistant *M. tuberculosis*.

In a study carried out by our research group, the antibacterial activity of *F. cernua* obtained with hexane, ether, ethanol, and metanol-chloroform mixture, at different dose, was evaluated on *Pseudomonas cichorii* (Pc), *Xanthomonas axonopodis* pv. phaseoli (Xap), and *Pectobacterium caratovora* subsp. atroseptic (Pca) (Peralta, 2006). All the extracts showed activity on Xap and Pc, however, none of them showed any inhibition effect when evaluated on Pca. *F. cernua* hexane extract at a 4000 $\mu l\ ml^{-1}$ ($P \leq 0.05$) concentration showed the highest inhibition on Xap (82.51%) and Pc (83.96%). The other extracts showed a lower activity (Table 1).

Extracts and		Inhibition of CFU ml^{-1}								
concentrations ($\mu l\ L^{-1}$)		Pc	A	B	Xap	A	B	Pca	A	B
Hexane	500	1.89		a	37.23		a	0		a
	1000	26.41	B	ab	51.55	C	ab	0	A	a
	2000	44.34		b	74.77		b	0		a
	4000	83.96		c	82.51		c	21.73		B
Diethyl ether	500	0		a	2.41		a	11.01		c
	1000	12.95	A	ab	0	B	a	5.29	A	b
	2000	17.27		b	2.08		a	0		a
	4000	47.48		c	63.44		b	3.96		Ab
Ethanol	500	18.7		a	0		a	4.58		a
	1000	20.87	A	ab	0	A	a	5.6	B	ab
	2000	43.04		b	1.47		a	13.49		c
	4000	52.17		c	14.73		b	9.92		B
Meth-Chlor	500	0		a	0		a	5.93		c
	1000	6.77	A	b	0	A	a	3.56	A	a
	2000	0		a	17.46		b	3.86		a
	4000	0		a	0		a	4.91		b

Table 1. Interactions of *F. cernua* extracts x concentrations of the CFU ml^{-1} inhibition precentage, of three bacteria at 24 h of incubation. Meth-Chlor= Methanol-Chloroform; Pc = *Pseudomonas cichorii*; Xcp = *Xanthomonas axenopodis* pv. Phaseoli; Pca = *Pectobacterium caratovora* subsp. Atroseptica; A: Extract, ** = $p \leq 0.05$; B: Concentration ** = $p \leq 0.05$.

8.3 Insecticide activity

The insecticide activity of the benzofuran 7-methoxy-2-isopropenyl-5-acetyl-2, 3-dihydrobenzofuran-3-ol-cinnamate proved its activity as a juvenile hormone causing anatomic malformation, juvenile characteristics retention and sterility in the insects treated from their second to fourth stages of development (Towers et al., 1975). The results were similar to those reported by Bowers (1971) with precosene. Termiticidal activity of hexane, diethyl ether and ethanol fractions was found by Tellez et al. (2001).

In lab studies carried out by our research team, when evaluating the bioinsecticide activity of crude extracts from *F. cernua* leaves extracted with solvents of variable polarity on three insect plagues of agronomical importance: *Sitophilus oryzae* (Linneaus), *Phthorimaea operculella* (Zeller), and *Brevicoryne brassicae* (Linnaeus), as well as the repellent or attraction effect on *Sitophilus oryzae* (Linneaus) (Martinez, 2006). The following results were obtained: extracts didn´t provoke mortality at 24 and 48 h ($P \leq 0.05$) on *Sitophilus oryzae* (Linneaus), *Phthorimaea operculella* (Zeller); mortality for cabbage plant louse (*B. brassicae*), by effect of all extracts was observed, although some required a higher concentration to kill beings 100%. The extract that presented insecticide effect potential against *B. brassicae* was hexane which had 100% mortality from concentration at 10,000 µl L^{-1} ($P \leq 0.05$) at 24 h. Besides, hexane fraction showed insectistatic effect when inciting repellency to *S. oryzae* at 5 and 45 days, in raffia sacks as well as in jute sacks (Fig. 6). The repellency effect incited by the hexane fraction may be due to the volatile substances borneol and camphor it contains.

8.4 Antioxidant activity

Salazar et al. (2008) in a study carried out in order to evaluate the antioxidant potential of six species of the Northeast of Mexico, reports that leaves, stem, root, and flowers of *F. cernua* possess antioxidant activity, due to the content of phenolic compounds showing up in the different parts of the plant. Stems and roots of this species report the highest contents of phenolic compounds.

8.5 Cytotoxic activity

Pure benzopyrans and benzofurans of *F. cernua* (Figs. 3 e and f) have been studied for cytotoxic activity using blood red cells and measuring the hemoglobin released on cell destruction. The benzopyrans were more active than benzofurans although no clear correlation between activity and structure has been obtained. The UV irradiated compounds showed higher cytotoxic activity than the non-irradiated ones (Towers et al., 1980).

Benzopyrans and benzofurans react with L-cystein (Towers et al., 1979) and the microbicide and cytotoxic activities may be associated with the alkyl formation capacity. Significant inhibition of radicle growth of *Amaranthus hypochondriacus* was reported by Mata et al. (2003). The crude extracts and their fractions are cytotoxic against five human breast cancer cell lines (Molina-Salinas et al., 2006).

9. Relevant achievements of *F. cernua* research

The crude extracts and fractions of different polarity inhibit mycelial development of *Rhizoctonia solani, Pythium* sp., *Fusarium oxysporum, Colletotrichum fragariae, Colletotrichum gloeosporioides* Penz, *Colletotrichum accutatum* Simmons, *Phytophthora infestans*, *Alternaria* sp., *Alternaría alternata*, *Penicillum digitatum*.

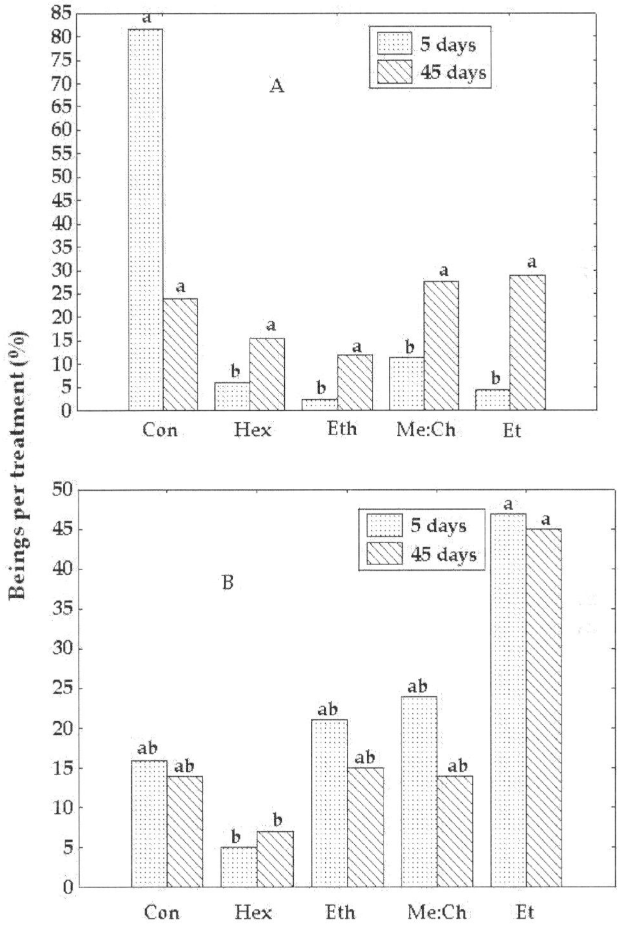

Fig. 6. Percentage of beings found in maize grains treated with four extracts of *Flourensia cernua* at 20,000 µl L^{-1} in (A) raffia sacks and (B) jute sacks. Con=Control; Hex=Hexane; Eth=Ether; Me:Ch=Methanol-Chloroform; Et= Ethanol.

Flourensia cernua extracts have a high bioinsecticide activity for controlling *B. brassicae* and insectistatic effect when provoking repellency of *S. oryzae* and high degree of antitermite activity.

The antibacterial activity from the hexane extract of the leaves against *M. tuberculosis* suggests that *F. cernua* could be an important source of non-polar compounds with bactericidal activity.

Crude extracts and its fractions are cytotoxic against five human breast cancer cell lines.

The antioxidant activity of this plant is considered to increase the fungicide activity of the extracts.

The institutional (UAAAN) research results as to the bactericidal and insecticide activity are scientifically reported for the first time.

10. Conclusions

The secondary metabolites of semidesert plants as a result of genetic, climatic, and soil factors, vary greatly from plants which develop under less extreme environmental conditions.

Flourensia cernua is a Mexican semidesert plant which proved to have the capacity to control *Rhizoctonia solani, Pythium* sp., *Fusarium oxysporum, Colletotrichum fragariae, Colletotrichum gloeosporioides* Penz, *Colletotrichum accutatum* Simmons, *Phytophthora infestans, Fusarium oxysporum, Alternaria* sp., *Alternaría alternata, Penicillum digitatum,* diseases which cause great worldwide losses in field and postharvest production of high commercial value cultivars as tomato, potato, apple, avocado, papaya, banana. As bioinsecticide it also controls the activity of *Brevicoryne brassicae,* which attacks different species of the *Cruciferae* family. Besides, it's a repellent for *Sitophilus oryzae* which attacks stored grains.

The hexane fraction of *F. cernua* leaf mainly contains monoterpenoids, while the ethanol fraction mainly contains sesquiterpenoids, volatile compounds. This extract also contains an unknown number of molecules next to the active principle.

Crude extracts and its fractions of this plant are cytotoxic against five human breast cancer cell lines

Phenolic compounds found in extracts could be responsible of this species antioxidant activity, besides, these extracts don´t show cytotoxicity.

Flourensia cernua phenolic compounds may be used as food additives with the purpose of preventing oxidation and in general, granting health benefits.

Taking into account the biological importance of this plant compounds, average and long term research must continue focusing on the following: 1. Isolation and identification of the plant active compounds; 2. *In vivo* evaluation of isolated extracts and compounds activity; 3. Product formulation for its industrial and pharmacological use.

Experimental results show a good correlation with the use of this plant in traditional Mexican medicine.

Because *F. cernua* is an endemic species, in the future it´s necessary to carry out a domestication for the commercial production of this species to be able to have enough vegetal material for its industrialization and commercialization.

11. References

Aregullin, M., & Rodríguez, E. (1983). Phytochemical studies of the genus *Flourensia* from the Chihuahuan desert. Department of Developmental and Cell Biology. University of California. Irvine, CA. 92717. Oct. *Proceedings, Second Chihuahuan Desert Symposium*

Arredondo, V. D. G. (1981). *Componentes de la vegetación del Rancho demostrativo Los Angeles.* Tesis licenciatura. Universidad Autónoma Agraria Antonio Narro, Saltillo Coahuila México

Bowers, W. S. 1971. Discovery of insect antiallatotropins. In: *Naturally occurring insecticides.* Jacobson, M., & Croby, D. G. Eds. Marcel Dekker Inc. New York. p. 394-405

Castillo, F., Hernández, D., Gallegos, G., Méndez, M., Rodríguez, R., Reyes, A & Aguilar, C. N. (2010). *In vitro* antifungal activity of plant extracts obtained with alterntive organic solvents against *Rhizoctonia solani* Kuhn. *Industrial Crops and Products* 32(2010) 324-328, ISSN: 0926-6690

Comisión Técnico *Consultiva para la Determinación de los Coeficientes de agostadero* (COTECOCA). (1979). Coahuila, Secretarla de Agricultura y Recursos Hidráulico, México. 255 pp

Correll, S. D., & Johnston, C. M. (1970). *Manual of the Vascular Plants of Texas.* Texas Research Foundation. Renner, Texas. p. 1879, 1981.

Domínguez, X. A., Escarria, S., & Butruille, D. (1973). The methyl-3,4'-kaempferol de *Cordia boissieri. Phytochemistry* 12, 724-725, ISSN: 0031-9422

Gamboa-Alvarado, R., Hernández-Castillo, F.D., Guerrero-Rodríguez E., Sánchez-Arizpe, A., & Lira-Saldívar, R.H. (2003). Inhibición del crecimiento micelial de *Rhizoctonia solani* Kuhn y *Phytophthora infestans* Mont (De Bary) con extractos vegetales metanólicos de hojasen (*Flourensia cernua* D.C.), mejorana (Origanum majorana L.) y trompetilla [*Bouvardia ternifolia* (Ca.) Schlecht]. *Revista Mexicana de Fitopatología.* 21, 13-18, ISSN: 0185-33-09

Guerrero, R.E. Solís, G.S., Hernández, C.F.D., Flores, O.A., Sandoval, L.V., & Jasso de Rodríguez, D. (2007). Actividad biológica in vitro de extractos de *Flourensia cernua* D.C en patógenos de postcosecha: *Alternaria alternata* (Fr.:Fr) Keiss., *Colletotrichum gloesporoides* (Penz.) Penz y Sacc. y *Penicillum digitatum* (Pers.:Fr.). *Revista Mexicana de Fitopatología.* 25, 48-53, ISSN: 0185-33-09

Jasso de Rodríguez, D., Hernández, CD., Angulo, S.J.L., Rodríguez, G.R., Villarreal, Q.J.A., & Lira, S.R.H. (2007). Antifungal activity *in vitro* of *Flourensia* spp. extracts on *Alternaria* sp., *Rhizotocnia solani,* and *Fusarium oxysporum. Industrial Crops and Products.* 25, 111-116, ISSN: 0926-6690

Kingston, D. G. I., Rao, M. M., Spittler, T. D., Pettersen, R. C., & Cuulen, D. L. (1975). Sesquiterpenes from *Flourensia cernua. Phytochemistry* 14,2033-2037, ISSN: 0031-9422

Martínez, M. (1993). *Las plantas medicinales de México.* Ed. Botas. México, D. F. p. 656

Martinez, Z.J. (2006). *Actividad de extractos vegetales* de *Flourensia cernua D.C. contra Brevcorine brassicae L y de Flourensia cernua D.C., Agave lechuguilla* Torr., *Azadirachta indica L., Argemone Mexicana* L. y *Larrea tridentata D.C. contra Sitophilus oryzae* L. Tesis de Licenciatura. Universidad Autónoma Agraria Antonio Narro, Saltillo Coahuila México

Mata, R., Bye, R., Linares, E., Macías, M., Rivero-Cruz, I., Pérez, O., & Timmerman, B. N. (2003). Phytotoxic compounds from *Flourensia cernua. Phytochemistry* 64,285-291, ISSN: 0031-9422

Molina-Salinas, G.M., Ramos-Guerra, M.C., Vargas-Villarreal, J., Mata-Cardenas, B.D, Becerril-Montes, P. & Said-Fernández, S. (2006). Bactericidal activity of organic extracts from *Flourensia cernua* DC against strains of *Mycobacterium tuberculosis. Archives of Medical Research,* 37, 45-49, ISSN: 0188-4409

Peralta, B.J.E. (2006). *Evaluación de la actividad de extractos de hojasen (Flourensia cernua D.C), in vitro en el control de las bacterias fitopatogenas Xanthomona campestris pv. phaseoli (Smith) Dye, Erwinia carotovora pv. atroseptica (Van Hall) Dye y Pseudomona cichorri (Swingle) Stapp.* Tesis de Licenciatura. Universidad Autónoma Agraria Antonio Narro, Saltillo Coahuila México

Saeedi-Ghomi, H., & Maldonado, R. (1982). *Potencial de la flora de las zonas áridas.* En: Ciencia y Desarrollo, (Ed. CONACYT) 47, 98-110. México

Salazar, R., Pozos, M.E., Cordero, P., Pérez, J., Salinas, M.C & Waksman, N. (2008). Determination of the Antioxidant Activity of Plants from Northeast Mexico. *Pharmaceutical Biology*, Vol.46, No. 3: 166-170, ISSN: 1388-0209

Sperry, D. E., Dollahite, J. W., Hoffman, G. O., & Camp, V. J. (1968). *Texas plants poisonous to livestock.* Texas A&M University. Texas Agriculture. Ext. Serv. USA. Bull. B- 1028, 59 pp

Tellez, M., Estelle, R., Frederickson, E., Powell, J., Wedge, D., Schrader, K., & Kobaisy, M. (2001). Extracts of *Flourensia cernua* (L.), Volatile constituents and antifungal, antialgal, antitermite bioactivities. *Journal of Chemical Ecology.* 27, 2263-2273, ISSN: 0098-0331

Towers, G. H. N., Macrae, W.D., Irwin, D. A. J. & Bisalputra,T. (1980). Membrane lesions in human erytrocytes induced the naturally occurring compounds alfa-terthienyl and phynylheptatryne. *Photobiochemistry and Photobiophysics.* 1, 309-318.

Towers, G.H.N., Picman, A. R., & Rodríguez, E. (1979). Formation of adducts of parthenin and related sesquiterpene lactones with cysteine and glutathion. *Chemico-Biological Interactions.* 28, 83-89.

Towers, G. H. N., Shan, G. F. Q., & Mitchell, J. C.(1975). Ultraviolet mediated antibiotic activity of thiophene compounds of *Tagetes. Phytochemistry* 14, 2295-2296, ISSN: 0031-9422

Valdes, S. I. (1988). *Efecto de la densidad de Flourensia cernua en la producción de forraje de pastizal mediano abierto.* Tesis profesional. Universidad Autónoma Agraria Antonio Narro, Saltillo Coahuila México

Vines, R.A. (1960). *Trees, shrubs and woody vines of the southwest.* Univ. of Texas, press. USA. 1104 pp

Witiak, D., Patel, D.,& Lin, Y. (1967). Nuclear magnetic resonance. Influence of substituents on the long-range spin-spin coupling constant between benzylic and ring protons in the orcinol series. *Journal American Chemistry Society.* 89, 1908-1911, ISSN: 0002-7863

Advances in Aerial Application Technologies and Decision Support for Integrated Pest Management

Ian M. McLeod[1], Christopher J. Lucarotti[2,3,*], Chris R. Hennigar[3], David A. MacLean[3], A. Gordon L. Holloway[4], Gerald A. Cormier[1] and David C. Davies[1]
[1]Forest Protection Limited, Fredericton International Airport, Lincoln,
[2]Natural Resources Canada, Canadian Forest Service - Atlantic Forestry Centre, Fredericton,
[3]Faculty of Forestry and Environmental Management and
[4]Department of Mechanical Engineering, The University of New Brunswick, Fredericton Canada

1. Introduction

The first aerial applications of a pesticide against forest insect pests in Canada used calcium arsenate to protect forest stands from defoliation by the spruce budworm (*Choristoneura fumiferana*). The program was first conducted in Nova Scotia in 1927, and then in Ontario in 1928 and 1929 (Randall, 1975). There were few developments during the 1930s, but World War II and the need to protect military personnel from mosquitoes helped establish aerial spray techniques using a variety of aircraft (Randall, 1975). Extensive outbreaks of spruce budworm populations in Canada were countered by aerial application of DDT (dichloro-diphenyl-trichloro-ethane) in 1944 in Ontario (Howse & Sippell, 1975), then in 1952 in Quebec (Blais et al., 1975) and New Brunswick (Miller & Kettela, 1975).

Insecticidal powders were first released from hoppers mounted directly on the aircraft, but liquid formulations of pesticides led to the development of gravity-flow, open-pipe systems, boom and nozzle systems and rotary atomizers variously distributed on the aircraft fuselage and wings (Randall, 1975). In the 1950s and 1960s, hydraulic boom and nozzle systems were used predominantly in forestry, but with research suggesting that 100-μm and smaller droplets are more effective against forest pest insects, rotary atomizers are now favoured (Weisner, 1995).

Aircraft guidance and direction to spray blocks was initially accomplished using topographic maps and features on the landscape. In the case of small blocks, helium-filled balloons, deployed from the ground, were sometimes used to mark the corners. Forest Protection Limited (FPL) (Lincoln, New Brunswick) was the first to use "spotter" aircraft that communicated information on block location and boundaries to spray pilots in flight (Flieger, 1964). Electronic guidance systems for the positioning of spray aircraft were first used in 1959 (Randall, 1975), and the availability of satellite-based, global positioning system (GPS) technologies onboard aircraft immensely improved aircraft guidance and navigation.

Since its incorporation in 1952, FPL has been a leader in the development of technologies to increase the efficacy, accuracy and safety of aerial application and fire control operations. In January 2007, FPL - in partnership with AG-NAV Inc., BioAtlantech, the Canadian Forest Service (CFS), Sylvar Technologies Inc., and the University of New Brunswick (UNB), received a 5-year grant from the Atlantic Innovation Fund (AIF) of the Atlantic Canada Opportunities Agency (ACOA) to further develop aerial application technologies and baculoviruses for use in controlling forest insect pests. In this review, we describe progress on the research that has led to the Accuair™ suite of aerial application technologies and services. For more detailed information on aerial application practices and technologies generally, readers are directed to articles referenced in this review (e.g., Payne, 1995; Weisner, 1995; Kilroy et al., 2003, Mierzejewski et al., 2007).

2. The development of Accuair™ technologies

2.1 Overview

Accuair™ is the brand for an integrated system that is capable of increasing aerial spray program efficiency. One component is the Accuair Aerial Management System (AMS) - an onboard guidance, navigation and control system that optimizes spray lines on a treatment block to compensate for changes in wind conditions and aircraft altitude at the time of spraying. The AMS uses spray droplet size information gathered from a second component, the Accuair Wind Tunnel. Droplet size is influenced by a host of factors whose effects cannot be easily predicted or generalized. A wind tunnel provides a means of measuring droplet sizes under a set of conditions that replicate actual operational situations. With these data from the wind tunnel, the AMS drift simulation software can develop more realistic predictions. Accuair™ Forest Protection Optimization System (ForPRO), the third component, is a planning software that uses forest stand information and risk factor overlays to identify areas within a region that are most in need of protection from forest insect pests.

2.2 Aerial management systems

The AMS that led to Accuair™ had its beginning in the early 1990s with the availability of inexpensive GPS and other sensing equipment that could be easily installed and used on spray aircraft. As early as 1991, the usefulness of newly available radar altimeters for determining of aircraft altitude above the ground and forest canopy was recognized (Mickle & Robinson, 1991). Knowledge of aircraft altitude above ground is a critical parameter in predicting spray drift, and such data are not provided by standard aircraft altimeters that use air pressure to determine height above sea level. A summary by Davies (1994) described trials of a system that made it possible to log flight data—including aircraft altitude and attitude, boom on/off functions, atomizer rotational speed, and application rate—that allowed managers to visualize these parameters on a computer interface, assess spray quality and make improvements. At the same time, Riley (1994) reported on work that had examined factors affecting the deposition and drift of pesticide sprays, including droplet size, atomizer configuration, influence of evaporation and product volatility, release height and atmospheric turbulence. Using the insights from these reports, Mickle (1998, 1999) used geographic information systems (GIS) and GPS-based guidance systems to improve the accuracy of aircraft positioning relative to the spray block. Wind data were also recorded on spotter aircraft using an Aircraft-Integrated Meteorological Measurement System (AIMMS-10; Aventech Research Inc., Barrie, ON) to obtain wind speed and direction at the altitude of the aircraft and at the time of spraying. In addition, a spray droplet dispersion model was used to determine optimal droplet size distributions to deliver a maximum number of droplets of pesticide to the target area for a variety of spray strategies. It was concluded that this practice delivered a more uniform application because it compensated for low upwind deposition as a result of wind-driven drift (Mickle, 1998, 1999). A second conclusion was that better efficacy was achieved if smaller droplets were used, as drifting sprays inherently give more uniform coverage of the target area. This second conclusion agreed with the report by Picot (1994) describing simulations of spray drift using a dispersion model where spray efficacy was evaluated based on delivering the maximum number of droplets to a block. It was found that increasing the small droplet content (<20 μm diameter) delivered the most uniform coverage over the target area.

In 2000, spray aircraft were equipped with an auto-flow system that automatically adjusted the flow rate of the control product to maintain a constant application rate as the aircraft speed changed during flight (Mickle, 2000). An auto-flow system takes information from the GPS system to determine aircraft speed relative to ground. This parameter is critical in determining application rate but is not available from standard aircraft airspeed indicators. A comparison was made on blocks sprayed with, and without, auto-flow and it was found that auto-flow technology yielded a significant reduction in application rate variability (Mickle, 2000).

The aircraft-mounted AIMMS-20 probe, an evolution of the AIMMS-10 system, was flown at different altitudes, and results were compared with SODAR (sonic detection and ranging) (Mickle, 2005). SODAR systems use sound waves to estimate the wind speed profile above ground. This study is of particular significance because it validated the ability of the AIMMS-20 probe to accurately measure wind speed and demonstrated that a single-point measurement of wind speed from a single location on the ground is insufficient to characterize the wind speed at altitude. In a project led by FPL, an AMS that incorporated real-time meteorology measurements and flight line offsetting was used on spray aircraft (Cormier, 2005; 2006; Mickle et al., 2007). The system was used to spray 8-ha triangular blocks (a difficult shape to treat), and a second set of 64-ha blocks was also sprayed with, and without, flight-line offsets. This early version of the AMS was the first operational system of its kind in agricultural/forest spraying, and results of this research demonstrated that offset spraying gave better deposit on the target area than spraying with no offset of flight lines in crosswind scenarios (Cormier, 2005, 2006; Mickle et al., 2007).

2.3 Wind tunnel

In parallel with AMS development, a wind tunnel facility was established to provide accurate droplet size information to the AMS. In 1975, the UNB Department of Chemical Engineering was invited by FPL to become involved in studies related to the prevention of conifer defoliation by the spruce budworm in the forests of New Brunswick. Professors Jules Picot and David Kristmanson constructed a wind tunnel specifically designed to measure spray droplet spectra incorporating then state-of-the-art particle size measurement systems. The 1-m diameter portable tunnel was completed in 1984. Newly emerging particle measuring systems were assessed, and in 1991, a Malvern Instruments Inc. (Malvern, UK) Fraunhofer laser diffraction spectrometer was purchased to determine droplet spectra from spray atomizers. Throughout this period, the work was coordinated with the New Brunswick Spray Efficacy Research Group (later SERG-International), which provided timely focus on the spray atomizer characterization requirements. In 2004, a partnership agreement was signed between UNB, FPL and CFS, whereby FPL became the caretaker of the wind tunnel in its new permanent location at the CFS-owned Acadia Research Forest in Noonan, NB. Extensive upgrades to the original tunnel and its new home were made at this time. In 2007, a new experimental and computational research program on aerial sprays was initiated by the UNB Department of Mechanical Engineering and FPL. The program was funded by the Natural Sciences and Engineering Research Council (NSERC) together with AIF funding to further improve the wind tunnel facility and to purchase a Sympatec (Clausthal-Zellerfeld, Germany) particle size analysis system. Since 2008, FPL and its research partners at UNB have been involved in the establishment of standards for the characterization of droplet spectra from atomizers for aerial spray applications.

2.4 Forest protection planning systems

Spruce budworm is the most widespread and economically important forest insect pest in eastern North America, affecting over 40 million hectares of forest at the peak of the last (1970s–1980s) outbreak. Repeated defoliation during budworm outbreaks, which typically last about 10 years, results in up to 90% growth reduction and 40-85% mortality in forests containing high quantities of spruce and balsam fir. As a result, mitigation measures such as insecticide spraying and salvage of impacted stands have been widely used during past infestations. Nevertheless, growth and mortality losses wreak havoc with forest management plans, creating considerable uncertainty about future forest conditions. Both theory and past experience imply that another spruce budworm outbreak is due across the northern forest regions of North America. The spruce budworm outbreaks of the 1970s and 1980s stimulated major insecticide spray programs as well as extensive salvage of vulnerable, dead and dying stands in some regions. It is vital that land managers responsible for forests understand potential consequences of the next spruce budworm outbreak on their wood supplies, land values and management plans. This provided the rationale for development of a forest <u>pro</u>tection <u>p</u>lanning <u>s</u>ystem or PROPS. PROPS is the software component of the Spruce Budworm Decision Support System (SBWDSS) used to assist with spruce budworm population management. The concept of the SBWDSS was developed by Erdle (1989), and the software application was developed by the CFS between 1992 and 1996. From 1996 to 1999, it was operationally implemented, on a cost-shared basis with industry and the provincial government, on all 6 million hectares of forest in New Brunswick (MacLean et al., 2001, 2002). The SBWDSS provides the conceptual basis for calculating marginal timber supply benefits (m^3/ha) of alternative foliage protection scenarios for each stand in a forest. PROPS software is used to implement these methods and has GIS tools to visually display SBW projected volume impacts on inventory at set time periods in the future or at the time of planned harvest; it facilitates manual spatial blocking of aerial bio-insecticide operations. PROPS allows users to determine effects of different foliage protection strategies on forest development and timber harvests (Erdle, 1989;

MacLean et al., 2000, 2002). PROPS was implemented for all forests in New Brunswick and for test areas in four other Canadian provinces.

3. Accuair™ – Components

3.1 Accuair™ aerial management system

Over the years, a variety of fixed-wing aircraft have been developed or adapted for use in aerial pesticide application programs (see Randall, 1975; Kilroy et al., 2003; Estey, 2004; Mierzejewski et al., 2007). Air Tractor Inc. (Olney, Texas) manufactures a number of single-engine aircraft that are used for pesticide application and in fire control operations. Currently, FPL owns and operates six AT-802F (Figs. 1A and B) that can be used both for aerial application and fire suppression and an AT-802F Fire Boss that is equipped with amphibious floats, allowing it to scoop surface water from lakes and rivers near a wildfire whereas other tankers must return to the airstrip for refilling with fire-retardants.

The current version of the AMS integrates several components that make real-time spray optimization possible. The AIMMS-20 probe measures and records air velocity and direction on the wing of the aircraft (Fig. 1C). Using data from the probe and information on aircraft position and orientation, as determined by GPS, the AIMMS system calculates wind speed at the altitude of the aircraft and the direction of the wind at this level relative to the ground. In addition to wind speed and direction, the altitude of the aircraft above the target area is another important parameter in predicting the distance that spray droplets will drift once they leave the atomizers. The altitude of the aircraft is measured using radar or laser altimeters that use radio waves or laser beams, respectively, to determine the altitude of the aircraft above ground level instead of using air pressure, which determines altitude above sea level.

A light bar (AG-NAV Inc., Newmarket, ON) (Fig. 1D), mounted on the exterior and in front of the cockpit, provides the pilot with information related to spray parameters, e.g., total number of swaths, current swath, total area to be sprayed, area sprayed, application and flow rates, ground speed, course deviation indicator and obstacle warning messages.

Two types of atomizers are routinely used for aerial application: hydraulic nozzles and rotary atomizers. With hydraulic nozzles, liquid is atomized as it is forced through a small orifice. The design of the orifice determines the shape of the spray, the size of drops and ultimate usefulness of the nozzle to a given application scenario. Rotary atomizers may be wind driven or electrically powered. During flight, wind flowing over the wing drives small propeller blades of wind-driven rotary nozzles that are attached to a cylindrical wire cage (Fig. 1E). Liquid enters the cage and is broken up into droplets by centrifugal force as it hits the rotating cylinder. The droplet spectrum, generated by rotary atomizers, is determined by the physical characteristics of the spray liquid, the liquid flow rate and the speed of the rotors. The revolutions per minute (rpm) of the rotary atomizer can be adjusted by changing the pitch of the propeller blades. Boom on/off functions and adjustment of flow rates from atomizers (e.g., to compensate for aircraft speed) are made automatically using the AG-FLOW system (AG-NAV Inc., Newmarket, ON) (Fig. 2). The required spray offset is calculated by a central, onboard processor based on droplet spectrum information (from wind tunnel data) and inputs from aircraft instrumentation including aircraft altitude, attitude and speed, relative to the ground and wind speed and direction at altitude (Fig. 2).

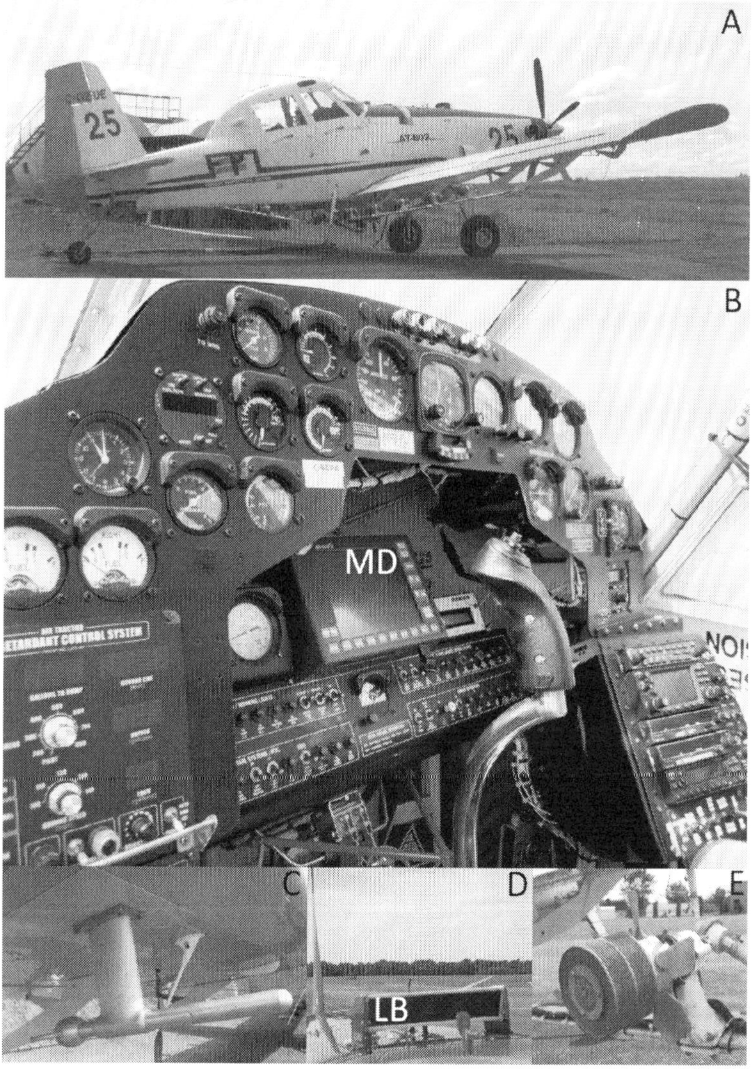

Fig. 1. Air Tractor Inc. AT-802F and components of the Accuair™ Aerial Management System. **A**. Forest Protection Limited (FPL) AT-802F aircraft. **B**. Cockpit instrument display of the FPL AT-802F shown in A. The AG-NAV differential GPS navigation and moving map display system (MD) is centrally located within the instrument display. **C**. The air-data probe is located on the underside of the wing near the wing-tip. **D**. The light bar (LB) is exterior to the cockpit on the nose cowling of the aircraft. **E**. A Micronair AU 4000 rotary atomizer.

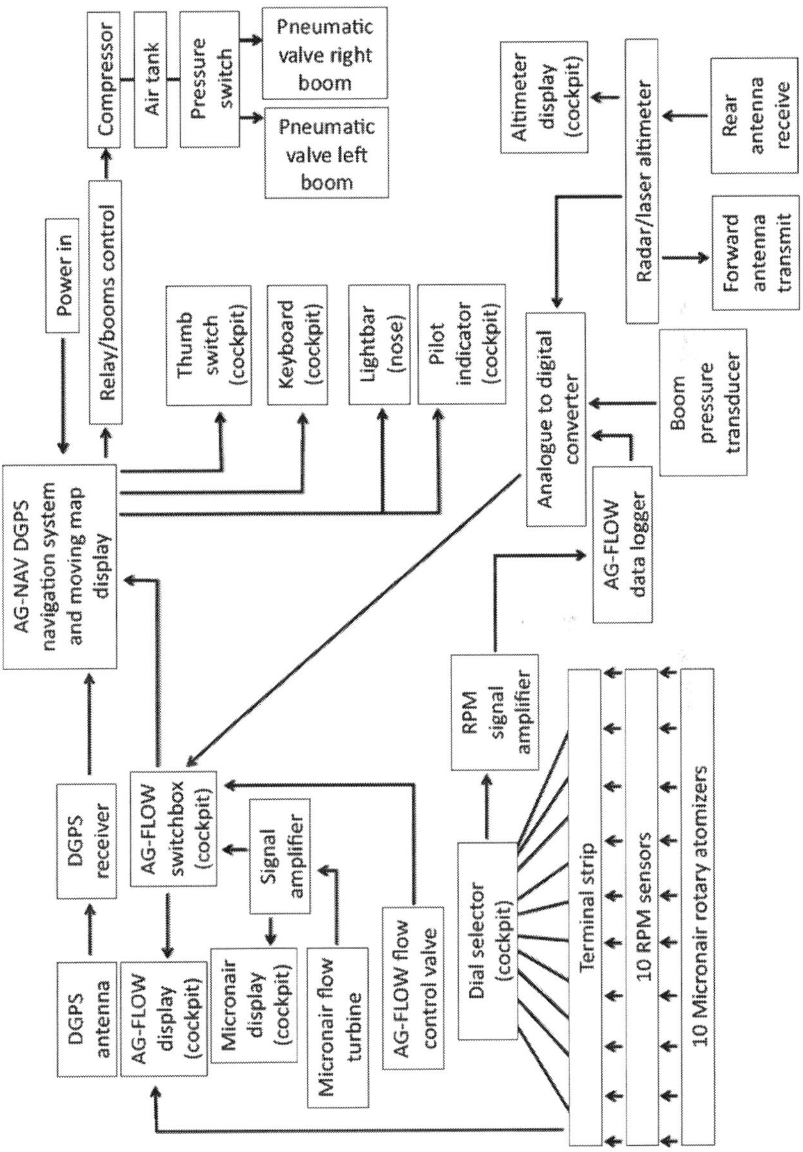

Fig. 2. Schematic diagram of the Accuair™ Aerial Management System (AMS).

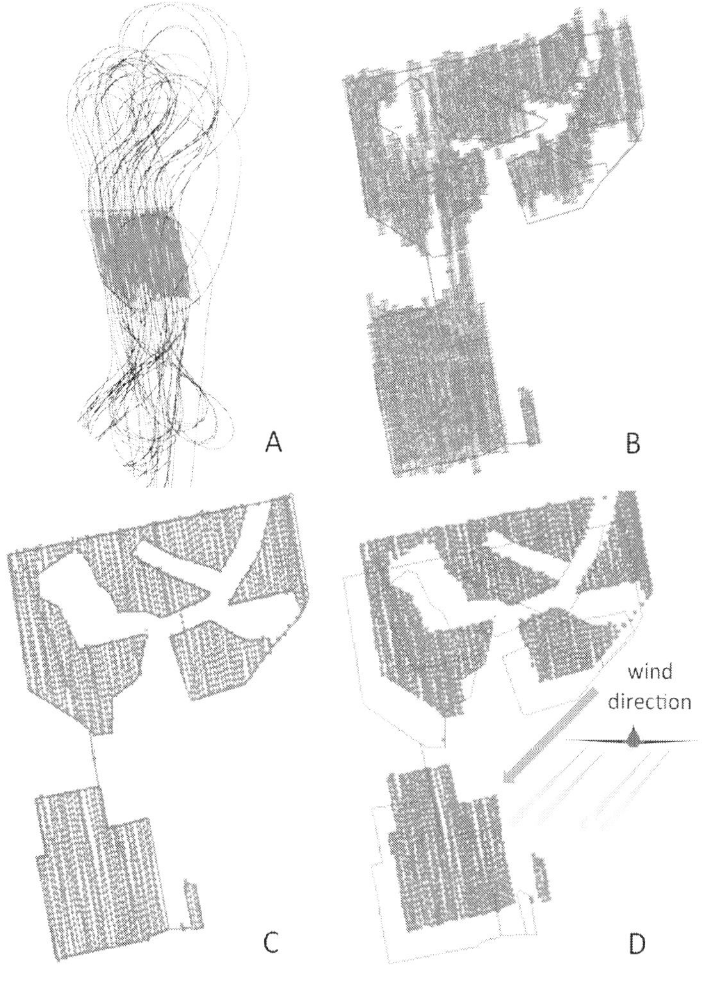

Fig. 3. Printouts of spray operation records. **A**. The boundaries of the block are input into the on-board computer from GPS data. Spray booms are automatically turned on as the aircraft enters the airspace over the block and turned off as the airplane exits. The black tracings are records of the aircraft flight path and the pink shows where the spray booms were turned on. **B** and **C**. Logs showing the accuracy of aerial applications when the booms were turned on (pink) and off manually by the pilot (**B**) and automatically by the GPS-directed autoboom controller (C) over an irregularly shaped block. **D**. Simulation of an optimized spray scenario over the same block as Figs. B and C, where spray lines have been off-set to allow ambient wind conditions to drift the atomized spray into the block.

As it flies through the air, an aircraft's propulsion system and wings generate a strong aerodynamic wake. This wake extends hundreds of meters behind the aircraft and entrains much of the fine component of a spray in long horizontal vortices. The size and shape of the aircraft wake depend on aircraft wingspan, velocity and weight. The wake downwashes (moves downward) toward the ground at a typical rate of 0.3–1.0 m/s (Payne, 1995), which is much faster than the settling rate of a fine spray (40-80 µm) in still air. The downwashing wake is one of the primary influences on transport of fine sprays downward into the forest canopy (Mickle & Rousseau, 1999). With the dissipation of the aircraft wake, spray droplet trajectories are influenced by wind speed, direction and turbulence. In turn, these are influenced by air temperature and humidity, solar radiation, time of day and characteristics of the ground cover (Mierzejewski et al., 2007). Using empirical data from numerous pesticide deposition and off-target drift studies, researchers with the United States Department of Agriculture Forest Service (USDA-FS) developed the AGDISP model for use by pesticide applicators to simulate and assess spray deposition and drift. The AGDISP model has been validated and is widely used within the aerial pesticide application industry worldwide (Mierzejewski et al., 2007).

In most aerial application scenarios, treatment blocks, buffer zones and exclusion zones are established by program managers and are then entered into the AMS computer as GPS coordinates (Fig. 3A). Aircraft spray along flight lines within the block boundaries, and the boundaries act as triggers to the auto-boom system. This type of system achieves much greater accuracy than manual pilot control (Figs. 3B and 3C). Spray drift can result in significant amounts of the control product being transported and deposited outside of the target area. Procedures to optimize spray deposit were developed where the control product is released outside or at the edge of the target block so it can drift into the block (Fig. 3D) resulting in greater amounts of product actually landing within the treatment area (Mickle & Rousseau, 1999; Cormier, 2005, 2006; Mickle et al., 2007).

3.2 Accuair™ wind tunnel

Spray testing is done in wind tunnels to obtain reliable control over wind speed and experimental conditions and to facilitate the use of bulky, vibration-sensitive laser drop-sizing equipment that cannot easily be used to measure sprays in an operational setting. A wind tunnel enables researchers to understand how fluids are atomized under real conditions and to measure the size spectra of spray droplets produced by various atomizers. This information is essential for accurate predictions using spray drift models and for optimizing the effectiveness of sprays using modern AMS. Spray drift models are also used to guide regulatory decisions and to determine the content of pesticide labels (i.e., instructions and restrictions for use), which are of interest to operators and manufacturers alike. Knowledge of spray droplet-size spectra and the ability to calibrate atomizers to produce desired droplet-size spectra enable more accurate estimates of spray drift, increase the accuracy of flight line offset predictions, increase spray efficacy, reduce costs and reduce the environmental impact of spray programs.

The Accuair™ Wind Tunnel (Fig. 4) is owned and operated by FPL and is one of only a few wind tunnels in the world capable of testing pesticide spray products. In order to simulate aerial application conditions, the Accuair™ Wind Tunnel was designed to produce a highly uniform, low-turbulence airstream within its 1-m diameter, 5-m long test section at speeds of up to 300 km/h. In the wind tunnel, sprays are measured to determine the characteristics of the droplet size spectrum. These data are vital to understanding the behaviour of droplets released into the environment and in predicting the amount of spray that will be deposited on a target. Fine sprays composed primarily of small droplets are more prone to drift but provide higher efficacy in many cases, whereas coarser sprays provide better drift control but sacrifice some efficacy (Weisner, 1995). Depending on the application and the type of pesticide product used, either efficacy or drift or both may be of primary concern. It is important, therefore, to know what sort of droplet spectrum a particular nozzle or atomizer will produce under actual spray conditions. Once the performance characteristics of a nozzle or atomizer are known, the information can then be used in computer drift simulations to determine the magnitude of flight line offsets and to classify the nozzle or atomizer according to its drift potential. Drift potential is used to determine the size of buffer zones that restrict how close operators may spray near sensitive areas. Manufacturers of atomizers are motivated by regulatory authorities to incorporate drift-reducing technologies in their products that spray program operators can then use to obtain reduced buffer zone requirements. Qualification as a drift-reducing technology requires wind tunnel testing and subsequent favourable comparison of data to those obtained for standardized or reference atomizers.

Fig. 4. The Accuair wind tunnel facility.

The measurement of droplet size spectrum is easily distorted by aerodynamic obstructions so measurements are typically made using laser drop-sizing instruments that pose no obstruction to the airflow or droplet trajectory in the wind tunnel. The most common form of expressing the droplet size distribution (DSD) for forestry and agricultural sprays is the volumetric DSD. This is represented in histograms where each histogram bar is assigned a droplet-size range, and height of the bar indicates the percentage of the emitted spray volume composed of droplets in that size range (volume fraction). The droplet size corresponding to the point where the cumulative sum of the volume fractions reaches 50% is the volumetric median drop diameter (VMD or D_{V50}). The D_{V50} provides a basic measure of spray droplet size but it is used along with other parameters such as D_{V10} and D_{V90}, which are the points at which the cumulative volume fraction sum reaches 10% and 90% of the total, respectively, indicating how widely the spray droplet sizes are distributed about the median.

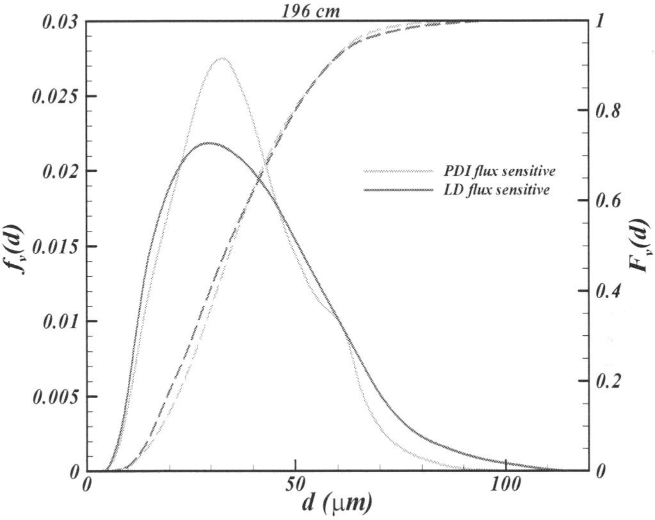

Fig. 5. Sample droplet size distribution (DSD) from the laser diffraction (LD) system and phase Doppler instrument (PDI).

The Accuair™ Wind Tunnel employs two instruments to characterize emitted droplet spectra. The laser diffraction (LD) system provides rapid measurements of a DSD across the entire plume that is based on droplet concentration and can be applied to opaque droplets, making it a good choice for characterization of pesticide spray products (Fig. 5). The phase Doppler (PD) system provides measurements of a DSD that are based on droplet flux and measurements of droplet velocity, which is vital for data interpretation. For the DSD results to be meaningful, sampling must take into account the flux (flow rate) of droplets through the measurement location because droplet flux rather than concentration is correlated to spray coverage and deposition The requirement to transform LD measurements to flux, based DSD poses a significant problem for spray characterization in wind tunnels where droplet velocity measurements are not available. The Accuair™ Wind Tunnel facility, with its complete set of instrumentation, has played a central role in developing guidelines to address issues related to LD spray measurement. These guidelines were compiled in a standard testing methodology submitted to the American Society for Testing and Materials (ASTM) in 2011.

Research into full physics computational fluid dynamic (CFD) modelling (Figs. 6 and 7) can provide a means of examining spray development beyond the wind tunnel environment. Typically, CFD models require validation and calibration using a set of measurements from limited numbers of test cases. For example, CFD models validated against measurements of turbulent droplet transport in the wind tunnel may be extended to simulations of the near wake of the aircraft (10-250 m downwind) and ultimately to predict transport far beyond the wake of the aircraft. These are long-term goals of the present computational research program of UNB and FPL.

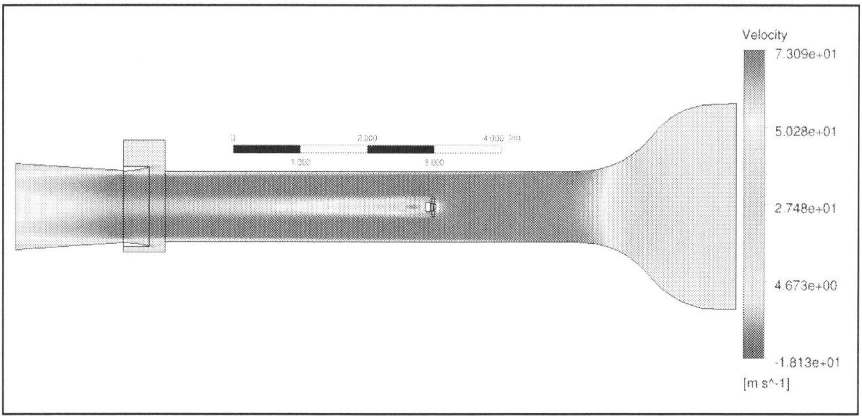

Fig. 6. Computational fluid dynamic (CFD) simulation of the axial velocity field in the wind tunnel.

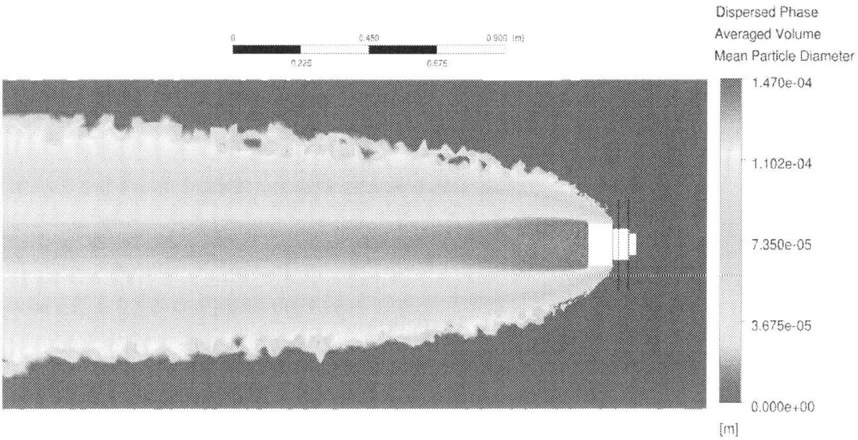

Fig. 7. Computational fluid dynamic (CFD) simulation of the dispersed droplet phase in the wind tunnel.

3.3 Accuair™ ForPRO

The Accuair™ Forest Protection Optimization System (ForPRO) is a new software package that allows integration between forest management planning and optimization models and underlying tree impact information derived from pest management decision support tools. ForPRO includes defoliation impact modelling data for a number of insect species, including spruce budworm, jack pine budworm (*Choristoneura pinus*), eastern hemlock looper (*Lambdina fiscellaria fiscellaria*) and balsam fir sawfly (*Neodiprion abietis*). Potential users of ForPRO include government agencies responsible for forest protection and maintenance of sustainable timber supplies, forest companies that conduct insecticide spray programs, and private landowners interested in quantifying vulnerability of their forest to insects and prioritizing ways to reduce future losses. Proactive assessment of vulnerability to, and implications of, current or possible future insect outbreaks is essential to develop and justify effective policies, response strategies and appropriate infrastructure before these events unfold. ForPRO can assist land managers in quantifying marginal benefits of protecting forest stands against insect defoliation (e.g., in terms of timber volume in m^3/ha or value as \$/ha). Protection cost:benefit analyses can be conducted using existing forest inventory and insect monitoring data in combination with forest management planning models to project the effects of foliage protection strategies on forest development and forest values.

ForPRO is composed of a number of specialized tools that allow users to simulate insect impacts on trees, stands, and forests. These tools leverage stand growth modelling capabilities from the FORUS Simulation Framework (FORUS Research, Fredericton, NB) and allow forest impact analyses to be conducted with existing strategic forest management optimization models (Remsoft Spatial Planning System; Remsoft, Inc. Fredericton, NB). These capabilities permit efficient exploration of cost-effective foliage protection or salvage scenarios.

ForPRO tools can be divided into those used in estimating stand impacts for strategic forest impact analysis (non-spatial tools) and those used for optimal spatial design of operational spray blocks (spatial tools). Non-spatial tools can be used to calibrate and implement SBWDSS. In 2009, ForPRO was used by the New Brunswick Department of Natural Resources (NBDNR) to estimate timber supply impacts on all Crown lands in New Brunswick for a variety of spruce budworm outbreak and foliage protection scenarios. These results were also used in the analyses of the effects of spruce budworm outbreaks on direct and indirect economic benefits from forests (Chang et al., 2012). The generalized framework of the SBWDSS has also been applied to insects other than spruce budworm. ForPRO was used to calibrate and predict timber supply impacts of balsam fir sawfly and hemlock looper defoliation scenarios in Newfoundland, Canada (Iqbal et al., 2011c), and NBDNR subsequently used ForPRO to help prioritize protection treatments for a localized balsam fir sawfly outbreak in New Brunswick.

Spatial tools include the 'Blocking Assistant' and ArcGIS extensions to allow forest protection priority maps (expected volume loss/area) to be spatially blocked for aerial application of biological insecticides. The Blocking Assistant uses meta-heuristic optimization algorithms and information about aircraft cost and flight constraints to help users search for blocking arrangements that increase the cost:benefit ratio of an aerial protection program.

The ForPRO insect decision support framework includes several software tools and steps to translate measured or predicted annual defoliation levels into estimated tree, stand and forest volume impacts. Some of these tools can be used to calculate tree growth reduction and survival from annual defoliation estimates for specific insect pest species or to classifying stands based on host tree species susceptibility to defoliation by certain insect species. Other tools do not include input of insect-related data and are used for such functions as averaging projected stand volume impacts according to stand-type classes to build the Stand Impact Matrix (MacLean et al., 2001).

ForPRO provides three software tools to assist with stand-level impact modelling and compilation for use in forest-level impact analyses.

3.3.1 Stand model multiplier builder

Modelling insect effects on stand dynamics within ForPRO is typically conducted using the stand development model, STAMAN (New Brunswick Growth and Yield Unit [NBGYU], 2004) to forecast defoliation impacts on merchantable volume (Erdle & MacLean, 1999), estimate salvageable volume (Hennigar et al., 2007) and determine changes to stand harvest timeframes. For spruce budworm and jack pine budworm, ForPRO facilitates modelling via the Stand Model Multiplier Builder. This tool converts user-defined estimates of annual defoliation by species into periodic 5-year mean estimates weighted by the proportion of total foliage mass by age class on a healthy balsam fir crown (MacLean et al., 2001), and then uses 5-year periodic defoliation and defoliation-impact relationships to calculate percentage tree growth or survival multiplier values relative to undefoliated conditions. ForPRO has pre-defined defoliation-damage multiplier files derived for spruce budworm (Erdle & MacLean, 1999) and jack pine budworm (Iqbal et al., 2011b), but users can develop their own damage multiplier files for other insects or adjust existing ones.

The FORUS Simulation Framework can be used to execute stand development models such as STAMAN with, or without, defoliation-impact multiplier commands for one or thousands of forest inventory plots. The Simulation Framework also provides a means to summarize tree-level projection output files (tree lists or stand tables) into other measures such as merchantable volume over time by species or species groups. Summarized reports can be output to a database table or text file. Although ForPRO, STAMAN and the FORUS Simulation Framework can all work together, they can also act as stand-alone programs.

3.3.2 Stand composition classifier

The Stand Composition Classifier tool provides a forest stand-type class link between stand impacts and forest inventory attributes. Classes are generally derived based on stand species composition, site, development stage (young, mature, old), canopy density, historical silviculture regime and geographic region. Similar classification logic is used to aggregate plot-level impact forecasts into stand-type classes in order to minimize projected volume impact variance within types. Typically, stand impacts vary the most by species composition and age because these attributes, in general, most significantly influence tree defoliation - damage relationships (MacLean et al, 2001; Hennigar et al., 2007, 2008; Iqbal et al., 2010, 2011a, 2011b).

The ForPRO Stand Composition Classifier tool facilitates the classification of percent tree species composition information from a GIS layer or yield table stored in a database into stand-type impact classes. The stand-type impact classes calculated for each record are written back to the same database in a single column, which can then be manually merged with other forest classification attributes common to both the inventory plots and forest estate model or GIS. The combination of criteria used to classify stands is left up to the user. The species classification algorithm and species information required also depend on the insect species of concern.

3.3.3 Stand impact matrix builder

Once the stand impact class link has been established between data sets, differences (percent reductions) between volume projections by tree species and plot with, and without, defoliation are calculated using the ForPRO Stand Impact Matrix Builder. The Stand Impact Matrix Builder tool also averages impact multipliers by grouping plots into stand-type impact classes. Model outputs or key performance indicators can be any stand measure (e.g., carbon, wildlife habitat index), but are typically timber volume over time by tree species. Using these data, the Stand Impact Matrix Builder quantifies relative differences between the base (undefoliated) and defoliated stand growth, by scenario, stand and time period:

%volume change = defoliation volume/base volume x 100

Once stand-level impacts are calculated, the Stand Impact Matrix Builder automatically averages impacts for each scenario and stand measure by stand impact class and period. Options allow for some degree of control of this averaging algorithm. If stand measures of dead volume are included with live volume estimates, additive (insect-caused) or total salvageable volume relative to base live volume yields can be calculated using the Stand Impact Matrix Builder. The resulting stand-type class, salvageable volume can be referenced against base-yield tables in the forest estate model to estimate forest-level salvageable volume available over time. The Stand Impact Matrix is essentially a large lookup table of time-dependent volume impact multipliers by defoliation scenario, tree species and stand impact class that can be readily linked to forest inventory records and yield tables stored in a relational database. Once these information sources are linked, calculations of projected volume impacts at time of planned harvest or forest inventory reduction caused by a defoliator.

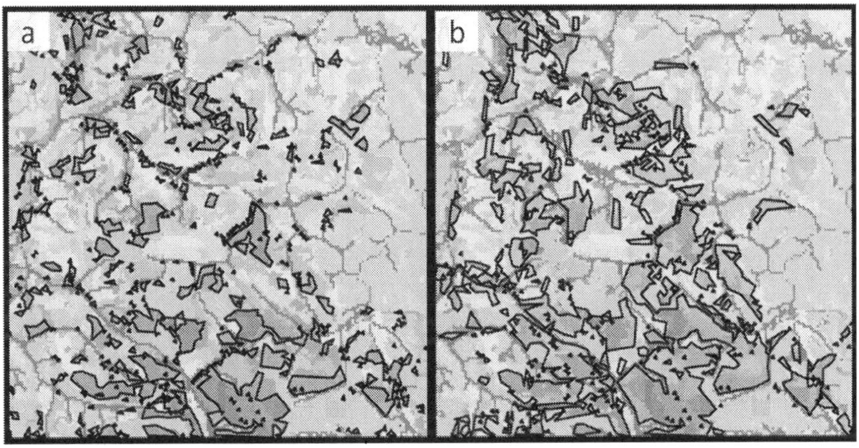

Fig. 8. Comparison of two $1 million aerial foliage protection programs scheduled for: a) highest volume loss protected first (non-spatial), and b) optimal spatial blocking of aerial operations to maximize the economic cost:benefit ratio of individual block flight plans. Each grid cell is 1 ha.

For land managers who use the Remsoft Spatial Planning System (RSPS; Remsoft, 2010), much of the database work described above can now be avoided by using the ForPRO plug-in for RSPS. The ForPRO plug-in allows yield tables in the RSPS forest estate model to be classified and linked to the Stand Impact Matrix with only minor changes to the RSPS model. This integration allows users to utilize the management optimization capabilities of RSPS with yield table impact information produced by ForPRO. The ForPRO framework allows for more efficient communication and integration between existing strategic forest planning optimization models and pest impact information. This can ultimately extend and accelerate implementation of the impact analysis methods used in the SBWDSS (Erdle, 1989; MacLean et al., 2001). ForPRO has also been expanded to allow for the automated spatial blocking of aircraft flight plans to optimize the net cost:benefit ratio of any foliage protection program (Fig. 8). The blocking algorithm blends strategic-level information on land value with industry-calibrated aircraft operational constraints and costs. A blocking cost algorithm that considers aircraft flight plan and spray constraints was refined with input from FPL and the Pest Management Branch of NBDNR.

4. Concluding remarks

The funding provided by the ACOA AIF and NSERC and the additional financial contributions of the research partners have allowed for the consolidation of a number of independent research efforts that has resulted in the Accuair™ suite of aerial application technologies and services (AMS, wind tunnel, and ForPRO). Additionally, these funds and this partnership have advanced the development of baculovirus-based biopesticides for use in forestry (not presented here). The results of this 5-year research project have established Canada as a leader in the development and practical use of aerial application technologies and decision support so that pesticides can be applied accurately, effectively and with minimal impact on the environment.

5. Acknowledgements

The financial support provided by the Atlantic Canada Opportunities Agency - Atlantic Innovation Fund, The Natural Sciences and Engineering Research Council of Canada, the Canadian Forest Service (Natural Resources Canada), the membership of SERG-International, AG-NAV Inc., BioAtlantech, Forest Protection Limited, Sylvar Technologies Inc. and the University of New Brunswick is gratefully acknowledged. A great number of people assisted in this project, and their contributions are also gratefully acknowledged. We thank C. Riley and E.G. Kettela for their reviews of an earlier draft of the manuscript.

6. References

Blais, J.R., Benoit, P., and Martineau, R. 1975. Aerial control operations against the spruce budworm in Quebec, In: *Aerial Control of Forest Insects in Canada*, M.L. Prebble (Editor), pp. 113-125. Department of the Environment, Ottawa, Canada.

Chang, W-Y., Lantz, V.A., Hennigar, C.R., and MacLean, D.A. 2011. Benefit-cost analysis of spruce budworm (*Choristoneura fumiferana* Clem.) control: incorporating market and non-market values. *Journal of Environmental Management* 93: 104-112.

Cormier, G. 2005. An aerial management system. *SERG-International / Canadian Forest Service Workshop: Forest pests and sustainable forest management in Canada 1994-2014*, p. 115. Fredericton, NB, Canada, February 7-10, 2005.

Cormier, G. 2006. Aerial application to PROPS polygons. *SERG-International / USDA Forest Service Workshop*, pp. 206-207, Banff, AB, Canada, February 13-16, 2006.

Davies, D. 1994. Some operational parameters for consideration in aerial pesticide applications and their potential for future development. *CAPCO Technical Session: Overview of Spray Drift and Deposit. Canadian Association of Pesticide Control Officials (CAPCO) Annual Meeting*, pp. 53-55, ISBN 0-662-22637-2, Ottawa, ON, Canada, October 25, 1994.

Erdle, T.A. 1989. *Concept and practice of integrated harvest and protection design in the management of eastern spruce fir forests*. Ph.D. Dissertation, University of New Brunswick, Fredericton, NB, Canada. 161 pp.

Erdle, T.A., and MacLean, D.A. 1999. Stand growth model calibration for use in forest pest impact assessment. *Forestry Chronicle* 75: 141-152.

Estey, R.H. 2004. Canadian use of aircraft for plant protection. *Phytoprotection* 85: 7-12.

Flieger, B.W. 1964. A new method for guiding spray aircraft. *Department of Forestry, Forest Entomology and Pathology Branch, Bi-monthly Progress Report* 20: 2-4.

Hennigar, C.R., MacLean, D.A., Porter, K.B., and Quiring D.T. 2007. Optimized insecticide application and harvest planning to reduce volume losses to spruce budworm. *Canadian Journal of Forest Research* 37: 1755-1769.

Hennigar, C.R., MacLean, D.A., Quiring, D.I., and Kershaw, J.A. Jr. 2008. Differences in spruce budworm defoliation among balsam fir and white, red, and black spruce. *Forest Science* 54: 158-166.

Howse, G.M. and Sippell, W.L. 1975. Aerial control operations against the spruce budworm in Ontario, In: *Aerial Control of Forest Insects in Canada*, M.L. Prebble (Editor), pp. 85-93, Department of the Environment, Ottawa, Canada.

Iqbal, J., MacLean, D.A., and Kershaw, J.A. 2010. Quantitative impacts of hemlock looper defoliation on growth and survival of balsam fir, black spruce and white birch in Newfoundland, Canada. *Forest Ecology and Management* 261: 1106-1114.

Iqbal, J., MacLean, D.A., and Kershaw, J.A. 2011a. Balsam fir sawfly defoliation effects on survival and growth quantified from permanent sample plots and dendrochronology *Forestry*. 84: 349-362

Iqbal, J., Amos-Binks L.J., and MacLean, D.A. 2011b. Meta-analysis of growth and mortality response of jack pine to jack pine budworm defoliation. *Canadian Journal of Forest Research* (under review).

Iqbal, J., Hennigar C.R., and MacLean, D.A. 2011c. Contrasting insecticide protection versus forest management approaches to reducing balsam fir sawfly and hemlock looper damage. *Agricultural and Forest Entomology* In press.

Kilroy, B., Karsky, D., and Thistle, H. 2003. Aerial application equipment guide 2003. *FHTET-2003, USDA-FS*, Morgantown, WV.

MacLean, D.A., K.B. Porter, W.E. MacKinnon, and K.P. Beaton. 2000. Spruce Budworm Decision Support System: lessons learned in development and implementation. *Computers and Electronics in Agriculture* 27: 293-314.

MacLean, D.A., Erdle, T.A., MacKinnon, W.E., Porter, K.B., Beaton, K.P., Cormier, G., Morehouse, S., and Budd, M. 2001. The spruce budworm decision support system: forest protection planning to sustain long-term wood supply. *Canadian Journal of Forest Research* 31: 1742-1757.

MacLean, D.A., Beaton, K.P., Porter, K.B., MacKinnon, W.E., and Budd, M.G. 2002. Potential wood supply losses to spruce budworm in New Brunswick estimated using the spruce budworm decision support system. *The Forestry Chronicle* 78: 739-750.

Mierzejewski, K., Reardon, R.C., Thistle, H., and Dubois, N.R. 2007. Conventional application equipment: aerial application. In: *Field Manual of Techniques in Invertebrate Pathology*, pp. 99-126, L.A. Lacey and H.K. Kaya (Editors), Springer, ISBN 978-1-4020-5931-5, Dordrecht, The Netherlands.

Mickle, R. 1998. Developing effective strategies for managing current and new epidemics of spruce budworm. *Report (unpublished) to Forest Protection Limited*, 100 pp. Lincoln, NB, Canada.

Mickle, R. 1999. Optimization trials for insecticide spraying into small blocks, Murdochville results, *Report (unpublished) to Forest Protection Limited*, 36 pp. Lincoln, NB, Canada.

Mickle, R. 2000. An evaluation of the Auto-Cal flow controller. *Report (unpublished) to Forest Protection Limited*, 15 pp. Lincoln, NB, Canada.

Mickle, R. 2005. Evaluation of the AIMMS-20 airborne meteorological package. *Report (unpublished) to Forest Protection Limited*, 19 pp. Lincoln, NB, Canada.

Mickle, R., and Robinson, A.G. 1991. An evaluation of a radar altimeter for forestry spraying. *Report (unpublished) to Forest Protection Limited*, 6 pp. Lincoln, NB, Canada.

Mickle, R.E. and Rousseau, G. 1999. Optimization trials for insecticide spraying into small blocks - Murdochville results. *Spray Efficacy Research Group (SERG) Report #1998/04*, 37 pp.

Mickle, R., Rousseau, G., and Cormier, G. 2007. Optimized aerial application to PROPS forest polygons. *SERG-International Workshop*, pp. 206-207, Quebec, QC, Canada, February 12-14, 2007.

Miller, C.A. and Kettela, E.G. 1975. Aerial control operations against the spruce budworm in New Brunswick. In: *Aerial Control of Forest Insects in Canada*, M.L. Prebble (Editor), pp. 94-112 Department of the Environment, Ottawa, Canada.

New Brunswick Growth and Yield Unit. 2004. *STAMAN 5.5 user's guide*, 15 pp, New Brunswick Department of Natural Resources, Fredericton, NB, Canada.

Payne, N.J. 1995. Spray dispersal, deposition, and assessment. In: *Forest Insect Pests in Canada*, J.A. Armstrong and W.G.H. Ives (Editors), pp. 465-478, Natural Resources Canada, Canadian Forest Service, ISBN 0-660-15945-7, Ottawa, Canada.

Picot, J. 1994. Deposit and drift in aerial spraying. *CAPCO Technical Session: Overview of Spray Drift and Deposit. Canadian Association of Pesticide Control Officials (CAPCO) Annual Meeting*, pp. 33-52, ISBN 0-662-22637-2, Ottawa, ON, Canada, October 25, 1994.

Randall, A.P. 1975. Application technology. In: *Aerial Control of Forest Insects in Canada*, M.L. Prebble (Editor), pp. 34-55, Department of the Environment, Ottawa, Canada.

Remsoft Inc. 2010. Remsoft Spatial Planning System. *Woodstock 2010.5 user guide*, Remsoft Inc., Fredericton, NB, Canada.

Riley, C. 1994. Factors affecting the deposition and drift of pesticide sprays. *CAPCO Technical Session: Overview of Spray Drift and Deposit. Canadian Association of Pesticide Control Officials (CAPCO) Annual Meeting*, pp. 13-24, ISBN 0-662-22637-2, Ottawa, ON. October 25, 1994.

Weisner, C.J. 1995. Review of the role of droplet size effects on spray efficacy. In: *Forest Insect Pests in Canada*, J.A. Armstrong and W.G.H. Ives (Editors), pp. 493-496, Natural Resources Canada, Canadian Forest Service, ISBN 0-660-15945-7, Ottawa, Canada.